Lecture Notes in Computer Science 3289

Commenced Publication in 1973
Founding and Former Series Editors:
Gerhard Goos, Juris Hartmanis, and Jan van Leeuwen

Shan Wang   Katsumi Tanaka
Shuigeng Zhou   Tok Wang Ling
Jihong Guan   Dongqing Yang
Fabio Grandi   Eleni Mangina
Il-Yeol Song   Heinrich C. Mayr (Eds.)

Conceptual Modeling for Advanced Application Domains

ER 2004 Workshops CoMoGIS, CoMWIM
ECDM, CoMoA, DGOV, and eCOMO
Shanghai, China, November 8-12, 2004
Proceedings

🐴 Springer

Volume Editors

Shan Wang
Renmin University of China, E-mail: swang@ruc.edu.cn

Katsumi Tanaka
Kyoto University, E-mail: tanaka@dl.kuis.kyoto-u.ac.jp

Shuigeng Zhou
Fudan University, E-mail: sgzhou@fudan.edu.cn

Tok Wang Ling
National University of Singapore, E-mail: lingtw@comp.nus.edu.sg

Jihong Guan
Wuhan University, E-mail: jhguan@whu.edu.cn

Dongqing Yang
Peking University, E-mail: dqyang@pku.edu.cn

Fabio Grandi
University of Bologna, E-mail: fgrandi@deis.unibo.it

Eleni Mangina
University College Dublin, E-mail: eleni.mangina@ucd.ie

Il-Yeol Song
Drexel University, E-mail: song@drexel.edu

Heinrich C. Mayr
Universität Klagenfurt, E-mail: heinrich@ifit.uni-klu.ac.at

Library of Congress Control Number: 2004114079

CR Subject Classification (1998): H.2, H.4, H.3, F.4.1, D.2, C.2.4, I.2

ISSN 0302-9743
ISBN 3-540-23722-4 Springer Berlin Heidelberg New York

Springer is a part of Springer Science+Business Media

springeronline.com

© Springer-Verlag Berlin Heidelberg 2004
Printed in Germany

Typesetting: Camera-ready by author, data conversion by Olgun Computergrafik
Printed on acid-free paper   SPIN: 11335238   06/3142   5 4 3 2 1 0

Preface

We would like to welcome you to the proceedings of the workshops held in conjunction with the 23rd International Conference on Conceptual Modeling (ER 2004).

The objective of the workshops associated with ER 2004 was to give participants the opportunity to present and discuss emerging hot topics related to conceptual modeling, thereby adding new perspectives to conceptual modeling. To meet the objective, we selected six workshops with topics on conceptual modeling for Web information integration, digital government, GIS, change management, e-business, and agent technology. Some of these workshops were brand new while others had been held in conjunction with ER conferences as many as five times. They are:

1. 1st International Workshop on Conceptual Model-Directed Web Information Integration and Mining (CoMWIM 2004)
2. 1st International Workshop on Digital Government: Systems and Technologies (DGOV 2004)
3. 1st International Workshop on Conceptual Modeling for GIS (CoMoGIS 2004)
4. 3rd International Workshop on Evolution and Change in Data Management (ECDM 2004)
5. 1st International Workshop on Conceptual Modelling for Agents (CoMoA 2004)
6. 5th International Workshop on Conceptual Modeling Approaches for E-Business (The Model-Driven Design Value Proposition) (eCOMO 2004)

These six workshops received 38, 16, 58, 21, 17, and 13 papers, respectively. Following the philosophy of the ER workshops, the selection of contributions was very carefully carried out in order to keep the excellent standard; 13, 6, 16, 8, 7, and 6 papers were accepted by the six workshops, respectively, with acceptance rates 34%, 37%, 27%, 38%, 41%, and 46%, respectively. In total, 163 workshop papers were received and 56 papers were accepted with an average acceptance rate of 34.4%. Paper presentations of the first five workshops took place on November 9, 2004, and the sixth workshop took place on November 10, 2004. Each workshop had 2 to 4 sessions with 3 or 4 paper presentations in each session.

This proceedings contains selected workshop papers presented in the workshop sessions. We are deeply indebted to the authors, the members of the program committees, and external reviewers, whose work resulted in this outstanding program. We would like to express our sincere appreciation for the hard work by individual workshop organizers in organizing their workshops, handling the paper submissions, reviewing, and selecting workshop papers to achieve a set of excellent programs.

At this juncture, we wish to remember the late Prof. Yahiko Kambayashi who passed away on February 5, 2004 at age 60, and was at that time a workshop co-chair of ER 2004. Many of us will remember him as a friend, a mentor, a leader, an educator, and our source of inspiration. We express our heartfelt condolence and our deepest sympathy to his family.

We hope all participants shared the recent advances in research findings and developments in these emerging hot topics, and enjoyed the programs of the workshops and the conference, and their stay in Shanghai.

November 2004   Tok Wang Ling
                                                            Shan Wang
                                                         Katsumi Tanaka

ER 2004 Conference Organization

Honorary Conference Chairs

Peter P. Chen   Louisiana State University, USA
Ruqian Lu   Fudan University, China

Conference Co-chairs

Aoying Zhou   Fudan University, China
Tok Wang Ling   National University of Singapore, Singapore

Program Committee Co-chairs

Paolo Atzeni   Università Roma Tre, Italy
Wesley Chu   University of California at Los Angeles, USA
Hongjun Lu   Univ. of Science and Technology of Hong Kong, China

Workshop Co-chairs

Shan Wang   Renmin University of China, China
Katsumi Tanaka   Kyoto University, Japan
Yahiko Kambayashi[1]   Kyoto University, Japan

Tutorial Co-chairs

Jianzhong Li   Harbin Institute of Technology, China
Stefano Spaccapietra   EPFL Lausanne, Switzerland

Panel Co-chairs

Chin-Chen Chang   Chung Cheng University, Taiwan, China
Erich Neuhold   IPSI, Fraunhofer, Germany

Industrial Co-chairs

Philip S. Yu   IBM T.J. Watson Research Center, USA
Jian Pei   Simon Fraser University, Canada
Jiansheng Feng   Shanghai Baosight Software Co., Ltd., China

---

[1] Prof. Yahiko Kambayashi died on February 5, 2004.

Demos and Posters Chair

Mong-Li Lee   National University of Singapore, Singapore
Gillian Dobbie   University of Auckland, New Zeeland

Publicity Chair

Qing Li   City University of Hong Kong, China

Publication Chair

Shuigeng Zhou   Fudan University, China

Coordinators

Zhongzhi Shi   ICT, Chinese Academy of Science, China
Yoshifumi Masunaga   Ochanomizu University, Japan
Elisa Bertino   Purdue University, USA
Carlo Zaniolo   University of California at Los Angeles, USA

Steering Committee Liaison

Arne Solvberg   Norwegian University of Sci. and Tech., Norway

Local Arrangements Chair

Shuigeng Zhou   Fudan University, China

Treasurer

Xueqing Gong   Fudan University, China

Registration

Xiaoling Wang   Fudan University, China

Webmasters

Kun Yue   Fudan University, China
Yizhong Wu   Fudan University, China
Zhimao Guo   Fudan University, China
Keping Zhao   Fudan University, China

ER 2004 Workshop Organization

ER 2004 Workshop Chairs

Shan Wang   Renmin University of China, China
Katsumi Tanaka   Kyoto University, Japan
Yahiko Kambayashi[2]   Kyoto University, Japan

CoMoGIS 2004 Program Chair

Jihong Guan   Wuhan University, China

CoMWIM 2004 Organization

Workshop Chairs

Dongqing Yang   Peking University, China
Tok Wang Ling   National University of Singapore, Singapore

Program Chairs

Tengjiao Wang   Peking University, China
Jian Pei   Simon Fraser University, Canada

ECDM 2004 Program Chair

Fabio Grandi   University of Bologna, Italy

CoMoA 2004 Program Chair

Eleni Mangina   University College Dublin, Ireland

DGOV 2004 Program Chairs

Il-Yeol Song   Drexel University, USA
Jin-Taek Jung   Hansung University, Korea

eCOMO 2004 Program Chairs

Heinrich C. Mayr   University of Klagenfurt, Austria
Willem-Jan van den Heuvel   Tilburg University, The Netherlands

---

[2] Prof. Yahiko Kambayashi died on February 5, 2004.

CoMoGIS 2004 Program Committee

Gennady Andrienko	Fraunhofer Institute AIS, Germany
Masatoshi Arikawa	University of Tokyo, Japan
Chaitan Baru	San Diego Supercomputer Center, USA
Michela Bertolotto	University College Dublin, Ireland
Patrice Boursier	University of La Rochelle, France
Stephane Bressan	National University of Singapore, Singapore
James Carswell	Dublin Institute of Technology, Ireland
Barbara Catania	University of Genoa, Italy
Jun Chen	National Geomatics Center of China, China
Christophe Claramunt	Naval Academy, France
Maria Luisa Damiani	University of Milan, Italy
Max Egenhofer	NCGIA, USA
Andrew Frank	Technical University of Vienna, Austria
Jianya Gong	Wuhan University, China
Michael Gould	University Jaume I, Spain
Giovanna Guerrini	University of Pisa, Italy
Erik Hoel	ESRI, USA
Bo Huang	University of Calgary, Canada
Christian Jensen	Aalborg University, Denmark
Ning Jing	National Univ. of Defense Technology, China
Menno-Jan Kraak	ITC, The Netherlands
Robert Laurini	INSA, France
Ki-Joune Li	Pusan National University, South Korea
Hui Lin	Chinese University of Hong Kong, China
Tok Wang Ling	National University of Singapore, Singapore
Dieter Pfoser	CTI, Greece
Philippe Rigaux	Univ. Paris Sud, France
Andrea Rodriguez	University of Concepcion, Chile
Wenzhong Shi	Hong Kong Polytechnic University, China
Ryosuke Shibasaki	University of Tokyo, Japan
Charles A. Shoniregun	University of East London, UK
Stefano Spaccapietra	EPFL, Switzerland
George Taylor	University of Glamorgan, UK
Nectaria Tryfona	CTI, Greece
Christelle Vangenot	EPFL, Switzerland
Nancy Wiegand	University of Wisconsin-Madison, USA
Stephan Winter	University of Melbourne, Australia
Kevin Xu	NICTA, Australia
Yong Xue	Chinese Academy of Sciences, China
Xiofang Zhou	University of Queensland, Australia
Shuigeng Zhou	Fudan University, China

CoMWIM 2004 Program Committee

Svet Braynov   University at Buffalo, USA
Ying Chen   IBM China Research Lab, China
Zhiyuan Chen   Microsoft Research, USA
David Cheung   University of Hong Kong, China
Jun Gao   Peking University, China
Joshua Huang   University of Hong Kong, China
Dongwon Lee   Pennsylvania State University, USA
Chen Li   University of California, Irvine, USA
Bing Liu   University of Illinois at Chicago, USA
Dong Liu   IBM China Research Lab, China
Mengchi Liu   Câleton University, Canada
Qiong Luo   Univ. of Science and Technology of Hong Kong, China
Zaiqing Nie   Microsoft Research, China
Il-Yeol Song   Drexel University, USA
Haixun Wang   IBM Research, USA
Wei Wang   Fudan University, China
Yuqing Wu   Indiana University, USA
Chunxiao Xing   Tsinghua University, China
Lianghuai Yang   National University of Singapore, Singapore
Qiang Yang   Univ. of Science and Technology of Hong Kong, China
Jeffrey Yu   Chinese University of Hong Kong, China
Senqiang Zhou   Simon Fraser University, Canada
Xiaofang Zhou   Queensland University, Astralia

ECDM 2004 Program Committee

Alessandro Artale	Free University of Bolzano-Bozen, Italy
Sourav Bhowmick	Nanyang Technical University, Singapore
Michael Boehlen	Free University of Bolzano-Bozen, Italy
Carlo Combi	University of Verona, Italy
Curtis Dyreson	Washington State University, USA
Kathleen Hornsby	University of Maine, USA
Michel Klein	Vrije Universiteit Amsterdam, The Netherlands
Richard McClatchey	University of the West of England, UK
Federica Mandreoli	University of Modena and Reggio Emilia, Italy
Erik Proper	University of Nijmegen, The Netherlands
Chris Rainsford	DSTO, Australia
Sudha Ram	University of Arizona, USA
John Roddick	Flinders University, Australia
Nandlal Sarda	IIT Bombay, India
Myra Spiliopoulou	Otto-von-Guericke-Universitaet Magdeburg, Germany
Tetsuo Tamai	University of Tokyo, Japan
Babis Theodoulidis	UMIST, UK
Carlo Zaniolo	University of California at Los Angeles, USA

CoMoA 2004 Program Committee

Javier Carbo	Univ. Carlos III of Madrid, Spain
Rem Collier	University College Dublin, Ireland
Kaiyu Dai	Fudan University, China
Maria Fasli	Essex University, UK
Ken Fernstrom	UCFV, Canada
Nathan Griffiths	University of Warwick, UK
Yinsheng Li	Fudan University, China
Eleni Mangina	University College Dublin, Ireland
Pavlos Moraitis	University of Cyprus, Cyprus
Iyad Rahwan	University of Melbourne, Australia
Chris Reed	University of Dundee, UK
Paolo Torroni	Università di Bologna, Italy
Shuigeng Zhou	Fudan University, China

DGOV 2004 Program Committee

Yigal Arens	Digital Government Research Center, USC, USA
Carlo Batini	University of Milano Bicocca, Italy
Athman Bouguettaya	Virginia Tech, USA
David Cheung	University of Hong Kong, China
Ahmed K. Elmagarmid	Purdue University, USA
Joachim Hammer	University of Florida, USA
Ralf Klischewski	University of Hamburg, Germany
Chan Koh	Seoul National University of Technology, Korea
William McIver Jr.	National Research Council of Canada, Canada
Enrico Nardelli	University of Roma Tor Vergara, Italy
Sudha Ram	University of Arizona, USA
Kathy Shelfer	Drexel University, USA
Maurizio Talamo	University of Roma Tor Vergara, Italy

eCOMO 2004 Program Committee

Fahim Akhter	Zayed University, United Arab Emirates
Boldur Barbat	Lucian Blaga University, Sibiu, Romania
Boualem Benatallah	Univ. of New South Wales, Australia
Anthony Bloesch	Microsoft Corporation, USA
Antonio di Leva	University of Torino, Italy
Vadim A. Ermolayev	Zaporozhye State Univ., Ukraine
Marcela Genero	Univ. of Castilla-La Mancha, Spain
Martin Glinz	University of Zurich, Switzerland
József Györkös	University of Maribor, Slovenia
Bill Karakostas	UMIST Manchester, UK
Roland Kaschek	Massey University, New Zealand
Christian Kop	University of Klagenfurt, Austria
Stephen Liddle	Brigham Young University, USA
Zakaria Maamar	Zayed University, United Arab Emirates
Norbert Mikula	Intel Labs, Hillsboro, USA
Oscar Pastor	University of Valencia, Spain
Barbara Pernici	Politecnico di Milano, Italy
Matti Rossi	Helsinki School of Economics, Finland
Michael Schrefl	University of Linz, Austria
Daniel Schwabe	PUC-Rio, Brazil
Il-Yeol Song	Drexel University, Philadelphia, USA
Bernhard Thalheim	BTU Cottbus, Germany
Jos van Hillegersberg	Erasmus University, The Netherlands
Ron Weber	University of Queensland, Australia
Carson Woo	UBC Vancouver, Canada
Jian Yang	Tilburg University, The Netherlands

ER 2004 Workshop External Referees

Jose Luis Ambite
Franco Arcieri
Mohammad S. Akram
Elena Camossi
Luo Chen
Federico Chesani
Junghoo Cho
Denise de Vries
Matteo Golfarelli

C.Q. Huang
Yunxiang Liu
Andrei Lopatenko
Paola Magillo
Hyun Jin Moon
Barbara Oliboni
Torben Bach Pedersen
Fabrizio Riguzzi
Maria Rita Scalas

Min Song
Yu Tang
Igor Timko
Alexis Tsoukias
Bas van Gils
Huayi Wu
Qi Yu
C.Y. Zhao
Xiaohua (Davis) Zhou

Co-organized by

Fudan University of China
National University of Singapore

In Cooperation with

Database Society of the China Computer Federation
ACM SIGMOD
ACM SIGMIS

Sponsored by

National Natural Science Foundation of China (NSFC)
ER Institute (ER Steering Committee)
K.C. Wong Education Foundation, Hong Kong

Supported by

IBM China Co., Ltd.
Shanghai Baosight Software Co., Ltd.
Digital Policy Management Association of Korea

Table of Contents

First International Workshop on Conceptual Model-Directed Web Information Integration and Mining (CoMWIM 2004)

Web Information Integration

Third International Workshop on Evolution and Change in Data Management (ECDM 2004)

First International Workshop on Conceptual Modeling for Agents (CoMoA 2004)

Conceptual Modelling of Agents

Agents' Applications and Conceptual Models

First International Workshop on Digital Government: Systems and Technologies (DGOV 2004)

Digital Government: Systems

Fifth International Workshop on Conceptual Modeling Approaches for E-Business (eCOMO 2004)

E-Business Systems Requirements Engineering

E-Business Processes and Infrastructure

First International Workshop on Conceptual Modeling for GIS (CoMoGIS 2004)

at the 23rd International Conference
on Conceptual Modeling (ER 2004)

Shanghai, China, November 9, 2004

Organized by

Jihong Guan

Preface to CoMoGIS 2004

The recent advances in remote sensing and GPS technologies have increased the production and collection of geo-referenced data, thus requiring the rapid development and the wide deployment of various geographic information systems. With the popularity of World Wide Web and the diversity of GISs on the Internet, geographic information can now be available via personal devices anytime and anywhere. Nowadays GIS is emerging as a common information infrastructure, which penetrates into more and more aspects of our society, and converges with most areas in the IT scenario, such as office automation, workflow, digital libraries, Web searching, virtual reality *etc*. This has given rise to new challenges to the development of conceptual models for GIS. Recently some new approaches have been applied in the development of geo-spatial systems to accommodate new users requirements.

CoMoGIS 2004 (**Co**nceptual **Mo**deling for **GIS**) was the first International Workshop held in conjunction with the 23rd International Conference on Conceptual Modeling (ER 2004) on 9th November in Shanghai, China. Its purpose is to bring together researchers, developers, users, and practitioners carrying out research and development in geographic information systems and conceptual modeling, and foster interdisciplinary discussions in all aspects of these two fields. It tries to provide a forum for original research contributions and practical experiences of conceptual modeling for GIS and highlight future trends in this area. Topics of interest cover but are not limited to theoretical, technical and practical issues of conceptual modeling for GIS.

The workshop attracted overwhelming response from 16 different countries distributed over Asian, North American, South American, Europe, Africa and Australia. We received a total of 58 submissions. A strong international program committee was organized, which consists of 41 scholars and experts from GIS and conceptual modeling areas. CoMoGIS 2004 adopted blind reviewing mechanism and each submission has at least three reviewers. Through careful review by the program committee, 16 papers were selected for the presentation and inclusion in the proceedings. As indicated by these numbers, CoMoGIS 2004 is quite selective. We grouped the sixteen accepted papers into four sessions, each of which consists of four papers: *geographical conceptual modeling*; *spatial storage, indexing and data consistency*; *spatial representation and spatial services*; *spatial querying and retrieval*.

In the session of *geographical conceptual modeling*, Qiang Han *et al* proposed a conceptual model to support multiple representations and topology management; Yvan Bédard *et al* summarized 16 years' research and experimentations in modeling geospatial databases with visual languages; Daniel T. Semwayo and Sonia Berman discussed how to represent ecological niches in a conceptual model; and Jugurta Lisboa Filho *et al* reported a CASE tool for geographic database design.

In the *spatial storage, indexing and data consistency* session, Yuzhen Li *et al* gave a method to store GML documents in spatial database. Two papers from Korea investigated moving objects indexing, where two new indexing structures are developed. Hae-Kyong Kang *et al* addressed topological consistency for collapse operation in multi-scale databases.

The *spatial representation and spatial services* session consists of three papers from China and the other one from Japan. Xiaoming Jin *et al* presented a method to symbolize mobile object trajectories with the support to motion data mining; Jingde Cheng and Yuichi Goto studied spatial knowledge representing and reasoning based on spatial relevant logic; Yu Tang *et al* discussed system architecture and service-based collaboration in spatial information grid; Yang An *et al* investigated Geo Web Services considering semantics.

In the *spatial querying and retrieval* session, Sai Sun *et al* proposed a scaleless data model for direct and progressive spatial query; José Poveda *et al* introduced a new quick point location algorithm. The other two papers, one from Chile and the other from Singapore, focus on spatial and retrieval in Heterogeneous Spatial Databases. Mariella Gutierrez and M. Andrea Rodriguez proposed the method to query heterogeneous spatial databases by using ontology with similarity functions; Shanzhen Yi *et al* addresses heterogeneous geographical information retrieval on semantic Web also using Ontology technology.

A successful workshop requires a lot of efforts from many people. In addition to the authors' contributions, the regular program committee work like reviewing and selecting papers, efforts are required that range from publicity to local arrangements. I'm grateful to all the authors contributing to the workshop. I appreciate the program committee members of the workshop who helped in reviewing the submissions and selecting such quality papers, which results in an excellent technical program. I thank the Workshop Co-chairs Prof. Shan Wang and Prof. Katsumi Tanaka for selecting this workshop, and for excellent coordination and co-operation during this period. I thank also ER2004 Conference Chair Prof. ToK Wang Ling for his valuable instructions on workshop organization. In the early stage of this workshop's organization, Dr. Michela Bertolotto from University College Dublin was also PC chair of CoMoGIS 2004. I appreciate very much Michela's contribution to CoMoGIS 2004. She is an excellent collaborator. Finally, I would like to thank the local organizing committee for its wonderful arrangement.

CoMoGIS 2004 is the first, but would not be the last of such event. I sincerely hope next year CoMoGIS 2005 will still stay with ER conference, and can attract more submissions and participants.

November 2004

Jihong Guan
Program Chair
CoMoGIS 2004

A Conceptual Model for Supporting Multiple Representations and Topology Management*

Qiang Han, Michela Bertolotto, and Joe Weakliam

Computer Science Department,
University College Dublin, Dublin, Ireland
{qiang.han,michela.bertolotto,joe.weakliam}@ucd.ie

Abstract. In this paper a joint topology-geometry model is proposed for dealing with multiple representations and topology management to support map generalization. This model offers a solution for efficiently managing both geometry and topology during the map generalization process. Both geometry-oriented generalization techniques and topology-oriented techniques are integrated within this model. Furthermore, by encoding vertical links in this model, the joint topology-geometry model provides support for hierarchical navigation and browsing across the different levels as well as for the proper reconstruction of maps at intermediate levels.

1   Introduction

Many applications using geographic data require access to that data at different levels of detail (LoD) for information browsing and analysis. In traditional geographic information systems (GIS) these requirements are met by storing a series of maps. These maps provide individual representations at different LoD while referring to the same real world phenomena. This approach suffers from scale limitations imposed by the source data, and presents the difficulty of maintaining consistency within the multiple versions of the maps.

One solution to the above problem is to store a single map housing the finest detail in a GIS database (Master Database). From this single Master Database, maps can be generated for any arbitrary LoD. This approach, however, remains a challenging area of research as current GIS lack methods for automated map generalization. Several different issues have been investigated with the aim of providing partial solutions to this problem (e.g. [11], [8], [13], [3] and [7]).

Focusing on spatial analysis, Rigaux and Scholl [11] discuss multiple representation issues from a data-modelling point of view, with emphasis placed on hierarchical structures. They describe how entity hierarchies can be modelled and queried by means of an object-oriented (OO) DBMS. In [8] a multi-scale GIS database, known as *GEODYSSEY,* uses object-oriented, deductive, and procedural programming techniques to minimize manual intervention and to efficiently retrieve spatial information while maintaining database integrity. Ruas [12] describes an OO framework that models geographic objects in terms of their topological and geometric primitives and

---

* The support of the Informatics Research Initiative of Enterprise Ireland is gratefully acknowledged.

S. Wang et al. (Eds.): ER Workshops 2004, LNCS 3289, pp. 5–16, 2004.

designs an environment defined by constraints. This research tries to abstract and formalize the knowledge of map generalization used by cartographers and represent this knowledge in an OO DBMS.

On an algorithmic level, several techniques have been developed that address issues of topological consistency that arise during map generalization. For example, algorithms ([4] and [13]) were developed to avoid topological inconsistencies arising during cartographic line simplification [5]. More recently, Bertolotto [1] showed that seven minimal atomic generalization operators maintain topological consistency in a pure topological model. These operators can be used to build consistent multiple representations of vector maps to support automated generalization [2]. This model can then be applied to the progressive transmission of spatial data across the network [3].

In this paper, we propose a new joint topology-geometry (JTG) model, supporting multiple representations and topology management during map generalization, which relies on these seven atomic topological operators. In this new model we investigate the issues of how to efficiently store vector maps at different levels of detail, how to effectively retrieve and manage topology information, how to control consistency, and how to integrate other existing map generalization techniques. Furthermore, this model can be implemented as an extension to a spatial DBMS (e.g. Oracle Spatial 9i) by offering both topology management and geometry modelling. The main contribution of this paper extends to both the cartographic and the GIS communities by providing new concepts and tools for the topological generalization and management of vector data. A prototype is being implemented that allows the evaluation of consistent topological generalization and the identification of issues that may arise when performing map generalization.

The remainder of this paper is organized as follows. In Section 2 we introduce the topological structure of vector maps as represented in TIGER/Line files. In Section 3 we describe the new joint topology-geometry (JTG) model dealing with multiple representations. Section 4 discusses issues concerning the management of topology information in this model. Section 5 describes our prototype. Section 6 outlines conclusions and future work.

2 A Structured Map: Topology

We represent the topology of vector maps by following the TIGER/line files [14] topological data structure. The primitive topological elements in TIGER/Line files are nodes, edges, common compound boundaries (CCBs) and faces. A node is a zero-dimensional object. An edge is bounded by two nodes: a start node and an end node. A CCB is bounded by two nodes and composed of a list of edges. A face is an atomic two-dimensional element. Its associated geometric object is a simply connected polygon. A geographic map given by TIGER/Line files contains three entity sets, i.e. a point/node set (P), a CCB/Edge (line) set (L), and a face (region) set (F).

Nodes can be isolated nodes. Non-isolated nodes can be either bounding nodes or shape nodes depending on whether they lie on the boundary of a CCB(s) or not. One or more CCBs can meet at every bounding node.

The order of nodes gives the direction of a CCB. The link from start node to end node gives the positive logical direction <*start_node, end_node*>, while the reverse, i.e. *<end_node, start_node>*, gives the negative direction. CCBs lie between two faces (left face and right face) and hold references to both of these faces. Each CCB contains references to all its contiguous CCBs. In describing a face, it is sufficient to refer to all directed CCBs forming its outer boundaries. All CCBs forming the outer boundary of a particular face contain a reference to the next (directed) CCB in the contiguous perimeter of that face on its left side. Faces can contain inner features. These include point features, line features, and holes or islands that lie inside a face.

3   A Joint Topology-Geometry Model

In this section, we propose a Joint Topology-Geometry model (the JTG model) supporting multiple representations and topology management. In this JTG model the Topology Model houses a structured vector map, while the Geometry Model is a geometry-oriented representation of the underlying topological structures inherent in the vector map. By wrapping topological structures and presenting them as geometries (i.e. point, polyline, polygon), the Geometry Model offers a common geometry interface to GIS applications. One advantage of this new JTG model is that the model offers two different views (i.e. Topology and Geometry) of the same spatial data. This characteristic allows techniques that were once developed to deal with either topology or geometry to be integrated together. Forming an important extension to this JTG model, a multiple-representation data structure encoding spatial objects at different LoD is proposed at the end of this section. We describe both components (i.e. Topology and Geometry) of the JTG model in the following sections.

3.1   Topology Model

In the JTG model, a geographic map contains objects belonging to one of three sets: P, L, and F (See Section 2). This map is structured as a collection of topological primitives together with topological relationships among them. Topological primitives comprise collections of objects from three sets. They include 0-dimensional objects, i.e. isolated nodes (*INode*) and non-isolated nodes (*UNode*), 1-dimensional objects, i.e. edges (*Edge*), and common compound boundaries (*CCB*), and 2-dimensional objects, i.e. faces (*Face*). It should be noted that a '*UNode*' distinguishes itself as either a shape node (*ShNode*) or a bounding node (*BNode*). The topology model explicitly represents the entities and a subset of their topological relationships [6].

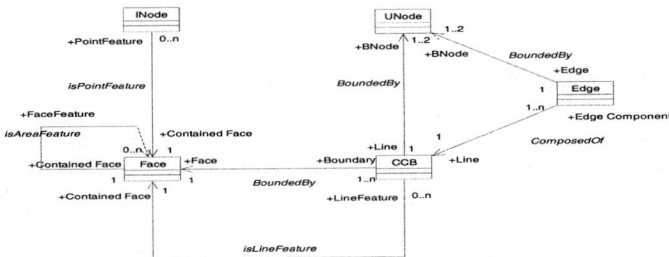

Fig. 3.1. Topology Model using UML

Table 3.1 gives the definitions of the relationships that are explicitly stored in the topology model. Fig. 3.1depicts this *Topology Model* using UML (Unified Modelling Language).

Table 3.1. Topological Relationships stored explicitly

Relationship	Definition
BoundedBy	This defines how an edge, a CCB or a face is composed, e.g. one edge is *bounded by* two nodes (start node, end node) while a face is *bounded by* a series of ordered CCBs.
ComposedOf	Every CCB is *composed of* a series of ordered edges.
IsPointFeature	Every isolated node must be associated with only one face. This point is defined as a point feature of that face by the relationship: *isPointFeature*.
IsLineFeature	Every *isolated polyline* (CCB) must be associated with only one face. This polyline is referred to as a line feature of that face by the relationship: *isLineFeatue*.
isAreaFeature	Every isolated face must be contained inside only one outer face. This inner face is defined as the area feature of that outer face. An area feature may be a hole or an island.

3.2 Geometry Model

In the *Geometry Model* the geometries are composed of ordered sequences of vertices. These vertices are connected by straight-line segments. In the following, geometry primitives are defined. These primitives include:

- *Point*: A point is defined by its unique spatial location (longitude, latitude).
- *Polyline*: A polyline is an ordered sequence of vertices that are connected by straight-line segments. A polyline cannot be self-intersecting but can be closed.
- *Primary Polygon:* A primary polygon is a simply connected polygon that does not have any cuts and punctures. It is bounded by an ordered sequence of polylines.

Combinations of these primitives generate more complex geometry types:

- *Point Cluster*: A point cluster is composed of a group of semantically relevant points. These points usually fall close to each other.
- *Compound Polygon*: A compound polygon is composed of an outer boundary and its interior. The interior may include polygons (primary or compound) or lower-dimensional features (point or polyline).

Fig. 3.2 depicts this geometry model using the object-oriented language UML. This *Geometry Model* represents a map as a collection of geometries.

3.3 The Joint Topology-Geometry (JTG) Model

In this section we describe a new conceptual model that integrates the *Geometry Model* together with the *Topology Model*. This methodology allows the GIS applications to utilize existing geometry-oriented generalization techniques (e.g. line simplification) with the support of an underlying *Topology Model*.

The structure of the JTG model is outlined in Fig. 3.3. In this model spatial data is organized and stored in the *Topology Model*. The *Geometry Model* forms the presen-

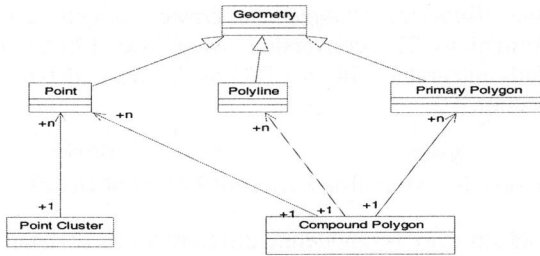

Fig. 3.2. Geometry Model using UML

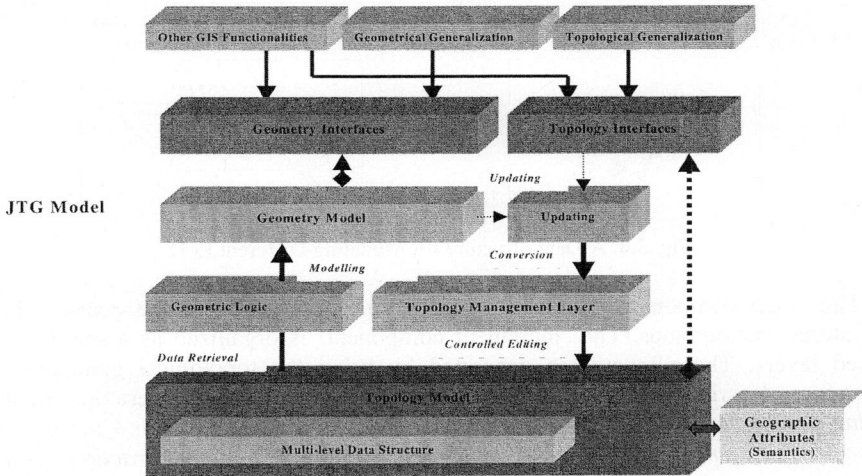

Fig. 3.3. Structure of the JTG Model

tation layer of the *Topology Model* (see Fig. 3.3). A geometric logic is a procedure for generating geometries by means of exploiting topological relationships. Geometry logics are used to generate geometrical representations of the topology model without physically storing geometries in this geometry model. From this presentation layer, different geometry-oriented algorithms can be easily applied to support map generalization. For example, line simplification algorithms can easily work with the cartographic lines presented in the *Geometry Model*. This geometry model also provides topological information as well as simply producing geometries, e.g. one CCB can encode information about its boundary nodes along with its ordered vertices. All connected geometry lines can be retrieved from these boundary nodes.

As depicted in Fig. 3.3, all updates applied to the JTG model ultimately affect the *Topology Model*. This structure ensures that only topologically consistent changes can occur in the model (further discussed in Section 4). This layered structure also enables the JTG model to control generalization behavior from different aspects, i.e. topological generalization is controlled through the topology interface while geometric generalization (e.g. line simplification) is controlled in the geometry interfaces. Interfaces implemented in the 'Topology Management Layer' convert geometric changes to topological transformations. This conversion layer allows different geometry-oriented algorithms to be fully integrated with this JTG model. More details of this mechanism are discussed in Section 4.2.

3.4 A Data Structure for Modeling Different Levels of Detail

In this section a data structure for modeling different levels of detail is described. This data structure forms an extension to the Topology Model of the JTG model. It is composed of two basic components (see Fig. 3.4): a Map Component and an Increment Component.

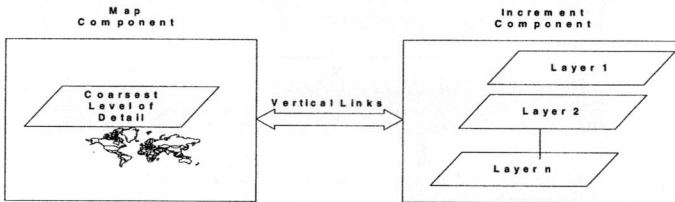

Fig. 3.4. A Data Structure for Modeling Different LoD

The 'map component' represents map entities appearing at the coarsest level for stored vector maps. The 'increment component' is organized as a series of ordered layers. These layers are generated by using the topological generalization operators described in [1]. These topological generalization operators include: point_abstraction, line_abstraction, line_to_point_contraction, line_merge, region_to_line_contraction, region_merge, and region_to_point_contraction. In this data structure a spatial object has its simplest representation in the 'map component'. Only increases in detail between two distinct representations at different LoD are stored in the 'increment component'.

Moreover, the map representation with the coarsest detail keeps a link to its progressive modifications stored in the layers contained in the increment component. These links, called vertical links, are introduced in [2]. Maintaining vertical links facilitates the reconstruction of a representation at an intermediate level of detail and it allows us to know what entities that are modified at subsequent levels and what ones are left unchanged.

In the JTG model the 'Vertical Links' are implemented as a sequence of 'function links'. The 'function link' is a unique generalization operator code along with a timestamp. This code describes what generalization operators have been applied to what spatial object. A sequence of function links represent all transformations applied to an object. The function link allows the system to reconstruct the spatial object by exploiting the information inherent in this route.

The timestamp is composed of a 'Layer ID' and a 'Transaction ID'. The 'Layer ID' shows what layer this spatial object belongs to. The 'Transaction ID' is a series of ordered values generated by the system to give the positional information of the operator during a generalization process. This positional information is important for reconstructing a spatial object and its topological relationships in the correct sequence. For example, when a complex polygon along with its line features is generalized to a single point in a less detailed level, the transaction for topological generalization operator *'line_abstraction'* applied to line features must be followed by the transaction of topological generalization operator *'region_to_point_contraction'* [1] applied to the region. When the reconstruction process takes place, the system can correctly recover the polygon before its line features using the *'Transaction ID'*. During this transformation, the 'Vertical Link' between this particular polygon and its simplified representation as a point, is interpreted as a sequence of function links, i.e. a link for *'line_abstraction'* followed by a link for *'region_to_point_contraction'*. The format of a function link is given as follows:

$$Function\ Link = Operator\ ID \mid Layer\ ID \mid Transaction\ ID$$

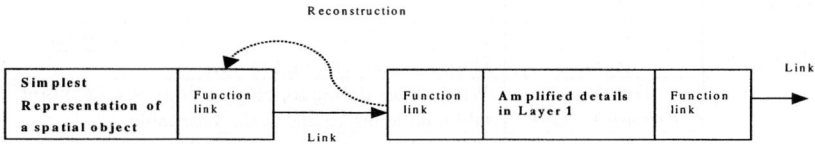

Fig. 3.5. A Spatial Object with Function Link

Fig. 3.5 shows how a spatial object is linked to its various multiple representations at different levels using this link. The function link is shared between a spatial object and its progressive modifications. In the JTG model, every topological primitive (i.e. point, node, edge, CCB, and face) uses function links to represent spatial objects at different LoD. These links help a spatial object find its intermediate representations.

4   Spatial Topology Management in a DBMS

In this section, a solution for achieving accurate topology management in a spatial DBMS is introduced (indicated as Topology Management Layer in Fig. 3.3). This solution makes use of topological metadata, controlled topological editing, and a toolbox for topology validation.

4.1   Using Metadata to Manage Topology

One alternative to extending the DBMS kernel with topological functionality is to create tables with metadata describing the topological structures in the DBMS. The metadata is a collection of data outlining structural knowledge of the topology stored in order for the DBMS to support topology management. For example, the names and definitions of the topological primitives must be accessible. These assist both the DBMS and applications to describe how topology information is stored in the database.

Definitions of generalization operators, measures for generalization operators, and topological relationships form essential parts of the metadata. These metadata enable systems to manage not only the storage of topology but also the maintenance of generalization for topological data. Table 4.1 shows a list of Metadata defined in the JTG model.

In order to achieve efficient topology management, we need to address the issue of inconsistency. Controlled Topological Editing is proposed as a solution to this issue in the JTG model.

Table 4.1. Metadata for Topology Management in the JTG Model

Name	Entries	Description
Structural Topology Metadata	Owner	Which user can create this Metadata
	Name	Name of this Topology Metadata
	Topology	Type of topology
	Spatial-Reference	Name of Spatial Reference System used by this map
	Tolerance	Numeric Tolerance level for measure or distance
	Unit	Unit, i.e. mm, cm, m, km, etc.
	Topology-Primitive-Table	What tables define topology primitives, i.e. point table, node table, line table, face table.
	Topology-Relation-Table	What tables define topology relationships, i.e. face boundary table, point feature table, line feature table, and face feature table.
Generalization Operator Metadata	Owner	Which user can create this Metadata
	Name	Name of generalization operator, e.g. line simplification.
	ID	A unique identification for this operator
	Parameters	What parameters this operator takes
	Output value	What is the output value, e.g. output value of line simplification is a more basic line.
	Measure	What measures are used to control this operator, e.g. minimum area of a region.
Measure Metadata	Owner	Which user can create this Metadata
	Name	Name of measure, e.g. minimum area of a region
	ID	A unique identification of this measure
	Input data types	What are the input data types of this measure, e.g. a region for minimum area
	Output data types	What are the output data types of this measure, e.g. double for minimum area
	Parameters	What parameters should be defined for this measure.

4.2 Controlled Topological Editing

Avoiding topological inconsistencies when updating spatial data is one essential requirement regarding topological management. In the JTG model, controlled topological editing is proposed as one solution for dealing with this problem. This mechanism is implemented in the *Topology Management Layer* of the JTG model where requests to update the data in the JTG model must comply with this mechanism. Topological updates involve three possible actions on spatial objects: *delete*, *modify*, and *replace*. Topology-preserving constraints are introduced for these three operators to prevent topological inconsistencies from arising. The following gives their definitions:

1. *Delete*: this removes a topology primitive from the topology layer without affecting the consistency of the overall topological structure. If inconsistency results from executing a 'delete' action then this operation terminates and returns an exception. In many cases 'delete' is accompanied by a deleting relationship, e.g. deleting a face by opening its boundary gives rise to a CCB(s), and this in turn will delete its associated interior relationships.
2. *Modify*: this updates the spatial and the non-spatial data associated with each spatial entity, e.g. when a node is moved to a new location, its spatial location (longitude, latitude) needs to be modified. If any topological violations occur during this modification (e.g. self-intersection), the operation is halted and an exception is returned.
3. *Replace*: this is composed of two phases: 1) deleting one or more entities, 2) inserting one or more new entities as a result of 1). In the context of map generalization, more complex spatial entities are always replaced by less complex spatial entities, e.g. a face is replaced by a line, etc. An entity of a higher dimension is always reduced to an entity of a lower dimension, e.g. a simple face might be replaced by a point when its area falls below a certain tolerance value, etc. If more complicated entities are involved after replacement or any topological violations occur, this replace operation is halted and an *exception* is returned.

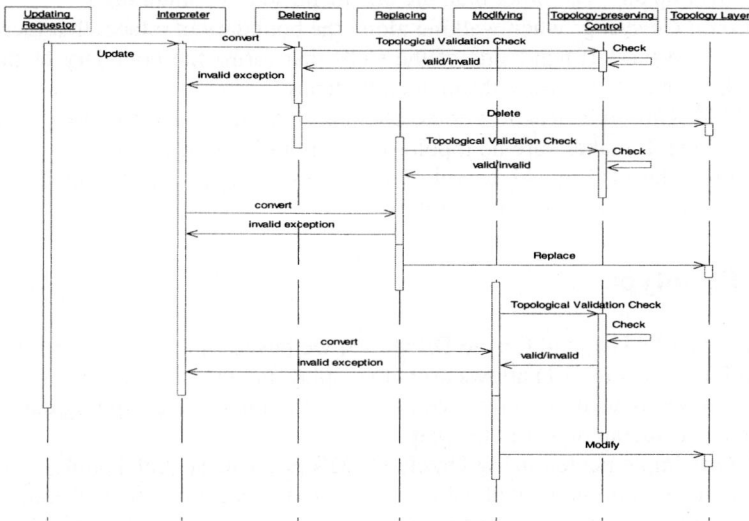

Fig. 4.1. Interactions between Controlled Topological Editing and Updating Request in UML

Fig. 4.1 illustrates how these three operators interact with other generalization operators (e.g. line smoothing) when performing topological updates. Generalization processes (Fig. 4.1) are composed of two phases: 1) a generalization operator (updating requestor) is the implementation of one specific generalization algorithm (e.g. line smoothing). This algorithm comprises a sequence of three possible actions (i.e. delete, modify and replace), together with associated algorithm logics. For example, line simplification can be interpreted as point selection logics together with a sequence of delete actions; 2) once a sequence of three possible actions is formed, each action is then applied to the topology layer unless topology inconsistency arises during one of these actions. This decomposition separates the generalization logics from the relevant data updates within the 2 phases. The separation allows generalization algorithms to focus solely on generalization as inconsistent update requests are rejected in Phase 2). On the other hand, this separation allows both geometry-oriented generalization algorithms and topology-oriented algorithms to be integrated in phase 1).

4.3 A Toolbox for Managing Topology Validation

In a DBMS, detecting and correcting inconsistencies is as important as avoiding inconsistencies. A toolbox for validating topology in the JTG model is proposed in the Topology Management Layer. This toolbox provides users with the facility to check the integrity of the topological structure in the JTG model. The following outlines the functionality provided by this toolbox:

- *Closure Checking*: each face must be closed by its boundary CCBs.
- *Containment Checking*: an isolated spatial object is defined as lying inside, and not touching, the outer boundary of its containing face. The isolated entity must also not touch any other isolated entities contained in that same face; the isolated entity cannot overlap any face other than the one identified as containing it.
- *Directional Checking:* when walking along the boundary of a face, that face should always be on the left-hand side of the CCB that forms the boundary of that face. This is known as "face-walk-through consistency".
- *Overall Validation Checking:* it is composed of the three-topology validation checks above. It makes sure no topology violations have occurred.

To evaluate the JTG model, a prototype system has been implemented. Section 5 outlines some basic features of this prototype.

5 The Prototype

Using OpenMap™ [10] and Oracle DBMS [9] we have implemented the JTG model. An OpenMap GUI (Fig. 5.1) allows users to display and interact with maps stored in a remote database, as well as to retrieve details of the underlying topological structure for any selected spatial object in the map.

The JTG is implemented using Oracle DBMS [9]. The Spatial Topology Management mechanism is implemented using Java Stored Procedures. Spatial indexing, i.e. R-tree indexing, is used to improve spatial relevant queries. Even though Oracle Spatial provides topological functionality, obtaining topological relationships between spatial objects is not efficient since topological relationships are not explicitly stored. On the other hand, Oracle (Spatial) 9i does not support spatial topology management. This may enable topological inconsistency to be introduced into the DBMS without being detected. The JTG model, forming an extension to the spatial model represented in Oracle Spatial DBMS, deals with these two difficulties by explicitly encoding topology information. In the JTG model topological inconsistency can be avoided by rejecting illegal actions and generating exception information. Furthermore, having the geometry model integrated with and conforming to the Oracle spatial model, the

JTG model fully benefits from the spatial functionality provided by Oracle Spatial. As a result, data encoded in the JTG model in Oracle DBMS is able to provide two different views: topology and geometry. The topology is used for advanced spatial analysis while the geometry is useful for displaying, indexing and metric computing. More importantly, in contrast to the spatial model in Oracle DBMS, the JTG model encodes the 7 atomic topological generalization operators that supports map generalization and stores the spatial objects generated by these operators at different LoD. This allows users to produce, query, download and display vector maps stored in the JTG model in a multiple-representation way.

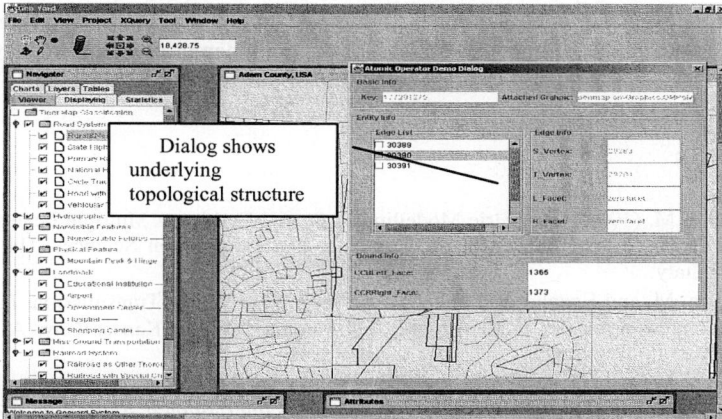

Fig. 5.1. GUI Prototype

6   Conclusion and Future Work

A joint topology-geometry model supporting multiple representations and topology management is presented in this paper. This JTG model proposes a solution to integrate and manage spatial information in a single DBMS by combining a Topology Model with a Geometry Model. This model places an emphasis on topological information. A data structure is proposed to encode spatial entities at different levels of detail. A prototype was implemented using Oracle 9i and OpenMap™. Extensive testing is currently on-going.

In contrast to the other solutions proposed for dealing with the multiple-representation spatial databases ([11], [8], [13], [3] and [7]), the JTG model provides not only a platform for the implementation of multiple-representation databases, but also offers a solution to the critical issue that arises when generalizing maps- how to balance between topology and geometry techniques. By integrating two different models (i.e. Topology and Geometry as described in Section 3), the JTG model offers a practical solution to this problem. In an application that avails of the JTG model functionality, the model itself ensures that only consistent topological transformations can occur. Moreover, it defines common interfaces for both Topology-oriented and Geometry-oriented generalization techniques to be integrated (see Section 3 and 4). This allows the system to benefit from both existing computational algorithms (e.g.

[5]) and new techniques developed for topological generalization (e.g. [1], [2]). Furthermore, this model is proposed for the purpose of progressive vector transmission scenario described in [2]. By encoding the concept of 'Vertical Links' ([2], [3]), a hierarchical data structure is designed to store the vertical links for every transformation caused by map generalization. This hierarchical data structure allows clients of the system to progressively reconstruct the original maps in full detail. By using both topology and geometry oriented techniques, the JTG model enables a generalized map to be stored at LoD and downloaded progressively.

Consistent topological generalization is supported by 7 atomic topological transformation operators [1]. Future work will involve extending our JTG model to support consistent topological generalization with the possible integration of other generalization methods (e.g. line simplification, etc). The implementation of progressive vector data transmission across the Internet is also one of our future perspectives.

References

1. Bertolotto, M. (1998) Geometric Modelling of Spatial Entities at Multiple Levels of Resolution. *Ph.D. Thesis*, Department of Computer and Information Sciences, University of Genova, Italy.
2. Bertolotto, M. and Egenhofer, M. J. (1999) Progressive Vector Transmission. In *7th ACM Symposium on Advances in Geographic Information Systems*, Kansas City, ACM Press, 152-157
3. Bertolotto, M. and Egenhofer, M.J. (2001) Progressive Transmission of Vector Map Data over the World Wide Web. In *GeoInformatica - An International Journal on Advances of Computer Science for Geographic Information Systems*, Kluwer Academic Publishers
4. de Berg, M., van Kreveld M. and Schirra S. (1995) A New Approach to Subdivision Simplification. In *ACSM/ASPRS Annual Convention, Autocarto 12*, Charlotte, NC, pp.79-88.
5. Douglas, D. and Peucker, T. (1973) Algorithms for the Reduction of the Number of Points Required to Represent its Digitalized Line or Caricature. In *The Canadian Cartographer*, 10, 112-122
6. Egenhofer, M. and Herring, J. (1990) A Mathematical Framework for the Definition of Topological Relationships. In K. Brassel and H. Kishimoto (Eds), *Fourth International Symposium on Spatial Data Handling*, Zurich, Switzerland, pp. 803-813
7. Han, Q. and Bertolotto, M. (2003) A Prototype for Progressive Vector Transmission within an Oracle Spatial Environment. In *Proceedings GISRUK 2003*, London, pp.189-193.
8. Jones, C. B., Kidner, D. B., Luo Q., Bundy, G. L. and Ware, J. M. (1996) Database Design for a Multi-Scale Spatial Information System. In *International Journal Geographical Information Systems* 10(8): 901-920.
9. Oracle Spatial. (2003) Spatial User's Guide and Reference, Oracle Documentation Library, http://download-east.oracle.com/docs/cd/B10501_01/appdev.920/a96630.pdf
10. OpenMap TM. (1998), BBN Technologies, http://openmap.bbn.com
11. Rigaux, P. and Scholl, M. (1994) Multiple Representation Modeling and Querying. In Proceedings of IGIS '94: Geographic Information Systems. *International Workshop on Advanced Research in Geographic Information Systems*. pp 59-69.
12. Ruas, A. (1998) OO-Constraint Modelling to Automate Urban Generalization Process. In Spatial *Data Handling '98 Conference Proceedings,* Vancouver, BC, Canada, July 1998, 225-235.
13. Saalfeld, A (1999) Topologically Consistent Line Simplification with the Douglas-Peucker Algorithm. In Cartography and GIS, Vol. 26 (1)
14. TIGER/Line. (2000) U.S. Census Bureau, http://www.census.gov/geo/www/tiger/

Modeling Geospatial Databases with Plug-Ins for Visual Languages: A Pragmatic Approach and the Impacts of 16 Years of Research and Experimentations on Perceptory

Yvan Bédard, Suzie Larrivée, Marie-Josée Proulx, and Martin Nadeau

Centre for Research in Geomatics, Dept of Geomatics Sciences
Laval University, Quebec City, Canada
Tel.: 1-418-656-2131
Fax: 1-418-656-7411
{yvan.bedard,suzie.larrivee,marie-josee.proulx,
martin.nadeau}@scg.ulaval.ca

Abstract. Modeling geospatial databases for GIS applications has always posed several challenges for system analysts, developers and their clients. Numerous improvements to modeling formalisms have been proposed by the research community over the last 15 years, most remaining within academia. This paper presents generic extensions (called Plug-Ins for Visual Languages or PVL) to facilitate spatial and temporal modeling of databases. For the first time, we explain its intrinsic relationship with an extended repository and how it has been influenced by pragmatic lessons learned from real life projects. We describe how we use PVLs with UML and how 16 years of fundamental research, diverse experimentations and feedbacks from users over the world shaped our approach. The final section presents Perceptory, a free repository-based UML+ PVL CASE developed to improve geospatial database modeling.

1 Introduction

Modeling geospatial databases for GIS applications has always posed several challenges for system analysts, system developers as well as for their clients whose involvement into the development of such a project is not a familiar endeavor. To help solving this problem, numerous software and database modeling techniques have been proposed over the last 25 years. Over the last 8 years, UML has emerged as a standard and has been widely adopted by industry and academia, including the geomatics community. Amongst the characteristics of UML, there is a formal way to extend it with stereotypes identified by a textual or pictogram notation [16]. Using pictograms to model databases isn't new. In the field of cartography, it was first proposed in 1989 by Bedard and Paquette [7] to help E/R models depicting the geometry of cartographic features. This first solution has been tested in several projects (National Topographic Database of Canada, Montmorency Forest, Urban database, etc.) and enhanced over time to become Modul-R [5, 6]. Modul-R was an E/R-based solution supported by Orion, the first CASE including spatial pictograms and automatic code

S. Wang et al. (Eds.): ER Workshops 2004, LNCS 3289, pp. 17–30, 2004.
© Springer-Verlag Berlin Heidelberg 2004

generation for commercial GIS [6,8]. In 1996, after several experimentations with real projects, an important shift was made to simplify the proposed solution while increasing its expressive power. Practical projects clearly indicated the need to better support unexpected complex geometric and temporal situations. On the other hand, Modul-R expressiveness was continuously underused by both practitioners and students in spite of previous training. Such common trend has psychological roots (ex. the concept of satisficing) and stems from projects constraints such as allowable time, allowable cost and client desire to see something running on his machine. In addition, many practitioners hesitated to abandon their familiar formalism to adopt Modul-R underlying MERISE E/R graphical notation. These practical experiences suggested realignment and led to improved pictograms, better balance of information between the schema and the dictionary, and formalism-independent PVLs (Plug-ins for Visual Languages). This was a major departure from other solutions which kept adding features to their unique graphical notation (sometimes proprietary) and didn't propose a formal repository to store textual information hardly suitable for graphical representation.

A first test was made in 1996-97 while designing the road network OO model for the government of British Columbia. In Summer 1997, the first version of the CASE tool Perceptory was running; it implemented the PVLs as UML stereotypes. In parallel, other researchers built slight variations of the pictograms, added extra ones or developed similar concepts [27, 28, 14, 25, 26, 17, 21, 29]. Fig. 1 from Filho and Lochpe [15] presents a historical overview of formalisms development.

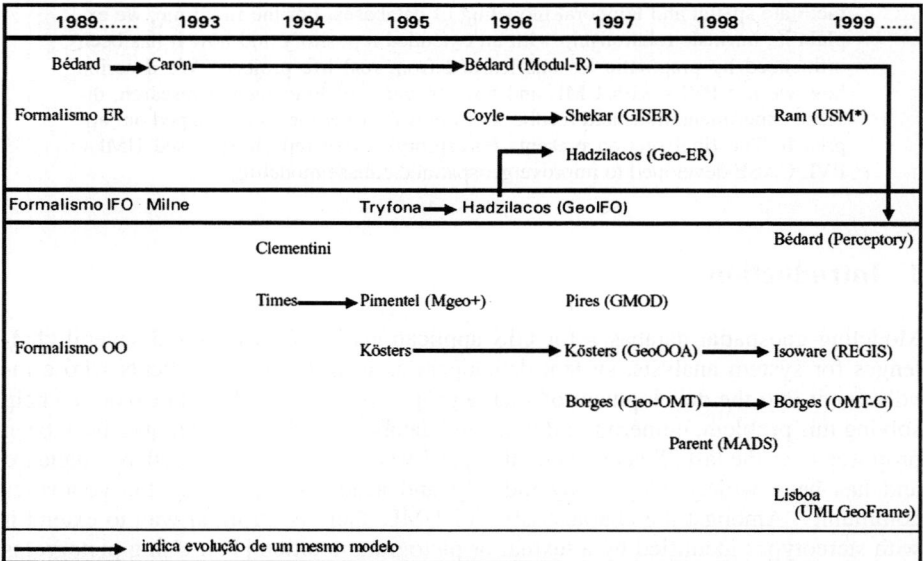

1989.......	1993	1994	1995	1996	1997	1998	1999 ......
Bédard ➞ Caron				➞ Bédard (Modul-R)			
Formalismo ER				Coyle ➞	Shekar (GISER)	Ram (USM*)	
					Hadzilacos (Geo-ER)		
Formalismo IFO Milne			Tryfona ➞ Hadzilacos (GeoIFO)				
		Clementini					Bédard (Perceptory)
		Times ➞ Pimentel (Mgeo+)			Pires (GMOD)		
Formalismo OO			Kösters ➞		➞ Kösters (GeoOOA) ➞	Isoware (REGIS)	
					Borges (Geo-OMT) ➞	Borges (OMT-G)	
					Parent (MADS)		
						Lisboa (UMLGeoFrame)	
➞ indica evolução de um mesmo modelo							

Fig. 1. Historical overview of spatial formalisms (published in 15).

It is beyond the scope of this paper to compare the different solutions, however the approach presented here differs from the others in two fundamental ways: a "plug-in approach" and a "fully-integrated schema + repository approach". The rational for the first difference is presented in [4]. The rational for the second differences is a reaction to the leading tendency to perceive "modeling" solely as a schema exercise while in fact it is not; a schema without clear and complete explanations of its content is meaningless and its robustness cannot be validated. We have searched for the most efficient balance between information best represented in the schema and information best represented in textual form [6]. This has allowed us to avoid building too many pictograms (as we tended in the early years, prior practical experiences). This has also allowed us to define spatial and temporal extensions to traditional data dictionaries, a work that is unique amongst the solutions proposed. The proposed balance between graphical notation extensions and object dictionary extensions results from 16 years of research and numerous lessons from database designers in several organizations and projects such as the Canada Road Network, Canada Hydrographic Network, National Topographic Database, Quebec Multi-Scale Topographic and Administrative Data Warehouse, M3Cat Georeferenced Digital Libraries, Quebec Public Health, and many more involving parcels, facilities, natural resources, archeology, olympic athletes training, etc. Over 1000 undergraduate and graduate students have worked with the developed solution as well as GIS professionals in continuing education courses in Canada, France, Switzerland, Tunisia, Mexico and World Bank projects for Burkina-Faso and Senegal. Perceptory is downloaded over 300 times per month and its web site (http://sirs.scg.ulaval.ca/perceptory) is visited thousands of times yearly. Perceptory is used in over 40 countries. The feedback from this large base of users and projects contributes to the improvement of PVLs and repository extensions.

In the next section, we present the modeling approach we developed over the years. Pragmatic lessons combined with studies in cognitive science and psychology have led us to adopt a symbiotic approach that takes into account humans' abilities and limitations to create and read models and texts (analysts, developers, users), the constraints that apply to real-life projects (time, cost), the needed simplicity vs expressive power of the formalism, the need to be compatible with international standards, and the possibility to use other tools to support modeling than Perceptory.

2 Pragmatic Lessons

Experience has shown that database designers have difficulty to create a first model that reflects "what" users need, i.e. the intended content of the database, without interfering with implementation issues ("how" to provide it). Not separating properly conceptual and implementation issues pollutes the analysis process with irrelevant details and often hides the intentions of the clients. The emphasis of several UML books on syntax and grammar lead to underestimating the importance of going incrementally from technology-independent database analysis to technology-specific development. In order to help analysts stay at the conceptual level and to facilitate communication with his clients, we make schemas using only the essentials of objects' semantics, geometry and world time. Accordingly, we use only the required subset of the UML object-class model and details are stored in the repository in natural language or codes as needed (detailed semantics, geometric acquisition rules, existence and evolution rules, minimum dimensions, resolution, extent, integrity constraints, derivation rules, ISO data types, etc.). In fact, experience has led our team to adapt an agile approach to database modeling (see [1], [9], [12], [18] and [22]) and to make conceptual schemas with a UML-Lite approach as opposed to [23]. These essentials include class, attribute without data type and visibility, operation without visibility, association with multiplicities (written consistently with two explicit numbers for min and max, no * nor single number) written next to its name to make reading easier, multiplicities for attributes, association class, generalization, aggregation, package, ad hoc stereotype, spatial properties of classes and attributes, temporal properties of classes and associations (existence) as well as temporal properties of attributes and geometries (evolution). Constraints on relationships are accepted as well as notes. On the other hand, since hundreds of spatial relationships between classes are possible in a schema, they are not included in the schema and are kept implicit (they are assumed to be derivable using GIS spatial analysis applied to the geometry of the classes) unless a user desires to put emphasis to some spatial associations for semantics reasons or integrity checking (ex. House IsOn 1,1 Lot). Spatial integrity constraints involve semantics and geometry (ex. Road cannot overlay Dam if Road Class is type 2 and Dam is shorter than 200m) and are typically kept outside of the schema since they are detailed information, they often vary according to attribute values and consequently may end up numbering over hundreds of times the number of classes. They are better managed in specialized tools such as CSory (Constraints-in-Space repositORY) which uses both the repository of Perceptory and the Erelate [13] and CRelate [11] ISO matrices to allow users to define their constraints based on geometry and semantics (objects and attributes). Doing so avoids polluting the schema and disturbing the focus of the model from clients' immediate objects of interest. Instead, an easily readable report is automatically produced. In fact, organizations don't create large models of spatial integrity constraints since it is practically impossible to cover a significant part of all cases, commercial software cannot check them automatically, and they rely on data providers for quality assurance. At best, they write the most important ones textually and CSory has been designed with this fact in mind. Such decision to explicitly exclude spatial integrity constraints that are not sensitive to the client from the database schema seems to be a unique approach among the above-mentioned solutions, but it likely is the only one based on real experiences with large topographic databases where strategies have to be defined to restrict constraints to a manageable number (ex. 400 out of 20000 possibilities for about 48 object classes [24].

Compatibility with international standards such as ISO/TC-211 and OGC is also commonly required [19, 20]. However, ISO jargon doesn't express all possible geometries (ex. alternate and facultative geometries) and they are not cognitively compatible with clients' conceptual view who assumes a topologically consistent world (ex. GMPoint vs TPNode, GMCurve vs TPEdge, GMSurface vs TPFace, Aggregate vs Multi). So, conceptual modeling is more efficient with PVLs. They can be mapped with standard ISO geometries automatically with Perceptory or with user assistance [10]. Doing so allows for automatic generation of code, reports, XML files and so on with ISO technical syntax and content (see lower left of Fig. 6). Similarly, there frequently is a linguistic need to support a single schema and dictionary (as well as reports) expressed in multiple languages (ex. French and English in Canada) in combination with a specific technical jargon (ex. French+OMG/UML, English+ISO; see upper-right of Fig. 6).

Another reality is the necessity for some analysts to continue working with their own formalism (sometimes, these are in-house variations of a known formalism) and

CASE tool they already have. This pragmatic requirement was behind the idea to develop a plug-in solution that can be integrated into any object-oriented and entity-relationship formalism, and to make the pictograms available in a special font (downloadable from Perceptory web site) which can be integrated into any software. PVLs can thus be installed and used in commercial CASE tools, word processing, DBMS, GIS, in-house software, etc. Extending CASE tools repository allow them to accept detailed information about the geometry of objects. We developed Perceptory to have a research tool we can modify at will for experimentation, to encourage teaching spatial database modeling by providing a free tool to academia worldwide and to help industry seeking for a more efficient technology specifically geared towards spatial databases. The next two sections present the PVL and Perceptory.

3   Using PVLs for Spatial Database Modeling

The PVL concept was introduced in 1999 by Bedard [4]. PVLs are simple but powerful sets of pictograms and grammar forming a graphical language used to depict any possible geometry. In database modeling, they are used for spatio-temporal properties of object classes, of attributes, of operations and of associations via association classes (see [4, 10] for more details). They offer a high level of abstraction by hiding the complexity inherent to the modeling of geometric and temporal primitives as well as implementation issues specific to each commercial GIS product or universal server. They encapsulate parts of the data structure dealing with geometric primitives, a technique called Sub-Model Substitution when introduced by Bedard and Paquette [7]. They are compatible with common practices to extend formalisms such as UML stereotypes (ex. relational stereotypes [23], multidimensional stereotypes [30]). We have developed two PVLs for 2D applications: spatial PVL and spatio-temporal PVL. Their equivalent 3D spatial PVLs are also available. The next sections present the use of PVLs in UML class diagrams for the modeling of 2D spatial databases, 2D multi-representations databases, 2D temporal and spatio-temporal databases, and finally 3D spatial databases.

3.1   Using PVL for 2D Spatial Database Modeling

This PVL was the first developed and it by far the most largely used as it suffices for most applications. It is composed of only three spatial pictograms and three special pictograms to keep the schema simple and to facilitate communication with customers. The geometric details are written in the repository. These pictograms are made of shapes (0D, 1D, 2D) within a 2D square which represents the 2D dimension of the system universe ($\boxed{\cdot}$ $\boxed{\diagup}$ $\boxed{\square}$). Simple point geometry pictogram can be used for object classes like "Fire hydrant", simple linear geometry for object classes like "Road segment centerline" and simple surface geometry for object classes like "Park". Using no pictogram indicates the objects of a class are not going to be represented on a map. All forms of pictograms that represent the object geometry are presented in Table 1.

Table 1. Possibles PVL forms for 2D spatial database modeling of object geometry.

Grammar	Notation	Description and examples
Simple geometry		
One pictogram with implicit 1,1 default multiplicity (not written).	Ex. [•] Ex. [▱ line] Ex. [◁]	Every instance is represented by one and only one simple geometry. Ex. A 'Fence' is represented by one and exactly one 1D geometry.
Aggregate geometry (AND)		
Simple aggregate has a simple pictogram with 1,N multiplicity.	Ex. [•]1,N Ex. [line]1,N	Each instance is represented by an aggregate of shape of *same dimension*. Ex. An 'Orchard' is an aggregate of points (not a polygon).
Complex aggregate has different shapes in the same pictogram where shape order has no incidence.	Ex. [◁ line] Ex. [• line] Ex. [• line ◁]	Each instance is represented by an aggregate of shape of *different dimensions*. Ex. 'Large rivers' are composed of lines and polygons. *An aggregate of aggregates is an aggregate (i.e. no need of 1,N multiplicity).*
Alternate geometry(XOR)		
Two or more pictograms on the same line.	Ex. [• line] Ex. [• line ◁] Ex. [line]1,N[◁]1,N Ex. [• ◁]1,N	Each instance is represented by a geometry or another but not both (the exclusive OR). The proposed geometries can be simple and/or aggregates. Ex. 'Building' can be represented by a point if it is smaller than a fixed value or by a surface if it is larger.
Multiple geometries		
Two or more pictograms on different lines	Ex. [•] [◁] Ex. [• ◁]1,N [◁]	Each instance is represented by each geometry but usually only one is used at a time. Ex. A 'Town' may have two geometries, a surface delimited by its boundary and a point representing downtown. *All geometry variations can be used within multiple geometries.*
Complicated geometry		
Exclamation mark pictogram *Rarely used as it is facultative*	[!]	When the analyst feels that although it is possible, expressing the geometry with previous pictograms becomes too complicated. Description is in the repository. Ex. Bus Network made of complex aggregations of roads (multiple lines) + bus stops (points or polygons) + parking lots (single or multiple polygons).
Any possible geometry		
Wildcard pictogram *Rarely needed*	[✶]	All geometries are possible, without restriction. Ex. Historical Feature as a point (ex. statue), line (ex. historic street), polygon (ex. park), polygon + line (ex. historic place with adjacent streets), set of polygons (ex. group of buildings), etc.
Temporarily undefined geometry		
Question mark pictogram. *Rarely used.*	[?]	The geometry is temporarily unknown. Will be replaced by meaningful pictograms.

Table 1. (Continued).

Grammar	Notation	Description and examples
Facultative geometry		
Add a minimum multiplicity of 0 and a maximum multiplicity after the pictogram.	Ex. [◁]0,N Ex. [• ✎]0,1 Ex. [✱]0,1	Only certain instances have a geometry. For example, all buildings are in the database, but only public ones have a position and shape and will appear on maps. *All the above geometry variations can be facultative.*
Derived geometry		
Italicize the pictogram (to remind the slash used in UML).	Ex. [• ╱] Ex. [• ✎ ╱] Ex. [• ╱✎] Ex. [• ╱✎]	The geometry is obtained from the processing of other geometries or attributes. Ex. Country polygons can be derived from state polygons. *All the above geometry variations can be derived, each pictogram independently from the other.*

For object classes, we place the pictograms on the left side of their name. When applied to attributes, pictograms indicate that the value of an attribute varies within the geometry of an object (an attribute with different values in different places for a single conceptual object). This avoids having to create at the conceptual level a new object class which would be meaningless for the client. For example, using [✎] next to the attribute "number of lanes" of the class "Road" (see Fig. 2) is cognitively compatible with the client perception of "number of lanes" as a characteristic of Road that varies spatially within an occurrence. Implementation can either create sub-segments or use GIS dynamic segmentation.

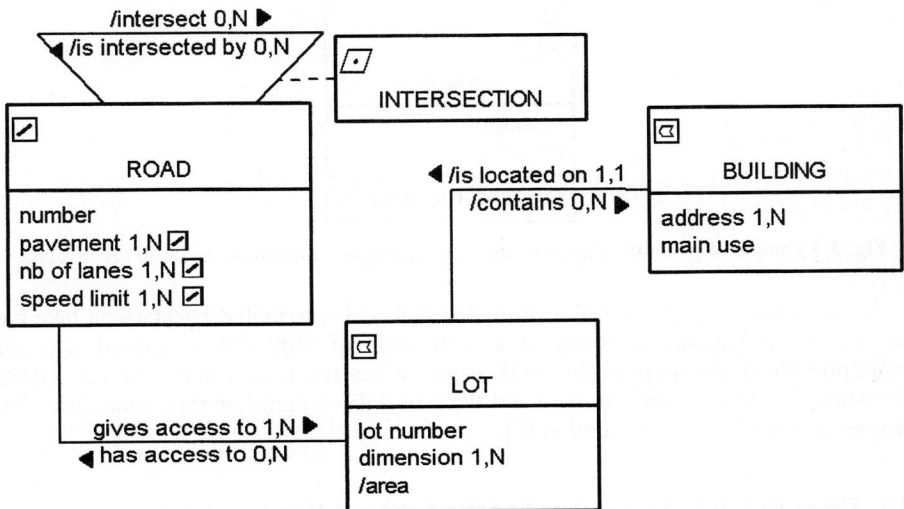

Fig. 2. Spatial schema showing 2D pictograms for object classes and attributes, two spatial associations derived from object geometries, one derived geometry for an association class, one spatially non-derivable association.

Concerning the modeling of spatial relationships, we include only spatial relationships that are semantically significant or sensitive for the customer to avoid overloading the diagram. In fact, spatial relationships of immediate interest to the user will often be the ones implemented explicitly, typically being derived from GIS processing and translated into foreign keys or tables during code generation. Such meaningful spatial relationships are modeled as derived associations (see UML "/" syntax in Fig. 2). Derivation details are explained in the repository.

Pictograms can also represent the result of a spatial process as for Intersection class in Fig. 2 where the derived geometry of the association class is the result of the spatial intersection association. Details of the processing are in the repository.

3.2   Using PVL for Multi-representations Database Modeling

The PVL already supports multiple geometries and is immediately usable for multi-representations databases. In addition, it also allows expressing generalization processes to build such multiple representations. Generalization processes are expressed as operations in OO models, i.e. operations on the geometry of an object class. Of the four possible generalization results presented hereafter, the last three lead to storing multiple representations:

- automatic on-the-fly generalization;
- automatic generalization storing the resulting geometry for performance reasons;
- semi-automatic generalization storing results to avoid repeating the operation;
- manual generalization and storing of the results.

One could add an additional way to create multiple representations: new acquisition.

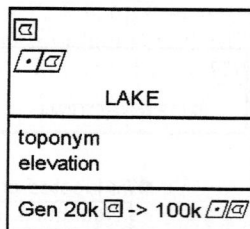

Fig. 3. Example of generalization operation and multiple representations of an object class.

In the previous example taken from the Quebec Topographic Database, a lake has an original polygonal geometry at a map scale of 20K and a derived alternate point/polygonal geometry at the 100K scale. When resulting geometries (ex. 100K) are stored in the database, we then add them *in italic* as multiple representations. Examples and details can be found in [3].

3.3   Using PVL for Temporal and Spatio-temporal Database Modeling

The spatio-temporal PVL is used to manage real world time (valid time, measured or simulated) and to indicate the user want to access it on-line (sometimes called "long transactions"), i.e. to build a temporal database. The spatio-temporal PVL is defined by 0D or time instant (⊘) and 1D or time interval (⊖) pictograms which are used for class existence, descriptive evolution and geometric evolution.

For a class existence, pictograms are located on the right side of its name. They illustrate if the life of every occurrence is considered instantaneous or durable. For example, ⊘ can be used for classes having instantaneous existence like "Accident", and ⊖ can used for classes having durable life like "Building". For evolution, we use ⊘ if attribute values or geometry are valid only for an instant and ⊖ if they are valid for longer periods. "Temperature" is an example of instant validity when new values are obtained every 15 minutes and the temporal granularity of this field or of the database also is 15 minutes. "Commercial value" is an example of attribute with durable values. Selecting between ⊘ or ⊖ depends on the temporal granularity defined into the repository for each class, attribute and geometry.

All the forms presented for 2D spatial pictograms can be applied to temporal pictograms. For example, complex aggregated temporality is represented by ⊖, simple aggregate by ⊘1,N, "any possible temporality" by ⊛, complicated temporality by ⊕ and unknown temporality by ⊘. Alternate temporality is represented by many pictograms with or without multiplicity on the same line ⊘⊖ while they are on different lines for multiple temporalities. No user interest into keeping trace of object existence leads to use no temporal pictogram next to the name of the class. Similarly, no user interest to keep trace of past/future attribute values and geometries lead to no temporal pictogram next to them.

The default multiplicity for class existence is 1,1 because objects have only one life (if well defined). The evolution of an object being the aggregate of its states, its default multiplicity is 1,N since pictograms indicate a client's interest to keep several states of the attribute or geometry it is applied to. Typically, most occurrences will have time to evolve during their life (the N multiplicity), but some won't with regard to a given attribute or geometry due to their short life or stability (the 1 multiplicity). Fig.4 shows a Building having a durable existence (one and only one life), durable states for its attribute "commercial value" and no need to keep track of past addresses.

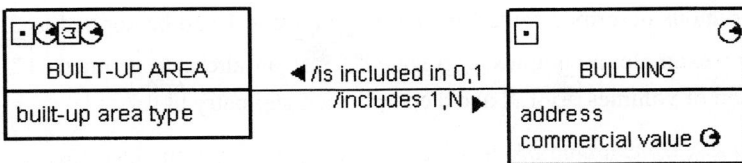

Fig. 4. Example of existence, descriptive and geometric evolutions.

Geometric evolution of objects involves shape modifications and/or displacements. A temporal pictogram is then placed at the right of the spatial pictogram (after the multiplicity if it has one). In the case of an alternate geometry, each spatial pictogram has its own temporal pictogram. Fig.4 illustrates the geometric evolution of the Built-Up Area of the Canadian National Topographic Database which occurrences are a point if smaller than 250 000m² or an area if greater. This area may evolve and the point move or become an area if the surface becomes greater than 250 000 m². These rules are indicated in specific fields of the repository.

3.4 PVL to Design 3D Spatial Database

Modeling 3D spatial databases is more complex than modeling 2D databases. Several objects of a 2D universe can be translated for a 3D universe simply by deriving their elevation from a DTM (Digital Terrain Model) or by extruding their height from an attribute. One may also have true 3D objects. To represent these nuances adequately, pictograms are made of shapes within a 3D cube instead of a 2D square in order to represent the three dimensions of the system universe. Second, to obtain the shape of the objects, we transpose each shape of the 2D spatial pictograms in a 3D space in a way preserving the ground trace (2D view) and giving an elevation or not to them. We obtain 6 pictograms as illustrated in Fig 5, where:

- a 0D object (⊡) in a 2D universe becomes a 0D object (⊡) in a 3D universe if it is an object without thickness or a vertical 1D object (⬕) if it has a height;
- a 1D object (⬈) in a 2D universe becomes a 1D object (⬓) in a 3D universe if it is an object without thickness or a vertical 2D object (⬓) if it has a constant or variable thickness;
- a 2D object (⬓) in a 2D universe becomes a 2D object (⬛) in a 3D universe if it is an object without thickness that embraces or not the digital terrain model or becomes a 3D object or volume (⬛) if it has a constant or variable thickness;

Objects of the reality	manhole	tree	road	wall	football field	building
2D Universe	⊡	⊡	⬈	⬈	⬓	⬓
3D Universe	⬕	⬕	⬓	⬓	⬛	⬛

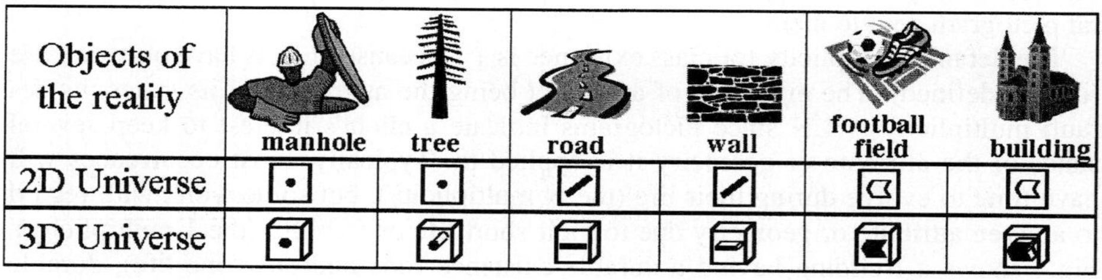

Fig. 5. Similarity between 2D and 3D pictograms to represent real world objects.

All variations described for 2D pictograms in Table 1 can be applied to 3D pictograms. For example, a complex aggregate ⬛, an alternate geometry ⬕, an aggregation of volumes ⬛ 1,N and a complicated geometry ⬓.

4 Perceptory

Perceptory is a free CASE tool developed as a MS-Visio stencil for spatial database modeling (although it may also serve for non-spatial databases). It is based on UML class diagram and the PVL identified earlier. The name *Perceptory* comes from *perception*, referring to the process of phenomenon representation, and *repository*, for knowledge documentation. It is supported by a rich web site [2]. This CASE tool has a rich multi-language (actually French, English and Spanish) and multi-standard repository (OMG-UML, ISO/TC 211 19115, ISO/TC 211 19110) where schema and forms can switch on-the-fly between languages as needed once the user has translated

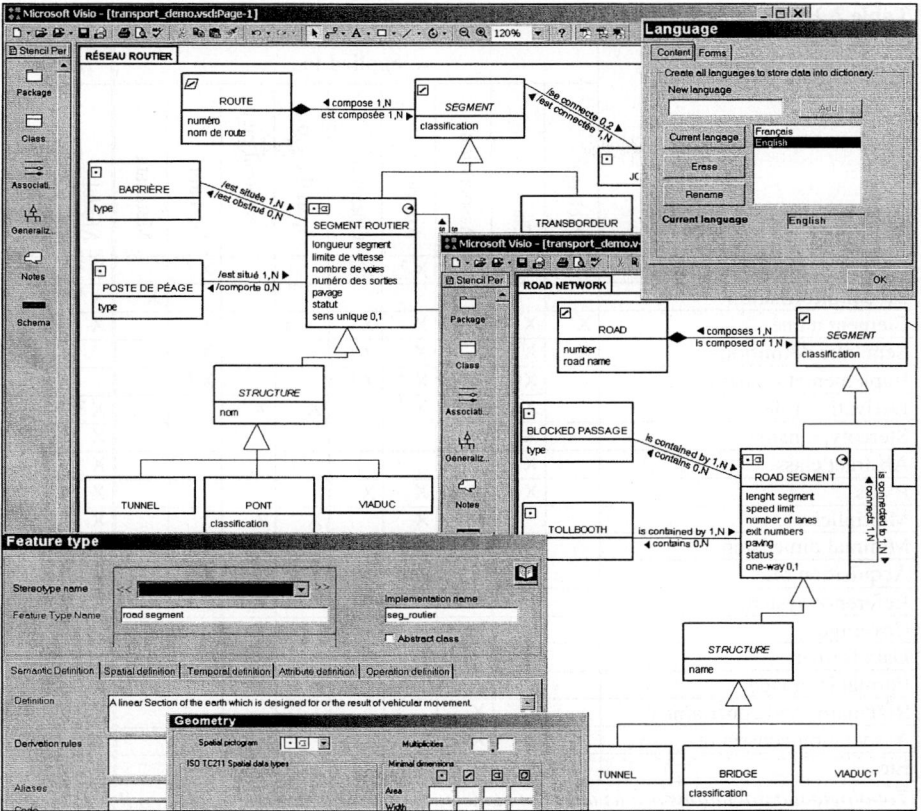

Fig. 6. Perceptory multistandard and multilanguage interfaces and schema viewing.

text fields. A same form can also be viewed in many languages simultaneously. Fig. 6 shows schemas in French and English synchronized thru the same repository. The larger form in the lower left shows ISO-TC211 19110 terms in blue for object definition while the smaller one shows the geometric characteristics of a class.

Perceptory has a utility to generate automatically the database code skeleton for commercial GIS and universal servers. SQL commands or database structures proper to each product are generated as well as log files and mappings between conceptual and implementation views. This process is user-assisted and allows the developer to select optimization strategies.

There is a report generator to print user-selected parts of the repository according to a user-defined content template or a predefined one (ex. standard ISO TC211 19110 content). Pictures or drawings that have been inserted into Perceptory repository during the design of the database are included (ex. to print topographic data acquisition specifications). The report can be produced in all the languages used in the repository in MS-Word and XML formats. The following table shows how Perceptory adheres to the philosophy we described earlier with regard to balancing information load between the schema and the repository.

Table 2. Balancing the information in the schema and in the repository for the main elements.

Modeling elements ...	... applied to...										Info in	
	Package	Class	Association	Attribute	Operation	Spatial picto	Temporal picto	Domain	Metadata	Model	Repository
UML notation (shape/text)	X³	X³	X^{1,3}	X³	X³			X²		X³	
Model information									X	X⁴	X
Element name	X	X	X	X	X			X		X	X
Semantic definition	X	X	X	X	X			X			X
Implementation name		X⁶	X⁶	X⁶	X⁶			X⁶			X⁶
Derivation rule		X	X	X		X	X			X⁵	X
Stereotype name		X								X	X
Abstract class		X								X	X
PVLs		X	X	X	X					X	X
Multiplicity			X	X		X	X			X	X
Minimal dimension						X	X				X
Acquisition rule						X	X				X
Reference system								X	X		X
Coverage								X	X		X
Data format				X		X	X				X
Formal language					X						X
Role name and constraint			X							X	X
Association constraint			X							X	X
Etc.											

[1] *Topological associations are usually not modeled*   [2] *Enumerated domains only*
[3] *Must be created in the schema*   [4] *Key elements only*
[5] *Via UML slash (/) or italic pictogram*   [6] *If entered by choice during analysis*

5   Conclusion

In a field such as database design which involves people, communication skills and technical knowledge, theoretical research cannot succeed without thorough experiences. Over the years, such experiences has alerted us that going after the theoretically most complete and rigorous language has to be balanced with human capabilities, tools functionalities and project constraints. Taking these considerations into account leads to a more complete research, one buildings on their symbiosis to deliver results in sync with their fundamental raison d'être: being used to improve spatial database design. The proposed solution has evolved over time and resulted in unique concepts: (1) PVLs made of pictograms that constitute a powerful yet simple language of their own and which may be used for several purposes and in different tools; (2) a global design balancing and testing both schema representations and textual representations in a repository; (3) a clear separation of the tasks involved in conceptual modeling from the ones involved in defining spatial integrity constraints, while keeping the two in sync. In particular, balancing the expressive power of PVL with that of textual explanations in a repository to describe the thousands of potential geometric and temporal characteristics combinations of an object is a challenge forever. So is balancing the readability and usability of PVLs with conceptual fidelity, robustness, completeness and, one must admit, some personal taste. All this research has also been influenced by pragmatic lessons which, on the overall, have led us to adopt a more agile approach to database design (in system development sense, see [1], [9], [12], [18], [22]).

Future research will involve a more global approach, a repository-based approach where Perceptory will become a UML front-end to enter and visualize database design components. Such expanded repository, called ISTory (Integrated Spatial and Temporal repositORY), will be the heart of a family of tools sharing common concepts: ontology engine, code generators, report generators, spatial integrity constraints, data cubes, catalogs, etc. ISTory metastructure will contain information to facilitate semantic interoperability, data quality analysis, transactional and analytical database development, language and standard translating, etc. The result of such endeavor, which builds on past and ongoing projects, will be made accessible as an XML file and provide a unique, powerful solution in sync with today's trends such as the semantic web and interoperability.

Aknowledgements

The authors wish to acknowledge the financial support of Canada Natural Sciences and Engineering Research Council, Laval University, and major testing partners such as the Ministry of Natural Resources Canada and the Ministère des Ressources Naturelles, de la Faune et des Parcs du Québec. We are thankful to the other project partners, to the users who send us feedbacks and requests, and to the three anonymous reviewers.

References

1. Ambler, S.: Agile Model-Driven Development with UML 2.0. Wiley & Sons, NY (2004)
2. Bédard, Y., Proulx, MJ.: Perceptory Web Site, http://sirs.scg.ulaval.ca/Perceptory/ (2004)
3. Bédard Y, Proulx MJ, Larrivée S, Bernier E: Modeling Multiple Representation into Spatial Datawarehouses: A UML-based Approach, ISPRS WG IV/3, Ottawa, July 8-12 (2002).
4. Bédard, Y.: Visual Modelling of Spatial Database towards Spatial PVL and UML, Geomatica, 53(2), (1999) 169-185
5. Bédard, Y., Caron, C., Maamar, Z., Moulin, B., Vallière, D.: Adapting Data Model for the Design of Spatio-Temporal Database. Comp. Env. and Urban Systems, 20(l) (1996) 19-41.
6. Bédard, Y., Pageau, J., Caron, C.: Spatial Data Modeling: The Modul-R Formalism and CASE Technology. ISPRS Symposium, Washington, August 1-14 (1992)
7. Bédard, Y., Paquette F.: Extending entity/relationship formalism for spatial information systems, AUTO-CARTO 9, April 2-7, Baltimore (1989) 818-827
8. Bédard, Y., Larrivée, S.: Développement des systèmes d'information à référence spatiale: vers l'utilisation d'ateliers de génie logiciel. CISM Journal ACSGC, 46(4) (1992) 423-433
9. Boehm, B., Turner, R.: Balancing Agility & Discipline. Addison-Wesley, NY (2004) 304 p.

10. Brodeur, J., Bédard, Y., Proulx, MJ.: Modelling Geospatial Application Database using UML-based Repositories Aligned with International Standards in Geomatics, ACMGIS, November 10-11, Washington DC, (2000) 36-46
11. Clementini, E., Di Felice, P., Van Oosterom, P.: A small set of formal topological relationship suitable for end users interaction. Third symposium on Large Spatial Database, No. 692, Singapore, Springer-Verlag, NY (1993) 277-295
12. Cockburn, A.:. Agile Software Development. Addison-Wesley, NY (2002) 278 p.
13. Egenhofer, M., Herring J.: Categorizing Binary Topological Relations Between Regions, Lines, and Points in Geographic Databases, Tech. Report, University of Maine (1990) 28 p.
14. Filho, J.L., Iochpe, C.: Specifying analysis patterns for geographic databases on the basis of a conceptual framework. ACMGIS, Vol. 7, Kansas City, USA. (1999) 7-13
15. Filho, J. L., Iochpe, C.: Um Estudo sobre Modelos Conceituais de Dados para Projeto de Bancos de Dados Geográficos, Revista IP-Informática Pública, 1(2) (1999) 37-90
16. Fowler, M.: UML 2.0, CampusPress (2004) 165 p.
17. Hadzilacos T., Tryfona N.: An Extended Entity-Relationship Model for Geographic Applications. SIGMOD Record, 26 (3) (1997)
18. Highsmith J.: Agile Software Development Ecosystems. Addison-Wesley, (2002) 448 p.
19. ISO/TC211 19110, Geographic information: Methodology for feature cataloguing (2004)
20. ISO/TC211, 19115, Geographic information: Metadata (2003) 140 p.
21. Kosters, G, Pagel, B., Six, H.: GIS-Application Development with GeoOOA. IJGIS, 11(4) (1997) 307-335
22. Larman, C.: Agile & Iterative Development. Addison-Wesley (2004) 342 p.
23. Naiburg EJ., Maksimchuk, RA.: UML for Database Design, Addison-Wesley (2001) 300 p.
24. Normand, P., Modélisation des contraintes d'intégrité spatiale : théorie et exemples d'application, Ms. Degree, Dept. Geomatics Sciences, University Laval, 1999
25. Parent, C, Spaccapietra, S, Zimanyi, E., Donini, P.: Modeling Spatial Data in the MADS Conceptual Model. Int. Symp. on Spatial Data Handling, Vancouver (1998) 138-150
26. Parent C, Spaccapietra, S., Zimanyi, E.: Spatio-Temporal Conceptual Models: Data Structures + Space + Time, 7th ACMGIS, GIS'99, Kansas City, (1999) 26-33
27. Shekhar, S., Vatsavai1, R.R., Chawla, S., Burk, T. E.: Spatial Pictogram Enhanced Conceptual Data Models and Their Translation to Logical Data Models, ISD'99, Lecture Notes in Computer Science, Vol 1737, Springer Verlag (1999) 77-104
28. Shekhar, S., Chawla, S.: Spatial Databases A Tour, Prentice Hall (2003) 262 p.
29. Tryfona, N. Price, R., Jensen, C.S. Conceptual Models for Spatio-temporal Applications. Spatio-Temporal Databases: The CHOROCHRONOS Approach 2003, (2003) 79-116
30. Priebe, T., Pernul, G.: Metadaten-gestützer Data-Warehouse-Entwurf mit ADAPTed UML, 5th Int.Tagung Wirtschaftsinformatik (WI 2001), 19.-21. September, Germany (2001)

Representing Ecological Niches in a Conceptual Model

Daniel T. Semwayo and Sonia Berman

Computer Science Department, University of Cape Town, Rondebosch 7701, South Africa
{dsemwayo,sonia}@cs.uct.ac.za

Abstract. The niche concept has been recognised as a fundamental notion in biological and business systems, amongst others. Existing data models do not have adequate primitives that faithfully represent ecological niches for geographical information systems. This paper shows the limitations of the status quo, manifested in semantic heterogeneity and the meaningless integration of ecological data sets. We advocate an ontological engineering approach to ecological data modelling and propose two new data modelling primitives, *Functional object granules* [FOGs] and *Functional Contexts,* which can be used to represent niches. We present our data modelling environment, OntoCrucible, and illustrate our model and methodology in the context of a case study involving African smallholder farmer upliftment.

1 Introduction

Recent literature [1, 2, 3] argue that traditional conceptual models, with primitives such as entities, attributes, relationships, aggregation and generalisation, are based on set theory which does not adequately cover all the real world phenomena that the data models attempt to abstract. Real world entities show a tendency to exist in nested ecological niches which determine their interactions and behaviour. We define ecology broadly after [4] as the study of the interaction of entities among themselves and with their environment. An ecological niche is a containment space bounded by spatial and/or temporal parameters favouring a specific group of members [1]. While set theoretic entity–relationship and object-oriented models have good modelling primitives for representing static and dynamic behaviour respectively [2], they fall short of fully representing ecological niches.

Problems associated with Geographical Information systems (GIS) include semantic heterogeneity and erroneous data integration [5, 6]. We argue that a conceptual model which represents niches, used with an ontological engineering methodology, increases the probability of detecting and resolving these problems. Semantic heterogeneity problems arise, particularly in multi-disciplinary projects, when different perspectives give rise to different interpretations and representations of the real world. It occurs when there is disagreement about the meaning of similar terms. For example the term farm refers to a bounded piece of land for the land surveyor or the tax man interested in rateable property. On the other hand, to a social scientist, a farm refers to an agricultural administrative unit with internal interactions between people, animals and crops.

S. Wang et al. (Eds.): ER Workshops 2004, LNCS 3289, pp. 31–42, 2004.

Ecological phenomena are observed and analyzed at different granular levels, e.g. catena, landscape and global. The methods for observing and the type of analyses employed are dependent on the scale of observation [7]. It is therefore important that data is collected at the correct granularity to obtain meaningful results. The integration of data from say average rainfall figures (obtained from 50 year data figures) with monthly soil moisture data will give a "result", but a misleading one. The two data sets are from two very different time horizons. Secondary users of such integrated data obtain erroneous information without being aware of it.

We contend that such problems are a result of conceptual models that fail to recognize and represent entities and relationships according to the niche in which they exist and interact. In this paper we show that representing niches means modeling concepts at the correct granularity and within appropriate contexts. We present new data modelling primitives aimed at addressing the niche representation gap inherent in conceptual models for geographical information systems. The proposed new primitives are the Functional Object Granule (FOG), and the Functional Context, which together enable niches to be identified and modelled. We further advocate a collaborative modelling approach based on ontological engineering, which uses ontological foundations (notions grounded in what exists in the real world) to construct and evaluate meaningful data models [8]. We argue that using these primitives, and building ontologies that correctly model context and granularity, will significantly reduce the problems outlined above.

This paper is organised as follows. In the next section we briefly outline related work, and in section 3 we introduce our OntoCrucible modelling environment. Sections 4 and 5 describe and discuss our approach to modelling niches for GIS, and we conclude with a summary of the paper.

2 Related Work

We provide a brief outline of related work in this section covering the domains of ontological engineering and ecological cognitive models.

2.1 Ontological Engineering

Ontological engineering provides the mechanism to design new primitives to enrich data models [8, 9, 10]. This paradigm is based on notions first studied in philosophy, and first incorporated in computer science in the field of artificial intelligence. From a philosophical perspective, ontology is concerned with the study of what exists as seen (observed and conceptualised) from a particular world view [11]. In artificial intelligence, an ontology refers to an engineering artifact, constituted by a specific vocabulary used to describe a certain reality, plus a set of explicit assumptions regarding the intended meaning of the vocabulary of words [10]. Put in another way, an ontology refers to a model with which the contents of the conceptualised world are defined, ordered and signified [11]. Comprehensive reviews of the development of philosophical metaphysics (which explores what exists as seen) is carried out in [1,10]. Metaphysical realists propose that all world phenomena can be divided into: *particulars* (individual things – e.g. you and me) and *universals* (repeatable phenomena – e.g. persons), with attributes (height, weight, etc), kinds (taxonomies), and relations (e.g. topological). This is the view we subscribe to in this paper.

2.2 Ecological Cognitive Models

Traditional ecological modelling focuses on identifying individual themes (soils, vegetation, rainfall, slope) and then attempting to integrate and analyse relationships between these thematic "layers" using tools like GIS. The layered approach, which is based on set theory [3, 2], puts emphasis on the intersection space of themes using spatial coincidence as the integration factor. For example, that a river runs through a farm is a worked out from a spatial overlay of two thematic layers of *rivers* and *farms*.

Current thinking in ecological modelling suggests a paradigm shift in developing cognitive models to represent ecological phenomena [1,4, 11,12]. The crux of the new wisdom is that ecological systems are not disjoint but are in essence connected wholes at different scales of operation, thus the set-theory-based approach is fraught with modelling inadequacies [13]. The suggestion that the functional niche concept be used as the fundamental unit of investigating ecological systems appears in [1, 12]. The advantage of this approach is that niches are space and time-scale dependent. This is closer to the "true" nature of ecological phenomena, also consistent with the view that entities and processes occur at a specific resolution, e.g. climate change (decades), denudation (hundreds of years), etc [14]. Niches are connected as part of a nested hierarchy. For example, swamps are populated by particular types of waterborne entities that thrive and interact in wet environments: fishes, frogs, birds, plant species, etc. Microbial interactions here occur at a different sub-niche and granularity from that of birds and frogs. Niches are not restricted to bio-physical phenomena however, they have been referred to in literature as; settings, environments, physical behavioural units [15] and spatial socio-economic units [1].

Aggregations, compositions and generalisation hierarchies have been used in ecological models. Composition and aggregation provide the best existing approximation to represent niches [2]. That the impact of the niche containment properties, its resolution, and the collective interaction of its members, affect the nature and behaviour of these members is well known and understood [1, 4, 12]. However this is not captured in conceptual models using aggregation or any other existing modelling primitive.

3   The OntoCrucible Integrated Modelling Environment

We propose a new model and methodology for GIS design, and have built an ontological engineering environment (OntoCrucible) for building, checking and normalizing such models. Our approach centres on appropriately modeling ecological niches where entities operate and interact at different resolution and within different contexts. The proposed new data modeling primitives should fill the gap created by the limitations of set theory based modeling and minimize inappropriate data collection and integration in GIS usage.

OntoCrucible is a collaborative ontological modeling environment based on the Poseidon UML case tool [16]. The OntoCrucible modelling environment is a best practices, project-based approach for ontology design to be used by a multi-disciplinary group under the guidance of an ontology engineer. It focuses on collaborative domain ontology building involving several stakeholders: project benefactors, domain specialists and ontological engineers. The collaborative approach is used to ensure ownership of the ontology by all, and not just by some system designer or domain specialist. The ontological design process bridges the gap between different conceptual models of the various stakeholders - an ontology after all is a negotiated explicit representation of a world view [8]. In what follows, we use an agricultural case study to illustrate how the proposed model and methodology are utilized. The case study is based on The Initiative for Development and Equity in African Agriculture (IDEAA) which aims to transform smallholder agriculture in Southern Africa [20]. The programme is jointly funded by the Kellogg and Rockefeller Foundations.

4   Applying Our Approach

4.1   Shortcomings of Class Diagram / ER Modelling

During a data modelling exercise the entities and relationships in figure 1 were identified for the production subsystem by the multi-disciplinary IDEAA project group. Figure 1 shows some of the classes and associations identified by the IDEAA team. The logical links in this data model assumes a "flat structure" of object interaction within the universe of discourse of the IDEAA case study – i.e. without any identification of niches. This is typical of ER and O-O data modeling approaches. However, virtually all world entities belong to or are contained in some ecosystem described in the literature as niches socio-economic units, or physical behavioural units [1]. Existing data modelling primitives do not adequately model this containment in space and time. While the domain of mereotopology provides primitives to capture the spatial components of containment of one entity by another, like a dam on a farm [3], these primitives do not adequately cover the niche meta-properties of context or granularity. We will expand on these meta-properties in the next section.

4.2   Our Approach

4.2.1   Modelling Entities in Niches
We now turn to the modeling steps of the OntoCrucible process using the IDEAA case study as illustration. The resultant ontology is then compared with the data model in figure 1.

We define a *functional context* (or *context*) as a subsystem or sphere of operation representing one node in a value chain for the domain being modeled. The first step of our methodology involves breaking down the project into subsystems by identifying the value chain and hence the different contexts in which activities occur. IDEAA identified a value chain for its project which comprised three subsystems (production, processing, marketing), as a first step. Prior to this, project participants decomposed their task into separate areas where they predicted innovation could occur, and con-

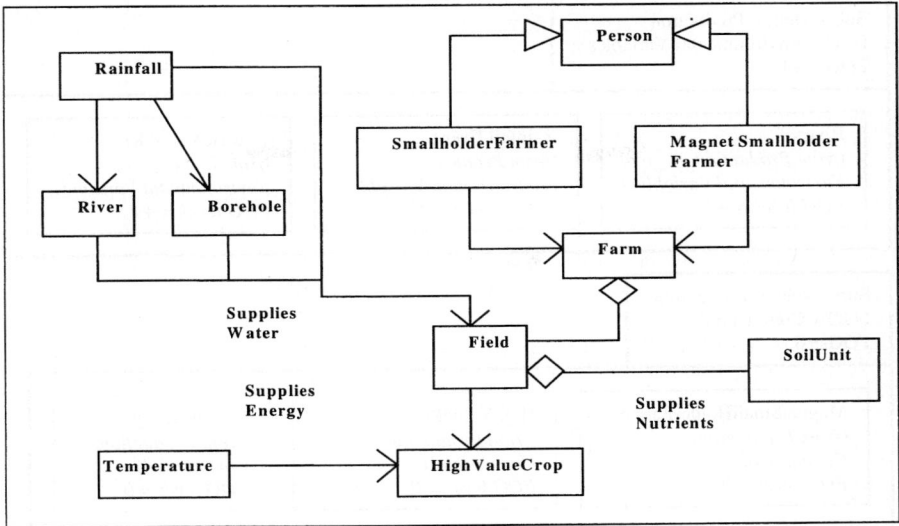

Fig. 1. Subset of classes and associations of the IDEAA production system

structed separate models for each of these, in an attempt to manage the size and complexity of the original goal (smallholder farmer upliftment). However, without the value chain identification, there was confusion as to how entities in different submodels affected each other. For the purposes of this paper we focus on the production sub-system.

Step 2 involves a brainstorming/mind-mapping approach to identify the relevant entities for each context, using cue cards or a white board. A facilitator, an ontological engineer, coordinates the process. In step 3, entities are grouped in FOGs (*functional object granules*). Each FOG is populated by entities at the same functional granular level. Granularity refers to the relative scale, level of detail, that characterizes an object or activity [17]. In other words, granularity refers to the relevant scale / resolution at which processes on the earth's surface operate. We define the *functional granularity* of a class as the level of interaction, impact, significance or influence of the object. It is constrained by the context within which objects are identified, their role, and the niche in which they operate. A FOG and a context together represent niches at different hierarchical levels, see figure 2.

The concept of class (as in object oriented modeling) is used to refer to concepts or universals in the ontological sense. Classes have properties (attributes) which induce distinctions among individuals. Meta-level properties induce distinctions among concepts (*universals*). A niche is a concept or universal with meta-properties context and granularity. In figure 2 e.g., an extract from our OntoCrucible environment in which the niche view has been selected, *"Production, Commercial farming FOG=0"* means the niche is Commercial farming, its context is Production and its relative granular level is 0, as distinct from the Subsistence farming niche *"Production, Subsistence farmingFOG =-1"* where context is Production and relative granular level is lower in a hierarchy of contextual niches.

Sub-system = Production
FOG = Environmental Variables
FOG = +1

River
(*from Production,*
Environmental Variables)
FOG level = +1

Rainfall Event
(*from Production,*
Environmental Variables)
FOG level = +1

ClimateVariables
(*from Production,*
Environmental Variables)
FOG level = +1

Sub-system = Production
FOG = Commercial
FOG = 0

MagnetSmallHolderFarmer
(*from Production,*
Commercial)
FOG level = 0

HighValuePlant
(*from Production,*
Commercial)
FOG level = 0

ContourRidge
(*from Production,*
Commercial)
FOG level = 0

CommercialFarm
(*from Production,*
Commercial)
FOG level = 0

ImprovedFertilisedField
(*from Production,*
Commercial)
FOG level = 0

GreenHouse
TemperatureRegime
(*from Production,*
Commercial)

Sub-system = Production
FOG = Subsistence
FOG = -1

SmallholderFarmer
(*from Production,*
Subsistence,)
FOG level = -1

SoilUnit
(*from Production,*
Subsistence,)
FOG level = -1

Field
(*from Production,*
Subsistence,)
FOG level = -1

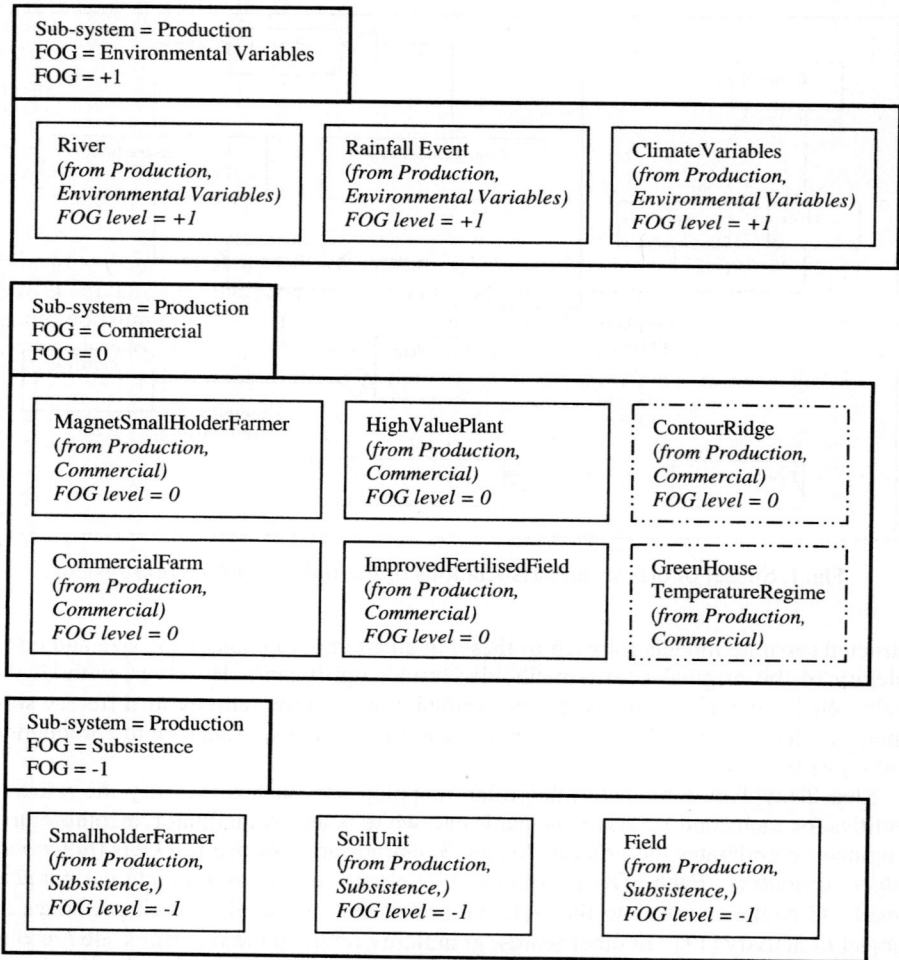

Fig. 2. Representing niches as functional contextual granules

Every subsystem is a separate context, and entities that appear in more than one will have different roles in each context/subsystem. A primary entity is chosen for a project (*MagnetSmallholderFarmer* in our example) and assigned a FOG level (circle of influence) of 0. In every subsystem/context, entities are then assigned FOG levels according to whether their granularity is at a higher or lower level of influence/impact. Within each subsystem (e.g. Production) functional object granules (FOGs) are defined, and then niches identified along with their granularity and context meta-properties.

As an example to illustrate the need to capture context in characterizing niches, consider typical disagreements on what a farm is. Is it a collection of fields, buildings, crops and people, or is it a land parcel with a cut-off minimum size? When is a property with agricultural activity a plot and when does is become a farm? In a Production context, the former is appropriate, while in a land-use context   the latter is suitable. Context and granularity (or FOGs) are useful tools for ensuring that a team converges on a single, agreed conceptual model or world view even when they come from different perspectives.

As another example: the term magnet smallholder farmer refers to a skilled and innovative individual engaged in commercial farming.   Class *magnet smallholder farmer* is placed in the same FOG as class *commercial farm,* and *high value crop.* What this means is that in our conceptualized view we expect the *magnet smallholder farmer* to be engaged in commercial agriculture, producing high value crops for commercial gain; we see this as his/her niche. A commercial farming niche has greater impact/influence/power than a smallholder farming niche, and must be represented at a higher FOG level.   The *smallholder farmer* (SF) cannot operate in the FOG *Production, Commercial Farming* and interact with other entities therein. This would be a violation of the interaction rules in the model, and represents an anomaly in the real world. In the real world, the SF does not enjoy the privileges of entities belonging to that niche.   Note that in comparison to figure 1 we are using more specific terms like *commercial farm* instead of simply "farm" to capture context and functional granularity (niche). We also note that each class maintains an indication of its niche, e.g. *(Commercial Production FOG = 0).*   That both SF and MSF "is-a" person is well represented in figure 1, using object oriented data modeling; however that these persons participate in two different ecological niches at two different resolutions is not represented.

If the IDEAA programme achieves its objectives then *smallholder farmer* will migrate to a higher FOG (higher socio-economic niche) and becomes a *magnet smallholder farmer.* In other words the person with role, *smallholder farmer*, would change state (say through skills transfer) and acquire the new role of *magnet smallholder farmer.* The extent of possible change of state of an object to a higher FOG is determined by its original state and context and the nature of its containment space and interactions (niche). A semi-skilled smallholder farmer in a stimulating and supportive environment is likely to progress faster than a non- skilled smallholder farmer in a non-supportive environment.   Thus we see that the modeling of niches helps in identifying environmental factors that affect progress/improvement, because of the hierarchical nature of niches.   This can be made explicit by considering intra- vs inter-niche relationships, as described below.

4.2.2 Modelling Relationships with Niches

The next step involves collaborators identifying relationships and addressing any inconsistencies that may arise in these relationships. Inconsistencies are used to detect missing entities and so present an opportunity to close gaps which may otherwise have gone unnoticed using ER and O-O modeling approaches. Consider figures 3 and 4.

In figure 3 the dashed lines indicate associations between classes at different granular leves e.g. *ClimateRegime* and *HighValuePlant.* This means the associations are irregular as they associate classes from unequal granular levels. To resolve this situation collaborators are prompted to "normalize" the relationships. This can be done by: (i) correcting any modeling error (e.g. entity in wrong FOG or incorrect relationship specified), or (ii) by replacing a participating entity by a more appropriate entity from the same aggregation hierarchy (introducing a new aggregation hierarchy if necessary, to explicitly distinguish the specific component/composite involved here) or (iii) by replacing a participating entity by a more appropriate entity from the same generalization hierarchy (introducing a new generalization hierarchy if necessary, to explicitly distinguish the specific type of entity involved in this relationship). Examples of how this may be achieved are shown in figure 4.

In fig. 4 for instance, a new class *GreenhouseRegime*, with a granular value of 0, is placed between *ClimateRegime* (FOG = 1) and *HighValuePlant* (FOG = 0), and the new generalization hierarchy *GreenhouseRegime* is-a *ClimateRegime* is introduced. What this means in reality is that we have used the normalization process to identify a potential intervention / innovation to facilitate interaction across niches – viz. to control the local temperature through the use of a greenhouse to reduce the risk to crop production. We also note in passing that the term *HighValueCrop* has been replaced by *HighValuePlant*. This occurred during the ontological engineering process when it was agreed through collaborative negotiation that the latter term was more ontologically correct because a *plant* only becomes a *crop* after harvesting. While other approaches like brainstorming workshops may arrive at a similar solution, OntoCrucible provides a methodological approach to creative innovation that greatly increases the possibility of such innovation.

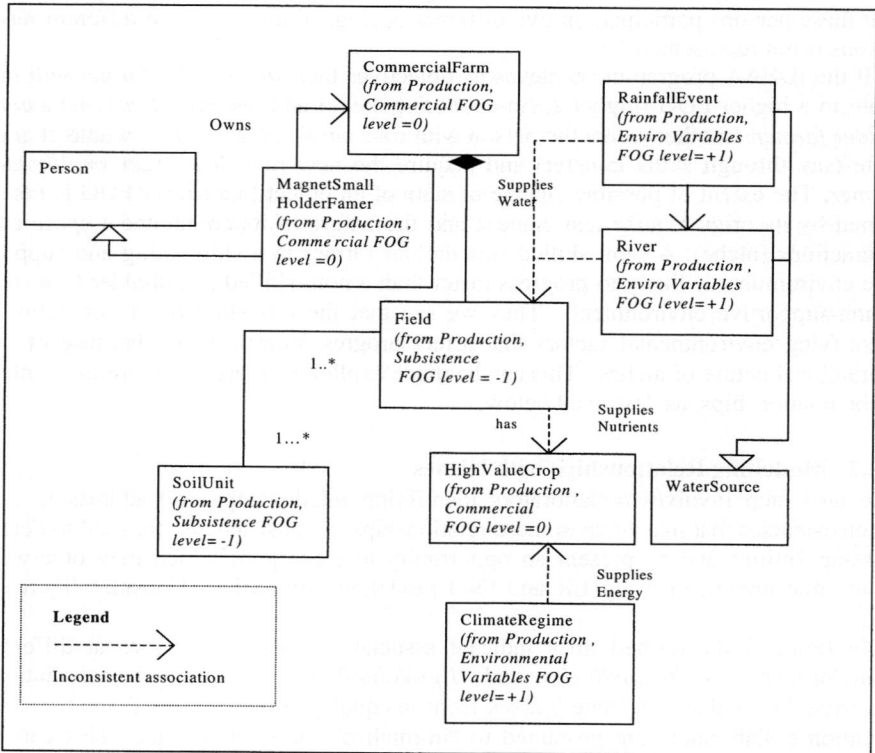

Fig. 3. Inconsistencies in cross FOG associations

4.2.3 Correctness Criteria

Having introduced new modeling primitives functional context and FOG, we now consider their relationship with other modeling primitives, viz. relationships, aggregation and generalization. Guarino and Welty [18] defined criteria for validating subsumption (generalization hierarchies). As an example, their first modeling constraint stipulates that anti-rigid classes cannot subsume rigid ones, where rigidity is an essentialness property - one that must be held by every member of the class. For example we cannot say person is-a student if person is rigid (every instance must be a person at all times) and student is anti-rigid (every instance of student may also be a nonstudent). In the same spirit, we define criteria for validating models/ontologies by means of the following axioms:

Axiom 1: *Classes from different FOG levels cannot interact directly in an association* (unless the association is "part-of" or "is-a") as each FOG represents a niche which favours particular members with particular influence/impact. A cross-FOG association needs to be normalized by replacing a participating entity, say X, by a more appropriate entity at the correct FOG level using either aggregation or generalization. That is, the relationship should instead involve an alternative entity Y drawn from the same aggregation hierarchy as X, or from the same generalization hierarchy as X, such that Y is at the correct granular level. If no such hierarchy or entity exists, then one should be explicitly introduced to model the situation where the anomaly arose.

An example of normalization using axiom 1 in figure 4, was the introduction of the generalization hierarchy *GreenhouseRegime* "is-a" *ClimateRegime*. Another example might be the introduction of an aggregation hierarchy such as Farmer "part-of" *FarmerOrganisation*, to model the relationship between farmers and higher level institutions such as suppliers or regional markets, which requires that a group of farmers get together in order to have sufficient influence/impact to interact with suppliers at their level.

Axiom 2: *In an aggregation hierarchy, if X is part-of Y, then Y cannot have a FOG level lower than that of X.* Since X is just one part of Y, the aggregate has at least the impact/influence of its X component (and possibly more).

As an example of axiom 2, if *MagnetSmallholderFarmer* has FOG level of 0 and is "part-of" a Consortium with FOG level −1, then either the aggregation or the FOG level is incorrect and must be remedied to reflect the real-world situation.

Axiom 3: *A class can have only one FOG level associated with it.* If X has been assigned two different FOG levels in two different contexts, this means that two different roles which X can play have been identified, and this needs to be modeled explicitly. To do so, a new (abstract) supertype Z must be introduced, along with X1 is-a Z (to represent the one role of X) and X2 is-a Z (to represent the other role of X). X1 and X2 can then have different FOG levels commensurate with the niches in which the object interacts in its different roles.

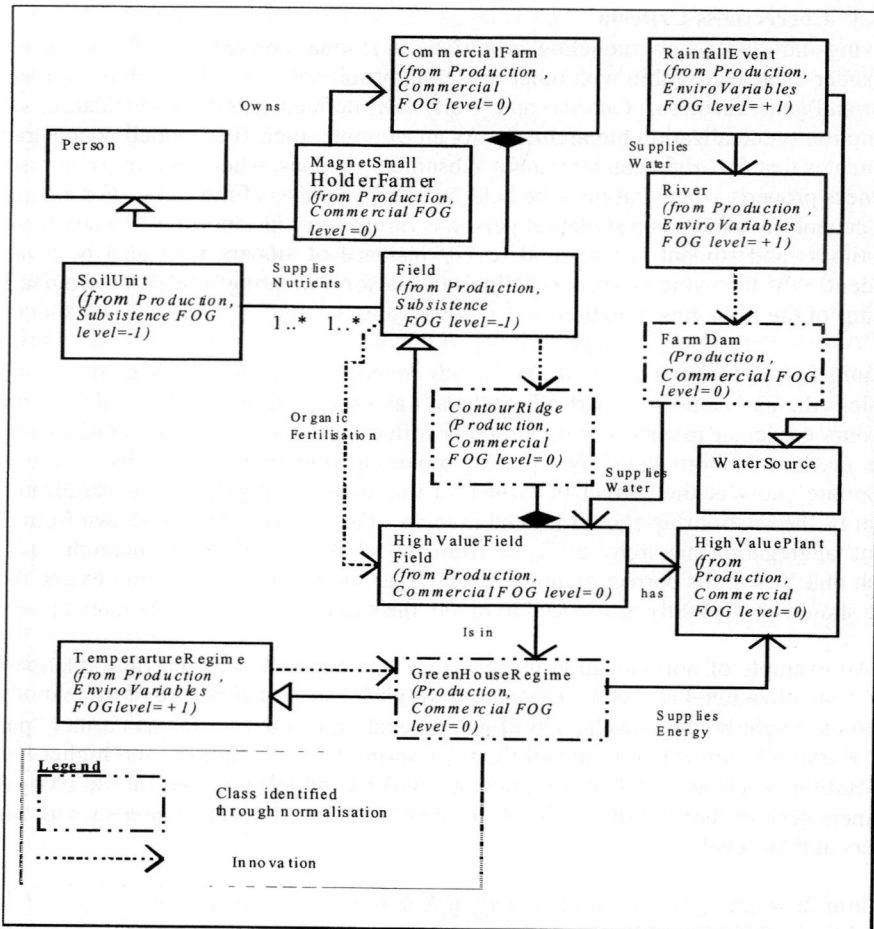

Fig. 4. Normalised FOG Associations

As an example of axiom 3, if *MagnetSmallholderFarmer* exists in the Commercial production niche with FOG level of 0 and in the Subsistence Production niche with FOG level of −1, then the role of that farmer in the latter context needs to be identified and made explicit - e.g. if the farmer is acting as mentor to subsistence farmers, then a new class FarmingMentor with FOG level of −1 is needed, and the taxonomy for Farmer extended to include MagnetSmallholderFarmer (FOG 0) is-a Farmer and FarmingMentor (FOG −1) is-a Farmer.

5   Discussion

Current conceptual models do not adequately represent niche meta-properties like granularity and context. E-R is good at representing static phenomena and O-O for representing dynamic phenomena [2]. We contribute to the domain of ontological modelling by proposing, in addition to existing primitives (entities, attributes, relationships, aggregation, generalisation, etc.), new primitives of context, functional granularity and niches. We argue in this paper that tackling and representing context and functional granularity to model niches in ecological systems avoids erroneous data integration, and also presents an opportunity for ecological innovation. Our proposed modeling environment, OntoCrucible [19], provides a collaborative ontological engineering tool that can be used to build consensus models and to "normalize" relationships between entities to correctly capture the interactions in real-world ecosystems. Workshops conducted [19] using other case studies indicate that a majority of GIS users find the model helpful and intuitive; future experimentation is scheduled to further test the validity and applicability of the modeling process and proposed primitives.

Ecological data models require additional primitives to adequately represent hierarchical niches that exist in the real world as a result of varying granularities of impact, time and space. We propose representing functional object granules, functional context and niches to capture additional semantics of entities and relationships. Context and granularity are meta-properties, they are not simply meta descriptions of entity attributes (*particulars or accidents*), but attributes of niches (*universals*) [1], thus they add to the range of modeling primitive available to design more cohesive ecological data models. Ontological engineering seeks to provide the nuts and bolts for building blocks of knowledge. In addition to the previously proposed criteria used to evaluate ecological ontologies (identity, rigidity, unity) [18], we propose criteria based on niches, to check for and correct inconsistencies in ecological ontologies.

We illustrated by means of examples how representation of niches offers several modeling benefits. It can help detect gaps and errors in a model, can facilitate the derivation of a shared conceptual view to avoid semantic heterogeneity and incorrect data collection/integration, and can identify areas where ecological innovation/intervention offers opportunities for meeting project goals.

Acknowledgements

This research programme was made possible through financial grants provided jointly by the South African National Research Fund and the Kellogg Foundation. The Kellogg Foundation dissertation grant is administered through the Academy for Educational Development Programme.

References

1. Smith B.: Objects and Their Environments; From Aristotle to Ecological Ontology., In Frank A, Raper J., Cheylan J-P (eds.): Life and Motion of Socio-Economic Units, Taylor Francis (2001) 79 – 97
2. Worboys M.F.: Modelling Changes and Events in Dynamic Spatial Systems With Reference to Socio-Economic Units. In Frank A, Raper J., Cheylan J-P (eds.): Life and Motion of Socio-Economic Units, Taylor Francis (2001) 129 – 136

3. Yuan M.: Representing Geographic Information to Support Queries About Life and Motion of Socio-Economic Units. In: Frank A, Raper J., Cheylan J-P (eds.): Life and Motion of Socio-Economic Units, Taylor Francis (2001) 217- 234
4. O'neill R.V., DeAngelis D.L. et al.: A hierarchical concept of ecosystems . Princeton, N.J: Princeton University Press, (c1986)
5. Vckovski, A.: Interoperable and distributed processing in GIS, Taylor & Francis, (1998)
6. Fonseca F., Egenhofer M, Davis C., Borges.: Ontologies and Knowledge Sharing in Urban GIS. In: Computer, Environment and Urban Systems 24 (3) (2000) 251-272
7. Cheng T., Molenaar M.: A process-oriented spatio-temporal data model to support physical environmental modeling. In: Proceedings of the 8th International symposium on spatial data handling (1997) 418-430
8. Gomez-Perez A.: Tutorial on Ontological Engineering: IJCAI'99
9. Maedche A., and Staab S.: Ontology Engineering Beyond the Modeling of Concepts and Relations. In: Proc. of the ECAI'2000 Workshop on Application of Ontologies and Problem-Solving Methods (2000)
10. Guarino N.: Formal Ontology, Conceptual Analysis and Knowledge Representation. In: Guarino and R. Poli (eds.): Special issue on Formal Ontology, Conceptual Analysis and Knowledge Representation, (2002)
11. Raper J.: Multidimensional geographic information science. Taylor Francis (2000)
12. Smith, B, Varci A.C.: The Formal Structure of Ecological Contexts, in Modeling and using Context. In: Bouquet P., Brezillon P., Serafini L. (eds.): Proceedings of the Second International and Interdisciplinary Conference, Berlin and Heiderberg:, Springer-Verlag, (1999) 339-350.
13. Schuurman N.: Reconciling Social Constructivism in GIS. ACME: An international E-journal for Critical Geographies.(2001)
14. Molenaar M.: An Introduction to the Theory of Spatial Object Modelling for GIS, London; Bristol, PA : Taylor & Francis (1998)
15. Raper J.: Defining Spatial Socio-Economic Units, Retrospective and Prospective In Frank A, Raper J., Cheylan J-P (eds.): Life and Motion of Socio-Economic Units, Taylor Francis (2001)
16. http://Poseidon.com/ Accessed 15 May 2004
17. http://whatistechtarget.com/ Accessed 15 May 2004
18. Guarino N., Welty C.: Evaluating Ontological Decisions with Ontoclean. In: Communications of the ACM, February vol 45. No.2 (2002) pp 61-65
19. Semwayo D.T.: OntoCrucible: A New Ontological Engineering Environment for Sensible and Meaningful Ecological Data Modelling, In: Proceedings of the SANTREN Conference on Challenges in Environmental Protection and Sustainable Use of Natural Resources. (2004) In print.
20. Whingiri E, et al,: IDEAA II Redesign Program Document (2001)

A CASE Tool for Geographic Database Design Supporting Analysis Patterns

Jugurta Lisboa Filho, Victor de Freitas Sodré, Jaudete Daltio,
Maurício Fidelis Rodrigues Júnior, and Valério Vilela

Department of Informatics, Federal University of Viçosa
36570-000, Viçosa, MG, Brazil
{jugurta,vsodre,jdaltio,mfrj,vvilela}@dpi.ufv.br

Abstract. This paper describes the development of an open source CASE tool, the ArgoCASEGEO, and its modular architecture. The ArgoCASEGEO allows the geographic database modelling based on the UML-GeoFrame conceptual model that is specific for Geographic Information Systems applications. The data dictionary associated to the modelled schema is stored as a XML/XMI document, aiming its use by other software. The ArgoCASEGEO follows a design methodology based on a reusable collection of analysis patterns. The analysis patterns collection is stored in a data base composing a Catalog. The catalog is attached to the ArgoCASEGEO. Thus, searching for existing analysis patterns will be an easier and efficient task.

Keywords: Geographic database design, Analysis Patterns, CASE tool.

1 Introduction

Geographic Databases (GeoDB) are collections of geographically referenced data, manipulated by Geographic Information System (GIS). In the Geoprocessing area, normally the user itself is the one who develops the GIS applications. Thus, redundancy and inconsistency are strong characteristics in the majority of the GeoDB, many times compromising the system reliability and, consequently, putting great public or private investments into risk. For that matter, the development of methodologies and tools that assist the GeoDB designers are essential to improve the quality of GIS applications.

A GeoDB should be designed following a database project methodology that includes conceptual, logical and physical design phases [5]. To elaborate the data schema in the conceptual phase, a data model must be chosen. Various models for GeoDB have been proposed in the past years as GeoOOA [12], MADS [19], OMT-G [2], UML+SpatialPVL [1] and UML-GeoFrame [14].

At this point, reuse mechanisms may help less experienced designers through instruments that allow software components reuse by patterns definitions. Analysis Patterns is a pattern category, which has been treated as a reuse instrument for requirement analysis and conceptual modelling [6], [7], [9], [11] and [20]. Analysis patterns permit reuse in a higher level than object-oriented class specialization because it makes possible to reuse a part of a data schema instead of a single class.

S. Wang et al. (Eds.): ER Workshops 2004, LNCS 3289, pp. 43–54, 2004.

Concluded the conceptual modelling, the next step - logical design - consists of the conceptual schema transformation into a data schema compatible with the data model of the GIS that will be used. This stage of a conceptual schema transformation into a logical-spatial schema, and its settlement in a GIS, can be made automatically by a CASE (Computer Aided Software Engineering) tool. Some of these conceptual models previously mentioned are supported by CASE tools, for example, Perceptory [1], REGIS [10], AIGLE [13] and Publisher Java MADS [19].

This paper describes the ArgoCASEGEO architecture, an open source CASE tool for GeoDB modelling that supports the UML-GeoFrame model [14]. The conceptual schema elaborated by this tool is stored in XML (eXtensible Markup Language) format, so can be easily accessed and used. This tool also provides an automatic generation module, able to generate data schemas to the most common formats usually found in commercial GIS. Moreover, the ArgoCASEGEO has a support for reuse based on analysis patterns [7] through the Analysis Patterns Module that implements a Manager and a Catalog. Further information about the using advantages of analysis patterns in GeoDB conceptual modeling can be seen in [15] and [16].

Section 2 presents the UML-GeoFrame Model whereas section 3 details the development of the ArgoCASEGEO tool, showing its architecture and describing each module. Finally, section 4 brings final considerations and future works.

2   The UML-GeoFrame Model

The conceptual modelling of GeoDB based on the UML-GeoFrame model [14] produces an easy understanding conceptual schema, improving the communication between designers and/or users. Besides being used in the database schema elaboration, the UML-GeoFrame model is appropriate to the analysis patterns specification.

The GeoFrame is a conceptual framework that supplies a basic class diagram to assist the designer on the first steps of the conceptual data modelling of a new GIS application. The mutual use of the UML class diagram and the GeoFrame allows the solution of the majority requirements of GIS applications modelling. A geographic conceptual schema built based on the UML-GeoFrame model includes, for example, the spatial aspects modelling of the geographic information and the difference between conventional objects and geographic objects/fields. The specification of these elements is made based on the stereotypes set shown in Figure 1.

The first stereotype set (Geographic Phenomenon and Conventional Object) is used to differ the two main object types belonging to a GeoDB. The Geographic Phenomenon class is specialized in Geographic Object (⬠) and Geographic Field (⬠) classes, according to two perception ways of the geographic phenomena, described by Goodchild [8]. Non-geographic Objects are modeled on traditional form and are identified through the stereotype (△).

The Geographic Object's Spatial Component and Geographic Field's Spatial Component stereotypes sets are used to model the phenomena spatial component according to object and field visions, respectively. The existence of multiple representations is modeled through the combination of two or more stereotypes on the same class. For example, a County class can have two abstraction ways of its spatial component, punctual and polygonal, that is specified by the stereotype pair (⊡⬓).

Fig. 1. Stereotypes of the UML-GeoFrame Model

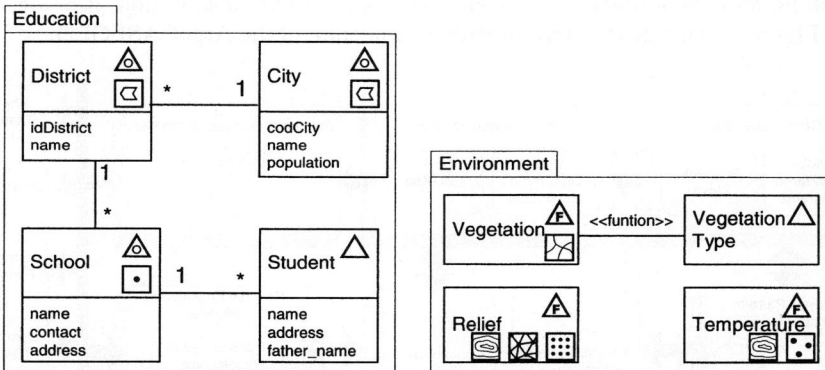

Fig. 2. An UML-GeoFrame schema example

Finally, the stereotype <<function>> is used to characterize a special type of association that occurs when modelling categorical fields. According to Chrisman [3], in a structure of categorical covering the spatial is classified in mutually exclusive categories, that is, a variable has a value of category type in all the points inside a region. Figure 2 exemplifies the UML-GeoFrame model use showing a class diagram containing two themes: Education and Environment.

The Education theme, modeled as a UML package, includes three geographic phenomena classes perceived in the object vision (District, City and School), and the Student class that is a non-geographic object. In the Environment theme, three classes of geographic phenomena perceived in the field vision are modeled, Vegetation, Relief and Temperature, each one with its different types of spatial representation. This theme still includes the Vegetation Type class, which is modeled as non-geographic object, being associated to the Vegetation class through the stereotype <<function>>, that is, each polygon is associated to a vegetation type.

3 The ArgoCASEGEO Tool

ArgoCASEGEO is a CASE tool whose goal is to give support to the GeoDB modelling based on the UML-GeoFrame model. The data schemas elaborated using this tool are stored in XMI (XML Metadata Interchange) format, a syntax for conceptual schema storage, in XML documents [18].

XMI combines the definition, validation and sharing document formats benefits of XML with the specification, distributed objects and business-oriented models documentation and construction benefits of the UML visual modelling language.

A CASE tool is primarily a graphical drawing software. To avoid a great programming effort in developing a new graphical drawing tool, some existing graphical softwares were selected to be used as starting point. This software must support the UML class diagram drawing and be extensible to support the stereotypes defined in the UML-GeoFrame model.

After analyzing some options, the ArgoUML publisher was chosen as the base tool. Thus, the ArgoCASEGEO was developed as an ArgoUML software extension, a modelling tool found over a use license and open source distribution, developed in Java. Figure 3 illustrates the five-module-architecture of the ArgoCASEGEO.

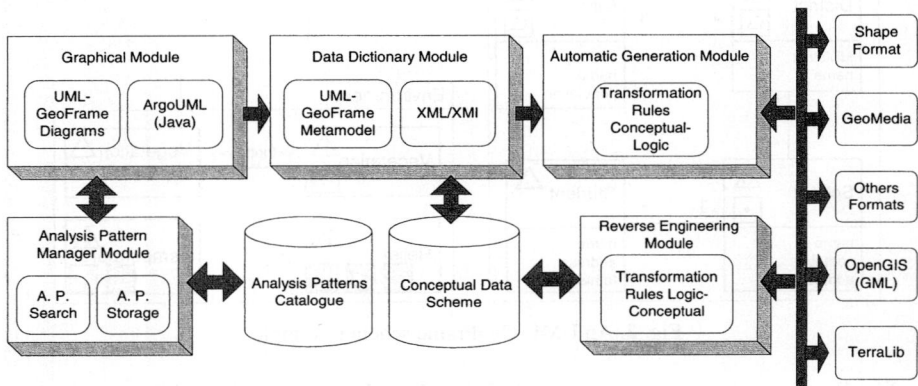

Fig. 3. The ArgoCASEGEO Tool Architecture

The Graphical Module allows the design of the conceptual schema, providing a set of constructors of the UML-GeoFrame model. The Data Dictionary Module stores the description of the diagram elements created by the designer. The Automatic Generation Module allows the transformation of the conceptual schema stored in the data dictionary into a logical schema corresponding to some models used in commercial GIS. The Analysis Patterns Catalog and its manager are defined in the Analysis Patterns Module. And finally, the Reverse Engineering Module, not yet implemented, will enable the designer to get conceptual schemas from existing GIS applications. The following sections describe these modules giving further details.

3.1 Graphical Module

The ArgoCASEGEO tool enables creation of diagrams that contain the constructors and stereotypes suggested by the UML-GeoFrame model. From this diagram the user can create its conceptual schema. An UML-GeoFrame conceptual schema supports three distinct class types: Geographic Object, Non-geographic Object and Geographic

Field. The existing fields in the implemented class have name, attributes, operations and symbols corresponding to the spatial representation type (stereotypes).

These classes can be related by relationships as generalization & specialization, aggregation, composition or association. In an association, the relationship name and the multiplicity of each class can be specified. The classes can be grouped to form a definitive theme, which is modeled by the UML's Package constructor. Figure 4 illustrates the ArgoCASEGEO environment.

An UML-GeoFrame data schema can be saved as a new Analysis Pattern, which can be (re) used in new data schema, composing thus, the Analysis Patterns Catalog. On the other hand, if the designer is starting a new project it would be interesting to look up in the catalog in order to find existing analysis patterns.

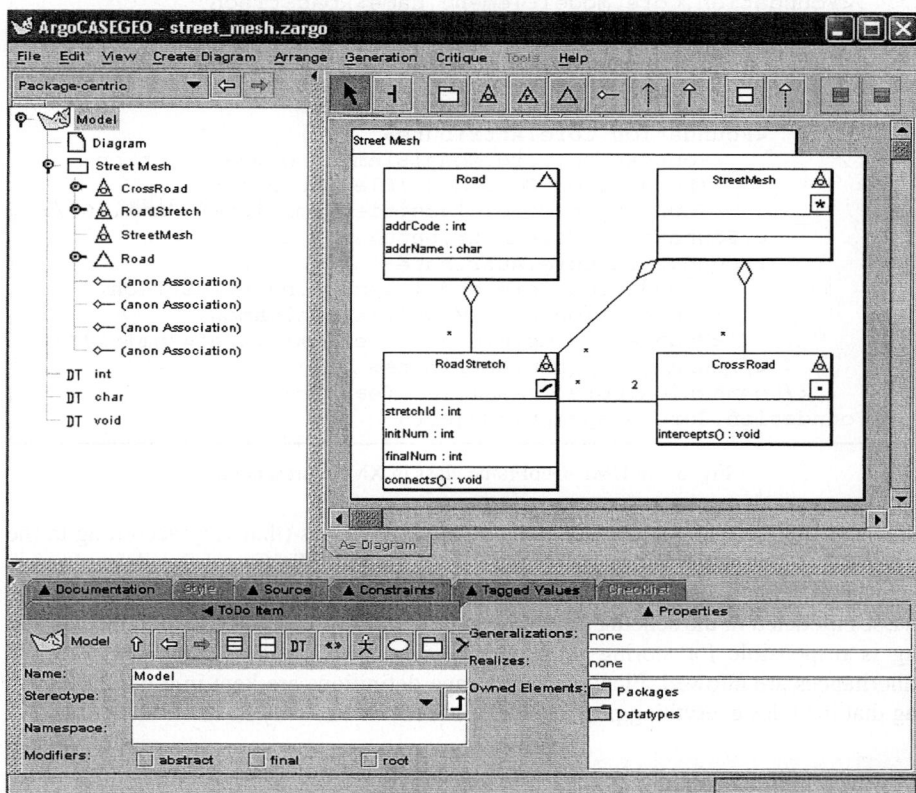

Fig. 4. The ArgoCASEGEO's graphical environment representing the analysis pattern Urban Street Mesh in the UML-GeoFrame model

3.2 Data Dictionary Module

The dictionary stores the data schema created by the user. A schema has two data types, the graphical data (drawing) and the semantic data (classes' names, attributes, associations' multiplicities, etc). The semantic data are stored in the data dictionary, while the graphical data are stored in an ArgoUML file. The data dictionary stores the conceptual schema in XMI format. Every class is delimited by a tag that contains the class name, its spatial representations and its features. The feature tag has two sub levels corresponding to the attributes and operations storage.

Figure 5 exemplifies the data dictionary to the Road Strech class (modeled in figure 4) whose spatial representation is line type, the direction and idStrech attributes. The types used in this definition, including the attribute type, parameters and operations' returned values are defined by the ArgoUML.

```
<Foundation.Core.GeographicObject>
    <Foundation.Core.ModelElement.name>RoadStrech
    </Foundation.Core.ModelElement.name>
        <Foundation.Core.GeneralizableElement.isLine
        xmi.value="true"/>
        <Foundation.Core.Classifier.feature>
            <Foundation.Core.Attribute>
                <Foundation.Core.ModelElement.name>direction
                </Foundation.Core.ModelElement.name>
                <Foundation.Core.Classifier xmi.idref="xmi.16"/>
            </Foundation.Core.Attribute>
            <Foundation.Core.Attribute>
                <Foundation.Core.ModelElement.name>idStrech
                </Foundation.Core.ModelElement.name>
                <Foundation.Core.Classifier xmi.idref="xmi.14"/>
            </Foundation.Core.Attribute>
        </Foundation.Core.Classifier.feature>
</Foundation.Core.GeographicObject>
```

Fig. 5. An UML-GeoFrame class in XMI representation

A specific tag that contains its name, related properties (that vary according to the type, association, aggregation or composition), its multiplicity and the classes' references that participate in the relationship, marks off the relationships between the classes modeled in the schema. From the generalizations definition vision, the internal tag is responsible for storing references to subclasses and super classes. Multiple inheritances are allowed. Finally, the package definitions are kept in a more external tag that includes everything previously described and has only its name as attribute.

3.3 Automatic Generation Module

After the conceptual modelling, the user needs to transform the elaborated schema into an effective implementation, characterizing a GIS application. As each GIS has its own data logical model, it is not possible to establish a single set of transformation rules to make the automatic generation of the logical-spatial schema. Thus, for each GIS the ArgoCASEGEO tool needs a specific Automatic Generation Module (AGM).

Two AGM have already been implemented in the ArgoCASEGEO tool. The first module transforms UML-GeoFrame schema to Shape format, used in the GIS Arc-

View [17]. A second AGM, described in the section below, transforms conceptual UML-GeoFrame schema into logical-spatial schema of the GIS GeoMedia.

3.3.1 GeoMedia Automatic Generation Module

The AGM-GeoMedia has as input the data dictionary identification that contains the conceptual schema to be transformed. To create a work environment in this software a connection with an existing database must be established. This connection will store the layers and the associated tables. For that matter, the AGM-GeoMedia creates a database Access (.mdb) to store all the elements generated by the automatic mapping.

For each element of the conceptual schema a specific transformation rule is applied. These rules are described as following.

Rule 1 – Packages: A package is formed by a set of interrelated classes. Therefore, a database is defined for each package in the conceptual schema that will store all the themes generated by the mapping related to it. The file name and its location are supplied by the user.

Rule 2 – Geographic Object Classes (⚑): Each class mapped as Geographic Object generates at least one layer inside the corresponding database, whose spatial representation is defined according to the representation's stereotype. The attributes defined in the class are mapped as fields of a relational table.

Rule 3 – Geographic Field Class (⚑): The GeoMedia is a vector software, therefore, it is not possible to carry through an automatic mapping of the phenomena that are modeled as field. However, according to Goodchild [8], the geographic fields' representations are simply aggregations of points, lines and polygons, connected to spatial characteristics. A field with spatial representation of Isolines type, for example, can be mapped in a layer of line type, having a value associated to each line.

According to this analysis, the program considers some mapping options to the designer. Beyond the suggested attributes, the attributes of each class are also added to the table. If these approaches are not useful, the designer can choose not applying them and the geographic fields are mapped similarly to non-geographic objects.

Rule 4 – Non-Geographic Object Classes (△): Each class modeled as Non-geographic Object generates directly one relational table. Objects are classified as Non-geographic exactly for having no spatial representation.

Rule 5 – Relationships: Relationships as association, aggregation and composition are made in accordance to the specified multiplicity. There are basically three types of multiplicities: one-to-one (1..1); one-to-many (1..*); and many-to-many (*..*).This mapping follows the same rules used to generate relational DBMS, which are well known [5]. Relationship of generalization-specialization type can also be translated using the same solutions applied in relational DBMS. Indeed, the spatial representation must be considered accordingly.

As an example, the AGM-GeoMedia will create the logical-spatial schema shown in Figure 6 taking the data dictionary shown in Figure 4 as input.

3.4 Analysis Patterns Manager Module

The idea of attaching an Analysis Patterns Catalog in a CASE tool is to help the database designer to find solutions that have already been used in similar GIS applications, which will improve the quality of the final database.

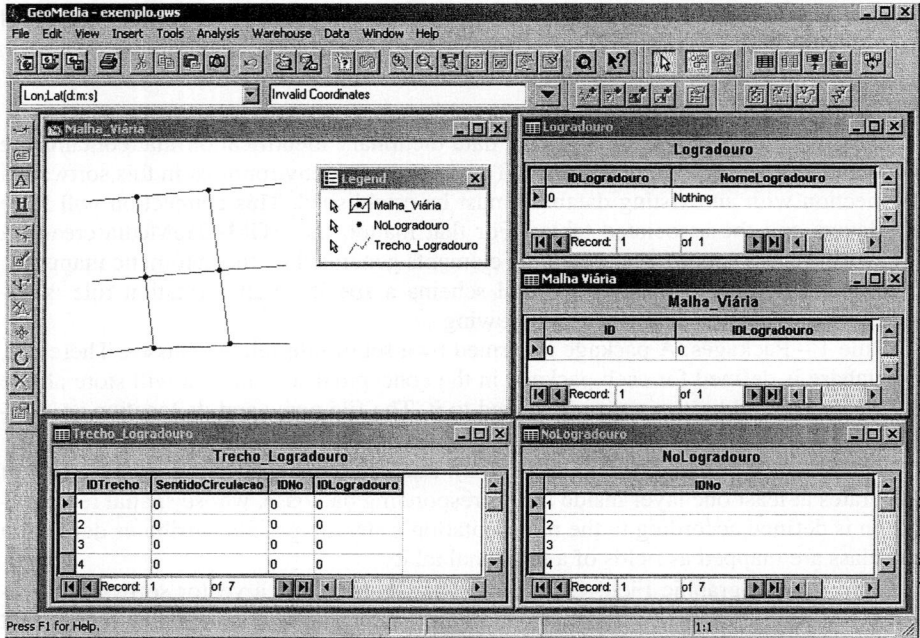

Fig. 6. Data schema generated automatically for the GeoMedia

An extensive collection of analysis patterns, that permits search in the existing patterns and their use in a new project that is under development, are kept organized by this module.

The Analysis Patterns Catalog and its Manager compose the Analysis Patterns Manager Module. The Catalog Manager deals with the Analysis Patterns Catalog and keeps its file system organized. The Catalog is a set of analysis patterns where each analysis pattern is stored without dependence on another one. In fact, they are grouped in a directory system, according to the pattern's theme. Therefore, different patterns proposing solutions for a specific class of problems are stored in distinct files in the same directory.

Besides the schema supplied as solution by the Analysis Patterns, its documentation is also stored so that the reasoning behind a solution can be searched and analyzed. The analysis patterns' documentation is stored in a XML file sharing the same name of the file that has the pattern modeled. Both files are kept in the same directory in order to make search an easier task.

In the ArgoCASEGEO tool the designer can add new patterns in the directory structure. The designer itself defines the patterns' themes hierarchy. Usually, there isn't an expressive number of analysis patterns available in an organization. Thus, designer groups can easily organize their patterns catalog in a simpler way. Analysis patterns can also be exchanged with partners from the same area. The tool also has import/export functions. Figure 7 shows an example of Analysis Patterns Directory.

```
♀ Analysis Patterns Catalogue
  ♀ 🗀 Environmental Theme
     ♦ 🗀 Hydrography
     ♦ 🗀 Vegetation
     ♦ 🗀 Temperature
     ♦ 🗀 Soil Type
     ♦ 🗀 Relief
  ♀ 🗀 Urban Theme
     ♦ 🗀 Urban Street Mesh
     ♦ 🗀 Urban Zoning
     ♦ 🗀 Street Traffic Network
     ♦ 🗀 Sewer System
     ♦ 🗀 Realty Cadastre
```

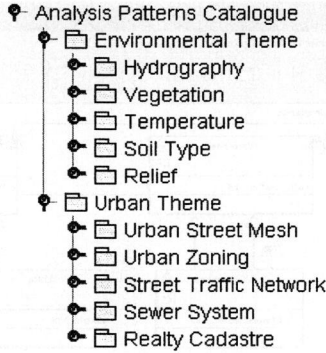

Fig. 7. An example of Analysis Patterns Directory

When the user needs to look for an analysis pattern, the manager builds a structure of packages, containing all the patterns recorded, in the Graphical Module. The Catalog Manager has a query mechanism which helps the designer to find analysis patterns by keywords that occur in the pattern's documentation.

An example of the Catalog's graphical environment can be seen in figure 8. On the left-hand side we can observe the directory hierarchy with the Hydrography pattern highlighted. All the pattern's components are listed below the theme's name. Opposite, the ArgoCASEGEO tool draws the schema related to the pattern. All the classes and relationships that form the schema can be identified and examined.

Putting the knowledge from the schema together with the information stored in the documentation we have enough means to understand and employ the pattern. As the documentation is stored in XML format, its recovering becomes easier and simpler to perform.

Figure 9 brings the Urban Street Mesh Pattern documentation source code. Every field found in the pattern template is kept in a special tag. As an example, the field Forces is represented by the pair of tags <forces> and </forces>. However, it might be interesting if we could store each force in a single pair of tags instead of keeping all the forces together. To solve this problem, a new pair of tag was created: <force> and </force>. Thus, to get the Forces data, it's only necessary to go through the <force></force> tags.

All the fields from the pattern template are put in a more external pair of tags: <documentation> </documentation>. This group of information constitutes the Pattern's documentation source code.

4   Conclusion

The CASE tool use during the GIS applications development makes creation time smaller, which, consequently, reduces cost. Moreover, the geographic databases quality increases.

The ArgoCASEGEO tool was implemented to assist the designer to develop its GIS applications with higher quality, following a design methodology based on a

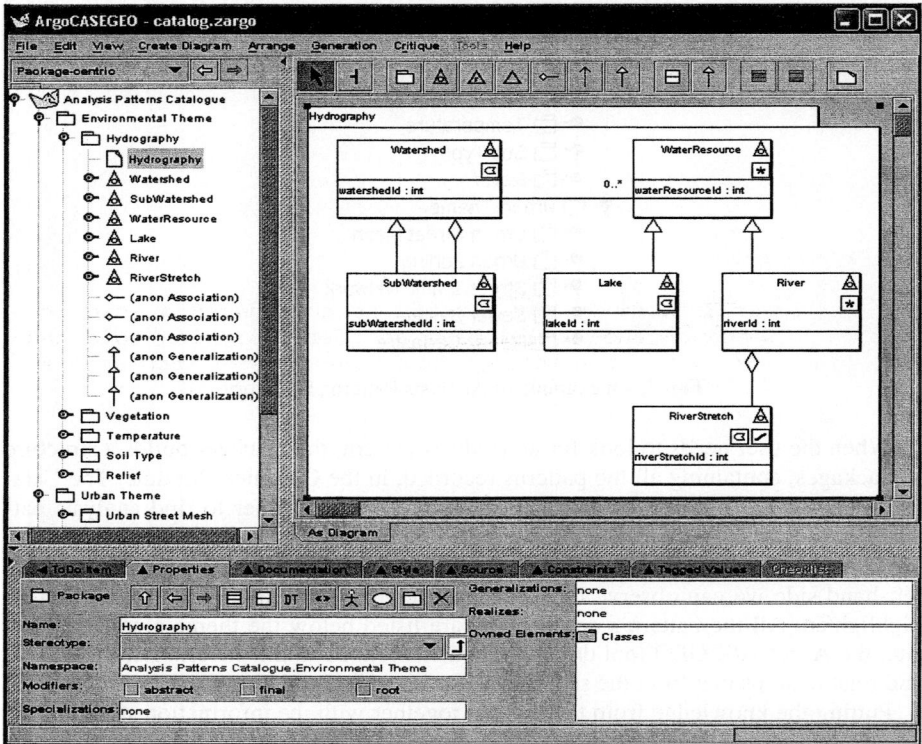

Fig. 8. The Analysis Patterns Catalog's graphical environment representing the analysis pattern Hydrography

conceptual model specific for geographic databases and on a reusable collection of analysis patterns. The documentation produced during the project (e.g.: conceptual schema and data dictionary) permits further references and visualization, which makes future system maintenance easier and the immediate generation of new versions of the application with the updates. The data dictionary storage in XML/XMI format allows the schema exchange and can be used by other applications, for example, analysis patterns discovery and search automatic tools. The Analysis Pattern Manager Module organizes all the patterns recorded into a directory architecture, which raises the efficiency while searching for a new analysis pattern to be used.

The Reverse Engineering Module development, the implementation of a new AGM for the OpenGIS feature model, and new analysis patterns mining and specification in different domains are futures works.

Acknowledgements

This work has been partially supported by CNPq (Brazilian National Research Council), the Brazilian governmental agency for scientific and technological development.

```xml
<?xml version="1.0" encoding="UTF-8"?>
<documentation>
  <problem>Which elements belong to a city's street mesh?
  </problem>
  <context>Every city in Brazil (and probably in the world)
          has shown the same organization pattern, which is
          structured by their pathways organization. The set
          of pathways stretches generates an urban street
          network.
  </context>
  <forces>
     <force>Each drive way stretch is considered a road in-
          stance and should have an identification code and
          a name. It normally should be divided into several
          segments as well.
     </force>
     <force>A road stretch is a pathway segment between two
          connections.
     </force>
     <force>The set formed by the connections (or terminal
          points) and road stretches create an urban street
          mesh.
     </force>
  </forces>
  <participants>The StreetMesh class is a geographic phe-
          nomenon represented by a complex spatial object
          (represented by the u symbol). In this class many
          attributes may be defined relating to the network
          as a whole. Road is a conventional class imple-
          mented normally as a table in a relational DBMS.
          Each road is made of several road stretches, which
          corresponds to a network arc. A road stretch may
          be connected to other stretches but this connec-
          tion is represented by the Crossroad class' in-
          stances, which are the network nodes. The network
          elements' manipulation operations may be imple-
          mented as classes' methods from StreetMesh, Road-
          Stretch and Crossroad depending on their function-
          ality.
  </participants>
  <related_patterns>The Urban Street Mesh uses the "State
          Across a Collection" pattern when modeling the
          Road and Road Stretch phenomena. Moreover, a new
          pattern project may be abstracted to create any
          network structure model made by nodes and arcs,
          whose topology's relationship among its elements
          is kept to make possible common network operations
          such as the shortest path calculation (a value for
          each is necessary), network navigation, distance
          between nodes, etc.
  </related_patterns>
</documentation>
```

Fig. 9. The Urban Street Mesh Pattern documentation source code

References

1. Bédard, Y.: Visual modelling of spatial databases towards spatial extensions and UML. Geomatica, v.53, n.2, (1999).
2. Borges, K. A. V.; Davis Jr, C. D. Laender, A.H.F.: OMT-G: an object-oriented data model for geographic applcations. GeoInformatica, v.5, n.3 (2001).
3. Chrisman, N.: Exploring Geographic Information Systems. John Wiley & Sons (1997).
4. Coad, P.: Object Models: Strategies, Patterns, and Applications. 2nd ed. New Jersey, Yourdon Press (1997).
5. Elmasri, R.; Navathe, S. B.: Fundamentals of Database Systems. Addison-Wesley (2000).
6. Fernandez, E. B.; Yuan, X.: An analysis pattern for reservation and use of reusable entities. Procs. of Workshop in the Conference of Pattern Language of Programs – Plop (1999).
7. Fowler, M.: Analysis Patterns: reusable object models. Menlo Park, Addison Wesley Longman (1997).
8. Goodchild, M. F.: Geographical data modelling, Computers & Geosciences, v.18, n.4 (1992).
9. Hay, D. C.: Data Model Patterns: conventions of thought. New York, Dorset House Publishing (1995).
10. Isoware: CASE-Toll REGIS. (2002). Available in http://www.isoware.de/.
11. Johannesson, P.; Wohed, P.: The deontic pattern – a framework for domain analysis in information systems design. Data & Knowledge Engineering, v.31 (1999).
12. Kösters, G. et al.: GIS-Application Development with GeoOOA. Int. Journ. GIS, v.11, n.4 (1997).
13. Lbath A., Pinet, F.: The Development and Customization of GIS-Based Applications and Web-Based GIS Applications with the CASE Tool AIGLE. In Proc. 8th ACM GIS (2000).
14. Lisboa Filho, J.; Iochpe, C.: Specifying analysis patterns for geographic databases on the basis of a conceptual framework. In Proc.7th ACM GIS, Kansas City (1999).
15. Lisboa Filho, J.; Iochpe, C.; Borges, K. A. V.: Analysis Patterns for GIS Data Schema Reuse on Urban Management Applications. CLEI Electronic Journal v.5, n.2 (2002).
16. Lisboa Filho, J.; Iochpe, C.; Beard, K.: Applying Analysis Patterns in the GIS Domain. In Proc. 10th Annual Colloquium of the SIRC, Dunedin, NZ (1998).
17. Lisboa Filho, J.; Pereira, M. A.: Desenvolvimento de uma ferramenta CASE para o Modelo UML-Geoframe com Suporte para Padrões de Análise. In the proceedings of the IV Simpósio Brasileiro de Geoinformática – GEOINFO`02, Caxambu (2002). (in Portuguese)
18. Object Management Group: Meta Objects Facility (MOF) Specification. (2000).
19. Parent, C. et al.: Spatio-temporal conceptual models: data structures + space + time. In Proc.7th ACM GIS, Kansas City (1999).
20. Wohed, P.: Tool support for reuse of analysis patterns – a case study. In: A. H. F. Laender, S. W. Liddle, V. C. Storey (eds): ER2000 Conference, LNCS 1920, 2000. Springer-Verlag Berlin Heidelberg (2000).

GML Storage: A Spatial Database Approach*

Yuzhen Li[1], Jun Li[1], and Shuigeng Zhou[2]

[1] School of Computer, Wuhan University, Wuhan, 430079, China
`yuzhenli@163.com`
[2] Dept. Computer Sci. and Eng., Fudan University, Shanghai, 200433, China
`sgzhou@fudan.edu.cn`

Abstract. GML was developed to standardize the representation of geographical data in XML, which makes the exchanging and sharing of geographical information easier. With the popularity of GML, more and more geographical
data is presented in GML. This arises the problem of how to efficiently store
GML data to facilitate its management and retrieval. This paper proposes an
approach to store GML documents into spatial databases (e.g. Oracle Spatial,
DB2 Spatial and PostGIS/PostgreSQL etc.). GML schema tree is first generated
based on the given GML schema, the generated schema tree is then mapping
into corresponding relational schema. All basic spatial objects are stored as values of the mapped tables' fields. Experiments are carried out to examine the
storage efficiency of the proposed approach.

1 Introduction

The Geography Markup Language (GML) is an XML based OGC specification for
the representation and exchanging of geographic information, including the geometry
as well as the properties of geographic features [1]. GML has been making a significant influential on the ability of organizations to share geographic information with
one another. By using GML, the users can deliver geographic information as distinct
features, and then control how they are displayed. Best of all, the users can view the
resulting maps using a standard browser - they don't need to purchase proprietary
client-side GIS software. Considering that GML is based on XML and other complementary protocols such as XML Schema, XLink, and Xpointer etc., it enables one to
leverage the whole world of XML technologies. GML data can be also easily mixed
with non-spatial data. By providing an open, standard, meta-language for geospatial
constructs, information can be shared across various domains like geology forestry,
tourism, archaeology, and telecommunications.

With the popularity of GML in industry, more and more geographical data will be
produced in GML format, which arises the problem of how to efficiently store GML
data to facilitate its management and retrieval. As the amount of GML data is rapidly
growing, it is natural to consider storing GML data into the databases, so that we can

---

* This work was supported by Hi-Tech Research and Development Program of China under
grant No.2002AA135340, and Open Research Fund Program of SKLSE and LIESMARS under grant No. SKL(4)003 and No. WKL(01)0303.

S. Wang et al. (Eds.): ER Workshops 2004, LNCS 3289, pp. 55–66, 2004.

make use of sophisticated functions available in traditional DMBS, such as concurrency control, crash recovery, scalability, and highly optimized query processors. Though GML is based on XML and XML storage has been extensively studied [2-9], XML storage techniques can't be straightforwardly transplanted to GML. First, traditional XML documents are mostly text-rich data, while GML exists mainly as data-rich documents. Second, GML documents contain topological and temporal information, which is unavailable in traditional XML documents. Last but not least, GML querying involves complex spatial operations and analysis. Therefore, new GML storage technology is necessary for efficiently supporting GML documents management and querying processing.

In this paper we propose an approach to store non-spatial and spatial data from GML documents into spatial databases. In our approach, GML schema tree is first generated based on the given GML schema, the generated schema tree is then mapping into corresponding relational schema. Finally, all basic spatial objects are stored as values of the mapped tables' fields. The major contributions of this paper include: an algorithm to generate GML schema tree from input GML document schema, and an algorithm to map GML schema tree to relational tables of spatial database. Expanding XQuery to query GML data with spatial properties is discussed and experiments are conducted to demonstrate the storage efficiency of the proposed approach.

The reminder of this paper is organized as follows. We first give an overview of related work in Section 2, and then introduce our approach to store GML data into spatial databases in Section 3. Particularly, system architecture and related algorithms as well as experimental results are presented. Finally, we conclude the paper and highlight future work in Section 4.

2 Related Work

Below we review related work from two aspects: XML storage and GML storage.

2.1 XML Storage Methods

The mainstream technology of XML storage is storing XML in relational databases. Currently, there are mainly two classes of approaches to mapping XML into RDBMS: model-based approaches and structure-based approaches [2]. The former uses a fixed database schema to store the structures of all XML documents. For example, Florescu and Kossmann[3] proposed six simple mapping strategies to map the relationship between nodes and edges of DOM into the relational schema based on the DOM model of XML document. The latter derives relational schemas from the structures of XML documents. For example, Deutsch et al.[4] presented a declarative language, STORED, which allows user to define mappings from the structure of semi-structured data to relational schemas; and Shanmugasundaram et al.[5] introduced two ways to map DTDs into relational schemas, namely SHARED and HYBRID respectively. HYBRID inlines all of the descents not attached by * in DTD graph into their parents, while SHARED maps the nodes with multiple parents (i.e., in-degree greater than 1) into independent relations. The different manners to deal with these nodes with in-degree greater than 1 consist of the only difference between SHARED and HYBRID. Comparing with the model based approaches, the structure based approaches, like the inlining techniques presented in [5], produce fewer document fragments, hence require less joins for path traversal, and subsequently outperform the model based approaches. Tian *et al.*[6] compared the performance of alternative mapping approaches, including relational and non-relational ones, and showed that structural mapping approach is the best one when DTD is available.

Recently, some articles addressed the problem of designing efficient XML storage schema by using user's query information. Bohannon *et al.* [7] presented a framework to design XML storage for a given application (defined by XML Schema, XML data statistics, and an XML query workload); Zheng *et al.* [8] studied storage schema selection based on the cost of given user queries. The major difference between their papers is the applied algorithms of schema selection. Xu *et al.* [9] developed a system that automatically adjusts storage schema according to the evolution of users' queries.

2.2 GML Storage Approaches

Although GML is gaining more and more attention of research community [], GML storage has not been properly studied. To the best of our knowledge, up to date only one research paper by J.E. Corcoles and P. Gonzalez [10] addresses this problem. In [10], the authors applied three XML storage approaches to GML storage and compared their performance: LegoDB, a structure-based approach, and two simple model-based approaches, i.e., Monet over Relational database and XParent. Their experiments show that the approaches obtaining good results with alphanumeric operators do not obtain good results when involving spatial operators. They also proposed a spatial query language over GML using extended SQL syntax [11]. However, such an approach requires implementation from scratch and does not conform to the current XML query languages.

Considering 1) GML documents contain geometry elements corresponding to spatial objects of various types, including Point, LineString, LinearRing, Polygon, MultiPoint, MultiLineString, MultiPolygon and GeometryCollection; 2) Spatial queries involve complex spatial operations and analysis, we argue that both native XML databases and XML-enabled databases are not suitable for GML storage, becasue these databases lack functionality to deal with complex spatial data types, and spatial operations. We have conducted some experiments of storing GML to native XML database (IPEDO), and found the storage efficiency is unacceptably low.

Different from the previous approaches, in this paper we store GML data into spatial databases (e.g. Oracle Spatial, DB2 Spatial and PostGIS/PostgreSQL etc.) so that we can re-use the functionality of spatial databases to handle spatial objects and carry out spatial operations and analysis. Furthermore, we introduce new method to mapping GML documents into spatial databases, which consists of two stages: the first stage is for generating schema tree of GML documents, and the second one is for mapping the generated schema tree to relational schema in spatial databases.

3 Storing GML in Spatial Database

3.1 System Architecture

We first present the system architecture for storing and querying GML based on spatial databases, and then introduce detailed techniques in later sections. Fig. 1 is the architecture, which consists of four main components: *Schema Tree Generator* (STG), *Mapping Rules Generator* (MRG), *Data Loader* (DL) and *Query Translator* (QT). STG's function is to generate GML *schema tree* by using the *schema tree generation algorithm* (see section 3.2), whose input is GML application schema. Fig.2 is an example of GML application schema. MRG is used to generate *mapping specification* by the GML *structure-mapping algorithm* (see section 3.3). DL takes as input a GML document, store it to the target SDBMS according to the *mapping specification* generated by MRG. QT translates a GML query into a corresponding SQL query that is ready to be submitted to the underlying SDBMS. The results returned are constructed and delivered to the user via QT.

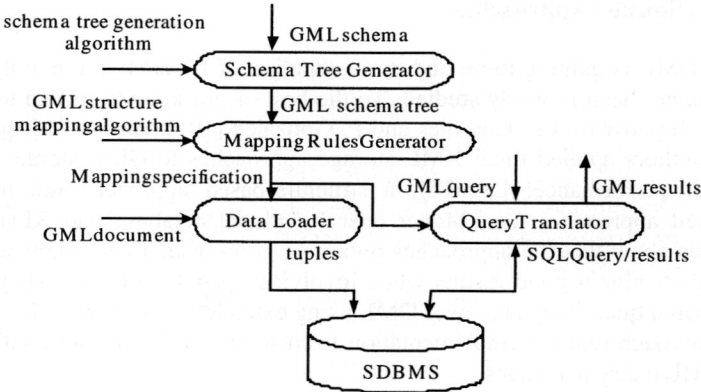

Fig. 1. The architecture of GML storage

In order to support spatial data types and spatial operations, we can convert GML data into Spatial DBMS. A spatial database is a collection of spatial data types, operators, indices, processing strategies, etc. A SDBMS is a software module that can work with an underlying DBMS, supports spatial data models, spatial ADTs, spatial query language from which these ADTs are callable, spatial indexing, algorithms for processing spatial operations, and domain specific rules for query optimization. Currently, there are some spatial databases, such as proprietary software ArcSDE, Oracle Spatial and DB2 Spatial, open source software PostGIS / PostgreSQL. In our implementation, we can choose Oracle Spatial. We store non-spatial geographic information in relational tables, but the spatial information in GML is mapped to Geometry column with geometry type. Oracle Spatial database has a special column called MDSYS.SDO_GEOMETRY, which is used to store geometry like point, line and polygon.

```
<?xml version="1.0" encoding="UTF-8" ?>
<!-- edited with XML Spy v4.4 U (http://www.xmlspy.com) by gml (207)    -->
- <schema xmlns="http://www.w3.org/2001/XMLSchema" xmlns:gml="http://www.opengis.net/gml"
    xmlns:xlink="http://www.w3.org/1999/xlink" xmlns:ex="http://www.opengis.net/examples"
    targetNamespace="http://www.opengis.net/examples" elementFormDefault="qualified" version="2.01">
    <import namespace="http://www.opengis.net/gml" schemaLocation="feature.xsd" />
    <element name="Map" type="ex:mapType" substitutionGroup="gml:_FeatureCollection" />
    <element name="Layer" type="ex:layerType" />
    <element name="Feature" type="ex:featureType" substitutionGroup="gml:_Feature" />
    <element name="SimpleProperty" type="ex:SimplePropertyType" />
    <element name="GeometryProperty" type="ex:GeometryPropertyType" />
    <element name="FloatProperty" type="ex:FloatPropertyType" />
    <element name="StringProperty" type="ex:StringPropertyType" />
  - <complexType name="mapType">
    - <complexContent>
      - <extension base="gml:AbstractFeatureCollectionType">
        - <sequence>
            <element ref="ex:Layer" />
          </sequence>
        </extension>
      </complexContent>
    </complexType>
  + <complexType name="layerType">
  + <complexType name="featureType">
  + <complexType name="GeometryPropertyType">
  - <complexType name="SimplePropertyType">
    - <sequence>
        <element ref="ex:FloatProperty" />
        <element ref="ex:StringProperty" />
```

Fig. 2. A sample of partial GML application schema

3.2 Schema Tree Generation

We can view GML schema as a tree. A *GML schema tree* is like this: the schema tag in the GML schema is the root of the tree. Other elements and its sub-elements or attributes are the node of the tree. The nodes containing other elements are the father nodes, and the nodes contained by any father node are child nodes.

Before generating the GML schema tree, the input GML schema is simplified. We propose four types of schema simplifying rules. For the convenience of rule description, we use the annotation in Table 1. Furthermore, we use "|" to represent *choice* tag of GML schema, and "," to represent *sequence* tag.

Table 1. Annotation used for Schema Simplification

Symbol	Meaning	XML schema example
A	'A' appears one time	<element type=A/>
A?	'A' appears one or zero time	<element type=A/ minOccurs=0/>
A*	'A' appears zero or more time(s)	<element type=A/ minOccurs=0 maxOccurs=unbounded/>
A+	'A' appears one or more time(s)	<element type=A/ minOccurs=0 maxOccurs=unbounded/>

Following are the four types of schema simplifying rules:
1) Decomposition rules: that is to convert nested form into flat form, i.e., making none operator contains binary operators "," "," and "|":
        (A1, A2)* -> A1*, A2*
        (A1, A2)? -> A1?, A2?
        (A1 | A2) -> A1?, A2?

2) Combination rules: that is to combine multiple unary operators to one single unary operator:

$$A** -> A*$$
$$A*? -> A*$$
$$A?* -> A*$$
$$A??-> A?$$

3) Grouping rules: that is to group sub-elements of similar name:

..., A*, ..., A*, ... -> A*, ...
..., A*, ..., A?, ... -> A*, ...
..., A?, ..., A*, ... -> A*, ...
..., A?, ..., A?, ... -> A*, ...
..., A, ..., A, ... -> A*, ...

4) All "+" operators are transformed to "*" operators.

A schema simplification example is shown in Fig. 3 and Fig. 4.

```
<complexType name="featureType">
<complexContent>
<extension base="gml:AbstractFeatureType">
<sequence
minOccurs="0" maxOccurs="unbounded">
<choice>
<element ref="gml:Point"/>
<element ref="gml:MultiPoint"/>
</choice>
</sequence>
<element name="POPULATION" type="float"/>
</extension>
</complexContent>
</complexType>
```

```
<complexType name="featureType">
<complexContent>
<extension base="gml:AbstractFeatureType">
<sequence>
<element ref="gml:Point" minOccurs="0"
maxOccurs="unbounded"/>
<element ref="gml:MultiPoint" minOccurs="0"
maxOccurs="unbounded"/>
</sequence>
<element name="POPULATION" type="float"/>
</extension>
<complexContent>
</complexType>
```

Fig. 3. A GML Schema sample before simplifying

Fig. 4. The GML Schema sample after simplifying

After simplifying the GML application schema, we start generating GML schema tree. From the root node of the GML schema, we traverse through every sub-node in pre-order to create the tree. In our algorithm, we determine the types of nodes by their attributes. If a node belongs to basic types, its sibling nodes will be searched, otherwise its child nodes will be searched. Here, the basic types refer to "string", "boolean", "float" and these listed in Table 2. The process of schema tree generation is shown in Fig. 5.

As an example, we use this method to generate schema tree of the GML application schema given in Fig. 2, the result schema tree is showed in Fig. 6. Here, the root node is map, the "*" indicates 1 to n relation between father node and child node.

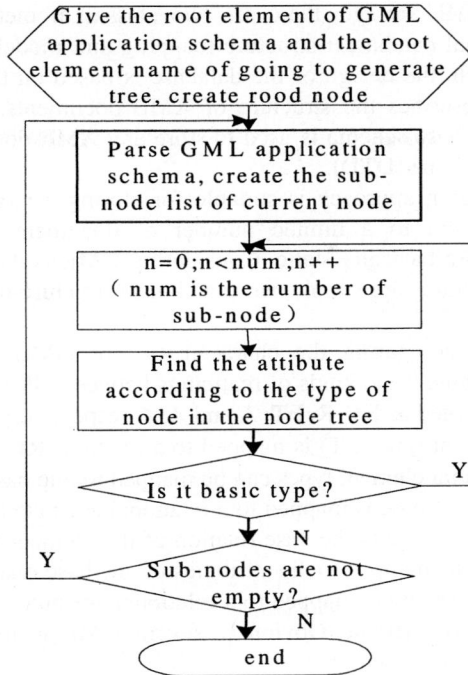

Fig. 5. The process of GML schema tree generation

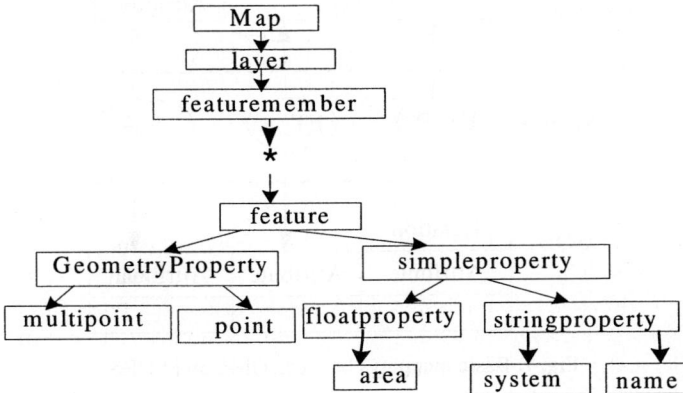

Fig. 6. Example of GML schema tree

3.3 Structure Mapping: From Schema Tree to Relational Table

In order to store a GML document into a spatial database, it is necessary to define a mapping mechanism between GML schema and database schema. That is to say, it is necessary to identify the relationship between GML elements/attributes and database tables/columns. In XML storage, there are two classes of methods to map XML documents to relational databases: structure-mapping and model-mapping. In structure-mapping approach, the design of the database is based on the understanding of XML schema that describes the structure of XML documents. In model-mapping approach, a fixed database schema is used to store any XML documents without the assistance of XML Schema (DTD).

The structure-mapping approach is suitable for storing a large number of XML documents that conform to a limited number of document structures, and the document structures are basically static. Considering GML is data-centric document and its document structures are simple, we can choose structure-mapping approach to store GML.

By analyzing and comparing the characteristics of GML schema and SDBS schema, we propose three basic kinds of mappings between GML schema and SDBS schema, which are denoted as ET_R, ET_A and A_A respectively:

(1) ET_R: An element type (ET) is mapped to a relation (R), which is called base relation. Note that several element types can be mapped to one base relation.

(2) ET_A: An element type is mapped to a relational attribute (A), whereby the relation of the attribute represents the base relation of the element type. Note that several element types can be mapped to the attributes of one base relation.

(3) A_A. A GML attribute is mapped to a relational attribute whose relation is the base relation of the GML attribute. Obviously, several GML attributes can be mapped to the attributes of one base relation.

The three kinds of mapping are illustrated in Fig. 7.

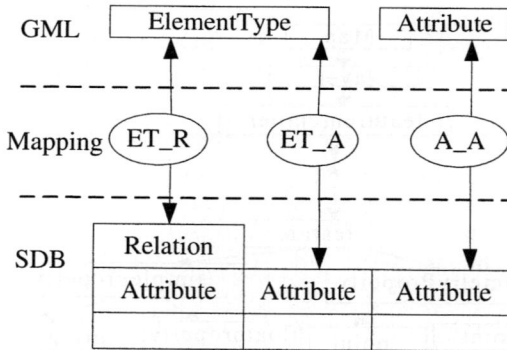

Fig. 7. Basic mappings between GML and SDBS

In what follows, we present the GMLSMA (GML structure-mapping algorithm) algorithm to map GML schema tree to relational schema based on the three kinds of mapping abovementioned. The algorithm's input is GML schema tree generated in section 3.2. GMLSMA is outlined as follows.

(1) For a node of zero in-degree (root node, isolated node), creating a table separately, the ET-R mapping is adopted.

(2) For a node directly below a "*", which means *n*:1 relationship exist between the node and its parent node, create a table separately.

(3) All remaining nodes are inlined as attributes of the relation corresponding to their closest ancestor nodes.

(4) Nodes of geometry types, including PointType, LineStringType, PolygonType, BoxType, MultiPointType, MultiLineStringType, MultiPolygonType, LinearRingType, are stored in a single column of their ancestor node's table.

The relationship between geometry types of GML documents and spatial database (take Oracle Spatial for example) is shown in Table 2. In the GML schema tree of Fig. 6, the types of nodes *multipoint* and *point* are multiPointType and PointType respectively, so they will be mapped into the spatial database (Oracle Spatial) as geometry objects of POINT and MULTIPOINT. Oracle is a well-known object-relational database. Although Oracle spatial can store geographical data in its relational tables, it allows the user to store and retrieve geographical objects such as point, polyline and polygon. It provides spatial access methods such as R-tree, and thus is able to perform spatial query and analysis. The users can issue any SQL query to Oracle and get results back in GML format.

Table 2. The geometry in GML schema and the geometry in Oracle Spatial

Geometric Type in GML	Geometric Type in Oracle Spatial (SDO_GTYPE)
PointType	POINT
LineStringType	LINE/CURVE
PolygonType/BoxType	POLYGON
MultiPointType	MULTIPOINT
MultiLineStringType	MULTILINE/MULTICURVE
MultiPolygonType/LinearRingType	MULTIPOLYON

Applying the GMLSMA algorithm above over the schema tree in Fig. 6, we get the corresponding spatial database schema shown in Table 3. Here, the *Map* node is the root node, so a separate table is created for it. The *Feature* node below "*", we also create a separate table for it. All the remaining nodes are inlined as attributes under the relation, which is created for their closest ancestor nodes. Table *feature* refers to table *Map* by the foreign key *feature.parentid*.

To examine the storage efficiency of our method, we conducted a set of preliminary experiments. The goal is to measure the time cost for storing GML documents of different size to spatial database (in this paper we use Oracle Spatial). The size of test GML documents varies from 2Mb to 45Mb. The test results are shown in Fig. 8. It can be seen that as the size of GML document grows, the time cost also increase linearly.

3.4 GML Query

The recent publication of XQuery standard by W3C offers a powerful standard query language for XML, however it lacks the support for spatial queries. There are some

Table 3. The structure of table in the spatial database

MAP table

map.layer.featuremember.ID	integer

feature table

ID	integer
feature.parentid	integer
feature.GeometryProperty.multipoint	MULTIPOINT
feature.GeometryProperty.point	POINT
feature.simpleproperty.floatproperty.area	float
feature.simpleproperty.stringproperty.system	string
feature.simpleproperty.stringproperty.name	string

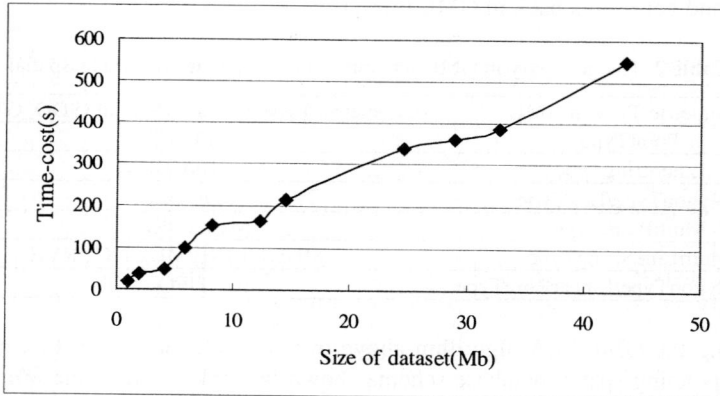

Fig. 8. The cost time of different size of GML documents

studies in extended XQuery language to support spatial queries over GML documents. The non-spatial queries can be directly answered with XQuery. However, spatial queries may involve a rich set of topological predicates and spatial analysis functions. Thus we have extended the XQuery to support spatial query operations.

In our system architecture of Fig. 1, GML queries submitted by users are first translated into spatial SQL queries and evaluated by the spatial database management system. The spatial database management system supports the query optimization methods and concurrency control for processing multiple queries simultaneously. Fig. 9 and Fig. 10 show an example of GML query and its equivalent SQL query.

```
FOR $c IN document("Country.xml")
//country,
RETURN
  <Country>
    <name> $c/name/text() </name>
    <area> $c/area </area>
    <distance> distance($c/point1,$c/point2) </distance>
  </Country>
```

Fig. 9. GML query with spatial operation

```
SELECT  C.name, C.area, distance(C.point1,C.point2) AS "distance"
FROM     County C
```

Fig. 10. The equivalent SQL query translated by Query Translator

4 Conclusion

In this paper we proposed a method to store GML documents into spatial database. With the proposed approach, GML schema tree is first generated based on the given GML schema, the generated schema tree is then mapping into corresponding relational schema. Finally, all basic spatial objects are stored as values of the mapped tables' fields. Spatial query can be submitted in XQuery-alike language with spatial functional extensions, and GML query is first translated into equivalent SQL query that is evaluated by spatial database management system.

In this paper, we carried out preliminary experiments to test the proposed method's storage efficiency. Considering that query-processing efficiency is more concerned by the users, in the future we will conduct extensive experiments to measure the query-processing efficiency based on the storage method. How to efficiently and accurately translate GML query to SQL query is another focus of our future work.

References

1. OpenGIS Consortium. GML Specifications. http://www.opengis.org/.
2. T. Shimura, M. Yoshikawa, and S. Uemura. Storage and Retrieval of XML Documents Using Object-Relational Databases. In International Conference on Database and Expert Systems Applications, pages 206–217, Florence, Italy, September 1999.
3. D. Florescu, D. Kossmann. Storing and Querying XML Data Using an RDBMS. Data Engineering Bulletin, Vol. 22, No. 3, 1999.
4. A. Deutsch, M. F. Fernandez, and D. Suciu. Storing Semi-structured Data with STORED. In Proceedings of the ACM SIGMOD International Conference on Management of Data, pages 431–442. ACM Press, 1999.
5. J. Shanmugasundaram, K. Tufte, G. He, C. Zhang, D.DeWitt, and J.Naughton. Relational Databases for Querying XML Documents: Limitations and Opportunities. In Proc. of the Int'l. Conf. On Very Large Data Bases, pages 302-314. 1999.

6. F. Tian, D. J. DeWitt, J. Chen, and C. Zhang, The Design and Performance Evaluation of Alternative XML Storage Strategies. SIGMOD Record special issue on "Data Management Issues in E-commerce", Vol.31, No. 1, March 2002.

7. P. Bohannon, J. Freire, P. Roy and J. Simeon. From XML Schema to Relations: A Cost-Based Approach to XML Storage. 18th International Conference on Data Engineering (ICDE2002), page 64, 2002.

8. S. Zheng, J. Wen and H. Lu, Cost-Driven Storage Schema Selection for XML, In: Proc. of 8th International Conference on Database Systems for Advanced Applications (DASFAA 2003), Kyoto, Japan, March, 2003. 26 – 28.

9. Z. Xu, Z. Guo, S. Zhou, and A. Zhou. Dynamic tuning of XML schema in VXMLR. In: *Proceedings of the 7th International Database Engineering and Application Symposium (IDEAS2003)*, pp. 76-86, IEEE CS, July 16-18, 2003, Hong Kong.

10. S. Shekhar, R. R. Vatsavai, N. Sahay, Thomas E. Burk, and Stephen Lime. WMS and GML based interoperable web mapping system. In Proceedings of the Ninth ACM International Symposium on Advances in Geographic Information Systems, ACMGIS. ACM Press, 2001.

11. Z. Guo, S. Zhou, Z. Xu, and A. Zhou. G2ST: A Novel Method to Transform GML to SVG. In: *Proceedings of ACM-GIS 2003*, ACM Press. November 2003.

12. J. Guan, S. Zhou, J. Chen, *et al*. Ontology-based GML schema matching for information integration. In: *Proceedings of 2nd IEEE International Conference on Machine Learning and Cybernetics*, vol.4, pp. 2240-2245, IEEE CS, Xi'an, China, November 2003.

13. W. Chung and H.-Y. Bae. A Specification of a Moving Objects Query Language over GML for Location-Based Services. In Advanced Web Technologies and Applications: 6th Asia-Pacific Web Conference, APWeb 2004, pages 788 – 793.

14. J. E. Corcoles, P. Gonzalez, Analysis of Different Approaches for Storing GML Documents, In Proceedings of ACM GIS2002, ACM Press, pages 11-16,2002.

15. J. E. Corcoles, P. Gonzalez, A Specification of a Spatial Query Language over GML, pages 112-117,2001.

TPKDB-Tree: An Index Structure for Efficient Retrieval of Future Positions of Moving Objects

Kyoung Soo Bok[1], Dong Min Seo[1], Seung Soo Shin[2], and Jae Soo Yoo[1]

[1] Department of Computer and Communication Engineering, Chungbuk National University
48 Gaesin-dong, Cheongju Chungbuk, Korea
{ksbok,dmseo,yjs}@netdb.chungbuk.ac.kr
[2] Department of Computer Engineering, Chungbuk National University
48 Gaesin-dong, Cheongju Chungbuk, Korea
shinss@chungbuk.ac.kr

Abstract. By continuous growing on wireless communication technology and mobile equipments, the need for storing and processing data of moving objects arises in a wide range of location-based applications. In this paper, we propose a new spatio-temporal index structure for moving objects, namely the TPKDB-tree, which supports efficient retrieval of future positions and reduces the update cost. The proposed index structure combines an assistant index structure that directly accesses to the current positions of moving objects with a spatio-temporal index structure that manages the future positions of moving objects. The internal node in our index structure keeps time parameters in order to support the future position retrieval and reduce the update cost. We also propose new update and split methods to improve search performance and space utilization. We, by various experimental evaluations, show that our index structure outperforms the existing index structure.

1   Introduction

With the rapid development of wireless communications and mobile equipments, many applications for location-based services have been emerged. In this type of applications, it is essential to efficiently manage the moving objects that change their positions over time. A moving object that has dynamic properties is represented by past, current and future positions in moving object databases. Many works on the efficient retrieval of the past trajectory and the current position of moving objects have been done. Recently, studies on a future position prediction of moving objects have been progressed as the importance of future position retrieval increases. In order to track and store moving objects that continuously move, new indexing structures are required to efficiently manage the consecutive changes of moving objects. However, traditional spatial index structures are not suitable for storing this continuously changing information because they suffer from numerous update operations. The traditional index structures typically manage the position data of static objects. For example, the R-tree [1], one of spatial index structures, requires a number of update

S. Wang et al. (Eds.): ER Workshops 2004, LNCS 3289, pp. 67–78, 2004.

operations to index the information on the positions of moving objects. Thus its performance is degraded greatly.

The spatio-temporal index structures based on a regular R-tree have been proposed to overcome the problems of spatial indexing structures. In order to efficiently process range queries and future position retrievals, [2] proposed the VCI-tree in which each node of a regular R-tree stores the maximum speed, v_{max}, of moving objects. And, [3] proposed the TPR-tree that uses the time-parameterized MBR(Minimum Bounding Rectangle) and a velocity vector to reduce the update cost of moving objects. However, those two approaches suffer from a high update cost and performance degradation when the insertion and delete operations of objects are issued continuously by the dynamic changes of moving objects because each update operation requires a round of the whole index structure.

We propose a new index structure, namely the TPKDB-tree, to efficiently process the current and future position retrieval of moving objects. The proposed index structure combines the KDB-tree[4] with an assistant index structure that directly accesses to the current positions of moving objects to reduce update cost. It stores the information on the current positions and velocities of moving objects to efficiently support the future position retrieval. It also stores the escape time which means the time when the objects escape from the region of a node. We also propose two novel strategies for node splitting to improve the space utilization and search performance.

The rest of this paper is organized as follows. In Section 2, we review related work. In Section 3, we propose a new indexing technique based on the KDB-tree to reduce the update cost and support efficient future position retrieval. In Section 4, we present the experimental results that compare our technique with an existing approach. Finally, conclusions and future work are discussed in Section 5.

2   Related Work

Recently, many works on indexing structure for moving objects have been progressed. In this section, we review the LUR-tree [5], Q+R-tree [6], VCI-tree [2], and TPR-tree [3] which have been proposed to support future position retrieval and reduce the update cost.

First, the LUR-tree reduces the update cost of indexing the current positions of moving objects [5]. In the traditional R-tree, the update of an object requires a deletion of its previous position and then an insertion of its new position with the itinerant of whole index structure. It requires a high cost of updating many MBRs to guarantee an optimum MBR. Therefore, if a new position of an object is in the MBR, the LUR-tree changes only the position of the object in the leaf node while avoiding unnecessary splits and merges. However, it does not support the future position retrieval and suffers from unsatisfactory search performance for moving objects due to many overlaps of nodes.

The Q+R-tree is a hybrid index structure which consists of both an R*-tree and a Quad-tree [6]. The motivation of this paper is as follows. First, fast-objects that require many update operations are managed in the Quad-tree because the update performance of a Quad-tree is better than that of an R-tree. Second, quasi-static objects that do not require many update operations are managed in the R^*-tree because the search performance of an R*-tree is better than that of a Quad-tree. As a result, this method improves overall performance of the index structure by improving the update and search performance, respectively. However, if most objects being indexed by the Quad-tree are skewed on any region, the height of the Quad-tree is increased and it becomes an unbalanced tree. Thus, the search performance of the Quad-tree is significantly declined. The update costs of the Quad-tree and the R^*-tree are also increased because of dynamic changes of moving objects.

The VCI-tree is a regular R-tree based index for moving objects [2]. It keeps an additional field in each node: v_{max}, which means the maximum speed of all objects covered by that node in the index, to reduce the update cost and support future position retrieval. Because the VCI-tree updates the positions of moving objects periodically, a query to moving objects cannot acquire the correct search results until the changed positions of moving objects are updated.

The TPR-tree is a time-parameterized R-tree which efficiently indexes the current and anticipated future positions of moving objects by using the velocity vectors and time-parameterized MBRs [3]. However, it does not guarantee an optimal MBR and an efficient search because the update of MBRs is performed under the velocities of moving objects and it has many overlaps of nodes.

As a consequence, the existing index structures based on the regular R-tree suffer from performance degradation due to the incessant index updates by the continuous position changes of moving objects. They also result in a high update cost for reorganizing MBRs in order to insert the positions of objects which are not in the previous node.

3   The TPKDB-Tree

3.1   Characteristics of Our Proposed Index Structure

The TPKDB-tree is a new indexing structure to minimize the update cost and support efficient future position retrieval. It is a hybrid index structure that combines the KDB-tree with an assistant index structure. Fig. 1 shows the structure of the TPKDB-tree. It is based on KDB-tree in order to reduce the update cost on MBR's change occurred by the insertion and deletion of objects, and improve the search performance, while the R-tree based approaches suffer from performance degradation due to the overlaps of nodes. An assistant index structure keeps pointers to the position information of moving objects in the leaf nodes of the KDB-tree in order to directly access to the current positions of objects. Therefore, the update cost of the TPKDB-tree is improved by quickly accessing to objects without a round of the whole index structure.

A moving object inserted in the TPKDB-tree is represented as (oid, t_{ref}, p_{ref}, v_{ref}), where oid is a unique id of an object, t_{ref} is an insertion time, and p_{ref} is a position of

Assistant Index   Index Structure based on the KDB-tree

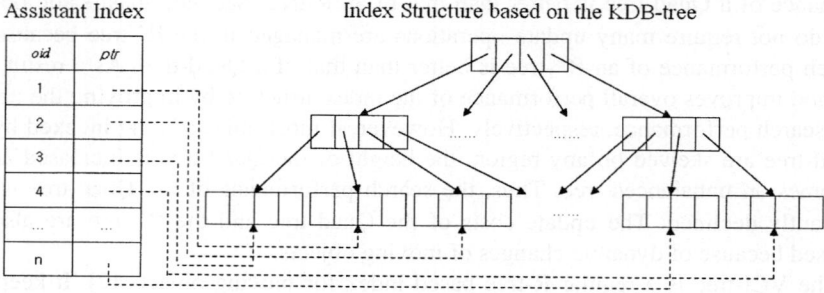

Fig. 1. The Index Structure of the TPKDB-tree

an object at t_{ref} p_{ref} is given by $(p_1, p_2, ..., p_n)$, which is a point in an n-dimensional space. v_{ref} is a moving speed of an object at time t_{ref} and is given by $<v_1, v_2, ..., v_n>$. An object's position at some future time t is given as follows.

$$p_t = p_{ref} + v_{ref}(t - t_{ref})$$ (1)

A leaf node of the TPKDB-tree consists of a header and entries for object's positions. The header of a leaf node is represented as $(RS, t_{upd}, v_{max}, t_{esc})$ and presents whether an object's update is dealt by an assistant index structure without accessing to parent nodes. The RS is the region information of a leaf node for managing objects and is given by $(RS_1, RS_2, ..., RS_n)$, where RS_i is given by $[p_i^-, p_i^+]$ and p_i^- and p_i^+ is the start position and end position of ith-dimensional region, respectively. The t_{upd} is the update time of a leaf node which indicates the update time of an object being included in the leaf node and having the oldest update time. The v_{max} is the maximum speed of objects covered by a leaf node and is given by $(v_{max1}, v_{max2}, ..., v_{maxn})$, where v_{maxi} is given by $[v_i^-, v_i^+]$ and v_i^- is the maximum speed of objects moving toward the left and v_i^+ is the maximum speed of objects moving toward the right in a ith-dimensional region. The t_{esc} is an escape time that a moving object escapes the RS of a leaf node, and takes the shortest escape time over objects covered by a leaf node in order to avoid the unnecessary node expansion for retrieving the future positions of objects. For example, if a future position retrieval is issued within some time t_{esc}, the regions of nodes need not be expanded unlike in the VCI-tree and TPR-tree. The entry of a leaf node that indicates the object's current position is represent as $(oid, t_{ref}, p_{ref}, v_{ref})$, and is the same as the inserted object's information.

An internal node of the TPKDB-tree consists of a header and entries for the headers of leaf nodes covered by that internal node. The header of an internal node is represented as $(RS, t_{upd}, v_{max}, t_{esc})$. The RS is region information for managing child nodes and t_{upd} is the update time of an internal node which indicates the oldest update time of objects being included in the leaf nodes covered by the internal node. The v_{max} is the maximum speed and t_{esc} is the shortest escape time of child nodes covered by the internal node. The entry of an internal node is represented as (RS, t_{upd}, v_{max}, t_{esc}, ptr_n) and is similar to the header of a leaf node. The ptr_i of an entry is a pointer that indicates a child node's address. A leaf node is organized with the information of objects, and an internal node is organized with the information of its child nodes.

The node structure of the TPKDB-tree is similar to the node structure of the TPR-tree. Fig. 2 represents leaf nodes N_1, N_2, and N_3 and an internal node that covers N_1, N_2, and N_3. In Fig. 2, t_{upd}=2 and t_{esc}=2 of the leaf node N_2 toward the west mean that the leaf node N_2 doesn't need to be expanded toward the west until query time 4. And t_{upd}=3 and t_{esc}=2 of the internal node toward the east mean that the internal node doesn't need to be expanded toward the east until query time 5.

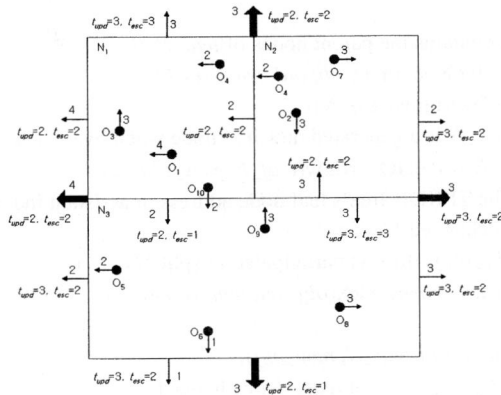

Fig. 2. Representation of the Nodes of the TPKDB-tree

3.2 Insertion

Insertion and deletion algorithms of the TPKDB-tree are very similar to those of the KDB-tree. First, the TPKDB-tree supports the updating algorithm of time-parameterized MBR information such as t_{upd}, t_{esc} and the regional information of the child node to retrieve the future positions of moving objects. The TPKDB-tree uses an assistant index to reduce the insertion and deletion costs. And it supports a new update method that avoids the update of the node if the node covering the new position of an object is same as the node covering the old position of the object and a moving pattern of the object is not changed. Thus, the TPKDB-tree adds the novel algorithms to the regular KDB-tree. Fig. 3 shows the insertion algorithm of the TPKDB-tree.

```
/* new_obj is the inserted object, new_obj.oid is the inserted object's id,
   root is the root node of TPKDB-tree */
Algorithm Insert(new_obj, root)
{
    /* pid is Page id including new_obj.oid of an assistant index */
    pid = SearchAssistantIndex(new_obj.oid);
    /* new_obj is already inserted in TPKDB-tree */
    if(pid)
        /* leaf_N is the leaf node of TPKDB-tree including new_obj.oid */
        leaf_N = ReadLeafNode(new_obj.oid, pid);
        /* Use only the assistant index */
        if(!Check_ChangeOfLeafNInform(leaf_N, new_obj))
            UpdateEntry(leaf_N, new_obj);
        /* Access root node */
        else
            /* stack contains the parent nodes of leaf_N */
            leaf_N=FindNode(new_obj.oid, root, stack);
            if(CheckOverFlow(leaf_N))
                /* new_N is a generated new leaf node when overflowed leaf_N is split*/
                new_N = TreatOverflow(leaf_N, new_obj, stack);
                /* The TPKDB-tree's leaf node address in assistant index's page
                   is changed */
                /* obj_set is objects participated in split */
                UpdateAssistantIndex(obj_set, leaf_N, new_N);
            Else
                UpdateEntry(leaf_N, new_obj);
                /* If t_ref, t_esc, v_max of leaf_N are changed,
                   those of its parent nodes are also changed */
                AdjustNode(stack, leaf_N);
                UpdateAssistantIndex(new_obj.oid, leaf_N);
    /* new_obj is first inserted in TPKDB-tree */
    else
        leaf_N = FindNode(new_obj.oid, root, stack);
        if(CheckOverFlow(leaf_N))
            new_N = TreatOverflow(leaf_N, new_obj, stack);
            UpdateAssistantIndex(new_obj.oid, leaf_N, obj_set);
        Else
            InsertEntry(leaf_N, new_obj);
            AdjustNode(stack, leaf_N);
            UpdateAssistantIndex(new_obj.oid leaf_N);
}
```

Fig. 3. Insert Algorithm

3.3   The Split Strategies of the TPKDB-Tree

Henrich et al. [7] proposes two methods for splitting a internal node in a KDB-tree. One is the forced splitting method that splits child nodes covered by that internal node. The other is the first division splitting method that splits an internal node on first split axis. They also present various experimental evaluations on the two split methods. The results of various experiments represent that the first division splitting is better than the forced splitting in terms of space utilization and search performance, since the forced splitting forces to split child nodes covered by that internal node. [8] proposes the data dependent split strategies and the distribution dependent split strategies for splitting the leaf node of a KDB-tree. The data dependent split strategies depend only on the objects stored in the node to be split. A typical example is to choose as the split position the average of all object coordinates with respect to a certain dimension. The distribution dependent split strategies choose a split dimension and a split position independently of the actual objects stored in the node to be split. A typical example is to split a cell into two cells with the same size along to a dimension.

However, the traditional split strategies of the KDB-tree are unsuitable because it does not reflect the moving object's attributes. If the distribution deviation of objects stored in the node is large, the space utilization of the KDB-tree might be decreased because the node's splitting is unequal. To improve space utilization, we propose an enhanced data dependent split (the EDD-split, for short) which is based on the data dependent split for the leaf node, and an enhanced first division split (the EFD-split, for short) which is based on the first division split for the internal node.

When a leaf node is overflowed, the EDD chooses the split positions of all axes by the data dependent split, where the split position is the average of all objects' coordinates with excluding ε that is some objects included in the minimum number of entries in a node. Thus, the space utilization of split nodes is improved. After split positions are chosen, the EDD split gets t_{upd}, t_{esc}, and v_{max} of a leaf node. And after the EDD split expands all leaf nodes until some time t (t is larger than all $t_{upd} + t_{esc}$ of leaf nodes), the split position of the axis that involves the smallest expansion of nodes is chosen as the split position of the overflowed leaf node. Fig. 4 is an example of the EDD split, where leaf nodes N_1, N_2, N_3 and N_4 of which ε is 2 are expanded until time 13 because the largest $t_{upd} + t_{esc}$ is time 13 among all t_{upd} values of nodes on the assumption that the insertion time of all objects is time 1. In the two-dimensional case, the expansion rate of a leaf node is given by formula 2.

$$N.ER = \frac{\sum_{i=x}^{y} \left(N_i.Length + (t_{sel} - N_i.t_{esc}^-) \times N_i.v^- + (t_{sel} - N_i.t_{esc}^+) \times N_i.v^+ \right)}{N_x.Length \times N_y.Length} \tag{2}$$

Because the expansion rates of N_1 and N_2 are smaller than those of N_3 and N_4, the split position of the leaf node including N_1 and N_2 becomes the split position of the leaf node being overflowed.

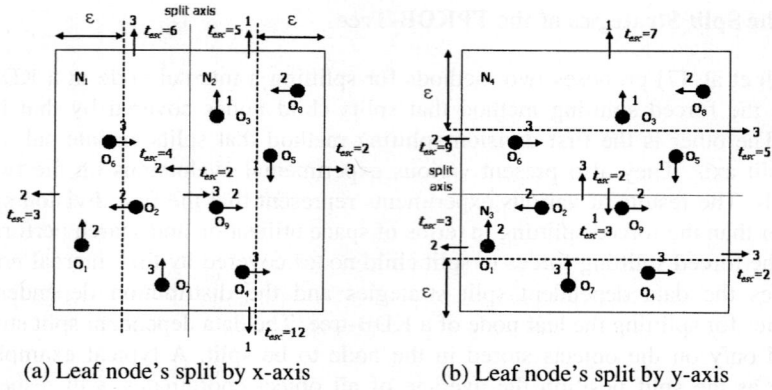

(a) Leaf node's split by x-axis   (b) Leaf node's split by y-axis

Fig. 4. The EDD Split of the TPKDB-tree

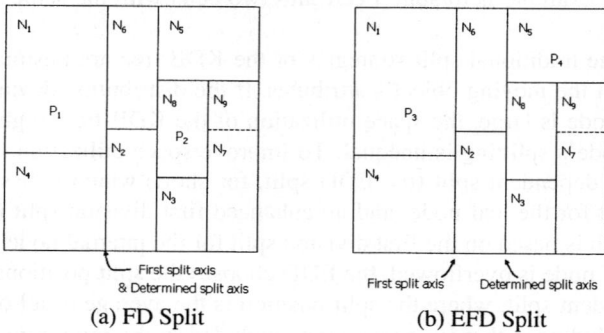

(a) FD Split   (b) EFD Split

Fig. 5. The EFD Split of the TPKDB-tree

To choose the split position of an internal node being overflowed, like Fig. 5, the EFD split does not choose the first split position of the internal node being overflowed but choose another split position with the better space utilization than the first split position. If there is no appropriate split position, the first split position is used.

3.4 Retrieval Technique on Future Positions of Objects

The TPKDB-tree supports not only current position retrieval but also future position retrieval. The query types of the TPKDB-tree are object query and range query. An object query is represented as $Q = (oid, t)$ and searches for the object with oid at some time t, which is a current or future time. Typical examples for such an object query are "Where is No. 29 bus now?" and "Where is No. 24 bus after ten minutes from now?". A range query is represented as $Q = (R, t)$ and searches for all objects covered by some region R which is given by $<R_1, R_2, ..., R_n>$ at some time t. Typical examples for such a range query are "Are there the taxies in front of the theater now?" and "Are there the buses in front of the school after three hours from now?".

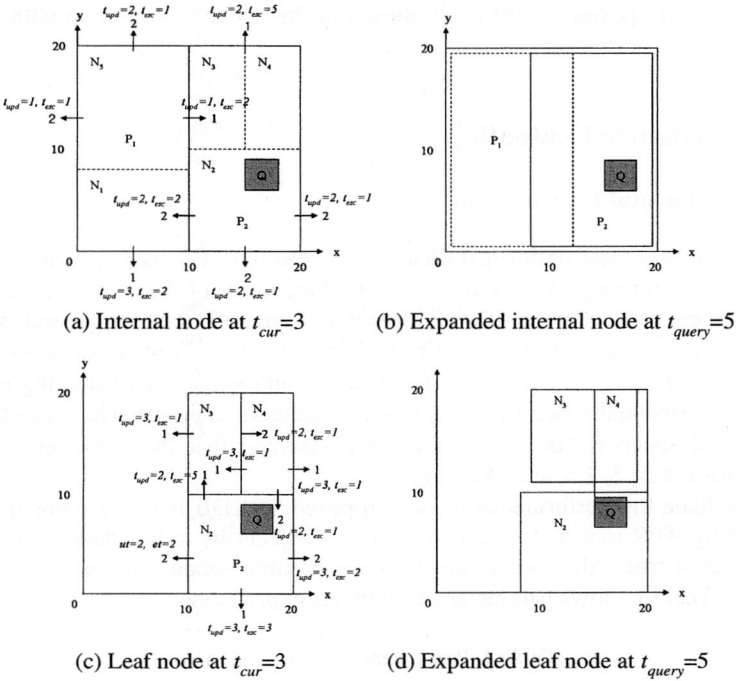

(a) Internal node at $t_{cur}=3$

(b) Expanded internal node at $t_{query}=5$

(c) Leaf node at $t_{cur}=3$

(d) Expanded leaf node at $t_{query}=5$

Fig. 6. The Future Position Retrieval of the TPKDB-tree

Fig. 6 shows the retrieval technique on future positions of moving objects in the TPKDB-tree. Fig. 6(a) shows the internal nodes of the TPKDB-tree at current time 3, where a range query is requested at time 5. In order to perform a requested range query, the internal nodes P_1 and P_2 are expanded by t_{upd}, t_{esc}, v_{max} of each node until query time 5. A node's expansion uses formula 3.

$$ES = \begin{cases} RS & , if \ t_{query} \leq (t_{upd} + t_{esc}) \\ RS + (t_{query} - (t_{upd} + t_{esc})) \times v_{max} & , else \end{cases} \tag{3}$$

ES is the region information of an expanded node to retrieve future positions at some query time. If $t_{query} \leq t_{upd} + t_{esc}$, node's expansion to perform future position retrieval is unnecessary. Therefore, ES is similar to RS. If $t_{query} > t_{upd} + t_{esc}$, each node is expanded until some query time. Each node is expanded to the whole region of the TPKDB-tree at most.

Fig. 6(b) shows each internal node's expansion until time 5. After internal nodes P_1 and P_2 are expanded, each leaf node covered by P_2 is only expanded because only P_2 is overlapped with the area of range query. Fig. 6(c) shows that each leaf node of P_2 is expanded by t_{upd}, t_{esc}, v_{max} of each leaf node until time 5. Fig. 6(d) shows that leaf nodes N_2 and N_4 are only overlapped in the range query. In conclusion, the retrieval of future position returns all objects in the nodes that overlap with the range query.

4   Performance Evaluation

4.1   Experimental Environment

Our experiments were performed on a 1.7GHz Pentium IV machine with 256Mbytes main memory running MS Windows 2000. Three kinds of data sets such as gaussian data set, skew data set and uniformly distributed synthetic data set generated by GiST [9] were used in our experiments. We configured the node size of our proposed index structure to 4Kbytes. We generate the positions and velocities of moving objects on the two-dimensional space by using random number generator. The size of the two-dimensional space is 1000×1000 km². We assumed that the velocities of moving objects are 1, 1.5, 2, 2.5, and 3km/min.

To evaluate the performance of our proposed TPKDB-tree, we compared it with the existing TPR-tree that organizes moving objects by using data split algorithm such as an R-tree. Also, we implemented a hashing scheme for an assistant index structure. Table 1 shows parameters used in our experiments.

Table 1. Parameters used in experiments

Parameter	Values Used
Number of inserted objects [integer]	10,000 ~ 100,000
Number of updated objects [integer]	500 ~ 100,000
Query interval length [minute units]	5 ~ 30
Query window size [% of the data space]	0.1 ~ 5

First, to measure insertion performance, we inserted 10,000~100,000 entries to index tree. Second, we measured the execution time of update operations by updating 500~10,000 entries in the constructed index tree. Also, to evaluate the search performance of future positions, we measured the execution time of 1000 range queries that search for the position of each moving object after 5~30 minutes, and then the future positions of objects after 20 minutes by expanding the area of range queries.

4.2   Experimental Result

To show the effectiveness of our approach compared to the existing one, we measured the search and building time of our proposed TPKDB-tree and the TPR-tree[3]. Fig. 7 shows the differences of insertion and updating times between the TPKDB-tree and the TPR-tree. This experimental result shows that the proposed TPKDB-tree improves the insertion performance by 300~550% and the updating performance by 200~300% over TPR-tree.

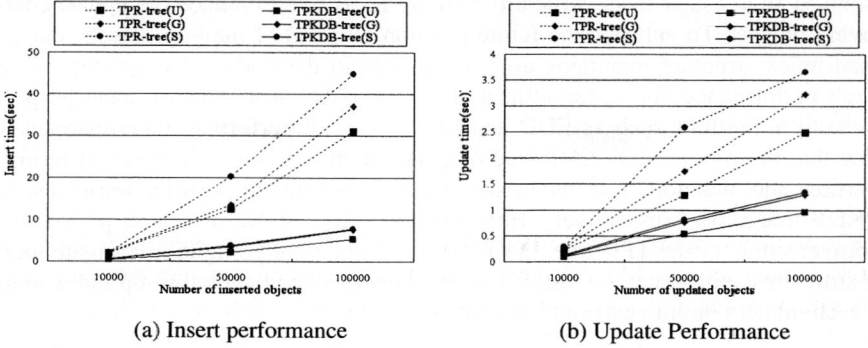

(a) Insert performance   (b) Update Performance

Fig. 7. Insert Performance and Update Performance of the TPKDB-tree and TPR-tree

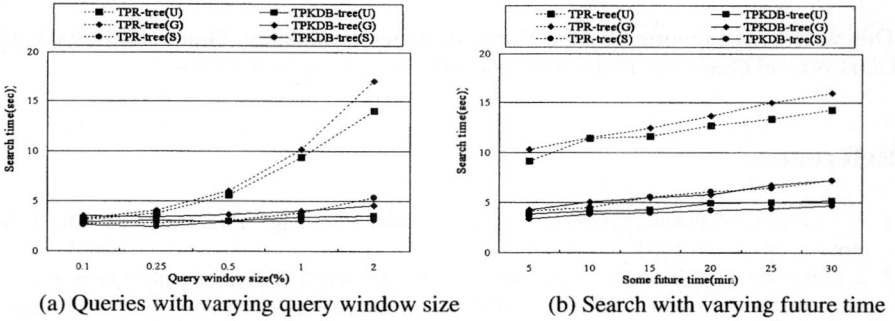

(a) Queries with varying query window size   (b) Search with varying future time

Fig. 8. Search Performance of the TPKDB-tree and TPR-tree

Fig. 8 shows the differences of execution time for range queries and some future position retrieval between the TPKDB-tree and the TPR-tree. This experimental result shows that the proposed TPKDB-tree improves search performance by about 100~400% for range queries and enhances about 100~300% of search performance for future position queries over TPR-tree.

Because the TPR-tree is based on an R-tree and performs frequent updates by moving objects, it requires high updating overhead. Therefore, overall performance of the TPR-tree degrades significantly. Our TPKDB-tree outperforms the TPR-tree because the TPKDB-tree uses enhanced space split methods to reduce the update cost and an assistant index structure to quickly access to the current positions of objects. Also, the TPKDB-tree searches faster than the TPR-tree because TPKDB-tree is based on a KDB-tree and has no nodes being overlapped.

5 Conclusion

In this paper, we have proposed a novel spatio-temporal indexing structure that supports efficient updating and the future position retrieval. The proposed TPKDB-tree combines an assistant index structure with the KDB-tree to reduce the update cost of moving objects. To enhance the future position retrieval of moving objects, our proposed index structure maintains time parameters in the node. Also, to improve the search performance and space utilization of the index structure, we have proposed new split algorithms such as EDD and EFD. Through performance comparison, we show that the proposed TPKDB-tree outperforms the existing TPR-tree in terms of insertion and deletion. For range queries and the future position retrievals, our TPKDB-tree shows respectively 100~400% and 100~300% of search performance improvement over the TPR-tree. For future works, we are going to investigate indexing structures appropriate for various types of queries and the bottom-up index updating technique to optimize the update cost.

Acknowledgement

This work was supported by the Korea Research Foundation Grant (KRF-2003-041-D00489) and Grant No. (R01-2003-000-10627-0) from the KOSEF.

References

1. A. Guttman, "A dynamic index structure for spatial searching", Proc. ACM SIGMOD Conference. pp.47–57, 1984
2. S. Prabhakar, Y. Xia, D.V. Kalashnikov, W.G. Aref, and S.E. Hambrusch, "Query indexing and velocity constrained indexing : scalable techniques for continuous queries on moving objects", IEEE Transactions on Computers, Vol. 51. No. 10. pp.1124–1140, 2002
3. S. Saltenis, C.S. Jensen, S.T. Leutenegger, and M.A. Lopez, "Indexing the Positions of Continuously Moving Objects", Proc. ACM SIGMOD Conference. pp.331–342, 2000
4. J.T. Robinson, "The K-D-B Tree : A Search Structure for Large Multidimensional Dynamic Indexes", Proc. ACM SIGMOD Conference. pp.10-18, 1981
5. D.S. Kwon, S.J. Lee, and S.H. Lee, "Indexing the Current Positions of Moving Objects Using the Lazy Update R-tree", Proc. International Conference on Mobile Data Management. pp.113–120, 2002
6. X. Yuni and S. Prabhakar. "Q+Rtree : Efficient Indexing for Moving Object Database", Proc. Eighth International Conference on Database Systems for Advanced Applications. pp. 175–182, 2003
7. A. Henrich, H.S. Six, and P. Widmayer, "The LSD-tree : Spatial Access to Multidimensional Point and Non-point Objects", Proc. International Conference on Very Large Data Bases. pp. 45–53, 1989
8. O. Ratko and Y. Byunggu, "Implementing KDB-Trees to support High-Dimensional Data", Proc. International Database Engineering and Applications Symposium. pp.58–67, 2001
9. J.M. Hellerstein, J.F. Naughton, and A. Pfeffer, "Generalized Search Trees for Database Systems", Proc. International Conference on Very Large Data Bases. pp.562–573, 1995

Moving Point Indexing for Future Location Query*

Hoang Do Thanh Tung, Young Jin Jung, Eung Jae Lee, and Keun Ho Ryu

Database Laboratory, Chungbuk National University, Korea
{tunghdt,yjjeong,eungjae,khryu}@dblab.chungbuk.ac.kr

Abstract. Nowadays, with a great number of mobile (moving) objects using wireless communication or GPS technology, monitoring systems having the capability of handling anticipated future queries of the mobile objects are literally important. To cater for these applications, the TPR-tree indexing method, which is able to support the future queries, is substantially feasible for the real world. The crucial idea of them is to make the bounding rectangle as a motion function of time. In this paper, we propose an extended method of the TPR-tree, named Horizon TPR-tree or HTPR-tree. The general idea that we make a new motion function of time is to reduce speed of growing sizes of MBR (minimum bounding rectangles). Smaller MBR alleviates overlap problem and increases the effect of query in R-tree based methods. During creating and updating index, motion function influences organizing operations, such as splitting and adjusting. Our experiments show that the HTPR-tree outperforms the TPR-tree for anticipated future queries. In addition, the idea of the proposed method is able to apply to not only the TPR-tree but also variants of the TPR-tree. The proposed method is applicable to vehicle-traffic network systems.

1 Introduction

Continued advances in wireless communications and GPS (Global Positioning System) technology enable people to get the location with velocities of mobile objects (e.g. car, plane, people, etc.) easily and in high accuracy. Therefore, many applications (e.g., traffic control, digital battlefield, tourist services, etc.) using the location information of moving object have been developed in these days. However, for efficiently access the voluminous moving object data, it needs to develop indexing method for answering ad hoc queries in the past, current as well as the future time. For example, an airplane supervisor wants to know "how many airplanes flied over Seoul in the past 25 minutes", or a policeman wants to know "how many vehicles will be passed the main street in the next 10 minutes".

In this paper, we will concentrate on the method that has the capability of dealing with anticipated future queries of moving objects. Generally, to support the future queries, databases store the current positions and velocities of moving objects as linear functions. Up to now, for processing current and future queries, several index-

---

* This work was supported by the RRC of MOST and KOSEF as well as University IT Research Center Project in Korea.

S. Wang et al. (Eds.): ER Workshops 2004, LNCS 3289, pp. 79–90, 2004.

ing methods have been proposed. One of them is the time-parameterized R-tree (TPR-tree) proposed by Simonas Saltenis [1]. It seems to be the most flexible and feasible with respect to real world. TPR-tree naturally extends the R*-tree, and employs time-parameterized rectangles termed *conservative bounding rectangles* instead of using truly minimum bounding rectangles. However, the defect is that the sizes of rectangles are minimum at current time point, but most likely not at later times. The lower bound of a conservative bounding rectangle is set to move with the minimum speed of the enclosed points, while the upper bound is set to move with the maximum speed of the enclosed points. This ensures that (conservative bounding) rectangles are indeed bounding moving points for all times considered. Obviously, the more those rectangles grow the more overlap probably occurs. Technically, Overlap crucially deteriorates query performance of any R-tree based method.

Until now, several methods have been proposed to improve the TPR-tree technique, referred in [1, 6, 7, 10]. Although those methods seemed actually to win better performance compared to the TPR-tree, most of them implicitly accept rapid growth of time-parameterized bounding rectangles (decisive point of TPR-tree idea) and undervalue Time Horizon parameter. In fact, this parameter, which consists of Update interval and Query interval, is the most important parameter of TPR-tree-based methods in most operators.

In this paper, we propose an extended indexing method of the TPR-tree for querying current and anticipated future positions of mobile objects, named Horizon TPR-tree (HTPR-tree). In order to reduce speed of growing sizes of MBR (minimum bounding rectangles), the general idea describes that the HTPR-tree adopts a new motion function for MBR, which is constrained to Time Horizon.

The rest of the paper is organized as follows. Section 2 reviews a number of important methods proposed to improve the TPR-tree. Section 3 gives a discussion of Time Horizon Heuristics in practical problem. Section 4 studies a problem of conservative bounding rectangle. Section 5 describes a new motion function for conservative bounding rectangle. Section 6 reports on performance experiments. We conclude in Section 7 by summarizing our work and suggesting further direction.

2   Related Work

Since Saltenis et al. [1] introduced the TPR-tree, a number of methods have been proposed to improve his method. In this part, we are reviewing some of these methods that have considered improvements to the TPR-tree in historical order.

In the paper [1], Saltenis et al. proposed the TPR-tree which modifies the R*-tree construction algorithms for managing moving objects. The main idea is to make the bounding rectangles as linear functions of time so that the enclosed moving objects will be in the same rectangles. This time-parameterized approach seems more practical than other previous methods because it is simple and flexible for high-dimension space.

In 2002, Saltenis et al. [6] proposed R^{EXP}-tree, which is an extension of the TPR-tree but handles moving objects with expiration times. With assumption about knowing the expiration times of all moving objects in advance, they improved search performance by a factor of two or more without sacrificing update performance. However, it seems not practical to know expiration times of all moving objects.

Yufei Tao proposed TPR*-tree [7], which modifies insertion/deletion algorithms to enhance performance of the TPR-tree. TPR*-tree uses swept regions (1) for insertion/deletion algorithm instead of the integral metric (2)

$$Cost(q) = \sum_{every\ node\ o} A_{SR}(o', q_T) \tag{1}$$

$$\int_{now}^{now+H} A(t)dt, \tag{2}$$

In the equation (2), A is the area of a MBR. His experiment showed better results compared to the TPR-tree. Despite that, not withstanding much improvement, it is important to stay mindful that the TPR*-tree still keeps using the conventional bounding rectangles.

Bin Lin and Jianwen Su [10] proposed a bulk-loading algorithm for TPR-tree for dealing with non-uniform datasets. Their paper presented a histogram-based bottom up algorithm (HBU) that uses histogram to refine tree structures for different distributions. Empirical studies showed that HBU outperforms the bulk-loading algorithm of the TPR-tree for all kinds of non-uniform datasets.

Consequently, the above methods achieved good successes of improving the TPR-tree, but most of them implicitly accept rapid growth of bounding rectangle that strongly affects quality of queries. In another word, query performance has not been optimized with these methods. The following sections illustrate how to optimize query performance.

3   Heuristics of Time Horizon Parameter

As mentioned in the previous sections, time Horizon (H) is inevitable parameter used in the majority of constructing operators of TPR-tree-based methods. At first, we revise the use of parameter H, and then discuss how we assume parameter H used in a traffic network system. At last, we propose how we use H parameter in our design.

In the TPR-tree method, there are three parameters that affect the indexing problem and the qualities of a TPR-tree. The first parameter is how far the future queries may reach, denoted W. The second parameter is the time interval during which an index will be used for querying, denoted U. The third is simply the sum of W and U, denoted H or Horizon. Figure 1 illustrates these parameters.

Intuitively, to have H value, we must estimate both parameters U and W in advance. Most methods, however, ignored this work. They normally assume an H value and allow queries for all times considered. However, these approaches have problems that (a) first, for more than one dimension space, the area of a bounding rectangle does not grow linearly with time so that with time interval longer than H, cost of query is unable to be estimated (b) second, it seems rather naïve to real system of which objects are moving constantly so that long time-interval queries must be extremely incorrect. Therefore, in our idea, we are supposed probably to estimate U and W values in order to increase the accuracy and performance of future queries.

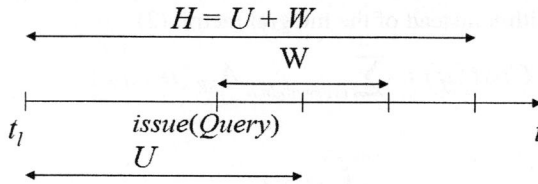

$$H = U + W$$

$$W$$

$$t_l \qquad issue(Query) \qquad t$$

$$U$$

Fig. 1. Time Horizon H, Index Usage Time U, and Querying Window W

As an idea, we assume our method is applied for a traffic network. For U value, we can estimate by either calculating total average interval between two updates (so that the value of U can be changed between adjacent updates) or imitating the frequency of traffic lights because most vehicles turn or stop under traffic light control. For W value, we can estimate by a value greater than or equal to time interval of the total average speed running on the longest road segment.

After all, to ensure the best performance and correctness, our method requires (a) H is always a sum of U and W (reference to experiment of [1], the best values of H lie between U/2 + W and U +W); (b) our indexing method invokes re-indexing (bulk loading algorithm) itself after U time units; (c) under no circumstances do queries anticipate values beyond a given W. Before going to our proposal, we are studying a problem in bounding rectangle of TPR-tree in the next section.

4   Problem Definition of TPR-Tree Bounding Rectangle

In this part, we define the problem in bounding rectangle of TPR-tree that is continuously growing and most likely not minimum in the anticipated future time.

Without loss of generality, we analyze the problem on an example for one axis. Given two objects A and B, in which V_A (velocity of A) is minimum velocity and V_B (velocity of B) is maximum velocity. Conservative bounding rectangle of two objects has the lower bound velocity V_L set to V_A and the upper bound velocity V_U set to V_B. We initially assume that,

At t=0, Distance (lower bound, upper bound) = D_{LU}, and

At t=0, Distance (A, B) = D_{AB} = D_{LU}.

Now, we exam two following situations

Situation 1: A stands before B. Figure 2 describes that the bounding interval is always tight in both (a) and (b) because the speed of each bound is equal to that of the objects nearest to it. Although (a) shows that its interval can grow very quickly, we still consider both situations (a) and (b) optimal because their bounding rectangles are always minimal over time.

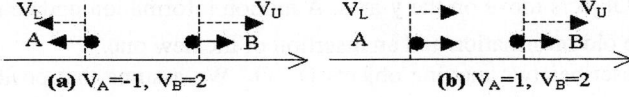

Fig. 2. A and B are (a) moving in the opposite direction (b) moving in the same direction

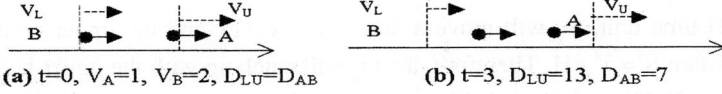

Fig. 3. A and B are moving on the same direction

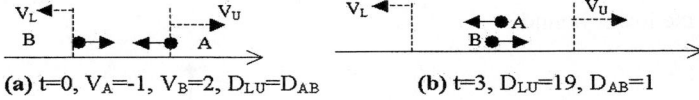

Fig. 4. A and B are moving toward each another

Situation 2: A stands after B. Figure 3 describes that A and B move in the same direction from timestamp 0 to 3. Analogously, Figure 4 describes that A and B move toward each other from timestamp 0 to 3. Nevertheless, both figures show that their bounding intervals are only tight at current time but no longer tight in the future because the speed of each bound is not equal to that of the objects nearest to it. Thus, in these situations, we consider the rectangle not optimal because the rectangle is not always minimal.

Now, we can answer, in such situation, bounding rectangles become worse in the future. In the next section we are designing a new motion function to deal with problem of the situation 2 to make bounding rectangles nearly to minimum.

5   New Motion Function for Bounding Rectangle

To solve the problem described in the previous section, we thought out a new motion function for conservative bounding rectangle. Using this function, the speed of the upper bound is reduced as much as possible and, analogously, the speed of the lower bound is increased as much as possible. As a result, the rectangles will be most likely minimal in the future. In the following, we are going to explain how we got this function.

5.1   Motion Function in One Dimension Space

Initially, we study objects moving on a 1-dimensional space. The location of each object is described as a linear function of time, namely the location $f_i(t)$ of the object O_i at time t is equal to $f_l + v_i(t - t_l)$, where v_i is the velocity of the object and f_l is its location at t_l. Objects move on the y-axis. A motion information update is treated as a deletion of the old information and an insertion of the new one.

Figure 5 describes two moving objects O_1, O_2. We assume that position P_u of the upper bound u is farther than position P_2 of O_2, which will move into the farthest position after H time units. V_2 is speed of O_2. The distant between O_2 and the upper bound is d= $P_u - P_2$, d>0.

After H time unit, O_2 will move a distant S= V_2*H, and the upper bound u will move a distant S'= V_u*H. Therefore, the O_2 will catch up with the upper bound when S= S' + d or V_2*H = V_u*H + d. Given V_2, inferably, O_2 always runs within the range of that rectangle as long as u run at speed V_u = (V_2×H - d)/H, V_u <V_2. Similarly, we can calculate the speed for the lower bound V_L = (V_L×H - d)/H, d=P_L-P_1, L is the position of the lower bound.

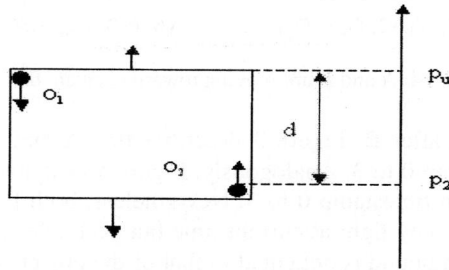

Fig. 5. Two objects O_1, O_2 moving on Y-axis

For generality, by considering objects moving furthest on each direction, we get the function for a number of objects greater than two.

$$v^{min} = \min\left\{ \frac{o.v \times H - L + o.x}{H} \Big| o \in node \right\}$$

$$v^{max} = \max\left\{ \frac{o.v \times H - u + o.x}{H} \Big| o \in node \right\} \tag{3}$$

5.2 Motion Function in Multi-dimension Space

In this part, we study objects moving in the $(x_1, x_2, \dots, x_d)$ d-dimensional space. The initial location of the object O is $(O.x_1, O.x_2, \dots, O.x_d)$ and its velocity is a vector v = $(O.v_1, O.v_2, \dots, O.v_d)$. The Object can move anywhere on the finite Terrain as linear functions of time $f_1(t), f_2(t), \dots, f_d(t)$ on each axis.

In fact, in d-dimensional space with a number of moving objects more than two objects, we need to choose objects, which will go farthest after H time for every dimension. Therefore, the bounds will reach the positions of objects that move farthest for each axis. This ensures that conservative bounding rectangles are indeed bounding for all times considered.

Therefore, we replace the general conventional formula (4)

$$v_i^{\min} = \min \left\{ o.v_i \,\middle|\, o \in node \right\}$$
$$v_i^{\max} = \max \left\{ o.v_i \,\middle|\, o \in node \right\}$$

(4)

with a new motion function to calculate new bound speeds as (5)

$$v_i^{\min} = \min \left\{ \frac{o.v_i \times H - l_i + o.x_i}{H} \,\middle|\, o \in node \right\}$$
$$v_i^{\max} = \max \left\{ \frac{o.v_i \times H - u_i + o.x_i}{H} \,\middle|\, o \in node \right\}$$

(5)

U_i is the upper bound's position of ith dimension; l_i is the lower bound's position of ith dimension.

We name the new motion function (5) Horizon function because it is constrained by H parameter. Before going to the next section, we illuminate our function by following example.

5.3   An Example for Bounding Rectangle Using New Motion Function

An example is used for comparison between two types of conservative bounding rectangle, one using our new motion function, the other using the conventional motion function of the TPR-tree. We assume that there are 4 moving cars with speeds (km/minute). The initial positions of 4 objects are a(-1,1), b(2,-2), c(-1,0), d(1,2). We are describing changes of the rectangle bounding 4 objects from timestamp 1 to 4 in Figure 6. In this example Horizon value is supposed to be 4 units.

First off by calculating the speeds of all bounds of the conventional conservative (dashed) bounding rectangle, we have

$$v_x^{\max} = \max(a_x, b_x, c_x, d_x) = 2$$
$$v_x^{\min} = \min(a_x, b_x, c_x, d_x) = -1$$
$$v_y^{\max} = \max(a_y, b_y, c_y, d_y) = 2$$
$$v_y^{\min} = \min(a_y, b_y, c_y, d_y) = -2$$

Now using the new function to calculate the new conservative (bold) bounding rectangle, we have

$$v_x^{\max} = \max(-1,1,0,1) = 1$$
$$v_x^{\min} = \min(-1,0,0,1) = -1$$
$$v_y^{\max} = \max(0.25, -1.25, 0, 1.25) = 1.25$$
$$v_y^{\min} = \min(0, -1.25, 0.75, 0) = -1.25$$

Intuitively, we can realize that the bounding rectangle using the new function is nearly minimal at all timestamp while the other bounding rectangle goes from bad to worse.

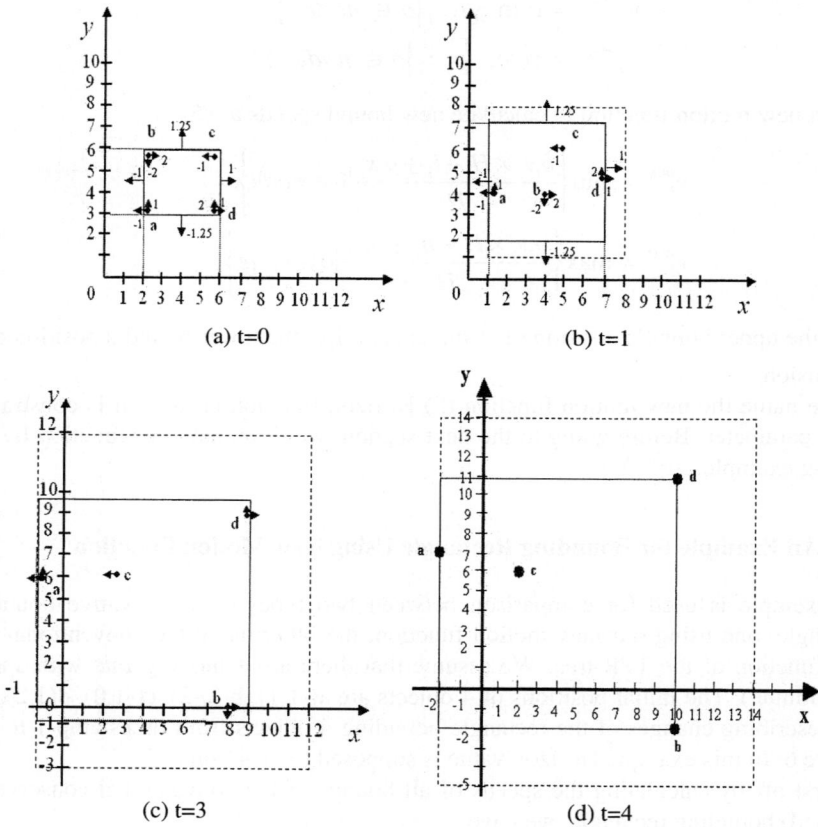

(a) t=0   (b) t=1

(c) t=3   (d) t=4

Fig. 6. Comparison between two kinds of bounding rectangle

In the next section, we are presenting our experimental results using the new function.

6 Performance Evaluation

The implementation of the TPR-tree and Horizon TPR-tree used in the experiments is performed on a Pentium III 500-MhZ machine with 256MB memory, running Windows 2000PRO. For all experiments, the disk page size is set to 1k bytes, and the maximum number of entries in a node is 27 for all indexes. Our goal is to compare query performance between the TPR-tree and the Horizon TPR-tree. Therefore, we don't show update experiments for our dataset. Our dataset simulates a traffic net-

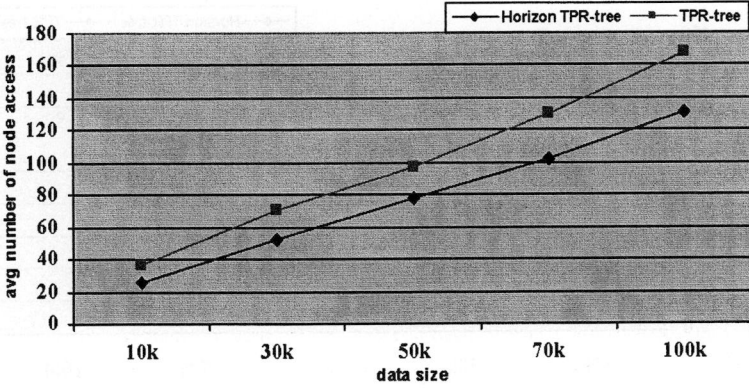

Fig. 7. Timestamp queries: $q_R len=100$, $q_V len=0$, $q_T len=1$

work area for mobile cars. Thus, we assume that moving cars/objects uniformly distributed in 2d area where each axis of the area is normalized to [0, 10000 meters], namely the region is 100km^2. Then, each object is associated with a MBR such that (a) object does not change spatial extents during its movement, (b) on each dimension, velocity value is in range [0, 50] per a time unit (a time unit is a twentieth minute or 3 seconds), namely a car can run at most of about 90km/h, velocity can be positive or negative as two opposite directions with equal probability. Each object can go to a target randomly selected within 2000 targets.

Each query q has three parameters: $q_R len$, $q_V len$, and $q_T len$, such that (a) its MBR q_R is a spare, with length $q_R len$, uniformly generated in the data space, (b) its VBR is $q_V = \{ -q_V len, q_V len, -q_V len, q_V len \}$, (c) its query interval is $q_T = [0, q_T len]$. Query cost is measured as the average number of node accesses in executing a workload of 200 queries with the same parameters. For each workload, the horizon parameters of the TPR-tree and Horizon TPR-tree are set to the corresponding $q_T len$. A query is generated as follows, on each spatial (velocity) dimension i ($1 \le i \le 2$), the starting point of its extent uniformly distributes in [0, 10000-$q_R len$].

6.1 Query Performance on Different Dataset Sizes

In order to study the deterioration of the indices with size of dataset, we measure the performance of the TPR-tree and Horizon TPR-tree, using the same query workload with sizes of dataset between 10000 objects and 100000 cars.

We took experiments for three kinds of query. With time horizon T=50 (about 2 minute 30 seconds), we first tested *Timestamp queries* (Figure 7) such that spatial extents are [100, 100] for every timestamp. And then we tested *Windows queries* (Figure 8) such that time interval is [0, 50] and the spatial extents are [100, 100]. Finally, we tested *Moving Window queries* (Figure 9) such that (a) spatial extents are

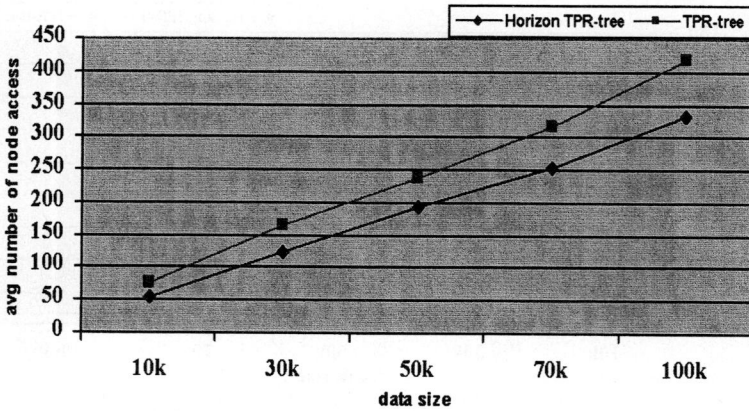

Fig. 8. Window queries: $q_R$len=100, $q_V$len=0, $q_T$len=50

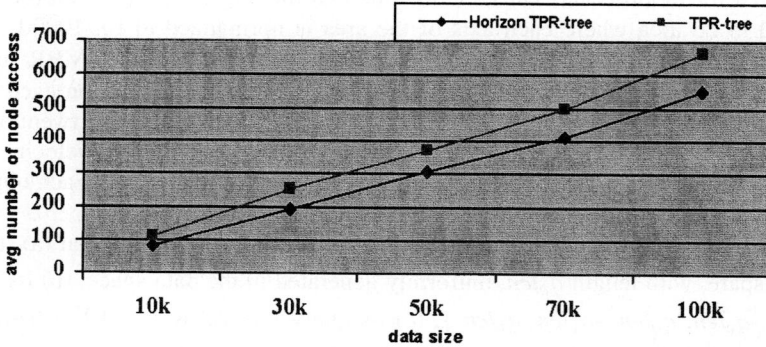

Fig. 9. Moving Window queries: $q_R$len=100, $q_V$len=10, $q_T$len=50

[100, 100] on each dimension, (b) velocities are {-10, 10, -10, 10} on each direction, (c) time interval is [0, 50]. For every experiment, we set spatial positions of queries randomly and operated 200 queries on a dataset to estimate query cost by average of node accesses. The experiments show that the query costs of the Horizon TPR-tree are always better than those of the TPR-tree. Moreover, HTPR-tree tends to increasingly better than TPR-tree when the number of objects increases.

6.2   Query Performance on Different Time Horizon Sizes

In order to study the deterioration of the indexes with size of Time Horizon, in this experiment, we tested fixed data size 10000 objects with different time Horizon sizes from 50 to 500 unit times (about 25 minutes). In this experiment, we only use *Moving Window query* because it is a general query type. Like previous experiment, we set Moving Window queries such that (a) spatial extents are [100, 100] on each dimen-

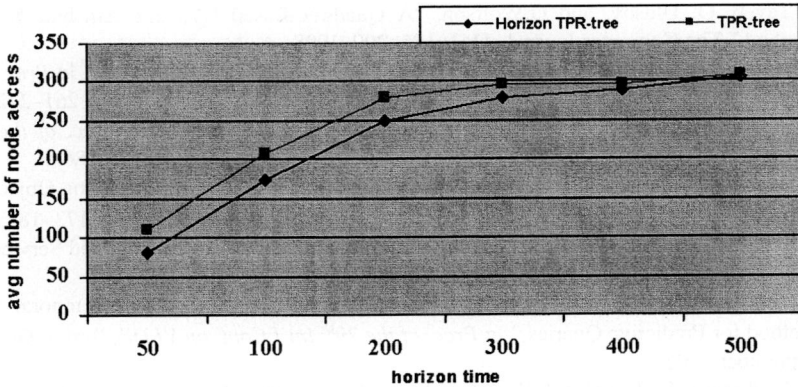

Fig. 10. Number of node access against different Time Horizon sizes

sion, (b) velocities are {-10, 10, -10, 10} on each direction, (c) time interval is [0, 50]. For every experiment, we set spatial positions of queries randomly and operated 200 queries on a dataset to estimate query cost by average of node accesses. Figure 10 shows that the Horizon TPR-tree is more efficient than the TPR-tree in short Time Horizons (near future queries). However it is not much better than the TPR-tree in very long Time Horizon (very far future queries). We believe, practically, very long time Horizon makes no sense for moving objects in real application. In this example of a traffic network, obviously, anticipating this traffic network for queries after 25 minutes is extremely useless.

7 Conclusion

In this paper, we proposed an extended indexing method of the TPR-tree called Horizon TPR-tree for processing anticipated future queries of moving objects. HTPR-tree treated an intractable problem of time-parameterized rectangle methods. The experimental results showed that the HTPR-tree outperforms the conventional TPR-tree in all future queries. In general idea, the new motion function (5) can apply not only to the TPR-tree but also to other TPR-tree variants, such as TPR*-tree. Also, we gave some discussion about time Horizon parameter and its partial parameters. In the future research, how to elaborately streamline the time parameters in real network system is needed.

References

1. S. Saltenis, C. S. Jensen, S. T. Leutenegger, and M. A. Lopez, "Indexing the Positions of Continuously Moving Objects," In *Proc. of the 2000 ACM SIGMOD Int'l Conf. on Management of Data*, Dallas, Texas, USA, May 2000.

2. J. Tayeb, O. Ulusoy, and O.Wolfson, "A Quadtree-Based Dynamic Attribute Indexing Method," The Computer Journal, 41(3):185–200, 1998.
3. G. Kollios, D. Gunopulos, and V. J. Tsotras, "On Indexing Mobile Objects," In *Proc. of the ACM Symp. on Principles of Database Systems (PODS1999)*, June 1999, pp. 261–272.
4. P. Agarwal, L. Arge, and J. Erickson, "Indexing Moving Points", In *Proc. of the ACM Symp. on Principles of Database Systems (PODS2000)*, May 2000, pp.175-186.
5. H. D. Chon, D. Agrawal, and A. E. Abbadi, "Storage and Retrieval of Moving Objects," In *Proc. of the 2nd Int'l Conf. on Mobile Data Management*, January 2001, pp. 173–184.
6. S. Saltenis and C. S. Jensen, "Indexing of Moving Objects for Location-Based Services," In *Proc. of the Int'l. Conf. On Data Engineering*, February 2002.
7. Y. Tao, D. Papadias, and J. Sun, "The TPR*-Tree: An Optimized Spatio-Temporal Access Method for Predictive Queries," In *Proc. of the 29th Int'l Conf. on VLDB*, Berlin, Germany, September 2003.
8. N. Beckmann, H.-P. Kriegel, R. Schneider, and B. Seeger, "The R*-tree: An Efficient and Robust Access Method For Points and Rectangles," In *Proc. of the 1990 ACM SIGMOD Int'l Conf. on Management of Data*, Atlantic City, New Jersey, USA, May 1990.
9. C. M. Procopiuc, P. K. Agarwal, and S. Har-Peled. "STAR-Tree: An Efficient Self-Adjusting Index for Moving Objects," In *Proc. of the 4th Intl Workshop on Algorithm Engineering and Experiments (ALENEX2002)*, January 2002, pp. 178–193.
10. Bin Lin, and Jianwen Su, "On Bulk Loading TPR-tree," In *Proc. of the 2004 IEEE Intl Conf. on Mobile Data Management (MDM'04)*, Berkeley, California, January 2004.
11. Mong Li Lee, Wynne Hsu, C. S. Jensen, Bin Cui, and Keng Lik Teo, "Supporting Frequent Updates in R-Trees: A Bottom-Up Approach," In *Proc. of the 29th Int'l Conf. on VLDB*, Berlin, Germany, September 2003.
12. D. Pfoser, and C. S. Jensen, "Indexing of Network-Constrained Moving Objects," In *Proc. of the 11th International Symp. on Advances in Geographic Information Systems*, New Orleans, LA, November 2003.
13. L. Speicys, C. S. Jensen, and A. Kligys, "Computational Data Modeling for Network-Constrained Moving Objects," In *Proc. of the 11th International Symp. on Advances in Geographic Information Systems*, New Orleans, LA, November 2003.
14. C. Hage, C. S. Jensen, T. B. Pedersen, L. Speicys, and I. Timko, "Integrated Data Management for Mobile Services in the Real World," In *Proc. of the 29th Int'l Conf. on VLDB*, Berlin, Germany, September 2003.
15. DongSeop Kwon, SangJun Lee, and SukHo Lee, "Indexing the Current Positions of Moving Objects Using the Lazy Update R-tree," In *Proc. of the 3rd Int'l. Conf. on Mobile Data Management (MDM2002)*, Singapore, January 2002, pp.113-120.

Topological Consistency for Collapse Operation in Multi-scale Databases

Hae-Kyong Kang[1], Tae-Wan Kim[2], and Ki-Joune Li[3]

[1] Department of GIS
Pusan National University, Pusan 609-735, South Korea
hkkang@pnu.edu
[2] Research Institute of Computer Information and Communication
Pusan National University, Pusan 609-735, South Korea
twkim@pnu.edu
[3] Department of Computer Science and Engineering
Pusan National University, Pusan 609-735, South Korea
lik@pnu.edu

Abstract. When we derive multi-scale databases from a source spatial database, the geometries and topological relations, which are a kind of constraints defined explicitly or implicitly in the source database, are transformed. It means that the derived databases should be checked to see if or not the constraints are respected during a derivation process. In this paper, we focus on the topological consistency between the source and derived databases, which is one of the important constraints to respect. In particular, we deal with the method of assessment of topological consistency, when 2-dimensional objects are collapsed to 1-dimensional ones. We introduce 8 topological relations between 2-dimensional objects and 19 topological relations between 1-dimensional and 2-dimensional objects. Then we propose four different strategies to convert the 8 topological relations in the source database to the 19 topological relations in the target database. A case study shows that these strategies guarantee the topological consistency between multi-scale databases.

1 Introduction

Multi-scale databases are a set of spatial databases on the same area with different scales. In general, small-scale databases can be derived from a larger-scale database by a generalization method[17]. During a generalization process, both geometries and relations such as topology in the source database are to be changed[11]. Although the relations should be respected, parts of them may not be identical. Thus, it is needed to assess if the derived relations are consistent or not.

The topological consistency can be checked in two ways. The first way is that if the topologies in source and derived databases are identical[16], the two databases are consistent as shown by figure 1-(1). In this figure, *Roads* in the source database figure 1-(1)(A) does not intersect with *Buildings*, while *Roads*

S. Wang et al. (Eds.): ER Workshops 2004, LNCS 3289, pp. 91–102, 2004.
© Springer-Verlag Berlin Heidelberg 2004

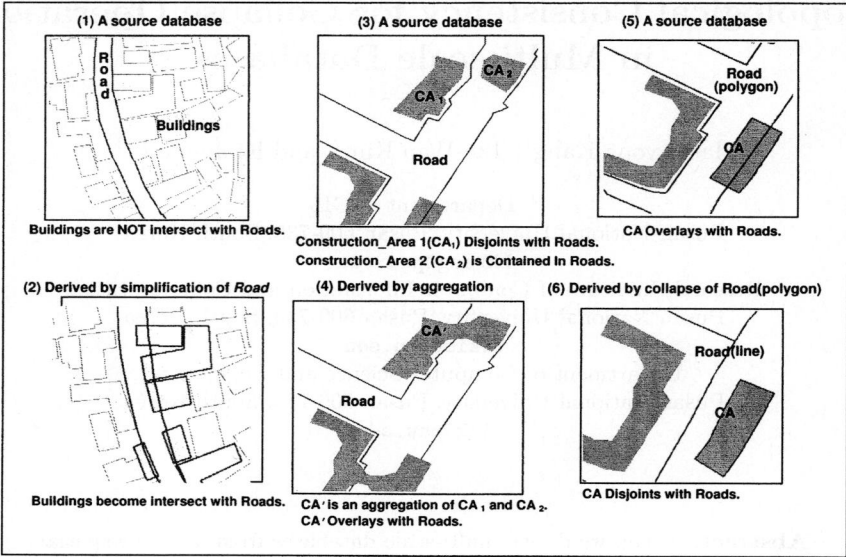

Fig. 1. Conversion of topological relations in multi-scale databases

in the derived database figure 1-(1)(B) intersects with *Buildings*. Consequently, the two databases are inconsistent and the intersected *Buildings* with *Roads* should be corrected. This approach can be applied when generalization operators change geometrical properties such as simplification. The second way is to check if or not topological relations in the derived database correctly correspond with the relations in the source database despite of their difference. In the figure 1-(3) and (4), for instance, *disjoint* in a source database can be converted into *overlay* in its derived database. Although the two relations are different, it may be a correct correspondence in a generalization process. While previous researches[4, 6, 8, 9, 13, 15] have focused on this field that the consistent conversion of topological relations between 2-dimensional objects, we consider the topological consistency in the case of the reduction of spatial dimension from 2-dimension to 1-dimension. Figure 1-(5) and (6) shows the motivation of our work.

The *overlap* relation between *Road* and *CA* in the source database becomes a *disjoint* relation in the derived database by a collapse operation. *Overlap* may be converted into another relations such as *meet* instead of *disjoint* depends on collapse algorithms. However, all of these conversions may not be acceptable. For instance, *disjoint* relations in the derived database is not acceptable since *Road(line)* and *CA* are intersect whereas *Road(polygon)* and *CA* in the source are not intersect. Thus, we must determine the possible conversions of relations to guarantee the consistency between the source and derived databases. The goal of our work is to discover the adequate topology conversions from the source database to the derived database by a *collapse* operator. In this paper, we propose four topology conversion strategies and give a comparison between these strategies.

The rest of this paper is organized as follows. In section 2, we will give a survey on the related work and explain the topological relations between two polygons and between a polygon and a line. Based on these topological relations, four topology conversion strategies will be proposed in section 3. A case study will be presented to compare these strategies in section 4. Finally, we will conclude this paper in section 5.

2   Related Work and Preliminaries

2.1   Related Work

We focus on the topological consistency on multi-scale databases. Thus our work is related to (i) topological relations, (ii) derivation or inference of new relations from existing relations in a source database, (iii) consistent conversion of topological relations on multi-scale databases. We will introduce a few novel researches on these themes briefly here and mention the limitation. Topological relations will be described in the section 2.2, separately.

• **Inference of spatial relations:** Relations in derived databases can be defined automatically by inference of relations in source databases. Thus we need to investigate the researches about inference of spatial relations. In [4], the inference of new spatial information has been formalized with properties of the binary topological relations such as symmetry, transitivity, converseness, and composition. For example of transitivity, *A contains B* and *B contains C* implies that *A meet C* is obviously inconsistent, therefore, {*contains, contains*} can not be transited to *meet*. Although this example shows part of the idea of [4], it proposed an assessment way for inferred spatial information. In [8], the transitivity of relations with set-inclusion was introduced to derive new topological relations. For instance, *A includes B* and *B intersect C* implies *A intersect C*. This approach can not be applied when relations are not represented by the set-inclusion. In [9], topology, cardinal directions like north, approximate distance like far and near, and temporal relations such as before and after were considered whereas previous researches were treated topological relations. The composition of different kinds of relations may have meaningful results since inferences can be made from not only isolated individual relations. For instance, *A is north from B* and *B contains C* implies that *A is north from C* and *A is disjoint with C*. During the derivation of a multi-scale database, spatial objects in a source database can be distorted by aggregation or collapse operators. However, these researches considered spatial objects which are not distorted unlike objects in multi-scale databases, and relations between regions(2-dimensional objects).

•**Consistent conversion of relations:** Relations in source database are, occasionally, converted to different but adequate ones on multi-scale databases. In this case, similarity or consistency between converted relations and its original ones needs to be evaluated. In [6], a boundary-boundary intersection was proposed to assess similarity of two relations on multi-scale representations.

The boundary-boundary intersection is part of 9-intersection model[5], and if boundary-boundary intersections of two relations are same each other, the two relations are same. The idea was developed based on the monotonicity assumption of a generalization, under any topological relations between objects must stay the same through consecutive representations or continuously decrease in complexity and detail. In [13], a systematic model was proposed to keep constraints that must hold with respect to other spatial objects when two objects are aggregated as shown in Figure 1-(3) and (4). This work can be a solution when a multi-scale database is derived by aggregation. However, we still need a solution for multi-scale databases derived by collapse from a source database.

2.2   Polygon-Polygon and Polygon-Line Topological Relations

The topological consistency is a core concept of our work. In order to explain it, we introduce the topological model [1, 2, 5, 7, 10, 14] in this section. The most expressive way to describe topological relations is *9-intersection model*. It describes binary topological relations between two point sets, A and B, by the set intersections of interior($\circ$), boundary(∂) and exterior($-$) of A with the interior($\circ$), boundary(∂), and exterior($-$) of B.

$$R(A, B) = \begin{pmatrix} A^\circ \cap B^\circ & A^\circ \cap \partial B & A^\circ \cap B^- \\ \partial A \cap B^\circ & \partial A \cap \partial B & \partial A \cap B^- \\ A^- \cap B^\circ & A^- \cap \partial B & A^- \cap B^- \end{pmatrix}$$

With 9-intersections being empty or non-empty, 512 topological relations are possible and 8 of which in figure 2-(1) can be realized for two polygons with connected boundaries if the objects are embedded in R^2. For lines and polygons, 19 distinct topological relations in figure 2-(2) are realized if lines are not branching and self-intersections and polygons are simple, connected and no holes in R^2 [1]. These categories of topological relations are an important basis of our work. Based on the categories, we will compare topological relations in source and derived databases respectively if they are consistent or not.

3   Consistent Correspondences Between Topological Relations

In this section, we introduce four approaches to correspond 8 topological relations in figure 2-(1) to 19 topological relations in figure 2-(2). We call the 8 polygon-polygon topological relations P-P relations and 19 polygon-line topological relations P-L relations for short. Four approaches are the matrix-comparison, the topology distance, the matrix-union, and hybrid approaches. The results of these correspondences are a partial set of 19 P-L relations, which is consistent with the set of P-P relations. In figure 3, we show the results of three approaches for comparisons. For example, *contain* in P-P relations corresponds to a R14 in P-L relations by a matrix-comparison and topology distance approach and R14, R15, R16, R17, R18, and R19 by the matrix-union approach. In the following, we describe how each of these approaches determines consistent relations.

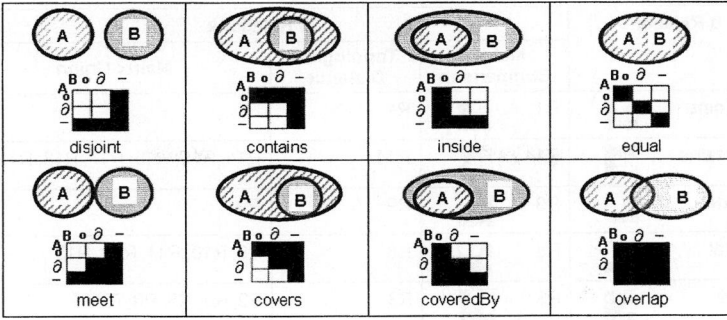

(1) 8 *P-P* Topological relations between polygons.

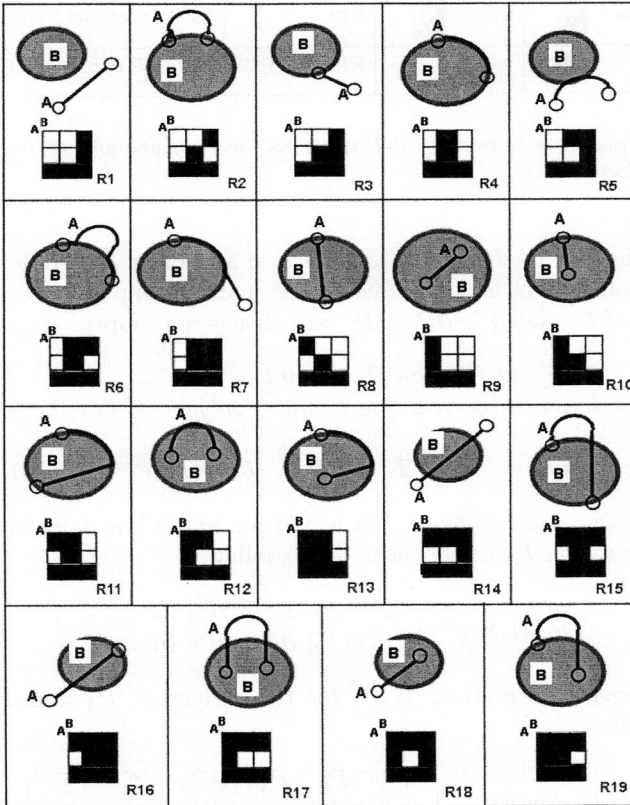

(2) 19 *P-L* Topological relations between polygons and lines.

Fig. 2. Topological Relations

3.1   Matrix-Comparison Approach

We represent a topological relation R as the 9-intersection matrix M_R. When M_Rs of two relations are equal, the two relations are the same. It is the same

8 Relations		19 Relations		
		Matrix-Comparison	Topological Distance	Matrix-Union
Disjoint		R1	R1	R1
Contains		R14	R14	R14, R15, R16, R17, R18, R19
Inside		R9	R9	R9
Equal		R8	R8	R8, R10, R11, R12, R13
Meet		R3	R3	R2, R3, R5, R6, R7
Covers		R16	R16	R14, R15, R16, R17, R18, R19
coveredBy		R10	R10	R8, R10, R11, R12, R13
Overlap		none	R16, R18, R19	R14, R15, R16, R17, R18, R19

Fig. 3. Corresponences between 8 *P-P relations for Polygons and 19 P-L relations for Polygon and Line*

with the basic idea of the matrix-comparison approach. The difference is that a 6-intersection matrix is considered in the matrix-comparison approach since intersections of L^- and P^o, ∂P or P^- are always not-empty.

Definition 1. *Line and Polygon Constraint*
A topological relation between a line L and a polygon P has the following constraint:

$$(L^- \cap P^\circ = \neg\emptyset) \wedge (L^- \cap \partial P = \neg\emptyset) \wedge (L^- \cap P^- = \neg\emptyset)$$

The 6-intersection matrix, $M_{R'_{1D}}$, for a polygon and a line L which is a simplification of a polygon P_A are constructed as follow:

$$M_{R'_{1D}(L,P_B)} = \begin{pmatrix} L^\circ \cap P_B^\circ & L^\circ \cap \partial P_B & L^\circ \cap P_B^- \\ \partial L \cap P_B^\circ & \partial L \cap \partial P_B & \partial L \cap P_B^- \end{pmatrix}$$

The 6-intersection matrix, $M_{R'_{2D}}$ for two polygons, P_A and P_B, are constructed as follow:

$$M_{R'_{2D}(P_A,P_B)} = \begin{pmatrix} P_A^\circ \cap P_B^\circ & P_A^\circ \cap \partial P_B & P_A^\circ \cap P_B^- \\ \partial P_A \cap P_B^\circ & \partial P_A \cap \partial P_B & \partial P_A \cap P_B^- \end{pmatrix}$$

Matrix-Comparison is that the two relations are consistent if and only if $M_{R'_{1D}}$ and $M_{R'_{2D}}$ are identical. For example, *equal* and R_8 are consistent since their 6-intersection matrix representations are the same.

$$M_{equal} \begin{pmatrix} 1 & 0 & 0 \\ 0 & 1 & 0 \end{pmatrix} = M_{R_8} \begin{pmatrix} 1 & 0 & 0 \\ 0 & 1 & 0 \end{pmatrix}$$

3.2   Topology Distance Approach

Similarity between two relations, R_1 and R_2, can be measured by a topology distance T_{R_1,R_2} as follows [3]:

$$T_{R_1,R_2} = \sum_{i=0}^{2} \sum_{j=0}^{2} \mid M_{R_1}[i,j] - M_{R_2}[i,j] \mid$$

The next example shows the calculation of a topology distance between *contains* and R_{16} in figure 2.

$$M_{contains} = \begin{pmatrix} 1 & 1 & 1 \\ 0 & 0 & 1 \\ 0 & 0 & 1 \end{pmatrix}, \quad M_{R_{16}} = \begin{pmatrix} 1 & 1 & 1 \\ 0 & 1 & 1 \\ 1 & 1 & 1 \end{pmatrix}, \quad M_{contains} - M_{R_{16}} = \begin{bmatrix} 0 & 0 & 0 \\ 0 & -1 & 0 \\ -1 & -1 & 0 \end{bmatrix}$$

$$\parallel M_{inside} - M_{contains} \parallel = 3$$

According to the topological distance approach, the correspondance of relations is shown figure 4. Smaller number of topology distances implies higher similarity of two relations. Therefore, the correspondent relations on the minimum topology distance should be selected as consistent relations.

Topology Distance / Relations	0	1	2	3	4	5	6	7	8
Disjoint	r1	r3, r5	r2, r7, r14	r6, r16, r18	r4, r8, r9, r15, r17	r10, r11, r12, r19	r13	X	x
CoveredBy	r10	r8, r9, r13	r11, r12, r19	r2, r4, r15, r17					
Cover	X	X	r16	r2, r4, r15, r17	r3, r6, r16, r18	r1, r7, r14, r15			
Overlap	x	r16, r18, r19	r7, r13, r14, r15, r17	r3, r5, r6, r10, r11, r12	r1, r2, r4, r8, r9				
Meet	r3	r1, r2, r7	r5, r6, r16	r4, r8, r14, r1 5	r10, r11, r18, r19	r9, r13, r17	r12		
Equal	X	X	r8	r10, r11	r2, r4, r9, r13, r15	r3, r6, r12, r16, r19	r1, r7, r14, r17	r5, r18	
Contain			r14	r5, r16, r18	r1, r7, r15, r17	r3, r6, r11, r12, r19	r2, r4, r8, r9, r13	r10	X
Inside	r9	r10, r12	r8, r13, r17	r11, r18, r19	r1, r2, r4, r14, r15	r3, r5, r6, r16,	R7		

Fig. 4. Topology Distance between 8 *P-P relations for polygons and 19 P-L relations for polygons and lines*

This approach has more expressive power than the matrix comparison approach since all of 8 *P-P* relations correspond to 19 *P-L* relations. But it has ambiguity in correspondences, that is has multiple corresponding relations in case of *overlap*. That is, *overlap* relations has three corresponding relations R_{16}, R_{18}, and R_{19}. Arbitrary selection of one of three possible relations does not still suggest satisfiable solution yet. That is, a relation whose distances are larger than another relations, can be a better relation in the context of generalization.

3.3   Matrix Union Approach

The main purpose of the matrix-union approach is to overcome the limitations of the topological distance approach, we described above, by proposing multiple consistent relations. Our purpose is to propose a method that finds consistent correspondences between P-P relations and P-L relations. A polygon could be collapsed to a line in generalization process. In other words, the interior and boundary of a polygon is collapsed to a line. Thus we can redefine a line as follows:

Definition 2. *Line(collapsed polygon)*
A line L is a subset of the union of the interior and the boundary of a polygon P. That is,

$$L \subseteq P^\circ \cup \partial P$$

Based on the definition 1 and 2, we compare 3-intersection matrixes of two relations. We call the 3-intersection matrix for a line and a polygon as $M_{R_{1D}}$ and for two polygons as $M_{R_{2D}}$. $M_{R_{1D}}$ and $M_{R_{2D}}$ are constructed as follow:

$$M_{R_{1D}(L,P_B)} = \left(L^{\circ \cup \partial} \cap B^\circ \quad L^{\circ \cup \partial} \cap \partial B \quad L^{\circ \cup \partial} \cap B^- \right)$$

$$M_{R_{2D}(P_A,P_B)} = \left((A^\circ \cup \partial A) \cap B^\circ \quad (A^\circ \cup \partial A) \cap \partial B \quad (A^\circ \cup \partial A) \cap B^- \right)$$

Matrix-Union is that the two relations are consistent if and only if $M_{R_{1D}}$ and $M_{R_{2D}}$ are identical. This approach allows multiple candidates when we generalize the P-P relation to the P-L relation. However, it may fail to suggest the most appropriate relation in some cases due to a information loss during a union process, unlike the matrix-comparison and the topological-distance approaches. In the next section, we suggest another approach, *the hybrid approach*, which resolve ambiguity in the matrix-union approach by proposing the way to select the best one.

3.4   Hybrid Approach: Matrix-Union and Topological Distance

For candidates relations found by the matrix-union approach, this approach applies topological distance to each of these relations and thus orders each of them. In generalization process, this approach suggests to select a relation among multiple candidate relations by this ordering. In figure 5, we show a set of consistent relations which are ordered by topological distance. Axis y denote the degree of consistency and axis x shows each of P-P relation. For example, the topological distance between *equal* and R_8 is 0 and thus this pair has the highest consistency among $\{R8, R10, R11, R13, R12\}$. But this fact does not mean we always choose R_8 when we generalize *equal* in $P - P$ relation to one of P-L relation. Since we do not know exact generalization process, e.g. threshold value to collapse polygons, criteria to select objects to be generalized, algorithms used to generalize, etc. Main purpose of the hybrid approach is to suggest a set of consistent correspondent relations, and is to reject relations which are not included in this set.

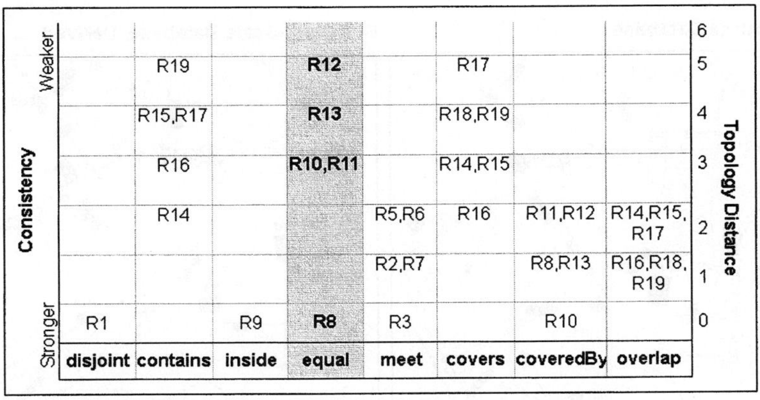

Consistency	disjoint	contains	inside	equal	meet	covers	coveredBy	overlap	Topology Distance
Weaker									6
		R19		**R12**		R17			5
		R15,R17		**R13**		R18,R19			4
		R16		**R10,R11**		R14,R15			3
		R14			R5,R6	R16	R11,R12	R14,R15, R17	2
					R2,R7		R8,R13	R16,R18, R19	1
Stronger	R1		R9	**R8**	R3	R10			0

Fig. 5. A Hybrid Approach(Matrix-Union and Topology Distance)

4   A Case Study

In the precedent section, we have proposed four approaches to define the consistent correspondences between 8 *P-P* topological relations and 19 *P-L* topological relations. These approaches, first, aims to detect erroneous or inconsistent conversion of topological relations on multi-scale databases and to let users correct the inconsistent relation conversions. On the second, they are useful in choosing the best topological relations among several possible topology conversions.

In this section, we present an example of application of the topology distance and matrix-union approaches to detect inconsistent topology conversions during a derivation of multi-scale database. For this example, we prepared a large-scale source database containing roads and road facilities, as shown by figure 6-A. Roads, which are 2-dimensional objects in the source database are collapsed into lines in the derived database. During the collapse, the topologies between roads and road facilities are converted as depicted in figure 6-B.

We can detect inconsistent relations with the consistent correspondences proposed in the previous section. Figure 7 shows consistent and inconsistent topological relations detected by the topological distance method and matrix-union method, respectively. By topological distance method and matrix-union method, we obtain different detections of inconsistent topological conversions. With small values of topological distance, we can allow only a small set of topologies in the derived database. For example, we observe that only 6 topological relations are considered as consistent when topological distance is 1, while 11 topological relation are detected as inconsistent relation on topology distance 2. It means that we enforce the topological consistency by giving a small value of topological distance. Figure 7-C shows the result of the matrix-union approach. The road facilities with circle indicate the difference between the results of topological distance method, when the topological distance is 1. They are considered as inconsistent by topological distance method, while they are consistent by matrix-union method. But in reality, it is obvious that they are not inconsistent unlike

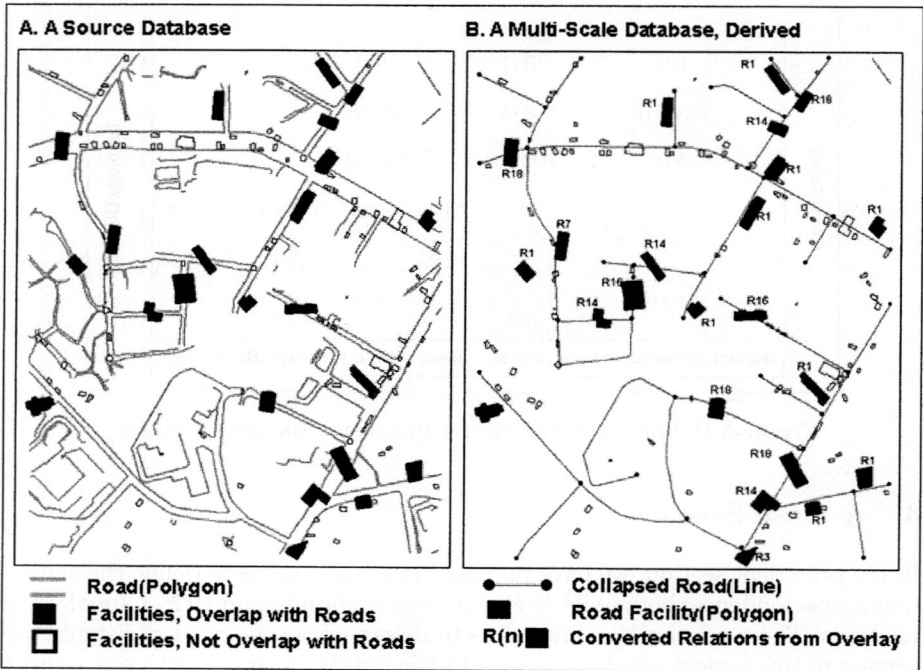

Fig. 6. A Source and A Multi-scale Databases

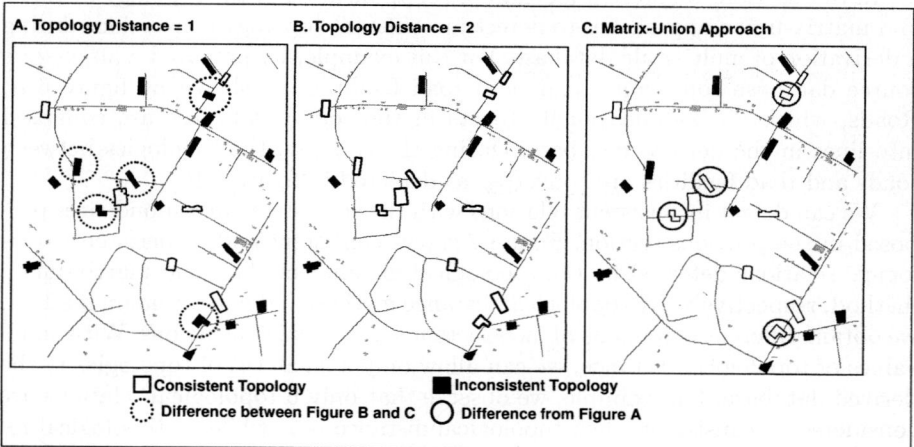

Fig. 7. Topology Distance and Matrix-Union Approaches

figure 7-A, and the results of matrix-union are correct as shown by figure 7-C since *overlap* topology between polygons must be still *overlap* after the collapse operation as well. On the contrary, when topological distance is given as 2, the results of two methods are almost similar as shown by figure 7-B and 7-C, except the road facilities with dotted circle. But there is no reason to consider it as inconsistent relation, since *overlap* relation can be converted to *meet* relation like the object of dotted circle in figure 7-B. It means that topological distance method is better than matrix-union method in this case.

From this case study, we conclude that none of the proposed methods exactly discovers inconsistent and consistent relations in derived database. The reason is that we cannot handle perfectly all cases in the real world by any method without intervention of users. And an appropriate method can be chosen by trial and errors of users, which satisfies the most requirements of users. For example, it is preferable to select the method detecting as many inconsistent relations as possible, in cases where topological relations are important.

5 Conclusion

When we generalize a source spatial database of a large scale to another database of a small scale, not only topological relations but also geometries are changed from the source database. Some topological conversions from a source database may be incorrect and we must detect erroneous conversions.

In this paper we proposed several methods to assess the consistency of topological conversions when polygonal object is collapsed to a line. Our methods are based on 8-topological relation model between polygon and polygon, and 19-topological relation model between polygon and line [1]. In particular, we defined the consistent topological correspondences between the source and derived topologies by four different approaches. With these correspondences, inconsistent or erroneous topology conversions can be detected and we can maintain the consistency between source and derived databases during generalization or updates on the source database.

We have considered only the case where 2-dimensional objects are collapsed to 1-dimensional one, and the rest types of collapse operations have not been fully investigated in this paper. For example, we should define a similar type of correspondences between topologies in source and derived databases, where 2-dimensional objects are collapsed to 0-dimensional ones. Future work therefore includes the completion of the consistent correspondences considering all cases of collapse operations.

References

1. M. J. Egenhofer and H. Herring, *Categorizing Binary Topological Relations Between Regions, Lines, and Points in Geographic Databases*, Technical Report, Department of Surveying Engineering, University of Maine, 1990.
2. M. J. Egenhofer, *Point-Set Topological Spatial Relations*, International Journal of Geographical Information Systems 5(2):161-174, 1991.
3. M. J. Egenhofer and K. K. Al-Taha, *Reasoning about Gradual Changes of Topological Relationships*, Theory and Methods of Spatio-Temporal Reasoning in Geographic Space, LNCS, VOL. 639, Springer-Verlag, 196-219, 1992.

4. M. J. Egenhofer, and J. sharma, *Assessing the Consistency of Complete and Incomplete Topological Information*, Geographical Systems, 1(1):47-68, 1993.
5. E. Clementini, J. Sharma, and M. J. Egenhofer, *Modeling Topological Spatial Relations: Strategies for Query Processing*, Computer and Graphics 18(6):815-822, 1994.
6. M. Egenhofer, *Evaluating Inconsistencies Among multiple Representations*, 6th international Symposium on Spatial Data Handling, 902-920, 1994.
7. M. J. Egenhofer, E. Clementini and P. Felice, *Topological relations between regions with holes*, International Journal of Geographical Information Systems 8(2):129-144, 1994.
8. M. J. Egenhofer, *Deriving the Composition of Binary Topological Relations*, Journal of Visual Languages and Computing, 5(2):133-149, 1994.
9. J. Sharma, D. M. Flewelling, and M. J. Egenhofer, *A Qualitative Spatial Reasoner*, 6th International Symposium on Spatial Data Handling, 665-681, 1994.
10. D. M. Mark, and M. J. Egenhofer, *Modeling Spatial Relations Between Lines and Regions:Combining Formal Mathematical Models and Human Subjects Testing*, Cartography and Geographical Information System, 21(3):195-212, 1995.
11. Muller J. C., Lagrange J. P. and Weibel R., *Data and Knowledge Modelling for Generalization in GIS and Generalization*, Taylor & Francis Inc., 73-90, 1995.
12. M. J. Egenhofer, *Consistency Revisited*, GeoInformatica, 1(4):323-325, 1997.
13. N. Tryfona and M. J. Egenhofer, *Consistency among Parts and Aggregates: A Computational Model.* Transactions in GIS, 1(3):189-206, 1997.
14. A.Rashid, B. M. Shariff and M. J. Egenhofer, *Natural-Language Spatial Relations Between Linear and Areal Objects: The Topology and Metric of English Language Terms*, International Journal of Geographical Information Science, 12(3):215-246, 1998.
15. J. A. C. Paiva, *Topological Consistency in Geographic Databases With Multiple Representations*, Ph. D. Thesis, University of Maine, 1998, http://library.umaine.edu/theses/pdf/paiva.pdf.
16. A. Belussi, M. Negri and G. Pelagatti, "An Integrity Constraints Driven System for Updating Spatila Databases", Proc. of the 8th ACMGIS, 121-128, 2000.
17. H. Kang, J. Moon and K. Li, *Data Update Across Multi-Scale Databases*, Proc. of the 12th International Conference on Geoinformatics, 2004.

Symbolization of Mobile Object Trajectories with the Support to Motion Data Mining

Xiaoming Jin, Jianmin Wang, and Jiaguang Sun

School of Software
Tsinghua University, Beijing, 100084, China
xmjin@mail.tsinghua.edu.cn

Abstract. Extraction and representation of the events in trajectory data enable us go beyond the primitive and quantitative values and focus on the high level knowledge. On the other hand, it enables the applications of vast off the shelf methods, which was originally designed for mining event sequences, to trajectory data. In this paper, the problem of symbolizing trajectory data is addressed. We first introduce a static symbolization method, in which *typical sub-trajectories* are generated automatically based on the data. For facilitating the data mining process on streaming trajectories, we also present an incremental method, which dynamically adjusts the *typical sub-trajectories* according to the most recent data characters. The performances of our approaches were evaluated on both real data and synthetic data. Experimental results justify the effectiveness of the proposed methods and the superiority of the incremental approach.

Keywords: Motion data mining, spatial trajectory, symbolization

1 Introduction

The recent advances in geographic data collection devices and location-aware technologies have increased the production and collection of spatial trajectories of moving objects [1, 2]. Well known examples include global positioning systems (GPS), remote sensors, mobile phones, vehicle navigation systems, animal mobility trackers, and wireless Internet clients. Briefly, trajectory of a moving object is a sequence of consecutive locations of the object in a multidimensional (generally two or three dimensional) space [3]. Fig. 1 shows a simple example of 2D trajectory data.

Recently, developments of data mining techniques on trajectory data have received growing interests in both research and industry fields [4, 5, 6]. For example, sequential rules with the format "if *movement A* then *movement B*" can be discovered by searching the historical trajectories generated by a vehicle user. Then based on the rules, personalized services can be provided according to the user's current movements in a vehicle navigation system. In another case, by analyzing the trajectories generated by all vehicles in a city, we can find some frequent patterns that reveal important passages, crossroads, highway, or other traffic facilities that are used frequently. The results of such analysis can be used for improving the management and maintenance of the traffic system.

S. Wang et al. (Eds.): ER Workshops 2004, LNCS 3289, pp. 103–113, 2004.
© Springer-Verlag Berlin Heidelberg 2004

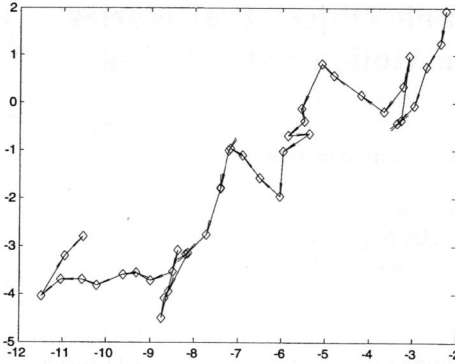

Fig. 1. An example of moving object trajectory in 2D space. Each diamond in the figure stands for a single location of the object at a certain time, and arrow represents a movement from one location to another.

In this paper, we address a novel problem, which is symbolization of trajectories. The problem is to map a trajectory into a symbolic sequence, in which each symbol represents the local movement at a time point or during a time period. For example, the trajectory of a car movement trajectory can be simply represented by a symbolic sequence in which each symbol indicates the shape of a quarter-hourly movement. Then the symbolic sequence can be examined for discovering rules such as "if a particular movement *A*, indicating a crossroads, occurs, then the car will go north straightly, say movement *B*, in the second quarter-hour."

Our research on this topic is mainly motivated by two ideas: First, there is a broad consensus that we are usually interested in high level representation of the data, rather than the primitive and quantitative values. Second, techniques for mining symbolic (event) sequences have been studies extensively in various application domains, such as click stream analysis, bio-informatics, and so on, and many sophisticated algorithms, models, and data structures were designed for handling symbolic sequences. Therefore, converting trajectory data into symbolic sequence could enable the applications of the vast off the shelf methods in mining trajectory data. Obviously, an appropriate and efficient method for extracting and representing the events in the trajectory data will achieve these goals, and whereupon, benefits the data mining tasks over trajectory data.

Symbolization is useful in mining spatial trajectory data. On the other hand, it can also help representing other data objects that seem different but are essentially with similar characters, such as features extracted from signature image and multiple attribute response curves in drug therapy.

Generally, symbolization can be viewed as an explaining process that classifies (or approximates) each individual "atomic" sub-trajectory into a *typical sub-trajectory* movements, e.g. "go north", "circumambulate", which is retrieved or defined beforehand. A simple and intuitive solution for generating the *typical sub-trajectory* is to choose it manually based on the domain expert's analysis and explanation. And then the "atomic" sub-trajectory at each time is represented by a simple nearest neighbor query through the given sets of *typical sub-trajectory*. Such ideas had been embedded into some ad hoc problem definitions, as well as data mining methods, implicitly or explicitly. However, it is usually extremely expensive to find and understand all candidate sub-trajectories in many real applications, whereupon it is very difficult, if not impossible to generate *typical sub-trajectories* by the above manual approach.

Another important aspect of mobile object trajectories is that the data are usually born with streaming property in many application domains. That is, the trajectories are frequently appended in the end over time. Since many factors that impact a moving object might be time varying, the patterns of the object movements are usually time varying, whereupon the *typical sub-trajectory* should also be adjusted correspondingly. In this case, it is usually difficult to apply a static approach from a practical point of view, where *typical sub-trajectories* are generated based on a snapshot of the trajectory, because it may fail to give a good representation of the new data. Alternatively, the *typical sub-trajectories* can be re-generated concurrently with each update. However, such re-generation will need all the data scanned, whereupon the time complexity might be extremely poor.

The above issues challenge the research on symbolizing trajectory data. Therefore, it is by no means trivial to consider this novel problem and develop effective and efficient methods for it. In this paper, the problem of symbolizing trajectory data is addressed. We first propose a symbolization method, in which *typical sub-trajectory* is generated automatically based on the data. For facilitating the data mining process on streaming trajectories, we also present an incremental symbolization method, which dynamically adjusts the *typical sub-trajectories* according to the most recent data characters, without scanning the whole date set. The performance of our approach is evaluated on both real trajectory data and synthetic data. Experimental results justify the effectiveness of the proposed methods and the superiority of the incremental approach.

2 Related Work

Data management and data mining on trajectory data had been studied in many applications [1, 2, 3, 4, 5]. Data mining applications can be found in [5, 6]. The most fundamental works in these contexts is on similarity measuring and indexing, e.g. [7, 8, 9, 10]. Related methods were developed mainly by extending the exist ones, e.g. Euclidean distance, Dynamic Time Warping (DTW), Longest Common Subsequence (LCSS), and multi-dimensional indexes. These works did not consider the symbolization process, but many ideas may serve as subroutines in the approach proposed in this paper.

In some previous work, unsupervised learning techniques were applied on the trajectories or similar data objects. For example, [11] presented an approach to clustering the experiences of an autonomous agent. Such methods were designed for purposes fundamentally different from ours, and were not further developed to fit the general data mining tasks.

Symbolization on time series data, as an important preprocessing subroutine, had been extensively studied for various data mining tasks [12,13,14], such as rule discovery, frequent pattern mining, prediction, and query by content, etc. A shape definition language was proposed in [15]. In [16], the time series was symbolized using cluster method for discovering rules with the format "if event A occurs, then event B occurs within time T." This method was then used in many applications that focus on mining time series [13,14]. [17] Claims that the method in [16] is meaningless if the step of sliding window is set to be 1. Actually, this problem could be solved by increasing the step of sliding windows. The above works deal with one-dimensional time series and the further extension for mining trajectory data have not been well considered. Our approach can be viewed as an expansion of the method introduced in [16] to facilitate the data mining process on trajectory data.

3 Problem Descriptions

A trajectory of a moving object is a sequence of consecutive locations in a multidimensional (generally two or three dimensional) space. In this paper, we only address the trajectories in 2D for clearness. A trajectory is denoted as: $T=T(1),...,T(N)$ and $T(n)=(Tx(n), Ty(n))$ stands for the location of the moving object at time n. $|T|=N$ denotes the length of T. The projection of T in x-axis and that in y-axis are represented as $Tx = Tx(1),..., Tx(N)$ and $Ty = Ty(1),..., Ty(N)$ respectively. The sub-trajectory $T(m), T(m+1),...,T(n)$ is denoted by $T[m,n]$.

As introduced in section 1, symbolization is a process to represent the object's behaviors at each individual time point. Then it is intuitive to first divide the trajectory to extract the sub-trajectories sequentially at various time points, and then to symbolize each extracted sub-trajectory individually by comparing it with a group of *typical sub-trajectories*. A *typical sub-trajectory* is a "prevalent" sub-trajectory that represents a typical form of sub-trajectory movements. Here we use the intuitive sliding window approach: Given trajectory T, window sub-trajectories of T are contiguous sub-trajectories $W_n=T[s(n),e(n)]$ or $(Wx_n, Wy_n)=(Tx[s(n),e(n)], Ty[s(n),e(n)])$ extracted sequentially by sliding the window through the trajectory, where $s(n)=nk-k+1$ and $e(n)=nk-k+w-1$ stand for the starting point and ending point of the n-th window respectively, n denotes the order number of the windows, parameter w controls the size of each window, and k controls the offset of positions of two consecutive windows.

Based on the above notions, the problem of symbolizing trajectory can be defined as follows: Given a trajectory T and all its window sub-trajectory W_n, symbolization is to convert T into a temporal sequence $S=(S(1),...,S(M))$, which is a ordered list of symbols where each symbol $S(n)$ comes from a predefined alphabet Σ and represents the movements in the n-th sub-trajectory W_n.

4 Symbolization Approach

The key idea of our symbolizing approach is to cluster all the sub-trajectories, and then each sub-trajectory is represented by the identifier of the cluster that contains it. The overall approach is formally illustrated as follows:

1) Extract all sub-trajectories W_n in trajectory S (W_n is defined in section 3).
2) Normalize each sub-trajectory W_n to W_n'.
3) Cluster all W_n' into sets $C_1, ... , C_H$, that is, $W_n' \in C_{j(n)}$.
4) For each cluster C_h, a unique symbol a_h from Σ is inducted.

5) The symbol sequence S is obtained by looking for each W_n' the cluster $C_{j(n)}$, and using the corresponding symbol $a_{j(n)}$ to represent the sub-trajectory at that point, i.e. $S = a_{j(1)}, a_{j(2)}, \ldots , a_{j(M)}, M = \lfloor N/k \rfloor$.

Though the overall strategy seems similar with that for time series data, the symbolization process for spatial trajectory is different in several important aspects, e.g. normalization process, similarity measurement, and cluster process.

The normalization step is applied for precise representation by removing the impacts of the absolute location value and the scaling factors in both dimensions. For time series $X=(X(1),\ldots,X(N))$, normalization can be easily done by $X'=(X-E(X))/D(X)$ where $E(X)$ is the mean of X and $D(X)$ is the standard deviation of X. However, this problem is somewhat less straightforward for trajectory data, because the scale of different dimension may differ. For example, given a trajectory T, it may be unadvisable to simply use the above strategy on Tx and Ty respectively, since this process zooms the movements in both dimensions into exactly the same scale, whereupon the shape of the original trajectory may be demolished. We normalize a sub-trajectory $W_n = (Wx_n, Wy_n)$, as follows:

$$Wx'_n = (Wx_n - E(Wx_n))/ \max(D(Wx_n), D(Wy_n))$$
$$Wy'_n = (Wy_n - E(Wy_n))/ \max(D(Wx_n), D(Wy_n))$$

This normalizing process zooms a sub-trajectory into a cell with unit size without changing its shape. The difference of the above two normalization processes is demonstrated in Fig. 2.

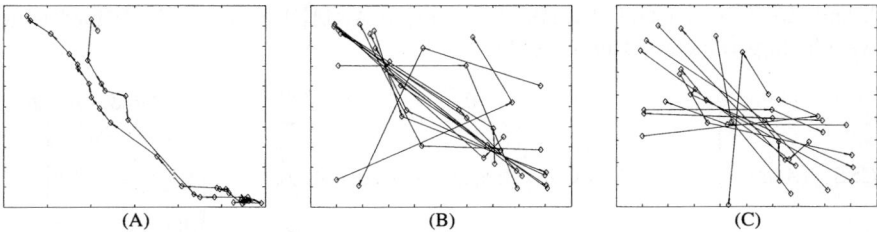

(A)   (B)   (C)

Fig. 2. Example trajectory (A) and the normalized sub-trajectories with the original shapes modified (B), and normalized sub-trajectories with the shape preserved (C).

In some applications, the information of interest is the relative movements that are irrelevant to its absolute directions, e.g. "go left" or "forward", rather than "go north" or "go along a meridian 45° counterclockwise from east." On such occasions, a rotating transform need to be applied based on the starting direction, i.e. each movement (represented by two consecutive locations) in a sub-trajectory is rotated to its relative direction to the last movement in the preceding sub-trajectory.

Clustering is the process of grouping a set of objects into classes of similar objects. Our symbolization approach has no constraint on the clustering algorithm, any common distance based clustering method could be used here, e.g. greedy method, recursive k-means, agglomerative and divisive hierarchical clustering, BIRCH, CUBE,

Chameleon, etc. We use greedy method as the cluster subroutine in presenting our method and experiments, because the time complexity is sound, it can be easily implemented, and it does a good job in supporting the incremental symbolization process. For each W_n, greedy method first finds the cluster center q such that the distance between W_n and q is minimal. If the distance is less than a predefined threshold d_{max}, W_n is added to the cluster whose center is q and the cluster center of q is regenerated as the point-wise average of all the sub-trajectory contained in it, otherwise a new cluster with center W_n is created.

In the cluster process, Euclidian distance is used as the distance measurement. Given two trajectories $A=(Ax(1),Ay(1)),\ldots, (Ax(|A|),Ay(|A|))$ and $B=(Bx(1),By(1)),\ldots, (Bx(|B|),By(|B|))$ with the same length M, the Euclidian distance between A and B is defined as:

$$D(A,B)=\sqrt{\sum_{i=1}^{M}\left((Ax(i)-Bx(i))^2+(Ay(i)-By(i))^2\right)}$$

Note that, other more sophisticated similarity measurements can also be applied in our method, such as *Dynamic Time Warping* (DTW) or *Longest Common Subsequence* (LCSS). DTW is defined as:

$$DTW(A,B)=Lp\left((Ax(|A|),Ay(|A|)),(Bx(|B|),By(|B|))\right)+\min\left\{\begin{array}{l}DTW(Head(A),B),\\ DTW(A,Head(B)),\\ DTW(Head(A),Head(B))\end{array}\right\}$$

where Lp stands for p-norm, i.e. $Lp(V1,V2)=(\sum_k|V1(K)-V2(K)|^p)^{1/p}$ for vector $V1$ and $V2$, and Head $(A) = (Ax(1),Ay(1)),\ldots, (Ax(|A|-1),Ay(|A|-1))$. LCSS is defined as follows: Given parameters δ and $0<\varepsilon<1$,

$$LCSS_{\delta,\varepsilon}(A,B)=\left\{\begin{array}{ll}0 & A=\phi\vee B=\phi\\ & |Ax(|A|)-Bx(|B|)|<\varepsilon\wedge\\ 1+LCSS_{\delta,\varepsilon}(Head(A),Head(B)) & |Ay(|A|)-By(|B|)|<\varepsilon\wedge\\ & \|A|-|B\|<\delta\\ \max(LCSS_{\delta,\varepsilon}(Head(A),B),LCSS_{\delta,\varepsilon}(A,Head(B))) & \text{otherwise}\end{array}\right.$$

Another important problem is the symbol mapping model, i.e. Σ. The symbol set is usually defined by a group of simple symbols, e.g. the lower case alphabet a...z. Since our symbolization approach generates the *typical sub-trajectories* automatically, the meaning of each symbol in Σ cannot be explained or defined a priori. However, after the whole trajectory has been symbolized, the meaning of symbol $a_h\in\Sigma$ can be extracted by manual reviews of the corresponding cluster center of C_h. Therefore, if meaningful representations are more favorable, a simple post-processing procedure can be applied by introducing a new alphabet Σ' that is generated by first replacing each symbol a_h in Σ by a meaningful one based on the manual explanation of the cluster C_h, and then rewriting the resulting symbolic sequence based on Σ'.

5 Incremental Symbolization Approach

Streaming trajectory is a trajectory with new data items generated and appended in the end frequently. On such occasions, the incremental symbolizing process can be formalized as follows: Whenever a trajectory T is updated to TU ($TU=T(1),\ldots,$ $T(N),U(1),\ldots,U(K)$ is the direct connection of $T=T(1),\ldots,T(N)$ and $U=U(1),\ldots,U(K)$), update the representing symbol sequence from S_T to $S_T S_U$, where S_T, S_U correspond to original trajectory T and the update U respectively. In this process, both *typical sub-trajectories* and symbol mapping model need to be dynamically updated with the data collections. Therefore, the static symbolization approaches are not applicable for streaming trajectories from a practical point of view as introduced in section 1.

The key idea of our dynamic symbolizing approach is that instead of using the static cluster centers as the *typical sub-trajectories*, we dynamically maintain the cluster information, and then the symbol sequence of new sub-trajectories is generated based on the up-to-date version of *typical sub-trajectories*.

Assume that, the method introduced in section 4 has already been applied on an initial trajectory, and the resulting cluster centers are saved as the initial *typical sub-trajectories*. Then, whenever the trajectory is updated, the cluster information is updated by introducing the new sub-trajectories and removing the "old" sub-trajectories with the generation time t such that $t < t_{now} - t_{max}$ where t_{now} denotes the current time and t_{max} is a predefined threshold *maximal time range*. Here it is assumed that the *typical sub-trajectories* will evolve with the updating process, whereupon the sub-trajectories that are generated too long ago will have minor effect on representing the current sub-trajectories.

The detailed method is illustrated as follows: First, extract all sub-trajectories W_n' contained in clusters and with the generation time $t < t_{now} - t_{max}$. Remove each W_n' from the corresponding clusters $C_{j(n)}$, and the cluster center $q_{j(n)}$ is re-computed as the point-wise average of all the sub-trajectories remained in $C_{j(n)}$. If $C_{j(n)}$ becomes empty after the deletion, remove it. Then extract and normalize all sub-trajectories contained in U, let the resulting sub-trajectories be W_m'. For each W_m', add it to a cluster $C_{j(m)}$ and generate symbol $a_{j(m)}$ for W_m'. The method used in this step is same as that introduced in section 4.

When a trajectory T is updated to TU, only the sub-trajectories in update part U or the "old" ones need to be considered. Then the total number of affected sub-trajectories is roughly $O(|U|+|U|)$. Each new sub-trajectory can be inserted into the proper cluster in $O(H)$ time where H is the number of clusters that is depended on predefined threshold d_{max}. And the update of *typical sub-trajectories* can be simply done by a weighted sum of the original *typical sub-trajectories* and the affected sub-trajectories. Therefore, the overall time complexity of the incremental symbolization approach is $O(H|U|+|U|)$. This time complexity, in our opinion, can meet the requirements of real applications.

6 Experimental Results

In this section, we present an empirical study of the proposed methods. The objectives of this study are: 1) to evaluate the effectiveness of the proposed method, and 2) to compare the performance of incremental symbolization method with the static one. In the experiments, two sets of data were used:

Real Data: The real data were a combination trajectory of a group of animal movements collected by satellite tracking. The whole trajectory consists of 238 locations (i.e. data points), each of which indicates the longitude and latitude of the observed object at a time point.

Synthetic Data: The real data is relatively small. To evaluate our approach on varying size data, we used synthetic data that were generated as $T(n)=L(n)+Rg$, where L was a trajectory that was generated based on the real data introduced above, but with varying size, each $L(n)$ was randomly selected around the relative position in the real trajectory. g was Gaussian noise with zero mean and unit variance. And parameter R controlled the noise-to-signal ratio.

The performance of a symbolization approach was evaluated by the average distance between original sub-trajectories and the corresponding *typical sub-trajectories*. That is, a representation that is more similar to the original sub-trajectory is a better symbolization result. Given the trajectory T, symbolization results is $S=a_{j(1)}, a_{j(2)}, \dots , a_{j(M)}$, then the performance of this symbolization process was evaluated as:

$$P(S)=\frac{1}{|S|}\sum_{m=1}^{|S|}D\left(C_{j(m)},W'_{m}\right)$$

where D is Euclidian distance, W_n' stands for the normalized sub-trajectories of T, and C stands for the cluster centers (detailed definitions can be found in section 4). It should be noted that the less the measuring result, the better the performance.

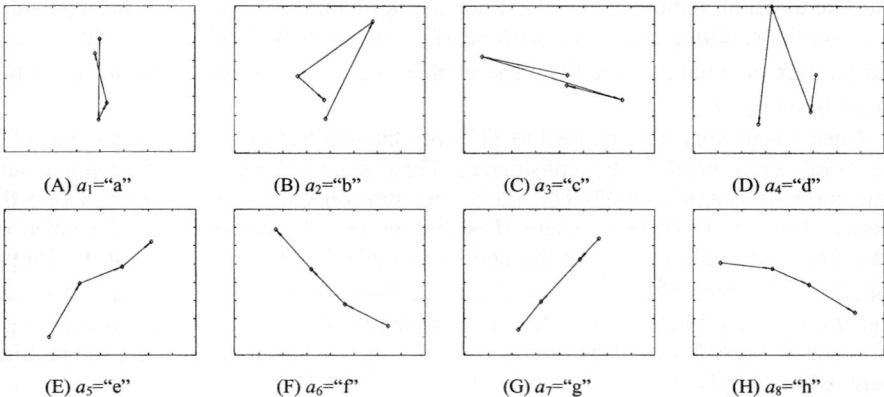

(A) a_1="a"   (B) a_2="b"   (C) a_3="c"   (D) a_4="d"

(E) a_5="e"   (F) a_6="f"   (G) a_7="g"   (H) a_8="h"

Fig. 3. *Typical sub-trajectories* and corresponding symbol representations, generated by static symbolization approach, on real data set.

Fig. 3 shows the experimental results on real data, which includes all the resulting *typical sub-trajectories* generated by the static approach. In the experiments, the width w and the step k of the sliding window were set to be 4 and 3 respectively, because this setting ($k=w-1$) can help fully avoiding the correlation between consecutive sub-trajectories without losing any information. The results indicated that there were 8 main typical movements in the trajectory, in which (E) - (H) represented straight motions roughly, (A), (B) represented round-motions in certain directions, and (C), (D) represented zigzag motions in two directions respectively.

By representing each sub-trajectory with a symbol indicating the corresponding *typical sub-trajectory*, the symbolization process gave us a concise and meaningful representation of the trajectory data, whereupon a data mining problem could be easily solved by simply applying an off the shelf symbol-sequence-oriented data mining approach. Recall from section 1 that such approaches can be easily found in many application domains. Since each symbol in the symbolic sequence represents the object's behaviors at an individual time point, the emphasis of this data mining process is on the high level representation of the data, rather than the quantitive values. In addition, the borrowed data mining method can be applied directly without any modification required. For example, we could apply a pattern mining method to discover frequent patterns with the form "a period of round-motions was always followed by a direct motion to the north-east in two hours."

In another group of experiments, synthetic data with various sizes and various amounts of noise were generated respectively for performance comparisons, and then both static symbolization approach and incremental one were used on the generated data alternatively. During these experiments, the window width and window step were also set to be 4 and 3 respectively. Since the static approach need all the data involved for re-computation, whereas in the incremental approach only the updated sub-trajectories need to be examined, the efficiency issue for the two approaches is quite obvious. Therefore, here we only give the results on effectiveness comparisons by considering the performance measure introduced above.

Fig. 4 shows the performances of the two approaches on trajectories with various lengths. By visual analysis, we could find that the incremental approach outperformed the static approach. This is partly because the static approach grouped many sub-trajectories into relatively incompact clusters, represented by out-dated *typical sub-trajectories*.

Fig. 5 shows the performances of the two approaches on trajectories with various amounts of noise involved, generated by varying the parameter R. This group of results also justifies the superiority of the incremental approach over the static one. In addition, the improvement in performance of the incremental approach decreased with the increase of the noise. Finally, the performances of the two approaches tended to be same. This is because the potential movement patterns implied by the original real data were influenced by the noise. When the amplitude of the noise was set to be large enough, the noise flooded the original real data, whereupon the generated data became a completely random one. For fully random data, both the static and the incremental approaches will perform like a meaningless random selection.

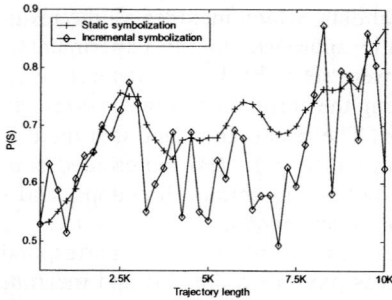

Fig. 4. Comparisons between the static and incremental symbolization approaches on trajectories with various lengths.

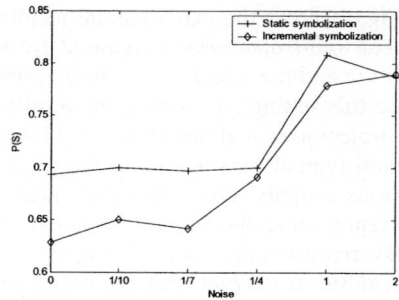

Fig. 5. Comparisons between the static and incremental symbolization approaches on trajectories with various amounts of noises.

7   Conclusions

In this paper, the problem of symbolizing trajectory data is addressed. We first introduce a static symbolization method, in which *typical sub-trajectories* are generated automatically based on the data. For facilitating the data mining process on streaming trajectories, we also present an incremental symbolization method, which dynamically adjusts the *typical sub-trajectories* to fit the up-to-date trajectory characters.

The performances of our approaches were evaluated on both real data and synthetic data. Experimental results justify the effectiveness of the proposed methods and show that the incremental approach outperformed the static one on trajectories with various lengths and various amounts of noise.

In future, we intend to generalize our symbolization approaches by introducing other similarity measurements and cluster models that are more sophisticated.

Acknowledgements

The work was partly supported by the NSFC 60373011 and the 973 Program 2002CB312000. We thank the anonymous reviewers for the valuable comments.

References

1. N. Priyantha, A. Miu, H. Balakrishnan, S. Teller. The cricket compass for context-aware mobile applications. In MOBICOM2001 Conference Proceedings, pages 1–14, 2001.
2. G. Chen, D. Kotz. Categorizing binary topological relations between regions, lines, and points in geographic databases. Technical Report TR2000-381. Dept. of Computer Science, Dartmouth College, 2000.
3. M. Vlachos, G. Kollios, Dimitrios Gunopulos. Discovering Similar Multidimensional Trajectories. In Proc. of the 18th International Conference on Data Engineering (ICDE'02). San Jose, California, 2002.
4. Y. Yanagisawa, J. Akahani, T. Satoh. Shape-based Similarity Query for Trajectory of Mobile Objects. In Proc. of the 4th International Conference on Mobile Data Management, pages 63 - 77. 2003.

5. C. S. Smyth. Mining mobile trajectories. H. J. Miller and J. Han (eds.) Geographic Data Mining and Knowledge Discovery, London: Taylor and Francis, 337-361. 2001.
6. G. Kollios, S. Sclaroff, M. Betke. Motion Mining: Discovering Spatio-Temporal Patterns in Databases of Human Motion. Workshop on Research Issues in Data Mining and Knowledge Discovery, DMKD 2001, Santa Barbara, CA, May 2001.
7. P. K. Agarwal, L. Arge, J. Erickson. Indexing moving points. In Proc. of the 19th ACM Symp. on Principles of Database Systems (PODS), pages 175–186, 2000.
8. S. Saltenis, C. Jensen, S. Leutenegger, and M. A. Lopez. Indexing the Positions of Continuously Moving Objects. In Proceedings of the ACM SIGMOD, pages 331–342, May 2000.
9. D. Pfoser, C. Jensen, and Y. Theodoridis. Novel Approaches in Query Processing for Moving Objects. In Proceedings of VLDB, Cairo Egypt, Sept. 2000.
10. M. Vlachos, M. Hadjieleftheriou, D. Gunopulos, E. Keogh. Indexing Multi-Dimensional Time-Series with Support for Multiple Distance Measures. In proc. of the 9th International Conference on Knowledge Discovery and Data Mining (KDD 2003). 2003.
11. T. Oates, M. Schmill, P. Cohen. A Method for Clustering the Experiences of a MobileRobot that Accords with Human Judgments. In Proc. of AAAI 2000.
12. Y. Zhu, D. Shasha. Fast approaches to simple problems in financial time series streams, Workshop on management and processing of data streams. 2003.
13. Z. Yao, L. Gao, X. S. Wang: Using triangle inequality to efficiently process continuous queries on high-dimensional streaming time series. In proc. of SSDBM 2003. 2003.
14. X. Jin, Y. Lu, C. Shi. Distribution discovery: local analysis of temporal rules. In proc. of the 6th Pacific-Asia Conf. on knowledge discovery and data mining (PAKDD 2002). 2002.
15. R. Agrawal, G. Psaila, E. Wimmers, M. Zaot. Querying shapes of histories. In proc. of the 21st international conference on very large database (VLDB'95). 1995.
16. G. Das, K. Lin, H. Mannila, G. Renganathan, P. Smyth. Rule discovery from time series. In proc. of the 4th International Conference on Knowledge Discovery and Data Mining (KDD 1998). 1998.
17. E. Keogh, J. Lin, W. Truppel. Clustering of time series subsequences is meaningless. In proc. of ICDM 2003. 2003.

Representing and Reasoning About Spatial Knowledge Based on Spatial Relevant Logic

Jingde Cheng and Yuichi Goto

Department of Information and Computer Sciences, Saitama University
Saitama, 338-8570, Japan
{cheng,gotoh}@aise.ics.saitama-u.ac.jp

Abstract. Almost all existing methodologies for representing and reasoning about spatial knowledge are somehow based on classical mathematical logic or its various conservative extensions. This approach, however, may be suitable to searching and describing a formal proof of a previously specified statement, under the condition that we have complete and consistent knowledge, but not necessarily suitable to forming a new concept and discovering a new statement, in particular, in the case that our knowledge is incomplete and inconsistent. This paper proposes a new approach: representing and reasoning about spatial knowledge based on spatial relevant logic.

1 Introduction

Spatial knowledge, i.e., shape, size, distance, orientation, relative position, connectivity, etc, plays an important role in our cognition and understanding of the world. There are many applications that need means for representing and reasoning about spatial knowledge, such as robotics, motion planning, machine vision, solid modeling, spatial database systems, geographic information systems, distributed systems, natural language understanding, etc.

In the area of geographic information systems, until now, almost all existing methodologies for representing and reasoning about spatial knowledge are somehow based on classical mathematical logic or its various conservative extensions [16]. This approach, however, may be suitable to searching and describing a formal proof of a previously specified statement, under the condition that we have complete and consistent knowledge, but not necessarily suitable to forming a new concept and discovering a new statement, in particular, in the case that our knowledge is incomplete and inconsistent. This is because the aim, nature, and role of classical mathematical logic is descriptive and non-predictive rather than prescriptive and predictive.

This paper proposes a new approach to spatial knowledge representation and reasoning: representing and reasoning about spatial knowledge with spatial relevant logic. The paper proposes a new family of logic, named "*spatial relevant logic*," and shows that it is a more hopeful candidate for the purpose.

S. Wang et al. (Eds.): ER Workshops 2004, LNCS 3289, pp. 114–126, 2004.
© Springer-Verlag Berlin Heidelberg 2004

2   Basic Notions

Reasoning is the *process* of drawing *new conclusions* from given premises, which are already known facts or previously assumed hypotheses to provide some *evidence* for the conclusions (Note that how to define the notion of 'new' formally and satisfactorily is still a difficult open problem until now). Therefore, reasoning is intrinsically ampliative, i.e. it has the function of enlarging or extending some things, or adding to what is already known or assumed. In general, a reasoning consists of a number of arguments (or inferences) *in some order*. An *argument* is a set of *statements* (or *declarative sentences*) of which one statement is intended as the *conclusion*, and one or more statements, called "*premises*," are intended to provide some evidence for the conclusion. An argument is a conclusion standing in relation to its supporting evidence. In an argument, a claim is being made that there is some sort of *evidential relation* between its premises and its conclusion: the conclusion is supposed to *follow from* the premises, or equivalently, the premises are supposed to *entail* the conclusion. Therefore, the correctness of an argument is a matter of the *connection* between its premises and its conclusion, and concerns the *strength* of the relation between them (Note that the correctness of an argument depends neither on whether the premises are really true or not, nor on whether the conclusion is really true or not). Thus, there are some fundamental questions: What is the criterion by which one can decide whether the conclusion of an argument or a reasoning really does follow from its premises or not? Is there the only one criterion, or are there many criteria? If there are many criteria, what are the intrinsic differences between them? It is logic that deals with the validity of argument and reasoning in general.

A *logically valid reasoning* is a reasoning such that its arguments are justified based on some *logical validity criterion* provided by a logic system in order to obtain correct conclusions (Note that here the term 'correct' does not necessarily mean 'true'.). Today, there are so many different logic systems motivated by various philosophical considerations. As a result, a reasoning may be valid on one logical validity criterion but invalid on another. For example, the *classical account of validity*, which is one of fundamental principles and assumptions underlying classical mathematical logic and its various conservative extensions, is defined in terms of *truth-preservation* (in some certain sense of truth) as: an argument is valid if and only if it is impossible for all its premises to be true while its conclusion is false. Therefore, a classically valid reasoning must be *truth-preserving*. On the other hand, for any correct argument in scientific reasoning as well as our everyday reasoning, its premises must somehow be *relevant* to its conclusion, and vice versa. The *relevant account of validity* is defined in terms of *relevance* as: for an argument to be valid there must be some connection of meaning, i.e. some relevance, between its premises and its conclusion. Obviously, the relevance between the premises and conclusion of an argument is not accounted for by the classical logical validity criterion, and therefore, a classically valid reasoning is not necessarily relevant.

Proving is the process of finding a justification for an explicitly specified statement from given premises, which are already known facts or previously assumed hypotheses to provide some *evidence* for the specified statement. A *proof* is a description of a found justification. A *logically valid proving* is a proving such that it is justified based on some logical validity criterion provided by a logic system in order to obtain a correct proof. The most intrinsic difference between reasoning and proving is that the former is intrinsically prescriptive and predictive while the latter is intrinsically descriptive and non-predictive. The purpose of reasoning is to find some new conclusion previously unknown or unrecognized, while the purpose of proving is to find a justification for some specified statement previously given. Proving has an explicitly given target as its goal while reasoning does not. Unfortunately, until now, many studies in Computer Science and Artificial Intelligence disciplines still confuse proving with reasoning.

Discovery is the process to find out or bring to light of that which was previously unknown. For any discovery, both the discovered thing and its truth must be unknown before the completion of discovery process. Since reasoning is the only way to draw new conclusions from given premises, there is no discovery process that does not invoke reasoning.

Logic is a special discipline which is considered to be the basis for all other sciences, and therefore, it is a science prior to all others, which contains the ideas and principles underlying all sciences [17, 24]. *Logic* deals with *what entails what* or *what follows from what*, and aims at determining which are the correct conclusions of a given set of premises, i.e. to determine which arguments are valid. Therefore, the most essential and central concept in logic is the logical consequence relation that relates a given set of premises to those conclusions, which validly follow from the premises.

In general, a *formal logic system* L consists of a formal language, called the *object language* and denoted by $F(L)$, which is the set of all *well-formed formulas* of L, and a *logical consequence relation*, denoted by meta-linguistic symbol $\vdash_L$, such that for $P \subseteq F(L)$ and $c \in F(L)$, $P \vdash_L c$ means that within the framework of L, c is a valid conclusion of premises P, i.e. c validly follows from P. For a formal logic system $(F(L), \vdash_L)$, a *logical theorem* t is a formula of L such that $\phi \vdash_L t$ where ϕ is the empty set. We use $Th(L)$ to denote the set of all logical theorems of L. $Th(L)$ is completely determined by the logical consequence relation $\vdash_L$. According to the representation of the logical consequence relation of a logic, the logic can be represented as a Hilbert style formal system, a Gentzen natural deduction system, a Gentzen sequent calculus system, or other type of formal system. A formal logic system L is said to be *explosive* if and only if $\{A, \neg A\} \vdash_L B$ for any two different formulas A and B; L is said to be *paraconsistent* if and only if it is not explosive.

Let $(F(L), \vdash_L)$ be a formal logic system and $P \subseteq F(L)$ be a non-empty set of *sentences* (i.e. *closed well-formed formulas*). A *formal theory* with premises P based on L, called a *L-theory with premises P* and denoted by $T_L(P)$, is defined as $T_L(P) =_{df} Th(L) \cup Th_L^e(P)$, and $Th_L^e(P) =_{df} \{et \mid P \vdash_L et \text{ and } et \notin Th(L)\}$ where $Th(L)$ and $Th_L^e(P)$ are called the *logical part* and the *empirical part* of the formal theory, respectively, and any element of $Th_L^e(P)$ is called an *empirical theorem* of the formal theory. A formal theory $T_L(P)$ is said to be *directly inconsistent* if and only if there exists a formula A of L such that both $A \in P$ and $\neg A \in P$ hold. A formal theory $T_L(P)$ is said to be *indirectly inconsistent* if and only if it is not directly inconsistent but there exists a formula A of L such that both $A \in T_L(P)$ and $\neg A \in T_L(P)$. A formal theory $T_L(P)$ is said to be *consistent* if and only if it is neither directly inconsistent nor indirectly inconsistent. A formal theory $T_L(P)$ is said to be *explosive* if and only if $A \in T_L(P)$ for arbitrary formula A of L; $T_L(P)$ is said to be *paraconsistent* if and only if it is not explosive. An explosive formal theory is not useful at all. Therefore, any meaningful formal theory should be paraconsistent. Note that if a formal logic system L is explosive, then any directly or indirectly inconsistent L-theory $T_L(P)$ must be explosive.

In the literature of mathematical, natural, social, and human sciences, it is probably difficult, if not impossible, to find a sentence form that is more generally used for describing various definitions, propositions, and theorems than the sentence form of 'if ... then ...'. In logic, a sentence in the form of 'if ... then ...' is usually called a *conditional proposition* or simply *conditional* which states that there exists a relation of sufficient condition between the 'if' part and the 'then' part of the sentence. Scientists always use conditionals in their descriptions of various definitions, propositions, and theorems to connect a concept, fact, situation or conclusion to its sufficient conditions. The major work of almost all scientists is to discover some sufficient condition relations between various phenomena, data, and laws in their research fields. Indeed, Russell 1903 has said, "Pure Mathematics is the class of all propositions of the form 'p implies q,' where p and q are propositions containing one or more variables, the same in the two propositions, and neither p nor q contains any constants except logical constants" [22].

In general, a conditional must concern two parts which are connected by the connective 'if ... then ...' and called the *antecedent* and the *consequent* of that conditional, respectively. The truth of a conditional depends not only on the truth of its antecedent and consequent but also, and more essentially, on a necessarily relevant and conditional relation between them. The notion of conditional plays the most essential role in reasoning because any reasoning form must invoke it, and therefore, it is historically always the most important subject studied in logic and is regarded as the heart of logic [1]. In fact, from the age of ancient Greece, the notion of conditional has been discussed by the ancients of Greek. For example, the extensional truth-functional definition of the notion of material implication was given by Philo of Megara in about 400 B.C. [18, 24].

When we study and use logic, the notion of conditional may appear in both the *object logic* (i.e. the logic we are studying) and the *meta-logic* (i.e. the logic we are using to study the object logic). In the object logic, there usually is a connective in its formal language to represent the notion of conditional, and the notion of conditional, usually represented by a meta-linguistic symbol, is also used for representing a logical consequence relation in its proof theory or model theory. On the other hand, in the meta-logic, the notion of conditional, usually in the form of natural language, is used for defining various meta-notions and describing various meta-theorems about the object logic.

From the viewpoint of object logic, there are two classes of conditionals. One class is empirical conditionals and the other class is logical conditionals. For a logic, a conditional is called an ***empirical conditional*** of the logic if its truth-value, in the sense of that logic, depends on the contents of its antecedent and consequent and therefore cannot be determined only by its abstract form (i.e. from the viewpoint of that logic, the relevant relation between the antecedent and the consequent of that conditional is regarded to be empirical); a conditional is called a ***logical conditional*** of the logic if its truth-value, in the sense of that logic, depends only on its abstract form but not on the contents of its antecedent and consequent, and therefore, it is considered to be universally true or false (i.e. from the viewpoint of that logic, the relevant relation between the antecedent and the consequent of that conditional is regarded to be logical). A logical conditional that is considered to be universally true, in the sense of that logic, is also called an ***entailment*** of that logic. Indeed, the most intrinsic difference between various different logic systems is to regard what class of conditionals as entailments, as Diaz pointed out: "The problem in modern logic can best be put as follows: can we give an explanation of those conditionals that represent an entailment relation?" [14].

For a formal logic system where the notion of conditional is represented by a primitive connective, say '$\Rightarrow$', a formula is called a ***zero degree formula*** if and only if there is no occurrence of $\Rightarrow$ in it; a formula of the form $A \Rightarrow B$ is called a ***first degree conditional*** if and only if both A and B are zero degree formulas; a formula A is called a ***first degree formula*** if and only if it satisfies the one of the following conditions: (1) A is a first degree conditional, (2) A is in the form +B (+ is a one-place connective such as negation and so on) where B is a first degree formula, and (3) A is in the form B*C (* is a non-implicational two-place connective such as conjunction or disjunction and so on) where both of B and C is first degree formulas, or one of B and C is a first degree formula and another is a zero degree formula. Let k be a natural number. A formula of the form $A \Rightarrow B$ is called a ***k^{th} degree conditional*** if and only if both A and B are $(k-1)^{th}$ degree formulas, or one of A and B is a $(k-1)^{th}$ degree formula and another is a jth ($j<k-1$) degree formula; a formula A is called a ***k^{th} degree formula*** if and only if it satisfies the one of the following conditions: (1) A is a kth degree conditional, (2) A is in the form +B where B is a kth degree formula, and (3) A is in the form B*C where both of B and C is k^{th} degree formulas, or one of B and C is a kth degree formula and another is a jth ($j<k$) degree formula.

Let $(F(L), \vdash_L)$ be a formal logic system and k be a natural number. The ***k^{th} degree fragment*** of L, denoted by $Th^k(L)$, is a set of logical theorems of L which is inductively defined as follows (in the terms of Hilbert style formal system): (1) if A is a jth ($j \leq k$) degree formula and an axiom of L, then $A \in Th^k(L)$, (2) if A is a jth ($j \leq k$) degree formula which is the result of applying an inference rule of L to some members of $Th^k(L)$, then $A \in Th^k(L)$, and (3) Nothing else are members of $Th^k(L)$, i.e. only those obtained from repeated applications of (1) and (2) are members of $Th^k(L)$. Obviously, the definition of the kth degree fragment of logic L is constructive. The kth degree fragment of logic L can be constructed in principle by a forward deduction based on the logic. Similarly, we can define k^{th} degree fragment of a L-theory with premises P.

3 Spatial Relevant Logics

The fundamental logic system to underlie representing and reasoning about spatial knowledge must satisfy the following at least four essential requirements: First, as a general logical criterion for the validity of reasoning as well as proving, the logic must be able to underlie relevant reasoning as well as truth-preserving reasoning in the sense of conditional. Second, in order to reason about new spatial knowledge, the logic must be able to underlie ampliative reasoning in the sense that the truth of con-clusion of the reasoning should be recognized after the completion of the reasoning process but not be invoked in deciding the truth of premises of the reasoning. Third, the logic must be able to underlie paracomplete and paraconsistent reasoning, because in any application area in the real world the completeness and the consistency are often not necessarily guaranteed. In particular, the principle of Explosion that every-thing follows from a contradiction cannot be accepted by the logic as a valid princi-ple. Finally, of course, the logic must be able to underlie spatial reasoning.

Classical mathematical logic adopts the classical account of validity as its criterion of logical validity and represents the notion of conditional, which is intrinsically in-tensional but not truth-functional, by the notion of material implication, which is intrinsically an extensional truth-function. As a result, a reasoning based on classical mathematical logic is not necessarily relevant, the classical truth-preserving property of a reasoning based on classical mathematical logic is meaningless in the sense of conditional, a reasoning based on classical mathematical logic must be circular and/or tautological but not ampliative, and reasoning under inconsistency is impossible within the framework of classical mathematical logic [1, 2, 6, 7, 9, 10, 15, 19, 21]. These facts are also true to those classical conservative extensions or non-classical alternatives of classical mathematical logic where the classical account of validity is adopted as the logical validity criterion and the notion of conditional is directly or indirectly represented by the material implication. Therefore, all these logics cannot satisfy the four essential requirements for the fundamental logic system to underlie representing and reasoning about spatial knowledge.

Some spatial logics was proposed in order to deal with geometric and topological entities, notions, relations, and properties, and therefore to underlie spatial reasoning, i.e. reasoning about those propositions and formulas whose truth-values may depend on a location [3-5, 11-13, 20, 23]. However, these existing spatial logics are classical conservative extensions of classical mathematical logic in the sense that they are based on the classical account of validity and they represent the notion of conditional directly or indirectly by the material implication. Therefore, these spatial logics can-not satisfy the first three of the four essential requirements for the fundamental logic system, even though that can underlie spatial reasoning in some certain sense.

On the other hand, relevant logics cannot underlie spatial reasoning, even though them can underlie relevant reasoning as well as truth-preserving reasoning in the sense of conditional, ampliative reasoning, paracomplete reasoning, and paraconsistent reasoning well [1, 2, 6, 7, 9, 10, 15, 19, 21].

Consequently, at present, there is no existing logic system which can satisfy all four essential requirements for the fundamental logic system. What we need is a suitable combination of spatial logics and relevant logics such that it can satisfy the all four essential requirements for the fundamental logic system.

We now propose a new family of relevant logic systems, named *Spatial Relevant Logic*, which can satisfy the all four essential requirements for the fundamental logic system to underlie representing and reasoning about geographic knowledge. The logics are obtained by introducing region connection predicates and axiom schemata of RCC [11, 20], point position predicates and axiom schemata, and point adjacency predicates and axiom schemata into strong relevant logics.

Let $\{r_1, r_2, r_3, ...\}$ be a countably infinite set of individual variables, called *region variables*. Atomic formulas of the form $C(r_1, r_2)$ are read as 'region r_1 connects with region r_2.' Let $\{p_1, p_2, p_3, ...\}$ be a countably infinite set of individual variables, called *point variables*. Atomic formulas of the form $I(p_1, r_1)$ are read as 'point p_1 is in region r_1.' Atomic formulas of the form $Arc(p_1, p_2)$ are read as 'points p_1, p_2 are adjacent such that there is an arc from point p_1 to point p_2, or more simply, points p_1 is adjacent to point p_2.' Note that an arc has a direction. Atomic formulas of the form $Path(p_1, p_2)$ are read as 'there is a directed path from point p_1 to point p_2.' Atomic formulas of the form $B(p_1, p_2, p_3)$ are read as 'points p_1, p_2, p_3 is on a straight line and point p_1 is between point p_2 and point p_3.' Note that here we use a many-sorted language.

The logical connectives, region connection predicates, point position predicates, point adjacency predicates, axiom schemata, and inference rules are as follows:

Primitive Logical Connectives:
$\Rightarrow$:   entailment
$\neg$:   negation
$\wedge$:   extensional conjunction

Defined Logical Connectives:
$\otimes$:   intensional conjunction, $A \otimes B =_{df} \neg(A \Rightarrow \neg B)$
$\oplus$:   intensional disjunction, $A \oplus B =_{df} \neg A \Rightarrow B$
$\Leftrightarrow$:   intensional equivalence, $A \Leftrightarrow B =_{df} (A \Rightarrow B) \otimes (B \Rightarrow A)$
$\vee$:   extensional disjunction, $A \vee B =_{df} \neg(\neg A \wedge \neg B)$
$\rightarrow$:   material implication, $A \rightarrow B =_{df} \neg(A \wedge \neg B)$ or $A \rightarrow B =_{df} \neg A \vee B$
$\leftrightarrow$:   extensional equivalence, $A \leftrightarrow B =_{df} (A \rightarrow B) \wedge (B \rightarrow A)$

Primitive Binary Predicate:
C :   connection   ($C(r_1, r_2)$ means 'r_1 connects with r_2')
I :   be in   ($I(p_1, r_1)$ means 'p_1 is in r_1')

Arc :	arc	(***Arc***(p_1, p_2) means 'p_1 is adjacent to p_2')
Path :	path	(***Path***(p_1, p_2) means 'there is a directed path from p_1 to p_2'),
B :	be between	(***B***(p_1, p_2, p_3) means 'p_1, p_2, p_3 are on a straight line and p_1 is between p_2 and p_3')

Defined Binary Predicates:

$DC(r_1, r_2) =_{df} \neg C(r_1, r_2)$ ($DC(r_1, r_2)$ means 'r_1 is disconnected from r_2')

$P(r_1, r_2) =_{df} \forall r_3(C(r_3, r_1) \Rightarrow C(r_3, r_2))$ ($P(r_1, r_2)$ means 'r_1 is a part of r_2')

$PP(r_1, r_2) =_{df} P(r_1, r_2) \wedge (\neg P(r_2, r_1))$ ($PP(r_1, r_2)$ means 'r_1 is a proper part of r_2')

$EQ(r_1, r_2) =_{df} P(r_1, r_2) \wedge P(r_2, r_1)$ ($EQ(r_1, r_2)$ means 'r_1 is identical with r_2')

$O(r_1, r_2) =_{df} \exists r_3(P(r_3, r_1) \wedge P(r_3, r_2))$ ($O(r_1, r_2)$ means 'r_1 overlaps r_2')

$DR(r_1, r_2) =_{df} \neg O(r_1, r_2)$ ($DR(r_1, r_2)$ means 'r_1 is discrete from r_2')

$PO(r_1, r_2) =_{df} O(r_1, r_2) \wedge (\neg P(r_1, r_2)) \wedge (\neg P(r_2, r_1))$

($PO(r_1, r_2)$ means 'r_1 partially overlaps r_2')

$EC(r_1, r_2) =_{df} C(r_1, r_2) \wedge (\neg O(r_1, r_2))$

($EC(r_1, r_2)$ means 'r_1 is externally connected to r_2')

$TPP(r_1, r_2) =_{df} PP(r_1, r_2) \wedge \exists r_3(EC(r_3, r_1) \wedge EC(r_3, r_2))$

($TPP(r_1, r_2)$ means 'r_1 is a tangential proper part of r_2')

$NTPP(r_1, r_2) =_{df} PP(r_1, r_2) \wedge (\neg \exists r_3(EC(r_3, r_1) \wedge EC(r_3, r_2)))$

($NTPP(r_1, r_2)$ means 'r_1 is a nontangential proper part of r_2')

Axiom Schemata:

E1	$A \Rightarrow A$
E2	$(A \Rightarrow B) \Rightarrow ((C \Rightarrow A) \Rightarrow (C \Rightarrow B))$
E2′	$(A \Rightarrow B) \Rightarrow ((B \Rightarrow C) \Rightarrow (A \Rightarrow C))$
E3	$(A \Rightarrow (A \Rightarrow B)) \Rightarrow (A \Rightarrow B)$
E3′	$(A \Rightarrow (B \Rightarrow C)) \Rightarrow ((A \Rightarrow B) \Rightarrow (A \Rightarrow C))$
E3″	$(A \Rightarrow B) \Rightarrow ((A \Rightarrow (B \Rightarrow C)) \Rightarrow (A \Rightarrow C))$
E4	$(A \Rightarrow ((B \Rightarrow C) \Rightarrow D)) \Rightarrow ((B \Rightarrow C) \Rightarrow (A \Rightarrow D))$
E4′	$(A \Rightarrow B) \Rightarrow (((A \Rightarrow B) \Rightarrow C) \Rightarrow C)$
E4″	$((A \Rightarrow A) \Rightarrow B) \Rightarrow B$
E4‴	$(A \Rightarrow B) \Rightarrow ((B \Rightarrow C) \Rightarrow (((A \Rightarrow C) \Rightarrow D) \Rightarrow D))$
E5	$(A \Rightarrow (B \Rightarrow C)) \Rightarrow (B \Rightarrow (A \Rightarrow C))$
E5′	$A \Rightarrow ((A \Rightarrow B) \Rightarrow B)$
N1	$(A \Rightarrow (\neg A)) \Rightarrow (\neg A)$
N2	$(A \Rightarrow (\neg B)) \Rightarrow (B \Rightarrow (\neg A))$
N3	$(\neg(\neg A)) \Rightarrow A$
C1	$(A \wedge B) \Rightarrow A$
C2	$(A \wedge B) \Rightarrow B$

C3	$((A{\Rightarrow}B){\wedge}(A{\Rightarrow}C)){\Rightarrow}(A{\Rightarrow}(B{\wedge}C))$
C4	$(LA{\wedge}LB){\Rightarrow}L(A{\wedge}B)$, where $LA =_{df} (A{\Rightarrow}A){\Rightarrow}A$
D1	$A{\Rightarrow}(A{\vee}B)$
D2	$B{\Rightarrow}(A{\vee}B)$
D3	$((A{\Rightarrow}C){\wedge}(B{\Rightarrow}C)){\Rightarrow}((A{\vee}B){\Rightarrow}C)$
DCD	$(A{\wedge}(B{\vee}C)){\Rightarrow}((A{\wedge}B){\vee}C)$
C5	$(A{\wedge}A){\Rightarrow}A$
C6	$(A{\wedge}B){\Rightarrow}(B{\wedge}A)$
C7	$((A{\Rightarrow}B){\wedge}(B{\Rightarrow}C)){\Rightarrow}(A{\Rightarrow}C)$
C8	$(A{\wedge}(A{\Rightarrow}B)){\Rightarrow}B$
C9	$\neg(A{\wedge}\neg A)$
C10	$A{\Rightarrow}(B{\Rightarrow}(A{\wedge}B))$
IQ1	$\forall x(A{\Rightarrow}B){\Rightarrow}(\forall xA{\Rightarrow}\forall xB)$
IQ2	$(\forall xA{\wedge}\forall xB){\Rightarrow}\forall x(A{\wedge}B)$
IQ3	$\forall xA{\Rightarrow}A[t/x]$ (if x may appear free in A and t is free for x in A, i.e., free variables of t do not occur bound in A)
IQ4	$\forall x(A{\Rightarrow}B){\Rightarrow}(A{\Rightarrow}\forall xB)$ (if x does not occur free in A)
IQ5	$\forall x_1 ... \forall x_n (((A{\Rightarrow}A){\Rightarrow}B){\Rightarrow}B)$ (n≥0)
RCC1	$\forall r_1\forall r_2(C(r_1, r_2){\Rightarrow}C(r_2, r_1))$
RCC2	$\forall r_1(C(r_1, r_1))$
PRCC1	$\forall p_1\forall r_1\forall r_2((I(p_1, r_1){\wedge}DC(r_1, r_2)){\Rightarrow}\neg I(p_1, r_2))$
PRCC2	$\forall p_1\forall r_1\forall r_2((I(p_1, r_1){\wedge}P(r_1, r_2)){\Rightarrow}I(p_1, r_2))$
PRCC3	$\forall p_1\forall r_1\forall r_2((I(p_1, r_1){\wedge}PP(r_1, r_2)){\Rightarrow}I(p_1, r_2))$
PRCC4	$\forall p_1\forall r_1\forall r_2((I(p_1, r_1){\wedge}EQ(r_1, r_2)){\Rightarrow}I(p_1, r_2))$
PRCC5	$\forall r_1\forall r_2(O(r_1, r_2){\Rightarrow}\exists p_1(I(p_1, r_1){\wedge}I(p_1, r_2)))$
PRCC6	$\forall p_1\forall r_1\forall r_2((I(p_1, r_1){\wedge}DR(r_1, r_2)){\Rightarrow}\neg I(p_1, r_2)$
PRCC7	$\forall r_1\forall r_2(PO(r_1, r_2){\Rightarrow}$ $\exists p_1(I(p_1, r_1){\wedge}I(p_1, r_2)){\wedge}\exists p_2(I(p_2, r_1){\wedge}\neg I(p_2, r_2)){\wedge}\exists p_3(\neg I(p_3, r_1){\wedge}I(p_3, r_2)))$
PRCC8	$\forall p_1\forall r_1\forall r_2((I(p_1, r_1){\wedge}EC(r_1, r_2)){\Rightarrow}\neg I(p_1, r_2))$
PRCC9	$\forall p_1\forall r_1\forall r_2((I(p_1, r_1){\wedge}TPP(r_1, r_2)){\Rightarrow}I(p_1, r_2))$
PRCC10	$\forall p_1\forall r_1\forall r_2((I(p_1, r_1){\wedge}NTPP(r_1, r_2)){\Rightarrow}I(p_1, r_2))$
PAC1	$\forall p_1\forall p_2(Arc(p_1, p_2){\Rightarrow}Path(p_1, p_2))$
PAC2	$\forall p_1\forall p_2\forall p_3((Path(p_1, p_2){\wedge}Path(p_2, p_3)){\Rightarrow}Path(p_1, p_3))$
PPC1	$\forall p_1\forall p_2(\neg B(p_1, p_1, p_2))$
PPC2	$\forall p_1\forall p_2\forall p_3(B(p_1, p_2, p_3){\Rightarrow}B(p_1, p_3, p_2))$
PPC3	$\forall p_1\forall p_2\forall p_3(B(p_1, p_2, p_3){\Rightarrow}\neg B(p_2, p_1, p_3))$
PPC4	$\forall p_1\forall p_2\forall p_3\forall p_4((B(p_1, p_2, p_3){\wedge}B(p_4, p_1, p_3)){\Rightarrow}B(p_4, p_2, p_3))$
PPC5	$\forall p_1\forall p_2\forall p_3\forall p_4((B(p_1, p_2, p_3){\wedge}B(p_4, p_1, p_2)){\Rightarrow}B(p_4, p_2, p_3))$

Inference Rules:

$\Rightarrow$E :   from A and A$\Rightarrow$B to infer B (Modus Ponens)

$\wedge$I :   from A and B to infer A$\wedge$B (Adjunction)

$\forall$I :   if A is an axiom, so is $\forall$xA (Generalization of axioms)

Thus, various relevant logic systems may now defined as follows, where we use 'A | B' to denote any choice of one from two axiom schemata A and B.

$\mathbf{T}_\Rightarrow =_{df} \{E1, E2, E2', E3 \mid E3''\} + \Rightarrow E$

$\mathbf{E}_\Rightarrow =_{df} \{E1, E2 \mid E2', E3 \mid E3', E4 \mid E4'\} + \Rightarrow E$

$\mathbf{E}_\Rightarrow =_{df} \{E2', E3, E4''\} + \Rightarrow E$

$\mathbf{E}_\Rightarrow =_{df} \{E1, E3, E4'''\} + \Rightarrow E$

$\mathbf{R}_\Rightarrow =_{df} \{E1, E2 \mid E2', E3 \mid E3', E5 \mid E5'\} + \Rightarrow E$

$\mathbf{T}_{\Rightarrow,\neg} =_{df} \mathbf{T}_\Rightarrow + \{N1, N2, N3\}$

$\mathbf{E}_{\Rightarrow,\neg} =_{df} \mathbf{E}_\Rightarrow + \{N1, N2, N3\}$

$\mathbf{R}_{\Rightarrow,\neg} =_{df} \mathbf{R}_\Rightarrow + \{N2, N3\}$

$\mathbf{T} =_{df} \mathbf{T}_{\Rightarrow,\neg} + \{C1{\sim}C3, D1{\sim}D3, DCD\} + \wedge I$

$\mathbf{E} =_{df} \mathbf{E}_{\Rightarrow,\neg} + \{C1{\sim}C4, D1{\sim}D3, DCD\} + \wedge I$

$\mathbf{R} =_{df} \mathbf{R}_{\Rightarrow,\neg} + \{C1{\sim}C3, D1{\sim}D3, DCD\} + \wedge I$

$\mathbf{Tc} =_{df} \mathbf{T}_{\Rightarrow,\neg} + \{C3, C5{\sim}C10\}$

$\mathbf{Ec} =_{df} \mathbf{E}_{\Rightarrow,\neg} + \{C3{\sim}C10\}$

$\mathbf{Rc} =_{df} \mathbf{R}_{\Rightarrow,\neg} + \{C3, C5{\sim}C10\}$

$\mathbf{TQ} =_{df} \mathbf{T} + \{IQ1{\sim}IQ5\} + \forall I$

$\mathbf{EQ} =_{df} \mathbf{E} + \{IQ1{\sim}IQ5\} + \forall I$

$\mathbf{RQ} =_{df} \mathbf{R} + \{IQ1{\sim}IQ5\} + \forall I$

$\mathbf{TcQ} =_{df} \mathbf{Tc} + \{IQ1{\sim}IQ5\} + \forall I$

$\mathbf{EcQ} =_{df} \mathbf{Ec} + \{IQ1{\sim}IQ5\} + \forall I$

$\mathbf{RcQ} =_{df} \mathbf{Rc} + \{IQ1{\sim}IQ5\} + \forall I$

Here, $\mathbf{T}_\Rightarrow$, $\mathbf{E}_\Rightarrow$, and $\mathbf{R}_\Rightarrow$ are the purely implicational fragments of $\mathbf{T}$, $\mathbf{E}$, and $\mathbf{R}$, respectively, and the relationship between $\mathbf{E}_\Rightarrow$ and $\mathbf{R}_\Rightarrow$ is known as $\mathbf{R}_\Rightarrow = \mathbf{E}_\Rightarrow + A{\Rightarrow}LA$; $\mathbf{T}_{\Rightarrow,\neg}$, $\mathbf{E}_{\Rightarrow,\neg}$, and $\mathbf{R}_{\Rightarrow,\neg}$ are the implication-negation fragments of $\mathbf{T}$, $\mathbf{E}$, and $\mathbf{R}$, respectively; $\mathbf{Tc}$, $\mathbf{Ec}$, $\mathbf{Rc}$, $\mathbf{TcQ}$, $\mathbf{EcQ}$, and $\mathbf{RcQ}$ are strong relevant (relevance) logics proposed by the present author. We can now obtain some spatial relevant logics as follows:

$\mathbf{RTcQ} =_{df} \mathbf{TcQ} + \{RCC1, RCC2, PRCC1{\sim}PRCC10, PAC1, PAC2, PPC1{\sim}PPC5\}$

$\mathbf{REcQ} =_{df} \mathbf{EcQ} + \{RCC1, RCC2, PRCC1{\sim}PRCC10, PAC1, PAC2, PPC1{\sim}PPC5\}$

$\mathbf{RRcQ} =_{df} \mathbf{RcQ} + \{RCC1, RCC2, PRCC1{\sim}PRCC10, PAC1, PAC2, PPC1{\sim}PPC5\}$

The spatial relevant logics have the following possible applications:

First, as conservative extensions of strong relevant logics satisfying the strong relevance principle, the spatial relevant logics can underlie relevant reasoning as well as truth-preserving reasoning in the sense of conditional, ampliative reasoning, paracomplete reasoning, and paraconsistent reasoning. Moreover, the logics can reasoning about relative relations among points as well as regions. Therefore, they can be used to represent and specify geographic relations in a geographic information system based on incomplete or even inconsistent information at first, and then reasoning (again, not proving) about the unspecified new geographic relations. This is a very useful way in modeling, designing, developing a geographic information system. Probably, at present, the family of spatial relevant logics is the only one to have such application.

Second, once we modeled a geographic information system and specified its desirable properties with the formal language of the spatial relevant logics, we can verify the properties based on the logics, even if there are some incompleteness and inconsistency.

For the first and second applications, an automated reasoning and verifying engine based on the spatial relevant logics is indispensable. Based on EnCal, an automated forward deduction system for general-purpose entailment calculus [8], we are developing an automated reasoning and verifying engine for the spatial relevant logics.

Third, the spatial relevant logics provide us with a foundation for constructing more powerful logic systems to deal with those issues on size, distance, and orientation in geographic information systems.

4   An Example

Here we give an example to show that the spatial relevant logic can underlie representing and reasoning about spatial knowledge. The example is to investigate what can be deduced based on a spatial relevant logic from a GIS about countries and their cities in the world, and what will happen when a logical theorem of classical mathematical logic, which is an implicational paradox, is added for deduction as a 'valid' logical theorem.

Let us suppose that our GIS includes the following knowledge:

$\forall r_1 \forall r_2((C(r_1, r_2) \wedge PP(r_2, \text{Europe})) \Rightarrow PP(r_1, \text{Europe}))$, $C(\text{France}, \text{Germany})$, $C(\text{France}, \text{Italy})$, $I(\text{Paris}, \text{France})$, $I(\text{Berlin}, \text{Germany})$, $I(\text{Rome}, \text{Italy})$, $PP(\text{France}, \text{Europe})$, $I(\text{Tokyo}, \text{Japan})$, and $I(\text{Beijing}, \text{China})$.

Based on 2nd degree fragment of **REcQ** and limit the degree of nested '$\wedge$' to 1, we deduced 226 formulas from the above GIS. All the 226 formulas are relevant to the GIS in the sense of conditional as well as facts. Some new facts in the deduced 226 formulas are: $I(\text{Paris}, \text{Europe})$, $PP(\text{Germany}, \text{Europe})$, $PP(\text{Italy}, \text{Europe})$, $I(\text{Berlin}, \text{Europe})$, and $I(\text{Rome}, \text{Europe})$.

On the other hands, besides 2nd degree fragment of **REcQ**, we added a logical theorem of classical mathematical logic '$(A \wedge B) \Rightarrow (A \Rightarrow B)$', which is an implicational paradox, as a 'valid' logical theorem in our deduction, but kept other things as the same as the above deduction. In this case, we deduced a lot of irrelevant conditionals such as I(Berlin, Germany)$\Rightarrow PP$(France, Europe), I(Tokyo, Japan)$\Rightarrow C$(France, Italy), PP(France, Europe)$\Rightarrow I$(Beijing, China) and so on. All these irrelevant conditionals were not included in the results of the above deduction using 2nd degree fragment of **REcQ** only.

5 Concluding Remarks

The work presented in this paper is our first step for establishing a theoretical framework of spatial relevant logic to underlie representing and reasoning about spatial knowledge and applying it to various applications where spatial reasoning plays an important role. There are many challenging theoretical and technical problems that have to be solved in order to apply the spatial relevant logic to practices in the real world.

References

1. Anderson, A.R., Belnap Jr., N.D.: Entailment: The Logic of Relevance and Necessity, Vol. I. Princeton University Press, Princeton (1975)
2. Anderson, A.R., Belnap Jr., N.D., Dunn, J. M.: Entailment: The Logic of Relevance and Necessity, Vol. II. Princeton University Press, Princeton (1992)
3. Caires, L., Cardelli, L.: A Spatial Logic for Concurrency (Part I). In: Kobayashi, N. and Pierce, B.C. (eds.): Theoretical Aspects of Computer Software. Lecture Notes in Computer Science, Vol. 2215. Springer-Verlag, Berlin Heidelberg New York (2001) 1-37
4. Caires, L., Cardelli, L.: A Spatial Logic for Concurrency (Part II). In: Lubos Brim, L., Jancar, P., Kretinsky, M., Kucera, A. (eds.): CONCUR 2002 - Concurrency Theory. Lecture Notes in Computer Science, Vol. 2421. Springer-Verlag, Berlin Heidelberg New York (2002) 209-225
5. Cardelli, L., Gardner, P., Ghelli, G.: A Spatial Logic for Querying Graphs. In: Widmayer, P., Triguero, F., Morales, R., Hennessy, M., Eidenbenz, S., Conejo, R. (eds.): Automata, Languages, and Programming. Lecture Notes in Computer Science, Vol 2380. Springer-Verlag, Berlin Heidelberg New York (2002) 597-610
6. Cheng, J.: Logical Tool of Knowledge Engineering: Using Entailment Logic rather than Mathematical Logic. In: Proc. ACM 19th Annual Computer Science Conference. ACM Press, New York (1991) 228-238
7. Cheng, J.: The Fundamental Role of Entailment in Knowledge Representation and Reasoning. Journal of Computing and Information, Vol. 2, No. 1, Special Issue: Proceedings of the 8th International Conference of Computing and Information. Waterloo (1996) 853-873
8. Cheng, J.: EnCal: An Automated Forward Deduction System for General-Purpose Entailment Calculus. In: Terashima, N., Altman, E. (eds.): Advanced IT Tools, Proc. IFIP World Conference on IT Tools, IFIP 96 – 14th World Computer Congress. Chapman & Hall, London Weinheim New York Tokyo Melbourne Madras (1996) 507-514
9. Cheng, J.: Relevant Reasoning as the Logical Basis of Knowledge Engineering. In: Cantu, F.J., Soto, R., Liebowitz, J., Sucar, E. (eds.): Application of Advanced Information Technologies, - 4th World Congress on Expert Systems, Vol. 1. Cognizant Communication Co. (1998) 449-457

10. Cheng, J.: A Strong Relevant Logic Model of Epistemic Processes in Scientific Discovery. In: Kawaguchi, E., Kangassalo, H., Jaakkola, H., Hamid, I.A. (eds.): Information Modelling and Knowledge Bases XI. IOS Press, Amsterdam Berlin Oxford Tokyo Washington DC (2000) 136-159

11. Cohn, A.G., Bennett, B., Gooday, J., Gotts, N.M.: RCC: A Calculus for Region based Qualitative Spatial Reasoning. GeoInformatica, Vol. 1 (1997) 275-316

12. Cohn, A.G., Bennett, B., Gooday, J., Gotts, N.M.: Representing and Reasoning with Qualitative Spatial Relations About Regions. In: Stock, O. (ed.): Spatial and Temporal Reasoning. Kluwer Academic, Dordrecht Boston London (1997) 97-134

13. Cohn, A.G., Hazarika, S.M.: Qualitative Spatial Representation and Reasoning: An Overview. Fundamenta Informaticae, Vol. 45 (2001) 1-29

14. Diaz, M.R.: Topics in the Logic of Relevance. Philosophia Verlag, Munchen (1981)

15. Dunn, J.M., Restall, G.: Relevance Logic. In: Gabbay, D., Guenthner, F. (eds.): Handbook of Philosophical Logic, 2nd edition, Vol. 6. Kluwer Academic, Dordrecht Boston London (2002) 1-128

16. Egenhofer M.J. and Golledge R.G. (eds.): Spatial and Temporal Reasoning in Geographic Information Systems. Oxford University Press, New York Oxford (1998)

17. Godel, K.: Russell's Mathematical Logic. In: Schilpp (ed.): The Philosophy of Bertrand Russell. Open Court Publishing Company, Chicago (1944)

18. Kneale, W., Kneale, M.: The Development of Logic. Oxford University Press, Oxford (1962)

19. Mares, E.D., Meyer, R.K.: Relevant Logics. In: Goble, L. (ed.): The Blackwell Guide to Philosophical Logic. Blackwell, Oxford (2001) 280-308

20. Randell, D., Cui, Z., Cohn, A.: A Spatial Logic Based on Regions and Connection. In: Proc. 3rd International Conference on Knowledge Representation and Reasoning. (1992) 165-176

21. Read, S.: Relevant Logic: A Philosophical Examination of Inference. Basil Blackwell, Oxford (1988)

22. Russell, B.: The Principles of Mathematics. 2nd edition. Cambridge University Press, Cambridge (1903, 1938). Norton Paperback edition. Norton, New York London (1996)

23. Stock, O. (Ed.): Spatial and Temporal Reasoning. Kluwer Academic, Dordrecht Boston London (1997)

24. Tarski, A.: Introduction to Logic and to the Methodology of the Deductive Sciences. 4th edition, Revised. Oxford University Press, Oxford (1994)

Towards System Architecture of Spatial Information Grid and Service-Based Collaboration*

Yu Tang, Luo Chen, and Ning Jing

School of Electronic Science and Engineering, National University of Defense Technology
Changsha, Hunan, P.R. China
yutang_nudt@163.com, {luochen,ningjing}@nudt.edu.cn

Abstract. Spatial information is any type of information that can be spatially referenced and it is widely used in many application domains. Traditional techniques can't solve the existing application problems of spatial information, e.g. enormous data, format heterogeneity, processing complexity, and wide distribution. Based on the novel technologies that can implement large-scale distributed resource sharing, i.e. grid, Web services and OpenGIS specifications, we propose a new service-oriented application grid named Spatial Information Grid (SIG). System architecture and service-based collaboration are two of the most important research issues of SIG. By building an open SIG architecture, some key theories and methods are discussed in detail to implement service-based collaboration of spatial information. Hence, a novel process model named Service/Resource Net (SRN) is proposed based on Petri net and graph theory, and a dynamic service selection model is presented to select the optimum service. The validity and usability of SIG are verified by an application example in accordance with SIG application flow.

1 Introduction

Spatial information is a kind of basic information resource that is used in a wide range of applications, such as planning, land management, environment evaluating and renewable energy resources analyzing. Generally, spatial information is defined as any type of information that can be spatially referenced [1], including remote sensing images, GPS data, digital maps, DEM data, 3-D spatial models and etc.

Spatial information has three outstanding features different from other types of information, which are spatial character, thematic character, and temporal character [1]. Traditional spatial information techniques can't solve the existing application problems of spatial information, such as enormous distributed data, complicated processing, heterogeneous system structures, and etc.

As new technologies to share distributed, sophisticated and heterogeneous resources, grid [2, 3] and Web services [4] form an open and standard information infrastructure to implement large-scale resources sharing. Hence grid, Web services together with OpenGIS specifications [5] built the technical foundation of sharing

---

* This work is supported in part by the National High Technology Research and Development 863 Program of China (Grant Nos. 2002AA134010, 2002AA134020, 2003AA135110).

S. Wang et al. (Eds.): ER Workshops 2004, LNCS 3289, pp. 127–138, 2004.

and integration of spatial information. Based on the three key technologies, we propose a new service-oriented application grid named Spatial Information Grid (SIG) [6] aiming to solve the application problems. The definition of SIG is as below:

Definition 1: Spatial Information Grid (SIG) is a spatial information infrastructure with the ability of providing services on demands, and implementing organizing, sharing, integration, and collaboration of enormous distributed spatial information.

SIG is a service-oriented application framework which defines mechanisms for creating, managing, and exchanging spatial information among entities called SIG services, and enables the integration and composition of services and resources across distributed, heterogeneous, dynamic environment. So service is one of the most important technical elements of SIG. According to the concepts defined by OGC [7], some definitions related to SIG service are given as follows:

Definition 2: SIG service is a collection of spatial operations which is accessible through an interface and provides spatial application functions.

Definition 3: Operation is specification of an interaction that can be requested from an object to effect behavior.

Definition 4: Interface is an implementation of operations including the syntax of the interaction for a given distributed computing technology.

SIG involves many research issues among which system architecture describes SIG structure and forms research foundation of SIG. On the other side, single SIG service can only support simple spatial information application, but most of current spatial information applications are so complex that they need involve effective composition of multiple individual SIG services, i.e. service-based collaboration, which is a critical step towards the development of interoperable spatial information applications. Hence, we put research emphasis on system architecture and service-based collaboration here.

The remainder of this paper is organized as follows. Section 2 gives a brief overview of related work. Section 3 is devoted to system architecture of SIG. Technical framework and some key issues of service-based collaboration are presented and discussed in Section 4. Section 5 illustrates application flow of SIG and presents a case example in our project. Conclusion is given in section 6.

2   Related Work

To date, many approaches are introduced and applied to implement sharing and integration of spatial information, in which grid technology, OpenGIS and Web services are most important three ones. Considering system architecture and service-based collaboration, we briefly overview the technologies closely related to our work.

There have been some studies on integrating grid technology with spatial information application. The Committee on Earth Observation Satellites (CEOS) started its research on a prototype system to share satellite data and spatial information in global areas from 2001. Another projects are ESG (Earth Grid System) supported by U.S. DOE and EOS (Earth Observation System, which is a sub-project of Europe Data-Grid project).

Furthermore, World Wide Web Consortium establishes Web Services Architecture [4] to define components and their relations of service framework. And Heather Kreger proposed WSCA (Web Services Conceptual Architecture) [8] to provide foundation of building and deploying Web services applications. As chief organizations of establishing open interoperation standards for spatial information, the OpenGIS Consortium (OGC) together with committee of ISO TC211 proposed a frame of OGC Web Services (OWS) which is an online spatial information service frame enables seamless integration of spatial information services [9].

On the other hand, as an effective method to implement service-based collaboration, web service composition has received much attention and several methods are proposed and investigated to describe and model process of web service composition. Recent emerging workflow projects focus on loosely coupled processes, such as eFlow [10], which is a process management system that supports adaptive and dynamic service composition. The industrial world has developed a number of XML-based standards to formalize the specification of Web services, their flow composition and execution, such as WSFL[11], XLANG[12] and BPEL4WS[13]. There is also a DARPA project DAML-S that has taken also the procedural approach to service composition [14]. Meanwhile, we see the use of a Petri-net-based algebra to model control flows for Web service composition in [15]. Moreover, service selection has become an important research issue of service composition. The main solution is based on service quality, e.g. in [16], the authors proposed a Web service QOS model and compute the QOS values of execution plans and execution paths to implement service selection and flow execution.

As pointed out above, we shortly outline some related technologies. In designing SIG system architecture of and implementing service-based collaboration, we took advantage of all research contributions above and extend them as follow:
- By combining OpenGIS specifications, grid technology and Web services synthetically, we establish open system architecture of SIG.
- We build technical framework of service-based collaboration, and propose SIG service protocol framework and SIG ontology (SIGonto).
- We introduce three new elements, i.e. service taxonomy, time and conditions, to form a new SIG service-based collaboration model (Service/Resource Net, SRN).
- To implement dynamic service selection in the execution process of SIG service-based collaboration, we propose a Service Instance Selection Model (SISM), and introduce an algorithm based on multiple attribute decision making method.

3   System Architecture of SIG

SIG is a complex spatial information system which may be used through whole application course of spatial information from information acquiring, storing to applying. SIG is mainly composed of following components: spatial information acquiring systems, storing systems, processing systems, multi-layer users, computing resources (PCs, high-performance servers etc), and application systems. In SIG, these components are integrated by SIG services. Figure 1 shows the constitution of SIG.

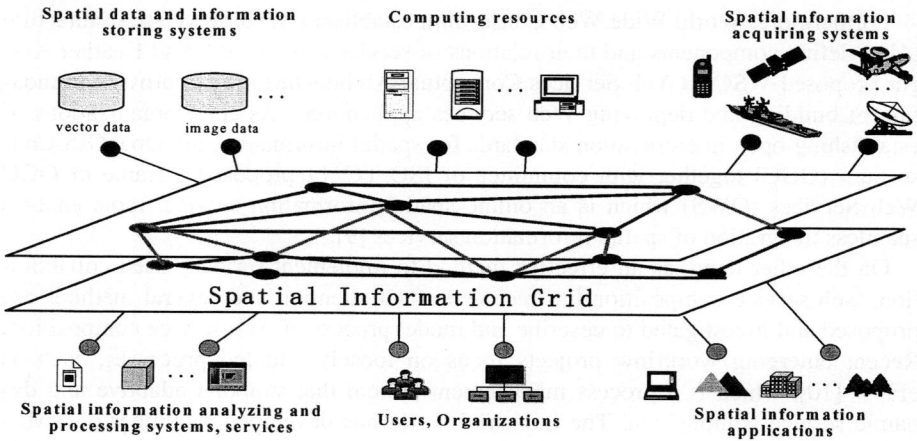

Fig. 1. The constitution of SIG

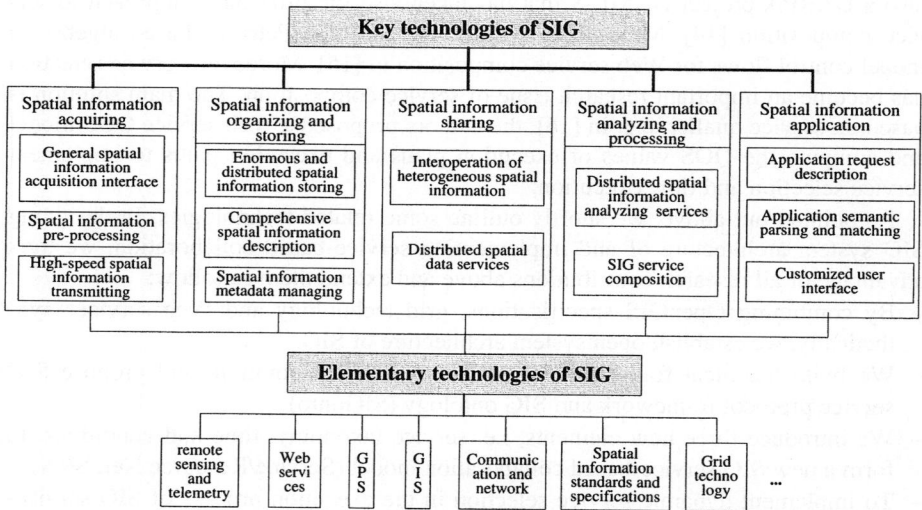

Fig. 2. The technical architecture of SIG

Corresponding to the constitution of SIG, we establish SIG system architecture to compartmentalize logic and technical layers, express the relations among these layers, and form a complete description of SIG. From the view of system theory, SIG system architecture can be presented from two aspects: application architecture and technical architecture. The application architecture of SIG is designed as seven-layer structure, which is called SIGOAA (SIG Open Application Architecture). The detailed presentation of SIGOAA is in [6].

Based on SIGOAA, we present technical architecture here by describing technical constitution, contents and framework of SIG to provide technical basis for designing and implementing SIG (shown in Figure 2).

4   Service-Based Collaboration of Spatial Information

Generally, an individual SIG service can only support simple spatial information application and can't accomplish complex task. But most of current applications often require wide linking and composition of multiple different SIG services to create new functionality web processes. By combining and composing SIG services, we can easily implement service-based collaboration of spatial information to support complex spatial information applications.

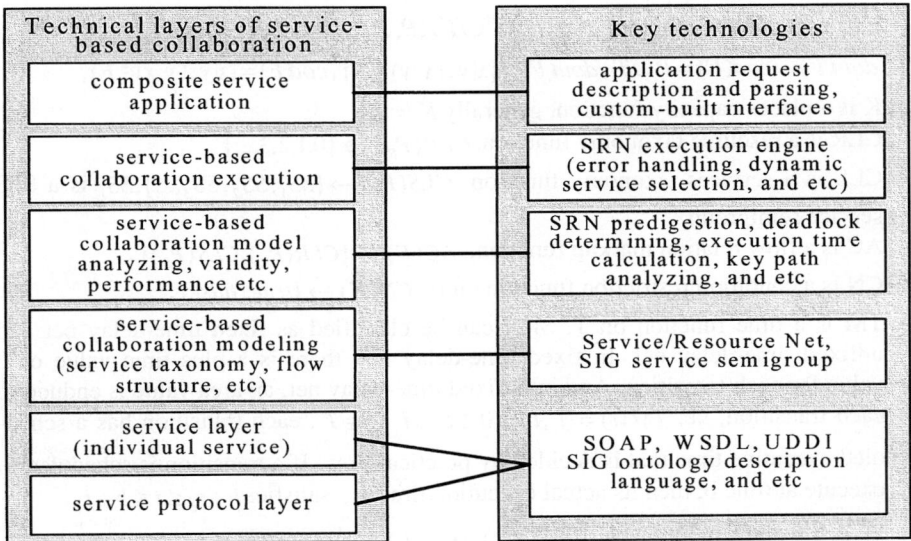

Technical layers of service-based collaboration	Key technologies
composite service application	application request description and parsing, custom-built interfaces
service-based collaboration execution	SRN execution engine (error handling, dynamic service selection, and etc)
service-based collaboration model analyzing, validity, performance etc.	SRN predigestion, deadlock determining, execution time calculation, key path analyzing, and etc
service-based collaboration modeling (service taxonomy, flow structure, etc)	Service/Resource Net, SIG service semigroup
service layer (individual service)	SOAP, WSDL, UDDI SIG ontology description language, and etc
service protocol layer	

Fig. 3. Technical framework of SIG service based collaboration

Service-based collaboration of spatial information includes six technical layers (shown in Figure 3). Based on application demands, we currently put research emphasis on service protocol framework, service semantics description[1], service-based collaboration modeling, and service dynamic selection.

4.1   Process Model of Service-Based Collaboration

The process of SIG service-based collaboration can be regarded as workflow. Workflow model is the foundation and precondition of workflow. As a practical and effective method and tool, Petri net is used widely in workflow modeling, and its basic definitions and concepts are in [17,18]. But basic Petri net can't accurately describe and model the workflow in which activity and resource flow run synchronously [17,18]. So we introduce three useful and additional elements to extend basic Petri

---

[1] The detailed presentation of service protocol framework and service semantics description can be referred in [6].

net, i.e. time, conditions, and service taxonomy. Hence we propose a new process model of service-based collaboration named Service/Resource Net (SRN).

Definition 5: Service/Resource Net (SRN) is an extended Petri net, i.e. a tuple $SRN = (P,T,F,K,CLR,CLS,AC,CN,TM,W,M_0)$ where:

- P is a finite set of places, $P = \{P_R, P_S\}, |P| = n$, $P_R = \{P_{R1}, P_{R2}, ..., P_{Rn}\}$ is a set of resource (data, information, etc), $P_S = \{P_{S1}, P_{S2}, ..., P_{Sn}\}$ is a set of SIG services,

- T is a finite set of transitions representing the activities,
  $T = \{T_1, T_2, ..., T_m\}, |T| = m$,

- F is a set of flow relation, $F \subseteq P \times T \cup T \times P, P \cap T = \varnothing, P \cup T \neq \varnothing$;
  $dom(F) \cup cod(F) = P \cup T, dom(F) = \{x | y : (x, y) \in F\}, cod(F) = \{x | y : (y, x) \in F\}$,

- K is a places capacity function, generally $K = \infty$,

- CLR is a resource taxonomy function, $CLR(P_R) \rightarrow \{0,1,2,3,...\}$,

- CLS is a services taxonomy function, $CLS(P_S) \rightarrow \{SS_1, SS_2, SS_3,...\}, SS_i$ is a SIG service semigroup [2].

- AC is a flow relation markup function, $AC(F) \rightarrow \{CLR(P_R), CLS(P_S)\}$.

- CN is a condition restriction function on F, $CN(F) \rightarrow \{true, false\}$.

- TM is a time function on T. SRN can be classified as fixed time-delay net and unfixed time-delay net. In fixed time-delay net, there is a plus time value or 0 value for each transition. And in unfixed time-delay net, an area value is endued to each transition, i.e. $TM(t) = [t_i, t_j], (0 \leq t_i \leq t_j), t \in T$, each transition has a scheduled executive time that is decided by practical flow. If a transition is scheduled to execute at time b, then its actual execution time (t) satisfies $b + t_i \leq t \leq b + t_j$.

- W is a weight function on F, $W(x, y) = (w(cl_1), w(cl_2), ..., w(cl_l)), w(cl_i) \rightarrow \{0,1,2,3,...\}$, $CL = CLR \cup CLS, cl \in CL, (x, y) \in F, l = |CL|$.

- M is a marking function (marking is an assignment of tokens to the places), $M = (m(p_1), m(p_2), ... m(p_n)), p_i \in P, m(p_i) = (|cl_1|, |cl_2|, ... |cl_l|), M_0$ is the initial marking.

SRN is a novel extended Petri net-based process model. To implement execution, simulation and verification of SRN, many elementary research issues should to be studied in detail, such as transition firing rules, running structures, characteristics of SRN (e.g. liveness, reachability etc), validity analyzing methods, performance evaluating algorithms, and so on. Further and detailed discussion of these issues can refer to our published paper, i.e. [19].

4.2 Dynamic Service Selection Model and Method

A well-known problem with execution of service composition is that finding and selecting the many appropriate services is difficult. In an execution process of service-based collaboration, there are many service roles (denoted in P_S) which are respectively specified by different functions in accordance with the service-based

---

[2] Service semigroup is a service taxonomy theory whose detailed presentation is in [19]

collaboration process logic. In actual application, there may be multiple service sets that can implement a same function, and each service set may involve lots of service. Hence, the course of dynamic service selection is divided into two steps:

1. For each service role of the given composition process, a service set should be selected from multiple service sets as the matching service set, i.e. efficient service set filtering;
2. To select the most suitable service from such service set as the optimum service instance to carry out corresponding function of the given service role, i.e. efficient service filtering.

At first, we propose Service Family (SF) as a new concept to implement selection of the matching service set. The concept of Service Family is as below.

Definition 6: Service Family is a set of SIG services; these SIG services are provided by different service providers, have same invocation interfaces, and can implement same functions, i.e. $SF = \{s_1, s_2, ..., s_n\}$, $s_i (i \in \{1, ..., n\})$ is a SIG service.

Considering definitions of SF and SIGonto [3], we invent a SF selection method based on semantics descriptions of function requests and services to implement matching of service role and SF. This method is to calculate semantic similarity degree of function request and service family, and the detailed algorithm is as follows:

---

step1: To generalize the concepts in SIGonto base, and form semantic vectors of function request and each SIG service.

step2: To calculate the central vector of each SF, i.e. $d = \sum_{i=1}^{n} c_i \Big/ n$, d denotes central vector of a SF, n is the number of SIG services included in such SF, c_i denotes vector of corresponding service in such SF.

step3: To calculate semantic similarity degree of function request vector and central vector of each SF, i.e. $sim(d_i, d_j)$:

$$sim(d_i, d_j) = \cos\theta = \frac{\sum_{k=1}^{M} W_{ik} \times W_{jk}}{\sqrt{(\sum_{k=1}^{M} W_{ik}^2)(\sum_{k=1}^{M} W_{jk}^2)}}$$

, where d_i is function request vector, d_j is central vector of the jth *SF*, M denotes dimensions of d_i and d_j, W_k denotes the kth dimension of d_i and d_j.

The angle between d_i and d_j s smaller, i.e. the value of $\cos\theta$ is bigger, the semantic similarity degree of function request and service family is higher. The value of $sim(d_i, d_j)$ is between 0 and 1, and we will select the SF which makes the value of $sim(d_i, d_j)$ be maximal as the matching SF of the given function request. For example, there exists a service role whose functional requirement is to query city map

---

[3] SIG ontology (SIGonto) is a formal, explicit specification of shared spatial conceptualization.

and analyze traffic status, and its function vector is expressed by $R = (a,b,c,d)$, where a denotes city, b denotes map querying, c denotes traffic data, d denotes data statistics. Given two service families, i.e. SF_1, SF_2 , their central vectors are $d_1 = (a,b,c,e)$ and $d_2 = (a,b,f,h)$, where e denotes data summing (data summing is a sub-concept of data statistics), f denotes geographic data, h denotes data querying (data querying is a sub-concept of data summing). By defining the matching weight between a concept and its sub-concept is 0.7, the matching weight between two concepts which have no relations is 0, and the matching weight between a concept and itself is 1, we translate the function vector and two central vectors into the following expressions: $R = (1,1,1,1), D_1 = (1,1,1,0.7), D_2 = (1,1,0,0.49)$.

Then, we can calculate similarity degree of R and D_1, D_2: $sim(R,D_1) = 0.99$, $sim(R,D_2) = 0.83$. For $sim(R,D_1) > sim(R,D_2)$, SF_1 is selected as the matching SF.

After selecting of the matching SF, we need to select an optimum service instance from the SF to carry out corresponding function of the given service role. Then we propose a Service Instance Selection Model (SISM) by defining a set of generic selection criteria (including service grade, reliability, invocation frequency etc) to implement optimum service instance selection.

Definition 7: Service Instance Selection Model is a tuple composed of six elements, i.e. $SISM = (D, T, C, IP, R, UE)$, where:

- Degree (D) denotes service grade. For a SIG service s, D(s) is determined by the status and attributes of its providers, e.g. given a service s_1 developed by a government department and a service s_2 provided by private organization, then $d(s_1) > d(s_2)$.
- Time (T) denotes the time interval for user to implement proposing request, invoking service, and getting the final processing results provided by the service. Given a service s, $T(s)$ is computed based on past service executions, i.e.

$$T(s) = \frac{\sum_{i=1}^{n} T_i(s)}{n}$$, where $T_i(s)$ is a past observation of the invocation and execution time of s, and n is the number of execution times observed in the past.
- Cost (C) denotes the cost of invoking a service, including fee etc.
- Invocation Probability (IP) denotes invocation frequency of a service. Given a SIG service s, the value of invocation probability is computed from historical data about past invocations using the expression $IP(s) = \frac{N(s)}{k}$, where $N(s)$ is the number of times that s has been invoked, and k is the total number of invocations.
- Reliability (R) denotes the reliability of service, i.e. the rate of invoking and executing service successfully. The reliability of service is influenced by the status of hardware, software and network. It is computed based on historical statistical data of service invocation, i.e. $R(s) = 1 - \frac{M(s)}{N(s)}$, where $M(s)$ denotes the number

of times that s has been unsuccessfully invoked and executed, $N(s)$ is the number of times that s has been invoked.

- User Evaluation (UE) denotes the user satisfied degree of a service. Its computing basis is the historical statistical evaluation data given by the users who invoked the

service, i.e. $UE(s) = \dfrac{\sum\limits_{i=1}^{n} UE_i(s)}{n}$, $UE_i(s)$ denotes the service evaluation of

the ith user, $UE(s) \rightarrow \{0,1,2,...,100\}$.

The SISM model provides computing elements for selection of the optimum service. So the process of optimum service selection can be concluded as a problem of Multiple Attribute Decision Making (MADM): given a set of projects, the decision-maker computes the criterion values of such projects based on some criterions, then compares the projects based on a decision making rule and criterion values to get the sort order of all projects, and select the optimum project [20]. The process of optimum service selection based on MADM is shown as Figure 4.

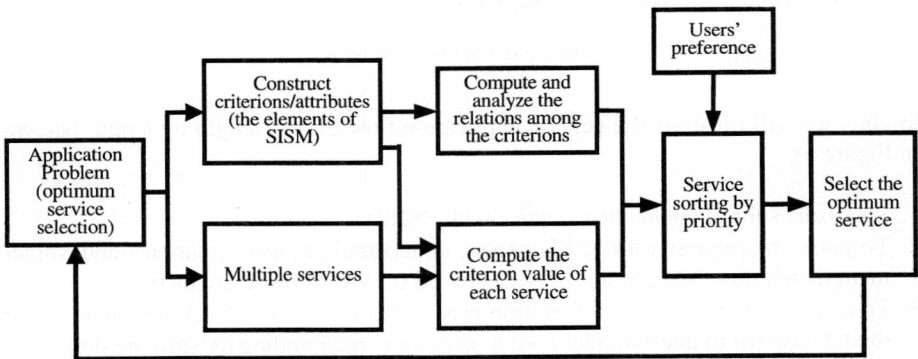

Fig. 4. The process of optimum service selection based on MADM

The six elements of SISM model construct the criterions. According to the methods of MADM, we combine the SIG services and criterions to establish a MADM matrix. Based on the matrix, we can compute the criterion value of each service and analyze the relations among the six criterions. Finally, we select the optimum service in term of users' preference and the sort order of the service. The detailed algorithm and rules is beyond the scope of this paper and will be presented in another paper.

The SISM model can be extended conveniently, and we will adjust and modify its parameters according to application requirements.

5 SIG Application Flow and Example

As presented above, SIG has many novel and interesting technical characteristics. Hence, SIG can be widely used in most spatial information applications. In the fol-

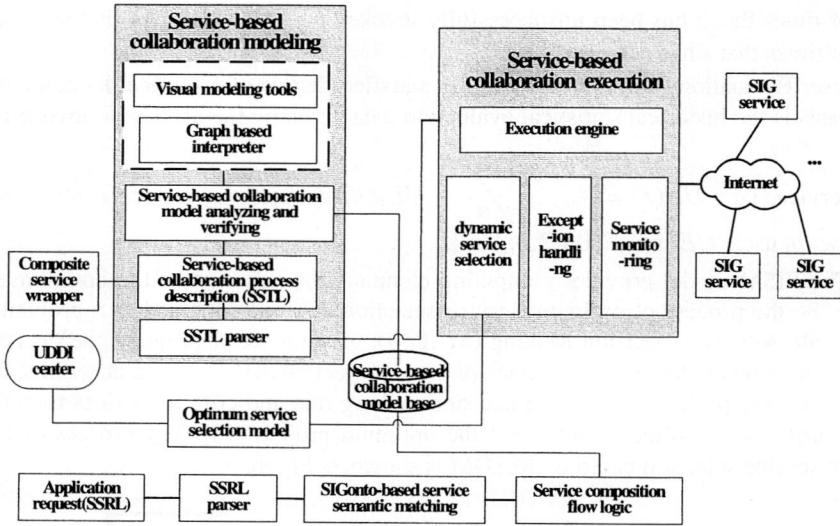

Fig. 5. SIG application flow

lowing, we will illustrate the general application flow of SIG length by length (shown in Figure 5).

1. SIG users submit application request with SSRL[4].
2. To parse the request with SSRL parser, understand request semantics, and implement matching of request and services based on SIGonto semantics rules.
3. To create service-based collaboration process logic and store such new model into model base (or to query model base to acquire corresponding existing model).
4. To analyze and verify the validity of such service-based collaboration model.
5. To use SSTL [5] to provide computer-interpreted description of service-based collaboration model (the composite service may register itself at UDDI center as a new individual SIG service).
6. SSTL parser parses process logic and sent it to execution engine to execute.
7. Service-based collaboration execution engine invokes corresponding SIG services orderly according to process logic, and many optimum service instances are selected based on SISM; meanwhile, some simple methods for exception handling and service monitoring are provided.
8. Return final results to users.

Furthermore, developers can use visual modeling tools to define service-based collaboration model, and graph-based interpreter will parse and understand the process

---

[4] SSRL (SIG Service Request Language) is a XML-based language that present application request.
[5] SSTL (SIG Service Task Language) is a XML-based language that describe the process structure of service-based collaboration.

logic. After being analyzed and verified, the process model is to be translated into SSTL file (shown in dashed rectangle) as well.

Corresponding to SIG application flow, we have implemented an experiment system based on SIG services and service-based collaboration. A case example of city environment evaluating is illustrated in Figure 6: (a) denotes the process of services discovery in SIG UDDI center; (b) is the district map; (c) shows the integration of data and map; (d) gives the evaluating results.

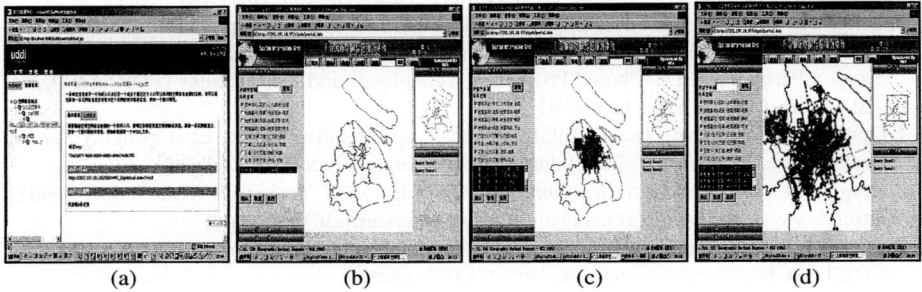

(a)   (b)   (c)   (d)

Fig. 6. SIG application example

From this application example, we learn that SIG can share and aggregate distributed spatial information and services from different departments and organizations all over the city, province, and country. With SIG services, we can easily implement service-based collaboration to provide powerful abilities of spatial information acquiring, sharing, processing, and applying.

6 Conclusion and Future Work

SIG is an innovative service-oriented spatial information system. In this paper, we put research emphasis on SIG system architecture and service-based collaboration. Hence, we introduce application and technical architecture of SIG, propose service-based collaboration framework, and discuss some key technologies which are vital in implementing service-based collaboration, i.e. a novel process model of service-based collaboration (SRN), the dynamic service selection model and the related method.

The research of SIG is at the initial stage. The architecture, concepts and key technologies of SIG need to be improved and extended. Our ongoing research includes enormous spatial data managing, more effective methods and technologies for service-based collaboration execution, new analysis and evaluating methods of SRN, and etc.

References

1. Longley, P., Goodchild, M., Maguire, D., Rhind, D. (eds.): Geographical Information Systems: Principles, Techniques, Applications and Management. John Wiley & Sons, New York (1999)

2. Foster, I., Kesselman, C. et al: The Physiology of the Grid: An Open Grid Services Architecture for Distributed Systems Integration. (2002).
http://www.gridforum.org/ogsi-wg/drafts/ogsa_draft2.9_2002-06-22.pdf
3. Tuecke, S., Czajkowski, K. et al : Grid Service Specification. Open Grid Service Infrastructure WG, Global Grid Forum, Draft 2. (2002). http://www.globus.org
4. World Wide Web Consortium: Web Services Architecture Draft 14 (2002).
http://www.w3.org/TR/2002/WD-ws-arch-20021114/
5. OpenGIS Consortium: The OpenGIS Abstract Specification Topic 0: Abstract Specification Overview, Version 4.0 (1999). http://www.opengis.org/docs/99-100r1.pdf
6. Tang, Y., Chen, L., He, K.T., Jing, N.: SIG: A Service-Oriented Application Grid for Spatial Information Sharing and Integrating. In Proceedings of the 2004 International Conference on Parallel and Distributed Processing Techniques and Applications (PDPTA'04), Nevada, USA, CSREA Press (2004) 915-922
7. OpenGIS Consortium: The OpenGIS Abstract Specification Topic 12: OpenGIS Service Architecture, Version 4.3. (2002). http://www.opengis.org/docs/02-112.pdf
8. Kreger, H.: Conceptual Architecture (WSCA 1.0). IBM Technical White Paper (2001).
http://www- 3.ibm.com/software/solutions/webservices/pdf/WSCA.pdf
9. OpenGIS Consortium: OGC Web Services Common Implementation Specification, Version 1.0 (2003). http://www.opengis.org/docs/03-088r1.pdf
10. Casati, F., Ilnicki, S., Jin, L., Shan, M.C.: An Open, Flexible, and Configurable System for E-Service Composition. Technical Report HPL-2000-41, HP Labs (2000)
11. Leymann, F.: Web Services Flow Language (WSFL). IBM (2001).
http://www.ibm.com/software/ solutions/webservices/pdf/WSFL.pdf
12. Thatte, S.: XLANG: Web Services for Business Process Design. Microsoft (2001).
http://www.gotdotnet.com/ team/xml_wsspecs/xlangc/default.htm.
13. Curbera, F., Goland, Y., Klein, J. et al : Bpel4ws white paper (2002).
http://www-3.ibm.com/software/ solutions/ webservices
14. Ankolekar, A., Burstein, M., Hobbs, J. et al: DAML-S: Web Service Description for the Semantic Web. In Proceedings of the 1st International Semantic Web Conference (ISWC) (2002)
15. Hamadi, R., Benatallah, B.: A Petri net-based model for web service composition. In Proceedings of the Fourteenth Australasian database conference on Database technologies, Adelaide, Australia (2003) 191-200
16. Zeng, L., Benatallah, B., Dumas, M.: Quality Driven Web Services Composition. In Proceedings of WWW2003
17. Murata, T.: Petri Nets: Properties, Analysis and Applications. Proceedings of the IEEE, 77(4) (1989) 541-580
18. Yuan, C.Y.: Petri Net Theory (in Chinese). Publishing House of Electronics Industry, Beijing, China (1998)
19. Tang, Y., Chen, L., He, K.T., Jing, N.: SRN: An Extended Petri-Net-Based Workflow Model for Web Services Composition. In Proceedings of the 2nd International Conference on Web Services (ICWS 2004), San Diego, CA, USA, IEEE Press (2004) 591-599
20. Yue, C.Y.: Decision Making Theory and Method (in Chinese). Science Press, Beijing, China (2003)

Geo Web Services Based on Semantic[*]

Yang An[1,2,3], Bo Zhao[1], and Fulin Bian[3]

[1] School of Computer, Wuhan University, 430079, Wuhan, China
{yangan,zhaobo}@whu.edu.cn
[2] State Key Lab of Software Engineering, Wuhan University, 430079, Wuhan, China
[3] Spatial Information and Digital Engineering Research Center, Wuhan University
430079, Wuhan, China
Flbian@wtusm.edu.cn

Abstract. The fast development of Web Services enables Geo spatial data and functionality to be shared in a distributed computing environment. However, Industry standards of Web Services, such as UDDI are not power enough for automated services discovery, binding and composition. The main inhibitor is the lack of semantics in the description of Web Services and discovery process. In this paper, pattern of Geo Web Services based on Ontology is proposed, which employs Ontology to advertise Geo Web Services and uses them for semantics-based discovery of relevant Geo Web Services. Composition of Geo Web Services for the requirement of complex task is also based on Ontology if it is necessary.

1 Introduction

Development in general Web Services technologies and the efforts of OGC in the areas of service categorization and interoperability of service interfaces have resulted in the evolution of Geographic Information Systems towards the Web Services model. Web Services are modular, self-contained, self-describing applications that are accessible over the Internet [1]. They constitute software modules that describe a collection of operations that are network-accessible through standardized XML messaging. Geo Web Services is such Internet applications, which use Geospatial data and related geo-functions to implement basic geo-processing task such as address matching, mapping and route planning, which allow application developers integrate GIS functionality into their application without having to build or host the functionality locally [2].

The core technology of Web Services is based on a series of XML-based open standard protocols. Among them, the industry standards for Web Services are described as follows: WSDL (Web Services Description Language), is an XML format for describing network services as a set of endpoints operating on messages con-

---

[*] The work was supported by National Natural Science Foundation of China under grant No. 60373019, and Hi-Tech Research and Development Program of China under grant No.2002AA135340, and Open Researches Fund Program of SKLSE and LIESMARS under grant No. SKL(4)003 and No. WKL(01)0303, and IBM Research Award.

S. Wang et al. (Eds.): ER Workshops 2004, LNCS 3289, pp. 139–147, 2004.

taining either document oriented or procedure-oriented information[3]; SOAP(Simple Object Access Protocol), provides a simple and lightweight mechanism for exchanging structured and typed information between peers in a decentralized, distributed environment using XML[4]; UDDI(Universal Description, Discovery and Integration) specification, offers users a unified and systematic way to find service providers through a centralized registry of services that is roughly equivalent to an automated online "phone directory" of Web Services[5]. Web services are described using WSDL definitions and advertised in UDDI registries. Discovery mechanism supported by UDDI provides an interface for keyword and taxonomy based searching, which is valid to web discovery based on syntax, while not be powerful enough for automated discovery in dynamic environment. If requesters and providers use different terms (or different languages) for referring to the same concepts or the same term for different concepts, this can cause some problems [6], such as: No match, services that might be useful for solving the requester's problem might not be found at all; Inappropriate match: Services that are found might not be useful for solving the requester's problem. Therefore, this makes it difficult for the requester to choose an appropriate term for his search or to judge whether the discovered services are indeed useful for his task. In GIS domain, geo-information technology standards developed by ISO/TC211 and OGC could not address the issue of semantic heterogeneity [7]. It is the goal of this paper to overcome the problems resulting from semantic heterogeneities in Geo Web Services.

2   Geo Web Services Pattern Based on Semantics

The main reason of problems described above is the lack of semantics in the service descriptions and discovery process, which makes UDDI less effective. Therefore explicit specified concepts rather than free text or simple taxonomies for describing Geo Web Services and user requirements are taken into account in this paper. Such explicit specifications of conceptualizations are called Ontologies.

Ontology is an explicit specification of a sharing conceptualization, which can be used in an integration task to describe the semantics of information sources and to make the contents explicit, also can be used for the identification and association of semantically corresponding information concepts [8]. Therefore, Geo Web Services pattern that we proposed in this paper employs Ontology to describe services and requestors' requirement.

Fig.1 depicts a general layered architecture that enables semantic interaction of Geo Web Services in dynamic environments. The architecture is described as follows:

- Network Layer: The Network Layer forms the lowest layer in the architecture and encapsulates standard network protocol (such as HTTP, SMTP and IIOP) and SOAP allowing for exchange of object descriptions by standard means. In addition, machine agents need the possibility to communicate at an adequate level using protocols such as Knowledge Query and Manipulation Language (KQML) or Agent Communication Language (ACL).

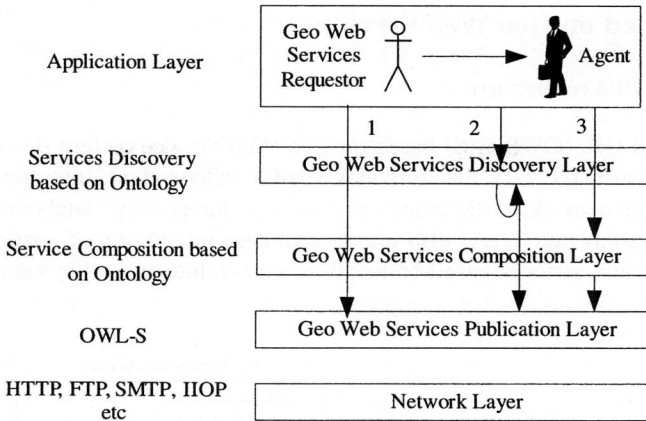

Application Layer

Geo Web Services Requestor — Agent

Services Discovery based on Ontology

Geo Web Services Discovery Layer

Service Composition based on Ontology

Geo Web Services Composition Layer

OWL-S

Geo Web Services Publication Layer

HTTP, FTP, SMTP, IIOP etc

Network Layer

Fig. 1. Geo Web Service Pattern

- Geo Web Services Publication Layer: The Geo Web Services providers create Geo Web Services and their services definition that are based on Ontology, and then publish the services with the Geo Web Services registry based on OWL-S (an Ontology of service). Once a Geo Web Service is published, a service requestor may find the services via the registry.
- Geo Web Services Discovery Layer: This layer is used to find the appropriate Geo Web Services for the requestor's task. Geo Web Services are advertised using Ontology, which is used in service matchmaking too. In this pattern, Geo Web Services accessing is distinguished in two possibilities. In the first one (arrow 1 in Fig.1), the requestor directly accesses the Geo Web Services when human user knows the location of the service description, which often is not the case. Alternatively (arrow 2 in Fig.1), the user or application interacts with the registry that helps requestor finding a service and retrieving the descriptions about how to invoke it. Because Agent may better find services and access them directly, Agent is employed to find and invoke Geo Web Services for the requirement of requestor. Therefore, protocols such as KQML or ACL described above in network layer are needed by Agents to communicate with each other.
- Geo Web Services Composition Layer: Geo Web Services composition can be interpreted as pipelining services in a dependent series to attain a customized, more complex request. And this layer is responsible for carrying out the process of managing the discovery and integration of services to construct a composite service (arrow 3 in Fig.1). The process model of the composite service is supplied as input to this layer.
- Application Layer: The application layer embodies any application and human user that request the Geo Web Services. The application layer encompasses different GUI facilities to display the result of Geo Web Services and provides the functionality to initiate a request for Geo Web Services.

3   GIS Based on Geo Web Services

3.1   GWS-GIS Architecture

We developed GIS (GWS-GIS) based on Geo Web Service pattern described above, which is a distributed GIS that provides vendor independent, interoperable framework to implement data retrieving, processing, integrating, analyzing, decision-making and visualization, and also allows seamless integration of spatial data processing system and services based on location. The architecture of system is shown in Fig.2.

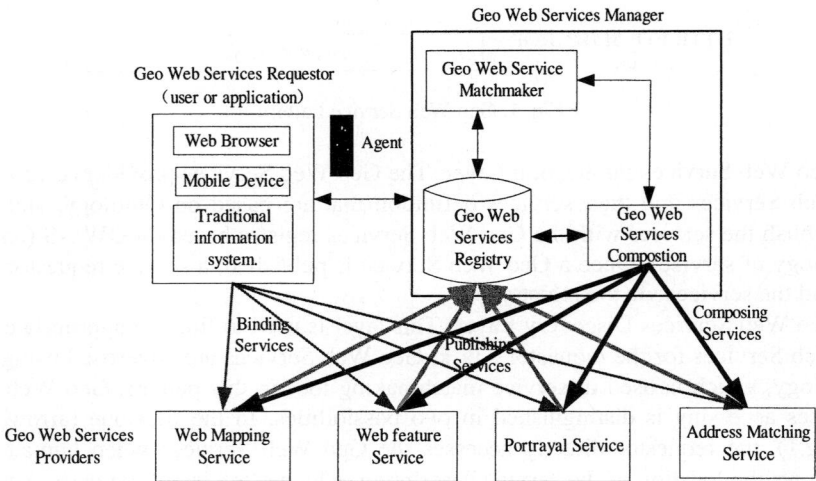

Fig. 2. Architecture of GWS-GIS

Geo Web Services can be grouped into two basic categories in GWS-GIS [9]:

(1)   Geo Data Services: offer customized data to users (such as OGC Web mapping, Web Coverage and Web Feature services). These services are tightly coupled with specific data sets.

(2)   Geo Processing Services: provide operations for processing or transforming data in a manner determined by user-specified parameters, which are not associated with specific datasets. Such services can provide generic processing functions such as projection/coordinate conversion, rasterization/vectorization, map overlay, imagery manipulation, or feature detection and imagery classification.

Once Geo Web Services are deployed, client applications can be built more flexibly by mixing and matching available services. Fig.2 shows a variety of Geo Web Services' requestors such as: Web browser, mobile devices and traditional information system.

The most important component in GWS-GIS is Geo Web Services Manager. It offers functions of Web Services discovery and Web Services composition. When user needs service to fulfill task, Agent sends the requirement to Geo Web Service Manager to find service. Some services will be composed when user's requirement is too complex to be fulfilled by one service.

3.2 Geo Web Services Discovery

Geo Web Services Discovery is the process of finding the appropriate service advertisements for a request description through Geo Web Services Matchmaker that is implemented based on matchmaking algorithm. Fig.3 shows the architecture of Geo Web Services Matchmaker.

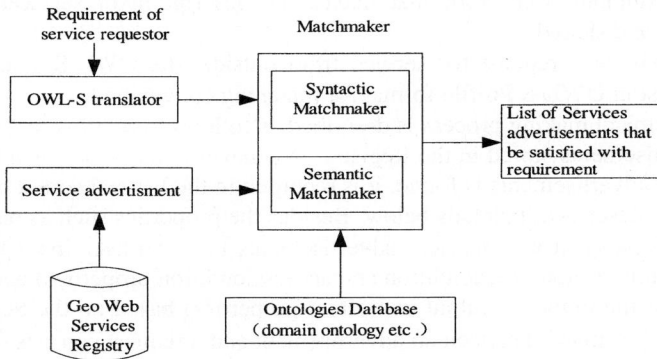

Fig. 3. Architecture of Geo Web Services Matchmaker

The service provider advertises services to Registry. When matchmaker receives the requirement description of requestor, the services advertisements in Registry will be sent to matchmaker for matching. Ontologies database contains Ontologies of general knowledge, GIS domain Ontologies and knowledge concept Ontologies, which are used to judge relation of related elements of two descriptions. The result of the matching is the list of advertisements that are satisfied with requirements.

In order to discover Geo Web Services based on semantics, OWL-S is employed to describe Geo Web Services in GWS-GIS. As part of the OWL (Web Ontology Language) [10] for Services, OWL-S is an ontology of services, which provides a set of basic concepts and relations for declaring and describing services, by utilizing the ontology structuring mechanisms provided by OWL.

An OWL-S Service is defines by three parts [11]: the ServiceProfile, which tells "what the service does"; the ServiceModel, which tells "how does the service works"; and the ServiceGrounding, which specifies "how is the service used". Generally speaking, the ServiceProfile gives the abstract description of services to a service-seeking agent to determine whether the service meets its need, while the ServiceModel and the ServiceGrounding provides the information for an agent to make use of a service. What publishes in GWS-GIS is attributions of ServiceProfile and

ServiceModel, such as: ServiceName, ServiceType, input (specifies the information that the service requires to proceed with the computation), output (specify the result of the operation of the service), precondition (present one or more logical conditions that should be satisfied prior to the service being requested) and effect (events that are caused by the successful execution of a service). These attributions are used to justify whether the services fulfill the requirements. And the elements used in OWL-S are annotated with concepts from domain Ontology in order to make the service description explicit. Domain Ontology provides an agreed-upon conceptualization of a certain part of the world, in this paper a small part of the domain of geography information, and presents the world in a machine interpretable way. And domain Ontology is written in OWL that enables the creation of Ontologies for any domain, and it is a particularly suitable framework that makes the description of services computer-interpretable and shared.

Upon receiving a request for service from outside, the OWL-S Translator constructs request in OWL-S Profile format for matching conveniently.

The main matchmaking process is described as follows: request is matched against all the advertisements stored in the Registry. Whenever a match between the requests of any of the advertisements is found, it is recorded in the list and sorted by similarity degree that is described in details below. Besides the properties such as name, type or provider, this paper groups services advertisements into two lists: input list (consisting of the union of input, precondition and accessCondition properties) and output list (consisting of the union of output and effect properties) based on the ServiceProfile of OWL-S [12]. A match between an advertisement and a request consists of the match of all the input/output of the request against the input/output of the advertisement. Output list of request and advertisement is determined for ReqOutputs and AdvOutputs respectively. The rule for output matching is: the output list ReqOutputs is compared with an advertisement in AdvOutputs, and a match is recognized if ReqOutputs$\subseteq$AdvOutputs. That means the match is recognized if and only if for each output of the request, there is a matching output in the advertisement.

The matchmaking between inputs is computed following the same algorithm, but with the order of the request and the advertisement reversed. Input list of request and advertisement is determined for ReqInputs and AdvInputs respectively. The rule for output matching is: the output list AdvInputs is compared with AdvOutputs, and a match is recognized if AdvInputs$\subseteq$ReqInputs. That means the match is recognized if and only if for each output of the advertisement, there is a matching output in the request.

In comparing process of related element between request and advertisements, "Similarity Degree" is computed. Similarity Degree is the concept that measures the degree of similarity of two elements, which captures the values from 0 to 1, and 1 shows the most accurate and restrictive matchmaking while 0 shows complete unmatchmaking. In this paper, the measure of Similarity Degree works at the syntactical and semantic levels.

In syntactic comparing level, the string matchmaking of two elements is taken into account. We propose a syntactic similarity measure for strings, the Syntactic Simlarity (SynSim), which comparing two strings $S_1.S_2$:

$$\text{SynSim}(S_1,S_2)=\text{Max}\left[\ 0\ ,\ \frac{\text{Min}\ (\ |S_1|\ ,|S_2|\ )-\text{ed}\ (S_1,\ S_2)}{\text{Min}\ (\ |S_1|\ ,\ |S_2|\ )}\right]$$

The ed is determined for edit distance that formulated by Levenshtein [13], which is a well-established method for weighting the difference between two strings. It measures the minimum number of token insertions, deletions, and substitutions required to transform one string into another using a dynamic programming algorithm.

In semantic comparing level, Similarity Degree is computed by considering synonymy, abbreviation, data type and namespace. Synonymy matching that described by Dsyn compares synonymy relation between two descriptions; abbreviation matching that described by Dabre compares names of two descriptions; data type matching that described by Dtype compares data type of element in two descriptions; namespace matching that described Dnsp compares namespace of element in two descriptions. After computing four matching degrees, Semantic Similairty (SemSim) could be defined by:

$$\text{SemSim}=\sqrt[4]{Dsyn\times Dabre\times Dtype\times Dnsp}$$

Based on the definition of Syntactic and Semantic Similarity, Similarity Degree (SM) is defined as follows:

SM= W • SynSim + (1- W) • SemSim

Where $W\in[0,1]$ determines the degree of influence of Syntactic Similarity and is determined by users.

After counting the similarity degree of related element between advertisements and requirements, all the matching services that SM exceed a given threshold are sorted by Similarity Degree. Then users could choice the most appropriate Geo Web Services.

3.3 Geo Web Services Composition

Geo Web Services composition is the process of making a combination of one or more Geo Web Services into users' applications by certain working flow to fulfill complex task.

In OWL-S, each service can be viewed as a process, and process model is used to control interoperation of services. One chief component of a process model is the process model (be defined as Process Ontology), which describes a service in terms of its component actions or processes, and enables planning, composition and agent/service interoperation. Process Ontology describes Web Services' inputs, outputs, precondition and component sub-process. OWL-S classified process into three groups: primitive, which has not any subprocess; undecomposable process, which can't be invoked for it is the abstraction of primitive process and composite process; composite process, which has additional properties called components to indicate the ordering and conditional execution of the subprocesses from which they are composed. Composition process employs some control structure to define how to receive

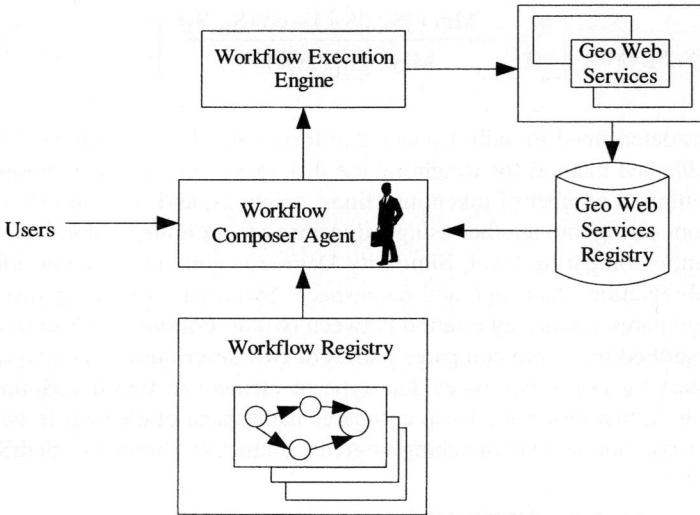

Fig. 4. Architecture for Geo Web Services composition

input and how to return output. Geo Web Services Composition is implemented based on them. Fig.4 depicts the architecture for Geo Web Services composition.

Each workflow that stored in the workflow repository is composed of one or more web services, which can be situated anywhere on the web. The OWL-S descriptions of the Geo Web Services are queried from the Registry, and the service matching that described above is applied. The workflow composer agent composes the executable workflow that comes from workflow repository based on composite process, and sends it into the workflow execution engine. Finally, the execution engine executes the workflow using the Geo Web Service instances.

4   Conclusions

In this paper, model of Geo Web Services based on Ontology is proposed, and GIS based on Geo Web Services is developed. This system employs Ontology to advertise Geo Web Services and uses them for semantics-based discovery of relevant Geo Web Services. Composition of Geo Web Services for the requirement of complex task is also based on Ontology if it is necessary.

In the future, we plan to consummate the prototype of GWS-GIS system, and also intend to research composition of Geo Web Services deeper.

References

1. Roy Jaideep, Ramanujan Anupama; Understanding Web Services; IEEE, IT Pro, November/December 2001
2. ISO/TC-211, OGC: Geogaphic information – Services (ISO/DIS 19119) v4.3.

3. Chinnici, R., Gudgin, M., Moreau, J. and Weerawarana, S. "Web Services Description Language (WSDL) Version 1.2 ", W3C Working Draft 24 January 2003, Available at http://www.w3.org/TR/2003/WD-wsdl12-20030124/
4. Don Box, David Ehnebuske, Gopal Kakivaya etc. Simple Object Access Protocol (SOAP) 1.1;World Wide Web Consortium, May 2000
5. Bellwood Thomas A.; UDDI – A Foundation for Web Services; XML Conference & Exposition 2001, December 2001
6. Bernard, L., U. Einspanier, M. Lutz and C. Portele (2003): Interoperability in GI Service Chains - The Way Forward, in: Gould, M., R. Laurini & S. Coulondre (ed.): 6th AGILE Conference on Geographic Information Science.
7. OGC: OpenGIS Web Services Architecture. OpenGIS Consortium, OpenGIS Discussion Paper OGC 03-025 (2003)
8. R.Fike, W.Pratt. Collaborative ontology construction for information integration. In: Technical Repor KSL-95-63, Stanford University Knowledge Systems Laboratory, 1995.
9. Nadine Alameh. Service Chaining of Interoperable Geographic Information Web Services
10. McGuinness, van Harmelen, eds. OWL Web Ontology Language Overview. W3C Recommendation 10 Feb 2004
11. OWL-S 1.0 Release. http://www.daml.org/services/owl-s/1.0/
12. Paolucci et al; Semantic Matching of Web Services Capabilities. In Proceedings of the 1st International Semantic Web Conference (ISWC2002)
13. I. V. Levenshtein. Binary Codes capable of correcting deletions, insertions, and reversals. Cybernetics and Control Theory, 10(8): 707–710, 1966.

A Scaleless Data Model for Direct and Progressive Spatial Query Processing

Sai Sun, Sham Prasher, and Xiaofang Zhou

School of Information Technology and Electrical Engineering
The University of Queensland, Australia
{sunsai,sham,zxf}@itee.uq.edu.au

Abstract. A progressive spatial query retrieves spatial data based on previous queries (e.g., to fetch data in a more restricted area with higher resolution). A direct query, on the other side, is defined as an isolated window query. A multi-resolution spatial database system should support both progressive queries and traditional direct queries. It is conceptually challenging to support both types of query at the same time, as direct queries favour location-based data clustering, whereas progressive queries require fragmented data clustered by resolutions. Two new scaleless data structures are proposed in this paper. Experimental results using both synthetic and real world datasets demonstrate that the query processing time based on the new multiresolution approaches is comparable and often better than multi-representation data structures for both types of queries.

1 Introduction

The driving force behind developing multi-resolution spatial database systems is to reduce the cost of spatial query processing by lowering the precision of results according to application requirements. This can lead a smaller amount of data used, thus minimizing both I/O and CPU costs of query processing. Different from conventional spatial databases which only have single resolution, a multi-resolution spatial database system is designed to support variable solutions such that the data is used at a proper resolution according to applications. Functionally, a multi-resolution system should support both direct and progressive spatial queries. A *direct query* is a spatial query whos results are not used as intermediate results for subsequent queries. This is a commonly (implicitly) assumed scenario for the traditional spatial query processing approach. The database system processes a query by only considering the query itself and uses a fixed resolution. A *progressive query*, on the other hand, considers a query as one in the context of a sequence of queries. In a spatial context this is typically a window query that is iteratively refined in both location and precision under some user's direction. An increase of precision generally accompanies a reduction in query region. Example applications include map navigations, interactive environmental analyses and spatial data mining algorithms, where users are initially not clear about

S. Wang et al. (Eds.): ER Workshops 2004, LNCS 3289, pp. 148–159, 2004.

what they are looking for and require several queries to refine their search. Progressive queries are similar to those techniques that allow partial query result viewing, such as skyline queries [7]. Here, users can initiate complex operations and stop execution during processing if the results produced at that time are 'good' enough. The advantage is that a rough-and-ready result could take a fraction of the usual time to execute a lengthy, possibly futile, 'complete' query. This approach is largely targeted at mobile digital systems which have limited resources but need quick responses. Providing partial query results also avoids long response times which are unacceptable in on-line and interactive systems [3]. Multi-resolution databases are well-suited for progressive queries because they use data structures which can be retrieved either in part or in their entirety. Progressive query processing is particularly useful for spatial applications because of the volatile and diverse nature of distributed systems, and because the nature of exploratory querying is implicitly imprecise [14].

However, until now little work has been done to support multi-resolution spatial query processing. Many current multi-resolution systems still incur extraneous data retrieval because they are designed for non-progressive queries. Supporting both query types is conceptually challenging, as direct queries favour location-based data clustering whereas progressive queries require fragmented data to be clustered by resolution. In this paper, we investigate the effects of existing spatial data storage schemes for their CPU and I/O costs in the context of supporting both direct and progressive queries. Two scaleless data structures, *Position-Map* (PM) and *Bit-Map* (BM), are proposed to improve direct as well as progressive query processing performance. A cost model is proposed to estimate the storage cost of the BM scheme, and an algorithm to find the optimal base level is presented. The effect of three clustering methods for both schemes (by location, by resolution and by both) has been investigated. Our new techniques are evaluated using both synthetic and real world datasets. Experimental results demonstrate that our approaches produce comparable, and often better, performance than the traditional methods for direct queries, and significant better results for progressive queries.

The rest of the paper is organized as follows. Section 2 gives a brief introduction to related work. After analysing the advantages and disadvantages of two extreme data structures, we propose our data structures in Section 3. Section 4 discusses the progressive cost model and presents experiment results. Section 5 contains concluding remarks.

2   Background

Polygon-based structures are the most common data structure used in spatial databases. A row under this kind of scheme may be structured as in the form of *Spatial_table(ObjectId, Geometry)*, where *Geometry* is a spatial object stored as a collection of vertices. Given n spatial objects we may get n tuples. In order to minimize seek-time, objects are clustered on disk according to their spatial location. The rationale behind this is that multiple objects in close spatial proximity are usually all needed together to answer queries. To further improve disk utilization data can often be compressed. The advantage of this technique is that the information of the exact geometry can be accessed directly without any extra step of object reconstruction. This technique is suitable for spatial databases in which objects are processed wholly. Because geometry elements are opaque to applications, for those operations which do not need full retention of geometry, this method causes a waste in I/O time and CPU time. Multi-resolution data structures are introduced to provide access to differently scaled representations of data, and preserve visual and topological constraints when changing scales. The multi-representation approach [16][4] pre-generates copies from the finest data at different resolution levels. In each copy, data are organized the same way as in conventional spatial database systems, i.e, based on geometry. Complex structures such as abstract cells [10] are used between different copies to measure approximation errors to maintain data coherence. This technique is, intuitively, superior in terms of data retrieval speed for single queries. However it suffers two drawbacks. First, the number of distinct resolutions made available is dictated by storage capacity. Second, progressive querying strategies cannot benefit from data replication. The total replication scheme of multi-representation defeats the purpose of progressively iterating queries as data in different resolutions are not associated, so duplicated data at a low resolution will be retrieved again when a higher resolution is required. The multi-resolution approach breaks data into fragments according to some algorithm, and then derives the proper version of the spatial data on-the-fly. It avoids the main problems of multi-representation but it costs extra CPU and I/O to derivate the data. In [6], objects are grouped into different layers according to their non-spatial attributes which are thought to be able to indicate scale. When processing a query, the non-spatial attributes are examined to check whether the corresponding objects need to be retrieved. The layered approach has drawbacks in that object must be retrieved or ignored as entirety. In [5], spatial objects are indexed under a BANG tree and stored with an explicit scale value. In a similar paper [15], objects are split into groups of vertices called vertex layers, which are also assigned a scale value and indexes as individual entities in a 3D R-tree. The two approaches allow the DBMS to deal with parts of a complex object and to approach different resolution levels by choosing different point sets. But the pre-computation of vertex layers is very complex and adding a dimension also inevitably increases the complexity of processing the index. In [9] we use z-values to encode data and adopts a Li-Openshaw algorithm for automatic map generalization. Spatial simplification can be done inside the database system. The most obvious advantage of this technique is its 'flexibility'. Each part of data may be accessed independently, thereby saving time, as there is no handling of unnecessary data. However, if a polygon has n vertexes, it will be presented as n tuples, which results in high overhead. Thus, although this method is suitable for spatial generalization, it is not good for complex spatial operations because it is difficult to compactly cluster highly fragmented data.

3   Scaleless Data Structures

In multi-resolution systems, a data object should be both fragmented and opaque. The fragmentation lets the system be scalable, which is necessary for progressive query processing, while opacity helps data clustering and reducing data reconstruction time. In order to balance these two extremes, we introduce two scaleless spatial data models in this section.

3.1   Position-Map Scheme

Under this scheme, each object is broken down into a set of levels $O_1, \ldots O_m$. A level contains vertexes having values within a given range, given as $[O_i^{min}, O_i^{max}]$. These ranges are disjoint (i.e., $O_i^{min} > O_{(i-1)}^{max}$ and $O_i^{max} < O_{(i+1)}^{min}$). Levels are stored with a respective delta and objected identifier. Geometry data is stored within an array object data type. As a result reconstructing an object's geometry at a certain resolution level requires that all levels of an equal or lesser delta value be joined. Thus $O_m = O_1 \oplus O_2 \oplus \ldots \oplus O_m$. To ensure correctness of reconstruction, the position of each vertex in O must be recorded explicitly. A simple way to achieve this is by incorporating vertex positions into the geometry array. A position map is used to store information regarding the position of points in a level. Bits corresponding to those stored within the representation are flagged thus requiring a single scan of the position map during reconstruction; the same complexity needed when explicitly storing point positions. For example, for an object of 10 vertexes with its 2th, 5th and 7th in level i, the data in O_i should be (x_2, y_2), (x_5, y_5) and (x_7, y_7), and PM should be the bit sequence '0100101000'.

Because several levels may be needed together to construct an object, levels are clustered together according to their associated `objectID`. Compared with a traditional single-resolution spatial data structure, a small overhead is incurred to split an object over several tuples. This scheme may, however, save data retrieval time for progressive queries, because only the data with high resolutions than that in previous queries need to be fetched. It must be noted that partial replication exists in this scheme. Next, we introduce the BM Scheme.

3.2   Bit-Map Scheme

Points can be encoded using z-values [11], such that the resolution of a point data (represented by a rectangle surrounding the point) can be reduced by truncating the right-most n bits of its z-values, where n is the resolution used for initial simplification [8]. At high resolutions an object will be represented by more z-values than at a low resolution. Hence some of these values will share a common prefix. Put another way, every point at a low resolution will exist in every higher resolution representation. By truncating z-values we can obtain a prefix to reduce their precision. However, under the PM scheme, points from lower resolution levels may be needed to construct a high-resolution object. Therefore some extra information must be stored so that z-valued can be reconstructed to the precision required by the user from a starting prefix (resolution).

An object is reconstructed by first selecting a base resolution from the dataset, at which all points are stored. Points are then truncated to the resolution length of this resolution and stored in a point array with duplicate points (after z-value truncation) removed. We select the resolution at least at which the total number of unique vertices ensures each object in the dataset has at least c points, where $c \geq 1$ (typically $c = 4$); when $c < 1$, the base levels of two objects may be equivalent, hence partially redundant. Second, for each subsequent resolution level a *bitmap* is constructed to map the extension of ancestors into descendants. Because the degree of extension is not known, we must also indicate which ancestors an extension applies to. As illustrated in figure 1, 3 bits are used to denote a single extension: the first two bits for the ordinate, and the third bit to indicate whether the mapping applies to the next ancestor point (0 - continue; 1 - next point). Each is stored with a resolution value. The original object is reconstructed by iteratively extending point levels using a bitmap of the next higher resolution.

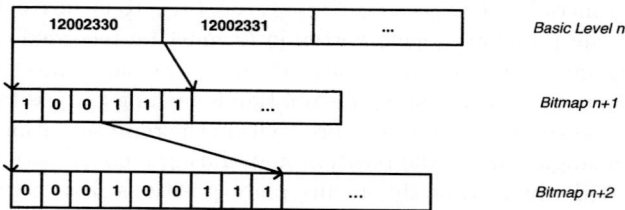

Fig. 1. Base level and bitmaps

In Figure1, for base level n, we have a point '12002330' and it expands to two new points in level $n + 1$ ('120023302' and '120023303'). '120023302' further expands to three new points in level $n + 2$. They are '1200233020', '1200233022' and '1200233023'. Note that if the change in the number of points between resolutions is small, multiple bitmaps can be compressed into a single bitmap. In this case, for k bitmaps compressed each extended point will require $2k + 1$ bits.

3.3   Storage Cost Model

We now provide a basic storage cost model for Bitmap-based scheme. Let n_r denote the number of all points whose resolution is less than or equal to r. Then, a sequence $\{n_0, n_1 \ldots, n_i, \ldots, n_m\}$, ($0 \leq i < m$), where m is the highest resolution level of the dataset, can be used to describe data distribution of the number of points in a different resolution. The resolution of the base level that stores a sequence of truncated z-values of vertices is denoted as i. Each z-value requires a bits of storage, i.e. $a = 2$. Finally the number of points within a certain resolution range $[r, r + 1]$ is denoted as $n_r = n_{r+1} - n_r$. Note n_r is never less than 0. Given this, we may determine that the storage of the traditional single-resolution scheme as:

$$S_o = n_m \times m \times a = (n_i + \sum_{r=i}^{m-1} \Delta n_r) \times m \times a \tag{1}$$

And, the storage of the bitmap method is:

$$S_{bit} = n_i \times i \times a + \sum_{r=i+1}^{m} n_r \times 3 = n_i \times i \times a + n_i \times (m-i) \times 3 + \sum_{r=i}^{m-1} (m-r) \times \Delta n_r \times 3 \tag{2}$$

When substituting S_o into equation 2, we have:

$$S_{bit} = S_o - n_i \times (m-i) \times (a-3) - m \times (a-3) \times \sum_{r=1}^{m-1} \Delta n_r - 3 \times \sum_{r=1}^{m-1} \Delta n_r \times r \tag{3}$$

We can see that storage of Bitmap method is dependant on i, m and n_r. Because m is determined by the precision of data set and n_r is decided by data distribution, the actual storage for S_{bit} is affected by the selection of i. Statistical analyses reveal that the increase of Δn_r satisfies some rules which is small on both ends and large in the middle, forming a bell curve [2], which can be approximated by a Gaussian distribution. Figure2(a) shows the curves of the data distributions for the synthetic and SEQUOIA datasets. We can regard the curve as a probability density function of a variable R, where $\frac{\Delta r \Delta n_r}{N}$ is the probability that R is located in the range $[r \; r + \Delta r]$. Following this, the Gaussian distribution approximation can be given as follows:

$$\Delta n_r = \frac{1}{\sqrt{2\pi}\sigma} e^{\frac{-(r-\mu)^2}{2\sigma^2}}, \; \mu = \frac{\sum_{r=1}^{m-1} \Delta n_r \times r}{\sum_{r=1}^{m-1} \Delta n_r}, \; \sigma = \sqrt{\sum_{r=1}^{m-1} (r-\mu)^2} \tag{4}$$

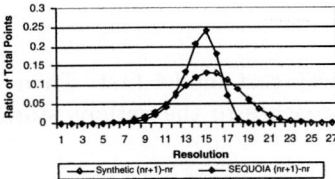

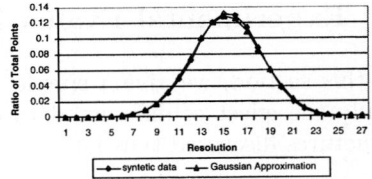

(a) Actual   (b) Gaussian Approximation

Fig. 2. Data Distributions

Thus μ is close to the level value that has the biggest Δn_r and σ is the standard deviation of the Gaussian distribution. Figure2(b) shows that the Gaussian approximation is a very good indication of the actual distribution of the synthetic dataset.

For simplification, we set $a = 3$ in equation 3, then:

$$S_{bit} = S_o - 3 \times \sum_{r=i}^{m-1} \Delta n_r \times r = S_o - 3 \times \sum_{r=i}^{m-1} \frac{1}{\sqrt{2\pi}\sigma} e^{\frac{-(r-\mu)^2}{2\sigma^2}} \times r \tag{5}$$

$$S_o - 3 \times \frac{\sigma}{\sqrt{2\pi}}(e^{\frac{-(m-1-\mu)^2}{2\sigma^2}} - e^{\frac{-(i-\mu)^2}{2\sigma^2}}), \qquad (i \le m - 1 \le \mu)$$

$$= S_o - 3 \times \frac{\sigma}{\sqrt{2\pi}}(2 - e^{\frac{-(m-1-\mu)^2}{2\sigma^2}} - e^{\frac{-(i-\mu)^2}{2\sigma^2}}), (i \le \mu \le m - 1) \qquad (6)$$

$$S_o - 3 \times \frac{\sigma}{\sqrt{2\pi}}(e^{\frac{-(i-\mu)^2}{2\sigma^2}} - e^{\frac{-(m-1-\mu)^2}{2\sigma^2}}) \qquad (\mu \le i \le m - 1)$$

We can see that as i decreases, less storage is needed, which generally will reduce disk I/O costs. Consequently, the time complexity of reconstruction is $O(\sum_{r=i}^{m-1} n_r)$. So a small i incurs a larger cost for object reconstruction. Thus, if i is too small the time used to reconstruct data may negate any performance advantage caused by less storage. To different systems, applications can find a trade-off point for i between this two factors according to above discussion. In our experiment for synthetic data, we chose $i = 10$, $m = 28$. If $a = 2$, the reduction of storage is 24,270,821 bits; if $a = 3$, the reduction is 63,508,107 bits. For the real data, we chose $i = 15$, $m = 22$. If $a = 2$, the reduction is 22,844,603 bits and if $a = 3$, the reduction is 59,481,354 bits.

In order to further improve the storage efficiency, we combine and compress the bit map of high resolution levels with small. Since at the highest several resolution levels, a small increase of the number of points will weaken the benefit of bit map and enhance the cost of reconstruction. An extreme case is that when level k and level $k+1$ contain the same number of points. Here each point in the bit map for level $k+1$ is associated with a flag bit that always set to be 0, which is a waste. If we can combine and compress the bit map of multiple continuous levels, say 3 levels, we can directly derive z-values of points for level $k+3$ from data of level k. Therefore each point in the bit map will contain 6 bits (for 3 levels) plus one flag bit.

4   Experimental Evaluation

In this section, we give a simple progressive query cost model first. Then we study the effect of different data clustering approaches to the various storage structures discussed thus far.

4.1   Progressive Query Cost Model

We defined a progressive query set Q as a series of queries in numerical order as $\langle Q_1, Q_2, \ldots, Q_n \rangle$. For Q_i, we use l_i to represent its resolution level, $S(Q_i)$ to represent the set of Object IDs in its query result, and A_i to represent its query area. Therefore, those parameters satisfy the following rules: $l_1 \le l_2 \le \ldots \le l_n$, $S(Q_1) \supseteq S(Q_2) \supseteq \ldots \supseteq S(Q_n)$, $A_1 \supseteq A_2 \supseteq \ldots \supseteq A_n$. Generally, the time cost $T(Q)$ for certain spatial queries is mainly composed of two parts: Data retrieval $D(Q)$ and Query processing $J(Q)$. The cost for data retrieval is primarily composed of I/O whilst for join processing more time is spent on CPU calculation. Hence, we consider the time needed for progressive query Q as: $T(Q) = T(Q_1) + T(Q_2) + \ldots + T(Q_n)$, where $T(Q_i) = D(Q_i) + J(Q_i)$, $i = 1 \ldots n$. Since queries are normally specified with a certain resolution level l and area A, we may incorporate l, A into $D(Q)$ and $J(Q)$, giving $D(Q, l, A)$ and $J(Q, l, A)$ respectively. Thus, for a multi-representation spatial database, the time cost of a progressive query Q would be:

$$T_M = D(Q_1, l_1, A_1) + J(Q_1, l_1, A_1) + ... + D(Q_n, l_n, A_n) + J(Q_n, l_n, A_n) \quad (7)$$

Here, $l_1 \leq l_2 \leq ... \leq l_n$ and $A_1 \supseteq A_2 \supseteq ... \supseteq A_n$

For a conventional spatial database with only a fixed resolution l_{fix}, in order to get the same query result, l_{fix} should be not less than l_n, and T would be:

$$T_S = D(Q, l_{fix}, A_1) + J(Q, l_{fix}, A_1) \quad (8)$$

Comparing equation 7 and 8, we can see that T_M is obvious less than T_S when $l_n \ll l_{fix}$, which is the case that users stop query according to partial query result. Two factors influence T_M greatly. First is the decrease in A_i. A larger reduction means better performance. In a worst case scenario $A_1 = A_2 = ... = A_n$. Second is the redundancy between the data of different resolution levels. Too much redundancy will result in noticeable overhead in accessing data. That is the reason why multi-representation is not suitable for progressive queries.

4.2   Performance Study

In the following sections we investigate the effect of four different data organization methods: single resolution (SR), multi-representation (MR), Position Map (PM) and Bit Map (BM), on performance over various aspects of the multi-resolution spatial join on a window. Window query is a basic geometric selection because spatial operations frequently involve location as a condition [1]. Spatial join is the most commonly used way to combine information from two or more spatial relations [13]. For MR all objects of a particular representation are clustered together. Objects are ordered according to a z-curve on the centroid of their MBRs. The same organization is used for objects in SR. For PM and BM, spatial data are clustered 1) based on location 2) based on resolution and 3) based on location then resolution (implemented by 3D z-value), termed PM_O, PM_R, and PM_3D; BM_O, BM_R, and BM_3D respectively. The aim of our tests is to determine the most efficient multi-resolution scheme overall for both direct and progressive queries. Performance is measured in both number of I/O accesses and time. While I/O is generally considered a more important factor, if seek time is high then overall response time may be higher than expected. We aim to minimize I/O cost while maintaining a low processing time.

Environment and Data. All tests were conducted in an Oracle database and JDK 1.3 environment on a PIII-800 256Mb system. Point arrays are stored using a VARRAY data structure of NUMBER types. A RAW binary data type is used to store each bitmaps. Disk pages are 4kb. Every point array and bitmap is associated with an object identifier and resolution value, r, in the same tuple. Standard B^+ Trees are placed on these fields. We used the California SEQUOIA polygon dataset [12] for join testing. The data was split into two layers of regular feature polygons, and polygon holes. Very large and very small objects were removed leaving 20,137 objects in layer A, and 5,798 in layer B (holes). In final dataset contains 2,635,065 of the original 3,207,350 data points; i.e. removed objects were treated as outliers. All layers are supplemented with an auxiliary table that stores MBR information for each object. Because the same MBR table is used for different schemes we do not elaborate on performance of the filter step.

Query Process. The join query on a window $w = \{x_{min}, y_{min}, x_{max}, y_{max}\}$ for a resolution r can be informally given as: 'Select objects from layer A and layer B that are inside w and that intersect at resolution r'. Let an object O, stored in p resolution-specific fragments, be the combination of all vertices stored in those fragments, or $O = \{O_1 \oplus ... \oplus O_p\}$ where $1 \leq r \leq p$. In a direct query the candidate set produced by the filter step, S, for some arbitrary window, is a set objects obtained as the joining of their respective r fragments. Every fragment of an object in S is retrieved contiguously before moving on to the next candidate. A progressive query is issued over a series of windows $\hat{W} = \{w_1, ..., w_n\}$ where $w_1 > ... > w_r$, corresponding to a series of result sets $\hat{S} = \{S_1, ..., S_r\}$ where $|S_1| \geq ... \geq |S_r|$. For each set S_k in $\hat{S}$ a single resolution-specific fragment O_k of each object is retrieved. After the first set is processed an arbitrary object is described as $O' = \{O_k\}$ where $k = 1$. Each set is processed iteratively, and new fragments are added to old fragments, until S_r where $O' = \{O_1 \oplus ... \oplus O_r\}$.

For reference the filter step comprises of a hash-join on the MBRs of objects from both layers within w and produces an unordered candidate set. Ordering this set is identical across all tested methods, hence we do not elaborate on it in the test results. We measure retrieval time as the total time required to fetch candidate object geometries, and bitmaps where applicable, from disk and perform object reconstruction.

Test Results. We compare the single-resolution and multi-resolution schemes under direct and progressive queries. In progressive querying an initial 100% window on the data space is issued at resolution 15, which gives data with a precision of 128m. The window is eventually refined to a 0.39% window with data at resolution 22, with 1m precision. The results of the direct query test (figure 3) show comparable performance of all schemes for smaller window sizes. We separate the variant approaches for the PM and BM schemes for easy visualization. The 3D variations, shown to perform best, are then compared with the other schemes. Since data loads are relative small this is expected. The PM_O variation has an overall worst performance because it does not consider resolution whereas both 3D clustering schemes prove more efficient than their lower-dimensional counterparts. PM also exhibits a larger time and I/O requirement than bitmap schemes, which is can be attributed to its higher storage overhead. Expectedly, MR has an ideal time performance, however incurs higher data retrieval than our compressed bitmaps. Note I/O reduction does not correlate directly to retrieval time due to the inherent variation in seek time with different clustering methods. At resolution 15 the multi-resolution schemes require 47% of the original data load. This figure is close to the I/O retrieval in BM_R and BM_3D schemes (45-44% of the original), though is noticeably lower than that seen in the MR (60%) and

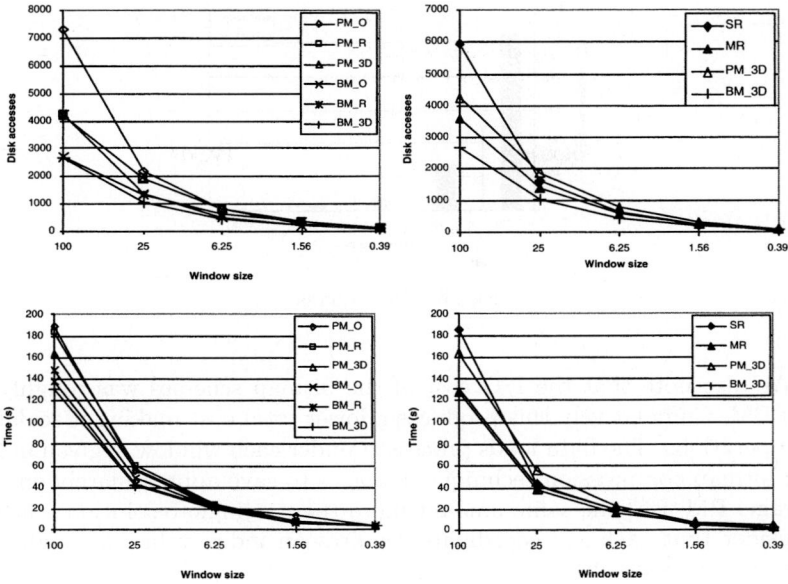

Fig. 3. Direct query test

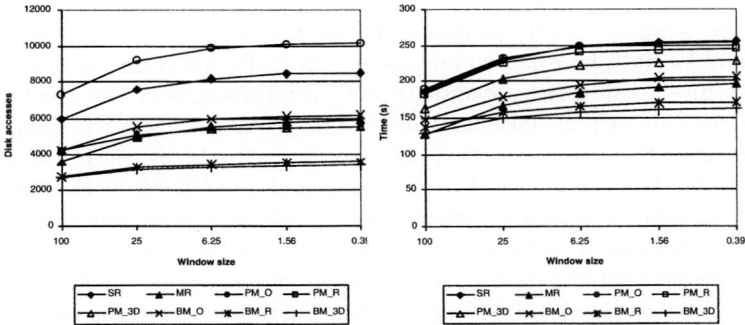

Fig. 4. Progressive query test

all PM variations (minimum 71%). Note the bitmap scheme has an I/O reduction slightly lower than the 47% reduction because compression in the bitmaps reduces overall storage by 30%. To illustrate storage requirements, under SR the data table size is 28.6Mb, 121.6Mb under MR, 38.9Mb in PM, and 19,9Mb in BM.

Figure 4 shows the total performance time and I/O costs for the progressive query calculated at each step of five iterations. The results show a clearer distinction between the single-resolution and multi-resolution methods. The bitmap approach performed comparably with MR while maintaining an overall lower I/O

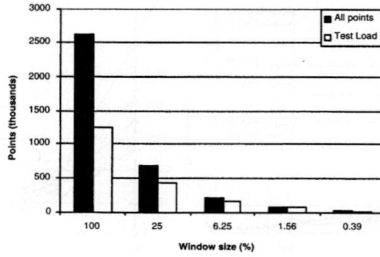

Fig. 5. Data Loads

cost. We also note that the I/O cost of the bitmap schemes was roughly 50% that of PM. Cumulatively this produces a lower total cost making it scalable to larger operations. The data loads processed under each window is given in figure 5. The bitmap compression technique manages to save approximately one third in storage. PM uses the same amount more than a single-resolution scheme. If we consider that each x,y coordinate requires an index value, a roughly 33% increase in size is justified.

5   Conclusions

In order to support direct queries and progressive queries at the same time, data should be both fragmented and opaque in multi-resolution systems. In this paper, we investigated the advantages and disadvantages of the common data structure used in current databases and the fragment data structure used in our previous work. To balance the two extreme data structures, we introduced two scaleless data structures - the Position-Map and the Bit-Map. Between them, the Bit-Map scheme is more efficient because this technique removes the replication existing in most current schemes. To estimate the storage of the Bit-Map scheme, we constructed a mathematical model and analysed how to find the optimal value for base level in Bit-Map scheme. Then, we discuss the cost model for progressive query. The aim of our tests is to determine the most efficient multi-resolution scheme overall for both direct and progressive queries. Our experiment results demonstrated that our approaches have comparable, often better, performance than traditional methods to direct queries, and superior performance under progressive queries.

Acknowledgment

The work reported in this paper has been partially supported by grant DP0345710 from the Australian Research Council.

References

1. A. Aboulnaga and J. R. F. Naughton. Estimation of the cost of spatial selections. In *ICDE*, pages 123–134, 2000.
2. P. J. Diggle. *Statistical analysis of spatial point patterns*. Oxford University Press, 2003.
3. J. Dittrich, B. Seeger, D. S. Taylor, and P. Widmayer. Progressive merge join: A generic and non-blocking sort-based join algorithm. In *VLDB*, pages 299–310, 2002.
4. A. U. Frank and S. Timpf. Multiple representations for cartographical objects in a multi-scale tree - an intelligent graphical zoom. *Computers and Graphics*, 18(6):823–829, 1994.
5. M. Horhammer and M. Freeston. Spatial indexing with a scale dimension. In *SSD*, pages 52–71, 1999.
6. R. Kanth, D. Agrawal, A. E. Abbadi, A. K. Singh, and T. R. Smith. Parallelizing multidimensional index structures. In *IEEE Symposium on Parallel and Distributed Processing*, pages 376–383, 1996.
7. D. Kossmann and F. Ramsak abd S. Rost. Shooting stars in the sky: An online algorithm for skyline queries. In *VLDB*, pages 275–286, 2002.
8. P Prasher. Perfect cartographic generalisation and visualisation. In *VDB*, 2002.
9. S. Prasher, X. Zhou, and M. Kitsuregawa. Dynamic multi-resolution spatial object derivation for mobile and WWW applications. *J. WWW*, 6(3):305–325, 2003.
10. E. Puppo and G. Dettori. Towards a formal model for multiresolution spatial maps. In *SSD*, pages 152–169, 1995.
11. H. Samet. *Applications of Spatial Data Structures*. Addison-Wesley, 1990.
12. M. Stonebraker. An overview of the sequoia 2000 project. *Digital Technical Journal of Digital Equipment Corporation*, 7(3):39–49, 1995.
13. C. Sun, D. Agrawal, and A. E. Abbadi. Selectivity estimation for spatial joins with geometric selections. In *EDBT*, pages 609–626, 2002.
14. R. Vijayshankar. Partial results for online query processing. In *SIGMOD*, pages 275–286, 2002.
15. S. Zhou and C. B. Jones. Design and implementation of multi-scale databases. In *SSTD*, pages 365–384, 2001.
16. S. Zhou and C. B. Jones. A multi-representation spatial data model. In *SSTD*, pages 33–51, 2003.

Querying Heterogeneous Spatial Databases: Combining an Ontology with Similarity Functions

Mariella Gutiérrez[1] and Andrea Rodríguez[2]

[1] School of Engineering, Universidad Católica de la Santísima Concepción
Caupolicán 490, Concepción, Chile
`marielag@ucsc.cl`
[2] Department of Computer Science, Universidad de Concepción
Edmundo Larenas 215, Concepción, Chile
`andrea@udec.cl`

Abstract. This paper uses a knowledge-based approach to querying heterogeneous spatial databases based on an ontology and conceptual and attribute similarities. The ontology, which may be independent of the databases, expands and filters a user query. Then, queries are translated into a formal specification of entity classes, which are compared against definitions in databases. This process is carried out by determining the conceptual similarity between entities in a user ontology and by comparing these entities in the ontology with entities in the conceptual models of databases. In addition, the specification of a query is done not only by identifying entity classes but also by considering constraints based on attribute values. The paper describes the system architecture and presents a case study with data from a forestry information system.

1 Introduction

This paper presents a system architecture for accessing information across heterogeneous spatial databases based on a user ontology and similarity functions. The focus of the paper is at the semantic level, where the ontological definitions of geographic features are independent of their geometric representations.

Studies that use an ontology for data integration require that databases subscribe to a common ontology, which is similar to subscribing to a shared schema at the schematic level. This common ontology is obtained by a single ontology or by the integration of multiple and independent ontologies [2, 17–19, 25, 29]. This work, in contrast, relaxes this strategy of using a common ontology, since it does not force databases either to subscribe to a common ontology or to have a complete semantic description of their information content. The approach of this work is to use semantic similarity measures to associate dynamically entities from different conceptualizations while maintening these conceptualizations independent [12].

This work follows and extends ideas from [12–14] that define similarity functions between ontologies and between ontologies and databases. Unlike these

S. Wang et al. (Eds.): ER Workshops 2004, LNCS 3289, pp. 160–171, 2004.

previous works, in this paper we define a mechanism that retrieves data from heterogeneous databases based on the identification not only of entity classes, but also of instances that are similar to a user request. This work assumes that each database has a conceptual schema. The use of the logical schema was explored in [14], but this approach has strong limitations respect to the description of the information content of a database. Conceptual schemas and the user ontology are expressed in OWL, a standard language for the definition of Ontologies in the Semantic Web [3, 15].

The organization of this paper is as follows. Section 2 reviews related work about querying heterogeneous databases. Section 3 describes the system architecture followed by Section 4 that addresses the description of the user ontology and conceptual schemas of databases. Section 5 adapts similarity functions of previous works [12–14] to evaluate similarity within the user ontology and between the ontology and conceptual schemas. A case study in the area of a forestry information system illustrates the access to databases in Section 6. Conclusions and future work are presented in Section 7.

2   Related Work
on Querying Heterogeneous Data Repositories

Many studies have treated the problem of accessing independent databases as a problem of solving heterogeneities among these databases. Focusing on semantic heterogeneities, studies have proposed the use of ontologies to specify queries and describe the content information of databases [2, 8, 18, 19]. In current ontology-based information systems, semantic matching has meant the agreement on the vocabulary used by different agents. This implies sharing the same conceptualization or agreeing to adopt a common conceptualization, which is usually the intersection of the original conceptualizations [10, 11]. Consequently, the general approach to handling semantic heterogeneity has been to map the local terms in a database onto a shared or common ontology. Most of these approaches use the terms interrelationships to determine semantic similarity between concepts [4–6]. Other approaches are measures based on graph matches and probabilistic measures that predict the probability that an instance of a concept in a differentiated ontology will satisfy a request [30].

In environments with multiple and independent information systems, however, each system may have its own conceptualization and, therefore, its own intended model or ontology. Nonetheless, if existing ontologies are well defined, their integration may reduce the cost of building a global ontology from scratch [2, 16]. Ontology integration is a complex problem, because concepts can overlap or definitions of concepts may be inconsistent across ontologies [27]. Some systematic approaches to handling ontology integration are composition algebras [21], lexical interrelations [2, 18, 19], mappings with mediator agents [22], inheritance from top-level ontologies [8], and semantic correspondence that relies on a common vocabulary for defining concepts across different concepts [25, 29]. All of these approaches are manual or semi-automatic, requiring some input from domain experts.

Applications that use an ontology-based access to information require associations between concepts in an ontology with data stored in information sources. Ontologies may relate to database schemas or single terms. A simple strategy for mapping ontologies onto databases is to translate the database structure into a language in which automatic reasoning is possible [1]. Another approach uses the ontology to further refine terms in the databases or database schema [25]. A structure enrichment combines the translation of data structure with the use of an ontology for enriching the definition of terms [16]. In the World Wide Web domain, the use of metadata adds semantics to an information source or databases. For these metadata, efforts have been made concerning the use of standards for expressing content information of data repositiories, such as the Resource Description Framework (RDF) and the Web Ontology Language (OWL) [28].

3   System Architecture

Main components of the proposed system include a *user ontology, conceptual schemas*, and *similarity functions* (Figure 1). An ontology describes concepts of terms in a user query; conceptual schemas describe the content of databases; and similarity functions compare concepts or descriptions at two different levels: (1) comparing entities within the user ontology for query validation and expansion, and (2) comparing entities in the ontology with entities in the conceptual schemas of databases.

An ontology allows users to express queries in their own terms according to their own conceptualizations without having to know the underlying modeling and representation of data in heterogeneous databases. Concepts used by the user in a query can be then compared in order to search not only for what the user has explicitly requested, but also for semantically similar terms (i.e., query expansion). These concepts are compared at the ontological level where there is a more complete description of the semantics of terms. The user can also select attributes of entities classes to constrain answers.

Since databases may have been designed without assuming the same user ontology, our system compares definitions in the user ontology with the content description of databases. This type of comparison differs from the one within a single ontology, since different levels of explicitness and formalization may affect the way definitions may be compared. Therefore, a second similarity evaluation compares ontological definitions with available components of conceptual schemas of databases. This comparison reduces the search space in each heterogeneous database to the set of entities that are semantically similar to the terms that belong to a user query.

4   Ontology and Conceptual Schemas

This work uses three basic components that define entity classes in an ontology of a spatial domain [12, 13]: (1) a set of synonym words (synset) that denotes an entity class, (2) a set of semantic interrelations among these entity classes, and

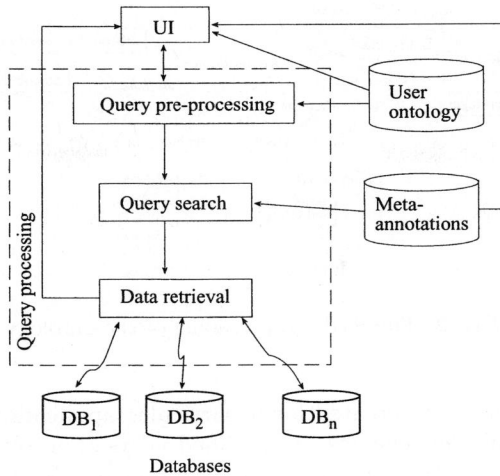

Fig. 1. System architecture

(3) a set of distinguishing features that characterize entity classes. This ontology, similar to a terminological ontology, supports information retrieval rather than query answering, which is typically done with an axiomatized ontology [24]. In this ontology, the use of a set of words to denote entity classes addresses polysemy and synonymy in the process of linking words to meaning. Polysemy occurs when the same word denotes more than one meaning, and synonymy occurs when different words denote the same or very similar entity classes [20]. Synonym sets attempt to capture more semantics than a single word that denotes an entity class.

Two semantic relations play an important role in the specification of ontologies: hyponymy, also called the is-a relation, and meronymy, which is a partial ordering of concept types by the part-whole relation [9, 23]. Properties that distinguish entity classes from the same superclass are called distinguishing features [24]. Usually, attributes describe different types of distinguishing features of a class. They provide the opportunity to capture details about classes, and their values describe the properties of individual objects (i.e., instances of a class). Unlike our previous work [12], this work does not distinguish between types of features, since such a distinction could only work at the ontological level, but not when comparing the ontological description of a user request with the description of entities in a traditional databases.

In order to be able to specify constraints in terms of attribute values, we complement the ontological definitions of entities' attributes by the description of the attributes' values, that is, values' domain and, if necessary, values' units. The definition of the ontology as a RDF graph model is presented in Figure 2, which was then expressed in the OWL language [3, 7, 15].

Conceptual schemas are rich enough to establish attributes and semantic interrelations between entity classes. Unlike the ontological definitions of en-

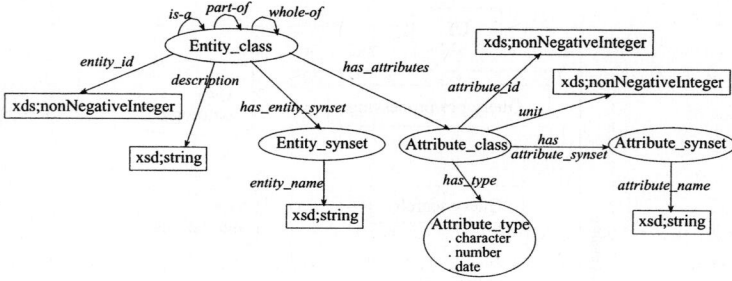

Fig. 2. The RDF graph model of the ontology

tity classes, databases are created for a particular application in a domain. In this context, we consider schemas as simplified views of ontological descriptions, which are expressed in the same way than the user ontology in the RDF graph model.

5 Similarity Functions

5.1 Query Expansion

In this work, the expansion of a query translates the query into a set of entity classes that are semantically related to terms in a user request. This expansion is accomplished by applying matching terms of a query with entity classes in an ontology and, then, by determining similarity between entity classes in the ontology.

Similarity is calculated as a function of common and different features $S_{exp}(a, b)$ (Equation 1) [12], which is based on the ratio model of a feature-matching process [26]. In $S(a, b)$, a and b are two entity classes, A and B correspond to the description sets of a and b (i.e., synonym set of features of entity classes)), and c is the the first class that subsumes a and b by the is-a or whole-of relation. The matching process determines the cardinality ($\|\|$) of the set intersection $(A \bigcap B)$ and the set difference (A/B).

$$S_{exp}(a, b) = \frac{|A \bigcap B|}{|A \bigcap B| + \alpha(a, b)|A/B| + (1 - \alpha(a, b))|B/A|}$$

where

$$\alpha(a, b) = \begin{cases} \frac{distance(a,c)}{distance(a,b)} & \text{if } distance(a, c) \leq distance(b, c) \\ 1 - \frac{distance(a,c)}{distance(a,b)} & \text{if } distance(a, c) > distance(b, c) \end{cases} \quad (1)$$

5.2 Mapping Ontology onto Databases

Mapping ontological definitions onto descriptions of a database can be achieved if the ontology representation and the conceptual schema of the database share some components. A natural way to exploit the full expressiveness of concept representations for a similarity evaluation is to compare each component in those representations. Thus, two different descriptions (i.e., ontologies, conceptual schemas) that have at least one common specification component can still be compared.

In this study we follow some of the results from the work by Rodríguez and Egenhofer [12] where they compare definitions in different ontologies. In our case, we make comparisons between a user ontology and databases' schemas, whereas in their work Rodríguez and Egenhofer compare entity classes across ontologies and use three independent similarity assessments: name-based similarity, neigborhood-based similarity, and attribute-based similarity.

Our work takes two steps to evaluation similarity between a query and stored data. The first step compares entity classes in the ontology with entities in databases based on name and neighborhood similarities. The similarity of names between an entity class in the ontology and an entity in a database's conceptual schema aims at exploiting the general agreement in the use of words and detects equivalent words that likely refer to the same entity class. This evaluation takes the maximum similarity between names, which are composed of one or more words. This similarity is evaluated as a simple matching process $Name(a, b)$ (Equation 2), where a_n is a name in the entity's synonym set in the user ontology and b_n is a name of an entity in a database's schema. Consider that a name may be composed of one or more words.

$$Name(a, b) = \frac{|a_n \bigcap b_n|}{|a_n \bigcap b_n| + |a_n/b_n| + |b_n/a_n|} \quad (2)$$

The similarity of neighborhoods involves semantic relations themselves as the subject of comparison. Since the types of semantic relations are known (e.g., is-a or part-whole relations), the interesting aspect of comparing semantic relations is whether target entities (i.e., entity classes that are the subject of comparison) are related to the same set of entity classes. If so, the entities may be semantically similar. Comparing semantic relations becomes a comparison between the semantic neighborhoods of entities $SN(a, b)$. The semantic neighborhood (N) of an entity a consists of those entities that are at a minimum distance from a, i.e., those entities that have a direct relationship with a (Equation 3).

$$SN(a, b) = \frac{|N(a) \bigcap_n N(b)|}{|N(a) \bigcap_n N(b)| + \delta(N(a), N(b)) + \delta(N(b), N(a))} \quad (3)$$

where

$$\delta(N(a), N(b)) = \begin{cases} |N(a)| - |N(a) \bigcap_n N(b)| & \text{if } |N(a)| > |N(a) \bigcap N(b)| \\ 0 & \text{otherwise} \end{cases}$$

The similarity evaluation of entities in semantic neighborhoods compares all entities in one neighborhood with entities in a second neighborhood in such a way that yields the maximum similarity between neighborhoods. The comparison between entities in semantic neighborhoods is based on name matching (Equation 4). In a extreme case, it is possible that the comparison between neighborhoods gives a larger value than the number of elements in one of the neighborhoods. This can happen when more than one entity in a neighborhood is similar to a single entity in another neighborhood. In such a case, $\delta()$ is considered to be equal to zero.

$$|N(a) \bigcap_n N(b)| = \left[\sum_{c \in N(a)} max_{d \in N(b)} Name(c,d) \right] \tag{4}$$

Combining name and neighborhood similarities, a global entity similarity is given by Equation 5, where ω_w and ω_n are the relative importance of name similarity and neighborhood similarity, respectively.

$$S_{entity}(a,b) = \omega_w \cdot Name(a,b) + \omega_n SN(a,b) \tag{5}$$

Unlike [12], this work uses the similarity evaluation between attributes as a subsequent evaluation that is only applied when entities are similar. In this sense, attribute similarity can be seen as a second filter and final discriminator between entities in the database, a discriminator that is only useful when the system has detected that the entities in the databases are similar to the entities in the user ontology. We do so, because feature similarity was found to be more useful for comparing definitions within a single ontology or for comparing semantically similar entities [12].

In our system two query cases are when the user has not specified attribute values as constraints or when the user has filtered the entities to be retrieved by attribute values. In the first case, comparing attributes at a semantic level could discriminate between similar entities by considering an attribute matching at the conceptual level, that is, matching of the class of attribute rather than between attribute values (Equation 6). In Equation 6, a_t is the set of synonym sets that refer to attributes in the entity class a of the user ontology and b_t is the set of terms that refers to attributes in the entity b of a database, respectively. Attribute correspondence is determined by considering a strict matching between terms, that is, a common attribute means that the term that refers to this attribute in the database was found as one of the terms in a synonym set of the attributes in the ontology.

$$S_{query}(a,b) = \begin{cases} \frac{|a_t \cap b_t|}{|a_t \cap b_t| + |b_t/a_t|} & \text{if } S_{entity}(a,b) \geq \tau \\ 0 & \text{otherwise} \end{cases} \tag{6}$$

We assume that the ontology is semantically richer than the conceptual schemas of the databases, so that entities in the ontology may have a more complete set of attributes. In this case, the set of attributes in the database is likely a subset of the set of attributes of the ontology, and attributes that are present in the database and not in the ontology reflect potential differences in the semantics of entities. Therefore, only attributes that are included in the databases' schemas and not in the ontology will affect the similarity assessment.

In the second query case, in order to be able to answer the query, attributes in the query specification should be part of the entities in databases. So, these attributes in the query specification q_t are necessary conditions of entities (Equation 7).

$$S_{query}(a, b) = \begin{cases} \frac{|a_t \cap b_t|}{|a_t \cap b_t| + |b_t/a_t|} & \text{if } (S_{entity}(a, b) \geq \tau) \wedge (q_t \subseteq b_t) \\ 0 & \text{otherwise} \end{cases} \quad (7)$$

6   Case Study: Forestry Management System

As a case study, we consider a forestry management domain. For such domain, an ad-hoc ontology in Spanish was created with definitions derived from available dictionary, thesaurus and glossaries used in agencies of resource management in Chile. A portion of the ontology translated to English is shown in Figure 3.

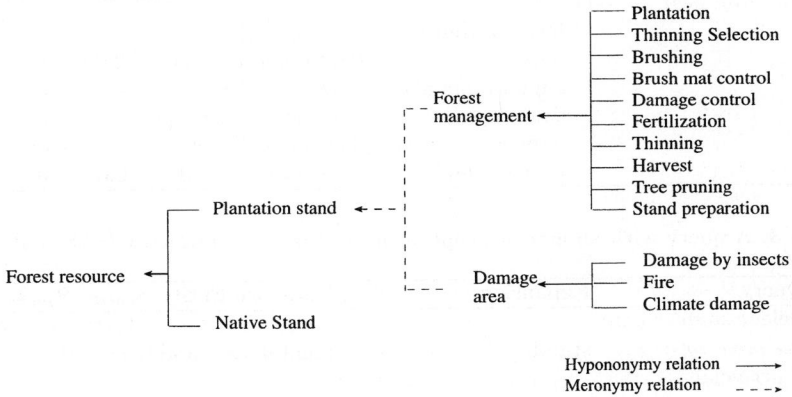

Fig. 3. A portion of an ontology for forest management

The case study uses a real forestry database. This database contains 100 entities, 27 of them related to forest management. As examples of how the system works, consider the following queries. The first case is a query without constraints based on attribute values (Table 1).

The original query is expanded to include seven different entity classes based on a threshold of $S_{exp} \geq 0.5$. For each of these entity classes, the system finds the most similar entity in the database based on the similarity S_{entity}. Only entities in the database with a similarity $S_{entity} \geq 0.5$ are considered in the final evaluation of query similarity S_{query}. Since this query does not consider attribute values as search criteria, S_{query} between an entity class in the ontology and an entity in the database is given by their common attributes (Equation 6).

A second case is a query with the specification of an attribute value that exists in a database (Table 2).

Unlike the first query, the second query uses a search criterion that is defined by an attribute value (i.e., *plantation year* > 2000). In this case, the system will

Table 1. A query without an attribute value specification

Query	Expansion	S_{exp}	Database Entity	S_{entity}	S_{query}
Select *stand*	stand	1	plantation stand	0.58	0.63
	forest resource	1	nonexistent	-	-
	damage	0.67	stand damage	0.62	0.28
	climate damage	0.62	nonexistent	-	-
	fire	0.62	nonexistent	-	-
	damage by insects	0.59	nonexistent	-	-
	native stand	0.56	native stand	0.5	0.5

Table 2. A query with an attribute specification that exists in the database

Query	Expansion	S_{exp}	Database Entity	S_{entity}	S_{query}
Select *stand* where *plantation year* > 2000	stand	1	plantation stand	0.58	0.63
	forest resource	1	nonexistent	-	-
	damage	0.67	stand damage	0.62	0
	climate damage	0.62	nonexistent	-	-
	fire	0.62	nonexistent	-	-
	damage by insects	0.59	nonexistent	-	-
	native stand	0.56	native stand	0.5	0

Table 3. A query with an attribute specification that does not exist in the database

Query	Expansion	S_{exp}	Database Entity	S_{entity}	S_{query}
Select *stand* where *resource type* = "artificial"	stand	1	plantation stand	0.58	0
	forest resource	1	nonexistent	-	-
	damage	0.67	stand damage	0.62	0
	climate damage	0.62	nonexistent	-	-
	fire	0.62	nonexistent	-	-
	damage by insects	0.59	nonexistent	-	-
	native Stand	0.56	native stand	0.5	0

check if this attribute exists in each of the entities that were found to be most similar to the entity classes in the expanded query. Only in the case that this attribute exists in the entity of the database, S_{query} is calculated by common attributes between the entity class in the ontology and the entity in the database; otherwise, S_{query} is equal to zero (Equation 7).

The last case is a query with the specification of an attribute value that do not exist in the database (Table 3).

The last query is an example where, although there are entities in the database that are similar to entity classes in the expanded query, none of these entities includes the attribute that is used as the search criterion. Consequently, S_{query} is always equal to zero.

7   Conclusions and Future Work

This paper describes a systems that queries heterogeneous spatial databases by using a user ontology and similarity functions that compare entities and instances. The advantages of this system are that databases can be independent of user ontologies and updates in both the user ontology and the databases will not affect the system. By using similarity functions, the system can dynamically associate user requests with entities stored in databases. Requirements of the system are a user ontology and conceptual schemas of databases described in OWL with two basic components: semantic relations (i.e., generalization and aggregation) and distinguishing features or attributes.

As future work, we plan to incorporate constraints or query conditions that combine types of entity classes (e.g., joins). Such types of queries impact the way the final similarity (S_{query}) between a query and a stored data is determined. We also expect to be able to express queries that use spatial criteria such as geographic windows and spatial relations. From the implementation point of view, we expect to have a fully running system where ontologies and conceptual schemas can be modified, and with a user friendly visualization of results.

Acknowledgment

This work has been partially funded by CONICYT, Chile, under grant Fondecyt 1030301. Mariella Gutierrez' research is also funded by Universidad Católica de la Santísima Concepción under grant DIN 03/2004.

We would like to thank FORESTAL BIOBIO S.A. for providing the real database for our case study. We also thank Dr. Max Egenhofer for his contribution to previous works published in [12, 13].

References

1. Y. Arens, C.-N. Hsu, and C. Knoblock. *Readings in Agents*, chapter Query Preprocessing in the SIMS Information Mediator, pages 82–90. Morgan Kaufmann, San Francisco, CA, 1997.
2. B. Bergamaschi, S. Castano, S. De Capitani di Vermercati, S. Montanari, and M. Vicini. An intelligent approach to information integration. In N. Guarino, editor, *First International Conference on Formal Ontology in Information Systems*, pages 253–268, Pisa, Italy, 1998. IOS Press.
3. J. Berner-Lee, J. Handler, and O. Lassila. The semantic web. *Scientific American*, 184(5):34–43, 2001.
4. Y. Bishr. Overcoming the semantic and other barriers to gis interoperability. *Int. J. Geographical Information Science*, 12(4):299–314, 1998.
5. M. Bright, A. Hurson, and S. Pakzad. Automated resolution of semantic heterogeneity in multidatabases. *ACM Transactions on Database Systems*, 19(2):212–253, 1994.

6. P. Fankhauser and E. Neuhold. Knowledge based integration of heterogeneous databases. In H. Hsiao, E. Neuhold, and R. Sacks-Davis, editors, *Database Semantics Conference on Interoperable Database Systems*, pages 155–175, Victoria, Australia, 1992. Elsevier Science Publishers.
7. D. Fensel and M. Musen. The semantic web: A new brain for humanity. *IEEE Intelligent Systems*, 16(1):24–25, 2001.
8. F. Fonseca, M. Egenhofer, P. Agouris, and C. Camara. Using ontologies for integrated information systems. *Transactions in GIS*, 6(3):231–257, 2002.
9. N. Guarino. Fornal ontology, conceptual analysis, and knowledge representation. *Int. Jounal on Human Computers Studies*, 43:625–640, 1995.
10. N. Guarino. *Information Extraction: A Multidisciplinary Approach to an Engineering Information Technology*, chapter Semantic Matching: Formal Ontological Distinctions for Information Organization, Extractionm and Integration, pages 139–170. Springer-Verlag, Francasi, Italy, 1997.
11. N. Guarino. *Formal Ontology in Information Systems*, chapter Formal Ontology in Information Systems, pages 3–15. IOS Press, Trento, Italy, 1998.
12. A. Rodríguez and M. Egenhofer. Determining semantic similarity among entity classes from different ontologies. *IEEE Transactions on Knowledge and Data Engineering*, 15(2):442–456, 2003.
13. A. Rodríguez and M. Egenhofer. Comparing geospatial entity classes: An asymmetric and context dependent similarity measure. *Int. Journal of Geographical Information Science*, 18(3):229–256, 2004.
14. A. Rodríguez and M. Varas. A knowledge-based approach to querying heterogeneous databases. In Z. Rás M.-S. Hacid, D. Zighed, and Y. Kodratoff, editors, *Foundations of Intelligent Systems. LNAI 2366*, pages 213–222. Springer-Verlag, 2002.
15. I. Horrocks and P. Patel-Schneider. Three thesis of representation in the semantic web. In *Proceeding of the 12th International Conference on WWW*, pages 39–47, 2003.
16. V. Kashyap and A. Sheth. *Cooperative Information Systems: Trends and Directions*, chapter Semantic Heterogeneity in Global Information Systems: The Role of Metadata, Context, and Ontologies, pages 139–178. Academic Press, London, UK, 1998.
17. M. Kavouras and M. Kokla. A method for formalization and integration of geographic categorizations. *Internationa Journal of Geographical Information Science*, 16(5):439–453, 2002.
18. E. Mena and A. Illarramendi. *Ontology-Based Query Processing for Global Information Systems*. Kluwer Academic Publishers, Norwell, MA, 2001.
19. E. Mena, A. Illarramendi, V. Kashyap, and A. Sheth. Observer: An approach for query processing in global information systems based on interoperation across pre-existing ontologies. *Distributed and Parallel Databases*, 8(2):223–271, 2000.
20. G. Miller. *WordNet: An Electrical Lexical Database*, chapter Nouns in WordNet, pages 23–46. The MIT Press, 1998.
21. P. Mitra and G. Wiederhold. *Handbook on Ontologies in Information Systems*, chapter An Ontology-Composition Algebra, pages 97–119. Springer, Berlin, 2003.
22. A. Preece, K.-J. hui, W. Gray, P. Marti, T. Bench-Capon, D. Jones, and Z. Cui. The kraft architecture for knowledge fusion and transformation. *Knowledge Based Systems*, 13(2-3):113–120, 2000.
23. J. Smith and D. Smith. Database abstractions: Aggregations and generalizations. *ACM Transactions on Database Systems*, 2:105–133, 1977.

24. J. Sowa. *Knowledge Representation: Logical, Philosophical and Computational Foundations*. Brook/Cole, Pacific Grove, CA, 2000.
25. H. Stuckenschmidt and H. Wache. Context modelling and transformations for semantic translation. In M. Bouzeghoub, M. Klusch, and U. Sattler, editors, *Knowledge Representation Meets Databases*, pages 115–126, Berlin, Germany, 2000.
26. A. Tversky. Features of similarity. *Psychological Review*, 84(4):327–352, 1977.
27. Visser, D. Jones, T. Bench-Capon, and M. Shave. *Formal Ontology in Information Systems*, chapter Assessing Heterogeneity by Classifying Onotlogy Mitmaches, pages 148–162. IOS Press, Trento, Pisa, 1998.
28. W3C. Semantic web, 2001.
29. H. Wache. Towards rule-based context transformation in mediators. In S. Conrad, H. Hasselbring, and G. Saake, editors, *International Workshop on Engineering Federated International Systems*, pages 107–122, Khlungsborn, Germany, 1999. Infix-Verlag.
30. P. Weinstein and P. Birmingham. Comparing concepts in differentiated ontologies. In *12th Workshop on Knowledge Adquisition, Modeling, and Management*, Banff, Canada, 1999.

Retrieval of Heterogeneous Geographical Information Using Concept Brokering and Upper Level Ontology

Shanzhen Yi[1], Bo Huang[2], and Weng Tat Chan[1]

[1] Department of Civil Engineering, National University of Singapore
10 kent Ridge Cresent, Singapore 119260
{cveys,cvecwt}@nus.edu.sg
[2] Department of Geomatics Engineering
University of Calgary, Calgary, AB, T2N 1N4, Canada
huang@geomatics.ucalgary.ca

Abstract. Different concept and attribute definitions for geographical features underlying information systems make it difficult to share geographical information. This paper provides concept brokering methods based on upper level ontology for retrieval of heterogeneous geographical information. The methods include design of upper level ontology using Concept Lattices. The user ontology is associated with domain ontology through upper level ontology by concept brokering rules. An example for the upper level ontology of transportation is given to illustrate the concept brokering methods.

1 Introduction

Geographical related information are produced and distributed in heterogeneous data sources. Integration and retrieval of geographical information from those heterogeneous depositories are still a challenge. The systems for the problem include Spatial Data Infrastructure, Digital Library [1] and Spatial Internet Marketplaces [2]. A core challenge is achieving interoperability in those large-scale distributed systems with a high degree of heterogeneity [2]. In order to implement interoperability at different levels, several standards have been developed, such as the Content Standard for Digital Geospatial Metadata [3], the International Standards Organization TC211[4], and the OpenGIS Consortium[5].

The new visions of web technologies, such as XML, semantic web [6] [7] and web services [8], provide the potential for the interoperability of geographical information over the web. The semantic web with the characteristics of ontology and agents, will promotes web-based GIS with new functions, such as information brokering, search agents, and semantic interoperability. Ontologies have shown their potential in global information systems, intelligent information integration and information brokering [9]. Therefore, their use is of high interest to geographic information interoperability.

The method for geographic information retrieval in this paper is based on upper level ontology and concept brokering. A client query is modeled as user ontology which is represented by concepts, attributes and relations. The user ontology is asso-

S. Wang et al. (Eds.): ER Workshops 2004, LNCS 3289, pp. 172–183, 2004.
© Springer-Verlag Berlin Heidelberg 2004

ciated with domain ontology through the concept brokering of upper level ontology. Predicates are used to represent the concepts, attributes and relations in upper level ontology, and the concept deriving method and concept brokering rules are developed. The results of client retrieval are the instances of concepts from domain ontologies with restricted concept relations and attributes. Finally we give an example to illustrate the building of upper level ontology and the XML query and inference for concept brokering.

In the remainder of the paper, section 2 discusses upper level ontology, user ontology, and domain ontology. Section 3 addresses our method of ontology design, concept brokering rules. Section 4 an example for query and inference is given. Section 5 concludes the paper.

2   User Ontology, Upper Level Ontology and Domain Ontology

Two systems can communicate only if they have conceptualizations overlap [10]. That indicates they should have common concepts to share. Ontology, defined as an explicit specification of conceptualization [11], is considered as an important step towards information exchange. From the view of formal ontology and information system, Guarino[10] uses the top level ontology to describe the general concepts like space, time, matter, object, event, and action, etc., which are independent of a particular problem or domain. An upper level ontology is limited to concepts that are meta, generic, abstract and philosophical, and therefore are general enough to address (at a high level) a broad range of domain areas [12].

2.1   Shared Upper Level Ontology

There are different forms and applications of upper level ontology, from lexical system to knowledge bases. WordNet [13] is a public domain on-line lexical reference system, which organizes nouns, verbs, adjectives and adverbs into hierarchical sets. Upper Cyc Ontology [14] is a large knowledge base and has an efficient inference engine, which processes users query with commonsense knowledge; IEEE Standard Upper Ontology [12] in computer science will enable computers to utilize ontology for applications such as information search and retrieval, automated inference, and natural language processing; SDTS Part 2 [15] provides general spatial entity types and their attributes for geographic data transformation. Upper level ontology has the following features: (1) It describes general-purpose and standard ontology of common concepts. (2) It is independent of particular problem or domain, provides a solid framework for more specialized application. (3) It is agreed upon and shared by all the application domain participated in. (4) It supports knowledge-based reasoning applications.

2.2   User Ontology and Domain Ontology

Support and use of multiple independently-developed ontologies are important for developing scalable information systems with multiple information producers and consumers. One challenging issue in supporting semantic interoperability is how to allow both users and providers to associate their ontologies with upper level ontology.

User ontology represents the view of a user, which describes the single user personalized requirement. The requirement may relate to multiple information sources. By contrast, domain ontology represents the conceptualization of specific information resources produced by provider. When a user retrievals information with his/her own view, the upper level ontology serves as the middle level for concept brokering which associates user queries with domain ontologies of related information resources. Domain ontology describes special conceptualization of the world for the information provider's domain.

3   Formalization of Upper Level Ontology and Concept Brokering

There are many different ways in which an ontology may explain a conceptualization and the corresponding context knowledge. The possibilities mentioned above range from a purely informal natural language up to strictly formal approaches. Different application scopes and motives have different methods. For upper level ontology of geographic information, one of method based on Concept Lattices and Formal Concept Analysis (FCA) has been explored in [16] [17].

In this section, we define the concept and attribute predicates to model upper level ontology using Concept Lattices, and also define concept brokering rules to associate user ontology with domain ontology through upper level ontology.

3.1   Concept Lattices

The Concept Lattices [18], as a mathematical theory, is the basic notions of FCA, which deals with concept structures and relationship between concepts. The following is the basic content of the Concept Lattices and FCA which we will employ for ontology design and concept brokering.

Definition 1. A formal context $K= (G, M, I)$ consists of two sets G and M and a relation I between G and M. The elements of G are called the objects and the element of M are called the attributes of the context. If the object g has the attribute m we write $gIm \in I$.

Definition 2. For a set $A \subseteq G$ of objects and a set $B \subseteq M$ of attributes, we define: $A' = \{m \in M \mid gIm$ for all $g \in A\}$(the set of attributes common to the concept objects in A).

$B' = \{g \in G \mid gIm$ for all $m \in B\}$ (the set of objects which have all attributes in B).

Proposition 1. If (G,M,I) is a context, A_1, A_2, $A_3 \subseteq G$ are sets of objects and B_1, B_2, $B_3 \subseteq M$ are sets of attributes, then

$$A_1 \subseteq A_2 => A_2' \subseteq A_1' \tag{1}$$

$$B_1 \subseteq B_2 => B_2' \subseteq B_1' \tag{2}$$

$$A \subseteq A \text{ '' and } A' = A''' \tag{3}$$

$$B \subseteq B \text{ '' and } B' = B''' \tag{4}$$

Definition 3. A formal concept of the context (G, M, I) is a pair (A, B) with $A \subseteq G$, $B \subseteq M$, $A' = B$ and $B' = A$. We call A the extent and B the intent of the concept (A, B). $\beta(G, M, I)$ denotes the set of all concepts of the context (G, M, I).

Definition 4. If (A_1, B_1) and (A_2, B_2) are concepts of a context, (A_1, B_1) is called a subconcept of (A_2, B_2), provided that $A_1 \subseteq A_2$, (which is equivalent to $B_2 \subseteq B_1$). In this case, (A_2, B_2) is a superconcept of (A_1, B_1), and we write $(A_1, B_1) \leq (A_2, B_2)$. The relation $\leq$ is called the hierarchical order of the concepts. The set of all concepts of (G, M, I) ordered in this way is denoted by $\beta(G, M, I)$, and is called the concept lattice of the context (G, M, I), for which infimum and supremum are described in [18].

Any concept has a unique name. The name is given by the label of the concept. In our method, for the concept (A, B), we use A as the label of the concept, and indicate concept class A has an attribute set B. If concept (A_1, B_1) is a sub-concept of (A_2, B_2), we say concept (A_2, B_2) has sub-concept (A_1, B_1), and represent as predicate $SubConcep$ (A_2, A_1). According to Proposition 1, if concept A_1 is sub-concept of A_2, then the attribute set A_1' is the super-set of A_2', $A_1' \supseteq A_2'$, represented as another predicate $SuperProperty(A_1', A_2')$.

3.2 Concept and Relationship

FCA provides the basic methods to analyze concept classes, attribute set and relationships between the concept and the attribute set. With concept classes and their hierarchy relations we can determine their attribute set and attribute relations, and vice versa. In order to represent them formally, we give the predicates to represent concept classes, hierarchy relations of concept classes, attribute set and hierarchy relations of attribute set in Table 1.

As Concept Lattices and FCA have defined the terms of *subconcept* and *supercIncept* for the relation between concepts, it is necessary to define predicates *SubProperty* and *SuperProperty* to represent the relation between attributes. Concept A has sub-concept B (which is equivalent to that A is super-concept of B) indicates attribute set of super-concept A is a subset of attribute set of sub-concept B. Therefore, super-concept A has a wider extension and a narrower intension than sub-concept B. We use P_1 to represent extension predicates, which includes *SameConcept*, *SuperCencept*,

Table 1. Predicates of concepts and attributes

Predicates	Description	Interpretation
Concept (A,M)	concept A has attribute set M	$A'=M$
SubConcept (A, B)	concept A has sub-concept B	$A \supseteq B$
SuperConcept (A, B)	concept A has super-concept B	$A \subseteq B$
SameConcept(A, B)	concept A is same with concept B	$A=B$
Attribute (A, m)	concept A has attribute element m, represented as A.m	$m \in A'$
SubProperty(M, N)	attribute set M has sub-property of attribute set N	$M \supseteq N$
SuperProperty(M, N)	attribute set M has super-property of attribute set N	$M \subseteq N$
SameProperty(M,N)	attribute set M is same with attribute set N	$M=N$

SubConcept, and P_2 to represent intension predicates which includes *SameProperty*, *SuperProperty*, *SubProperty*.

3.3 Deriving Rule

The motive of concept deriving rule is to provide the connection and brokering relationship between user ontology, upper level ontology and domain ontology. In Concept Lattices, the attribute set of a super-concept is the common attributes of all its sub-concepts (described in Definition 2). Super-concept can be built by including sub-concepts through excluding the private attributes in the sub-concepts. When we build a concept system, we can generalize the sub-concepts by removing their private attributes and retaining their common attributes to form a super-concept. Therefore, the super-concept can be derived from its sub-concepts by excluding private attributes. For example, the concept *avenue* is described as a broad ground roadway in urban area with passage for vehicle, the concept *road* is described as a ground roadway. *Avenue* is a kind of *road*. When we query a kind of road with specified attributes which match the attributes of *avenue*, the query can be resulted from *avenue*. In Concept Lattices, the instance of a sub-concept is a kind of the instance of its super-concept, and a super-concept is constructed by the common attributes of its sub-concepts. We call this bottom-up method of conceptualization as generalization. The generalized super-concept can be derived from the special sub-concepts. This case is the inverse of inheritance in object oriented design, which has top-down method for sub-classes to inherit from super-classes. If the attribute set of concept class A includes the attribute set of concept class B (which is equivalent to *SubConcept (B, A)*), we say concept class B can be derived from A, i.e. $A' \supseteq B' \Rightarrow B \supseteq A \Rightarrow B := A$. The concept deriving rule (represented by "$:=$") has following characteristics:

Reflexive: $A:=A$ $(A:=B$ and $B:=A \Rightarrow A=B)$, A can be derived from itself.

Anti-symmetric: $A:=B$ and $A \neq B \Rightarrow not\ B:=A$, A is derived from B, and A is not equal B, than we can not hold $B:=A$.

Transitive: $A:=B$ and $B:=C \Rightarrow A:=C$, A is derived from B, and B is derived from C, than we can hold $A:=C$.

If super-concept B has sub-concept set $\{A_1, A_2, \dots A_n\}$:

$$B \quad \overset{n}{\underset{t=1}{\bigcup}} A_t \tag{5}$$

B can be derived from any sub-concept: $B := A_1 \vee A_2 \vee \ldots \vee A_n$. The attributes of B is common to all of its sub-concepts. The attributes of B can be derived from same-named attributes of its sub-concepts. $B.m := A_1.m \vee A_2.m \vee \ldots \vee A_n.m$. $(B.m \in \{m \mid m \in B'$ and $m \in A_i' \}$; $A_i.m \in \{ m \mid m \in A_i'$ and $m \in B' \}$, $i = 1 \ldots n$).

A concept of user ontology is associated with a concept in the upper level ontology by restricting concept relation and attributes. If the associated concept in upper level ontology has the sub-concepts, we indicate the concept in the user ontology can be derived from the sub-concepts in upper level ontology.

3.4 The Rules of Concept Brokering

User ontology represents user's query requirement, which is specified by restricting extension of concepts and the intension of attributes of upper level ontology. The inquiry is derived from the domain ontology through the concept of upper level ontology. The procedure has three aspects. First, user's query is built by binding variables of concepts and attributes to upper level ontology with restrictions. Second, the bound concepts in upper level ontology are derived from other sub-concepts in upper level ontology. Third, the concepts and attributes in the upper level ontology are associated with domain ontology of information resources by using deriving rule.

Given user's ontology *Concept (x, y)*, variable x is bound with concept in upper level ontology through extension predicates P_1, and variable y is bound with attributes in upper level ontology through intension predicates P_2, represented as: $P_1(x, g_1$) or $P_2 (y, g_1')$, where g_1 and g_1' are concept and attribute set in upper level ontology. Upper level ontology *Concept (g_1, g_1')* or its sub-concepts, for example the *Concept (g_2, g_2')*, are associated with domain ontology *Concept (v, w)*, represented as $P_1 (g_1, v)$ and $P_2 (g_1, w)$, or $P_1 (g_2, v)$ and $P_2 (g_2, w)$. Through the concept brokering of upper level ontology, the user's ontology *Concept (x, y)* can be derived from domain ontology *Concept (v, w)*. User has no context knowledge about domain ontology and the query result can be derived from multiple information resources.

In Concept Lattices, the implication between super-concepts and corresponding sub-concepts are dual direction. In the definition of upper level ontology, we adopted the dual direction. That implies if concept a has super-concept b, than concept b has sub-concept a.

However, from user ontology to shared ontology and from shared ontology to domain ontology, we defined single direction derived rule. That indicates if concept a in user ontology has super-concept b in upper level ontology, it does not hold that concept b has sub-concept a. The implication relations are represented in Fig. 1 by arrow and diamond respectively.

3.4.1 The Rules from User Ontology to Upper Level Ontology

The user ontology is represented as a query for the new concept which satisfies restricted conditions for concepts and attributes in upper level ontology. Given user ontology *Concept (X, X')* and the restricted concept condition *SuperConcept (X, G)* (G is concept in upper level ontology), we can get the restricted attribute condition *SubProperty (X', G')*, i.e. $X' \supseteq G'$. It indicates the set of attributes X' includes G'. For any concept g in upper level ontology, if it satisfies the two conditions: (1) g is subconcept of G (i.e. *SuperConcept (g, G)* or Subproperty (g',G'), i.e. $g' \supseteq G'$), and (2) the set of attributes g' includes the set of attributs X' (i.e. $X' \subseteq g'$)(X' is specified by user), according to deriving rule, concept X can be derived from concept g. The concept brokering rules mapping from user ontology to upper level ontology are shown in the left of Fig. 1. The example of user ontology and upper level ontology is shown in the right of Fig. 1.

Rules from user ontology to upper level ontology	Example	
SuperConcept (X, G) =>SubProperty (X' , G') => $X' \supseteq G'$	User ontology	Upper level ontology
$\forall g$ (SuperConcept (g, G) $\wedge X' \subseteq g'$) => $g' \supseteq G' \wedge X' \subseteq g'$ => $g' \supseteq X' \supseteq G'$ => $g \subseteq X \subseteq G$ =>X: =g => X.m:=g.m (X.m $\in$ {m| m $\in$ X' and m $\in$ g' }) =>Attribute(X,m):=Attribute(g,m)	 X, X' Concept(X, X'), SuperConcept(X, G) $X' \subseteq g'$	G$_2$, G$_2'$ G$_1$, G$_1'$ G, G' Concept(G, G') Concept(G$_1$, G$_1'$) Concept(G$_2$, G$_2'$) SuperConcept (G$_1$, G), SuperConcept (G$_2$, G).

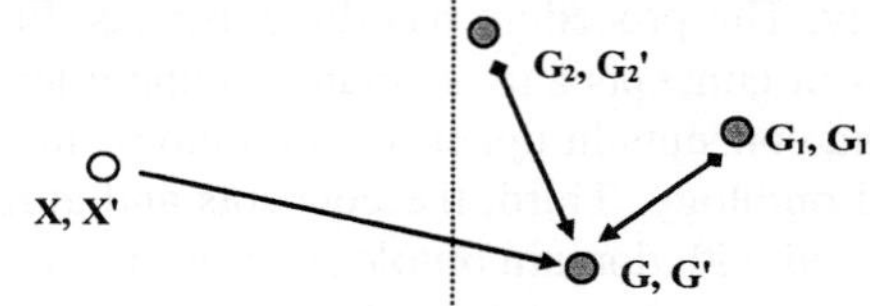

Fig. 1. The brokering rules from user ontology to upper level ontology

The user ontology *Concept(X, X')* is represented with empty circle, and has the single direction relation with the *Concept (G, G')* in the upper level ontology. The facts in the example of the upper level ontology include *Concept (G, G')*, *Concept (G$_1$, G$_1'$)*, *Concept (G$_2$, G$_2'$)*, *SuperConcept(G$_1$,G)* and *SuperConcept(G$_2$, G)*. The brokering rules are represented as for any sub-concept g of G (represented as $\forall g$ *(SuperConcept(g, G))* in the left of Fig. 1), for example, G_1 or G_2, if the attribute set X' is sub-property of g' (represented as $X' \subseteq g'$), according to the definition in deriving rule, X can be derived from g. And the attributes in X' are derived from the same-named attributes in g' (represented as *X.m $\in$ {m\| m $\in$ X' and m $\in$ g' }; g.m $\in$ {m \| m $\in$ g' and m $\in$ X' }*).

The above query is translated to determine which concepts from the upper level ontology can be derived for X and which attributes can be derived for attributes X'. The conditions are concept relation *SuperConcept(X, G)* and the restricted attribute set X'. G_1 and G_2 are the sub-concepts of G, if attribute set G_1' is the super-property of X', then X can be derived from G_1.

If user application ontology has the restricted concept condition *SameConcept (X, G_1)*, the derivation of X is simply represented as $X:=G_1$, and $X.m:=G_1.m$, where $X.m \in \{ m \mid m \in X'$ and $m \in G_1' \}$ and $G_1.m \in \{ m \mid m \in G_1'$ and $m \in X' \}$.

In the upper level ontology, the implication between *SuperConcept()* and *SubConcept()* is dual direction (shown in Fig. 2). If concept G_1 has super-concept G, represented as *SuperConcept(G_1,G)*, then G has sub-concept G_1, represented as *SubConcept (G,G_1)*.

Rules in upper level ontology	Example
$\forall g$ (SuperConcept(g,G)<==> SubConcept (G, g)) =>SubProperty (g', G')<==>SuperProperty (G', g') =>G:=g =>G.m:=g.m (G.m$\in$ {m| m $\in$ G' and m$\in$g'}; g.m$\in$ {m | m$\in$g' and m$\in$ G' })	Upper level ontology 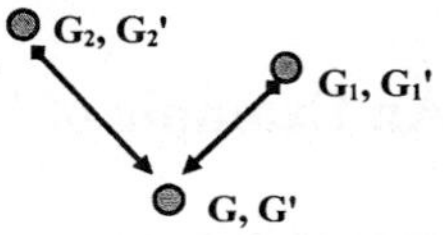 G_2, G_2' G_1, G_1' G, G' Concept(G, G') Concept(G_1, G_1'), Concept(G_2,G_2') SuperConcept((G_1,G) SuperConcept((G_2,G)

Fig. 2. The rules in upper level ontology

3.4.2   The Rules from Upper Level Ontology to Domain Ontology

Domain ontology provides conceptualization for information sources. Different formalizations of concepts and instances exist in multiple heterogeneous information sources which complicates the design of concept brokering.

Upper level ontology includes generalized concepts and non-private attributes from multiple domain ontologies. By contrast, domain ontology includes more detail concepts and private attributes. The predicates *SubConcept( )* and *SameConcept( )* are used for the rules from upper level ontology to domain ontology. *SubConcept(a,b)* implies a can be derived from b, and the attributes of a are derived from the same-named attributes of b. The rules for the upper level ontology to domain ontology are shown in the left of Fig. 3.

Given a concept G_1 in the upper level ontology, for any concept v in domain ontology, the predicate *SubConcpet(G_1, v)* implicates G_1 can be derived from v. G_1 can be derived from multiple concepts of different domain ontologies. For example in right of Fig. 3, the G_1 can be derived from V_1 and V_2 in domain ontology.

As we has assumed in the beginning of Section 3.4, if user ontology concept X is derived from upper level ontology concept G_1, $X:=G_1$, and upper level ontology concept G_1 is derived from domain concepts V_1 or V_2, $G_1:=V_1$ or $G_1:=V_2$, then we have $X:=V_1$ or $X:=V_2$. Therefore, the upper level ontology concepts G and G_1 act as the concept broker to derive user ontology concept X from (multiple) domain ontologies.

Rules from upper level ontology to domain ontology	Example	
	Upper level ontology	Domain ontology
$\forall v$ (SubConcept (G_1, v) => $G_1 \supseteq v$) => $G_1 := v$ => SuperProperty (G_1', v') => $G_1' \subseteq v'$ => $G_1.m := v.m$ ($G_1.m \in \{m \mid m \in G_1'$ and $m \in v'\}$; $v.m \in \{m \mid m \in v'$ and $m \in G_1'\}$) => Attribute $(G_1, m) :=$ Attribute(v, m).	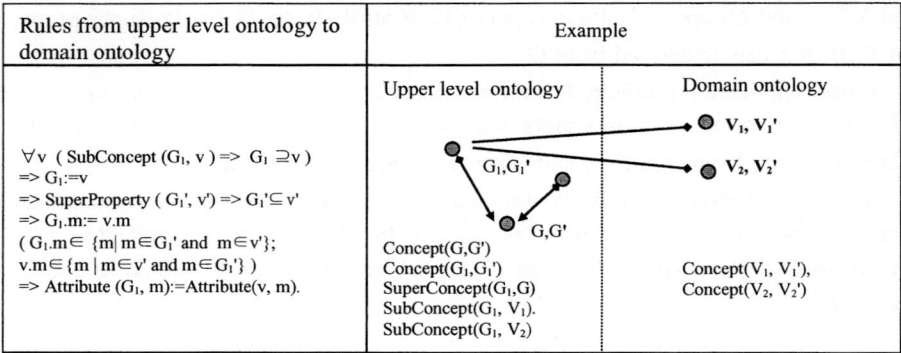 Concept(G,G') Concept(G_1,G_1') SuperConcept(G_1,G) SubConcept(G_1, V_1). SubConcept(G_1, V_2)	Concept(V_1, V_1'), Concept(V_2, V_2')

Fig. 3. The brokering rules from upper level ontology to domain ontology

4   An Example of Upper Level Ontology

We give an example of the concept *road*, its partial sub-concepts and attributes (categorical attributes and quantitative attributes) by referring to SDTS spatial features [15] and Transportation Spatial Data Dictionary [19]. We use Concept Lattices to design the upper level ontology of *road*. A cross table of *road* context is show in Table 2. The first column represents concepts, the first row attributes, and the cross X in column i and row j the concept in column i which has the attribute in row j.

In any ontology, concepts have to be given unique names. We introduce the function label (A) that returns a label for concept A. The label is the unique name chosen in the course of design. In any particular ontology, concepts are always identified by their label. For example, concept *road* has a label *1* as its identifier. The names for attributes are similar.

Table 2. The cross table of road context

		a	b	c	d	e	f	g	h	i	j
1	Road	X									X
2	Alley	X	X					X			X
3	Avenue	X		X					X	X	X
4	Expressway	X			X				X	X	X
5	Freeway	X					X		X	X	X
6	Highway	X			X				X	X	X
7	Footpath	X						X			X

a: ground roadway;
b: narrow ground roadway, between buildings.
c: broad ground roadway, in urban area
d: partial controlled access
e: full controlled access
f: full access
g: passage of foot
h: passage for vehicles
i: volume
j: geometry

The labeled line diagram of the concept lattices *road* is shown in Fig. 4. The shadowed circle represents a concept and its attributes in Table 2. The empty circle represents a combination of attributes for no-entity concepts. The concept *road* is labeled by *1* with attribute set *{a ,j}* (represented as *aj*), which is super-concept of the others, and it can be derived from any sub-concepts (represented as *1:= 2 ∨ 3 ∨ 4 ∨ 5 ∨ 6 ∨ 7*). The attributes *a* and *j* of concept *road* are common to all of its sub-concepts.

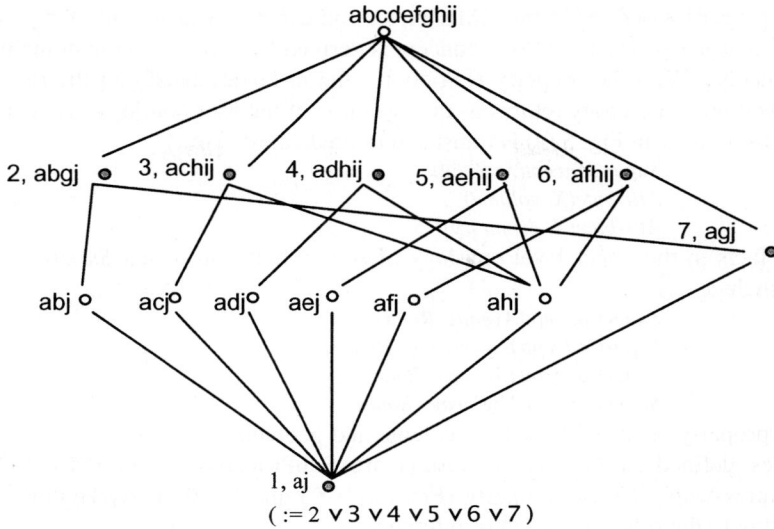

Fig. 4. Concept lattices of road context

4.1 User Query Based on Upper Level Ontology

A user query, for example, searching the roads whose average volume of traffic flow is more than 150 units/min and returning their locations. The query expression of user task ontology based on the upper level ontology is represented by XML shown in Fig. 5(a).

```
<s:Concept about="X" >
    <s:SuperConcept resource=" Road" />
    <s:attribute resource=" volume">
        <restriction  morethen="150" >
            <MeasurementUnit ="units/min"/>
        </restriction>
    </s:attribute>
    <s:attribute resource="geometry" />
</s:Concept>
<retrun  X.geometry />
```

(a)

```
<s:Concept about="X ">
    <s:SuperConcept resource="Road" />
    <s:attribute resource=" geometry">
        <restriction across >
            < s:Concept about="Y">
                <s:SuperConcept resource=" landuse" />
                <s:attribute resource=" landusetype">
                    <restriction value="agriculture"/>
                </s:attribute >
                <s:attribute resource="geometry" />
            </s:Concept>
            <retrun Y.geometry />
        </restriction>
    </s:attribute >
</s:Concept>
<return X.name />
```

(b)

Fig. 5. Two user queries

Attribute restriction is added in query expression for the element of *attribute*. The restriction includes an operator, attribute value, and the measurement unit of the value. In the end of the *concept* element, the *return* element represents specified attribute will be returned.

The properties defined in the XML expression can be translated into the predicates defined in Table 1. User ontology concept is derived from one or more domain ontology concepts. With the property restrictions, the instances satisfying the restrictions from the domain ontology returns as the instances of the user ontology concept.

The user query in Fig. 5 (a) is translated to predicates:

$$SuperConcept(X, Road)$$
$$Attribute(X, volume)$$
$$Attribute(X, geometry)$$

The facts in the upper level ontology of *road* which can match *SuperConcept(X, Road)* include:

$$SuperConcept(Avenue, Road)$$
$$SuperConcept(Expressway, Road)$$
$$SuperConcept(Freeway, Road)$$
$$SuperConcept(Highway, Road)$$

The property set X' of X include volume and geometry. According to Table 2 and the rules defined in Fig. 1, the results are SubProperty(Avenue', X'); SubProperty(Expressway', X'); SubProperty (Freeway', X') and SubProperty(Highway', X '). According to the rules, the result of inference:

$$X:= Avenue \lor Expressway \lor Freeway \lor Highway$$

The upper level ontology concepts are associated with domain ontology, for example, in the XML annotation of the upper level ontology of *road*, the resource *Highway* has a property *subconcept*: *<s:subconcept   rdf:resource:="http://domain ontology/ schema #v" />*. Therefore, there is *X:=http: //domain ontology/schema#v*.

As an example of query with join of two concepts X and Y, searching the roads which location is across the area of landuse with "agriculture" may be represented based on the expression shown in Fig. 5(b). The attribute *geometry* is the shape of *road*, and the *inside* is spatial operator, which determines whether a geometry is inside the other geometry.

5 Conclusion

This paper has presented how to design and represent the upper level ontology to support the concept brokering and inference between user ontology and domain ontology. The design methods of upper level ontology using FCA and Concept Lattices have been examined that it is suitable for concept brokering for the hierarchic geographic information. Considering the characteristics of Concept Lattices, the concept deriving rule is introduced to obtain the instances of super-concept from its sub-concept, upon which, the brokering rules of the upper level ontology are then built. User ontology concepts are associated with the provider's domain ontology concepts by the brokering rules and inference.

The example of upper level ontology of road is given to illustrate the query and inference among user ontology, upper level ontology and domain ontology. Future work includes the extension of concept analysis with geometric attributes in the design of upper level ontology, and the further refinement of the approach and architecture to implement such a concept brokering.

References

1. Andresen, D., Carver, L., Dolin, R., Fisher, C., Frew, J., Goodchild, M., Ibarra, O., Kothurl, R., Larsgaard, M., Manjunath, B., Nebert, D., Simpson, J., Smith, T., Yang, T. and Zheng, Q., 1995. The WWW Prototype of the Alexandra Digital Library. In *Proceedings of ISDL'95: International Symposium on Digital Libraries (Japan)*, *http://www.dl.ulis.ac.jp/ISDL95/*
2. Abel, D. J., Grade, V.J., Taylor, K.L., and Zhou, X.F, 1999, SMARTL: Towards Spatial Internet Marketplaces. *GeoInformatica*, **3**, 141-164.
3. FGDC, 1998. Content Standard for Digital Geospatial Metadata, http://www.fgdc.gov/metadata/metadata.html
4. ISO/TC 211, 1999. Geographic Information/Geomatics, http://www.isotc211.org/scope.htm#scope.
5. OGC, 2001. Open GIS Consortium, http://www.opengis.org
6. Berners-Lee, J., Hendler, J. and Lassila, O., 2001, The Semantic Web. *Scientific American*, 184, 34-43
7. Yi, S., Zhou, L., Xing, C., Liu, Q. and Zhang, Y., Semantic and interoperable webGIS, in Proceedings of Second International Conference on Web Information Systems Engineering (WISE'01) Volume 2, Kyoto, Japan, December 03 - 06, 2001.
8. W3C WS 2002. Web Services, http://www.w3.org/2002/ws/
9. Kashyap, V. and Sheth, A., 2000, Information Brokering Across Heterogeneous Digital Data: A Metadata-based Approach (Boston/Dordrecht/London: Kluer Academic Publishers).
10. Guarino, N. 1998, Formal ontology and information Systems. *In Proceedings of FOIS'98, Trento,Italy,*(Amsterdam, ISO Press), pp.3-15.
11. Gruber, T.R., 1993, Towards Principles for the Design of Ontologies Used for Knowledge Sharing. Technical Report KSL 93-04, Knowledge Systems Laboratory, Stanford University.
12. SUO, 2002. Standard Upper Ontology Study Group, http://suo.ieee.org/
13. WordNet, 1998. A Logical Database for English, Cognitive Science Laboratory, Princeton University, http://www.cogsci.princeton.edu/~wn
14. CYCORP,1997. Upper Cyc Ontology, CYCORP Inc., http://www.cyc.com
15. USGS, SDTS Part2 1997. Spatial Data Transfer Standard (SDTS)-Part2, Spatial Features, http://mcmcweb.er.usgs.gov/sdts/SDTS/-standard-oct91/part2.html
16. Kokla, M. and Kavouras, M. 2001, Fusion of top-level and geographical domain ontologies based on context formation and complementarity. *International Journal of* Geographical Information Science, 15, 679-687
17. Kavouras, M. and Kokla M., 2002, A method for the formalization and integration of geographical categorizations. *International Journal of Geographical Information Science*, 16, 439-453.
18. Ganter, B., and Wille, R., 1999, Formal Concept Analysis, mathematical foundations (Berlin: Springer-Verlag).
19. FGDC, Transportation spatial data dictionary, 1997, http://www.bts.gov/gis/fgdc/pubs/tsdd.html

A New Quick Point Location Algorithm

José Poveda[1], Michael Gould[1], and Arlindo Oliveira[2]

[1] Departamento de Lenguajes y Sistemas Informáticos
Universitat Jaume I, Castellón, Spain
{albalade,gould}@uji.es
[2] Departamento de Engenharia Informática, Instituto Superior Técnico
Universidade Técnica de Lisboa, Lisboa, Portugal
aml@inesc-id.pt

Abstract. We present a new quick algorithm for the solution of the well-known point location problem and for the more specific problem of point-in-polygon determination. Previous approaches to this problem are presented in the first sections of this paper. In the remainder of the paper, we present a new quick location algorithm based on a quaternary partition of the space, as well as its associated cost and data structures.

1 Introduction

In this paper we present a new quick algorithm for the general solution of the point location problem[1] and the specific solution of the point-in-polygon problem [1] on which we will focus our attention. Some of the most efficient solutions for the point-in-polygon problem reduce the solution to that of other fundamental problems in computational geometry, such as computing the triangulation of a polygon or computing a trapezoidal partition of a polygon to solve, then, in an efficient way, the point-location problem for that trapezoidal partition. Two different methods for solving the point-in-polygon problem have become popular: counting ray-crossings and computing "winding" numbers. Both algorithms lead to solutions with a less-than-attractive cost of O(n), however the first one is significantly better than the second [2]. An implementation comparison by Haines [3] shows the second to be more than twenty times slower.

Methods of polygon triangulation include greedy algorithms [2], convex hull differences by Tor and Middleditch [4] and horizontal decompositions [5]. Regarding solutions for the point-location problem, there is a long history in computational geometry. Early results are surveyed by Preparata and Shamos [1]. Of all the methods suggested for the problem, four basically different approaches lead to optimal O(log(n)) search time, and O(n) storage solutions. These are the *chain method* by Edelsbrunner et al. [6], which is based on segment trees and fractional cascading, the *triangulation refinement method* by Kirkpatrick [7], and the *randomised incremental method* by Mulmuley [8]. Recent research has focused on dynamic point location,

---

[1] The well-known point-location problem could be more appropriately called point-inclusion problem.

S. Wang et al. (Eds.): ER Workshops 2004, LNCS 3289, pp. 184–194, 2004.

where the subdivision can be modified by adding and deleting edges. A survey of dynamic point location is given by Chiang and Tamassia [9]. Extension beyond two dimensions of the point-location problem is still an open research topic.

2 Point Location Problems in Simple Cases

Several algorithms are available to determine whether a point lies within a polygon or not. One of the more universal algorithms is counting ray crossings, which is suitable for any simple polygons with or without holes. The basic idea of the counting ray crossings algorithm is as follows:

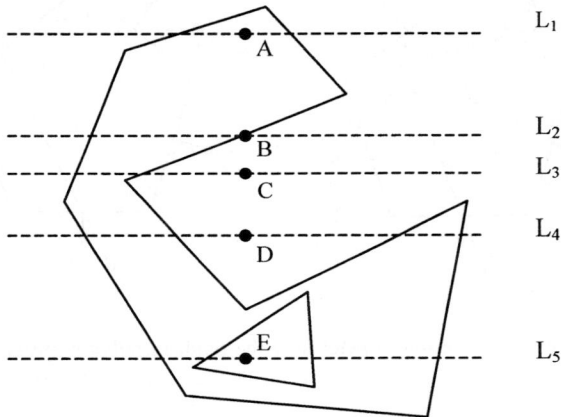

Fig. 1. Ray-crossings algorithm

Let L be the scan-line crossing the given point, Lcross is the number of left-edge ray crossings and Rcross is the number of right-edge ray crossings. First, we need to determine whether the given point is a vertex of the polygon or not; if not, we calculate Lcross and Rcross. If the parities of Lcross and Rcross are different (Fig.1, case B) then this point lies on the edge of the polygon. Otherwise, if Lcross (or Rcross) is an odd number, (Fig.1 case A) the point must lie inside the polygon. If Lcross (or Rcross) is an even number (Fig 1, cases C, D, E) the point must lie outside the polygon.

When the polygon is degraded to a triangle, the above-mentioned problem can be solved by the following algorithm shown in Fig. 2. We assume the three vertices A, B and C are oriented counterclockwise. If point P lies inside ABC, then it must lie to the left of all three vectors AB, BC and CA. Otherwise, p must lie outside ABC. This algorithm may also be applied to any convex polygon, in principle without holes, if the vertices of the polygon are ordered counterclockwise or clockwise.

An equivalent algorithm is based on the inner product among the vectors defined by the viewpoint of the query point and the normal vectors of the edges. This algorithm can be used for convex polygons, and extended for the degenerate case of convex polygons with holes in its interior. The main criterion of this algorithm is based

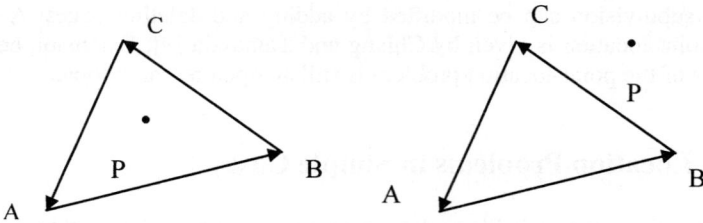

Fig. 2. Conditions for the location of P. (a) P lies to the left of all vectors AB, BC an CD. (b) P lies to the right of vector BC

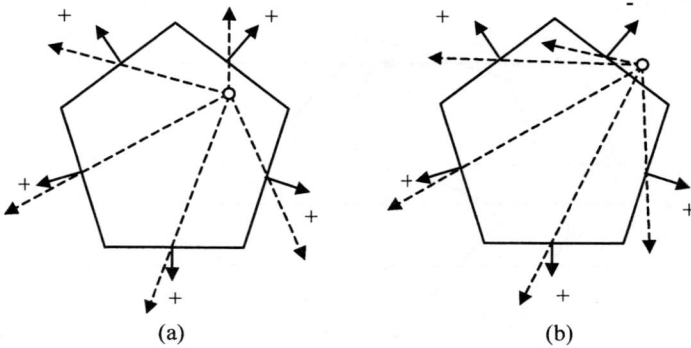

(a)   (b)

Fig. 3. Inner product as a criterion to determine the point-in-polygon problem for the case of convex polygons

on the analysis of the sign of the inner products among the normal vectors of the edges of the polygon, with the unitary vector whose direction is determined by the query point and any point of the edge (normally the middle point of the edge is taken).

If all the inner products are positive, the query point lies in the interior of the polygon, whereas if all the products are positive except one that is null, the point *could* lie on the edge whose associated product is null. If all the products are positive except two that are null, the point lies in the vertex associated with the edges whose inner products are null. In any other case the point is outside the polygon.

3   A Quick Point Location Algorithm

The method we present here is based on a recursive quaternary partition of the space by means of a quadtree data structure. In essence it is not very different from the methods presented previously, though it does have some important characteristics. On one hand the following method uses a "Bounding Quad-tree Volume, (BQV)" in seeking to obtain O(log n) to solve the point location problem as the best methods do, but on the other hand it tries to resolve commonly encountered problems with some adaptive characteristics to the particular problem.

Given P as a partition of the space, as for example a polygon, the objective of the BQV method is to define a set of two-dimensional intervals, V_i, E_i, I_i, O_i in order to define a partition Q of all the space, in such a way that for any query point, we return an interval of the partition Q in which the query point is located. If the query point is in a box V_i or E_i we can calculate with just a simple operation--in fact an inner product--the location of the point regarding the whole polygon P and for points located in an interval I_i or O_i we resolve directly with a simple comparison. Additionally, in order to define an optimum method we should be able to index this set of intervals with at least O(log n).

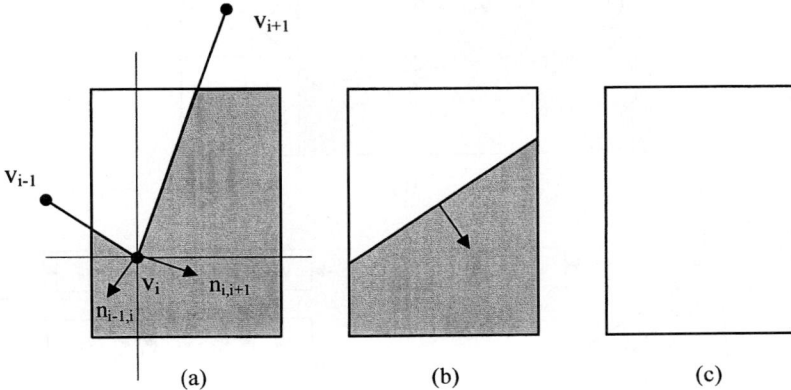

(a)   (b)   (c)

Fig. 4. The different types of intervals defined to cover the space; (a) interval V_i, just with a vertex of P, (b) interval E_i, just with an edge of P, (c) interval I_i, wholly inside the polygon, or interval O_i, wholly outside the polygon

Although the sizes of the intervals shown in Fig. 4 have been represented with the same size, this factor depends on the partition of the space created by the quadtree. Note also, that for the case of polygons we define and order the vertices of the polygon, for the sake of simplicity of implementation. For the case of traversing polygons it is easier to use an array as a storage data structure instead of typical data structures for graphs in the case of a general partition.

Given a polygon P, the first step is to create a recursive quaternary partition of the space for the identification of the V_i intervals. To do that we create a quadtree associated to the space in which we embed the polygon with the simple condition that any node must have associated either one vertex or zero. In the creation of the quadtree, every node of the tree has a list associated with the points inside of its associated area as shown in figure 5. Once the tree has been created these lists are not useful and can be deleted.

For the identification of the intervals E_i, once the V_i intervals have been identified we proceed to calculate the intervals E_i between each two consecutive V_i intervals as shown in Fig. 6. The property we will use for making this calculus is twofold: on one hand we will calculate the intersection point between the intervals and the edge defined by two consecutive vertices and, on the other hand, we will use neighbourhood properties between intervals.

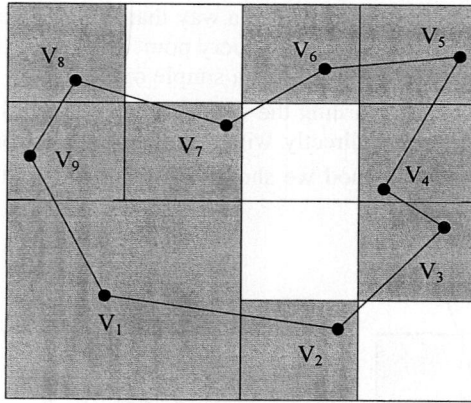

(a)

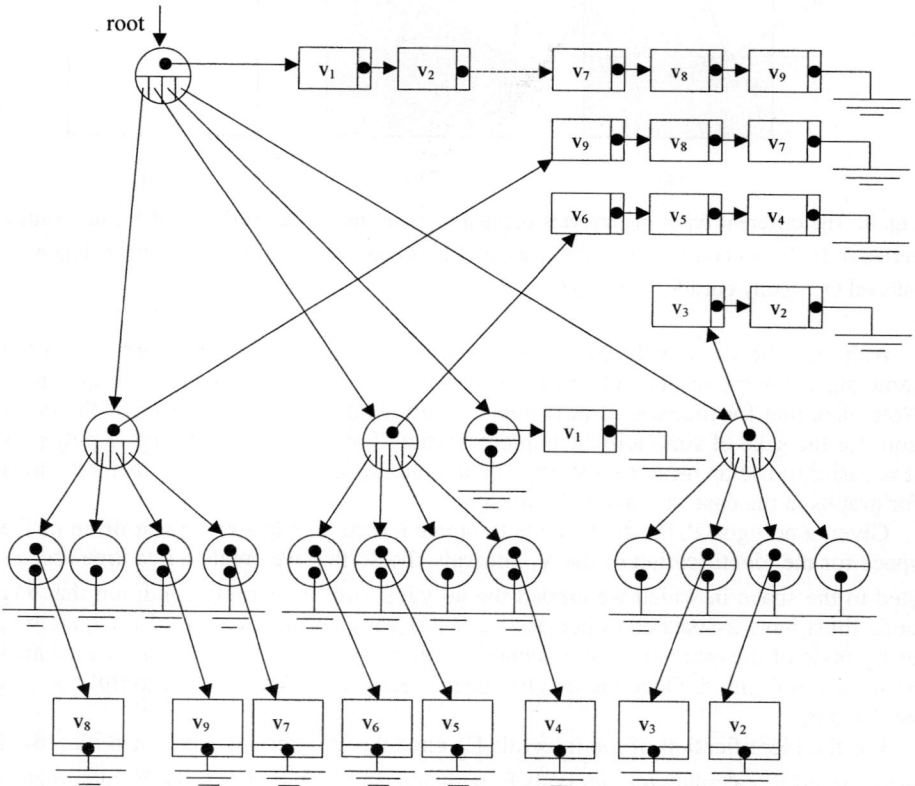

(b)

Fig. 5. Creation of a quaternary partition with identification of intervals V_i. (a) Each Vi interval identifies in the creation of the quad-tree is marked as V_i interval, in the figure shadowed intervals. (b) Intermediate data structure for the creation of the quaternary tree

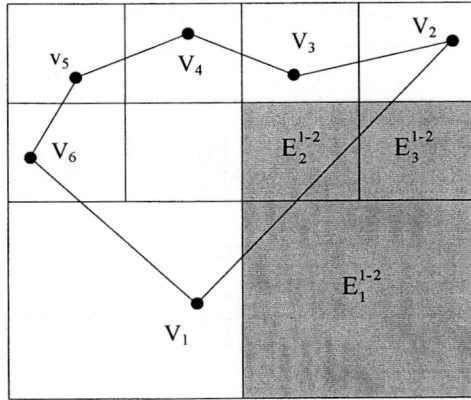

Fig. 6. Identification of E_i intervals of the space partition for a polygon P

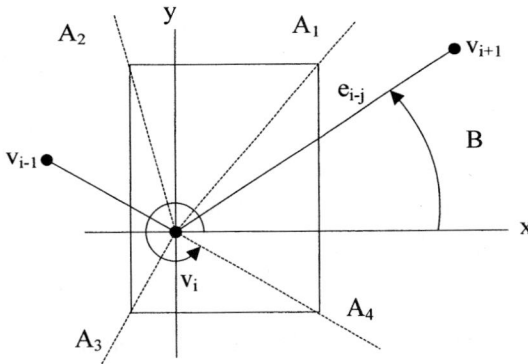

Fig. 7. Space and interval V_i are divided in four areas by the angles A_1, A_2, A_3 and A_4

For the calculus of the successive intersection points among edges and the intervals we need to calculate previously the angle B defined between the edge $e_{i\text{-}j}$ and the x axis. Note that B is an invariant for each E_i interval. After obtaining each intersection point we identify the neighbour to that common intersection point as the following E_i interval. The process continues until reaching the interval V_{i+1}.

For the calculus of the intersection points we start from a V_i interval. For this calculus we will define a reference system centred on the same vertex as shown in Fig. 7. For knowing at which edge will be the intersection point we will need to calculate the angles A_i as shown in Fig. 7.

In order to identify the intersection point of the edge V_i-V_{i+1} with the sides of the interval V_i, we firstly identify the intersection side by comparing the angle B with the angles A_i. The angles A_1, A_2, A_3, A_4 and B are given by the following formulas, in which we have taken into account the negative sense of the axis y as it is usual in graphics libraries.

Table 1. Calculus of the angle B

	$\Delta y > 0$	$\Delta y < 0$
$\Delta x > 0$	$B = 2\pi - \text{atan} \dfrac{|\Delta y|}{|\Delta x|}$	$B = \text{atan} \dfrac{|\Delta y|}{|\Delta x|}$
$\Delta x < 0$	$B = \pi - \text{atan} \dfrac{|\Delta y|}{|\Delta x|}$	$B = \pi + \text{atan} \dfrac{|\Delta y|}{|\Delta x|}$

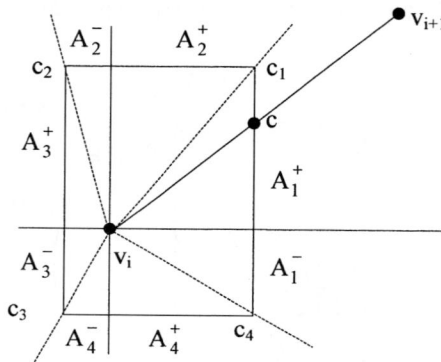

Fig. 8. Splitting of the sides of the interval V_i

Defining Δx and Δy as the differences between the coordinate points of the vertices V_i, V_{i+1}, given by

$$\Delta x = v_{i+1}.x - v_i.x$$
$$\Delta y = v_{i+1}.y - v_i.y$$

once we have identified the side of the interval in which the intersection point c lies, we are prepared for the particular calculus of the intersection point c. Taking into consideration the angles A_i, we identify for each side of the interval V_i a positive part A_i^+ as well as a negative one A_i^-, as indicated in Figure 8.

Once identified the side in which the intersection point lies, we explicitly obtain the coordinates of the intersection point c. For example, if B is lower than A_1 and greater or equal to 0 – which means that c lies over A_1^+ as shown in Fig. 8 – we calculate the coordinates of the intersection point c; c.x and c.y, as follows[2]

---

[2] The detailed calculus as well as an implementation of the more relevant algorithms can be found in [13].

$$c.x = c_1.x$$
$$c.y = v_i - \tan(B)(c_1.x - v_i.x)$$

After obtaining the intersection point c_i, the following step is to find the interval E_i neighbour to V_i on the side of the intersection point c_i. Assuming δ is the length of the minimum side of any interval of the space partition Q that we have created with the quadtree structure, then we can ensure that the point q defined in Table 2 is inside of the neighbour E_i. This fact allows us to locate the neighbour interval E_i by simply traversing the quadtree in order to locate the interval in which that point q is contained.

Table 2. Calculus of a point q, inner to the interval E_i neighbour to V_i

Side	q.x	q.y	Side	q.x	q.y
A_1^+	c.x+d/2	c.y	A_3^+	c.x-d/2	c.y
A_1^-	c.x+d/2	c.y	A_3^-	c.x-d/2	c.y
A_2^+	c.x	c.y-d/2	A_4^+	c.x	c.y+d/2
A_2^-	c.x	c.y-d/2	A_4^-	c.x	c.y+d/2

If the interval E_i found, is V_{i+1} the calculus process for the intervals E_i that cover the edge V_iV_{i+1} has finished, and there is no any interval E_i between the vertices V_i and V_{i+1}, which is the case of vertices V_5V_6 or V_6V_1 in Fig. 6. If the interval found is not a vertex V_i we mark this interval in the quadtree as interval E_i and we proceed in a similar way to calculate the neighbour of this new interval, which shares an intersection point. For the calculus of neighbour intervals from an interval E_i we consider in each E_i interval, two angles A_1 and A_2, and four possible sides in which the intersection point can lie (see Fig. 9). The reference system now, as is shown in Fig. 9, is centred on the same intersection point.

At the beginning of the process, previous to the identification of the E_i intervals, in the identification process of V_i all the intervals of the spatial partition are created and labelled as outer intervals to the polygon if they do not correspond to a V_i interval, which in practice means to do nothing.

Once the process for the identification of the V_i intervals and E_i intervals has completed, for the labelling of inner intervals to the polygon P we must define what is inside and what is outside of P, by defining an inner point of the polygon and after that by means of a seed algorithm we will be able to fill the inner space to the polygon labelling all the inner intervals. After the process defined what we have, as result, is a partition of the space in which we can find four types of intervals; V_i if the interval contains only one and only one vertex of the polygon, E_i if the interval is cut by any edge of the polygon, I_i if the interval is wholly inside the polygon and finally O_i if the interval is wholly outside the polygon.

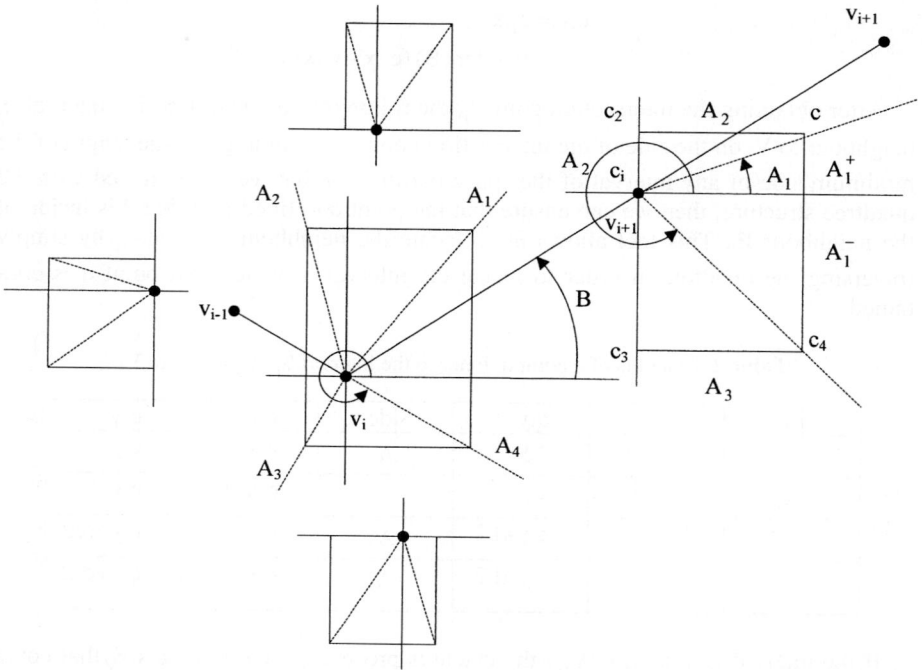

Fig. 9. Definition of reference systems for the calculus of E_i intervals

4   Computational Cost

The cost of BQV algorithm is determined by the access to the interval in which the query point is located, plus a constant cost for each interval that in principle we ignore. As the intervals are indexed with a quaternary tree, in the average case in which we suppose a uniform spatial distribution of the vertex of the polygon over the initial space, the average cost of BQV method corresponds to the average height of the tree, log(m), where m is the number of intervals, always greater for a polygon but in the same order than the number of vertices for a graph.

However, surprisingly we find that there is no a direct relationship between the number of vertices of the polygon and the computational cost required to solve by the BQV method. This fact has to do with the adaptability presented by the BQV method. Let us calculate the worst case of the cost of the BQV method and we shall see that it corresponds to the depth of the tree. Then we will compare for the average case, the cost for a regular polygon inside a circle of a given radius.

The depth of a quadtree for a set of points P in the plane is as maximum log(D/d)+3/2, where d is the minimum distance between any two points of P and D is the length of the initial square that contains P.

In a quadtree the length of the side associated to each node is half of the length of the parent. In general can be said that the length of the associated square to a node of depth i, will be $D/2^i$. On the other hand, the maximum distance between two inner points to a square is as maximum the distance of the diagonal, given by

$$\frac{D\sqrt{2}}{2^i}$$

for the area associated to the node of depth i. Take into consideration that an inner node of the quadtree has as minimum two points in its interior and the minimum distance between two point is d, at level i the inner nodes must satisfy

$$\frac{D\sqrt{2}}{2^i} \geq d$$

which implies

$$i \leq \log\left(\frac{D\sqrt{2}}{d}\right) = \log(D/d) + 1/2$$

and from the fact that the depth of a quadtree is exactly the maximum of the inner nodes plus one, we conclude that as maximum the depth of the quadtree will be

$$\log(D/d) + 1/2 + 1 = \log(D/d) + 3/2$$

and therefore the cost of the BQV algorithm, in the worst case, will be as maximum

$$\log(D/d) + 3/2$$

From the previous relationship we observe that if d tends toward zero the maximum cost tends to infinity, the more relevant cost on which we must focus attention is the average cost, that can be calculated as follows

$$\bar{c} = \sum_i p(Q_i) c(Q_i)$$

In general it can be proven that for a closed polygon with a number n of vertices this cost is finite and corresponds to a small value. So, for example for an regular polygon built inside a circle of radius 300 pixels, the average cost with a number of vertices over 10000, is on the order of just three comparisons.

5 Conclusions

We have presented a new algorithm for the solution of the well-known point location problem and for the more specific problem of point-in-polygon determination. Far from the O(log n) of other algorithms presented in the literature, we have obtained an extraordinary cost that does not depend directly of the number of vertices of the polygon for the average case.

The worst case of the BQV method depends on the factor D/d in which the minimum distance between the vertices of the polygon is crucial, though for the average case can be demonstrated that the worst cases do not contribute relevantly. Although the focus of this paper has been centred on polygons, for simplicity reasons, the BQV method can be extended in the same way to any spatial partition.

Regarding the storage and the building cost, the costs obtained are higher than the optimum O(n) obtained for other algorithms for particular cases seen in early sections of this paper, but are lower than the $O(n^2)$ of other algorithms as for example the slab method. The BQV method does not offer an implementation as simple as other particular algorithms seen in this paper, but the use of the BQV algorithm could be suitable to solve the point-in-polygon problem in cases of high frequency queries and typical real-time constraint situations.

Acknowledgements

This work was partially supported by European Union projects Esprit 25029 (GeoIndex) and IST-2001-37724 (ACE-GIS). We would like to thank in particular Dr. Antonio Rito da Silva from INESC-ID, Universidade Técnica de Lisboa for the unconditional support given during the development of this work.

References

1. Preparata, Franco and Shamos, Michael. "Computational geometry, an introduction", Springer-Verlag 1985.
2. O'Rourke, Joseph. "Computational geometry in C", Cambridge University Press 2001.
3. Haines, Eric.. "Point in polygon strategies". In Paul Heckbert, editor, Graphics Gems IV, pages 24-46. Academic Press, Boston, 1994.
4. Tor, S. B: and Middleditch, A. E.. "Convex decomposition of simple polygons". ACM Trans. on Graphics, 3(4):244-265, 1984.
5. Seidel, R.. "A simple and fast incremental randomised algorithm for computing trapezoidal decomposition and for triangulating polygons", Comput. Geom. Theory Appl. 1991.
6. Edelsbrunner, H., Guibas, L. J. and Stolfi, J., "Optimal point location in a monotone subdivision", SIAM J. Comput.1986.
7. Kirkpatrick, D. G., "Optimal search in planar subdivisions", SIAM J. Comput., 1983.
8. Mulmuley, K. and Schwarzkopf, O.. " Randomized algorithms", in J. E. Goodman and J. O'Rourke, eds., Handbook of Discrete and Computational Geometry, CRC Press LLC, Boca Raton, 1997.
9. Chiang, Yi-Jen, Tamassia, Roberto. "Dynamization of the trapezoid method for planar point location". Seventh annual symposium on computational geometry. ACM Press, New York, NY, USA, 1991.
10. de Berg, M., van Kreveld M., Overmars, M., Schwarzkopf, O.. "Computational Geometry, Algorithms and applications", Springer-Verlag 2000, Second Edition.
11. Dobkin, D. P. and Lipton, R. J. "Multidimensional searching problems", SIAM J. Comput., 1976.
12. Kirkpatrick, D. G., Klawe, M. M. and Tarjan R. E. "Polygon triangulation in O(n log log n) time with simple data structures", In Proc. 6th Annu. ACM Sympos. Comput. Geom., 1990.
13. Poveda, J.. "Contributions to the Generation, Visualisation and Storage of Digital Terrain Models: New Algorithms and Spatial Data Structures", PhD. dissertation, Universitat Jaume I, Castellon Spain. 2004.

First International Workshop on Conceptual Model-Directed Web Information Integration and Mining (CoMWIM 2004)

at the 23rd International Conference
on Conceptual Modeling (ER 2004)

Shanghai, China, November 9, 2004

Organized by

Dongqing Yang and Tok Wang Ling

Preface to CoMWIM 2004

As Internet accesses become ubiquitous, information retrieval and knowledge acquisition from WWW have become more and more important. One critical challenge is that both the contents and the schema/structures of the data sources on the web are highly dynamic. The attempts to provide high quality services to various users by integrating and mining information from the web stimulate a lot of interesting research problems.

In conjunction with the 23rd International Conference on Conceptual Modeling (ER2004), the International Workshop on Conceptual Model-directed Web Information Integration and Mining (CoMWIM 2004) brought together researchers and experienced practitioners from academia and industry to address the challenges and opportunities for conceptual model-directed web information integration and mining. The CoMWIM Workshop provided an international forum for the sharing of original research results and practical development experiences.

We received 39 valid paper submissions coming from different counties and regions. After careful review of the submissions by members of the international pro-gram committee, 13 papers were accepted for presentation at the workshop and included in the proceedings. The CoMWIM 2004 workshop organized 4 sessions around important subtopics such as Web Information Integration, Web Information Mining, conceptual Model for Web Information, Web Information Systems and Web Services, etc.

We would like to thank the many people who volunteered their time to help making the workshop a success. We thank ER Conference Co-chair Professor Tok Wang Ling and ER Workshop Co-chairs Shan Wang for their help and important instructions. We would also like to thank the COMWIM Steering Co-Chairs Professor Xingui He and Professor Hongjun Lu for their kind support. We thank professor Shuigeng Zhou serving as COMWIM local Chair. Special thanks go to all the PC members: Svet Braynov, Ying Chen, Zhiyuan Chen, David Cheung, Joshua Huang, Dongwon Lee, Chen Li, Bing Liu, Dong Liu, Mengchi Liu, Qiong Luo, Zaiqing Nie, Il-Yeol Song, Haixun Wang, Wei Wang, Yuqing Wu, Chunxiao Xing, Lianghuai Yang, Qiang Yang, Jeffrey Yu, Senqiang Zhou and Xiaofang Zhou. Finally, we would also like to thank all the speakers and presenters at the workshop, all the authors who submitted papers to the workshop, and all the participants to the workshop, for their engaged and fruitful contributions.

This workshop was also supported by the NKBRSF of China (973) under grant No.G1999032705.

November 2004   Dongqing Yang, Shiwei Tang,
                                           Jian Pei, Tengjiao Wang, and Jun Gao

Checking Semantic Integrity Constraints
on Integrated Web Documents

Franz Weitl and Burkhard Freitag

University of Passau, Germany
{weitl,freitag}@fmi.uni-passau.de

Abstract. A conceptual framework for the specification and verification of constraints on the content and narrative structure of documents is proposed. As a specification formalism we define $\mathsf{CTL}_{\mathcal{DL}}$ which is a new version of the temporal logic CTL extended with description logic concepts. In contrast to existing solutions our approach allows for the integration of ontologies to achieve interoperability and abstraction from implementation aspects of documents. This makes it specifically suitable for the integration of heterogenous and distributed information resources in the semantic web.

1 Introduction

A large amount of information in the web is available in the form of web documents. A web document is a coherent collection of web pages presenting information or knowledge about a specific matter to certain target groups.

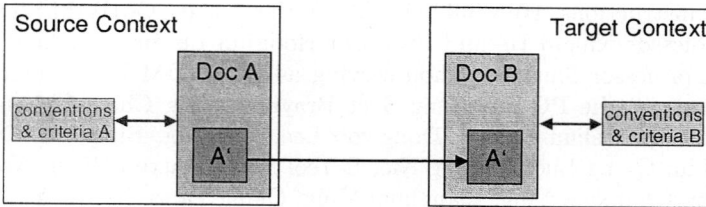

Fig. 1. A document fragment is integrated into a new document: does the integrated document (still) conform to the set of conventions & criteria of the new context?

When integrating external resources into a document, its overall *semantic integrity* needs to be maintained (Fig. 1) where *semantic integrity* is understood as the conformance of a document to *conventions and criteria* which apply in a certain *authoring context*. For instance, it could be required in a scientific setting that terminology and symbols should be used consistently within the document, or that a document has to conform to a specific global structure (introduction - problem description - solution - discussion - conclusion) or to a

S. Wang et al. (Eds.): ER Workshops 2004, LNCS 3289, pp. 198–209, 2004.

certain argumentation style ("a new concept should be well defined before it is discussed"). Especially when document fragments are assembled dynamically or changed frequently as e.g. in the domain of eLearning, it is highly desirable to have support in ensuring the document's overall integrity [16].

In this paper we present an approach to representing relevant types of conventions and criteria in a *formal specification* and verifying the specification against a *model* representing relevant aspects of the document's content and structure. In contrast to existing methods of hypermedia verification our approach allows for the adoption of terminological background knowledge represented in formal ontologies. This increases the expressivity and interoperability of the specification formalism. Secondly, "semantic" properties are represented and specified independently from implementation aspects such as the hypertext/frameset structure or the document format. This makes our approach particularly suitable for integrating heterogeneous and distributed web resources.

The rest of the paper is structured as follows: first we will summarize existing approaches in hypermedia verification and argue why more has to be done to be able to check the semantic integrity of a document. Subsequently we will give a short overview of our approach followed by a formal definition of the involved data structures and algorithms. We conclude with a summary and outlook.

2   State of the Art in Hypermedia Verification

A hypertext verification method based on temporal logic has been developed by Stotts, Furuta et al. [25, 26]. A hypertext is modelled as a finite state machine (browsing automata). Its states represent hypertext nodes, its transitions hypertext links. Local properties of nodes such as the availability of certain buttons, menu options, and content fragments are represented by boolean state variables. The temporal logic HTL*, an extension of CTL* [11], is used as a specification language for global properties of browsing histories such as reachability of certain pages from other pages and availability of certain functions on pages. In [24] the method has been enhanced for the verification of quasi-parallel browsing activities in framesets.

A similar goal is pursued in the MCWeb project [7]. Framesets are modelled in an hierarchical graph of web nodes and links. Properties of single WebNodes are modelled using boolean predicates. A restricted version of the $\mu - calculus$ [15] (*constructive μ-calculus*) has been developed to be able to verify *partly known* models of web sites. Instead of specifying properties which should hold within a frameset model, the model is queried explicitly for counter examples violating desired properties.

Dong [8] has used the $\mu - calculus$ for specification and verification of hypermedia application designs composed of different design patterns of object oriented programming. Knowledge about design patterns are represented in Prolog. Desired properties such as "whenever a link is followed the showpage method is eventually invoked" are specified using the $\mu - calculus$ and verified against the model of the design using the model checker XMC.

Santos et al propose an approach to verification of hypermedia *presentations* [23, 21, 22]. The goal is to discover possible inconsistencies in interactive, synchronized presentations of multimedia content such as multiple access on exclusive resources (audio) and deadlock situations. The nested context model is used to model relevant aspects of an interactive multimedia presentation. The specification language RT-LOTOS (Real Time Language of Temporal Ordering Specification) [6] is used to specify desired properties of the interactive presentation process.

Especially the latter two approaches aim at the *technical correctness* of the design or the implementation of a hypermedia document. Technical correctness, however, does not imply that the document's content and structure makes sense to a *human* recipient.

We extend the approach of Stotts and Furuta 1) towards the specification and verification of properties of the document's content and narrative structure rather than its hypertext or frameset structure, and 2) towards the adoption of terminological knowledge in ontologies to be able to reason about document fragments from different sources with potentially different content models and document formats.

3   General Approach

3.1   Knowledge Representation

We assume that the required knowledge about the document is made explicit by semantic tagging using XML or external *metadata*. The metadata is provided manually by the author or extracted automatically by appropriate analysis techniques [12, 18]. Based on metadata and a *mapping specification*, a *semantic model* is constructed containing assertions of properties of document fragments w.r.t. general background knowledge in *ontologies* (Fig. 2). The semantic model serves as the desired abstraction from implementation aspects such as hypertext structure, document formats, or metadata schemata.

The following aspects need to be represented in the semantic model:

Document Structure: Documents are usually structured in distinct units such as chapters, sections, paragraphs, definitions, examples, tables, lists, images, interactive objects, etc. In the domain of eLearning, various models and formats for structuring information have been proposed [20, 17, 14, 16]. A general model of structural units is represented in the *structure ontology*. The semantic model contains assertions on the structural function of single document fragments.

Topics: Documents have a certain domain of discourse. Knowledge about the entities and relationships between the entities of the domain is represented in a domain ontology [19, 16]. The semantic model of the document contains assertions about which topics are covered in a particular document fragment.

Further Domain Specific Knowledge: In the case of learning documents *objectives*, *learning outcomes* and *competencies* play a major role. Various pedagogic models exist in order to classify objectives and competencies [5, 10]. In other domains different aspects may be relevant for checking the document's integrity. These aspects can be represented in further ontologies and used for verification.

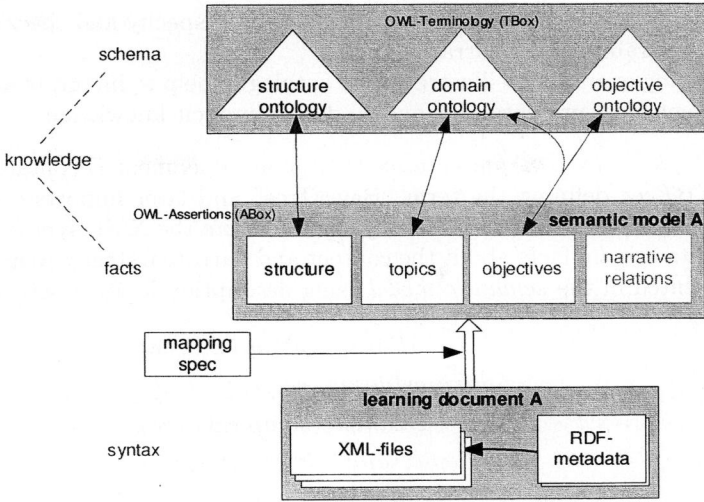

Fig. 2. Representation of knowledge about an integrated document using ontologies

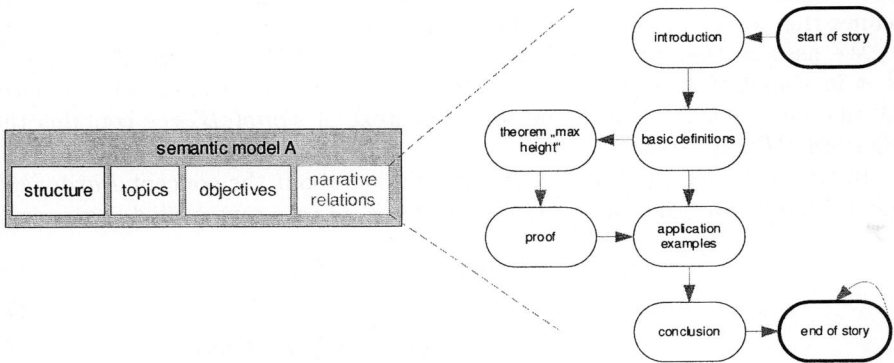

Fig. 3. Narrative structure of a document having a short narrative path and an extended path with additional information

Narrative Structure: The *narrative structure* of a document determines the sensible orders of accessing its fragments in standard situations. It guides the reader through the content along coherent *narrative paths*. Complex documents such as manuals, text books, or web based trainings may contain several alternative narrative paths suitable for different target groups or information demands.

We choose a directed graph of *narrative relations* between document fragments as an appropriate way to represent the document's narrative structure (Fig. 3). Since the main purpose of the narrative structure is to guide the reader through the content, it has to be guaranteed that the document makes sense along narrative paths. Therefore, it is appropriate to specify and check semantic integrity constraints along narrative paths.

In following example we illustrate how ontologies help to integrate knowledge from different local presentations and to derive implicit knowledge.

Example 1 (Knowledge Representation). Assume a document D contains a paragraph $defBTree$ defining the term "BayerTree" and later imports a fragment $sampleBTree$ with an example of "BayerTree". From the XML syntax or external metadata certain facts about the content and narrative structure are derived and represented in the *semantic model* using description logics assertions [4]:

$$Paragraph(defBTree) \tag{1}$$

$$defines(defBTree, bayertree) \tag{2}$$

$$Example(sampleBTree) \tag{3}$$

$$hasTopic(sampleBTree, b{-}tree) \tag{4}$$

$$followedBy(defBTree, sampleBTree) \tag{5}$$

The assertions (1) and (3) describe the fragments $defBTree$ and $sampleBTree$ in terms of the *structure ontology*. The assertions (2) and (4) describe the fragments in terms of the the *domain ontology*. The role assertion (5) followedBy defines the *narrative relationship* between the two fragments.

We assume that in the *mapping specification* of the context of document D it is stated that $bayertree$ is a $BTree$-topic: $BTree(bayertree)$. Similarly the mapping specification of the source context of $sampleBTree$ contains the assertion $BTree(b{-}tree)$.

In the *structure ontology* further information about the concept *Paragraph* and *Example*, and the roles *defines* and *hasTopic* are represented, e.g.

$$Definition \doteq Paragraph \sqcap \exists defines.Topic \tag{6}$$

$$Example \doteq Paragraph \sqcap \exists exemplifies.Topic \tag{7}$$

$$exemplifies \sqsubseteq hasTopic \tag{8}$$

$$defines \sqsubseteq hasTopic \tag{9}$$

In the *domain ontology* further information about topics and relationships of topics are represented, e.g.

$$BTree \sqsubseteq Tree \tag{10}$$

$$BTree \sqsubseteq IndexStructure \tag{11}$$

$$IndexStructure \sqsubseteq \exists usedIn.DataBases \tag{12}$$

Let KB be a knowledge base containing the semantic model, the mapping specification, and the structure and domain ontology as sketched above. We now can reason in an uniform way about the two fragments and infer, for instance, that fragment $defBTree$ is a $Definition$ (13) and that both fragments deal with something used in data bases (14,15):

$$KB \models (definition \sqcap \exists hasTopic.BTree)(defBTree) \qquad (13)$$

$$KB \models (\exists hasTopic \exists usedIn.DataBases)(defBTree) \qquad (14)$$

$$KB \models (\exists hasTopic \exists usedIn.DataBases)(sampleBTree) \qquad (15)$$

3.2 Representing Conventions and Criteria

For expressing semantic constraints on the narrative structure of documents, we enhance the branching time temporal logic CTL (computation tree logic [11]) to $\mathsf{CTL}_{\mathcal{DL}}$ such that concept descriptions instead of atomic propositions can be used for local constraints. We formally define the syntax and semantics of $\mathsf{CTL}_{\mathcal{DL}}$ in section 4.2. For now we give some sample specifications and state their intuitive meaning.

Example 2 ($\mathsf{CTL}_{\mathcal{DL}}$ *Specifications on Narrative Structures).* "On every narrative path in the document the reader will learn something about index structures"

$$\mathsf{AF}\ \exists hasTopic.IndexStructure \qquad (16)$$

"There is at least one narrative path which contains an application of b-trees in the context of data bases"

$$\mathsf{EF}\ \exists hasTopic.DataBases \sqcap \exists applies.BTree \qquad (17)$$

"Whenever a fragment is formal or discusses a central topic an illustration should be provided in the sequel"

$$\mathsf{AG}(Formal \sqcap \exists hasTopic.CentralTopic \Rightarrow Illustration \vee \mathsf{AX}\ Illustration) \quad (18)$$

"The document starts with an introduction followed by some presentations, each offering an optional exercise with least two tasks and ending with a summary"

$$Intro \wedge \mathsf{AF}\ Presentation \wedge \mathsf{AG}(Presentation \Rightarrow$$
$$\mathsf{EX}\ Exercise \sqcup \exists^{\geq 2} contains.Task \wedge \mathsf{AF}\ Summary) \qquad (19)$$

4 Formal Framework

4.1 Representation of Knowledge About the Documents Content

In the sequel we formally define the *narrative structure* of a document. The narrative structure represents relevant aspects of the document's content and structure in a semantic model.

Definition 1 (Narrative Structure).
A **narrative structure** *is a tuple* $NS = (F, ID, m, id_0, id_e, NR, SM)$ *where*

- *F is a non empty finite set of fragments.*
- *ID is a set of individuals such that $\#ID = \#F$.*

- $m : ID \rightarrow F$ is a bijective function mapping individuals onto fragments.
- $id_0 \in ID$ is the representation of a start fragment of all narrative paths.
- $id_e \in ID$ is the representation of an end fragment of all narrative paths.
- $NR \subseteq ID \times ID$ is a representation of narrative relations between fragments of a document such that $(id_e, id_e) \in NR$, and $\forall id \in ID \; \exists (id_0, .., id, .., id_e) \in \{(id_0, id_1, ..., id_n) \mid \forall i \in \{0..n-1\} : (id_i, id_{i+1}) \in NR\}$, i.e. for each fragment id there is a narrative path leading from id_0 over id to id_e. $(a, b) \in NR$ expresses that fragment $m(b)$ should not be read before fragment $m(a)$.
- $SM \supseteq NR$ is a $\mathcal{DL}$-ABox containing a set of assertions on ID (semantic model).

The individuals in ID represent the actual document fragments F. We have defined ID to represent exactly the fragments of a given document. We also assume to have knowledge about the concrete mapping m of individuals on document fragments.

We further assume a document to have a distinct start and end fragment id_0 and id_e. The start fragment is the entry point of the document (associated with the document's URL). The end fragment is the fragment from where the document is left. id_0 and id_e may represent "pseudo fragments" which do not have any actual content. We require the narrative structure to be coherent in the sense that every document fragment can be reached on some narrative path from id_0 to id_e. Non coherent documents can be broken down into coherent parts which are treated separately. For technical reasons we require $(id_e, id_e) \in NR$ such that NR is a left total relation, i.e., $\forall i \in ID \; \exists j \in ID : (i, j) \in NR$.

4.2   Specification of Semantic Properties by CTL$_{\mathcal{DL}}$

We now define formally the syntax and semantics of the branching time temporal logic CTL over a description logic language $\mathcal{DL}$ which we suggest as a formalism for expressing semantic integrity constraints on documents.

Definition 2 (syntax of CTL$_{\mathcal{DL}}$).
Let $\mathcal{DL}$ be a decidable description logic. Let $\mathbf{C}_{\mathcal{DL}}$ be the set of concept descriptions as defined by the syntax of $\mathcal{DL}$. The set of CTL$_{\mathcal{DL}}$ formulae is the set of state formulae generated by the following rules:
(S1) each concept description $C \in \mathbf{C}_{\mathcal{DL}}$ is a state formula;
(S2) if p, q are state formulae then so are $p \wedge q, \neg_{ctl} p$;
(S3) if p is a path formula then $\mathsf{E}p, \mathsf{A}p$ are state formulae;
(P0) if p, q are state formulae then $\mathsf{X}p, p \; \mathsf{U} \; q$ are path formulae.

A CTL$_{\mathcal{DL}}$ formula is interpreted w.r.t a structure $M = (S, R, KB)$ such that

- S is a nonempty set of *individuals* representing states.
- $R \subseteq S \times S$ is a *left total* binary relation on S.
- $KB = TBox \cup ABox$ is a $\mathcal{DL}$ knowledge base.

The *semantics* of CTL$_{\mathcal{DL}}$ defines when a formula $p \in$ CTL$_{\mathcal{DL}}$ is *true* in a structure $M = (S, R, KB)$ at state $s_0 \in S$, in symbols: $M, s_0 \models p$. For the definition of $\models$ we define a *fullpath* in S as a sequence $x = (s_0, s_1, s_2, ...)$ such that $\forall i \in \mathbb{N} : (s_i, s_{i+1}) \in R$. $M, x \models p$ denotes that a path formula p is *true* in the structure M for the fullpath x.

Definition 3 (semantics of $\mathsf{CTL}_{\mathcal{DL}}$).

Let $\models_{\mathcal{DL}}$ be the entailment operator of $\mathcal{DL}$ as defined in [3]. We define $\models$ inductively as follows:

(S1) $M, s_0 \models C$ *iff* $KB \models_{\mathcal{DL}} C(s_0)$

(S2) $M, s_0 \models p \wedge q$ *iff* $M, s_0 \models p$ *and* $M, s_0 \models q$

   $M, s_0 \models \neg_{ctl} p$ *iff not* $M, s_0 \models p$

(S3) $M, s_0 \models \mathsf{E}p$ *iff* $\exists x = (s_0, s_1, s_2, ...)$ *in* M *such that* $M, x \models p$,

   $M, s_0 \models \mathsf{A}p$ *iff* $\forall x = (s_0, s_1, s_2, ...)$ *in* $M : M, x \models p$

(P0) $M, x \models \mathsf{X}p$ *iff* $M, s_1 \models p$

   $M, x \models p \;\mathsf{U}\; q$ *iff* $\exists i \in \mathbb{N} \; [M, s_i \models q$ *and* $\forall j \in \mathbb{N}(j < i$ *implies* $M, s_j \models p)]$

Remark 1 (Syntax and Semantics of $\mathsf{CTL}_{\mathcal{DL}}$). Some comments on the definitions above are in order:

- In rules (S2) above we denote the negation of a state formula $\neg_{ctl} p$ with an index to distinguish it from the negation used in $\mathcal{DL}$ concept descriptions. In fact the two negations have a different semantics: a state formula $\neg C$ holds in M, s_0 iff for all models $\mathcal{I}$ of KB it can be proven true that $s_0^{\mathcal{I}} \notin C^{\mathcal{I}}$. In contrast, $\neg_{ctl} C$ holds in M, s_0 iff it cannot be proven true that $s_0^{\mathcal{I}} \in C^{\mathcal{I}}$ for all models $\mathcal{I}$ of KB (negation as failure).
- We use the following abbreviations as defined in [11]: $p \vee q$, $p \Rightarrow q$, *true*, *false*, $\mathsf{F}p$ ("eventually p"), $\mathsf{G}p$ ("always p"), $p \;\mathsf{B}\; q$ ("p before q") .
- The binding precedence of connectives is: the connectives of the sublanguage $\mathcal{DL}$ ($\forall, \exists, \neg, \sqcap, \sqcup$) have highest binding power, followed by the path operators A, E, temporal operators $\mathsf{F}, \mathsf{G}, \mathsf{X}, \mathsf{U}, \mathsf{B}$, and finally $\neg_{ctl}, \wedge, \vee, \Rightarrow$.
- In contrast to existing temporally extended description logics [1, 2, 13], $\mathsf{CTL}_{\mathcal{DL}}$ has a very modular structure: a $\mathsf{CTL}_{\mathcal{DL}}$ formula p is a second order formula over a $\mathcal{DL}$ knowledge base KB. More precisely, p is interpreted on reified individuals themselves representing objects of a domain which in general is not known (open world assumption). In our application we assume a certain set of individuals ID and relationships NR to be known completely such that the more efficient closed world reasoning (model checking) can be applied for that part of knowledge. The modular structure of $\mathsf{CTL}_{\mathcal{DL}}$ serves well for integrating open and closed world reasoning within one framework as we will see later on.

Proposition 1 (relationship between $\mathsf{CTL}_{\mathcal{DL}}$ and CTL).

Let $\models_{\mathsf{CTL}_{\mathcal{DL}}}$ be the entailment operator as defined in Definition 3 and $\models_{\mathsf{CTL}}$ the entailment operator defined by the standard CTL semantics [11]. Every $\mathsf{CTL}_{\mathcal{DL}}$ formula p, interpreted over (S, R, KB), can be interpreted as a CTL formula over a structure (S, R, L) such that $(S, R, KB), s_0 \models_{\mathsf{CTL}_{\mathcal{DL}}} p \Leftrightarrow (S, R, L), s_0 \models_{\mathsf{CTL}} p$.

Proof. Let p be a $\mathsf{CTL}_{\mathcal{DL}}$ formula. Let $M = (S, R, KB)$ be a structure as defined above. Let $\mathbf{C}_p$ denote the set of concept descriptions used in p. Let L be a labeling

$$L : S \to \mathcal{P}(\mathbf{C}_p) \text{ with } L(s) := \{C \in \mathbf{C}_p \mid KB \models_{\mathcal{DL}} C(s)\}$$

Such a labeling exists because we assumed $\mathcal{DL}$ to be decidable.

By interpreting all concept descriptions $C \in \mathbf{C}_p$ as atomic propositions we can now interpret p in terms of standard CTL semantics over the structure $M = (S, R, L)$. The semantics of CTL is identical to the semantics of $\mathsf{CTL}_{\mathcal{DL}}$ except for rule (S1) which in the case of CTL reads:

(S1') $M, s_0 \models C$ iff $C \in L(s_0)$.

Let $C \in \mathbf{C}_p$ be a concept description in p. Then

$$\begin{aligned}
(S, R, KB), s_0 \models_{\mathsf{CTL}_{\mathcal{DL}}} C &\Leftrightarrow KB \models_{\mathcal{DL}} C(s_0) &&\text{(definition of } \models_{\mathsf{CTL}_{\mathcal{DL}}}) \\
&\Leftrightarrow C \in L(s_0) &&\text{(definition of } L) \\
&\Leftrightarrow (S, R, L), s_0 \models_{\mathsf{CTL}} C &&\text{(definition of } \models_{\mathsf{CTL}})
\end{aligned}$$

Since $\models_{\mathsf{CTL}_{\mathcal{DL}}}$ and $\models_{\mathsf{CTL}}$ coincide in the cases (S2, S3, P0) the equivalence $(S, R, KB), s_0 \models_{\mathsf{CTL}_{\mathcal{DL}}} p \Leftrightarrow (S, R, L), s_0 \models_{\mathsf{CTL}} p$ follows by induction on the structure of p.

Relevant for our application is the model checking problem of $\mathsf{CTL}_{\mathcal{DL}}$:

Definition 4 (model checking problem of $\mathsf{CTL}_{\mathcal{DL}}$).
*Let M be a finite structure (S, R, KB) as defined in Definition 3 and let $p \in \mathsf{CTL}_{\mathcal{DL}}$ be a formula. The **model checking problem** of $\mathsf{CTL}_{\mathcal{DL}}$ is to decide for all $s \in S$, if $M, s \models p$.*

Proposition 2 (complexity of the model checking problem).
The model checking problem for $\mathsf{CTL}_{\mathcal{AL}}$ is ExpTime-*hard.*

Proof. Since model checking a $\mathsf{CTL}_{\mathcal{AL}}$ formula involves deciding the logical implication $KB \models_{\mathcal{AL}} C(a)$, and logical implications w.r.t. a knowledge base with general concept inclusion axioms is ExpTime-hard for $\mathcal{AL}$ (see [9]), the model checking problem for $\mathsf{CTL}_{\mathcal{AL}}$ is at least ExpTime-hard.

Proposition 1 implies that the model checking problem of $\mathsf{CTL}_{\mathcal{AL}}$ w.r.t a structure (S, R, KB) is equivalent to the CTL model checking problem w.r.t. a structure (S, R, L). Since S and p are assumed to be finite, (S, R, KB) can be transformed into a structure (S, R, L) in polynomially many steps (see proof of proposition 1). In each step the logical implication of an assertion w.r.t. an $\mathcal{AL}$ knowledge base must be decided which is known to be ExpTime-hard [9]. Hence the transformation of (S, R, KB) into (S, R, L) is ExpTime-hard. The model checking problem of CTL is known to be in deterministic polynomial time [11]. Thus the model checking problem for $\mathsf{CTL}_{\mathcal{AL}}$ can be reduced to a polynomial problem in ExpTime. As a result the model checking problem of $\mathsf{CTL}_{\mathcal{AL}}$ is at most ExpTime-hard.

Remark 2. For languages $\mathcal{DL}$ at least as expressive as $\mathcal{AL}$ the complexity of the model checking problem for $\mathsf{CTL}_{\mathcal{DL}}$ is determined by the complexity of deciding logical implication in $\mathcal{DL}$.

5   Verification of Narrative Structures

5.1   Definition of the Verification Problem

Let $NS = (F, ID, m, id_0, id_e, NR, SM)$ be a narrative structure (see Definition 1) of a web document. Let $B = Ont \cup MS$ be a $\mathcal{DL}$ knowledge base such that Ont represents background knowledge in the structure ontology, domain ontology, and other ontologies, and MS represents a mapping specification of SM on Ont. Let p be a $\mathsf{CTL}_{\mathcal{DL}}$ formula. We define the verification problem of NS, p, B as follows:

Definition 5 (satisfaction of a specification).
A narrative structure $NS = (F, ID, m, id_0, id_e, NR, SM)$ **satisfies** p *under* B *(in symbols* $NS \models_B p$*) iff* $(ID, NR, SM \cup B), id_0 \models p$.

Remark 3. Since for a given narrative structure NS $(ID, NR, SM \cup B)$ is finite and NR is left total, the verification problem of NS, p, B is a model checking problem restricted to state id_0.

5.2   Abstract Algorithm for Model Checking $\mathsf{CTL}_{\mathcal{DL}}$

Given decision procedures for instance checking w.r.t. a $\mathcal{DL}$ knowledge base $(KB \vdash_{\mathcal{DL}} C(a))$ and for model checking a CTL formula p $(M, s_0 \vdash_{\mathsf{CTL}} p)$ a straight forward algorithm for the verification of a narrative structure $NS = (F, ID, m, id_0, id_e, NR, SM)$ is:

```
function verify(ID, NR, SM, B, id_0, p):boolean {
    KB := SM ∪ B
    C_p := {C | C concept description in p}
    for each id ∈ ID {      #construct a labeling L
        L[id] := {}
        for each C ∈ C_p
            if KB ⊢_DL C(id) then L[id] := L[id] ∪ {C}
    }
    return (ID, NR, L), id_0 ⊢_CTL p
}
```

Proposition 1 and its proof imply that *verify* is sound and complete given that $\vdash_{\mathcal{DL}}$ and $\vdash_{\mathsf{CTL}}$ are sound and complete. Let $n = \#ID$ be the number of fragments in NS and $m = \#\mathbf{C}_p$ the number of concept descriptions in p. The time complexity of *verify* can be calculated as $n \cdot m \cdot Time(\vdash_{\mathcal{DL}}) + Time(\vdash_{\mathsf{CTL}})$. In the case of $\mathcal{DL} = \mathcal{AL}$ there is a procedure $\vdash_{\mathcal{AL}} \in \text{ExpTime}$ and a procedure $\vdash_{\mathsf{CTL}} \in \mathbf{P}$ such that *verify* $\in \text{ExpTime}$ which is optimal (Proposition 2).

6   Conclusion

When integrating fragments from different sources into a web document the integrity of the integrated document in terms of content and narrative structure becomes an issue. We argued that existing approaches to hypermedia verification are not sufficient to check for "semantic" integrity violations in web documents. We proposed $\mathsf{CTL}_{\mathcal{DL}}$ as an expressive and efficient means for the specification of integrity constraints and showed how $\mathsf{CTL}_{\mathcal{DL}}$ formulae can be evaluated on semantic document models w.r.t background knowledge in ontologies.

In future work we will look at possibilities to enhance our conceptual framework in terms of efficiency, expressivity and practicability. In particular, we will evaluate our approach on a large body of existing eLearning documents.

References

1. A. Artale and E. Franconi. Temporal description logics. In L. Vila, P. van Beek, M. Boddy, M. Fisher, D. Gabbay, A. Galton, and R. Morris, editors, *Handbook of Time and Temporal Reasoning in Artificial Intelligence*. MIT Press, 2000.
2. A. Artale and E. Franconi. A survey of temporal extensions of description logics. *Annals of Mathematics and Artificial Intelligence (AMAI)*, 30(1-4):171–210, 2001.
3. F. Baader, D. Calvanese, D. McGuinness, D. Nardi, and P. Patel-Schneider, editors. *The Description Logic Handbook - Theory, Implementation and Applications*. Cambridge University Press, 2003.
4. F. Baader and W. Nutt. Basic description logics. In *[3]*, chapter 2, pages 47 – 100. 2003.
5. B. S. Bloom, M. B. Engelhart, E. J. Furst, W. H. Hill, and D. R. Krathwohl. Taxonomy of Educational Objectives. The Classification of Educational Goals. In *Handbook I. Cognitive Domain*. Longman, New York, 1956.
6. J.-P. Courtiat and B. O. C.A.S. Santos, C. Lohr. Experience with RT-LOTOS, a Temporal Extension of the LOTOS Formal Description Technique. *Invited Paper in Computer Communications*, 2000.
7. L. de Alfaro. Model checking the world wide web. In *CAV 01: 13th Conference on Computer Aided Verification*. Springer-Verlag, 2001.
8. J. Dong. Model checking the composition of hypermedia design components. In *Proceedings of the 2000 conference of the Centre for Advanced Studies on Collaborative research*, Mississauga, Ontario, Canada, 2000. IBM Press.
9. F. M. Donini. Complexity of reasoning. In *[3]*, chapter 3, pages 101 – 141. 2003.
10. H. L. Dreyfus and S. E. Dreyfus. Künstliche Intelligenz. Von den Grenzen der Denkmaschine und dem Wert der Intuition. Hamburg, 1988.
11. E. Emerson. Temporal and modal logic. In J. van Leeuwen, editor, *Handbook of theoretical Computer Science: Formal Models and Semantics*, pages 996–1072. Elsevier, 1990.
12. S. Handschuh, S. Staab, and F. Ciravegna. S-CREAM - Semi-automatic CREAtion of Metadata. In *Proceedings of the European Conference on Knowledge Acquisition and Management - EKAW-2002, Lecture Notes in Computer Science*, Madrid, Spain, Oct. 2002. Springer-Verlag.
13. I. Hodkinson, F. Wolter, and M. Zakharyaschev. Decidable and undecidable fragments of first-order branching temporal logics. In *Proceedings of the 17th Annual IEEE Symposium on Logic in Computer Science (LICS 2002)*, pages 393–402, 2002.
14. M. Kohlhase. OMDoc: Towards an Internet Standard for the Administration, Distribution and Teaching of mathematical Knowledge. In *Proceedings of "Artificial Intelligence and Symbolic Computation*. Springer LNAI, 2000.

15. D. Kozen. Results on the propositional mu-calculus. In *Theoretical Computer Science*, volume 27, pages 333–355. 1983.
16. B. Krieg-Brückner et al. Multimedia instruction in safe and secure systems. *LNCS, Recent Trends in Algebraic Development Techniques*, 2755:82–117, 2003.
17. U. Lucke, D. Tavangarian, and D. Voigt. Multidimensional Educational Multimedia with $< ML >^3$. In *Proc. of the E-Learn Conference*, Phoenix, Arizona, USA, 2003.
18. S. Mukherjee, G. Yang, and I. V. Ramakrishnan. Automatic annotation of content-rich HTML documents: Structural and semantic analysis. In *Second International Semantic Web Conference (ISWC)*, Sanibel Island, Florida, Oct. 2003.
19. I. Niles and A. Pease. Towards a standard upper ontology. In C. Welty and B. Smith, editors, *Proceedings of the 2nd International Conference on Formal Ontology in Information Systems (FOIS-2001)*, Ogunquit, Maine, 2001.
20. C. Süß and B. Freitag. LMML - the learning material markup language framework. In *International Worshop ICL*, Villach, Austria, Sept. 2002.
21. C. A. S. Santos, J.-P. Courtiat, L. F. G. Soares, and G. L. De Souza. Formal specification and verification of hypermedia documents based on the nested context model. In *Proceedings of the Conference on Multimedia Modeling*, 1998.
22. C. A. S. Santos, P. N. M. Sampaio, and J. P. Courtiat. Revisiting the concept of hypermedia document consistency. In *Proceedings of the seventh ACM international conference on Multimedia (Part 2)*, pages 183–186, Orlando, Florida, United States, 1999. ACM Press.
23. C. A. S. Santos, L. F. G. Soares, G. L. de Souza, and J.-P. Courtiat. Design methodology and formal validation of hypermedia documents. In *Proceedings of the sixth ACM international conference on Multimedia*, pages 39–48, Bristol, United Kingdom, 1998. ACM Press.
24. D. Stotts and J. Navon. Model checking cobweb protocols for verification of HTML frames behavior. In *Proceedings of the eleventh international conference on World Wide Web*, pages 182–190, Honolulu, Hawaii, USA, 2002. ACM Press.
25. P. Stotts, R. Furuta, and J. Ruiz. Hyperdocuments as automata: Trace-based browsing property verification. In D. Lucarella, J. Nanard, M. Nanard, and P. Paolini, editors, *Proceedings of the ACM Conference on Hypertext (ECHT'92)*, pages 272–281. ACM Press, 1992.
26. P. D. Stotts, R. Furuta, and C. R. Cabarrus. Hyperdocuments as automata: Verification of trace-based browsing properties by model checking. *Information Systems*, 16(1):1–30, 1998.

X2S: Translating XPath into Efficient SQL Queries*

Jun Gao, Dongqing Yang, and Yunfeng Liu

The School of Electronic Engineering and Computer Science
Peking University
100871 Beijing, China
{gaojun,dqyang,yfliu}@db.pku.edu.cn

Abstract. Translating XPath into SQL queries is necessary if XML documents
are stored in relational databases. Although most of the current methods can
translate XPath into equivalent SQL queries, they seldom focus on the efficiency of the translated SQL queries. This paper proposes a method called X2S
to handle the problem. X2S is based on the extended relational encoding, which
can deal with the operators in XPath efficiently. X2S also exploits DTD to rewrite the XPath logically before the SQL translation. The goal of the rewriting
is to eliminate the wildcards of the XPath, to remove the ancestor/descendant
relationship and to shorten the length of the XPath. An estimation cost model is
used to choose one with the lowest cost from the candidate SQL statements.
The experimental results demonstrate that the SQL queries generated by X2S
outperform those from other methods significantly.

1 Introduction

The rapid emergence of XML as a standard for data exchange and representation has
led to much interest in the management of massive XML documents. Using mature
relational databases is a competitive choice for XML documents management because of their scalability, transaction management and other features. However, the
problems of manipulating XML data in a relational database have to be tackled before
this alternative becomes feasible. In this paper, we examine one basic problem,
namely, translating XML queries into SQL queries.

Suppose a set of XML documents are decomposed and stored in a relational database. When a user issues an XPath query, the system needs to generate the corresponding equivalent SQL on the underlying relational tables. This translation is a
challenging problem due to the semantic mismatch of the XML data model and the
relational data model. For instance, we can not map the ancestor/descendant axis or
wildcards in the XPath into SQL relational query directly.

XPath translation algorithm can be evaluated by two metrics [6]: the supported
features of XPath and the efficiency of the translated SQL queries. Although some of

---

* Supported by the National High-Tech Research and Development Plan of China under Grant
No.2002AA4Z3440; the National Grand Fundamental Research 973 Program of China under
Grant No.G1999032705.

S. Wang et al. (Eds.): ER Workshops 2004, LNCS 3289, pp. 210–222, 2004.

the current methods can translate an XPath query with full features into equivalent SQL queries correctly, the translated SQL queries suffer a serious efficiency problem especially when the XML query supports some kind of features.

1.1 Related Work

The relational storage schema for XML has a direct impact on the XPath translation. We review XPath translation methods according to the different XML storage strategies, namely, the schema oblivious method and the schema based method [6].

With a schema oblivious XML storage mechanism, such as the edge tables [2], or the pre/post encoding of XML documents [3, 4, 5], the complete strategy to translate XPath into SQL queries has been presented. In particular, Dehaan et al. [3] proposes a method to translate an XQuery into SQL queries based on the pre/post encoding. Although XQuery poses more challenges than XPath, the main drawback of the method is the low efficiency of the translated SQL. How to enhance relational database kernel to support tree structure data naturally is discussed in [4]. It also relies on pre/post encoding to capture the tree structure, but it focuses on physical optimization of basic operations [4]. In summary, the efficiency of the translated SQL on the schema oblivious XML storage schema is relatively low. For example, it needs to make union of all elements related relations together to handle the wildcards {*}.

The schema constraints in XML can not only be used to generate a reasonable relation schema, but also lead to high efficiency of the generated SQL. In particular, Krishnamurthy et al. [5] deals with XPath translation into SQL in the presence of recursive DTD in a XML publishing scenario. The translated SQL may include recursive statements when the XPath contains {//} and DTD is recursive, while the recursive SQL can only be supported by some commercial databases. The efficiency of translated SQL is studied recently [7]. The method employs constraints in DTD to eliminate some element nodes in the XPath so as to reduce the size of the generated SQL. However, the method cannot support the recursive DTD and can only run on the prefix part of the XPath (from the root element to the first element with predicates).

Earlier proposals and techniques for XML query translation leave much space to optimize the generated SQL queries. For instance, we can enrich the relational encoding for XML element nodes to deal with the operators in XPath more efficiently. DTD structural constraints can be used to optimize the generated SQL from all parts of the XPath rather than the prefix of the XPath with the support of the enriched database relational encoding.

1.2 Contributions

In order to address the XPath translation problem and overcome the limitations of the current methods, we propose a method called X2S to translate an XPath query into highly efficient SQL queries. We discuss the XPath translation problem in the XML storage scenario, where the XML data is decomposed into and queried by the

RDBMS. Specifically, the problem in this paper is formulated as follows: given an XML documents D with DTD (may be recursive), how to store XML documents into the relational database and translate an XPath on the XML documents into the equivalent efficient SQL on the underlying relational schema?

The contributions of our paper are summarized as follows:

- We propose a relational extended encoding (pre, post, parent) to capture the structural relation of nodes in XML document. Basic XPath-to-SQL translation template has been proposed on the extended relational encoding.
- We propose a method called X2S to generate efficient SQL queries. X2S exploits the constraints in DTD to rewrite XPath before the SQL is generated. The goal of rewriting includes the elimination of the wildcards in the XPath, the removal of the ancestor/descendant relationship from the XPath and the reduction of the total length of the XPath. We generate different candidate SQL plans from the logical equivalent XPath expressions. An estimated cost model is used to select one with the lowest cost.
- We have conducted performance studies on a commercial database server and our results show that the SQL queries generated by X2S outperform those by other methods.

The remainder of this paper is organized as follows. In section 2, we introduce some background information, including the XPath queries handled and the XPath tree patterns. In section 3, we describe the encoding for XML documents used in the XPath-to-SQL translation. In Section 4, we present X2S method to translate XPath into efficient SQL queries. Section 5 illustrates the performance study. Section 6 concludes the whole paper.

2   Preliminaries

This section defines the class of XPath queries handled, and introduces the XPath query tree pattern.

2.1   The Class of XPath Queries Handled

XML can be regarded as a node labeled directed tree. XPath has been proposed by W3C as a practical language for selecting nodes from XML [9] and is widely used in other XML query languages like XQuery, XSLT [10,11], etc. XPath supports a number of powerful features. Only a few of the features are commonly used in the applications. As discussed in [1,4,5,7], the operators in this paper include '/' for the parent/child axis; '//' for the ancestor/descendant axis; '*' for wildcards; '[]' for a branch condition.

2.2   XPath Query Tree Patterns

Given an XPath query with operators {/, //, *, []}, we can construct an equivalent tree pattern [14]. The evaluation of XPath queries can be regarded as the tree matching from the XPath tree pattern to the XML document tree. Informally, a branch node in the tree denotes the operator '[]' in XPath, an edge with a solid line denotes an operator '/' in XPath, and an

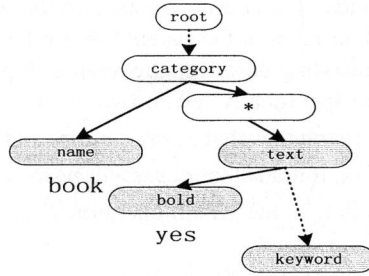

Fig. 1. The sample XPath tree pattern

edge with a dotted line denotes an operator '//' in Xpath.

We illustrate an XPath tree pattern in Fig. 1 for XPath *p=//category[name=book]* */*/text[/bold=yes]//keyword*. The node in the tree with predicates, e.g., the *name* or *bold* node, is called the *condition node*. The root node of query results, e.g., the *keyword* node, is called *the return node*. The node with the outgoing degrees larger than 1, e.g., the *category* node, is called *the branching node*. The path from the root node to the return node is called *the main path*. The paths other than main path are called the *predicate path*.

Definition 1: An XPath segment S in the XPath tree pattern $P=(N,E)$ is defined as a path $(n_1,..n_k)$ in P, where $(n_i, n_{i+1}) \in E$ $(1<i<k-1)$, n_1 and n_k are branching nodes, or leaf nodes, or root node, and as for $n_i(1<i<k)$, n_i is not a branching node.

Intuitively, the XPath segment is the longest path in the XPath tree pattern without a branching node in the middle of the path. Taking Fig. 1 as an example, the path *"/category/name"* or *"/category/*/text"* is an XPath segment.

The goal of this paper is to translate XPath into SQL. The translation process can be regarded as SQL annotating on the XPath tree pattern. The final tree pattern can be defined as:

Definition 2: *XPath tree pattern with SQL annotation.* An XPath tree pattern with SQL annotation is 4-element tuple $P=(N, E, S, M)$, where N is a node set, E is an edge set, S is a set of SQL queries, and M is a set of (s,n) tuples with $s \in S$ and $n \in N$. For each $s \in S$, there is $m \in M$, where $m=(s,n)$, $n \in N$.

3   An Extended Encoding of XML Document

The XPath-to-SQL translation is based on the mapping from the XML schema to the relational schema. Consequently, the relational schema has a great impact on the efficiency of the query translation. Previous work has proposed to encode the structural relationship of XML elements into relational schema. In this work, we extend this encoding further in order to handle the XPath operators $\{/\}$ and $\{//\}$ efficiently.

Two kinds of encoding are used in the XPath translation work [1,3,4,5]. The *parent* encoding records the parent-child relationship between two XML elements. The *pre/post* encoding describes the relationships among an element n and all descendants of n. More specifically, given two elements e_1 and e_2, if e_1 is an ancestor of e_2, the encoding guarantees that $e_1.pre< e_2.pre$ and $e_1.post> e_2.post$. In addition, if e_1 is a left sibling of e_2, it holds that $e_2.pre> e_1.post$. As the result, the parent encoding facilitates the translation of the XPath operator '/' to SQL and the pre/post encoding works for that of '//'.

Putting these two kinds of encoding together for our XPath-to-SQL translation, we propose an extended encoding with three fields (*Pre*, *Post*, *Parent*) for each element. Both *Pre* and *Post* are candidate keys of an element table. *Parent* records the *Pre* field of the parent element. If there are multiple XML documents in the database, we add *DocID* to distinguish the documents.

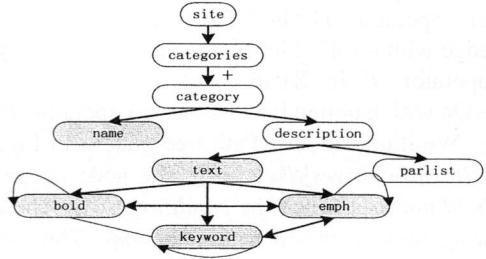

Fig. 2. A fragment of the auction DTD

Example 1: Given a fragment of a DTD named auction [5], we illustrate our extended encoding together with the relational storage schema of the XML documents that conform to the auction DTD.

The grey nodes in the DTD graph (see Fig. 2) can be the leaf nodes in the corresponding XML documents. A directed edge from node n_1 to n_2 denotes that n_2 is a child element of n_1. An edge with a "+" sign from node n_1 to n_2 indicates that n_1 has at least one child elements n_2. Note that the DTD is recursive since there are cycles in the DTD graph.

Let us first take the basic fragmented XML storage strategy in which each types of element is converted into a relational table. In this strategy, documents conforming to the auction DTD fragment in Fig. 2 are converted into relations named *site, categories, category, name, description, text, parlist, bold, keyword* and *emph*. Each relation contains three basic fields (*pre, post, parent*) as our encoding. In the remainder of this paper, we assume this kind of mapping from the XML data model to the relational model.

Note that we can extend the encodings in some ways. The value of the element can be supported by extending the encoding with "*value*" fields. In addition, the XML storage relation schema can be further optimized by an inline method. For example, we can create a relation *description*₁ in replacement of the three relations named *description, text, parlist*. The *parent* field of *text* or *parlist* is removed safely from relation *description₁* since they are already encoded in *description.pre*. The extension of the encoding does not change the translation process in the following.

The extended encoding can be implemented by scanning an XML document only once. Let us describe the encoding generation with the SAX [15] XML parser and a stack. On an event of the *startDocument* of SAX parser, we initialize an empty stack and set a counter *GUI(Global Unique Identifier)* to be 0. On an event of the *startElement*, we increase *GUI* and push *GUI* and the current element into the stack. On an event of the *endElement*, we increase *GUI* and assign the *Post* encoding of the current element with the value of *GUI*. We pop up the top layer from the stack. The *Pre* encoding of the current element is assigned with the *GUI* value in the popped layer. *Parent* information is assigned with the *GUI* value in the current top layer. Hence, we fill out the three fields (*Pre*, *Post*, *Parent*) of a tuple and insert the tuple into the corresponding relation.

4   XPath-to-SQL Translation

We first give an overview of X2S and then describe the translation process in more details.

4.1   Overview of X2S

X2S annotates SQL on the tree pattern until all nodes have been processed. Since the return node is the root node of query results, we translate the main path of the XPath in a top-down fashion. The other paths in the tree pattern can be regarded to impose some restrictions on the nodes in the main path. Hence, we translate other paths in a bottom-up fashion.

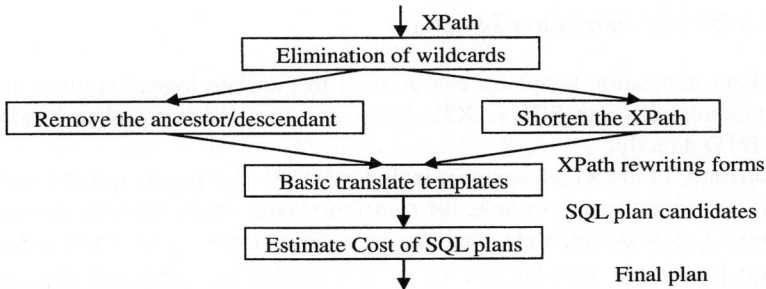

Fig. 3. The process of the XPath segment translation in X2S

X2S works at the granularity of an XPath segment. The translation process can be described in Fig. 3. We make logical rewriting on this XPath segment before it is translated if DTD is available. We eliminate wildcards {*} first, and then we shorten the total size of the XPath or to reduce the occurrence of {//}. We generate the candidate SQL plans from the different form of the logical equivalent XPath expressions. An estimation cost model can be used to choose the SQL statement with the lowest cost as the final SQL statement for the XPath segment.

Taking the XPath segment $S=/category/*//text$ and DTD in Fig. 2 as an example, we exploit DTD to rewrite this segment into different equivalent XPath expressions. First, we can determine that {*} in S represents element *description*. Second, the ancestor/descendant relationship can be replaced by the parent/child relationship safely, hence we get another segment expression $S_1=/category/description/text$. Due to the fact that all paths from *category* node to *text* node satisfy the XPath segment expression, we can get another equivalent expression $S_2=/category//text$. We obtain different SQL plans from S_1 and S_2. The final SQL is chosen with the lowest cost based on an estimation cost model. The final SQL statement may take the following form:

Example 2: The translated SQL from the XPath in Fig. 1:

View *category*: Select *category.** from *category*, *name* where *name.parent= category.pre* and *name.text='book'*;

View *text*: Select *text.** from *text, bold, category,description* Where *text.parent=description.pre and descryption.parent=category.pre* and *bold.text='yes'* and *bold.parent=text.pre;*

View *keyword*: Select *keyword.** from *text, keyword* Where *text.pre<keyword.pre* and *text.post>keyword.post.*

The final view *keyword* contains the request information of the original XPath. We can notice the roles of the (*pre, post, parent*) encoding and DTD in X2S. The *pre/post* encoding can be used to deal with the {//} operator; the *parent* encoding can be used to deal with the {/} operator; DTD can be used to rewrite the XPath into logical equivalent XPath expressions.

4.2 Basic XPath Translation Templates

Basic XPath translation templates can be used to translate logical equivalent XPath segment expressions into SQL in X2S. They are also used to translate XPath query without DTD directly.

The operator in the XPath segment includes {//,/,*}. Taking translation strategy in a bottom-up fashion as an example, the *translation context* of current node n includes the relation n_r related with node n and the view definition c_v on the child node c of n. The templates for SQL translation n_c on the current node can be described as follows:

As for the wildcards {*} labeled node, the relation n_r can be determined by the SQL statement in the form "select * from t union ...", where t is a relation mapped from any possible element in XML;

In the case of the parent/child relationship between node n and node c, we generate the SQL n_c in the form of "select $n_r.*$ from n_r, c_v where $c_v.parent= n_r.pre$";

In the case of the ancestor/descendant relationship between node n and node c, we generate the SQL n_c in the form of "select n_r.* from n_r, c_v where $c_v.pre> n_r.pre$ and $c_v.post< n_r.post$".

We iteratively apply the templates on the XPath segment until all the element nodes have been processed. The SQL annotated on the last processed node contains all the information of this XPath segment.

4.3 XPath Translation in the Presence of DTD

Although we can generate SQL from the XPath segment according to the templates above, the preliminary experimental results indicate three kinds of operations lead to the low efficiency in the generated SQL: the view containing all possible relations introduced by wildcards in the XPath; the range query join introduced by the ancestor/descendant relationship on pre/post encoding; the number of multi views join operations related with the size of the XPath. In the following, we aim at the improving the efficiency of the generated SQL by handle these problems with DTD.

4.3.1 The Production of the DTD Automata and XPath Automata

The first problem is how to exploit the constraints in the DTD to remove the wildcards and the ancestor/descendant relation in the XPath. Since tree automata are the fundamental device to present both XPath and DTD, we define the production operation of the automata to generate the optimization form of the XPath and DTD. The construction of tree automata from XPath or DTD is trivial. We only notice that the wildcards in the XPath is mapped into the transition rule which accepts the 'any' element.

Given XPath tree automata $PT=(Q_1,\Sigma_1,\delta_1,F_1,S_1)$ and DTD tree automata $DT=(Q_2,\Sigma_2,\delta_2,F_2, S_2)$, the production of XPath tree automata and DTD tree automata can be defined by $T_3=(Q_3,\Sigma_3,\delta_3,F_3,S_3)$, where $S_3=S_1\times S_2$, $Q_3=Q_1\times Q_2$, $F_3= F_1\times Q_2$, $\delta_3=\delta_1\times\delta_2$, where $\delta_1\times\delta_2=\{(q_{11},q_{21})\text{-}r_a\rightarrow(q_{12},q_{22})|(q_{11}\text{-}q_a\rightarrow q_{12})\in \delta_1,(q_{21}\text{-}d_a\rightarrow(f(q_{21'},....,q_{2m'}))\in \delta_2$, q_{22} is an state in state expression $f(q_{21'},....q_{2m'})$, $r_a=q_a\cap d_a\}$. The production is termed as $PT\times DT$. We omit the proof for the space reasons.

4.3.2 Elimination of the Wildcards and the Ancestor/Descendant Relationship

Due to the fact the elements in any transition rule in the automata from DTD are explicit, the transition rule in the production of the automata from XPath and automata from DTD will not accept "any" element according to the production construction rule. At the same time, the relation of elements in connected transition rules can be regarded as the parent/child relation. In the automata production process, we can build a mapping M from the elements in the XPath to the transition rules in the production. The mapping can be used to remove of wildcards and ancestor/descendant relationship in the XPath segment.

With the mapping M, we can locate the transition rules corresponding with "wildcards" element. We then replace the wildcards with the element(s) in the transition rule(s). In this way, we can eliminate the wildcards in the XPath.

Without loss of the generality, we assume that XPath expression takes the form of $p="a//b"$. We locate transition rule r_1 and transition rule r_2 mapped from element a and element b respectively. We then generate the language L accepted by transition rules sequence T from transition rule r_1 to r_2 in the case that there is no transition cycle in T. The XPath expression p can be replaced by L. We omit the proof for the space reason.

4.3.3 Shorten the Length of the XPath

Given XPath $p_1=/n_1/.../n_k$, the number of the join operations in the final SQL is related with the length of the XPath, which has great impact on the efficiency of the final SQL.

We observe that when any path from node n_1 to node n_k in the XML document satisfies expression p_1, p_1 can be converted into $p_2=/n_1//n_k$, which can be translated into SQL with one join opration.

The basic problem behind is XPath equivalence in the presence of DTD. Taking the above example, if p_1 and p_2 are equivalent with DTD, we can replace p_1 with p_2, which is shorter than p_1, hence to reduce the join operations in the final SQL.

As for the XPath with N nodes, we can shorten the XPath after the $O(N^2)$ equivalence test. As [16] shows, the equivalence of XPath $\{/,//,*\}$ in the presence of DTD can be detected in PTIME. Hence the XPath shortening can be implemented efficiently.

4.3.4 Query Cost Estimation Model

Notice that the reduction of the range query joins and the shortening of the total size are contradict in some way. Suppose that an XPath segment $s_1="/a/b/c=9"$, we may get another equivalent XPath segment $s_2="/a//c=9"$ in the presence of DTD. We translate XPath segment s_1 and s_2 into q_1 and q_2 respectively. Although the join operations in the SQL q_2 are less than that in the query q_1, q_2 contains a range query join while q_1 does not.

We can generate a SQL with the templates discussed above on the different logically equivalent XPath expressions. Next we select one SQL with the lowest cost based on the estimated cost of possible operation in the final SQL:

$C_{rangejoin}$ is the cost of range query join based on the pre/post encoding. The estimated cost for range join of relation r_1 and relation r_2 is denoted as $MIN(|r_1|, |r_2|)*\log(MAX(|r_1|,|r_2|))$, where $|r_1|$ is the size of the relation r_1.

$C_{pointjoin}$ is the cost of point query join based on the parent encoding. The estimated cost for point join of relation r_1 and relation r_2 is denoted as $Max(|r_1|, |r_2|)$;

C_{union} is the cost of union operation. The estimated cost of union of relation r_1 and relation r_2 is denoted as $(|r_1|+|r_2|)$;

C_{select} is the cost of selection operation. The cost for select operation of selection is denoted as $|r_1|$.

Given a query, we can estimate the query cost in a linear time. It is something like the cost based optimization strategy in RDBMS. Someone wonder why not let DBMS determine the estimated cost. This is because that only one SQL statement will be sent to the backend of DBMS.

4.3.5   XPath Segment Translation Algorithms with DTD

In this section, we combine the techniques above to translate an XPath segment into efficient SQL queries.

Procedure 1: translateXPathSegment(S,D)

   Input: the XPath fragment S, DTD D

   Output: the SQL q for the fragment

   1. Make a production operation on automata from XPath segment S and DTD D and create mapping M from the XPath to the production;
   2. Determine the explicit element(s) for wildcards in XPath segment S with mapping M and rewrite S into S_1;
   3. Rewrite S_1 into S_2 to remove the ancestor/descendant relationship with mapping M if the corresponding transition rules sequence does not contain a cycle;
   4. Shorten S_1 into S_3 with XPath equivalence test;
   5. Generate SQL q_1 from S_2 and q_2 from S_3;
   6. Select the query with the lower cost from q_1 and q_2 with the estimated cost model;

4.4   XPath Translation Algorithms with DTD

After the SQL generation for each segment of XPath, the next step is to combine all the queries into one for the entire XPath tree pattern. As the discussion above, the translation on the main path takes a top-down fashion, while the translation on the predicate paths takes a bottom-up fashion. The whole process can be described as follows:

Procedure 2: Translate the entire XPath tree pattern

   Input: XPath p, DTD D

   Output: the equivalent SQL for p

   1. Create an XPath tree pattern T for p;
   2. Decompose XPath tree pattern T into a set of XPath segment;
   3. Let S be XPath segment containing the root element;
   4. While (S!=null)
   5.   {Translate XPath segment S with Proc 1 in a top-down fashion;

6.   If end node n in XPath segment S is a branching node, then generate SQLs on the predicate paths with Proc 1 in a bottom up way;

7.   Merge the SQLs from the predicate paths and main path into SQL s_n, which is annotated on node n;

8. S is the next non-annotated XPath segment}

9.   return the SQL annotated on the return node of XPath tree pattern T.

Step 6 generates the SQL on the predicate path in the XML tree pattern. The translated SQL contains all conditions in predicate paths. Step 7 merges SQL queries from different predicate paths with main path by the join operation. The last node processed is the return node in the main path, which reflexes all predicate expressions in the XPath tree pattern.

5   Performance Study

We used Xmark [12] to generate XML documents with factor from 0.2 to 1.2 with each increase step 0.2. The size of the generated XML documents ranges from 23.3M(17,132 nodes) to 139.3M (100,456 nodes). We decompose XML documentts into DB2 V8.1 for Windows 2000 running on a Dell Optiplex GX260 with P4 2GHz CPU and 512MB RAM. Each type of elements is converted into a relation with three fields (*pre*, *post*, *parent*). Two unique indices were created on the pre and post fields. A cluster index was created on the parent field.

The XPath test set is generated by path generator [13]. We set different possibility ratio of operators {[],*,//} as {5%,5%,5%} and {20%,20%,20%}, and generate the different XPath sets with size of 100. The maximal length of the XPath is 5. We implement three translation methods. The basic method implements translation according to the basic templates in section 4.2, which can be treated as the schema oblivious translation method. The prefix method makes the optimization on the prefix of the XPath with DTD, which takes the similar idea as that in [7]. The X2S is the method described in this paper. The results can be illustrated as in Fig. 4.

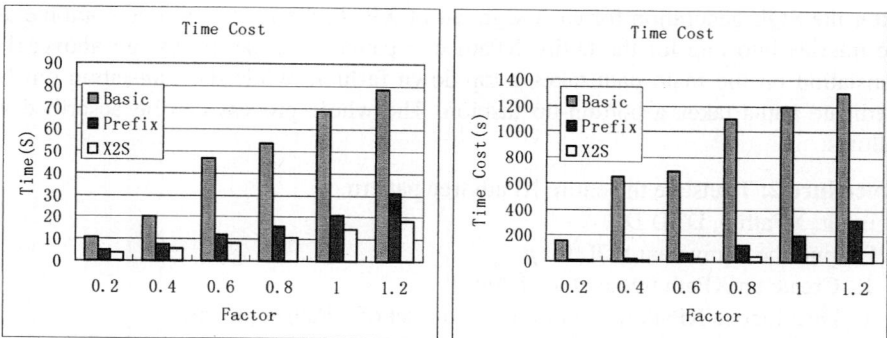

Fig. 4. The efficiency of translated SQL from XPath with {//,*,[]} possibility ratio as {5%,5%,5%} and {20%,20%,20%}

The X-axis denotes the factor used in the XMark and Y-axis denotes the time cost of the whole query set in Fig.4. We notice that the evaluation cost of XPath with a higher ratio of {//,*,[]} is much higher than that of XPath with a lower ratio, which is demonstrated by comparing the left figure with the right figure.

We observe that the efficiency of the SQL generated from X2S is much higher than that of others with the increase the possibility ratio of {//,*,[]}. This is because X2S makes optimization on all XPath segments, and prefix method makes optimization only on the prefix of the XPath.

We also notice that the difference between the SQL statements from prefix and X2S increases with the increase of {//,[],*} possibility ratio in the query set. This is due to that the prefix works on the nearly entire XPath when the {[]} ratio is lower.

6 Conclusion

In this paper, we propose a method called X2S to translate XPath into efficient equivalent SQL queries when RDBMS is used to host XML documents. X2S is based on the extended relational encoding and can exploit DTD to further improve the efficiency of the generated SQL. The experiments running on top of the commercial database demonstrate the effectiveness of our method.

There are a number of issues for the future research. The trade off between the space cost introduced by *pre/post/parent* encoding and efficiency improvement will be further studied. Another interesting, but more challenging issue is to translate XQuery to SQL based on the *pre/post/parent* encoding.

Acknowledgement

I would like to thank Prof Luo Qiong for the invitation to visit HKUST and very helpful discussions on this paper.

References

1. D.Dehaan, D.Toman, M.P.Consens, M.Tamer Ozsu. A Comprehensive XQuery to SQL Translation Using Dynamic Interval Encoding. In Proc. of SIGMOD, 2003, pp 623-634.
2. D.Florescu,D.Kossman, A Performance Evaluation of Alternative Mapping Schemes for Storing XML Data in a Relational Database. In Rapport de Recherche No.3680 INRIA, Rocquencourt, France, May 1999.
3. J.Shanmugasundaram, H.Gang, K.Tufte, C.Zhang, D.Dewitt, J.Naughton, Relational Database for Querying XML Documents: Limitation and Opportunities. In Proc. of VLDB, 1999, pp 302-314.
4. T.Grust, M.Keulen, J.Teubner. Teach a relational DBMS to watch its step. In Proc. of VLDB, 2003, pp 524-535.

5. R.Krishnamurthy, R.Kaushik, J.Naughton, V.Chakaravarthy. Recursive XML Schema, Recursive XML Queries and Relational Storage: XML-to-SQL Query translation. In Proc. of ICDE, 2004.
6. R.Krishnamurthy, R.Kaushik, J.Naughton: XML-SQL Query Translation Literature: The State of the Art and Open Problems. In Proc. of Xsym, 2003, pp 1-18.
7. R.Krishnamurthy, R.Kaushik, J.Naughton: Efficient XML-to-SQL Query Translation: Where to Add the Intelligence? In Proc. of VLDB, 2004.
8. J.Shanmugasundaram, J.Kiernan, E.Shekita, C.Fan and J.Funderburk. Querying XML Views of Relational Data. In Proc. of VLDB, 2001, pp 261-270.
9. J.Clark. XML Path language(XPath), 1999. Available from the W3C, http://www.w3.org/TR/XPath.
10. World Wide Web Consortium. XSL Transformations (XSLT) version 1.0. http://www.w3.org/TR/XLST, November 1999, W3C Recommendation.
11. World Wide Web Consortium. Xquery version 1.0. http://www.w3.org/TR/xquery, January 2001, W3C working draft.
12. XMark www.xml-benchmark.org
13. C.Chan, P.Felber, M.Garofalakis and R.Rastogi. Efficient Filtering of XML Document With XPath Expressions. In Proc. of ICDE, 2002, pp 235-244.
14. G.Miklau and D.suciu. Containment and Equivalence for an XPath Fragment. In Proc.of PODS, 2002, pp 65-76.
15. SAX, http://xml.apache.org
16. F.Neven, T.Schwentick: XPath Containment in the Presence of Disjunction, DTDs, and Variables. In Proc. of ICDT, 2003, pp 315-329.

On Aggregation Operators
for Fuzzy Information Sources

Ai-Ping Li and Quan-Yuan Wu

613#, School of Computer, National University of Defense Technology
Changsha 410073, P.R. China
apli@x263.net

Abstract. During the development of fuzzy information systems, the aggregation fuzzy information sources should be considered. Aggregation operators are the key technique that solve this problem. By assign 0.5 to the generalized triangular norm identy, the harmonic triangular norm operators and strict harmonic triangular norm operators is defined. The properties of the aggregation operators is discussed. We prove the continuous harmonic triangular norm operator does not exist. The operators when the information sources have weights are also given. Four type detail (strict) harmonic triangular norm operators are given. Finally, two measures associated with the operators are defined, one is marginal measure and the other is sensitive measure. Based on the measure, the relationships of those operators and classical operators are evaluated and compared. The result shows that the operators offer a new method to aggregation of fuzzy information sources in fuzzy environments.

1 Introduction

Information fusion techniques and aggregation operators are commonly applied into several fields of fuzzy information systems. Because different fields imply different requirements, a large number of aggregation operators exist today. For example, triangular norm (T-norms)[1] operator, triangular conorm (T-conorms)[1] operator, weighted mean(WM)[2] operator, Ordered Weighted Averaging(OWA)[3] operator, Ordered Weighted Geometric(OWG)[4] operator, Weighted OWA (WOWA)[5] operator and so on.

For T-norm and T-conorm are emphasized particularly on local detail information source, it may lose the global information. WM operator consider the global information, it will lose the local source information. The OWA operator (operator defined by Yager defined in Ref. 3; see also other similar operators in Ref. 4,5) generalized minimum and maximum and WM operators through permutation the truth-value of information source. But it doesn't solve considering the source global and local information at the same time. Besides the common relationship of fuzzy information sources, which is conjunction, disjunction and weighted average, there is an important weak logic relation[8]. Because the existent operators have not different aggregation method when the information source are all trend to true or all trend to false. For example, there is such situation, when the two source are trend to true, the aggregated value is not less than either of the source truth value; otherwise, when the two source

S. Wang et al. (Eds.): ER Workshops 2004, LNCS 3289, pp. 223–233, 2004.

are trend to false, the aggregated value is not greater than either of the source truth value either. When one source is trend to true and the other is trend to false, the aggregation value is between them, that is weak logic relation[8].

When define the identy of generalized triangular norm to 0 or 1, we can get T-norms or T-conorms. We find when define the identy of generalized triangular norm to 0.5, we can get a new kind of aggregation operators, we named it to harmonic triangular norm. It can aggregate the source when they have weak logic relation, and it can solve considering both the information sources' global property and local property. If strengthen the monotonic in the definition, we defined strict harmonic triangular norm.

2   Definition and Properties of Harmonic Triangular Norm

2.1   Two-Dimensional Operator

Definition 1[1]. (Generalized Triangular Norm) A Mapping T: $[0,1]\times[0,1]\rightarrow[0,1]$ is a generalized triangular norm iff it satisfies:

(t.1) $T(0,0) = 0, T(1,1) = 1$

(t.2) $T(x,y) = T(y,x)$

(t.3) $(x \leq x', y \leq y') \Rightarrow T(x,y) \leq T(x',y')$

(t.4) $T(T(x,y),z) = T(x,T(y,z))$.

If the generalized triangular norm T can satisfy:

(t.1') $T(x,1) = x$

then T is triangular norm(T-norm for short); If the generalized triangular norm T can satisfy:

(t.1") $T(x,0) = x$

then T is triangular conorm(T-conorm for short).

Obviously, (t.1') or (t.1") => (t.1).

When the generalized triangular norm, T-norm and T-conorm operator is continuous function, we call them continuous generalized triangular norm, continuous T-norm and continuous T-conorm operator respectively.

Definition 2. (Harmonic Triangular Norm) A Mapping T: $[0,1]\times[0,1]\rightarrow[0,1]$ is a generalized triangular norm, when it still satisfies:

(t.5) $T(x,0.5) = x$,

we call T is Harmonic Triangular Norm(H-norm for short).

Proposition 1. *H-norm T: $[0,1]\times[0,1]\rightarrow[0,1]$ satisfies:*

$T(x,y) \geq x$ and $T(x,y) \geq y$, if $x,y \in (0.5,1]$;

(t.6) $T(x,y) \leq x$ and $T(x,y) \leq y$, if $x,y \in [0,0.5)$;

$T(x,y) \geq x$ and $T(x,y) \leq y$, if $x \in [0,0.5), y \in (0.5,1]$.

Proof. For T is associative(t.2) and non-decreasing(t.3), then:

if $x,y \in (0.5,1]$, then

$T(x,y) \geq T(x,0.5) = x$ and $T(x,y) \geq T(0.5,y) = y$;

if $x,y \in [0,0.5)$, then

$T(x, y) \le T(x, 0.5) = x$ and $T(x, y) \le T(0.5, y) = y$ ；

if $x \in [0, 0.5), y \in (0.5, 1]$, then

$T(x, y) \ge T(x, 0.5) = x$ and $T(x, y) \le T(0.5, y) = y$. ∎

If the relation in (t.6) is strict, we can get strict harmonic triangular norm definition.

Definition 3. (Strict Harmonic Triangular Norm) *A Mapping T: $[0,1] \times [0,1] \to [0,1]$ is a H-norm, when it still satisfies:*

$T(x, y) > \max(x, y)$, if $x, y \in (0.5, 1)$;

(t.6') $T(x, y) < \min(x, y)$, if $x, y \in (0, 0.5)$;

$T(x, y) > x$ and $T(x, y) < y$, if $x \in (0, 0.5), y \in (0.5, 1)$.

we call T is Strict Harmonic Triangular Norm(Strict H-norm for short).

The semantic explanation of t.6 is as follows:

1) $T(x, 0.5) = x$, Any fuzzy value aggregate with 0.5(unknown true or false) is itself;

2) $T(x, y) \ge x$ and $T(x, y) \ge y$, if $x, y \in (0.5, 1]$, The aggregate value of two fuzzy value is greater than any of them if both trend to true;

3) $T(x, y) \le x$ and $T(x, y) \le y$, if $x, y \in [0, 0.5)$, The aggregate value of two fuzzy value is less than any of them if both trend to false;

4) $T(x, y) \ge x$ and $T(x, y) \le y$, if $x \in [0, 0.5), y \in (0.5, 1]$, The aggregate value of two fuzzy value is between them if one is trend to true and the other is trend to false.

Obviously, H-norm satisfies the weak logic relation in Ref. 8.

Theorem 1. *Continuous H-norm does not exist.*

To prove this theorem we need to prove the following lemma first.

Lemma 1. Any continuous generalized triangular norm can be expressed as follows: for $\forall x, y \in [0, 1]$,

$$T(x, y) = \begin{cases} T_\vee(x, y), & \text{if } (x, y) \in [0, \lambda] , \\ T_\wedge(x, y), & \text{if } (x, y) \in (\lambda, 1] , \\ \lambda , & \text{otherwise.} \end{cases}$$

where $\lambda = T(0, 1), T_\vee, T_\wedge$ is continuous T-conorm and T-norm.

Proof. When T is continuous T-conorm or T-norm, the lemma obviously holds. When T is continuous generalized triangular norm and $\lambda \ne 0$ or 1, we construct T-conorm $T_\vee$ and T-norm $T_\wedge$ as follows:

$$T_\vee(x, y) = \begin{cases} T(x, y), & \text{if } x, y \in [0, \lambda] , \\ \max(x, y), & \text{otherwise.} \end{cases}$$

$$T_\wedge(x, y) = \begin{cases} T(x, y), & \text{if } x, y \in (\lambda, 1] , \\ \min(x, y), & \text{otherwise.} \end{cases}$$

Let's prove $T_\vee$ is continuous T-conorm. It is obviously $T_\vee$ is generalized triangular norm and continuous. We need prove the identy of $T_\vee$ is zero. When $x \in [\lambda, 1]$, $T_\vee(x, 0) = \max(x, 0) = x$; when $x \in [0, \lambda]$, $T_\vee(x, 0) = T(x, 0)$. Let $f(x) = T(x, 0)$, then $f(0) = T(0, 0) = 0$, $f(\lambda) = T(\lambda, 0) = T(T(0, 1), 0) = T(T(0, 0), 1) = T(0, 1) = \lambda$. Thus $f(x)$ is continuous function, which both range of values and range of definition are in $[0,\lambda]$. From intermediate value theorem we can get $\exists \zeta \in [0,\lambda]$, $T(x,0) = T(T(\zeta,0),0)$ $= T(\zeta,T(0,0)) = T(\zeta,0) = x$, so the identy of $T_\vee$ is zero. Similarly, we can prove $T_\wedge$ is continuous T-norm.

Now let's prove $\forall x \in [0,\lambda], \forall y \in [\lambda,1]$, $T(x,y) = \lambda$. In fact, $= T(0,T(0,1)) = T(0,\lambda)$ $= \lambda \le T(x,y)$ $\lambda = T(0,1) = T(T(0,0),1) \le T(\lambda,1) = T(T(0,1),1) = T(0,T(1,1)) = T(0,1) = \lambda$.

So, Lemma 1 holds.

Using Lemma 1 we can prove Theorem 1 in the following way.

Suppose there exists continuous H-norm $\tilde{T}$. For $\tilde{T}$ is generalized triangular norm, so it can be expressed as follows:

$$\tilde{T}(x,y) = \begin{cases} T_\vee(x,y), & \text{if } (x,y) \in [0,\lambda], \\ T_\wedge(x,y), & \text{if } (x,y) \in (\lambda,1], \\ \lambda, & \text{otherwise.} \end{cases}$$

where $\lambda = T(0,1), T_\vee, T_\wedge$ is continuous T-conorm and T-norm.

If $\lambda = 0$ or 1, then $\tilde{T}$ is T-conorm or T-norm, obviously does not hold.

If $\lambda \ne 0$ or 1, then when $(x,y) \in [0,\lambda]$, $\tilde{T}(x,y) = T_\vee(x,y)$, $\forall \xi \in (0,\lambda]$ then $\tilde{T}(\xi,0) = T_\vee(\xi,0) = \xi > 0$, this conflicts with Proposition 1. So Continuous H-norm does not exist. ∎

2.2 Multi-dimensional H-Norm Operator

To satisfy more than two fuzzy information sources, we need multi-dimensional aggregation operator, let's introduce multi-dimensional Harmonic Triangular Norm.

First, let us introduce partial order relationship" $\le$ " in $[0,1]^m$:
$\forall X = (x_1, x_2, \cdots, x_m), \forall Y = (y_1, y_2, \cdots, y_m) \in [0,1]^m$, $X \le Y$ iff $(\forall j)(x_j \le y_j), j \in 1, \cdots, m$.

Four conversions, $p_i, q_i, r_i, \sigma_{ij} : [0,1]^m \to [0,1]^m$:

$p_i(X) \square (1, \cdots, 1, x_i, 1, \cdots, 1)$

$q_i(X) \square (0, \cdots, 0, x_i, 0, \cdots, 0)$

$r_i(X) \square (0.5, \cdots, 0.5, x_i, 0.5, \cdots, 0.5)$

$\sigma_{ij}(X) = \sigma_{ij}(x_1, \cdots, x_i, \cdots, x_j, \cdots, x_m) \square (x_1, \cdots, x_j, \cdots, x_i, \cdots, x_m)$

Definition 4. (m-Dimensional Generalized Triangular Norm) *A Mapping* $T_m : [0,1]^m \to [0,1]$ *is a m-dimensional generalized triangular norm iff it satisfies:*

(T.1) $T_m(0, \cdots, 0) = 0$, $T_m(1, \cdots, 1) = 1$

(T.2) $T_m(\sigma_{ij}(X)) = T_m(X)$

(T.3) $X \le Y \Rightarrow T_m(X) \le T_m(Y)$

(T.4)
$\forall X = (x_1, \cdots, x_m), \forall Y = (y_m, \cdots, x_{2m-1})$, $T_m((T_m(X), x_{m+1}, \cdots, x_{2m-1}) = T_m(x_1, \cdots, x_{m-1}, T_m(Y))$.

Obviously T_2 is a 2-dimensional generalized triangular norm. If T_m still satisfies:

(T.1') $T_m(p_i(X)) = x_i$

then T_m is a m-dimensional triangular norm(m-dimensional T-norm for short);

If T_m still satisfies:

(T.1'') $T_m(q_i(X)) = x_i$

then T_m is a m-dimensional triangular conorm(m-dimensional T-conorm for short).

When the m-dimensional generalized triangular norm, m-dimensional T-norm and m-dimensional T-conorm operator is continuous function, we call them continuous m-dimensional generalized triangular norm, continuous m-dimensional T-norm and continuous m-dimensional T-conorm operator respectively.

Definition 5. (m-Dimensional Harmonic Triangular Norm) *A Mapping $T_m : [0,1]^m \to [0,1]$ is a m-dimensional generalized triangular norm, when it still satisfies:*

(T.5) $T_m(r_i(X)) = x_i$,

we call Tm is m-dimensional Harmonic Triangular Norm(m-dimensional H-norm for short).

Similarly in proposition 1, we can prove the follow proposition.

Proposition 2. *m-dimensional H-norm $T_m : [0,1]^m \to [0,1]$ satisfies:*

When $r_i(X) \neq X$;

$T_m(x_1, \cdots, x_m) \geq \max_i(x_i)$ if $\min_i(x_i) \geq 0.5$;

(T.6) $T_m(x_1, \cdots, x_m) \leq \min_i(x_i)$ if $\max_i(x_i) \leq 0.5$;

$T_m(x_1, \cdots, x_m) \geq x_i$ if only one $x_i \in (0, 0.5)$;

$T_m(x_1, \cdots, x_m) \leq x_j$ if only one $x_j \in (0.5, 1)$.

Similarly in definition 3, we can get strict definition of m-dimensional H-norm.

Definition 6.(Strict m-dimensional H-norm) *A Mapping $T_m : [0,1]^m \to [0,1]$ is a m-dimensional H-norm, when it still satisfies:*

When $r_i(X) \neq X$;

$T_m(x_1, \cdots, x_m) > \max_i(x_i)$ if $\min_i(x_i) \geq 0.5$ and $\max_i(x_i) < 1$;

(T.6') $T_m(x_1, \cdots, x_m) < \min_i(x_i)$ if $\max_i(x_i) \leq 0.5$ and $\min_i(x_i) > 0$;

$T_m(x_1, \cdots, x_m) > x_i$ if only one $x_i \in (0, 0.5)$;

$T_m(x_1, \cdots, x_m) < x_j$ if only one $x_j \in (0.5, 1)$.

we call T_m is m-dimensional Strict Harmonic Triangular Norm(Strict m-dimensional H-norm for short).

Additionally, if m-dimensional H-norm Tm still satisfies:

(T.7) *Exist multi-dimensional H-norm*

$T_r, T_{m-r}, T_2, 2 \leq m < r, T_m(X) = T_2(T_r(x_1, \cdots, x_r), T_{m-r}(x_{r+1}, \cdots, x_m))$,

then m-dimensional H-norm T_m is decomposed;

If T_m still satisfies: Exist H-norm T2,

(T.8) $T_m(X) = T_2(T_2(\cdots T_2(T_2(x_1, x_2), x_3) \cdots, x_{m-1}), x_m)$,

then m-dimensional H-norm T_m is totally decomposed.

Certainly, we can get an m-dimensional H-norm from a H-norm T2. Like the proof in Theorem 1, we can prove the next theorem.

Theorem 2. *Continuous multi-dimensional H-norm does not exist.*

2.3 Weighted H-Norm Operator

For expressing importance of individual information sources, we will discuss weighted H-norm. Generically, the lager the weight, the more important is the source on determining the aggregated value.

Definition 7. *A vector* $v = (w_1, w_2, \cdots, w_m)$ *is a weighting vector of dimension N if and only if* $w_i \in [0,1]$ *and* $\sum_i w_i = 1$.

Definition 8. (Weighted m-dimensional H-norm) *A Mapping* $T_m : [0,1]^m \to [0,1]$ *is a m-dimensional H-norm, v is a weighting vector, then:*

$T_m^w(x_1, \cdots, x_m) = T_m(x_1^w, \cdots, x_m^w)$, *where*

$$
x_i^w = \begin{cases} 0.5 + (x_i - 0.5) \times \dfrac{w_i}{w_{max}}, & if \ x_i \geq 0.5, \\[2mm] 0.5 - (0.5 - x_i) \times \dfrac{w_i}{w_{max}}, & if \ x_i < 0.5. \end{cases} \quad , w_{max} = \max_i \{w_i\}
$$

is a weighted m-dimensional H-norm.

3 Four Type of m-Dimensional H-Norm

In this section we will give the detail H-norm operator, and explain the method of construct H-norm operators. For H-norm is the instance of multi-dimensional H-norm, we will discuss m-dimensional H-norm.

Theorem 3. $\forall X = (x_1, \cdots, x_m) \in [0,1]^m$, *A Mapping* $T_m^{min} : [0,1]^m \to [0,1]$:

$$
T_m^{min}(X) = \begin{cases} x_i, & if \ X = r_i(X); \\[1mm] \max_i(x_i) + \lambda \cdot (1 - \max_i(x_i)), & if \ X \in [0.5,1]^m; \\[1mm] \eta \cdot \min_i(x_i), & if \ X \in [0,0.5]^m; \\[1mm] \min_i(x_i), & otherwise. \end{cases}
$$

Where $\lambda, \eta \in \{0,1\}, i \in \{1, \cdots, m\}$.

Then T_m^{min} *is m-dimensional H-norm.*

For shorten the paper, the proof is omitted, the following proof of the theorem also is omitted.

Obviously, Operator T_m^{min} has these properties:

1) $T_2^{\min}(0,1) = 0$.

2) $T_m^{\min}$ is totally decomposed.

From Theorem 3, when assign $0,1$ to λ, η respectively; we can get four m-dimensional H-norms.

Theorem 4. $\forall X = (x_1, \cdots, x_m) \in [0,1]^m$, *A Mapping* $T_m^{\max} : [0,1]^m \rightarrow [0,1]$:

$$
T_m^{\max}(X) = \begin{cases}
x_i, & \text{if } X = r_i(X); \\
\max_i(x_i) + \lambda \cdot (1 - \max_i(x_i)), & \text{if } X \in [0.5,1]^m; \\
\eta \cdot \min_i(x_i), & \text{if } X \in [0,0.5]^m; \\
\max_i(x_i), & \text{otherwise.}
\end{cases}
$$

Where $\lambda, \eta \in \{0,1\}, i \in \{1, \cdots, m\}$.

Then $T_m^{\max}$ *is m-dimensional H-norm.*

As the same, Operator $T_m^{\max}$ has these properties:

1) $T_2^{\max}(0,1) = 1$.

2) $T_m^{\max}$ is totally decomposed.

From Theorem 4, when assign $0,1$ to λ, η respectively; we also can get four m-dimensional H-norms.

Now let's discuss strict m-dimensional H-norms.

Theorem 5. $\forall X = (x_1, \cdots, x_m) \in [0,1]^m$, *A Mapping* $T_m^{\Pi} : [0,1]^m \rightarrow [0,1]$:

$$
T_m^{\Pi}(X) = \begin{cases}
\lambda, & \text{If } \min_i(x_i) = 0 \text{ and } \max_i(x_i) = 1; \\
\dfrac{\prod\limits_{i=1}^{m} x_i}{\prod\limits_{i=1}^{m} x_i + \prod\limits_{i=1}^{m}(1 - x_i)}, & \text{Otherwise.}
\end{cases}
$$

Where $\lambda \in \{0,1\}$. *Then* T_m^{Π} *is Strict m-dimensional H-norm.*

Obviously, Operator T_m^{Π} has these properties:

1) $T_2^{\Pi}(0,1) = 0$ or 1.

2) T_m^{Π} is totally decomposed.

From Theorem 5, when assign $0,1$ to λ respectively; we can get two m-dimensional H-norms.

For the operators defined in Theorem 3,4,5, $T_2(0,1) = 0$ or $T_2(0,1) = 0$, i.e. False (truth-value is 0) and True(truth-value is 1) aggregated value is 0 or 1. Although they can be applied in many fields, there maybe such demand, $T_2(0,1) = 0.5$, i.e.

False (truth-value is 0) and True (truth-value is 1) aggregated value is 0.5 (unknown). Now let's focus on this H-norm operator.

First define some notations. When $\exists l, g \in \{1, \cdots, m\}$, $x_l \leq 0.5, x_g > 0.5$, define conversion $s : [0,1]^m \rightarrow [0,1]^m$:

$s(x_1, \cdots, x_m) = (x_1 ', \cdots, x_{k-1} ', x_k ', \cdots, x_m ')$, Satisfies:

i) $x_i ' \leq x_{i+1} ', i \in \{1, 2, \cdots, m-1\}$;

ii) $x_k ' > 0.5$, and $x_{k-1} ' \leq 0.5$.

Table 1. Comparison of each operator with classical operator

	min	max	Mean	$T^{\min}$				$T^{\max}$				T^{Π}		T^{*}
				$\lambda=0,$ $\eta=0$	$\lambda=0,$ $\eta=1$	$\lambda=1,$ $\eta=0$	$\lambda=1,$ $\eta=1$	$\lambda=0,$ $\eta=0$	$\lambda=0,$ $\eta=1$	$\lambda=1,$ $\eta=0$	$\lambda=1,$ $\eta=1$	$\lambda=0$	$\lambda=1$	
(0,0.2)	0	0.2	0.1	0	0	0	0	0	0	0	0	0	0	0
(0,0.5)	0	0.5	0.25	0	0	0	0	0	0	0	0	0	0	0
(0,0.7)	0	0.7	0.35	0	0	0	0	0.7	0.7	0.7	0.7	0	0	0
(0,1)	0	1	0.5	0	0	0	0	1	1	1	1	0	1	0.5
(0.2,0.3)	0.2	0.3	0.25	0	0.2	0	0.2	0	0.2	0	0.2	0.08	0.08	0.12
(0.2,0.5)	0.2	0.5	0.35	0.2	0.2	0.2	0.2	0.2	0.2	0.2	0.2	0.2	0.2	0.2
(0.2,0.8)	0.2	0.8	0.5	0.2	0.2	0.2	0.2	0.8	0.8	0.8	0.8	0.5	0.5	0.5
(0.3,0.9)	0.3	0.9	0.6	0.3	0.3	0.3	0.3	0.9	0.9	0.9	0.9	0.75	0.75	0.83
(0.2,1)	0.2	1	0.6	0.2	0.2	0.2	0.2	1	1	1	1	1	1	1
(0.5,0.7)	0.5	0.7	0.65	0.7	0.7	0.7	0.7	0.7	0.7	0.7	0.7	0.7	0.7	0.7
(0.5,1)	0.5	1	0.75	1	1	1	1	1	1	1	1	1	1	1
(0.7,0.8)	0.7	0.8	0.75	0.8	0.8	1	1	0.8	0.8	1	1	0.9	0.9	0.88
(0.7,1)	0.7	1	0.85	1	1	1	1	1	1	1	1	1	1	1

Theorem 6. $\forall X = (x_1, \cdots, x_m) \in [0,1]^m$, *A Mapping*

$$T_m^{\Pi} : [0,1]^m \rightarrow [0,1] : T_m^{*}(X) = \begin{cases} 1 - 2^{m-1} \cdot \prod_{i=1}^{m} (1 - x_i); & \text{If } X \in [0.5, 1]^m; \\ 2^{m-1} \cdot \prod_{i=1}^{m} x_i; & \text{If } X \in [0, 0.5]^m; \\ \dfrac{\varphi + \gamma - 1}{1 - \min(1 - 2\varphi, 2\gamma - 1)} + \dfrac{1}{2}; & \text{Otherwise.} \end{cases}$$

Where, $\varphi = 2^{k-2} \cdot \prod_{i=1}^{k-1} x_i '$, $\gamma = 1 - 2^{m-k} \cdot \prod_{i=k}^{m} (1 - x_i ') + 1$,

$X ' = s(X) = (x_1 ', \cdots, x_{k-1} ', x_k ', \cdots, x_m ')$, $x_k ' > 0.5$, $x_{k-1} ' \leq 0.5$. *Specially, when* $\varphi = 0$, $\gamma = 1$, *let* $T_m^{*}(X) = 0.5$.

Then T_m^{*} *is Strict m-dimensional H-norm.*

Obviously, Operator T_m^{*} has these properties:

1) $T_2^*(x, y) = 0.5$, when $x + y = 1$, specially, $T_2^*(0,1) = 0.5$.

2) T_m^* is totally decomposed.

We will compare the results of these 2-dimensional operators with minimum, maximum and mean operator (WM, weighted vector $v=(0.5,0.5)$) in $[0,1]^2$.

As shown in the table 1, minimum and maximum operator emphasize particularly on the local information of the sources, and that mean operator emphasize particularly on the global information of the sources. H-norms emphasize the local and global information of the sources at the same time. T^{min} and T^{max} can be considered as the different forms of minimum and maximum after they consider the global information of source. Strict H-norms have fairly good results after they consider both local and global information of the fuzzy sources. Thus, different H-norms can adapt to different applications in fuzzy information aggregation, can be have a good use in the situation when classical aggregation operator can't fit the demand of application.

4   Measures and Evaluation of Operators

In this section, we will define two measures of fuzzy aggregation operators, so we can choose appropriate operators in the applications.

Generally, if two aggregation operators T^1 and T^2 , if at point (x_0, y_0) there is $|T^1(x_0, y_0) - 0.5| > |T^2(x_0, y_0) - 0.5|$, we consider the marginality of T^1 is greater than T^2 at (x_0, y_0) .

Because generalized triangular norm is non-decreasing and bounded at $[0,1]^m$, it can be definite integrated at $[0,1]^m$ based on integration theorem.

Definition 9.(Marginal Measure) *The Marginal Measure of m-dimensional generalized triangular norm* $T^m : [0,1]^m \rightarrow [0,1]$ *is:*

$$M(T_m) = \iint_{x_i} {}_0^1 | T^m x_1, \cdots, x_m) - 0.5 | \, dx_i .$$

The meaning of above expression is the convolution integral of the absolute difference about T^m and 0.5 at each $x_i \in [0,1]$. Specially, when m=2, it is:

$$M(T_2) = \int_0^1 \int_0^1 | T^2(x, y) - 0.5 | dx dy .$$

Obviously, the greater the marginal measure of a operator is, the closer it is near by 0 or 1. The marginal measure of above operators in shown in Table 2.

Table 2. Comparison of each operator's marginal measure

	min	max	mean	T^{min}				T^{max}				T^{II}		T^*
				$\lambda=0,$ $\eta=0$	$\lambda=0,$ $\eta=1$	$\lambda=1,$ $\eta=0$	$\lambda=1,$ $\eta=1$	$\lambda=0,$ $\eta=0$	$\lambda=0,$ $\eta=1$	$\lambda=1,$ $\eta=0$	$\lambda=1,$ $\eta=1$	$\lambda=0$	$\lambda=1$	
$M(T)$	0.25	0.25	0.167	0.33	0.375	0.29	0.33	0.33	0.375	0.29	0.33	0.307	0.307	0.3125

Table 3. Comparison of each operator's sensitive measure

	min	max	mean	$T^{\min}$				$T^{\max}$				T^{Π}		T^*
				$\lambda=0,$ $\eta=0$	$\lambda=0,$ $\eta=1$	$\lambda=1,$ $\eta=0$	$\lambda=1,$ $\eta=1$	$\lambda=0,$ $\eta=0$	$\lambda=0,$ $\eta=1$	$\lambda=1,$ $\eta=0$	$\lambda=1,$ $\eta=1$	$\lambda=0$	$\lambda=1$	
$M(T)$	-0.17	-0.17	0	-0.21	-0.25	-0.29	-0.21	-0.17	-0.25	-0.17	-0.21	-0.141	-0.141	-0.146

As shown in Table 2, the marginal measure of every H-norms is greater than the measure of minimum, maximum and mean. The marginal measure of $T^{\min}$ and $T^{\max}$ is the maximum 0.375 when λ, η be assigned 0 or 1 respectively. And that two strict H-norms, the measure of T^* is a little greater than T^{Π}, the lowest is the measure of mean operator. In fact, the marginal measure is the measure of the operator's degree of considering local information of sources.

Next, we will introduce another measure, sensitive measure. When any of source changed, the aggregation value of mean operator will change, that is, its' considering global information of sources is very good. It is not the same with minimum and maximum operator. We hope the aggregation value change with the change of each information source. It will smoothly change from 0 to 1 in the interval $[0,1]^m$. In fact, the sensitive measure is the measure of the operator's degree of considering global information of sources.

Definition 10.(Sensitive Measure) *The Marginal Measure of m-dimensional generalized triangular norm* $T^m : [0,1]^m \to [0,1]$ *is:*

$$\mathbb{S}(T_m) = -\oiint_{x_i} {}_0^1 | T^m(x_1, \cdots, x_m) - \frac{1}{m} \sum_{i=1}^{m} x_i | \, dx_i .$$

The meaning of above expression is the negative convolution integral of the absolute difference about T^m and $\frac{1}{m} \sum_{i=1}^{m} x_i$ at each $x_i \in [0,1]$. Specially, when $m=2$, it is:

$$\mathbb{S}(T_2) = -\int_0^1 \int_0^1 | T^2(x, y) - \frac{1}{2}(x + y) | dx dy$$

Obviously, the greater the sensitive measure of the operator, the closer it is near by average plane, i.e. the degree of global information of sources is better.

As shown in Table 3, the sensitive measure of mean is the greatest, which accord with the actual situation. The measure of $T^{\min}$ and $T^{\max}$ is smaller, not very good at the degree of considering global information of sources, that is accord with these graph is not smooth. Two strict H-norms, T^* and T^{Π}, the sensitive measure is greater than minimum and maximum operators.

Compared with Table 2, we can see the marginal measure and sensitive measure is a pair of contrary measure, the bigger is one, the another will be smaller. Take into account two measures, the marginal measure and sensitive measure of T^* and T^{Π} are both better than minimum and maximum operators. So, T^* and T^{Π} are fairly good fuzzy aggregation operators.

5 Conclusions

This paper introduce a new kind of fuzzy aggregation operators in multicriteria decision environments. It consider both local and global information of sources, fetch up the lack of classical operators. It gives different forms when the sources are all trend to false or trend to true, solves the problem when the sources have weak logic relations. We give the definition and properties of H-norms and strict H-norms, also think about the weighted operators and multi-dimension operators. Then give four type of detail forms of multi-dimensional H-norms and strict H-norms. Then the measure and evolution of aggregation operators is introduced. The result shows the operators given in this article is fairly good. The H-norms can be used in multicriteria decision systems, fuzzy recognition systems, fuzzy control systems, fuzzy expert systems and multi-agent systems, it will have preferable theoretical and applied sense in the future.

References

1. Erich PK et al. Triangular norms, Kluwer Academic Publishers, 2000.
2. Josep DF et al. Median-Based aggregation operators for prototype construction in ordinal scales. Int J Intell Syst 2003; 18:633-655.
3. Yager RR. On ordered weighted averaging aggregation operators in multicriteria decision making. IEEE Trans Syst Man Cybernet 1988,18:183-190.
4. Chiclana F, Herrera F, Herrea-Viedma E. The ordered weighted geometric operator: Properties and application. In: Proc 8th Int Conf on Information Processing and Management of Uncertainty in Knowledge-based System. Madrid, Spain; 2000. pp 985-991.
5. Torra V. Negation functions based semantics for ordered linguistic labels. Int J Intell Syst 1996;11:975-988.
6. Kacprzyk J. Group decision making with a fuzzy linguistic majority. Fuzzy Sets Syst 1986;18:105-118.
7. Kacpryzk J, Fedrizzi M, Nurmi H. Group decision making and consensus under fuzzy preference and fuzzy majority. Fuzzy Sets Syst 1992;49:21-31.
8. An Shi-Hu. On the weak norms of the compound fuzzy proposition. Chinese Journal of Computers, 2001, 24(20): 1071-1076
9. Marichal, J-L. Aggregation operators for multicriteria decision aid.. PhD dissertation, Faculte des Sciences, Universite de Liege, Liege, Belgium, 1999.
10. Mayor G, Torrens J. On a class of operators for expert systems. Int J Intel Syst 1993;8: 771-778.
11. Godo L, Torra V. On aggregation operators for ordinal qualitative information. IEEE trans Fuzzy Syst 2000;8:143-154.
12. Yager RR. Induced aggregation operators. Fuzzy Sets Syst 2003. In press.
13. Herra F, Herrera-Viedma E, Chiclana F. Multiperson decision making based on multiplicative preference relations. Eur J Oper Res 2001;129:372-385.
14. Yager RR,Filer DP. Induced ordered weighted averaging operators. IEEE trans Syst Man Cybern 1999;29:141-150.

CL^2: A Multi-dimensional Clustering Approach in Sensor Networks*

Xiuli Ma[1,2], Shuangfeng Li[1], Dongqing Yang[1], and Shiwei Tang[1,2]

[1] School of Electronics Engineering and Computer Science, Peking University
Beijing, China, 100871
{xlma,sfli,ydq}@db.pku.edu.cn, tsw@pku.edu.cn
[2] National Laboratory on Machine Perception, Peking University
Beijing, China, 100871

Abstract. Sensor networks are among the fastest growing technologies that have the potential of changing our lives drastically. These collaborative, dynamic and distributed computing and communicating systems are generating large amounts of data continuously. Finding useful patterns in large sensor data sets is a tempting however challenging task. In this paper, a clustering approach, CL^2, *CLuster and CLique*, is proposed. CL^2 can not only identify clusters in a multi-dimensional sensor dataset, discover the overall distribution patterns of the dataset, but also can be used for partitioning the sensor nodes into subgroups for task subdivision or energy management. CL^2's time efficiency, and accuracy of mining are evaluated through several experiments. A theoretic analysis of the algorithm is also presented.

1 Introduction

Recent technology advances have witnessed the emergence of small, battery-powered, wireless sensor nodes equipped with sensing, computation, and communication capabilities [2][12]. Each sensor is capable of detecting ambient conditions such as temperature, sound, or the presence of certain objects. A network of sensors can gather meteorological variables such as temperature and pressure in order to forecast harsh natural phenomena. These sensor nodes create exciting opportunities for large-scale, data-intensive measurement, surveillance, and analytical applications. For example, there have been initial applications in wild life habitat surveillance, battlefield troop coordination, and traffic monitoring. In order for these applications to become ubiquitous in the few years to come, it is essential to analyze and mine the sensor readings in real-time, react the analytical result and the discovered patterns to the world to make intelligent decisions.

Clustering is the process of grouping a set of physical or abstract objects into classes of similar objects [7]. By clustering sensor data, one can gain insight into the overall distribution patterns and interesting correlations among sensory attributes. Clustering analysis can also serve as a preprocessing step for other analysis or mining

* This work is supported by the NKBRSF of China (973) under Grant No.G1999032705, the National '863' High-Tech Program of China under Grant No.2002AA4Z3440.

S. Wang et al. (Eds.): ER Workshops 2004, LNCS 3289, pp. 234–245, 2004.

operations on the detected clusters, or performance tuning of the sensor networks, such as sampling rate adjusting, prolonged network lifetime, scalability, and load balancing.

The context of the sensor networks makes the clustering problem more challenging [5][11]. First, sensors have limited resource which the mining needs to conserve whenever possible [13]. Second, there are several intrinsic differences between sensor based data sources and standard database sources [8]. In sensor networks application, data are commonly viewed as potential infinite time ordered data streams rather than finite data sets stored on disk. This view challenges fundamental assumptions commonly made in the context of data mining algorithms. Data from many different streams have to be examined dynamically and combined to detect the potential patterns. In such an environment, our goal is to process as much of the data as possible in a decentralized fashion, so as to avoid unnecessary communication and computation effort.

In fact, as clustering sensor nodes is an effective technique for achieving the network management goals, there have been considerable efforts for clustering nodes in ad-hoc sensor networks [3][6][14]. However, most of the previous researches have focused on *network topology and protocol*. The goal of the optimal use of sensor's energy is reached there by minimizing the total distance between sensor nodes and heads. The potentially useful distribution patterns hidden in sensor data have been rarely cared. The essential operation in sensor node clustering has been to select a set of cluster heads among the nodes in the network, and cluster the rest of the nodes with these heads. Cluster heads are responsible for intra-cluster coordination and inter-cluster communication.

In this paper, a new method will be proposed to do clustering from *sensor data*. Cluster sensor data (which is CLuster, the first 'CL' in CL^2), and then identify the equivalent subsets of sensor nodes on some attributes (which is CLique, the second 'CL' in CL^2). Nodes within the same equivalent subset are equivalent in some sensory attributes, thus one representative node can be selected to be the head of this subset. Each representative can answer all kind of queries about the particular sensory attributes of the other nodes in the same subset. The other nodes can remain asleep, and the communication energy is also saved. Thus, not only the sensor data distribution pattern can be discovered, but also the network lifetime is prolonged.

The remainder of this paper is organized as follows. Section 2 states the problem addressed in this work. Section 3 briefly surveys related work. Section 4 presents the CL^2 approach and argues that it satisfies our goal. Section 5 shows its effectiveness via simulations. Section 6 gives concluding remarks and directions for future work.

2 Problem Statement

Assume that N nodes are dispersed in a field to detect M attributes. Our goal is to answer the question of "which sensor nodes are similar on which sensory attribute(s)?" In fact, that would be to identify a set of possibly overlapping clusters. Each cluster has *two sets*: nodes set and attributes set. For each cluster, for each attribute in its attribute set, the nodes within its nodes set have similar values. What we want is to get the insight of the data distribution pattern of the physical world.

Another goal is to further adjust the network according to the resulted clusters. Such as optimal cooperation of nodes within one same cluster, task assignment among the nodes, selection of representative nodes for the cluster. The essence is minimization of the total energy dissipation of the sensors. However, this sub-goal is not focused on in this paper because of length limits. That will be explored in another paper.

Before clear definition of our problem, some terminologies [15] are briefly reviewed.

Definition 1. A *bipartite graph* $G = (U, V, E)$ has two distinct vertex sets U and V, and an edge set $E=\{(u, v) \mid u \in U$ and $v \in V\}$. A complete bipartite sub-graph $I \times T$ is called a *bipartite clique,* and is denoted as $K_{i,t}$, where $|I|=i$, $|T|=t$ and $I \subseteq U$, $T \subseteq V$.

Definition 2. Let P be a set. A *pseudoequivalence relation* on P is a binary relation $\equiv$ such that for all $X, Y \in P$, the relation is:

1. Reflexive: $X \equiv X$.
2. Symmetric: $X \equiv Y$ implies $Y \equiv X$.

The pseudoequivalence relation partitions the set P into possibly overlapping subsets called *pseudoequivalence class.*

The input database for our whole clustering task is essential a set of N sensor nodes with their sensor data for M attributes. Our goal is to identify a set of clusters such that:

1) Each node v_i, where $1 \leq i \leq N$, is mapped to at least one cluster c_j, where $1 \leq j \leq N_c$, and N_c is the number of clusters.

2) Each cluster consists of a set of sensor nodes V_c, and a set of sensory attributes A_c. For each attribute a_i in A_c, the values of all the nodes in V_c are similar. (Thus for each attribute, the value from one node can represent those from all the other nodes) For each cluster, all the nodes in its nodes set V_c are pseudoequivalent on its set of attributes A_c.

3) For each cluster, both the set of nodes and the set of attributes are *maximal*. That is, there can not be any set V_b of nodes and set A_b of attributes, satisfying $V_c \subseteq V_b$ and $A_c \subseteq A_b$ while nodes in V_b are pseudoequivalent on A_b, or V_b on A_c, or V_c on A_b.

The goal can be reached this way: First, cluster the nodes by each attribute separately; Second, construct the bipartite graph, with the nodes set and the clusters set obtained for all the attributes as the two distinct vertex sets. Third, discover all the bipartite cliques in the bipartite graph. The approach will be explored in detail in section 4.

3 Related Works

Although the clustering problem has been widely researched in the database, data mining and statistics communities [7][16][4], no work on the kind of clustering defined in section 2 has been seen.

There have also been some clustering methods in sensor networks [3][6][14]. However, existing sensor clustering methods mainly concerns about the distance among nodes and towards the network protocol. They do not care sensor data.

Recently, the clustering problem has also been studied in the context of the data stream environment [1][10]. However, these works cannot be directly adopted here.

Some other mining problems can also be related to discovering cliques [15]. The problem of enumerating all (maximal) frequent itemsets corresponds to the task of enumerating all (maximal) constrained bipartite cliques. However, discovering cliques in this paper is not the same as that in general bipartite graph. Both the items and the transactions vertex set in the bipartite graph of [15] are homogeneous, whereas in our paper the clusters in cluster set are not, some clusters are for one attribute, some other clusters are for other attributes.

4 Multi-dimensional Clustering

As mentioned above, the whole clustering process can be summarized as: First, cluster the nodes by each attribute separately; Second, construct the bipartite graph, with the nodes set and the clusters set as the two distinct vertex sets. Third, discover all the bipartite cliques in the bipartite graph. The bipartite cliques are the clusters desired. Actually, this is a two-phase algorithm. The first phase is mainly clustering the nodes along every "sensory attribute", i.e. "dimension". The second phase is discovering bipartite cliques. Example 1 is used for explanation.

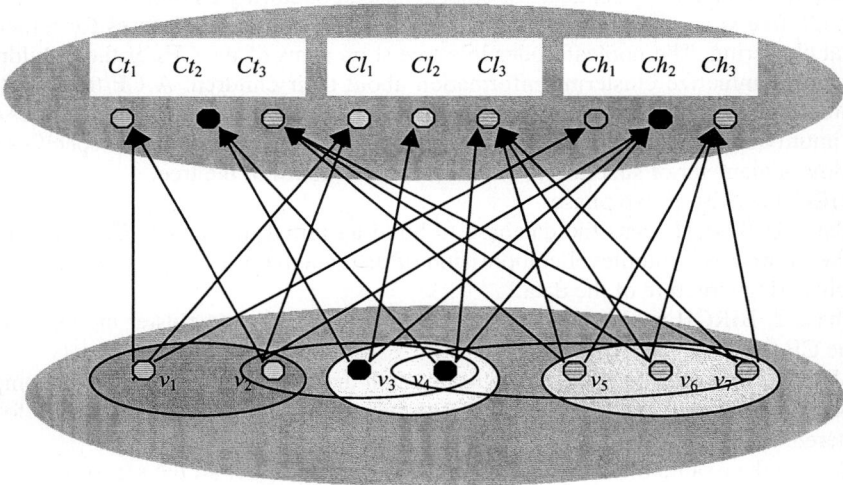

Fig. 1. Multi-dimensional clustering

Example1. Figure 1 illustrates multi-dimensional clustering. v_1-v_7 are seven sensor nodes. Ct_1- Ct_3 means the three clusters of temperature values. Cl_1- Cl_3 means the three clusters of light values. Ch_1- Ch_3 means the three clusters of humidity values. These clusters (cliques) can be discovered: (1) v_1 and v_2 are similar on temperature and light; (2) v_3 and v_4 on temperature and humidity; (3) v_5, v_6 and v_7 on all the three attributes: temperature, light and humidity; (4) v_4, v_5, v_6 and v_7 on light; (5) v_2, v_3 and v_4 on humidity; (6) v_3 on light; (7) v_1 on humidity.

4.1 Clustering

BIRCH [16] is adopted to cluster the sensor nodes along each attribute. BIRCH is an integrated hierarchical clustering method. The most direct reason why BIRCH is adopted is that, BIRCH is effective for incremental and dynamic clustering of incoming objects. The two concepts of *clustering features* (CF) and *clustering feature tree* (CF tree), are at the core of BIRCH's incremental clustering.

A CF is a triple summarizing information about subclusters. Given N objects $\{o_i\}$ in a subcluster, then the *CF* of the subcluster is defined as $CF = (N, \overrightarrow{LS}, SS)$, where N is the number of objects in the subcluster, $\overrightarrow{LS}$ is the linear sum on N objects (i.e., $\sum_{i=1}^{N} \overrightarrow{o_i}$), and SS is the square sum of objects (i.e., $\sum_{i=1}^{N} \overrightarrow{o_i}^2$). A *CF* is essentially a summary of the statistics for the given subcluster: the zeroth, first, and second moments of the subcluster from a statistical point of view. It registers crucial measurements for computing clusters and utilizes storage efficiently since it summarizes the information about the subclusters of objects instead of storing all objects.

A CF tree is a height-balanced tree that stores the clustering features for a hierarchical clustering. The nonleaf nodes in a tree store sums of the CFs of their children, and thus summarize clustering information about their children. A CF tree has two parameters: branching factor, B, and threshold, T. The branching factor specifies the maximum number of children per nonleaf node. The threshold parameter specifies the maximum diameter of subclusters stored at the leaf nodes of the tree.

BIRCH consist of two phase:

Phase 1: BIRCH scans the database to build an initial in-memory CF tree, which can be viewed as a multilevel compression of the data that tries to preserve the inherent clustering structure of the data.

Phase 2: BIRCH applies a (selected) clustering algorithm to cluster the leaf nodes of the CF tree.

BIRCH tries to produce the best clusters with the available resources. The computation complexity of the algorithm is $O(n)$, where n is the number of objects to be clustered.

Review the example 1. After clustering along temperature, all the sensor nodes are clustered into three clusters Ct_1, Ct_2, Ct_3. After clustering along light, the sensor nodes are clustered into another three clusters Cl_1, Cl_2, Cl_3. After clustering along humidity, the sensor nodes are clustered into another three clusters Ch_1, Ch_2, Ch_3.

In this clustering process, we note that some constraint must be imposed to the clusters. The diameter of each final cluster means the maximal dissimilarity between two nodes within the cluster. If it is not constrained, the accuracy (fidelity) of sensing later after the whole clustering will surely be impaired. So we should give a *threshold to the diameter* of each final cluster, such as 2 for temperature, which means two nodes with temperature difference greater than 2 cannot be clustered into one cluster.

After clustering along each attribute by BIRCH, the bipartite graph can be constructed. View each cluster generated above as a vertex, then all the clusters constitute a vertex set. All the sensor nodes constitute the other vertex set. For each sensor node, connections between it and the clusters of its sensory attributes constitute the edge incident with the sensor node vertex and the cluster vertex. Thus the bipartite graph is formed.

The following requirement must be met:

For each attribute in the attributes set we consider, any node must belong to one and only one cluster of that attribute.

4.2 Discovering Bipartite Cliques

The bipartite clique discovery problem here is not the general problem in graph theory. Clusters in the cluster vertex set, are not homogeneous. Some clusters belong to one temperature attribute; while some other clusters belong to other different attributes. However, our final goal is, "which sensor nodes are similar on which attributes". The goal is to discover all the maximal bipartite cliques. The algorithm for maximal clique generation in [15] cannot be used, because here it is bipartite clique to be discovered, not clique. Obviously, although a bipartite graph is a graph, a bipartite clique is *not* a clique, because there are no edges among the nodes in one vertex set.

Before our clique algorithm, the following terms need to be defined.

Definition 3. Let $G = (V, A, E)$ be a bipartite graph constructed from nodes set V and attributes set A. V_S is a set of nodes, A_S is a set of attributes, and $V_S \subseteq V$, $A_S \subseteq A$. *Nodes in V_S are equivalent according to A_S,* means that for each attribute a_i in A_S, (where $1 \leq i \leq n$, n is the number of attributes in A_S.), for any nodes v_i, v_j in V_S, where $i \neq j$, and $1 \leq i, j \leq m$, m is the number of nodes in V_S, $v_i.a_i$ and $v_j.a_i$ are similar, that is, fall into one same cluster of a_i.

Define *an equivalent relation r on the nodes set V_S according to attributes set A_S* as follows:

$$\forall \ v_i, v_j \in V_S, \ v_i \equiv_r v_j \Leftrightarrow v_i, v_j \text{ are incident with the same set of clusters according}$$

to attributes set A_S. That is, according to a set of attributes, two nodes are in the same equivalent class if they are similar for each attribute in this attributes set. (They fall into one same cluster for one attribute) We call *r* an attribute-set-based equivalent relation. Given an attributes set A_S, the equivalence relation partition the nodes set V_S into disjoint subsets, *equivalent classes*.

Definition 4. Given V_S, a set of nodes, and A_S, a set of attributes. If nodes in V_S are equivalent according to A_S, V_S is defined to be *a maximal equivalent class for A_S,* if there does not exist any set $V_S{}'$ of nodes such that $V_S \subset V_S{}'$ and $V_S{}'$ are equivalent according to A_S also.

Definition 5. Given an attribute set A_S, all the maximal equivalent class for A_S constitute the attributes set A_S's *maximal equivalent classes set,* denoted by $MEC(A_S)$.

Definition 6. The set of nodes incident with one cluster is called this cluster's *covering set.*

There exists the following property in bipartite graph.

Property 1. *Anti-monotone* of an equivalent relation r on the nodes set V_S according to attributes set A_S. If the nodes set V_S are not equivalent according to attributes set A_S, there can not be any set V_b of nodes and set A_b of attributes, satisfying $V_S \subseteq V_b$ and $A_S \subseteq A_b$, while nodes in V_b are equivalent according to A_b, or V_b are equivalent according to A_c, or V_c are equivalent according to A_b.

This is an important property, which forms the core of our Clique algorithm. Start from fewer attributes and these attributes' maximal equivalent classes set, extend the attributes set level by level, decompose the nodes set meanwhile, till all maximal bipartite cliques are discovered. The algorithm is as follows. We extend along the attributes set, as there are fewer attributes compared with nodes.

Algorithm 1 Clique

```
Input: bipartite graph G consisting of sensor nodes set and sen-
sory attributes set A and its clusters set.

Output: set of all bipartite cliques, R.(R is a set of <at-
trset,nodeset> pairs)

Method:

(1)  Let P(A) denote the set of all nonempty subsets of attrib-
utes set A;

(2)  Sort the subsets in P(A) according to their sizes and put
them into ListP;

(3)  While (ListP is not empty)

(4)  {

(5)       Get one element AttrSet out from ListP;

(6)       If AttrSet has only one attribute a

(7)       {

(8)          MEC(AttrSet) = Ø;

(9)          for each cluster c_i of a  {

(10)              insert (c_i.coveringset) into MEC(AttrSet);

(11)              insert(<Attrset,c_i.coveringset>)into R; }

(12)       }

(13)      else

(14)      {

(15)          choose any attribute b from AttrSet;

(16)          d =(AttrSet - b); //a direct subset of AttrSet;

(17)          for each set x in MEC(d)
```

```
(18)                    for each cluster c_i of b

(19)                        {

(20)                            M = x ∩ c_i.coveringset;

(21)                            if |M|>0  {

(22)                                insert (M) into MEC(AttrSet);

(23)                                if(|M|>1){

(24)                                    insert <Attrset,M> into R;

(25)                                    for each direct subset w of AttrSet

(26)                                        if(<w,M> IN R)delete <w,M> from R;}}

(27)                        } // end for cluster c_i of b

(28)        } // end else

(29) } //end while
```

Each attribute set and one class in its maximal equivalent classes set constitute a maximal bipartite clique.

We extend the maximal equivalent set along the attribute set. We first need to sort the attributes set according to their size. For a set of more than two attributes, extend its maximal equivalent set from its direct subset's maximal equivalent sets (from (13)), if one of its direct subset's maximal equivalent sets fall into the same cluster, extend the attribute set; if they fall into different clusters, decompose the original equivalent class. Next, an example is given to illustrate the process of the whole algorithm.

Example 2. The same scenario of sensor nodes and sensory attributes in example 1 is adopted. Clique first gets the list of all nonempty subsets of attributes set: $\{\{t\}, \{l\}, \{h\}, \{tl\}, \{lh\}, \{th\}, \{tlh\}\}$. Then, each element is addressed. For $\{t\}$, $\{l\}$ and $\{h\}$, just put the covering set of the clusters of the only attribute into its MEC and the corresponding pairs into result set R. Thus, $MEC(t)=\{\{v_1v_2\},\{v_3v_4\},\{v_5v_6v_7\}\}$, $MEC(l)=\{\{v_1v_2\},\{v_3\},\{v_4v_5v_6v_7\}\}$, $MEC(h)=\{\{v_1\},\{v_2v_3v_4\},\{v_5v_6v_7\}\}$. $R=\{<\{t\},\{v_1v_2\}>, \quad <\{t\},\{v_3v_4\}>, \quad <\{t\},\{v_5v_6v_7\}>, \quad <\{l\},\{v_1v_2\}>, \quad <\{l\},\{v_3\}>, <\{l\},\{v_4v_5v_6v_7\}>, <\{h\},\{v_1\}>, <\{h\},\{v_2v_3v_4\}>, <\{h\},\{v_5v_6v_7\}>\}$. As for those attributes set with more than one attribute: such as $\{tl\}$, choose one attribute, such as t, then expand the remaining attributes set $\{l\}$ by finding the common nodes set between $MEC(l)$ and t's covering sets. For element $\{v_1v_2\}$ in $MEC(l)$, the intersection gets $\{v_1v_2\}$. Then the expansion succeeds – put $\{v_1v_2\}$ into $MEC(tl)$ and $<\{tl\},\{v_1v_2\}>$ into R, and delete $<\{t\},\{v_1v_2\}>$ and $<\{l\},\{v_1v_2\}>$ from R. For $\{v_3\}$, we get $<\{tl\}, \{v_3\}>$. These resulted pairs with only one node have not been put into the result set, as they don't contribute much in real scenario. However, in order for the completeness of MEC, we remain them in MEC. Then deal with $\{v_4v_5v_6v_7\}$ of $MEC(l)$, resulting with putting $\{v_5v_6v_7\}$ into $MEC(tl)$ and $<\{tl\}, \{v_5v_6v_7\}>$ into R, deleting $<\{t\},\{v_5v_6v_7\}>$

from R. The same process happens for $\{lh\}$, $\{th\}$ and $\{tlh\}$. Finally, we get $R=\{<\{tl\},\{v_1v_2\}>,$ $<\{th\},\{v_3v_4\}>,$ $<\{tlh\},\{v_5v_6v_7\}>,$ $<\{l\},\{v_4v_5v_6v_7\}>,$ $<\{h\},\{v_2v_3v_4\}>, <\{l\},\{v_3\}>, <\{h\},\{v_1\}>\}.$

We summarize the whole multi-dimensional clustering algorithm CL^2 as follows.

Algorithm 2 CL^2

Input: sensor nodes set V, and sensor data for V on sensory attributes set A

Output: All maximal cliques of sensor nodes and attributes

Method:

(1) $ClusterSet = \varnothing$;

(2)   For each attribute a in A

(3)   {

(4) CS_a = the set of clusters discovered by clustering sensor data on a by BIRCH;

(5) $ClusterSet = ClusterSet \cup CS_a$;

(6)   }

(7)   Construct bipartite graph G using V and $ClusterSet$;

(8)   Clique(G);

(9)   Return all the maximal cliques;

CL^2 is a two-phase algorithm: first clustering, then discovering cliques. More detailed analysis about CL^2 and Clique are in next section.

5 Performance Study

Let us now examine the efficiency of the whole multi-dimensional clustering algorithm. The problem is transformed into two phases: one is the general clustering process by BIRCH; the other is the bipartite clique discovery process.

For BIRCH, it tries to produce the best clusters with the available resources. Given a limited amount of main memory, an important consideration is to minimize the time required for I/O. BIRCH applies a multiphase clustering technique: a single scan of the data set yields a basic good clustering, and one or more additional scans can (optionally) be used to further improve the quality. The computation complexity of the algorithm is $O(n)$, where n is the number of objects to be clustered. Experiments have shown the linear scalability of the algorithm with respect to the number of nodes, and good quality of clustering of the data.

Assume there are k sensory attributes, and for the i-th attribute, there are n_i cluster, where $i \leq k$.

For Clique, the cycle in line (3) run 2^k times. The cycle in line (17) and (18) altogether run at most $\Pi\, n_i$ times. The cycle in line (25) runs at most $(k-1)$ times. Thus, the complexity of the algorithm Clique is $2^k \bullet \Pi\, n_i \bullet (k-1) \leq 2^k \bullet n_{max}^k \bullet (k-1) = (2n_i)^k \bullet (k-1)$. Note that k is the number of sensory attributes, which cannot be too big, generally not exceed 10. n_{max} is the maximum among all the n_i $(i \leq k)$.

All the experiments are performed on a 733-MHz Pentium PC machine with 512M main memory. All the programs are written in Microsoft/Visual C++ 6.0. We simulate 1000 sensor nodes, each with three sensory attributes, temperature, humidity, light. Generally, the number of clusters for each attribute is five. Then we perform the experiments of efficiency of our algorithm vs. node number, attribute number and the number of clusters.

The sensor data are randomly generated. We start the analysis with the number of sensor nodes to be 100, and then tune it to 500, 1000, 1500, 2000, and 3000. Clearly, the scalability of the whole algorithm is good. Figure 2 shows, when the number of sensor nodes is 3000, CL^2 will cost more than 1.5 seconds. However, this is reasonable for real life usage. Clique only account for a small part of the runtime. Clustering takes almost all the time of CL^2. The linear scalability of CL^2 to the number of sensors clustered has been analyzed above. According to the above theoretical analysis, the algorithm Clique is efficient. Even in real deployment scenario, when we want to improve the accuracy of sensing by increase the cluster number and decrease the diameter of clusters, the cliques can still be discovered efficiently. As you see in Figure 3 and 4, the scalability of our algorithm to the number of clusters is really satisfying. In Figure 3 and 4, scalability experiments just care about 2000 sensor nodes, as that is enough in real life usage.

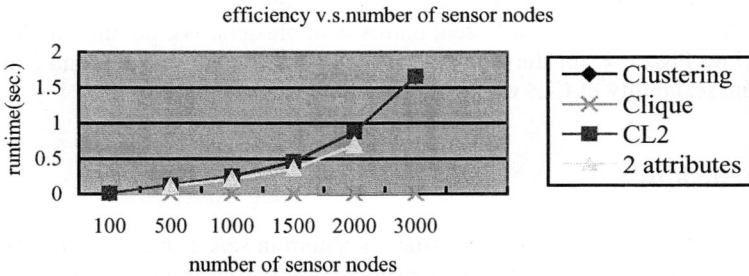

Fig. 2. Scalability of CL^2 vs. number of sensors

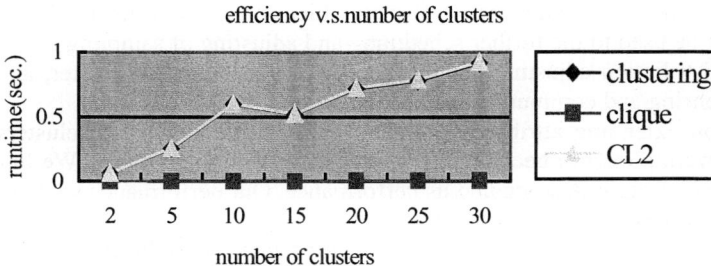

Fig. 3. Scalability of CL^2 vs. number of clusters (number of nodes: 1000, number of attributes: 3)

efficiency v.s.number of clusters

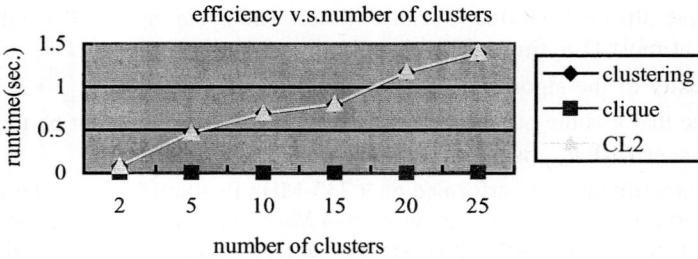

Fig. 4. Scalability of CL^2 vs. number of clusters (number of nodes: 2000, number of attributes: 3)

efficiency v.s.number of nodes

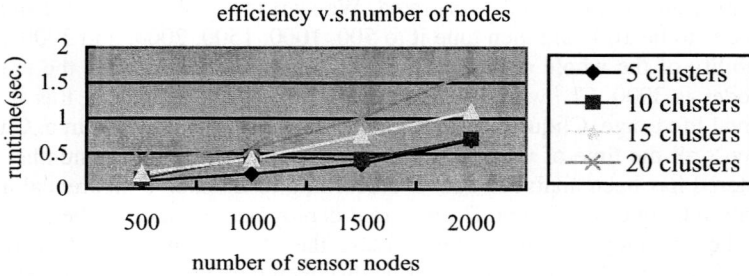

Fig. 5. Scalability of CL^2 vs. number of sensor nodes (number of attributes: 2)

Figure 5 is for explanation of number of attributes. In Figure 2-4, the numbers of attributes are all 3. In figure 5, we illustrate the scalability of the whole CL^2 algorithm vs. number of nodes for four different numbers of clusters. We put the line for 5 clusters also into Figure 2, for the convenience of comparing 3 attributes and 2 attributes easily. The scalability of CL^2 vs. the number of clusters is good.

6 Conclusions

An efficient multi-dimensional clustering algorithm in sensor networks has been proposed. Through clustering along each attribute, a bipartite graph is constructed from sensor nodes set and clusters set. Through discovering the bipartite cliques in the bipartite graph, we get the insight into the overall sensor data distribution pattern, such as, which sensor nodes are similar on which sensory attributes. The discovered pattern can be used to the further scheduling and adjusting of sampling task allocation [9], task distribution adjusting among the sensor nodes in each cluster, and energy-efficient sensing and communication model.

Based on extending attributes combination in pseudoequivalent clusters step by step, discovering cliques becomes a dynamic programming process. We have implemented the CL^2 algorithm, studied its performance. Our performance study shows that the method mines sensor data efficiently.

There are a lot of interesting research issues related to multi-dimensional clustering, including further study and implementation of it when the dataset is evolving in context of data streams, distributed operation of this mining process, energy-aware clustering, integrating the mining with the network topology, and other interesting problems encountered in sensor networks. We are taking efforts towards these problems.

References

1. C.C.Aggarwal, J.Han, J.Wang, P.S.Yu. A Framework for Clustering Evolving Data Streams. VLDB2003.
2. I.F.Akyildiz, W.Su, Y.Sankarasubramaniam, and E.Cayirci. Wireless sensor networks: a survey. Computer Networks, Vol.38, No.4, pp.393-422, 2002.
3. S.Bandyopadhyay and E.J.Coyle. An Energy Efficient Hierarchical Clustering Algorithm for Wireless Sensor Networks. IEEE INFOCOM 2003.
4. M.Ester, H.Kriegel, J.Sander, M.Wimmer, and X.Xu. Incremental clustering for mining in a data warehousing environment. VLDB 1998.
5. M.J.Franklin. Challenges in Ubiquitous Data Management. Informatics 2001.
6. S.Ghiasi, A.Srivastava, X.Yang, and M.Sarrafzadeh. Optimal Energy Aware Clustering in Sensor Networks. Sensors 2002, 2, pp.258-269.
7. J.Han and M.Kamber. Data Mining – Concepts and Techniques. Morgan Kaufmann Publishers,2001.
8. S.Madden and M.J.Franklin. Fjording the stream: An architecture for queries over streaming sensor data. ICDE 2002.
9. S. Madden, M.J.Franklin, J.M.Hellerstein, and W.Hong. The Design of an Acquisitional Query Processor For Sensor Networks. SIGMOD 2003.
10. N.H.Park and W.S.Lee. Statistical Grid-based Clustering over Data Streams. SIGMOD Record, Vol.33, No.1, March 2004.
11. F.Perich, A.Joshi, T.Finin, and Y.Yesha. On data management in pervasive computing environments. IEEE Transactions on Knowledge and Data Engineering. Vol. 16, No. 5, May 2004.
12. G.J.Pottie and W.J.Kaiser. Wireless Integrated Network Sensors. Communications of the ACM. Vol.43, No.5, pp.51-58, May 2000.
13. Y.Yao, J. Gehrke. Query processing for sensor networks, CIDR 2003.
14. O.Younis and S.Fahmy. Distributed Clustering in Ad-hoc Sensor Networks: A Hybrid, Energy-efficient Approach. IEEE INFOCOM 2004.
15. M.J.Zaki. Scalable Algorithms for Association Mining. In IEEE Transactions on Knowledge and Data Engineering, vol.12, No.3, May/June 2000.
16. T.Zhang, R.Ramakrishnan and M.Livny. BIRCH: An efficient data clustering method for very large databases. SIGMOD 1996.

TCMiner: A High Performance Data Mining System for Multi-dimensional Data Analysis of Traditional Chinese Medicine Prescriptions*

Chuan Li[1], Changjie Tang[1], Jing Peng[1], Jianjun Hu[1], Lingming Zeng[1], Xiaoxiong Yin[1], Yongguang Jiang[2], and Juan Liu[2]

[1] The Data Base and Knowledge Engineering Lab (DBKE)
Computer School of Sichuan University
{lichuan,tangchangjie}@cs.scu.edu.cn
[2] Chengdu University of Traditional Chinese Medicine
{cdtcm,liujuan0}@163.com

Abstract. This paper introduces the architecture and algorithms of TCMiner: a high performance data mining system for multi-dimensional data analysis of Traditional Chinese Medicine prescriptions. The system has the following competing advantages: (1) High Performance (2) Multi-dimensional Data Analysis Capability (3) High Flexibility (4) Powerful Interoperability (5) Special Optimization for TCM. This data mining system can work as a powerful assistant for TCM experts by conducting Traditional Chinese Medicine Data Mining such as Computer-Aided Medicine Pairing Analysis, Medicine Syndrome Correlation, Quality and Flavor Trend Analysis, and Principal Components Analysis and Prescriptions Reduction etc.

1 Introduction

Traditional Chinese medicine (TCM) has a long therapeutic history of thousands of years and the therapeutic value of which, especially on chronic diseases, has been winning wider and wider acknowledgement in the World [1]. In addition, the TCM seems to have made enormous strides forward after China's entry into the World Trade Organization (WTO), as large sums of capital investment become available to spur technical innovation. The World Health Organization (WHO) has also been keen to pursue the development of TCM in recent years.

However, despite its existence and continued use over many centuries, and its popularity and extensive use during the last decades, its chemical background and formula synergic effects are still a mystery at least in theoretical sense because of its complex physiochemical [2]. Newly developed techniques, such as data mining or

---

* This Paper was supported by Grant from National Science Foundation of China (60073046), Specialized Research Fund for Doctoral Program provided by the Ministry of Education (SRFDP 20020610007), and the grant from the State Administration of Traditional Chinese Medicine (SATCM 2003JP40). LI Chuan, PENG Jing, HU Jianjun are Ph. D Candidates at DB&KE Lab, Sichuan University. Jiang Yongguang is a Professor at Chengdu University of Traditional Chinese Medicine. And TANG Changjie is the associate author.

S. Wang et al. (Eds.): ER Workshops 2004, LNCS 3289, pp. 246–257, 2004.

knowledge discovery in database (KDD) which aim at discovering interesting patterns or knowledge from large scale of data by nontrivial approaches, provide us with a very promising means and a hopeful opportunity to do research on the TCM data.

Designing a data mining system for analysis of TCM data has been considered by both the TCM and data mining trade for quite a long period but due to lack of mutual understanding and the true complexity of the problem itself the work seems complicated and challenging. The Data Base and Knowledge Engineering Lab (DBKE) at Computer School of Sichuan University in Chengdu has been working on the development of TCMiner (Traditional Chinese Medicine Miner) in collaboration with Chengdu University of Traditional Chinese Medicine under a research grant from State Administration of Traditional Chinese Medicine (SATCM) to investigate new methods of multi-dimensional data analysis of Traditional Chinese Medicine prescriptions. This system can provide knowledge discovery and data mining capabilities for TCM data values as well as for categorical items, revealing the regularities of TCM paring and indicating the relationship of prescriptions, medicine and syndromes.

TCMiner has the following distinguishing features: (1) High Performance: Algorithms implemented in the system are all the leading algorithms in their respective domains. (2) Multi-dimensional Data Analysis Capabilities: The system has multi-dimensional analysis capabilities. E.g. it can discover Medicine Pairing regularities in consideration of digitalized Quality, Flavor, Channel tropism etc. in form of multi-dimensional frequent patterns or association rules (3) High Flexibility: In order to realize the highest flexibility of the system to meet the different requirements of different users, multiple engines were adopted. (4) Powerful Interoperability: By the highly interactive visual and friendly user interface, the system has wonderful interoperability. (5) Special Optimization for TCM: TCMiner optimizes the system architecture and algorithms according to characteristics of TCM data in the following aspects: (a) Standardization of discrete attributes such as Quality, Flavor, Channel tropism (b) TCM Knowledge Base Constructions for heuristic search (c) The system implements General Trend Evaluation and Contrast of Quality, Flavor, and Channel tropism, which proves to be of high value in general analysis of TCM prescriptions. (d) Construction of TCM Data Warehouse (e) Multi-dimensional Prescriptions Structure Analysis.

The remaining of the paper is organized as follows. Section 2 introduces the methodological knowledge in TCM data analysis. Section 3 covers the special designing issues concerning the particular TCM data process and system implementation. Section 4 details the system architecture of TCMiner. Section 5 presents the high-level performance of the fundamental algorithms based on which TCMiner engines are implemented. Section 6 discusses the probable future directions of TCMiner's later versions. And section 7 concludes the paper.

2 Methodology of TCM Data Analysis

TCM therapy theory is a very broad and deeply philosophy in that it contains not only the theoretical explanations of how TCM works but also comprises the particular prescriptions, medicines, herbs and laws. Traditional Chinese Medicine Prescriptions are the original records of detailed procedures and ways in which theses herbs are put to use. What implied are the Paring Regularities of TCM, actually represented in form of the correlations among Prescriptions, Medicines, and Syndromes. The major objective of the TCM Prescription Analysis Research is to reveal these corresponding associations by means of the latest data processing and analysis technology – Data Mining.

2.1 A Glance at TCM Prescription

Before we go into the complicated technical details, let's have a glance at a typical TCM prescription shown below (extracted from TCM professional materials). The prescription consists of two medicines/herbs: Corktree bark and Atractylodes rhizome with their corresponding Quality, Flavor, Channel tropism, etc. shown to the right. For detailed explanation of the terms and notations, please refer to reference [3].

	1st layer	2nd layer	3rd layer	
1st level	Components:	Corktree bark	Quality	Cold
			Flavor	Bitter
			Channel tropism	Kidney/Bladder
			Function	Resolving heat/ Eliminating dampness/ Consolidating Yin
			Dose	15g
		Atractylodes rhizome	Quality	Warm
			Flavor	Spicy and Bitter
			Channel tropism	Spleen/Stomach
			Function	Eliminating dampness
			Dose	15g
2nd level	Function:	Resolving heat/ Eliminating dampness		
	Indications:	Pathogenesis	downward flow of damp-heat	
		Syndrome	Flaccidity syndrome/Arthragia syndrome /Dermatophytosis /Leukorrhea /Eczema	
3rd level		Pulse ⋮	soft and floating pulse	
	Add/Reduce:		add Chaenomeles fruit /Dioscorea septemloba	
4th level	Derived Prescription	San Miao Powder /Si Miao Pill/Qi wei cang bai Powder/Qing re sheng shi Decoction		

Fig. 1. A sample TCM prescription

2.2 Technological Routines

Since Prescriptions Paring Regularities contain two indispensable factors: Medicine Paring Regularities and Prescription-Syndrome Matching Regularities, through continuous and ardent discussions with TCM experts, we are resolved to adopt the following general line: Start with Prescription Structure Analysis and Medicine-Syndrome Analysis concentrating on the medicine components data, and then conduct research on Prescription-Syndrome Matching Regularities regarding "Prescription-Medicine-Syndrome" as a whole. The research routine is outlined as follows:

As the first step, we investigate the Prescription Structure Analysis method with the following sub-objectives and solutions: (1) basic prescription finding: first, track down the principal medicine components in the prescriptions; second, sort and rate

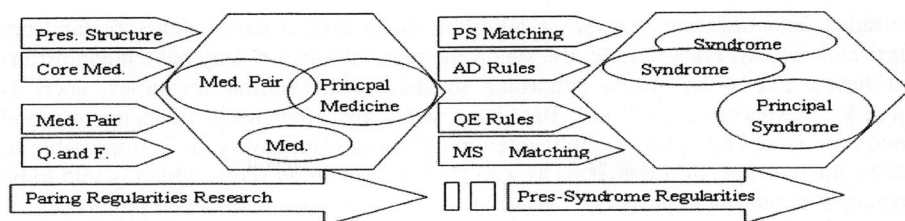

Fig. 2. The general technique routine of TCM prescription analysis

these medicines according to their importance; third, cored with principal medicine, summarize the structure of fundamental and derived prescriptions; finally, propose the initial formation of a basic prescription to the TCM professionals for evaluation (2) medicines group selection: first, try to find a group of medicines which are often used together in the same kind of prescriptions; second, refine the medicine pairs and groups through testing samples; third, ascertain the medicine pairs and groups through special "Medicine-Syndrome Matching" (3) Medicine-Syndrome Correlation Analysis: the law of the medicine usage for a certain disease or corresponding syndromes in form of multi-dimensional representation; the different medicine usage laws for different parts of the body with different Quality and Flavor.

In the second stage, we concentrate on the Prescription-Syndrome Correlation Analysis with the following summarized analysis contents: (1) Organization characteristics analysis of categorical prescriptions or related syndromes including: (a) the Recognition of major symptom (single symptom, symptoms group, and symptoms herd); (b) the analysis and judgment of combining patterns cored by principal syndrome. (2) The Analysis of Syndrome Data: (a) the analysis of Cold/Hot Trend, Weak/Strong Trend, Corresponding Body Parts, and Pathogenesis and characteristics of the Syndromes; (b) The correlations of related syndromes; (c) the significance and frequency of the syndromes. (3) The Syndromes Analysis through prescriptions and medicines: (a) the deductive and verifying analysis through the Quality, Flavor, Channel tropism, function and efficacy prescriptions and medicines; (b)the verification and deduction through the associations of the efficacy, clinical effects and pathogenesis of prescriptions, esp. basic prescriptions. (4) The methods of Construction of the Syndrome Knowledge Base.

Due to space limitation, we do not give further descriptive illustrations here. Interested readers shall contact us directly.

3   Design Issues

The TCM data that need to be analyzed are got through incipient handy input about 2 years ago in Microsoft Access format where 1355 Spleen-stomach prescriptions were screened out from Dictionary of Traditional Chinese Medicine [4] for sample data of analysis. However, most of these data fields are text attributed including some fundamental fields, such as medicine effect, Quality, Flavor, Channel tropism, etc. which make the analyses intangible and unfeasible regarding that most analysis methods are designed for continuous values. In addition, many textual expressions are far from clear and regular. For example, the syndrome item "tiredness" may have many different terms across the whole syndrome database with similar meanings, such as "weary", "exhaustiveness", etc. But these terms are often used independently and have corresponding specific applicable environment respectively. It is impossible to define the concepts of these terms as well as their governing scope which is said to be "perceptible but not expressible", however, to analyze these data in computer fashion, every trivial must be strictly formulated. Even among the rare countable fields, such as Quality, Flavor, Channel tropism, etc. which could be easily converted into categorical numbers, such expression still lag behind what the domain professional really want out of them because the categorical representations did not cover all their implications. Moreover, the TCM data sampled by the task is only a small portion of the huge amount of TCM data. What the system needs to face is large scale data bases containing thousands and thousands of prescriptions. So the designing and implementing of highly efficient and scalable algorithms remains the core of the whole project at least from the angle of the development process of this system though some may accentuate the system's righteousness while ignore its performance. At last, to best satisfy the users' requirements, the system must be designed highly flexible and have good interoperability. Based on the abovementioned considerations, the following issues have to be addressed:

3.1 Data Standardization and Regularity

There are a lot of ambiguities and irregularities in the original prescriptions data, mainly comprising the following two kind of occurrences: (1) Semantic ambiguities, i.e. some traditional definitions, terms or expressions need to be reanalyzed and positioned so as to take on a clearer and more sensible meaning. (2) Expression irregularity, i.e. many different expressions can be used to imply or indicate the same definition (namely, multiple terms one concept). To solve this problem, we adopt the following scheme to regulate and integrate the source data. For the one-term-multiple-concepts case, we eliminate the term from alphabet and reserve only the expressions for detailed concepts. And as to the multiple-terms-one-concept case, we only reserve the definition for the uniformed concept. By doing this, we unify the concepts' space and avoid the consequences caused by the inconsistency of the original data.

3.2 Handling Categorical Data

Most data analysis algorithms are designed to run on continuous data, however a large portion of the source TCM data is categorical. It then becomes a problem to measure the value of these categorical items. The key issue is to develop a scientific and practical data assessment scheme or system which transforms the categorical data items into continuous values. In the development of TCMiner, the establishment of such a system can be divided into two steps: firstly, try to preset the upper and lower bound of target continuous value scope and its corresponding mapping function with according to given constraints of the TCM knowledge; secondly, suppose the above digitalization scheme to be true and train the scheme by conducting sample test on prepared testing data; finally, test the trained scheme by validating on prepared testing data. The training and testing method can be justified by the introduction of highly precise and fast GEP approaches.

3.3 Adoption of High Performance Algorithms

Traditional Chinese Medicine has a treatment history of more than 3000 years and there have accumulated thousands of thousands of ancient prescriptions spanning across hundreds of TCM works. On the other hand, most valuable patterns are obtained through deeply and sophisticated analyses involving different aspects (Prescriptions, Medicines, Syndromes, etc.). To mine complicated knowledge through such a huge amount of data, the algorithms adopted must have high processing performance and scalability as well.

3.4 Flexibility and Interoperability Considerations

The very significant requirements for a data mining system are to provide enough flexibility and scalability. When we construct the system, one of the key points in our minds is to make it highly generalized so that it can be flexible enough to adapt itself to the diverse needs of users. Another key point is how to provide multiple data analysis technologies, i.e. mining engines so that users can select the appropriate one according to the characteristics of the data to be analyzed and the way in which the extracted information will be used for different TCM data analysis tasks involve different datasets and different datasets may have different formats in case of wide existence of legacy database, text dataset, excel forms established at different times.

4   System Features

In the development of TCMiner, we use the latest object Oriented techniques to achieve a high degree of portability, accessibility and maintainability. The implementation in Delphi allows the system to have high developing speed and run on multiple operating systems including Microsoft Windows 98, 2000 and XP Linux, and UNIX.

4.1   Process Flow

The TCMiner system architecture can be viewed as processes on a data stream in that the mining tasks are broken down into a series of subtasks with results from each step passed to the next. Since existing TCM prescriptions may be saved in different databases and have different formats and usually there are noisy data and inconsistencies, the first step toward knowledge is to clean and integrate the data. In this process, TCMiner finishes the following tasks: filling out vacant values, smoothing away noise data, removing the inconsistency and reducing the task irrelevant data. The second step is to construct TCM data warehouse, which is of great essence to the multidimensional data analysis. In TCMiner, four data marts are enclosed in the TCM data

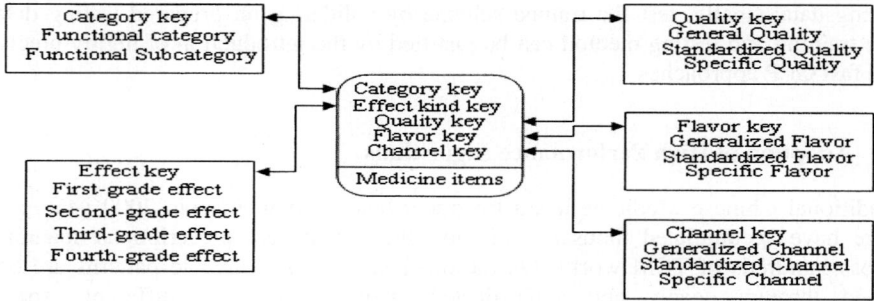

Fig. 3. Star schema of Prescription Data Mart

warehouse and their subjects are the prescription, medicine, syndrome and pathogenesis respectively. Due to limited space, we only give the star schema of Prescription Mart shown below.

The third step is execution of the essential tasks performed by the Multi-Dimension Analysis Engine or the OLAP Engine mainly on the TCM data warehouses. Finally, the patterns are presented to the Patterns Evaluation Module to make the last selection. And only the rules passing the filtering would have the opportunity to show on the end user's screen.

4.2 Major Components

The Multi-Dimensional Data Analysis Engine is responsible for most multi-dimensional data analysis tasks. For example, it can accomplish requests such as "What medicines and herbs are the building blocks of a typical spleen-stomach prescription treating menopause with the Quality to be cold and Flavor to be light bitter?" The Multi-dimensional Data Analysis Engine interprets a mining task specification which provides the details about each specified operation and their executing sequence and put to practice.

The OLAP Engine is responsible for providing a simple, summarized and generalized view of data as a whole through rolling up, drilling down, slicing, dicing, and pivoting operations according to user-specified conditions. The OLAP Engine works in corporation with the Multi-dimensional Data Analysis Engine at times when the tasks involve processing on both aspects.

Validation Tester works as a safeguard to check the user's request to see if there is grammatical mistakes, internal contradictions, and semantic errors or if the request is out of the system's processing capacity. If one of the abovementioned things happens, the request will be regarded as invalid and returned to the user with detailed situation description. Otherwise, the request will be added to the tasks queue of Task Scheduler with specific description of resource requirements. And the Task Scheduler will do the subsequent post-processing.

Task scheduler is in charge of maintaining and scheduling management of the task list. It examines task list and determines which task should be executed. The scheduling strategy can be customized to adapt to different system requirements.

Knowledge base works as a heuristic guide for TCM knowledge mining. The knowledge base embedded a considerable set of TCM knowledge rules containing the correlations of the prescriptions with certain Quality, Flavor, Channel tropism and their corresponding medicines and their respective syndromes in form of rules.

Patterns Evaluator reexamines the resulting patterns in the following four aspects: (1) availability: the resulting patterns should be validated according to corresponding rules of the Knowledge Base (2) size: association mining usually generates too many rules which makes it difficult for the users to pick valuable ones. Pattern Evaluator checks the resulting rules' size to see if it is relatively too large, and returns the result for refinery mining. (3) user-specified template: TCMiner supports user-specified template and extra constraint in associations mining. (4) Grammatical and consistency test: this process finish the formulary examination of the resulting patterns.

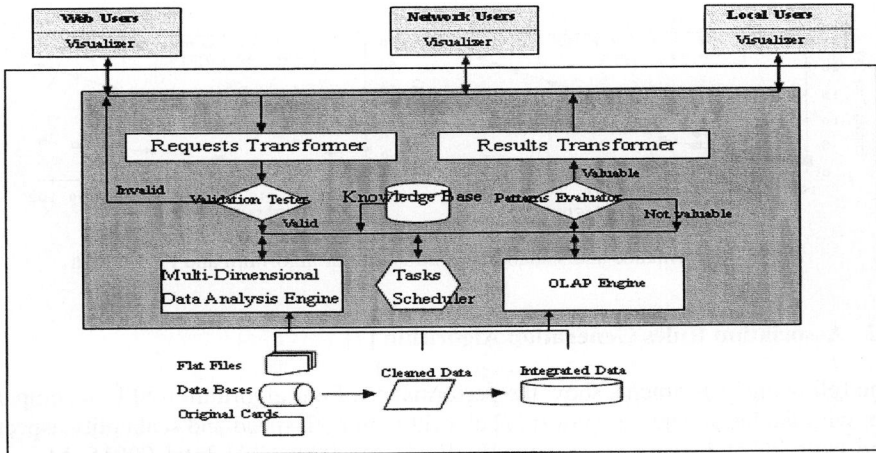

Fig. 4. General architecture of TCMiner

5 Performance Study of Implemented Algorithms

In this section, we present the performance study of the major algorithms implemented in TCMiner over a variety of datasets. All Algorithms are implemented in C++. Readers are recommended to refer to paper [5]-[9] for detailed design thought, pseudo code, example of these algorithms, which are our progresses at the previous stage. Due to limitation of space, we only demonstrate performances of them. All DataSets involved are the standard testing data which can be downloaded through websites [10]-[11] including T10K, T100K, T100K-800K, Connect-4, Pumsb, and Mushroom.

5.1 Frequent Patterns Mining Algorithm [5]

The following charts contrast the performance of our algorithm with the almost best efficient algorithm FP-Growth [12]. The experiments are done on a Windows 98 machine with CPU and memory to be Intel 450M and 128M respectively. As is shown in the first two figures, in both synthetic and real world dataset (dense datasets) our algorithm outperforms FP-Growth greatly and sharply. The subsequent two graphs of experiments on T10K-T100K datasets ensures that our algorithm has better scalability in both time and space perspectives.

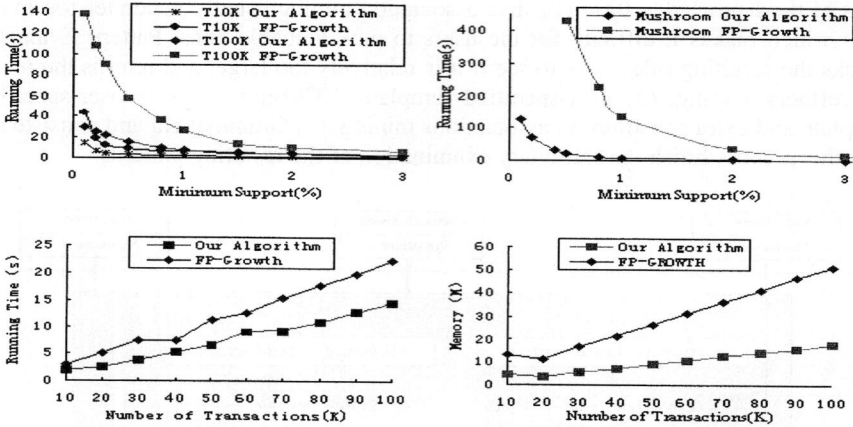

Fig. 5. Performance and scalability contrast of our algorithm and FP-Growth

5.2 Association Rules Generation Algorithm [7]

The following experiments show the performance of our algorithm – SPF in comparison with the latest version Apriori [7] algorithm in both speed and scalability aspects. The experimental environment is: OS Windows 2000, CPU Intel 900M, Memory 128M. This group of experiments shows that our algorithm runs far faster than Apriori algorithm and has far better excellent scalability.

5.3 Frequent Closed Patterns and Rules Mining Algorithm [8]

In paper [8], we propose an efficient algorithm for frequent closed itemsets mining. In addition, we have extended FCIS into CI_RULES, which can highly efficiently produce frequent closed itemsets rules. The following experiments were done on a 233MHz PC with 128MB of memory, running Windows 2000 to compare both performance and scalability of our algorithm and CLOSET on Pumsb and Connect-4. Please note that the time axes are all in logarithmic scale. The performance data of CLOSET and CHARM is extracted from paper [14].

These experiments show: (1) our algorithm outperforms CLOSET with at least an order of magnitude on real datasets; (2) our algorithm has a wonderful scalability; (3) our algorithm can highly efficiently produce frequent closed itemsets rules.

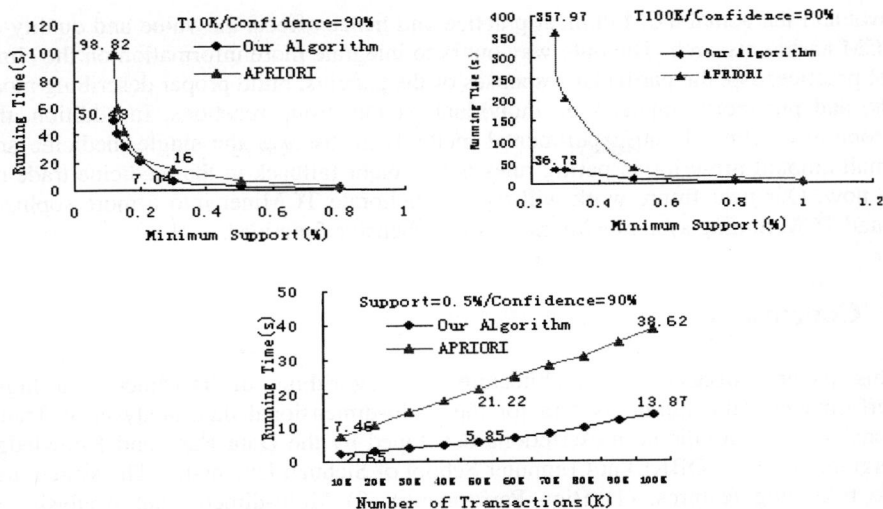

Fig. 6. Performance and scalability contrast of our algorithm and latest APRIORI

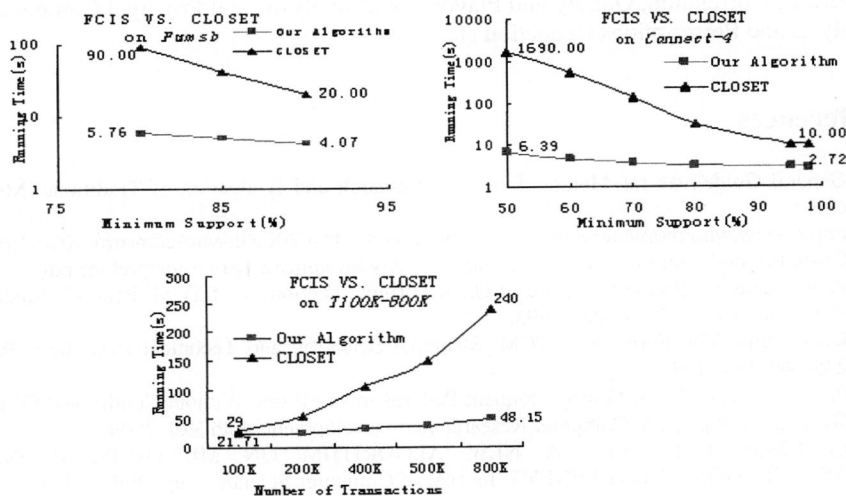

Fig. 7. Performance and scalability contrast of our algorithm and CLOSET

6 Future Directions

Although TCMiner has achieved inspiring results in both computer and TCM sides, some problems still exist. One noticeable problem is that for most theoretical prescription research the clinical application oriented issues are always ignored to some extent, which leads to the separation of research of TCM prescription paring laws and the actual TCM clinical practice and hence affects the value and quality of TCM paring research. The only way out is to integrate more information on the clinical practice, esp. the particular responses of the patients, build proper describing models, and put more emphasis on individual patient group reactions. In addition, the adoption of clinical and experimental methods to discover the single medicine and small amount prescription paring rules is the major fallback in the medicine trade up to now. Our near future work will try to ameliorate TCMiner into a more sophisticated TCM mining system with more comprehensive functions.

7 Conclusion

This paper introduces the architecture and algorithms of TCMiner – a high-performance data mining system for the multi-dimensional data analysis of Traditional Chinese Medicine prescriptions developed by the Data Base and Knowledge Engineering Lab (DBKE) at Computer School of Sichuan University. The system has the following features: (1) High Performance (2) Multi-dimensional Analysis (3) Flexibility (4) Interoperability (5) Optimization for TCM. This data mining system can work as a powerful assistant for TCM experts by conducting Traditional Chinese Medicine Data Mining such as Computer Aided Medicine Pairing Analysis, Medicine Syndrome Correlation, Quality and Flavor Trend Analysis, and Principal Components Analysis and Prescriptions Reduction etc.

References

1. General Guidelines for Methodologies on Research and Evaluation of Traditional Medicine,
   http://www.who.int/medicines/library/trm/who-edm-trm-2000-1/who-edm-trm-2000-1.pdf
2. Guste Editors' Notes on the special issue, http://www.sinica.edu.tw/~jds/preface.pdf
3. Peng Huairen. The First Volume of Great Formula Dictionary of TCM. People's Medical Publishing House. December 1993.
4. Duan Fujin. The Formula of TCM. Shanghai Scientific and Technical Publishers. Page 248-249, June 1995
5. FAN Ming, LI Chuan, Mining Frequent Patterns in an FP-tree Without Conditional FP-tree Generation, Journal of Computer Research and Development, 40th Vol. 2004
6. LI Chuan, FAN Ming, A NEW ALGORITHM ON MULTI-DIMENSIONAL ASSOCIATION RULES MINING, Journal of Computer Science, Aug. 29th Vol. A Complement, page 1-4, 2002
7. LI Chuan, FAN Ming, GENERATING ASSOCIATION RULES BASED ON THREADED FREQUENT PATTERN TREE, Journal of Computer Engineering and Application, 4th Vol., 2004
8. FAN Ming, LI Chuan, A Fast Algorithm for Mining Frequent Closed ItemSets, submitted to ICDM'04
9. LI Chuan, FAN Ming Research on Single-dimensional Association Mining, Full Paper Data Base of Wanfang Network
10. http://www.ics.uci.edu/~mlearn/MLRepository.html
11. http://www.almaden.ibm.com/cs/quest/demos.html

12. J. Han, J. Pei and Y. Yin. Mining frequent patterns without candidate generation. Proc. 2000 ACM-SIGMOD Intl. Conf. on Management of Data, pages 1-12. May 2000.
13. R. Agrawal and R. Srikant. Fast algorithms for Mining association rules. Proc. 1994 Int'l Conf. on Very Large Data Bases, pages 487-499, Sept. 1994.
14. Jian Pei, Jiawei Han, and Runying Mao. CLOSET: An Efficient Algorithm for Mining Frequent Closed Itemsets. Proc. 2000 ACM-SIGMOD Int. 2000 ACM SIGMOD Intl. Conference on Management of Data. page 8-10

Association Rule Mining Based on the Multiple-Dimensional Item Attributes*

Yunfeng Duan[1], Tang Shiwei[1], Yang Dongqing[1], and Meina Song[2]

[1] Peking University, Beijing, 100871, China
duanyunfeng@chinamobile.com
[2] Beijing University of Posts and Telecommunications, Beijing, 100876, China
mnsong@bupt.edu.cn

Abstract. The association rule mining is an important topic in recent data mining research. In this paper, a new association rule mining method based on the multiple-dimensional item attributes is proposed through the Market Basket Analysis. The corresponding average weight support is defined, and the AWMAR algorithm is described in detail. Finally, the performance study and results analysis of the improved algorithm is presented. AWMAR algorithm is effective for mining the association rules with acceptable running time.

1 Introduction

The association rule mining, introduced in [1], is widely used to derive meaningful rules that are statistically related in the underlying data[2,3,4,5].The main problem of association rule analysis is that there are so many possible rules, whose orders of magnitude always are billions or even trillions. It is obvious that such a vast number of rules cannot be processed by inspecting each one in turn. Efficient algorithms are needed to find the interested rules without missing important facts.

The Apriori algorithm, which is a classical method proposed by R.Agrawal[6], is the best known algorithm for association rule induction. This algorithm will generate the frequent itemsets (often called large itemsets) through computing the support and comparing it with the user-specified minimum support (i.e., minimum occurrence in a given percentage of all transactions). And finally the association rules are determined.

Currently no multiple-dimensional item attribute is considered for the current association analysis. For example, let's analyze the following cases for the retail market transaction. Customer A bought a top-level video player, two earphones and three CDs. Customer B bought one earphone and one CD. Customer C bought 10 CDs and 20 batteries. If we use the traditional association rules analysis, only the rules of item category can be found. No other item attributes, such as the item number and price is considered. So we can only draw the conclusion that the relationship existed between the earphone and the CD.

But if the number of the bought item is considered, the association between the CDs and the battery also existed. Because the total number of bought item for cus-

---

* This work is supported by the NKBRSF of China (973) under grant No.G1999032705, the National '863' High-Tech Program of China under grant No. 2002AA4Z3440.

S. Wang et al. (Eds.): ER Workshops 2004, LNCS 3289, pp. 258–265, 2004.

tomer A and customer B is three earphones and four CDs. The purchase pattern for customer C is worthy to be analyzed because of the number of bought item. And to the market manager, he will pay his attention on how to induce the customer to buy the top-grade video player for chasing the highest profit margin.

Then the problem will be brought forward during the association rules analysis. That is how to integrate multiple-dimensional item attributes for association rules mining. In [7] and [8], the concept of weighted item is introduced. The value of the weight is between 0 and 1. But the weight is roughly decided according to the normalized item value. Applied to the foregoing case, the normalized weight is 0.9 for the video player, 0.2 for the earphone and 0.1 for the CD. The intending weight is not fine enough to include all item attributes information in evidence. In order to mine the interested association rules among the items with multiple-dimensional attributes, item weights should be defined based on attributes. In this paper, we focus on developing and analyzing an AWMAR (Average Weight Mining Association Rules) algorithm to address the above problem. The algorithm description and performance study are also presented to show the validity and efficiency.

Without the loss of generality, in the following of this paper, it is convenient for us to explain the Association Rule Mining based on the two-dimensional item attributes, which are the item number and the item price.

The organization of the rest of the paper is as follows. In Section 2, we give a formal support definition with two-dimensional item attributes. In Section 3, we present the corresponding algorithm for mining association rules. In Section 4, we study the performance and the example analysis results of the improved algorithm. We'll conclude this paper with a summary in section 5.

2 Association Rules Based on the Two-Dimensional Item Attributes

When the multiple-dimensional item attributes is taken into account, the key problem is how to define the support. The frequent itemset is generated according to its support. And the association rules are found based on the confidence. The reason for introducing the concept of two-dimensional is to bring the importance of item number and item price into determining the final association rules. And the item contribution is related to its relative transaction revenue proportion to the total amount. Aiming at each item, both the item number and the item price attribute is considered representing the item transaction revenue. The total amount of each product is introduced to achieve the average weight combined the two attributes for one item.

Let's suppose $I = (i_1, i_2, \ldots, i_k)$ is the itemset and $T = (t_1, t_2, \ldots, t_m)$ is the transaction set. Each item can be represented as $t_i = \{(i_{i1}, n_{i1}, p_{i1}), (i_{i2}, n_{i2}, p_{i2}), \ldots, (i_{ik}, n_{ik}, p_{ik})\}$. n_{ik} is the number of item i_k for transaction t_i; p_{ik} is the price of item i_k for transaction t_i. $sum(i_j)$ is the total amount of item i_j, which equals to the product of the number and price of item i_j. For any q-itemset which is named X, the definition of its average weight named Ave-Weight is

$$\text{Ave - Weight}(X) = \frac{1}{q} \left(\sum_{i_j \in X}^{j} \left(\frac{\sum_{i_{jq} \in X}^{q} n_{jq} P_{jq}}{sum(i_j)} \right) \right).$$

(1)

The intuitionistic explanation of the foregoing definition is that the average weight of each item, which is named Ave-Weight, is the average value for all items contained

in itemset X. And $\left(\sum_{i_j \in X}^{j} \left(\dfrac{\sum_{i_{jq} \in X}^{q} n_{jq} P_{jq}}{sum(i_j)} \right) \right)$ is the proportion of amount for each item

contained in itemset X to total amount.

The support of q-itemset can be defined as

$$Sup_{\text{Ave-Weight}}(X) = \text{Ave - Weight}(X) \times Sup(X) .$$

(2)

Where $Sup(X)$ is the original support for item X, which is defined by the percentage of those transactions in all transactions under consideration which contain X[6,9,10]. The minimum support input by users is represented by $\min_{sup}$.

It can be observed from the above definition that three parts are included based on the two-dimensional support. The first part is the term for normal support, which aims at the occurrence frequency of item X relative to all transactions. The second part aims at the influence from number attribute of the item. The third part considers the effect from price attribute of the item. Through the integrated average weight, two-dimensional attributes will be combined to determine the final association rules.

For item X, if $Sup_{\text{Ave-Weight}}(X) >= \min_{sup}$, then X is named as Average-Weight-Frequency itemset. The followed two basic properties provide a basis for the Apriori algorithm. The first property is that if item X is frequent itemset then any subset of X are all frequent itemsets. The second property is that if X is not frequent itemset then any subset of X are not frequent itemsets. It's obvious that the above two properties don't come into existence. Thus a new algorithm is needed to generate frequent itemset, which will be discussed in the next section.

3   Algorithm Description

The process to generate the Average-Weight-Frequency itemset is similar to horizontal weight with generality proposed in [7]. Consequently the requirements of average weight are met through improving the MWAL (Mining Weight Association Rule).

Theorem 1. Supposing item X is an Average-Weight-Frequency itemset, for Y, which is any subset of X, if $Sup_{\text{Ave-Weight}}(Y) \geq Sup_{\text{Ave-Weight}}(X)$, then Y is also an Average-Weight-Frequency itemset.

Proof. Because Y belongs to X and any subsets will be contained by other itemsets, then $Sup(Y) \geq Sup(X)$. According to $Sup_{\text{Ave-Weight}}(Y) \geq Sup_{\text{Ave-Weight}}(X)$, $Sup_{\text{Ave-Weight}}(Y) =$ Ave-Weight(Y)×Sup(Y) $\geq$ Ave-Weight(X)×Sup(X) $\geq \min_{\text{sup}}$. Thus we draw the conclusion that Y is also an Average-Weight-Frequency itemset.

Theorem 2. Item X is supposed to be an Average-Weight-Frequency itemset and $X = X_1 \cup X_2 \cup \cdots \cup X_m$, $Ave\text{-}Weight(X_1) \leq Ave\text{-}Weight(X_2) \leq \cdots \leq Ave\text{-}Weight(X_m)$, for any subset Y of item X, if $Y = X_i \cup X_{i+1} \cup \cdots \cup X_m (i \geq 1)$ comes into existence, then subset Y is also an Average-Weight-Frequency itemset.

Proof. Because of $0 \leq Ave\text{-}Weight(X_1) \leq Ave\text{-}Weight(X_2) \leq \cdots \leq Ave\text{-}Weight(X_m)$ ≤ 1 and $Ave\text{-}Weight(X)$ is the average weight of the item attributes, then $Ave\text{-}Weight(Y) \geq Ave\text{-}Weight(X)$. Because Y is a subset of X, then we can learn from the theorem 1 that Y is also an Average-Weight-Frequency itemset.

Through improving the MWAL algorithm proposed in [7], the AWMAR algorithm based on the two-dimensional item attributes can be described in detail as follows.

```
Algorithm AWMAR
Begin
      C₁ = Generate_C₁(D);
      L₁ = {c∈ C₁•Sup_ave-weight(c)>=min_sup};
      k = 1;
      while (•L_k•> 0 ) do
         k++;
         C_k = Generate_C_k(C_k-1);
         C_k = Prune(C_k);
         If c∈ C_k  then
         Ave-Weight(c)= Generate_Average_Weight(c)
         L_k= { c∈ C_k•Supave-weight(c)>=min_sup};
         L = ∪k L_k;
         R = Generate_Rule(L);
   End
```

From the function named Generate_C_1, 1-itemset C_1 is produced. The original support Sup() is computed for all 1-itemsets firstly. Secondly the function named Generate_ Average_Weight () is called to compute the average weight. Thirdly the item in 1-itemset C_1 is sorted according to Ave-Weight() and the Average-Weight-Frequency itemsets are saved into L_1.

In order to prepare the precondition of average weight support, the function Generate_ Average_Weight () is responsible for the average weight computation. For each item, the total revenue amount is computed and the average weight Ave-Weight() is generated. This will decrease the complexity to produce the average weight support.

The function Generate_C_k will generate k-itemset using the k-1 candidate itemsets. Supposing that $C_k = L_k \cup S_k, L_k \cap S_k = \phi$ and L_k is the k-itemset, according to theorem 2, any k+1-average-weight-frequent itemset must be a low-order superset of k- average-weight-frequent itemset. The method to generate C_k is given as follows.

```
Begin
insert into Cₖ
        select b.itemᵢ , a.item₁, a.item₂, … , a.itemₖ₋₁
        from Lₖ₋₁ a , Sₖ₋₁ b
        where b.itemᵢ < a.item₁ ;
End
```

The useless itemsets are pruned from C_k by the function Prune(). At first, the original support of all itemsets in C_k is computed. For example, if the original support is smaller than the minimum support min_{sup} defined by users, then the itemset is pruned. And then the $Sup_{ave\text{-}weight}$ () is processed.

The method for Generate_Rule to generate the association rules can refer to [11].

4   Performance Study and Results Analysis

4.1   Performance Study

In this section, in order to analyze the performance of the AWMAR and the Apriori algorithms, we used a real supermarket retail database. The one-week sale data are considered, which include 235 transactions and 430 items. And we performed our experiments on PC with 1.8GHz processor and with 512MB of main memory under Microsoft Windows XP operating system.

Figure.1 presents the AWMAR algorithm running time performance compared with Apriori. Four groups of experiment were accomplished using the same database with different minimum support. The main differences between two algorithms are produced by the number of candidate itemsets. With the minimum support increase, the running time for both decreases for less candidate itemsets. Under the higher user-specified minimum supports, performance of AWMAR is close to that of Apriori. It is also duo to less candidate itemsets.

From Figure.1, we can observe that the performance of AWMAR algorithm is acceptable compared with Apriori algorithm.

Figure.2 is the result for the number of association rules under the condition that minimum support is 0.4.When the number of transactions increases, more association rules is found.

4.2   Results Analysis

To explain the difference between AWMAR and Apriori in a simple way, the followed example is analyzed.At first, let's assume the retail product include 5 kinds which is represented by {A, B, C, D, E} and 6 transactions occurred. The detailed instance is listed in Table 1.

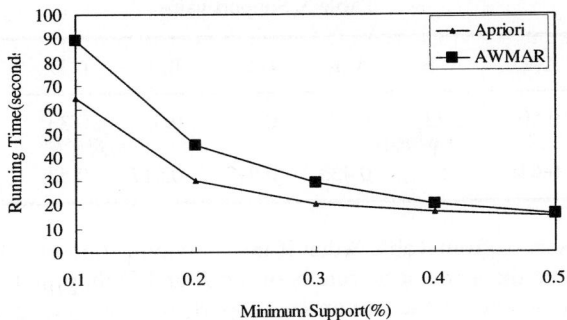

Fig. 1. AWMAR Performance 1(Running time)

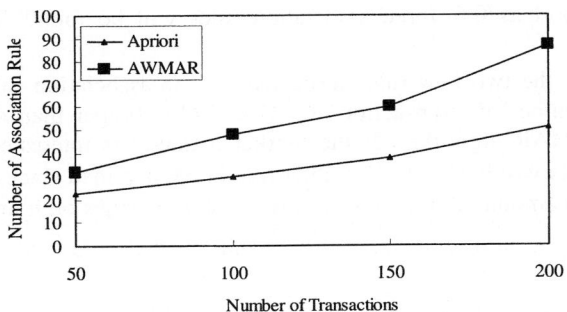

Fig. 2. AWMAR Performance 2(Association rules)

Table 1. Transaction instance

Sequence Number	Product Name, Number, Price
1	(A,1,10),(D,2,1)
2	(A,2,10),(B,1,5),(D,1,1)
3	(C,1,2),(D,2,1),(E,1,50)
4	(A,1,10),(B,3,5),(C,5,2)
5	(A,1,10),(D,2,1),(E,2,50)
6	(B,6,5),(C,7,2),(D,2,1)

Table 2. Total number and revenue for each products

Product name	A	B	C	D	E
Total number	5	10	13	9	3
Total Revenue	50	50	26	9	150

The total product number and transaction revenue is summarized in Table 2.what is showed in Table 3 are the values for original support and the average weight support.

Table 3. Support value

Association Rules	A, D	A, B	A, E	B,C	B, D	C,D	C,E	D,E
Original Support	0.50	0.33	0.17	0.33	0.33	0.33	0.17	0.33
Ave- Weight Support	0.678	0.5	0.433	0.912	0.517	0.53	0.205	0.722

It can be observed from Table 3 that if the minimum support is 0.5 the Apriori algorithm will mine the association rule between A and D through the original support definition. If we decrease the minimum support, the association rules that will be found are A→B, A→D, B→C, B→D, C →D, D→E. There are so many irrelevant association rules involved.

In the two-dimensional AWMAR algorithm improved in this paper, if we set the minimum support as 0.6, three association rules will be found. They are A→D, B→C, D→E.

Let's review the two new rules respectively. The association rule B→C is found because the number of transactions for B and C is bigger than others who is not found by AWMAR algorithm. If the market manager is interested on the product number, this rule will help him. The other rule D→E is mined since the price of product E is highest among all products. So the association rules on it are more important than others.

5   Conclusions and the Future Work

From the relative performance study and results analysis in section 4, we can draw a conclusion that the improved AWMAR algorithm based on two-dimensional item attributes is effective for mining the relative rules to market manager who care the specific product attributes with acceptable running time. In this paper, the experiment data is from real world. Thus how to define the "product price" to differentiate the item importance is not discussed. And it is our research direction for the future work.

References

1. Agrawal, R., Imielinski, T., Swami, A.:Mining Association Rules Betewwn Sets of Items in Large Databaseds. Proc.ACM SIGMOD Conf. management of Data.(1993)207–216
2. Rastogi,R., Kyuseok, S.: Mining optimized association rules with categorical and numeric attributes. IEEE Transactions on Knowledge and Data Engineering, Vol. 14 (2002)29–50
3. Xin,Y., Ju, S.G.: Mining conditional hybrid-dimension association rules on the basis of multi-dimensional transaction database. 2003 International Conference on Machine Learning and Cybernetics, Vol.1(2003)216–221
4. Li,T.R., Niu,Y.Q., Ma,J., Xu,Y.: Generalized association rule base mining and its algorithm. PDCAT'2003. Proceedings of the Fourth International Conference on Parallel and Distributed Computing Applications and Technologies(2003)919–922

5. Chiu,K.-S.Y., Luk,R.-W.P., Chan,K.-C.C., Chung, K.-F.L.: Market-basket analysis with principal component analysis: an exploration. 2002 IEEE International Conference on Systems, Man and Cybernetics, Vol.3(2002) 6–9
6. Agrawal, R., Imielienski, T., Swami, A.: Mining Association Rules between Sets of Items in Large Databases. In Proc. Conf. on Management of Data, New York(2001)207–216
7. Lu,S.F., Hu, H.P.: Mining Weighted Association Rules. Mini-Micro Systems(2001)
8. Cai,C.H., Ada Fu,W.C., Cheng,C.H., Kwong, W.W.: Mining Association Rules with Weighted Items. In Proc. Of the Int' 1 Database Engineering and Applications Symposium(1998)68–77
9. Chan, M.K., Fu, A., Man, H.W.: Mining fuzzy association rules in databases. in SIGMOD record , Vol.27(1998)41-46
10. Takeshi, F., Yasuhiko M.,Shinichi,M.: Data mining using two-dimensional optimized association rules. In proceedings of ACM SIGMOD(1996)13–23
11. Agrawal, R., Srikant,S.: Fast algorithms for mining association.In Proc. Of the 20th Int's 1 Conf. on Very Large Database(1994)487–499

OWL-Based Semantic Conflicts Detection and Resolution for Data Interoperability

Changqing Li and Tok Wang Ling

Department of Computer Science, School of Computing
National University of Singapore
{lichangq,lingtw}@comp.nus.edu.sg

Abstract. Ontology provides sharing knowledge among different data sources which will help to clarify the semantics of information. OWL is being promoted as a standard for web ontology language, thus in the future a considerable number of ontologies will be created based on OWL. Therefore automatically detecting semantic conflicts based on OWL will greatly expedite the step to achieve semantic interoperability, and will greatly reduce the manual work to detect semantic conflicts. In this paper, we summarize seven cases based on OWL in which semantic conflicts are easily to be encountered, and for each case, we give a rule to resolve the conflict. Via a series of examples, we show how this method works. Based on the seven cases, we propose a semantic conflicts detection and resolution algorithm. This aims at providing a semiautomatic method for semantic uniform data interoperability and integration.

1 Introduction

Data interoperability and integration is a long-standing challenge to the database research community. Merge of databases has been intensively studied in the past two decades, and some important architectures for data interoperability and integration have been proposed. These architectures include the early federated database system (FDBS) designed by Sheth and Larson [14], the later mediator systems proposed by Wiederhold [16]. Now the eXtensible Markup Language (XML) [20] as a standard of World Wide Web Consortium (W3C) has provided a syntactic platform for data representation and exchange on the Internet.

Past works have efficiently resolve the structure conflicts [7, 17], constraint conflicts [8], and schematic conflicts [4, 5] at syntactic level, however semantic interoperability is becoming more and more important since we must capture the semantics for the heterogeneous schemas and data before we could integrate them syntactically. To achieve semantic interoperability, we must firstly detect what kind of semantic conflicts may be encountered in data integration or interoperability. Ontology provides sharing knowledge among different data sources which will help to clarify the semantics of information. OWL (Web Ontology Language) [15] is being promoted as a standard for web ontology language, thus in the future, a considerable number of ontologies will be created based on OWL. Therefore in this paper, we show how to use OWL to detect the semantic conflicts in data interoperability and integration.

S. Wang et al. (Eds.): ER Workshops 2004, LNCS 3289, pp. 266–277, 2004.

The rest of the paper is organized as follows. In Section 2, we introduce the ontology languages; and in Section 2.1, we list some primitives of RDF, RDFS and OWL which provide a preliminary for the semantic conflicts detection in Section 3; moreover, we introduce the motivation of this paper in Section 2.2. In section 3, we summarize seven cases based on OWL in which semantic conflicts are easily to be encountered, also an algorithm is designed in this section. Related works and comparison are presented in Section 4. Finally we conclude the research in Section 5.

2 Preliminaries and Motivation

An ontology defines the basic terms and relations comprising the vocabulary of a topic area, as well as the rules for combining terms and relations to define extensions to the vocabulary [12]. Thus different applications can understand each other when referring to the sharing ontologies.

The Simple HTML Ontological Extensions (SHOE) [10] extends HTML with machine-readable knowledge annotated. Although SHOE has XML version, it is not based on the Resource Description Framework (RDF) [6] and RDF Schema (RDFS) [1].

RDF is a recommendation of W3C for Semantic Web [24], and it defines a simple model to describe relationships among resources in terms of properties and values. RDF does not define many particular primitives to annotate data, thus RDFS, the primitive description language of RDF, was created to provide appropriate primitives.

The Ontology Inference Layer (OIL) [23] from the On-To-Knowledge Project is the first ontology representation language that extends RDFS with additional language primitives not yet presented in RDFS.

The DARPA Agent Markup Language (DAML) [22] funded by the US Defense Advanced Research Projects (DARPA) aims at developing a language to facilitate the semantic concepts and relationships understood by machines. The DAML language is also based on the syntax of RDF and the primitives of RDFS.

DAML+OIL [18, 19] is the latest extension of DAML which has some important features of OIL imported. Presently, DAML+OIL is evolving as OWL (Web Ontology Language) [15]. OWL is almost the same as DAML+OIL, but some primitives of DAML+OIL are renamed in OWL for easier understanding.

In Section 2.1, we list the primitives of RDF, RDFS and OWL, which are used in Section 3 for the semantic conflicts detection. In Section 2.2, we introduce the motivation of this paper.

2.1 RDF, RDFS and OWL Primitives

In this section, the primitives of RDF, RDFS and OWL are explained so that the concepts in Section 3 could be understood more easily. In order to represent some concepts and relationships that cannot be described by the existing OWL primitives for semantic conflicts detection, we extend OWL by adding some new primitives such as "orderingProperty", "overlap". The namespace [21] "eowl" is used to refer to the extended primitives in the Extensions of OWL (EOWL).

Table 1. Part primitives of RDF, RDFS, OWL and EOWL

Category	Primitives	Comment
RDF	rdf:parseType="Resource"	To indicate the range of a property comprises several values
	rdf:Bag	An unordered collection (set) of members
	rdf:Seq	An ordered collection (set) of members
RDFS	rdfs:label	To provide a human-readable version of a resource
OWL	owl:equivalentClass	To indicate the equivalence between two classes
	owl:euqivalentProperty	To indicate the equivalence between two properties
	owl:sameIndividualAs	To indicate the equivalence between two instances
	owl:disjointWith	To indicate that two classes are disjoint
	owl:differentFrom	To indicate one instance is different from another instance
	owl:intersectionOf	A new class is the intersection of a list (set) of existing classes
	owl:unionOf	A new class is the union of a list (set) of existing classes
	owl:complementOf	A new class is the complement of a existing class
	owl:inverseOf	To define one property is the inverse of another property
EOWL	eowl:orderingProperty	To indicate that a property is order sensitive
	eowl:overlap	To indicate that two classes are overlap
	eowl:properSubClassOf	Class A is a sub class of B, but A is not equivalent to B
	eowl:properSubProperty-Of	Property A is a sub class of B, but A is not equivalent to B

In Table 1, we list the primitives of RDF, RDFS, OWL and EOWL and we explain the meaning of each primitive in the "Comment" column. The "rdf", "rdfs", "owl" and "eowl" in Table 1 are all XML namespaces [21] which refer to the languages RDF, RDFS, OWL and the Extensions of OWL (EOWL); the words after the colon following the namespaces, such as "ID" in "rdf:ID", are primitives defined in the corresponding language, such as RDF.

We will show how to use these primitives to detect semantic conflicts for data interoperability and integration in Section 3.

2.2 Motivation

Semantic interoperability is becoming more important than before. To achieve semantic interoperability, we must firstly detect what semantic conflicts may be encountered. But the current semantic conflicts detection depends mostly on human intervention which is an impedance for automatic semantic interoperability. Ontologies play an important role in providing clear semantics, and OWL is being promoted to be a standard of web ontology language, then in the future, a considerable number of ontologies will be created based on OWL. Thus if we could detect semantic conflicts based on OWL, that will greatly expedite the step of achieving semantic interoperability.

3   OWL-Based Semantic Conflicts Detection and Resolution

We use three steps to achieve data interoperability and integration:

Step 1. Detect possible semantic conflicts by examining ontologies.

Step 2. Resolve semantic conflicts detected at Step 1.

Step 3. Syntactically achieve the data interoperability after the semantics of the source data are clear.

Step 3 can be finished by applying syntactic integration methods, such as Mediator [16], XML [20]. Here we mainly consider Step 1 and Step 2 in this section.

3.1   OWL-Based Semantic Conflicts Cases

Seven semantic conflict cases based on OWL ontology language are summarized in this section. And in order to resolve the semantic conflicts, we propose the conflict resolution rules for each case. Examples are used to illustrate how to detect and resolve semantic conflicts.

A.   Name Conflicts

Case 1. Semantic conflicts may exist if one concept is defined the same as another concept (Synonyms). The OWL primitives "owl:equivalentClass", "owl:equivalent-Property" and "owl:sameInvidualAs" are indicators to detect this case.

Conflict Resolution Rule 1. If synonym conflicts are detected, different attribute names with the same semantics need to be translated to the same name for smooth data interoperability.

Example 1. Consider two distributed data warehouses, one is used to analyze the United States market, and the other is used to analyze the China market. The data can be analyzed at different region levels, such as country, state, city and district, or at different time levels, such as year, quarter, month and day. In order to analyze the global market, the data warehouses for the United States, China and many other countries need to be integrated. But some synonyms may exist in the integration, for example, the United States uses country, state, city etc. levels to analyze data, but China uses country, province, city etc. levels to analyze data, therefore in the integration, we must know that state and province are just at the same level. To detect this semantic conflict, we search "State" in the region ontology shown in Figure 1, and we find "Province" is defined equivalent to "State" using the OWL primitive "owl:equivalentClass". Then "State" and "Province" may encounter synonym conflict in the integration. To resolve this conflict, one name needs to be changed.

Remark 1. Other region concepts such as country, city could be similarly defined in the region ontology, but in Figure 1, we only show the equivalence between "State" and "Province" since only these two concepts are used in Example 1. The examples in later sections may also need this assumption.

```
<owl:Class rdf:ID="Province">
  <rdfs:label>Province</rdfs:label>
  <owl:equivalentClass
rdf:resource="#State"/>
</owl:Class>
```

Fig. 1. Detection of synonym conflicts

```
<rdf:Property rdf:ID="gregorianDate">
  <rdfs:label>date</rdfs:label>
</rdf:Property>
<rdf:Property rdf:ID="chineseDate">
  <rdfs:label>date</rdfs:label>
</rdf:Property>
```

Fig. 2. Detection of homonym conflicts

Remark 2. The first character of a class or instance ID should be in uppercase, such as "Province", "State" in Figure 1. But the first character of a property ID should be in lowercase, such as "gregorianDate", "chineseDate" in Figure 2.

Here we simply explain the definitions in Figure 1. Figure 1 is just an XML file, but it is different from a common XML file. The element name such as "owl:Class" and attribute name such as "rdf:ID" are not arbitrary strings defined as we like; they come from certain ontology languages such as "OWL", "RDF" (see Section 2.1 for the primitives of RDF, RDFS and OWL), thus the elements and attributes such as "Class" and "ID" have namespaces such as "owl" and "rdf" before. Namespaces "owl" and "rdf" are short terms to refer to the URIs which specify the definitions of ontology language OWL and RDF. The hash mark "#" in Figure 1 means the concept after the hash mark is just a reference but not a definition. For example, "Province" is defined equivalent to "#State"; here "State" is referred by "Province", but the definition of "State" is not shown in Figure 1.

Case 2. Semantic conflicts may exist if concepts with different IDs have the same labels (Homonyms). RDFS primitive "rdfs:label" is an indicator to detect this case.

Conflict Resolution Rule 2. If homonym conflicts are detected, the same name attributes should be renamed to different names for smooth data interoperability.

Example 2. Consider two database schemas Person(ID, name, sex, birthDate) and Person(ID, name, gender, birthDate). Both of the two person schemas use "birthDate" as an attribute. But when searching the time ontology, Figure 2 shows that both "gregorianDate" and "chineseDate" use "date" as their labels ("rdfs:label"), however, "gregorianDate" and "chineseDate" have different meanings. Thus when we integrate the two "birthDate" of person schemas, they may encounter homonym conflicts as one birth date may refer to the Gregorian date, but the other may refer to the Chinese date. Then it may be wrong if we just simply merge the data of "birthDate" of the two person schemas together. We need to rename the "birthDate" in the two person schemas to "gregorianBirthDate" and "chineseBirthDate" for clearer semantics.

B. Order Sensitive Conflicts

Case 3. Semantic conflicts may exist if one concept or property is defined to be order sensitive (Order sensitive). EOWL primitive "eowl:orderingProperty" and RDF primitive "rdf:Seq" are indicators to detect this case.

Conflict Resolution Rule 3. If order sensitive conflicts are detected, we need to adjust the member sequence according to the same criterion for smooth data interoperability.

Example 3. Consider the highest three scores of courses. The highest three scores of course A are listed as "90, 95, 100" at ascending order, but the highest three scores of course B are listed as "98, 95, 93" at descending order. By searching ontologies, we found that the "highestThreeScores" is defined as an "eowl:orderingProperty" (Figure 3) which means this property is order sensitive, then in the interoperability of the highest three scores for course A and course B, we are reminded that order sensitive conflicts may be encountered. The sequences of the highest three scores for course A and B should be adjusted both to ascending order or descending order.

```
<eowl:orderingProperty
          rdf:ID="highestThreeScores">
   <rdfs:label>highest three scores of a course
   </rdfs:label>
   <rdfs:domain rdf:resource="#Course"/>
   <rdfs:range rdf:resource="xsd#integer"/>
</eowl:orderingProperty>
```

```
<owl:DatatypeProperty rdf:ID="price">
   <rdfs:domain rdf:resource="#Product">
   <rdfs:range rdf:parseType="Resource">
      <rdf:value/>
      <currency:CurrencyUnit/>
   </rdfs:range>
</owl:DatatypeProperty>
```

Fig. 3. Detection of order sensitive conflicts   **Fig. 4.** Detection of scaling conflicts

C. Scaling Conflicts

Case 4. Semantic conflicts may exist if the value of a data type property comprises both value and unit (Scaling). RDF primitive "rdf:parseType="Resource"" and OWL primitive "owl:DatatypeProperty" are indicators for this case.

Conflict Resolution Rule 4. If scaling conflicts are detected, the value should be translated to refer to the same unit for smooth data interoperability.

Example 4. Consider two database schemas Product(ID, Price) and Product(ID, Price) for interoperability; although both of these two schemas use "Price" as an attribute, one price may refer to the US dollars, while the other may refer to the Singapore dollars. Figure 4 shows some concepts about a currency ontology; "price" is defined as a data type property using "owl:DatatypeProperty"; and the range of this data type property "price" is a complex type indicated by "rdf:parseType="Resource"" which includes value and currency unit. Based on the definition of price, we know in the interoperability of the two product schemas, the price may refer to currencies of different countries, which is a scaling conflict. We need to translate the price to refer to the same currency unit.

D. Whole and Part Conflicts

Case 5. Semantic conflicts may exist if one concept is completely contained in another concept (Whole and part). EOWL primitives "eowl:properSubClassOf", "eowl:properSubPropertyOf" and OWL primitive "owl:unionOf" are indicators to detect this case.

Conflict Resolution Rule 5. If whole and part conflicts are detected, the whole attributes should be divided into part attributes or the part attributes should be combined together to whole attributes for smooth data interoperability.

Example 5. Consider schemas Person(ID, name) and Person(ID, surname, given-Name). In Figure 5, "surname" is defined as a proper sub property of "name"; similarly "givenName" can be defined as a proper sub property of "name" (not shown in Figure 5). "eowl:properSubClassOf" has clearer semantics than "rdfs:subClassOf" because "rdfs:subClassOf" is ambiguous with two meanings: "eowl:properSubClassOf" and "owl:equivalentClass". The whole and part conflicts are detected when examining the "eowl:properSubPropertyOf" between "surname" and "name". We need to divide the whole attribute "name" to the part attributes "surname" and "givenName" or combine the part attributes "surname" and "givenName" together in the correct sequence to form the whole attribute "name".

```
<rdf:Property rdf:ID="surname">
  <eowl:properSubPropertyOf
        rdf:resource="#name">
</rdf:Property>
```

```
<owl:Class rdf:ID="ResearchAssistant">
  <eowl:overlap
        rdf:resource="#GraduateStudent"/>
</owl:Class>
```

Fig. 5. Detection of whole and part conflicts   **Fig. 6.** Detection of partial similarity conflicts

Furthermore, if a class A is defined as the "unionOf" another two classes B and C, and A is not equal to B or C, then whole and part conflicts may exist when integrating A and B, or A and C.

E. Partial Similarity Conflicts

Case 6. Semantic conflicts may exist if two concepts are overlapped (Partial similarity). EOWL primitive "eowl:overlap" and OWL primitive "intersectionOf" are indicators to detect this case.

Conflict Resolution Rule 6. If partial similarity conflicts are detected, the overlap part should be separated before integration.

Example 6. Consider the integration of the ResearchAssistant schema and GraduateStudent schema. The relationship between research assistant and graduate student is overlap because some research assistants are also graduate students, but not all research assistants are graduate students, and not all graduate students are research assistants. The "eowl:overlap" in Figure 6 will help to detect this semantic conflict. After integration, there should be three schemas: the first one is Research Assistant but not Graduate Student, the second one is Graduate Student but not Research Assistant, and the third one is both Research Assistant and Graduate Student.

F. Swap Conflicts

Case 7. Semantic conflicts may exist if the domain of a property in the first ontology is the range of the same property in the second ontology, and the range of the property in the first ontology is the domain of the same property in the second ontology (Swap).

Conflict Resolution Rule 7. If swap conflicts are detected, context restrictions (see Example 7) should be added to the schema for smooth data interoperability.

Example 7. Continued from Example 1, we analyze data at different region levels. In China, county is contained in city (city has larger area), but in US, city is contained in county (county has larger area). This semantic conflict must be known before we can analyze data at the city and county levels. Suppose in the region ontology, a property "containedIn" is defined to represent the relationship between a smaller area and a larger area, for example, "State" is "containedIn" "Country". The China ontology and the US ontology are two specific ontologies of the region ontology. From Figure 7a, we can see "County" is defined contained in "City" using the property "region:containedIn". However in the US ontology shown in Figure 7b, "City" is defined contained in "County" using the same property "region:containedIn". The domain ("County") of property "region:containedIn" in the China ontology is just the range of the same property "region:containedIn" in the US ontology, and the range ("City") of property "region:containedIn" in the China ontology is just the domain of the same property "region:containedIn" in the US ontology. This is a swap conflict if we do not care the context. We can add "China." or "US." before "City" and "County" for smooth data interoperability.

```
<owl:Class rdf:ID="County">
  <region:containedIn
        rdf:resource="#City"/>
</owl:Class>
```

```
<owl:Class rdf:ID="City">
  <region:containedIn
        rdf:resource="#County"/>
</owl:Class>
```

Fig. 7a. Detection of swap conflicts (the relationship between city and county in the China ontology)

Fig. 7b. Detection of swap conflicts (the relationship between city and county in the US ontology)

G. Other Cases

If one class is defined "owl:disjointWith" another class, or the members of a class are defined pairwise different using the OWL primitive "owl:AllDifferent", or one class is defined as the sub class of the "owl:complementOf" of another class, then the two classes or instances are different, and they may not encounter semantic conflicts in the data interoperability. For example, class "FemalePerson" is defined as a sub class of the "complementOf" class "MalePerson", then "FemalePerson" is disjoint with "MalePerson", thus no semantic conflicts may be encountered in the interoperability of "FemalePerson" and "MalePerson".

However, we focus on how to detect semantic conflicts, but not on the cases that semantic problems may not be encountered. Thus we will not take care of these cases that two classes or instances are disjoint.

Other primitives of OWL, such as "owl:allValuesFrom", "owl:someValuesFrom", "owl:hasValue", "owl:TransitiveProperty", "owl:SymmetricProperty", may not easily lead to semantic conflicts, thus we do not present them in detail.

3.2 Semantic Conflicts Detection and Resolution Algorithm

In this section, we summarize Case 1 to 7 presented in Section 3.1 into an algorithm for semantic conflicts detection. The algorithm will automatically output the possible semantic conflicts and the possible resolution method.

Algorithm *Semantic conflicts detection and resolution*
Input: Data for interoperability
Output: Possible semantic conflicts and resolving method

Main program:
> *Step 1*: Search the input data in the ontologies. The classes, properties and instances related to these input data will be returned. Analyze these returned ontology information for data conflicts detection.
>
> *Step 2*: Call procedure **Conflict_Cases**(O) (O represent classes, properties or instances related to the data for interoperability)
>
> *Step 3*: Output the possible semantic conflicts

Procedure: **Conflict_Cases**(Object: O)
> *Switch* (O){
>> **Case**(class, property or instance in O is "owl:equivalentClass", "owl:equivalentProperty" or "owl:sameInvidualAs" another class, property or instance in O):
>>> **Return** "Synonym conflicts may be encountered; rename different names to the same name";
>>
>> **Case**(the "rdfs:label" value in one class, property or instance in O is the same as the "rdfs:label" content in other classes, property or instance in O; and they have different IDs):
>>> **Return** "Homonym conflicts may be encountered; rename the same name to different names";
>>
>> **Case**(the property of O is defined using "eowl:orderingProperty" or "rdf:Seq"):
>>> **Return** "Order sensitive conflict may be encountered; adjust the member sequence";
>>
>> **Case**(the range of a "DatatypeProperty" in O is marked with "rdf:parseType="Resource""):
>>> **Return** "Scaling conflicts may be encountered; translate the values to refer to the same unit";
>>
>> **Case**(class or property in O is defined as "eowl:properSubClassOf", "eowl:properSubPropertyOf" another class or property in O; or one class in O is defined as the "owl:unionOf" other classes in O):
>>> **Return** "Whole and part conflicts may be encountered; divide whole concept or combine part concepts";
>>
>> **Case**(class or property in O is defined as "eowl:overlap" another class or property in O; or one class in O is defined as the "owl:intersectionOf" other classes in O):

> ***Return*** "Partial similarity conflicts may be encountered; separate the over-
> lap part";
> ***Case***(the domain and range of the same property in two ontologies are re-
> versed):
> ***Return*** "Swap conflicts may be encountered; add context restrictions";
> ***Case else:***
> ***Return*** NOTHING;
> }

The algorithm uses the main program to search ontologies and output the possible semantic conflicts and the corresponding resolution method.

There may be a lot of "owl:equivalentClass", "owl:equivalentProperty" etc. in the ontologies, but in the algorithm, we search the related concepts firstly, then we analyze whether there are "owl:equivalentClass", "owl:equivalentProperty" etc. among these related concepts which will improve the performance of this algorithm.

This algorithm can reminder the user that certain semantic conflicts may be encountered. But it is not to say that certain conflicts must happen in the real data interoperability.

The comparison between our algorithm and other conflict detection and resolution methods is presented in the next section.

4   Related Works and Comparison

Semantic interoperability means the meaningful exchange of information [3], and it is becoming more and more important for the federated database architecture. To achieve semantic interoperability, semantic conflicts must be firstly detected and resolved. Semantic conflicts have been summarized by many researchers [11, 13], and these conflicts can be categorized as naming conflicts, scaling conflicts, and confounding conflicts [3]. Semantic naming conflicts include the synonym and homonym conflicts. Scaling conflicts (domain mismatch conflicts) involve the use of different units for the same concept. Confounding conflicts mean that equating concepts are actually different.

[9] is a data mining method to discover and reconcile semantic conflicts for data interoperability. It uses a correlation analysis to detect the relevance among the attributes of schemas and it uses regression analysis to detect the quantitative relationship. In [9], an example is conducted which is about the "hotel rate after tax" and the "hotel rate before tax". [9] uses correlation analysis in statistics to detect that "rate after tax" and "rate before tax" are highly correlated, thus they may be the same or very similar concepts in semantics, then [9] needs human intervention to decide whether they are indeed the same or similar in semantics. But this method may detect the semantics irrelevant concepts to be highly correlated, for example, the weight and height of persons are highly correlated, but weight and height obviously have different semantics. Furthermore, if the attributes are all from the same schema, they are more possible to be detected correlated, then much human intervention is needed to decide whether all these attributes are indeed related in semantics. In addition, this

method could only detect the synonym and homonym conflicts, i.e., things seeming different are in fact the same or very similar (highly correlated), and things with the same name are in fact different (not correlated).

Furthermore, compared with the work in [2] which used XML Topic Maps (XTM) [25] for data interoperability, the method in this paper has the following merit: XTM method uses "occurrence" to link back the distributed data when discrepancies happen, but the method in this paper could help to resolve the discrepancies.

Ontology is an important technique to capture data semantics. By analyzing the latest web ontology language OWL, we show how to detect semantic conflicts. Our method not only can cover the basic semantic conflicts such as name conflicts and scaling conflicts, but it also can detect order sensitive conflicts, whole and part conflicts, partial similarity conflicts, and swap conflicts.

5   Conclusion

Ontologies are becoming more important in today's semantic data interoperability and semantic web. OWL is an ontology language designed mainly for semantic web, but in this paper, we try to find how OWL can be used to detect semantic conflicts in data interoperability. Firstly, we extend OWL with several primitives which have clearer semantics, then we summarize seven cases based on OWL in which semantic conflicts are easily to be encountered, furthermore the conflict resolution rules for each case are presented. Following these cases, we propose an algorithm to detect and resolve the semantic conflicts. In the future, OWL will be frequently used to build ontologies, and this paper provides a computer-aid approach to detect and resolve semantic conflicts for smooth data interoperability.

References

1. Dan Brickley and R.V. Guha. Resource Description Framework (RDF) Schema Specification 1.0, W3C Candidate Recommendation 27 March 2000. http://www.w3.org/TR/rdf-schema/
2. Robert M. Bruckner, Tok Wang Ling, Oscar Mangisengi, A. Min Tjoa: A Framework for a Multidimensional OLAP Model using Topic Maps. WISE (2) 2001: pp109-118
3. Cheng Hian Goh: Representing and Reasoning about Semantic Conflicts in Heterogeneous Information Sources. Phd, MIT, 1997.
4. Won Kim, Jungyun Seo: Classifying Schematic and Data Heterogeneity in Multidatabase Systems. IEEE Computer 24(12): pp12-18 (1991)
5. Ravi Krishnamurthy, Witold Litwin, William Kent: Language Features for Interoperability of Databases with Schematic Discrepancies. SIGMOD Conference 1991: pp40-49
6. Ora Lassila and Ralph R. Swick: Resource description framework (RDF). http://www.w3c.org/TR/WD-rdf-syntax
7. Mong-Li Lee, Tok Wang Ling: Resolving Structural Conflicts in the Integration of Entity Relationship Schemas. OOER 1995: pp424-433
8. Mong-Li Lee, Tok Wang Ling: Resolving Constraint Conflicts in the Integration of Entity-Relationship Schemas. ER 1997: pp394-407

9. Hongjun Lu, Weiguo Fan, Cheng Hian Goh, Stuart E. Madnick, David Wai-Lok Cheung: Discovering and Reconciling Semantic Conflicts: A Data Mining Perspective. DS-7 1997: pp409-427
10. Sean Luke and Jeff Heflin: SHOE Specification 1.01. http://www.cs.umd.edu/projects/plus/SHOE/spec.html
11. Channah F. Naiman, Aris M. Ouksel: A classification of semantic conflicts in heterogeneous database systems. J. of Organizational Computing 5(2): pp167-193 (1995)
12. Robert Neches, Richard Fikes, Timothy W. Finin, Thomas R. Gruber, Ramesh Patil, Ted E. Senator, William R. Swartout: Enabling Technology for Knowledge Sharing. AI Magazine 12(3): pp36-56 (1991)
13. Amit P. Sheth, Vipul Kashyap: So Far (Schematically) yet So Near (Semantically). DS-5 1992: pp283-312
14. Amit P. Sheth, James A. Larson: Federated Database Systems for Managing Distributed, Heterogeneous, and Autonomous Databases. ACM Comput. Surv. 22(3): pp183-236 (1990)
15. Frank van Harmelen, Jim Hendler, Ian Horrocks, Deborah L. McGuinness, Peter F. Patel-Schneider and Lynn Andrea Stein. OWL Web Ontology Language Reference. http://www.w3.org/TR/owl-ref/
16. Gio Wiederhold: Mediators in the Architecture of Future Information Systems. IEEE Computer 25(3): pp38-49 (1992)
17. Xia Yang, Mong-Li Lee, Tok Wang Ling: Resolving Structural Conflicts in the Integration of XML Schemas: A Semantic Approach. ER 2003: pp520-533
18. DAML+OIL Definition. http://www.daml.org/2001/03/daml+oil
19. DAML+OIL (March 2001) Reference Description, W3C Note. http://www.w3.org/TR/daml+oil-reference
20. Extensible Markup Language (XML) 1.0 (Second Edition), W3C Recommendation. http://www.w3.org/TR/2000/REC-xml-20001006
21. Namespaces in XML, World Wide Web Consortium 14-January-1999. http://www.w3.org/TR/REC-xml-names/
22. The DARPA Agent Markup Language Homepage. http://daml.semanticweb.org/
23. The Ontology Inference Layer OIL Homepage. http://www.ontoknowledge.org/oil/TR/oil.long.html
24. The SemanticWeb Homepage. http://www.semanticweb.org
25. XML Topic Maps (XTM) 1.0. http://www.topicmaps.org/xtm/index.html

From Ontology to Relational Databases

Anuradha Gali[1], Cindy X. Chen[1],
Kajal T. Claypool[1], and Rosario Uceda-Sosa[2]

[1] Department of Computer Science, University of Massachusetts
Lowell, MA 01854, USA
{agali,cchen,kajal}@cs.uml.edu
[2] IBM T. J. Watson Research Center, Hawthorne, NY 10532, USA
rosariou@us.ibm.com

Abstract. The semantic web envisions a World Wide Web in which
data is described with rich semantics and applications can pose complex queries. Ontologies, a cornerstone of the semantic web, have gained
wide popularity as a model of information in a given domain that can
be used for many purposes, including enterprise integration, database
design, information retrieval and information interchange on the World
Wide Web. Much of the current focus on ontologies has been on the
development of languages such as DAML+OIL and OWL that enable
the creation of ontologies and provide extensive semantics for Web data,
and on answering intensional queries, that is, queries about the structure of an ontology. However, it is almost certain that the many of the
semantic web queries will be extensional and to flourish, the semantic
web will need to accommodate the huge amounts of existing data that is
described by the ontologies and the applications that operate on them.
Given the established record of relational databases to store and query
large amounts of data, in this paper we present a set of techniques to
provide a lossless mapping of an OWL ontology to a relational schema
and the corresponding instances to data. We present preliminary experiments that compare the efficiency of the mapping techniques in terms
of query performance.

1 Introduction

The Semantic Web brings closer to realization the possibility of semantically
organized data repositories, or *ontologies*, throughout the Internet that can be
used for intelligent information searching both by humans and computational
agents. Ontologies are human readable, comprehensive, sharable and formal
which means that they are expressed in a language that has well-defined semantics. Ontologies are important to application integration solutions because
they provide a shared and common understanding of data that exist within an
application integration problem domain. Ontologies also facilitate communication between people and information systems. However, while today there is
an un-precedented wealth of information available on the Web, to fully realize

S. Wang et al. (Eds.): ER Workshops 2004, LNCS 3289, pp. 278–289, 2004.

the power of ontologies and to enable *efficient* and *flexible* information gathering, persistent storage of ontologies and its subsequent retrieval is of paramount importance.

There are three possible approaches to store the ontology data and execute queries on that data. One is to build a special purpose database system which will be tailored to store and retrieve ontology specific data. The second is to use an object oriented database system exploiting the rich data modeling capabilities of OODBMS. Given the popularity of SQL and the efficiency of executing these queries, a complimentary approach is to use relational database systems. In this approach the OWL data is mapped into tables of a relational schema and the queries posed in an ontology language are translated into SQL queries. Relational database systems are very mature and scale very well, and they have the additional advantage that in a relational database, ontology data and the traditional structured data can co-exist making it possible to build applications that involve both kinds of data. While there are approaches that focus on translating XML documents into corresponding relational tables and data [9, 3], to the best of our knowledge ours is the first to focus on the persistent relational storage of OWL ontologies.

The rest of the paper is organized as follows: Section 2 gives an overview of OWL. Section 3 describes briefly the related work in terms of the different mapping schemes available for storing the XML documents in the relational database. The algorithm for translating the given OWL file into relational schema is explained in Section 4. Section 5 describes the preliminary performance comparison of the different approaches. We conclude in Section 6.

2   Background: Ontologies and OWL

Ontology is the key enabling technology for the semantic web. The main purpose of an ontology is to enable communication between computer systems in a way that is independent of the individual system technologies, information architectures and application domain. The key ingredients that make up an ontology are a vocabulary of basic terms, semantic interconnections, simple rules of inference and some logic for a particular topic.

Ontologies may vary not only in their content, but also in their structure, implementation, level of description, conceptual scope, instantiation and language specification. Ontologies are not all built the same way. A number of possible languages can be used, including general logic programming languages such as Prolog. More common, however, are languages that have evolved specifically to support ontology construction. The Open Knowledge Base Connectivity (OKBC) model and languages like KIF (and its emerging successor CL – Common Logic) are examples that have become the bases of other ontology languages. There are also several languages based on a form of logic thought to be especially computable, known as description logics. These include Loom and DAML+OIL, which are currently being evolved into the Web Ontology Language (OWL) standard. When comparing ontology languages, what is given up for computability and simplicity is usually language expressiveness, which many developers disagree since they feel that a good language should have a sound underlying semantics.[1] gives a good overview of all the relevant technologies in Ontology.

OWL. OWL [6] is a language for defining and instantiating Web ontologies. An OWL ontology may include descriptions of classes, properties and their instances. Given an ontology, the OWL formal semantics specify how to derive its logical consequences, i.e. facts not literally present in the ontology, but entailed by the semantics. These entailments may be based on a single document or multiple distributed documents that have been combined using defined OWL mechanisms. The OWL language provides three increasingly expressive sub-languages designed for use by specific communities of implementers and users.

OWL Lite supports those users primarily needing a classification hierarchy and simple constraint features. For example, while OWL Lite supports cardinality constraints, it only permits cardinality values of 0 or 1. It is typically simpler to provide tool support for OWL Lite than its more expressive relatives, and provide a quick migration path for thesauri and other taxonomies.

OWL DL supports those users who want the maximum expressiveness without losing computational completeness (all entailments are guaranteed to be computed) and decidability (all computations will finish in finite time) of reasoning systems. OWL DL includes all OWL language constructs with restrictions such as type separation (a class can not also be an individual or property, a property can not also be an individual or class). OWL DL was designed to support the existing Description Logic business segment and has desirable computational properties for reasoning systems.

OWL Full is meant for users who want maximum expressiveness and the syntactic freedom of RDF with no computational guarantees. For example, in OWL Full a class can be treated simultaneously as a collection of individuals and as an individual in its own right. Another significant difference from OWL DL is that a owl:DatatypeProperty can be marked as an owl:InverseFunctionalProperty. OWL Full allows an ontology to augment the meaning of the pre-defined (RDF or OWL) vocabulary. It is unlikely that any reasoning software will be able to support every feature of OWL Full.

3   Related Work – XML Mapping Approaches

Given the similarities between XML and OWL, in this section we give a brief overview of the mapping approaches that have been proposed in literature for XML documents.

Edge Approach. The simplest scheme proposed by Florescu et al. [3] is to store all attributes in a single table called the *Edge* table. The Edge table records the object identities (oids) of the source and target objects of the attribute, the name of the attribute, a flag that indicates whether the attribute is an inter-object reference or points to a value, and an ordinal number used to recover all attributes of an object in the right order and to carry out updates if the object has several attributes with the same name. The Edge table, therefore, has the following structure:

`Edge(source, ordinal, name, flag, target)`

The key of the Edge table is `source, ordinal`. Figure 2 shows how the Edge table would be populated for the example XML graph.

Attribute Approach. Attribute mapping scheme groups all attributes with the same name into one table. Conceptually, this approach corresponds to a horizontal partitioning of the Edge table used in the first approach, using name as the partitioning attribute. Thus, as many Attribute tables are created as for different attribute names in the XML document, and each Attribute table has the following structure:

`Aname(source, ordinal, flag, target)`

The key of the Attribute table is `source, ordinal`, and all the fields have the same meaning as in the Edge approach.

Universal Approach. Universal table approach stores all the attributes of an XML document. This corresponds to a Universal table with separate columns for all the attribute names that occur in the XML document. Conception ally, this Universal table corresponds to the result of an outer join of all Attribute tables. The structure of the Universal table is as follows, if n1, n2,...,nk are the attribute names in the XML document, then

```
Universal(source, ordinaln1, flagn1, targetn1, ordinaln2,
          flagn2, targetn2, . . . , ordinalnk, flagnk, targetnk)
```

The Universal table has many fields which are set to null, and it also has a great deal of redundancy.

Normalized Universal Approach. The UnivNorm approach, is a variant of the Universal table approach. The difference between the UnivNorm and Universal approach is that multi-valued attributes are stored in separate Over-flow tables in the UnivNorm approach. For each attribute name that occurs in the XML document a separate Overflow table is established, following the principle of the Attribute approach. This way, there is only one row per XML object in the UnivNorm table and that table is normalized. The structure of the UnivNorm table and the Overflow tables is as follows, if n1; : : : ; nk are all the attribute names in the XML document, then

```
UnivNorm(source, ordinaln1, flagn1, targetn1, ordinaln2, flagn2,
         targetn2, . . . , targetnk)
OverflowOwn1(source, ordinal, flag, target)
OverflowOwnk(source, ordinal, flag, target)
```

The key of the UnivNorm table is `source`. The key of the Overflow table is `fsource, ordinalg`.

Basic Inlining Approach. Basic Inlining Approach converts the XML DTD into relations. Basic inlines as many descendants of an element as possible into a single relation. Basic creates relations for every element because an XML document can be rooted at any element in a DTD. Basic approach is highly inefficient since it creates a large number of relations.

Shared Inlining Approach. Shared Inlining tries to avoid the drawbacks of basic by ensuring that an element node is represented in exactly one relation. Shared identifies the element nodes that are represented in multiple relations in Basic and shares them by creating separate relation for these elements. In shared, relations are created for all elements in the DTD graph whose nodes have an in-degree greater than one. Nodes with an in-degree of one are inlined. Element nodes having an in-degree of zero are also made into separate relations, because they are not reachable from any other node. Elements below a "*" node are made into separate relations.

Hybrid Inlining Approach. The hybrid inlining technique is the same as shared except that it inlines some elements that are not inlined in shared. Hybrid additionally inlines elements with in-degree greater than one that are not recursive or reached through a "*" node.

4   Mapping OWL to Relational Schema

Semantic web is increasingly gaining popularity and importance with an increasing number of people using ontologies to represent their information. To augment the growth of ontology, efficient persistent storage of ontology concepts and data is essential. We now examine relational databases as an effective persistent store for ontologies. While there is research on storing XML documents in relational databases, there has not been much work done on the persistent storage of ontologies, in particular of ontologies described by OWL.

In this section, we present an approach to map OWL to relational schema. The goal of our work is to devise an approach that will result in fewer joins when a user issues a query since join is an expensive operation. Since OWL is based on XML Syntax, but defines a lot of constraints on classes and provides inference, the approaches to store XML documents in relational databases are not fully applicable. We need to preserve all the constraint information while mapping OWL into relational schemas.

4.1   System Architecture

Our OWL to relational schema mapping system contains three parts – Ontology Modeler, Document Manager and Ontology Resource Manager.

`Ontology Modeler` takes OWL documents as input and creates an ontology model that is similar to DOM [2] for XML documents. While parsing the OWL documents, all the constraints will also be detected and recorded. An appropriate

OWL reasoner, depending upon the input document if its OWL Lite, OWL full or OWL DL, is selected. Currently, all the OWL documents are stored in main memory for parsing.

Document Manager helps in processing and handling of OWL documents. It uses Jena [4] to assist importing of OWL documents. This is important because if we do not do this, once the imports have been processed, it would be impossible to know where a statement came from. Document Manager helps in processing and handling multiple OWL documents. It builds the union of the imported documents and creates new models for the imported OWL documents.

Ontology Reasoner provides methods for listing, getting and setting the RDF [8] types of a resource. The `rdf:type` property defines many entailment rules in the semantic models of the various ontology languages.

4.2   OWL2DB Algorithm

The algorithm we developed to map an OWL document into relational table is given in Algorithm 1.

Algorithm 1 Mapping OWL to Relational Schema

1: The given OWL file is first parsed for Root classes that returns an iterator over all the root classes.
2: Iterate through the root classes to determine the depth of the descendants.
3: The showclass method is called that keeps track of the depth of the subclasses from the root and when the depth of the graph is 3 then a relation table is created for all the subclasses above it with attribute names/instances as column names.
4: The above step is repeated starting from the last subclass and traversing the graph and collecting the list of all the table names and column names of subclasses that are a the depth of three from the starting node.
5: The depth of the Union/complement/disjoint subclasses on the same level are the same.
6: Separate relations are created for classes that are disjoint from each other.
7: Once the information on tablenames and corresponding attributes are collected, we need to collect the values of the instances to populate the database. The graph is parsed again to collect the values that are stored in a data structure.
8: Finally, a database connection is established and all the tables are created and then the values are inserted. Inlining the root is handled the same way as a parent node.

Figure 1 shows an example OWL document. Table 1, 2, 3, 4 represent the relational tables into which the OWL document was mapped.

```
<rdf:RDF xmlns:xsi="http://www.w3.org/2001/XMLSchema-
         instance" xsi:noNamespaceSchemaLocation=
   "C:\usr\Repository\iris1.0\data\schema\IRISData.xsd"
         Name="FurnitureOntology">
<owl:Ontology rdf:about="">
```

```
<rdfs:comment>An example of OWL Ontology</rdfs:comment>
<owl:priorVersion rdf:resource="C:\usr\Repository\iris1.0
                \data\schema\IRISData.xsd"/>
<owl:imports rdf:resource="http://www.w3.org/2001/XMLSchema-
        instance"/>
<rdfs:label>Furnite Ontology</rdfs:label>

<!----Main Classes----->
<owl:Class rdf:ID="OntologyDependency"/>
   <rdfs:label
     xml:lang="en">ShoppingCatalogOntology</rdfs:label>
 </owl:Class>

<owl:Class rdf:ID="Relation"/>
   <rdfs:label>Relation/rdfs:label>
</owl:Class>
<owl:Class rdf:ID="ProtoNode"/>
   <rdfs:label>ProtoNode</rdfs:label>
</owl:Class>
<owl:Class rdf:ID="AtomicNode"/>
   <rdfs:label>Atomic Node</rdfs:label>
</owl:Class>

 <!----Sub Classes----->
    <owl:Class rdf:ID="Name"/>
      <rdfs:subClassOf rdf:resource="#Relation" />
      <rdfs:label xml:lang="en">Name</rdfs:label>
    </owl:Class>
    <owl:Class rdf:ID="Arity"/>
      <rdfs:subClassOf rdf:resource="#Relation" />
      <rdfs:label xml:lang="en">Arity</rdfs:label>
    </owl:Class>
    <owl:Class rdf:ID="Reflexive"/>
      <rdfs:subClassOf rdf:resource="#Relation" />
      <rdfs:label xml:lang="en">Reflexive</rdfs:label>
    </owl:Class>
    <owl:Class rdf:ID="Name"/>
        <rdfs:label>Name</rdfs:label>
        <rdfs:subClassOf>
        <owl:Restriction>
            <owl:onProperty rdf:resource="#ProtoNode"/>
            <owl:DatatypeProperty rdf:ID="string">
        </owl:Restriction>
        </rdfs:subClassOf>
    </owl:Class>
```

```
<owl:Class rdf:ID="IsAbstract"/>
      <rdfs:label>IsAbstract</rdfs:label>
      <rdfs:subClassOf>
      <owl:Restriction>
            <owl:onProperty rdf:resource="#ProtoNode"/>
            <owl:DatatypeProperty rdf:ID="string">
      </owl:Restriction>
      </rdfs:subClassOf>
</owl:Class>
- <owl:Class rdf:ID="ProtoNode"/>
      <rdfs:subClassOf>
      <owl:Restriction>
            <owl:onProperty rdf:resource="#ProtoNode"/>
      </owl:Restriction>
      </rdfs:subClassOf>
    </owl:Class>
      - <owl:Class rdf:ID="Name"/>
            <rdfs:label>Name</rdfs:label>
            <rdfs:subClassOf rdf:resource="#ProtoNode" />
            <owl:Restriction>
                 <owl:DatatypeProperty rdf:ID="string">
            </owl:Restriction>
        </owl:Class>
        - <owl:Class rdf:ID="constraintNexus"/>
              <rdfs:label>Constraint Nexus</rdfs:label>
              <rdfs:subClassOf>
              <owl:Restriction>
                  <owl:onProperty
                       rdf:resource="#ProtoNode"/>
              </owl:Restriction>
              </rdfs:subClassOf>
            </owl:Class>

              - <owl:Class rdf:ID="Relation"/>
                    <rdfs:label>Relation</rdfs:label>
                    <rdfs:subClassOf rdf:
                        resource="#constraintNexus" />
                    <owl:Restriction>
                            <owl:DatatypeProperty
                                 rdf:ID="string">
                            <rdfs:domain rdf:resource=
                                   "#ProtoNode"/>
                    </owl:Restriction>
                </owl:Class>
```

Fig. 1. Sample OWL document

Table 1. Furniture Table

Name	OntologyDependency
Furniture Ontology	Shopping Cart

Table 2. Relation Table

Name	Arity	Reflexive	Symmetric
SPATIAL RELATION	2	true	true
HAS-OWNER	2	false	false

Table 3. Proto Table

ProtoName	IsAbstract	childProtoName
Shopping Catalog Base	true	Physical Object

Table 4. Child Proto Table

ChildProtoName	IsAbstract	SemIndp	ConstraintRelation	ConstraintTarget
Physical Object	true	false		IS-A ShoppingCatalogBase

5   Performance

To test the performance of our system, we generated two OWL files which are wide and deeply nested based on `wine.owl` and `food.owl` [7]. The OWL file was parsed and separate relations were created for all children nodes at a depth of 2. The nodes with an in-degree less than 2 were inlined. The experiment was repeated, creating relations for child nodes that were at depth 3, 4, 5, ..., 10, etc. After each mapping, queries were executed to determine the depth-level at which the inlining should occur for optimal performance. Based on the result, we have decided that the optimal performance is obtained when the nodes with an in-degree of 3 to 5 are inlined based on the depth of the OWL document. When the inlining increases more than that then the performance is close to the performance of Universal Table approach.

Experimental evaluations were conducted on all the different mapping schemes discussed in Section 3 and also for the OWL to relations mapping. The database used for the performance evaluation is IBM DB2 [5]. The parameters considered were time taken and number of joins needed to perform a simple select or update query. Separate databases were created for each mapping schemes. The same set of Queries were issued on all the databases.and the results are shown in Figure 2, 3, 4, and 5.

The queries that were used are as follows.

Query 1 *A simple select query to display all the information in the table.*

Select hasColor,hasLocation, hasBody, hasSugar from Wine

Query 2 *A delete query.*

Delete hasSugar from all the Wines that have hasColor=Red

Query 3 *An update query.*

Update the hasBody for all the Wines located in Sauterine Region

Query 4 *A select query with joins.*

List all the Wines made in Bordeaux (several location)

The performance evaluation showed in Figures 2, 3, 4, 5[1] clearly indicate that Basic Inlining, Universal and Normalized Universal approach perform worst among the other methods. It also proves that OWL2DB performs pretty close to shared and attribute mapping approaches that are known to be the optimal mapping approaches from XML to Relations. OWL2DB is therefore a clearly an optimal mapping from OWL to Relations.

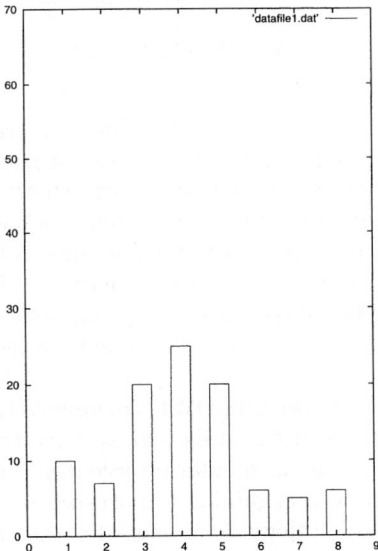

Fig. 2. Simple Select Query Q1

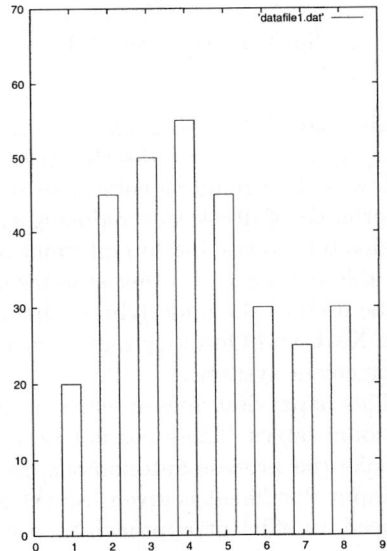

Fig. 3. Update Query Q2

6   Conclusion

Ontology has more expressive power than XML but XML is richer in defining structures and grammar. Ontology and XML can be used to describe semi-structured natural language texts. Ontology is about sharing and consensual

---

[1] In these figures, 1,2,3,4,5,6,7,8 of the X-Axis stands for the Edge, Attribute, Universal, UnivNorm, Basic, Shared, Hybrid, OWL2DB mapping schemes.

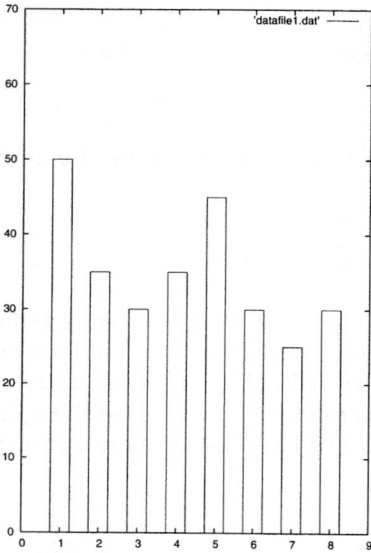

Fig. 4. Delete Query Q3

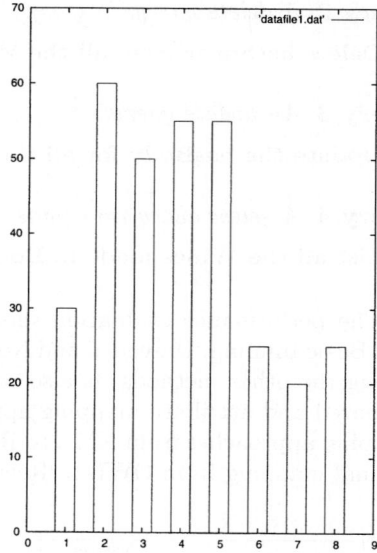

Fig. 5. Join Query Q4

terminology and XML is used to accomplish that. Ontology provides a domain theory while XML provides the structure of data container. Ontology is a powerful new tool for building enterprise and web-based information systems that hold the promise of drastically reducing system construction and maintenance costs. It helps businesses develop information systems that are reusable, maintainable, and able to talk with other systems sharing similar kinds of information. XML can be used as the underlying technology for Ontology since it is pretty mature. Both XML and Ontology together can pave the way for the dynamic knowledge management systems.

This paper describes an efficient approach to mapping OWL documents into relational tables. There has not been much work in this area and we have tried to make the process autonomous, that is there is no human intervention once the input document is given. As the semantic web is gaining importance lately, we need an efficient approach to store all the ontology documents for the ease of querying and this paper introduces a first attempt in this direction.

References

1. John Davies, Dieter Feasel, and Frank Van Harmelen. *Towards the Semantic Web: Ontology Driven Knowledge Management.* John Wiley snd Sons Inc, 2003.
2. Document Object Model (DOM) Level 3 Core Specification. *http://www.w3.org/TR/2004/REC-DOM-Level-3-Core-20040407/.*
3. Daniela Florescu and Donald Kossman. Performance evaluation of alternate mapping schemes for stoing xml data in a relational database. *Technique Report*, 1999.
4. Jena Documentation. *http://jena.sourceforge.net/ontology/index.html.*

5. David Martineau. *DB2 Universal Database for Application Development*. John Wiley snd Sons Inc, 1998.
6. OWL Web Ontology Language. *http://www.w3.org/TR/owl-guide/*.
7. OWL Web Ontology Language Reference. *http://www.w3c.org/TR/owl-ref/*.
8. RDF Primer. *http://www.w3c.org/TR/2004/REC-rdf-primer-20040210/*.
9. Jayavel Shanmugasundaram, Kristin Tufte, Gang He, Chun Zhang, David DeWitt, and Jeffrey Naughton. Relational databases for querying xml documents: Limitations and opportunities. *Proceedings of 25 VLDB Conference*, 1999.

A New Dynamic Indexing Structure
for Searching Time-Series Patterns

Zi-Yu Lin[1], Yong-Sheng Xue[1], and Xiao-Hua Lv[2]

[1] Department of Computer Science of Xiamen University
361005 Xiamen, China
cainiu@263.net, ysxue@jingxian.xmu.edu.cn
[2] College of Information and Electron of Zhejiang University of Sciences
310018 Hangzhou, China
bowen95@sohu.com

Abstract. We target at the growing topic of representing and searching time-series data. A new MABI (Moving Average Based Indexing) technique is proposed to improve the performance of the similarity searching in large time-series databases. Notions of Moving average and Euclidean distances are introduced to represent the time-series data and to measure the distance between two series. Based on the distance reducing rate relation theorem, the MABI technique has the ability to prune the unqualified sequences out quickly in similarity searches and to restrict the search to a much smaller range, compare to the data in question. Finally the paper reports some results of the experiment on a stock price data set, and shows the good performance of MABI method.

1 Introduction

Time-series data are of growing importance in the field of database application, more and more researchers are devoting themselves to developing powerful and useful data mining tools based on time-series database. Nowadays, we can see the application of these tools in many fields such as medicine, finance, meteorology, etc. In these application fields, large amounts of data are constantly added to systems, for example, MACHO project, an astronomy database, contains more than 0.5 TB data and receives a large amount of new data everyday [2].For this reason, there must be effective methods for solving problems such as representation and indexing of time-series.

Here we propose a new index mechanism—MABI (Moving Average Based Indexing), which is not only easy to implement but also has desirable performance. It has great capability in pruning those unqualified time-series and can also be applied in other fields

The remainder of this paper is organized as follows: Section 2 discusses relative work in this field. Section 3 gives some preliminary knowledge for our method. Section 4 presents MABI method in detail, which involves some relating definitions, useful theorems and algorithms. Section 5 gives performance analysis. Section 6 presents our conclusions.

S. Wang et al. (Eds.): ER Workshops 2004, LNCS 3289, pp. 290–301, 2004.

2 Related Work

R-tree is a well-known and wildly-used indexing mechanism. At the same time, some new indexing methods with good performance are also proposed in recent years. In [9], the authors proposed EMDF (Extended Multidimensional Dynamic index File) as indexing structure for time-series patterns. EMDF is a multidimensional spatial file to store and retrieve time-series patterns represented as multi-key records and is used for preselection process which select out a small set of candidate frames from the entire set, through this way, speedup in the pattern matching process is achieved. However, EMDF exists some disadvantages, for example, it has to maintain pointers for dynamic range tables which is very large in this method. In order to avoid these defection and enhance the performance, a new improved dynamic indexing structure for efficient handling of time-series was presented in [10]—TIP-index (TIme-series Patterns index). As far as structure is concerned, these two methods are similar. A TIP-index consists of frame identifier blocks, a dynamic region table and hierarchical directory blocks. However unlike EMDF, TIP-index only has s single dynamic region table which represents all dimensions. In the aspects of inserting operation and search, TIP-index can be more efficient than EMDF, furthermore, the former can achieve balanced performance for different types of data with special value distribution, however, the latter performs poorly when encountering such circumstances. In [11], the authors gave a novel method called STB-index (Shape-To-Vector-index), in which, index is formed by creating bins that contain time series subsequences of approximately the same length, for each bin, they can quickly calculate a lower-bound on the distance between a given query and the most similar element of the bin, this bound allows them to search the bins in best first order and to prune some bins from search space without having to examining the contents. Some other methods include: vantage-point-tree index in [12], using maximum and minimum to create index in [13], multilevel hierarchical structure based on feature of time series in [14], etc. There are also some well-known indexing mechanism such as KD-trees[4],K-D-B-trees[5] and Quad-trees[6], however, they cannot be applied in the field of time-series, because time-series databases are different from traditional databases in that the former must keep not only the value but also the time when the value is produced.

3 Preliminary

3.1 Representation of Data

For clarity we will refer to 'raw', unprocessed temporal data as time-series, and a piecewise representation of a time-series as a sequence. At the same time, a "series" will denote a time-series or a sequence. An effective representation of data is the prerequisite for constructing a dynamic indexing mechanism with desirable performance. Here we introduce a well-known and widely-used representation method in time-series field.

PAA (Piecewise Aggregate Approximation) is a method easy to understand [3]. A time series C of length n can be represented in N dimensional space by a vector $\overline{C} = \overline{c}_1, \overline{c}_2, .., \overline{c}_N$, and we assume that N is a factor of n. The ith element of $\overline{C}$ is calculated as follows:

$$\overline{c}_i = \frac{N}{n} \sum_{\frac{n}{N}(i-1)+1}^{\frac{n}{N}i} c_j \qquad (1)$$

This simple technique is surprisingly competitive with the more sophisticated transforms, so we take PAA as our method of data representation.

3.2 Moving Average

Moving average data are widely used in stock data analysis. Their primary use is to smooth out short term fluctuations and depict the underlying trend of a stock. The computation is simple, now we give the definition:

Definition 1. For a series $\overrightarrow{x} = [x_t]$ $(t=0,1,\ldots,n-1)$, which contains n data points, its M-l moving average is computed as follows: the mean is computed for an l-point-wide window placed over the end of the series; this will give the moving average for point $n - \lfloor l/2 \rfloor$; the subsequent values are obtained by stepping the window through the beginning of the series, one point at a time. This will produce a moving average of length $n-l+1$; We can use $M\text{-}l(\overrightarrow{x})$ to represent the M-l moving average of $\overrightarrow{x}$.

3.3 Euclidean Distance

In the field of time series, Euclidean distance is a method used to measure the distance between two series. For two series, $\overrightarrow{x} = \{x_0, x_1, \ldots, x_{n-1}\}$, $\overrightarrow{y} = \{y_0, y_1, \ldots, y_{n-1}\}$, the Euclidean distance between them is computed as follows:

$$D(\overrightarrow{x}, \overrightarrow{y}) = \left(\sum_{i=0}^{n-1} (x_i - y_i)^2 \right)^{\frac{1}{2}} \qquad (2)$$

We usually use Euclidean distance to determine whether two series are similar or not. If $D(\overrightarrow{x}, \overrightarrow{y}) < \varepsilon$ (ε is a threshold set by users), then we may say that the two series are similar.

3.4 Two Novel Theorems Based on Moving Average

Now we give two very important theorems.

Theorem 1. For two sequences $\vec{x} = \{x_0, x_1, \ldots, x_{n-1}\}$ and $\vec{y} = \{y_0, y_1, \ldots, y_{n-1}\}$, where $1 \leq l \leq n$, there M-l moving average are M-l ($\vec{x}$) and M-l ($\vec{y}$) respectively, then they must satisfy the following formula:

$$D(M\text{-}l(\vec{x}), M\text{-}l(\vec{y})) \leq D(\vec{x}, \vec{y})$$

The conditions for $D(M\text{-}l(\vec{x}), M\text{-}l(\vec{y})) = D(\vec{x}, \vec{y})$ are as follows:

    (1) $l = 1$;

    (2) $l > 1$ and $\vec{x}$ is identical to $\vec{y}$;

In other circumstances, $D(M\text{-}l(\vec{x}), M\text{-}l(\vec{y})) < D(\vec{x}, \vec{y})$.

Here we call Theorem 1 "distance reducing theorem". Due to the restriction of the pages, we would not give the proof here.

According to Theorem 1, we can get another useful theorem.

Theorem 2. Under the same similarity criterion, if two sequences are similar, then their M-l moving average are also similar.

Proof: It is easy to prove. Let $\vec{x} = \{x_0, x_1, \ldots, x_{n-1}\}$ and $\vec{y} = \{y_0, y_1, \ldots, y_{n-1}\}$, where $1 \leq l \leq n$, if they are similar, then there exists $D(\vec{x}, \vec{y}) < \varepsilon$, according to Theorem 1, $D(M\text{-}l(\vec{x}), M\text{-}l(\vec{y})) \leq D(\vec{x}, \vec{y})$, then $D(M\text{-}l(\vec{x}), M\text{-}l(\vec{y})) < \varepsilon$, so $M\text{-}l(\vec{x})$ and $M\text{-}l(\vec{y})$ are similar. ■

Definition 2. For a sequence s, its M_l *moving average is* $M\_l(s)$, Euclidean distance between s and horizontal axis is represented as Ds, Euclidean distance between $M\_l(s)$ and horizontal axis is D_M, DRR (Distance Reducing Rate) r between s and $M\_l(s)$ is defined as: $r = (Ds - D_M)/Ds$.

Here suppose there are two sequences s_1 and s_2, their M_l *moving average are* $M\_l(s_1)$ and $M\_l(s_2)$, Euclidean distances between the four sequences and horizontal axis are Ds_1, Ds_2, D_{M1} and D_{M2} respectively. Let r_1 denotes DRR (Distance Reducing Rate) between s_1 and $M\_l(s_1)$, r_2 denotes DRR between s_2 and $M\_l(s_2)$. We now give the most important theorem, it is the foundation of MABI method.

Theorem 3. For two sequences s_1 and s_2, if $Ds_1 > \varepsilon$, then they must satisfy the equa-

tion: $\dfrac{Ds_1 \times r_1 - 2\varepsilon}{Ds_1 + \varepsilon} < r_2 < \dfrac{Ds_1 \times r_1 + 2\varepsilon}{Ds_1 - \varepsilon}$

Proof: Suppose the distance between s_1 and s_2 is less than ε, namely, $D\ (s_1, s_2) < \varepsilon$, according to triangular inequality theorem [15], we can get that $Ds_1 - Ds_2 < D\ (s_1, s_2)$ and $Ds_2 - Ds_1 < D\ (s_1, s_2)$, namely, $|\ Ds_1 - Ds_2\ | < D\ (s_1, s_2)$, so we can get $|\ Ds_1 - Ds_2\ | < \varepsilon$; According to Deduction 1, $M\_1\ (s_1)$ and $M\_1\ (s_2)$ are also similar, so we get that $|\ D_{M1} - D_{M2}\ | < \varepsilon$, then $|\ Ds_1 - Ds_2\ | + |\ D_{M1} - D_{M2}\ | < 2\ \varepsilon$, according to the knowledge about absolute value we know that $|a| - |b| \le |a-b| \le |a| + |b|$, so we get that $|\ (Ds_1 - Ds_2\ ) - (\ D_{M1} - D_{M2})| \le |\ Ds_1 - Ds_2\ | + |\ D_{M1} - D_{M2}\ |$, so, $|\ (Ds_1 - Ds_2\ ) - (\ D_{M1} - D_{M2})| < 2\ \varepsilon$, then, $|\ (Ds_1 - D_{M1}\ ) - (\ Ds_2 - D_{M2})| < 2\ \varepsilon$, namely, $|Ds_1 \times r_1 - Ds_2 \times r_2| < 2\ \varepsilon$, because $Ds_2 = Ds_1 \pm \varepsilon$, then we have $|Ds_1 \times r_1 - (Ds_1 \pm \varepsilon) \times r_2| < 2\ \varepsilon$, when $Ds_1 \le \varepsilon$, $|Ds_1 \times r_1 - (Ds_1 \pm \varepsilon) \times r_2| < 2\ \varepsilon$ must be true for all the values of distance reducing rate, when

$Ds_1 > \varepsilon$, then we get: $\dfrac{Ds_1 \times r_1 - 2\varepsilon}{Ds_1 \pm \varepsilon} < r_2 < \dfrac{Ds_1 \times r_1 + 2\varepsilon}{Ds_1 \pm \varepsilon}$,so we can get:

$\dfrac{Ds_1 \times r_1 - 2\varepsilon}{Ds_1 + \varepsilon} < r_2 < \dfrac{Ds_1 \times r_1 + 2\varepsilon}{Ds_1 - \varepsilon}$. ∎

Here we call Theorem 3 "DRR relation theorem", with which we can prune quickly those unqualified sequences when searching for similar sequences of a given query sequence, because for a query *q*, if a sequence *s* is similar to *q*, it is a prerequisite that *s* and *q* satisfy DRR relation theorem.

4   MABI Method

MABI method makes full use of DRR (Distance Reducing Rate) relation theorem presented above and can prune those unqualified sequences quickly, through this way, the range of search is confined to very small area, which leads to desirable performance enhancement.

4.1   Building MABI Index Tree

In MABI index tree, the nodes are classified into leaf nodes and non-leaf nodes, moreover, non-leaf nodes include a root node and many mid nodes. The structures of leaf nodes and non-leaf nodes are different. The structure of non-leaf node is as Figure 4.1, in which, *range_node* records the DRR range that the node represents, *value* records the information of DRR range, *node_pointer* points to its son node whose *range_node* is equal to *value*. For a leaf node, it contains only two items: *range_node* and *table_pointer*, *table_pointer* points to information table *T*. *T* includes the following information: r, *address* and *euclidean*, where r is a DRR, *address* denotes where a sequence is stored, *euclidean* denotes the Euclidean distance between a sequence and horizontal axis. The maximum number of records that T can contain is p, and these records are arrayed in order according to r. If r of a sequence is covered by the *range_node* of a leaf node, then the information of the sequence will be recorded in T of the leaf node. When the number of records that a leaf node t contains reaches p, the leaf node t will automatically split its DRR range into k son DRR ranges of equal length and create k corresponding leaf nodes for these k son DRR ranges. Then the node t will put all the information of itself into the k new leaf nodes, after this, node t will change itself into a non-leaf node and record the information about its son nodes.

range_node	
range_node[1]	value
	node_pointer
range_node[2]	value
	node_pointer
......	......
	......
range_node[11]	value
	node_pointer

Fig. 4.1. The structure of non-leaf node of MABI index tree.

In order to build index tree from a time-series, we should first transform the time-series into a sequence with PAA algorithm, then we design a b-point-wide window placed at the beginning of the sequence, let the window move ahead along the sequence one point at a time, through this way we can get many subsequences of equal length, which will be used to build the index tree. The following algorithm in Figure 4.2 is used to insert a new subsequence s into a index tree.

4.2 Search Similar Sequences in MABI Index Tree

Here we presume that query q is a sequence got from a time-series with PAA algorithm. The search process involves three steps. In the first step, We will first calculate r_{q1}, which is a DRR between q and M_l (q), then compute *range*, which is a DRR

range, $range = (\dfrac{Dq \times r_{q1} - 2\varepsilon}{Dq + \varepsilon}, \dfrac{Dq \times r_{q1} + 2\varepsilon}{Dq - \varepsilon})$, after this, we can find all leaf

nodes whose *range_node* is covered by *range* (here if upper limit or lower limit of *range_node* is covered by *range*, then we call *range_node* is covered by *range*), and then select from T of these nodes those records whose r is covered by *range* and put them into a set S. In the second step, we use the condition $\mid Ds_1 - Ds_2 \mid < \varepsilon$ to continue pruning those unqualified elements from S. In the last step, we use condition D $(s_1,$ $s_2) < \varepsilon$ to get the final results. The algorithm is Figure 4.3.

```
Algorithm InsertSubsequence (s, root)
Input: s; root;
begin
    pointer:=root;//locate the pointer to root
    compute Ds and r of s;
  move along the tree according to r and reach leaf node u whose
  range_node covers r;
  add r to table T of node u;
  if the number of DRR that T contains reaches p then
    split the range_node of u into k parts of equal length;
  create new k leaf nodes for the k parts;
    move the information in T of u into the corresponding T of
    the k leaf nodes;
      recreate node u as a non-leaf node and record in it the in-
    formation of its k son nodes;
  endif;
end;
```

Fig. 4.2. Algorithm to insert a subsequence into MABI index tree.

Among the three steps, we use DRR relation theorem to prune many unqualified sequences in the first step, and use triangular inequality theorem [15] to do pruning job in the second step. For two sequences s_1 and s_2, according to triangular inequality theorem, we have $|Ds_1-Ds_2| < D(s_1,s_2)$, so if $\varepsilon < |Ds_1-Ds_2|$, then we surely have $\varepsilon < D(s_1,s_2)$, and s_1 and s_2 have no probability to be similar. The reason we design the second step is that the condition of $|Ds_1-Ds_2| < \varepsilon$ is more strict than the condition of $|Ds_1-Ds_2| + |D_{M1}-D_{M2}| < 2\varepsilon$ (which is the condition we use in the second step). Because D_{M1} and D_{M2} are the moving average of Ds_1 and Ds_2 respectively, so we have $|D_{M1}-D_{M2}| < \varepsilon_0$, where ε_0 is smaller than ε, so we say that $|Ds_1-Ds_2| < \varepsilon$ is more strict than $|Ds_1-Ds_2| + |D_{M1}-D_{M2}| < 2\varepsilon$. But we have to point out that the first pruning process is the critical aspect in enhancing query performance.

5 Performance Analysis

We do many experiments to get desirable results to support our new theory. Our experiments are executed on PC with one 1.3GHz CPU and 256M RAM running WIN XP. We download from internet the data of all the stocks in American stock market from 1980 to 2003 (*http://www.macd.cn*), which contains 1,228,764 records, then we get 10 time-series, each one contains 100,000 data points. We first transform the 10 time-series into 10 sequences of 10,000-point length with PAA algorithm. Here we let $w=200$, $b=200$, $k=5$, $p=500$, where w is the length of query q, b is the width of the window placed on the sequences, k and p are defined in section 4.1. Finally we get 9801 DRRs from every sequence. In order to show the pruning capability of DRR relation theorem, we select one sequence as an example. For this

```
Algorithm  Search(q,root,  ε);
Input: q; root; ε
Output: sequences similar to q
begin
  //the beginning of the first step
  initialize a new queue Q;
  put root into Q;
  calculate M_2 (q), Dq ,range and rq1;
  get an element from Q;
  pointer:= the element got from Q;
  while the node pointer points to is not a leaf node do
    for each pointer.range_node[i] do
      if range_node[i].value is covered by range then
        put range_node[i].node_pointer into Q;
      endif;
    endfor;
    get an element from Q;
    pointer:= the element got from Q;
  endwhile;
  while Q is not null do
    if pointer.range_node ⊆ range then
      put into S all records in T;
      //T belongs to the node that pointer
      //table_pointer points to;
    else // pointer.range_node are not completely
      //covered by range
      put into S those records in T whose r is covered by range;
      // T belongs to the node that pointer.
      //table_pointer points to;
    endif;
    get an element from Q;
    pointer:= the element got from Q;
  endwhile;
  //the end of the first step
  //the beginning of the second step
  for each s∈ S do
    if |Dq-s.Euclidean| > ε then delete s from S;
    endif;
  endfor;
  //the end of the second step
  //the beginning of the third step
  for each s∈ S do
  //let ls is the sequence that s.address points to
    if  the Euclidean distance between ls and q is more than ε
    then delete s from S;
    endif;
  endfor;
  //the end of the third step, get the result
end;
```

Fig. 4.3. Algorithm to search similar sequences.

chosen sequence, its DRRs are distributed in the range of $0.9{\times}10^{-2}{\sim}1.8{\times}10^{-2}$ (as Figure 5.1 shows), then we get a subsequence of 200-point length as query sequence (as Figure 5.2 shows) and calculate its DRR, the result is $r=1.628{\times}10^{-2}$, also we calculate the Euclidean distance between q and horizontal axis, the result is $Dq=101.79$, then we let $\varepsilon = Dq/f$ and endow f with different values to see how the performance of DRR relation theorem in pruning unqualified sequences changes. The experiment results are shown in Figure 5.3.

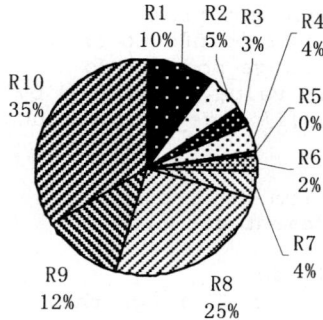

Fig. 5.1. Distribution of DRRs of the chosen sequence. The ranges are as follows: R1: $(0{\sim}1.0){\times}10^{-2}$, R2: $(1.0{\sim}1.1){\times}10^{-2}$, R3: $(1.1{\sim}1.2){\times}10^{-2}$, R4: $(1.2{\sim}1.3){\times}10^{-2}$, R5: $(1.3{\sim}1.4){\times}10^{-2}$, R6: $(1.4{\sim}1.5){\times}10^{-2}$, R7: $(1.5{\sim}1.6){\times}10^{-2}$, R8: $(1.6{\sim}1.7){\times}10^{-2}$, R9: $(1.7{\sim}1.8){\times}10^{-2}$, R10: $(1.8{\sim}100){\times}10^{-2}$.

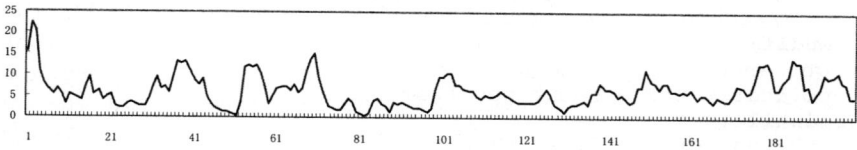

Fig. 5.2. The chosen query q of 200-point length.

We give in Figure 5.5 the pruning rate curve formed by changing the value of f. From Figure 5.5, we can draw the conclusion that, the larger the value of f is, the stronger the pruning capability is. At the same time, with the "expanding" of f, the extent of increase in pruning capability decreases, namely, when f is relative small, we can gain a great increase in pruning capability if we increase f a little. But when f reaches to certain value ($f = 3000$, for example), we can only achieve relative little enhancement from the increase of f.

In order to show the performance of MABI, we make it compete with sequential scanning optimized with Branch and Bound evaluation. To enable the results more reasonable and persuasive, we do 10 tests. In each test, we build index tree from one of the 10 sequences, and select randomly 9 200-point-long query sequences from the other 9 sequences, then we do search work with the index tree and the 9 query sequences. Finally we adopt the average of these results as our final results, which is as Figure 5.4.

f	1	500	1,000	2,000
pruning number	0	3551	5488	6538
pruning rate	0%	36.2%	56.0%	66.7%
range $(\times 10^{-2})$	-99.19~ 100.81	1.226~ 2.024	1.427~ 1.826	1.527~ 1.727

f	5,000	10,000	20,000	100,000
pruning number	8791	9333	9521	9728
pruning rate	89.7%	95.2%	97.1%	99.3%
range $(\times 10^{-2})$	1.588~1.668	1.608~1.648	1.618~1.638	1.626~1.630

Fig. 5.3. Performance of DRR relation theorem when f has different values.

f	length of sequence	sequential scanning (second)	MABI (second)	indexing speedup
10	100,000	980.86	287.94	5.41
500	100,000	737.84	179.96	4.10
2,000	100,000	727.67	86.38	8.42
10,000	100,000	725.19	28.43	25.51

Fig. 5.4. Performance contrast between MABI and Sequential scanning.

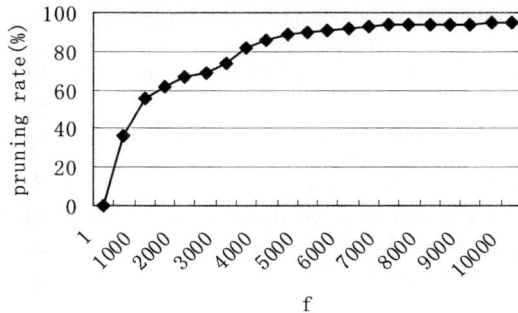

Fig. 5.5. Pruning rate curve.

Now we discuss a problem that whether MABI is an effective index mechanism. As presented in [8], there are several highly desirable properties for any indexing scheme:

- It should be much faster than sequential scanning.
- The method should require little space overhead.
- The method should be able to handle queries of various lengths.

- The method should be dynamic, which is to say it should allow insertions and deletions without requiring the index to be rebuilt.
- It should be correct, i.e. there should be no false dismissals (although probabilistic matching may be acceptable in some domains).

Our index mechanism has all the above properties except the third one. For the moment, we take sequential scanning as our method to deal with query shorter than w, and take the *long_query_searh* algorithm proposed in [11] to process query longer than w. In the future, we will continue our work to gain a better method to deal with this problem.

6 Conclusion

In this paper, we propose two novel theorems: distance reducing theorem and DRR relation theorem, based on which we propose MABI index mechanism, which is easy to implement and with little space overhead and desirable performance. However there exists a problem in our method, MABI can not effectively process queries of different length for the moment. We will do more work to perfect our method, at the same time, we will apply the two novel theorems proposed here in other fields.

References

1. Welch. D. & Quinn. P (1999) http://wwwmacho.mcmaster.ca/Project/Overview/status.html
2. Chan, K. & Fu, W. (1999). Efficient time series matching by wavelets. Proceedings of the 15th IEEE International Conference on Data Engineering.
3. Keogh, E,. Chakrabarti, K,. Pazzani, M. & Mehrotra (2000) Dimensionality reduction for fast similarity search in large time series databases. Journal of Knowledge and Information Systems.
4. Bently, J.L.(1975). Multidimensional binary search trees used for associative searching. Comm. ACM, Vol.18, No.9, 1975.
5. Robinson, J.T.(1981). The K-D-B-tree : A search structure for large multidimensional dynamic indexes. Proc. of Intl. Conf. on Management of Data, ACM SIGMOD, 1981.
6. Finkel, R. A. and Bently, J.L.(1974). Quad Trees: A data structure for retrieval on composite keys. Acta Informatica 4,1974.
7. Agrawal, R., Faloutsos, C., & Swami, A.(1993). Efficient similarity search in sequence databases. Proceedings of the 4th Conference on Foundations of Data Organization and Algorithms.
8. Faloutsos, C., & Lin, K. (1995). Fastmap: A fast algorithm for indexing, data-mining and visualization of traditional and multimedia datasets. In Proc. ACM SIGMOD Conf., pp 163-174.
9. Kim, Y. I., Ryu, K. H. and Park, Y.(1994). Algorithms of improved multidimensional dynamic index for the time-series pattern. Proc. of the 1st KIPS Spring Conf., Vol. 1, No. 1, 1994.
10. Kim, Y.I., Park, Y., Chun, J.(1996).A dynamic indexing structure for searching time-series patterns. Proceedings of 20th International Computer Software and Applications Conference, 21-23 Aug., 1996. Pages:270–275.

11. Keogh, E.J., Pazzani, M.J.(1999). An indexing scheme for fast similarity search in large time series databases. Proceedings of the 11th International Conference on Scientific and Statistical Database Management, 28-30 Jul, 1999. Pages:56-57.
12. Bozkaya, T. and Ozsoyoglu, M. Indexing large metric spaces for similarity search queries. ACM Transactions on Database Systems, Volume 24, Issue 3, September, 1999. Page: 361 - 404.
13. Park, S., Kim, S. and Chu, W.W.(2001).Segment-based approach for subsequence searches in sequence databases. Proceedings of the 2001 ACM symposium on Applied computing, 2001.Page: 248–252.
14. Li, C.S., Yu, P.S., Castelli, V.(1998). A Framework For Mining Sequence Database at Multiple Abstraction Levels. In Proceedings of the 1998 ACM 7th international conference on Information and knowledge management, 1998. Pages:267–272.
15. Shasha, D., & Wang, T. L., (1990). New techniques for best-match retrieval. ACM Transactions on Information Systems, Vol. 8, No 2 April 1990, pp. 140-158.

Automatic Location and Separation of Records: A Case Study in the Genealogical Domain

Troy Walker and David W. Embley

Department of Computer Science
Brigham Young University, Provo, Utah 84602, USA
{troywalk,embley}@cs.byu.edu

Abstract. Locating specific chunks (records) of information within documents on the web is an interesting and nontrivial problem. If the problem of locating and separating records can be solved well, the longstanding problem of grouping extracted values into appropriate relationships in a record structure can be more easily resolved. Our solution is a hybrid of two well established techniques: (1) ontology-based extraction [ECJ+99] and (2) vector space modeling [SM83]. To show that the technique has merit, we apply it to the particularly challenging task of locating and separating records for genealogical web documents, which tend to vary considerably in layout and format. Experiments we have conducted show this technique yields an average of 92% recall and 93% precision for locating and separating genealogical records in web documents.

1   Introduction

When looking for information on the web, the challenge is usually not scarcity of data; rather, it is locating the specific data we want. Searching for genealogical information is a prime example[1]. In March 2003, a search for "Walker Genealogy" on Google returned 199,000 documents; just one year later, the same search returned 338,000 documents. Rather than requiring a human to sift through this mountain of data, it would be helpful to have a software agent that could locate and extract desired information automatically.

There are, however, many obstacles to the ideal situation in which software agents return to their owners only the data they want at the time they desire it. One large obstacle a software agent must overcome is to be able to deal with the format of each page and translate it to its own representation[2]. The web presents information in a variety of formats. Even within the HTML standard, information is presented in many ways. We can classify HTML documents by

---

[1] Genealogy is also a popular example. We choose it, however, not so much because of its popularity, but because of the challenge it provides for locating, separating, and extracting records.

[2] Ontologies on the semantic web promise to make this easy, but the need to automatically or at least semi-automatically transform regular web documents into semantic web documents leaves us with the same format-recognition and translation problems.

S. Wang et al. (Eds.): ER Workshops 2004, LNCS 3289, pp. 302–313, 2004.
© Springer-Verlag Berlin Heidelberg 2004

the way they present data. For our genealogy application we classify HTML documents as follows.

- *Single-Record Document*: contains a record for one person (e.g. Figure 1).
- *Multiple-Record Document*: contains many records in a list (e.g. Figure 2).
- *Complex Multiple-Record Document*: contains many records organized in some sort of chart, usually an ancestral chart (e.g. Figure 3).

As humans view these pages, they effortlessly adapt to these and other presentation formats. How can we endow software agents with this same ability?

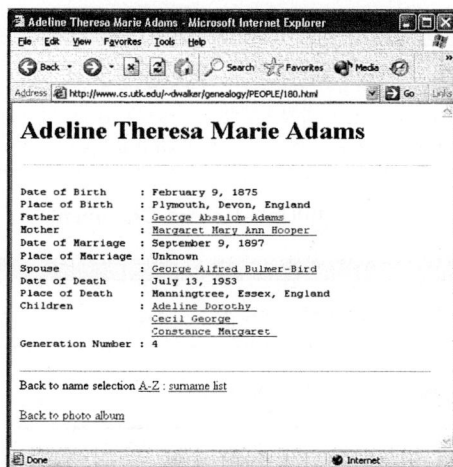

Fig. 1. Single-Record Document

Although we do not claim to be able to endow software agents with human-like format-reading ability, we do offer as our contribution in this paper a way to solve the problem of locating and separating the individual records so that the related information can be extracted as a unit. We base our solution on data extraction ontologies [ECJ+99] and on vector space modeling [SM83]. We [EJN99] and others [BLP01] have previously offered partial solutions to the problem of record separation. These solutions, however, depend on a document having multiple records consistently separated by some HTML tag such as <hr>. For many domains, genealogy being one good example, this assumption fails. The assumption holds for the web page in Figure 2, but fails for the web page in Figure 3 because there is no HTML tag to separate the record groups. Surprisingly, it also fails for the web page in Figure 1 because it assumes multiple records and attempts to incorrectly split the information into subcomponent groups. Recognizing the difference between a page with a large amount of genealogical information about one person and a page with a small amount of information for a few people is not as easy as it at first sounds.

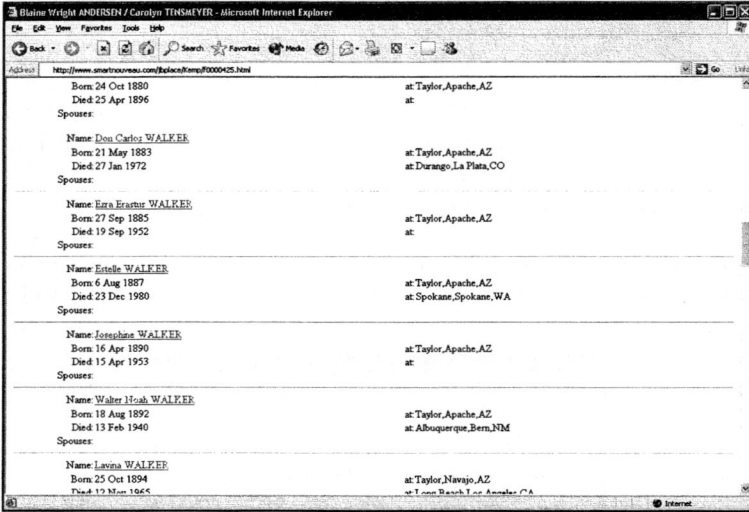

Fig. 2. Multiple-Record Document

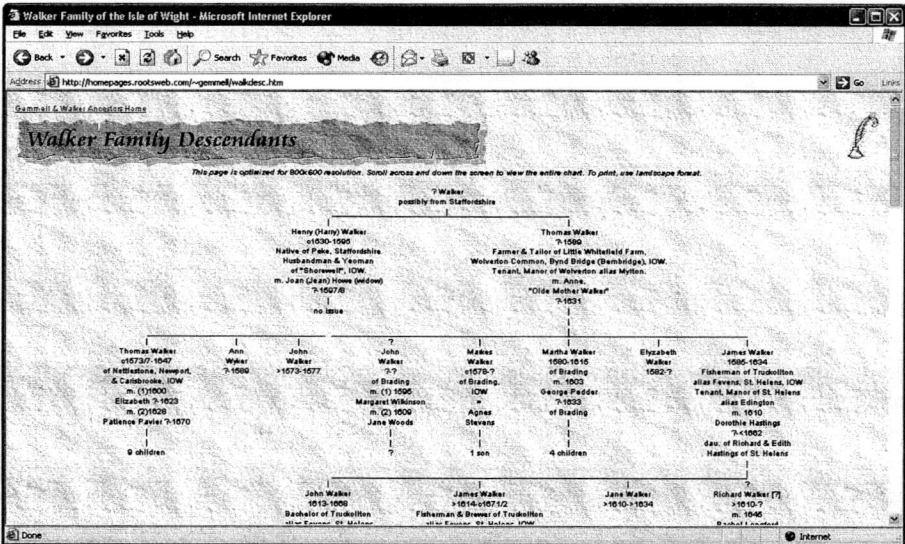

Fig. 3. Complex Multiple-Record Document

We present our contribution as follows. In Section 2 we explain what an extraction ontology is and argue that the approach is likely to be best among all extraction techniques for domains like genealogy where there are a huge number of continually changing web pages with a wide variety of structural formats. In Section 3 we show how to use information from extraction ontologies together with the cosine measure of vector space modeling to locate and separate records. In Section 4 we present and discuss experimental results, and in Section 5 we make concluding remarks.

2   Extraction Ontologies

Currently, the most prevalent approach to data extraction from the web is by using page-specific wrappers. (See [LRNdST02] for a survey.) Since page-specific wrappers extract data based on page layout clues, they are sensitive to changes in formatting. Because wrappers are tedious to write and must be written for each new page as well as every time a page changes, researchers have focused on semi-automatic generation of wrappers[3]. Even with semi-automatically generated wrappers, the use of page-specific wrappers requires a substantial amount of work to produce and maintain, especially for an application like genealogy where pages change often, and a substantial number of new pages continually appear. Extraction tools based on natural language processing avoid the problems of page-specific wrappers. They need clues from parsed sentences, however, to identify data of interest, which does not help for applications like genealogy where most web pages use a terse assortment of terms and fragments instead of complete sentences.

Unlike page-specific wrappers, ontology-based extraction tools are resilient to page changes and need no training or adaptation to work with new pages. Rather than using a generated page-specific wrapper for each site related to an application domain, they use an extraction ontology that wraps all pages related to a domain. For applications like genealogy, this is ideal because the effort[4] to create the ontology can be amortized over thousands (even millions) of sites.

An extraction ontology is an augmented conceptual model of the information pertaining to a narrow domain of interest. (See [ECJ+99] for a detailed explanation.) We represent the conceptual model of our ontologies in OSM (Object-oriented System Model) as described in [EKW92]. In particular, we use a sub-model of OSM, the Object Relationship Model, which models objects, relationships among objects, and constraints. Constraints of particular interest are participation constraints, which consist of a *minimum*, optional *average*, and *maximum* number of times an object participates in a relationship.

We augment the conceptual model with data frames [Emb80], which are descriptions of the data types to which objects adhere. An extraction ontology has

---

[3] There are at least 39 commercial and non-commercial wrapper generators in existence [KT02].

[4] Although the amount of effort required to create an extraction ontology is sometimes criticized, it is usually no more than the effort required to create training data for machine-learned wrapper generators. It does, however, require more expertise. Anecdotely, we asked 24 students, who have taken the extraction and integration course at BYU, to report the number of hours it took them to create extraction ontologies for a new domain. Projects included, for example, pharmacuetical drugs, jewelry, TV's, and camp grounds. All reported taking somewhere in the neighborhood of about two dozen hours or less.

a data frame for each object set. A data frame contains recognizers to identify data that belongs to an object set. These recognizers consist of extended regular expressions that match values, the context typically surrounding values, and keywords usually located close to a value. Our matchers support macro substitution and the inclusion of lexicons. This helps extraction-ontology engineers keep their regular expressions manageable.

3   Record Location and Separation

In order to extract the information related to each person in a genealogical document, the computer needs to separate the document into records. Each record should contain information on only one person. This greatly simplifies the task of selecting values and linking them together as objects and relationships. We divide web pages into three categories based on how they present information: single-record documents (Figure 1), simple multiple-record documents (Figure 2), and complex multiple-record documents (Figure 3).

Two clues aid in locating and separating records: structure and content. A well-designed page has a format that assists readers in distinguishing records. Human readers can also guess where record divisions occur based on the data they contain – when they see a second name and another birth date in a genealogical document, they know they are reading about a new person. Previous approaches [EJN99,BLP01] primarily use structural clues to locate and separate records. Straightforward structural clues, however, may either not exist (single-record documents) or may require extensive heuristics to recognize (complex multiple-record documents. On the other hand, it is also likely to be difficult to create an accurate record separator based purely on content. In a domain where data always appears in a certain order or all data is always present, it may be straightforward, but in genealogy, this is not the case. Missing data, imperfect matchers, and unpredictable order combine to make this approach infeasible. We thus adopt a hybrid approach.

Vector Space Modeling (VSM) comes from the field of information retrieval [SM83]. A set of features from a document makes up the values in a vector from which useful cosine and magnitude measures are derived. For record location and separation, we use the object sets from the extraction ontology as dimensions in vector space.

First, we create a vector that represents a prototypical record of genealogical data. We call this vector the *Ontology Vector (OV)*. To create the OV, we use the average value given in participation constraints. We do not include dimensions for all object sets; instead, we include dimensions only for those object sets most closely related to the object set of primary interest (e.g. *Person* for genealogy). These object sets give us the information most helpful for locating and separating instances of the primary object set while more indirectly related object sets have more of a potential for ambiguity and conflicts or for being completely unrelated. Specifically, we create the OV by following relationships from the primary object set until we encounter an object set with a keyword matcher or a value matcher. The dimensions and dimension values for the OV for our genealogy ontology are ($<Gender$, 0.8>, $<Name$, 0.99>, $<Birth$, 0.95>, $<Death$, 0.9>, $<Christening$, 0.6>, $<Burial$, 0.5>, $<Marriage$, 0.6>, $<Relationship$, 3.0>, $<RelationName$, 3.0>), or, when we fix the number and order of dimensions, simply (0.8, 0.99, 0.95, 0.9, 0.6, 0.5, 0.6, 3.0, 3.0).

For each candidate record, another vector contains the number of matches found in that candidate record's portion of the document. We call this vector the *Subtree Vector* (*SV*) because it gives the number of value matches for each dimension object set found within a subtree in the document's DOM tree (Document Object Module tree). Thus, for example, if in a subtree we find 50 gender references, 100 names, 100 birth dates, 100 death dates, 20 christening dates, 40 burial references, 60 marriages, and 250 pairs of relationship references and names of relatives, our SV for this subtree would be (50, 100, 100, 100, 20, 40, 60, 250, 250).

We judge each SV by its cosine score relative to the OV and by its magnitude score. The cosine of the acute angle between any two vectors in n-space is the dot product of the two vectors divided by the product of their magnitudes. This provides a reliable method of determining the similarity of SV to OV. Cosine measures close to one show that the subtree likely contains data that relates to the ontology, i.e. has the proportion of values expected by the ontology. The magnitude of SV divided by the magnitude of OV yields a rough estimate of the number of records in SV, which turns out to be accurate enough to decide whether to split a subtree into multiple records. Because large values along any dimension skew these measures, object-set dimensions that are expected to have an average of many more values than other object-set dimensions have too much weight in these measures. We therefore normalize all vectors, including the OV, prior to finding VSM scores by dividing each value in the vectors by its corresponding average in the OV. Thus our sample vector (50, 100, 100, 100, 20, 40, 60, 250, 250) becomes (63, 101, 105, 111, 33, 80, 100, 83, 83) and its cosine score is computed with respect to the normalized OV, (1,1,1,1,1,1,1,1,1).

Our method of record location and separation starts at the root of the tree and evaluates the subtree rooted at each node. If its magnitude measure is less than an empirically determined threshold value, we accept it as a record. If not, we split the subtree using the heuristics of [EJN99] to find a separator tag. In cases where these heuristics fail to find a separator (usually when a node has fewer than four child nodes), we simply use the subtrees that are children of the current node. We then use a technique from [EX00] to recombine these subtrees where appropriate. This technique combines pairs of subtrees if the combination has a better cosine measure with the OV than either of the subtrees alone. Finally, we discard the subtrees with low cosine scores (empirically determined to be less than 0.6), and we repeat the process with the remaining subtrees.

As an example, Figure 4 shows part of the DOM tree for the document in Figure 2 as well as the corresponding SVs, cosine scores, and magnitude scores. At the first level, there are only two children: `<!DOCTYPE>` and `<html>`. `<!DOCTYPE>` has a low cosine score (zero), so we discard it; `<html>` has a comparatively high

	Subtree Vector	Cosine	Magnitude
<!DOCTYPE...>	(0.0,0.0,0.0,0.0,0.0,0.0,0.0,0.0,0.0,0.0)	0.00	0.00
<html>	(0.0,150.5,93.7,84.4,0.0,0.0,80.0,7.7,7.7)	0.67	70.77
<head> ... </head>	(0.0,1.0,0.0,0.0,0.0,0.0,0.0,0.0,0.0,0.0)	0.33	0.34
<body>	(0.0,150.5,93.7,84.4,0.0,0.0,80.0,7.7,7.7)	0.67	70.53
<div> ...header... </div>	(0.0,0.0,0.0,0.0,0.0,0.0,0.0,0.0,0.0,0.0)	0.00	0.00
<div> ...	(0.0,147.5,92.6,84.4,0.0,0.0,80.0,7.7,7.7)	0.67	69.90
<td>	(0.0,1.0,1.1,1.1,0.0,0.0,0.0,0.0,0.0,0.0)	0.58	0.61

Fig. 4. DOM Tree Selections and Vector Scores

cosine score so we keep it. Since <html> has a comparatively large magnitude, we split it into <head> and <body>. We discard <head> because its cosine score is comparatively low. We split <body> because its magnitude is comparatively high. Because <body> has many child nodes, we find a separating tag, <div>, and divide the children accordingly. We continue processing with the second <div> tag since its cosine score is comparatively high. After repeating this process and dividing based on table rows and then table cells, we eventually start finding individual records. The <td> in the example is a table cell that happens to contain one record. Note that its magnitude score is considerably lower than the magnitude of the scores of the subtrees we split. Other table cells in this document actually contain multiple records and need to be split.

As we implemented this algorithm, we encountered three problems that limited its effectiveness. First, differences between our ontology and the schema used within documents often caused cosine measures to be too low. Second, the over-richness of data particularly in single-record documents often caused magnitude measures to be too high. Third, when separating simple multiple-record documents, our algorithm was sometimes outperformed by the old method ([EJN99]) because it did not take advantage of simple patterns in the data. We introduced refinements into our algorithm to cope with these problems.

Schema Differences. Not everyone agrees on what attributes should describe a person in genealogy. We designed our genealogy ontology to hold typical data we were likely to find on the web. It contains many attributes such as *Gender*, *Burial*, and *Christening* for which most web sites do not include data. The fact that these attributes do not appear on a page does not mean that the page is not about genealogy. It does affect the cosine scores of the document, however, and can cause valid records to have cosine scores below the threshold resulting in valid records being discarded. In Figure 4, for example, the untrimmed score of even our largest cosine scores are not as high as we would like them to be. If we detect these differences in schema at the page level and compensate for them, we can more accurately find the records within the page. We do this by pruning dimensions in our vector space. In order to detect which object sets do have matches in a document's schema, we count the object set matches and prune any dimension with no matches. Since a few erroneous matches are possible, we also prune dimensions with counts less than 5% of the average count, weighted according to the participation constraints.

Over-Richness of Data. Single-record documents tend to include more complete information than multiple-record documents. It is not unusual to find seven name matches in one single-record document. Single-record documents may also repeat information or have multiple instances of one keyword. Magnitude measures for records in single-record documents are therefore much higher than measures for multiple-record documents. To overcome this, we programmed our record locator and separator to require a higher magnitude to split a document than that needed to split at the highest level within a document.

Missed Simple Patterns. Simple multiple-record documents are distinguished by a simple list pattern. On some simple multiple-record documents, our old technique ([EJN99]) is able to produce more correct records than our new technique. In any document, some records contain more or fewer details than others. Sometimes our matchers do not accept all the valid data such as when names are incomplete or contain name components not in our lexicon. At times, this variation is enough to cause our record separator to erroneously discard or split a valid record. We can take advantage of the pattern in a simple multiple-record document if we can detect the pattern. We do so when we split a subtree by counting the records and computing the ratio of records with sufficient cosine scores and low enough magnitudes to be a single record to the total number of records. If there are more than three records and at least two-thirds of them are single records, we consider all of them to be single records. As a further refinement, we eliminate headers and footers by discarding records with low cosine scores at the head and tail of the list.

4   Experimental Results

While implementing our system, we used a few example documents to debug our code. Once our system was ready, we gathered 16 additional documents to test our algorithm and made further refinements. When our system performed adequately on this tuning set, we were confident it would perform well on any genealogical page from the web. We gathered test documents by searching the web for common surnames and genealogy. To ensure stability and reproducibility throughout our test and to reduce load on the web hosts, we created a local cache of our test pages. When collecting pages, we found that there were about three generators commonly used for genealogical web pages. Since we were interested in evaluating our system on a wide variety of sources, we only included a few pages generated by each program. Further, some pages contained close to a hundred records while others contained just one. To reduce skewing for record counts, we trimmed long documents to between ten and twenty records. We were careful to preserve any footers that might exist on each document.

We divided our test documents into three groups: single-record documents, simple multiple-record documents, and complex multiple-record documents. To test record location and separation, we compared the records produced to records that should have been produced. Because of the nested nature of genealogical data, this was not always simple. A name by itself in some contexts could be considered to be a record while in other contexts the name may just be the name of a relative within a valid record. As a general rule, we considered information about a relative to be a distinct record if it contained more than just a name.

Single-Record Documents. We tested 21 single-record documents. As can be seen in Table 1, our record separator correctly handled most of these documents resulting in 90% recall and 73% precision. This success is due to the refinement we made to compensate for over-richness of data. In two cases, data was still rich enough to overwhelm even this refinement. Attempting to split these records, the system destroyed the relationships within these records and produced several incorrect records, which explains the relatively low precision. We could increase the threshold to cover more of these cases, but raising it too much would cause multiple-record documents not to be split. Thus, our refinement worked fairly well, but since it is not just a simple matter of finding a proper balance, it is clear that a different approach is needed to produce even better results.

Simple Multiple-Record Documents. Table 2 shows the results of our experiments on simple multiple-record documents. By using our refinement for exploiting patterns in simple documents, we were able to correctly process seven more documents than we would have otherwise and achieved 95% precision and 93% recall. In some documents, we lost the first record because it did not have a high enough cosine score and was misinterpreted as part of the header. In one case less than two thirds of the records were acceptable as single records, so the algorithm did not detect that it should have treated it as a simple pattern that could be better handled in other ways.

Complex Multiple-Record Documents. Since most of the genealogical documents on the web fall into this category, performance on complex multiple-record documents for our application area is critical. Table 3 shows our results. Considering the difficulty of the task, 92% recall and precision should be seen as a very good result. The most common problem we encountered stems from conflicting matches. As currently programmed, our system has no way of knowing whether a name is a member of the *Name* object set or the *RelationName* object set and must consider it a potential match for both object sets. This becomes a problem when recombining fragments of a record. Figure 5 shows two records produced from *complex4* that should have been recombined. The first record has matches for *Name*, *Birth*, and *Marriage*. The second has matches for *Name*, *Relationship*, and *Death*. Although the names in the second record are really names of relatives, they prevented our system from recombining these

Table 1. Single-Record Document Results

	records	returned	correct	precision	recall
single1	1	1	1	100.00%	100.00%
single2	1	1	1	100.00%	100.00%
single3	1	1	1	100.00%	100.00%
single4	1	1	1	100.00%	100.00%
single5	1	1	1	100.00%	100.00%
single6	1	1	1	100.00%	100.00%
single7	1	1	1	100.00%	100.00%
single8	1	4	0	0.00%	0.00%
single9	1	1	1	100.00%	100.00%
single10	1	1	1	100.00%	100.00%
single11	1	1	1	100.00%	100.00%
single12	1	3	0	0.00%	0.00%
single13	1	1	1	100.00%	100.00%
single14	1	1	1	100.00%	100.00%
single15	1	1	1	100.00%	100.00%
single16	1	1	1	100.00%	100.00%
single17	1	1	1	100.00%	100.00%
single18	1	1	1	100.00%	100.00%
single19	1	1	1	100.00%	100.00%
single20	1	1	1	100.00%	100.00%
single21	1	1	1	100.00%	100.00%
Total	21	26	19	73.08%	90.48%

Table 2. Multiple-Record Document Results

	records	returned	correct	precision	recall
simple1	19	20	19	95.00%	100.00%
simple2	19	17	17	100.00%	89.47%
simple3	11	11	11	100.00%	100.00%
simple4	9	9	9	100.00%	100.00%
simple5	12	13	11	84.62%	91.67%
simple6	12	11	10	90.91%	83.33%
simple7	14	10	10	100.00%	71.43%
simple8	5	7	5	71.43%	100.00%
simple9	14	14	14	100.00%	100.00%
simple10	15	15	15	100.00%	100.00%
Total	130	127	121	95.28%	93.08%

two records. Since this problem prevented the correct record from being returned and created two incorrect records, it affected both the precision and the recall of our record separator. Another problem arose in *complex18*. Since the document only contained four records, its magnitude measure was low enough that it appeared to be a single-record document. Our record separator did not attempt to split it so we lost three records.

Table 3. Complex Multiple-record Document Results

	records	returned	missed	extra	correct	precision	recall
complex1	10	10	0	0	10	100.00%	100.00%
complex2	15	15	0	0	15	100.00%	100.00%
complex3	12	12	0	0	12	100.00%	100.00%
complex4	7	9	1	3	6	66.67%	85.71%
complex5	16	15	1	0	15	100.00%	93.75%
complex6	15	16	2	3	13	81.25%	86.67%
complex7	13	12	1	0	12	100.00%	92.31%
complex8	10	10	0	0	10	100.00%	100.00%
complex9	19	20	1	2	18	90.00%	94.74%
complex10	10	10	1	1	9	90.00%	90.00%
complex11	15	11	4	0	11	100.00%	73.33%
complex12	15	15	0	0	15	100.00%	100.00%
complex13	11	11	0	0	11	100.00%	100.00%
complex14	16	18	1	3	15	83.33%	93.75%
complex15	8	8	2	2	6	75.00%	75.00%
complex16	8	9	0	1	8	88.89%	100.00%
complex17	10	11	0	0	11	100.00%	110.00%
complex18	4	1	3	0	1	100.00%	25.00%
complex19	8	11	0	3	8	72.73%	100.00%
complex20	16	13	4	1	12	92.31%	75.00%
Total	238	237	21	19	218	91.98%	91.60%

> BROWN, Edwin, Born 18 Apr 1899 in Somerset, Kentucky, Married 1928
> Ruth V. Rosenburg dau. of Johan N. and Anna Marie Eriksson
> Rosenberg, he died 4 Aug 1960 in Toledo, Ohio at age 61

Fig. 5. Split Record

5 Conclusion

Based on vector space modeling, we implemented a technique to accurately locate and separate records even when these records have a complex layout such as is found in genealogy web pages. The technique we developed achieved an average of 93% precision and 92% recall in our experiments over a collection of single-record documents, multiple-record documents, and complex multiple-record documents. We added this technique to our system for ontology-based extraction.

Acknowledgements

This material is based upon work supported by the National Science Foundation under grant No. IIS-0083127.

References

[BLP01]   D. Buttler, L. Liu, and Calton Pu. A fully automated object extraction system for the world wide web. In *Proceedings of the 21st International Conference on Distributed Computing Systems (ICDC'01)*, Mesa, Arizona, April 2001.

[ECJ+99]   D.W. Embley, D.M. Campbell, Y.S. Jiang, S.W. Liddle, D.W. Lonsdale, Y.-K. Ng, and R.D. Smith. Conceptual-model-based data extraction from multiple-record web pages. *Data & Knowledge Engineering*, 31(3):227–251, November 1999.

[EJN99]   D.W. Embley, Y.S. Jiang, and Y.-K. Ng. Record-boundary discovery in web documents. In *Proceedings of the 1999 ACM SIGMOD International Conference on Management of Data (SIGMOD'99)*, pages 467–478, Philadelphia, Pennsylvania, 31 May - 3 June 1999.

[EKW92]   D.W. Embley, B.D. Kurtz, and S.N. Woodfield. *Object-oriented Systems Analysis: A Model-Driven Approach*. Prentice Hall, Englewood Cliffs, New Jersey, 1992.

[Emb80]   D.W. Embley. Programming with data frames for everyday data items. In *Proceedings of the 1980 National Computer Conference*, pages 301–305, Anaheim, California, May 1980.

[EX00]   D.W. Embley and L. Xu. Record location and reconfiguration in unstructured multiple-record web documents. In *Proceedings of the Third International Workshop on the Web and Databases (WebDB2000)*, pages 123–128, Dallas, Texas, May 2000.

[KT02]   S. Kuhlins and R. Tredwell. Toolkits for generating wrappers – a survey of software toolkits for automated data extraction from websites. In M. Aksit, M. Mezini, and R. Unland, editors, *Objects, Components, Architectures, Services, and Applications for a Networked World – Proceedings of the 2002 International NetObjectDays Conference*, pages 184–198, Erfurt, Germany, October 2002.

[LRNdST02] A.H.F. Laender, B.A. Ribeiro-Neto, A.S. da Silva, and J.S. Teixeira. A brief survey of web data extraction tools. *SIGMOD Record*, 31(2):84–93, June 2002.

[SM83]   G. Salton and M.J. McGill. *Introduction to Modern Information Retrieval*. McGraw-Hill, New York, 1983.

Research and Implementation of CORBA Web Services*

Jianqiang Hu, Bin Zhou, Yan Jia, and Peng Zou

School of Computer, National University of Defense Technology
410073 Changsha, China
jqhucn@hotmail.com

Abstract. CORBA (Common Object Request Broker Architecture) is one of the popular technologies that can be used to construct enterprise level application systems. However, none of conventional middleware technologies will dominate the software industry alone; therefore interoperability becomes one of the most crucial issues. Web Services is a kind of new Web application shape which shields from the existence of different middleware platforms and programming language abstractions. This paper firstly presents two models for wrapping CORBA objects to Web Services, and then analyzes some key technologies such as scalable architecture, SOAP/IIOP protocol datatype mapping, unified service providing infrastructure etc. Finally, it gives some tests and performance comparison on top of **StarWebService** that is developed by our team.

1 Introduction

CORBA is one of the popular technologies that can be used to construct enterprise computing platform. It enables the interconnection, intercommunication and interoperability of distributed application software in enterprise. Sharing arrangements are typically relatively static and restricted to occur within a single organization. The primary form of interaction is client/server mode, rather than the coordinated use of multiple resources. Moreover, CORBA cannot host next-generation grid applications [1].

With the emergence and rapid development of Internet, organizations of all sizes are moving their main businesses to the Internet/Web for more automation, efficient business processes, and global visibility. The distributed application is now open, loose-coupled and could be accessed publicly. Thus the hypothesis, made by the conventional middleware for relative close and static application environments, is broken. Problems arise when these components must interoperate or be integrated into a single executable process since it is hard to assemble semantically compatible and interoperable components based on multiple middleware platforms [2, 3].

In face of this challenge, Web Services are new kinds of distribution middleware services and components that complement and enhance existing conventional mid-

---

* This work was supported by National Natural Science Foundation of China under the grant No. 90104020 and National 863 High Technology Plan of China under the grant No. 2002AA6040 and 2003AA5410.

S. Wang et al. (Eds.): ER Workshops 2004, LNCS 3289, pp. 314–322, 2004.

dleware, which not only raise higher level of abstraction but also simplify distributed application development. They are Internet-based, distributed modular applications which can provide standard interfaces as well as communication protocols aiming at efficient and effective service integration. Many businesses on the Web benefit from Web service's loose-coupled, Service-Oriented, and HTTP-based attributes. Moreover, Web Services have gained a considerable momentum as paradigms for supporting both Business-to-Business (B2B) interaction and Grid computing collaboration.

The rest of the paper is structured as follows. Section 2 describes static and dynamic models of CORBA Web Services [4, 5], which wraps CORBA objects to Web Services. Section 3 analyzes some key technologies such as scalable architecture, SOAP/IIOP protocol datatype mapping, unified service providing infrastructure etc; Section 4 gives some tests and performance comparison on top of **StarWebService** that is developed by our team; Section 5 describes related research; and finally Section 6 concludes the paper and provides a brief overview for our future research.

2   CORBA Web Services

Web Services supports direct XML message-based interactions with other services and applications via Internet-based protocols like SOAP. CORBA is the use of the standard protocol IIOP allowing CORBA-based programs to be portable across vendors, operating systems, and programming languages. Therefore, it is necessary to have a bridge between SOAP server and the remote CORBA objects. The bridge allows static and dynamic model for wrapping CORBA objects to Web Services. Static model is based on compile time knowledge of interface specification of CORBA objects. This specification is formulated in IDL and is compiled into a CORBA stub. In the dynamic model, the object and its interface (methods, parameters, and types) are detected at run-time. Through the dynamic invocation interface (DII), the bridge dynamically constructs CORBA invocations and provides argument values from SOAP message.

2.1   Static Model

If the bridge holding stub wherein CORBA objects rarely changes, static model is enough. Static model architecture is illustrated in Fig.1. The bridge is responsible for: (1) loading CORBA stubs and invoking the CORBA objects; (2) Converting the result from the CORBA server into a SOAP envelope.

Several distinct advantages will be brought by static model:
1. Supporting multiple implementations with the same interface specification. In any case, SOAP server is unaware of CORBA implementations.
2. A CORBA stub is tight-couple with interfaces of a CORBA object.
3. Not need to construct CORBA request dynamically because CORBA stub is unchanged.

The static model provides an efficient, reliable way of wrapping CORBA objects to Web Services. However, as SOAP server is tight-couple with CORBA server, once the interfaces change, the bridge is sure to need some new CORBA stubs which are generated by IDL Compiler.

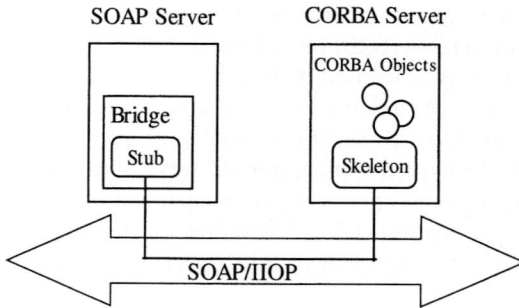

Fig. 1. A static model for CORBA Web Services. Its bridge is responsible for loading CORBA stubs.

2.2 Dynamic Model

In the dynamic model, the bridge translates SOAP request to CORBA invocation dynamically. Fig.2 shows relation between the bridge and CORBA objects. It has to adapt to unpredictable changes in CORBA server, and the static model may be too restrictive.

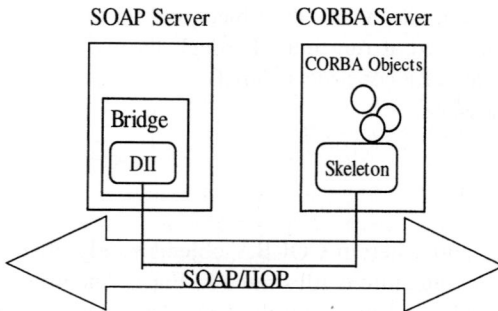

Fig. 2. A dynamic model for CORBA Web Services. Its bridge is responsible for translating SOAP requests to CORBA invocations dynamically via CORBA DII.

The bridge is responsible for: (1) binding CORBA objects dynamically; (2) checking the consistence of invocation methods; (3) processing parameters and mapping SOAP/IIOP protocol between SOAP Server and CORBA Server; (4) converting the result from the CORBA server into a SOAP envelope.

Compared with the former static model, it has some new characteristics:

1. Searching interface information from metadata repository in run-time.
2. Supporting multiple models of binding CORBA objects.
3. Constructing CORBA request dynamically from SOAP message.

This model provides a flexible and loose-coupled way of wrapping CORBA objects to Web Services. Dynamic model is more complex and difficult to implement than static model.

3 Key Technologies

CORBA Web Services has a scalable architecture to support seamless interoperability with CORBA objects. Based static and dynamic model, we demonstrated a prototype system **StarWebService** complying with CORBA Web Services specification [5, 6] and CORBA technology standard [7]. When designed and implemented, it solves several key technologies such as scalable architecture, SOAP/IIOP protocol datatype mapping, unified service providing infrastructure etc.

3.1 A Scalable and Configurable Architecture

CORBA Web Services has a series of Handlers are each invoked in order. SOAP message path is shown in the following Fig.3. The small cylinders represent Handlers and the larger, enclosing cylinders represent Chains. A Chain is a Handler consisting of a sequence of Handlers which are invoked [9]. This architecture may contain transport and global Chains and must contain a Provider, which is a Handler responsible for implementing the actual back end logic of the service. This Provider is a bridge which constructs CORBA invocations and receives the response from CORBA server.

1. Transport Chains is responsible for configuring various transport protocols, such as HTTP, SMTP, message-oriented middleware, etc.
2. Global Chains perform URL mapping, authenticate user whether or not satisfies one of the allowed roles of the target services.
3. Service Chains is a special Handler whose job is to actually perform whatever service-specific operations are necessary to send CORBA request and receive CORBA response. The Provider uses the SOAP RPC convention for determining the method to call, and makes sure the types of the incoming-encoded arguments match the types required parameters for constructing CORBA request.

CORBA Web Services gives a flexible mechanism to insert extensions into the architecture for custom header and SOAP body processing. It supports to configure all Handlers and set all kinds of the monitoring and controlling Handlers of the state of back services.

3.2 SOAP/IIOP Protocol Datatype Mapping

Web services may use SOAP transport protocol, while CORBA server receives the request over IIOP (Internet Inter-ORB Protocol) protocol. Thus the Provider is responsible for translating messages between these different protocols. Datatype mapping is one of the most crucial issues when SOAP and IIOP protocols are being translated.

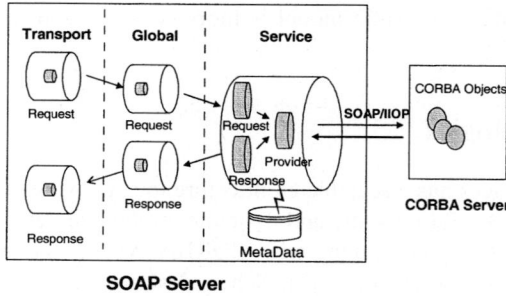

Fig. 3. An architecture for CORBA Web Services. It consists of transport chains, global chains and service chains.

Fig. 4. Data mapping rule between Source Protocol and Target Protocol.

Fig.4 shows datatype mapping rule between Source Protocol and Target Protocol based on a TypePair tuple. A tuple has four elements as the following form: <<*SourceProtocolType, TargetProtocolType>, CodeFactory, DecodeFactory>*, where *CodeFactory* and *DecodeFactory* are responsible for creating and managing specific Code (Decode) classes. Mapping rule's query is executed with the key element TypePair. An example for mapping SOAP *int* to IIOP *short* is as follows:

<<*int, short>, int2shortCodeFactory, short2intDecodeFactory>*

1. Querying Source2TypePair or Target2TypePair hash tables to determine TypePair *<int, short>*.
2. Finding *CodeFactory* and *DecodeFactory* element from TypePair2CodeFactory and TypePair2DecodeFactory hash tables with key TypePair.
3. Loading corresponding Code (Decode) class to serialize (deserialize) SOAP *int* and IIOP *short*. Finally it completes to map SOAP *int* to IIOP *short*.

3.3 Unified Service Providing Infrastructure

CORBA Web Services provides a scalable and customizable Unified Service Providing Infrastructure (USPI) to integrate conventional middleware, such as CORBA/CCM, EJB, DCOM and Enterprise Application Integration in general.

USPI uses plugins to simplify the extension with new service providers (see Fig.5). Each plugin defines unified interfaces that contain operations for accessing back services. *Invoke* method is responsible for locating and interacting with back services; *initServiceDesc* method allows initialization of service description including service interface information. CORBAProvider is a plugin, which retrieves CORBA objects reference and invokes CORBA server via *invoke* interface and loads service information from metadata repository via *initServiceDesc* interface.

It is convenient for programmers to implement plugins for legacy systems that do not adhere to a particular middleware model, similar to CORBAProvider.

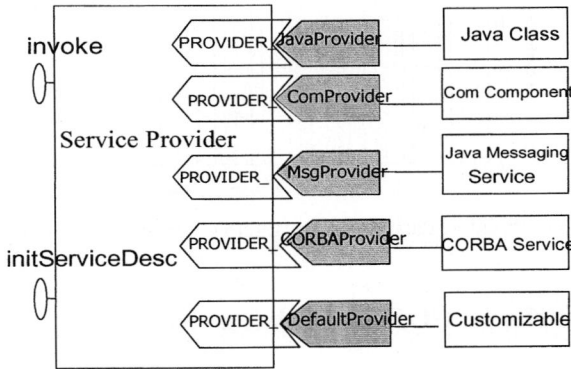

Fig. 5. Unified Service Providing Infrastructure for CORBA Web Services.

4 Performance Evaluations

This section is concerned with the implementation performance and the cost of wrapping CORBA objects. The results are directly affected by the environment the implementation is developed in. The tests run on the implementation with the following conditions: P4 2.6G/MEM 512M/Win 2000/JDK 1.4/**StarWebService** 2.0/**JStarBus** 3.0. **JStarBus** is a middleware platform complying with CORBA technology standard. JStarBus histogram shows the time CORBA client takes to invoke CORBA server via DII; StarWebService histogram shows the time SOAP client spends invoking CORBA server based on dynamic model. Every experiment is repeated 5000 times and all the results are the averages in order to avoid errors caused by randomness.

We are interested in CORBA objects with primitive datatypes in their implementation. Fig.6 presents the time under some environments with ten primitive datatypes. It can be seen that the time of **JStarBus** is rather low, and the maximum is 27.4ms in *float* datatype, while the minimum is just 22.3ms in *short* datatype. Meanwhile, the time of **StarWebService** has the same trend with JStarBus, and the maximum is 123.1ms in *float* datatype, while the minimum is about 110.7ms *short* datatype. The rate between **StarWebService** and **JStarBus** time is about 4.56. The reason for this phenomenon is that CORBAProvider needs to translate SOAP/IIOP protocol and construct CORBA request from SOAP message.

Of course, further work needs to be done the approach to optimize programs consequently to improve wrapping performance.

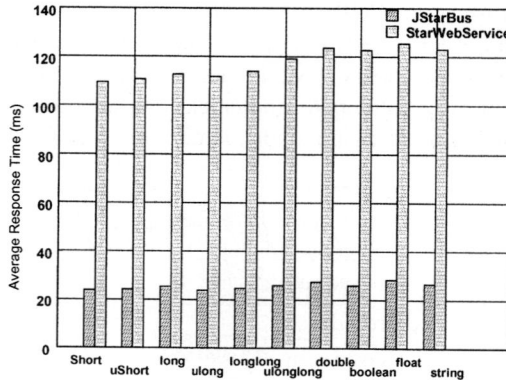

Fig. 6. Primitive datatypes invocation vs. average response time.

5 Related Works on CORBA Web Services

There has been much work done in the field of CORBA Web Services.

Open-source Axis project [9] publishes Axis v1.1 and. provides a scalable architecture. It presents- at the same time as us- a static model very similar to the one described in this paper except that no analytical results are shown and no test results are reported. It does not support dynamic model and bad system tools.

CapeClear [8] proposes to deal with Web Services integrating CORBA and gives perfect tools. CapeClear's platform offers an engine and dispatches SOAP requests to CORBA server based on static model. It has not a unified service providing infrastructure and adheres to INOA Orbix [11].

StarWebService supports static and dynamic models to bridge CORBA server.

To evaluate performance of these platforms, we write down average response time via SOAP client. Fig.7 shows how average response time is affected by the number of request number each platform. Apache Axis achieves low response time in curve S_1 and is hardly affected by request number. **StarWebService** displays a stable trend (between 49 to 50 ms) based on static model in curve S_2. Curve S_3 shows high and unstable time consumption because CapeClear engine loads too graphic tools. Therefore, **StarWebService** is efficient and robust.

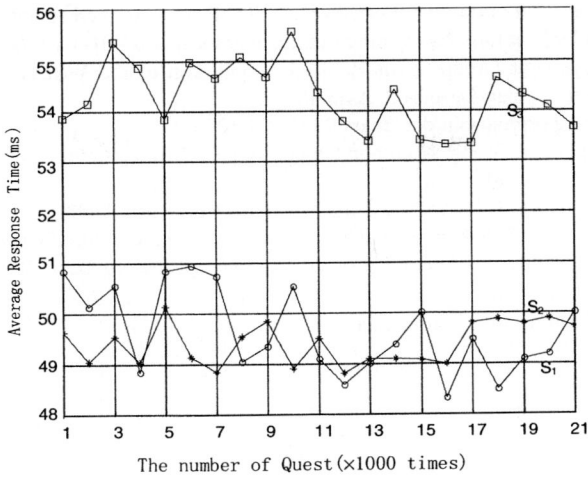

Fig. 7. Average response time vs. request number.

6 Conclusions and Future Work

Web Services are a promising technology that can be seen as the technology ground work facilitating a service-oriented architecture. Obviously enterprises will still use a variety of different middleware technologies and Web Services can bridge the existing boundaries. The major contributions of the paper are as follows: (1) present static and dynamic models to wrap CORBA objects as Web Services; (2) solve three key technologies in CORBA Web Services: scalable architecture, SOAP/IIOP protocol datatype mapping and unified service providing infrastructure. We have implemented the prototype system **StarWebService** and our test results illustrate the effectiveness of the proposed models and architecture. In the future, we will investigate how to include other desirable system properties such as fault tolerance and load balancing into **StarWebService**.

References

1. Aniruddha S. Gokhale, Balachandran N.: GriT: A CORBA-based GRID Middleware Architecture. In Proceedings of the 36th Hawaii International Conference on System Sciences (HICSS) (2003)
2. Joao Paulo A. Almeida, Luis Ferreria Pries, Marten J. van Sinderen: Web Services and Sealmess Interoperability. Available at
   https://doc.telin.nl/dscgi/ds.py/Get/File-31638/almeida-eoows-2003.pdf (2003)
3. Arne Koschel, Klaus Hofmann zur Linden: Interoperability of Standards Web Services & .Net, EJB and CORBA.(2003)
4. Object Management Group: CORBA Web Services, initial Joint Submission. Available at
   http://www.omg.org/technology/documents/spec_catalog.htm (2001)

5. Object Management Group: CORBA-WSDL/SOAP final adopted specification, ptc/03-01-14, Jan. Available at http://www.omg.org/cgi-bin/doc?ptc/03-01-14 (2003)
6. Object Management Group: Joint Revised Submission to the WSDL-SOAP to CORBA Interworking RFP, mars/03-03-03. Available at http://www.omg.org/cgi-bin/doc?mars/03-03-03. (2003)
7. Object Management Group: Common Object Request Broker Architecture: Core Specification, Version 3.0. Available at http://www.omg.org/technology/documents/spec_catalog.htm (2002)
8. CapeClear software 5.0. Available at http://www.capeclear.com/products/download/ (2004)
9. Apache <Web Services/> Axis 1.1 project. http://ws.apache.org/axis/ (2003)
10. WebServices.Org™. http://www.webservices.org
11. IONA software. http://www.iona.com

Grammar Based Interface Processing in Web Service Composition*

XiaoLing Wang[1] and Chi Cheung Shing[2]

[1] Department of Computer Science and Engineering
Fudan University, Shanghai, China
wxling@fudan.edu.cn
[2] Department of Computer Science
Hong Kong University of Science and Technology
Clear Water Bay, Kowloon, Hong Kong
scc@cs.ust.hk

Abstract. Web services have emerged as a new computing paradigm for information integration and sharing. The interfaces of Web services can evolve to adapt to the dynamic and autonomic nature of business workflows. As such, it is difficult to define interface rules between any couple of collaborative services. How to cope with the information exchange among such services is an important and challenging problem in the realization of service composition. This paper studies the problem and presents grammar based method by taking advantage of the XML documents exchanged among the services. Preliminary experiments showed that the grammar based method is promising in terms of efficiency and robustness under the dynamic Web environment.

1 Introduction

The Web is a common platform for enterprises to communicate with their partners, interact with their back-end systems, and perform electronic commerce transactions [2, 18]. Recently, Web services have emerging as a new computing paradigm that enables services providers to make their services available to other Web entities (the service consumer) [13,14]. Because of the dynamic and autonomic nature of business processes, services providing the same function can have different interfaces. On the other hand, web service interfaces, including input and output, can change dynamically. Under this autonomic environment, it is difficult to define interface rules between any couple of services in service composition.

Interface matching in Web service composition [7] is an interesting and challenging problem, which has not been formally addressed in previous works. This paper presents a solution for this problem by representing input and output items with XML [1] element, and then interface change is reflected in the XML elements. Therefore, the key operation is the efficient transformation of an XML document that satisfies one schema to a document satisfying another schema. Though XML transformation is

---

* This work was performed while the author was a visiting scholar at HongKong University of Science and Technology.

S. Wang et al. (Eds.): ER Workshops 2004, LNCS 3289, pp. 323–334, 2004.
© Springer-Verlag Berlin Heidelberg 2004

not a new topic, XSLT, and XQuery can implement XML transformation from one schema to another, but they all require users to write specific scripts. Under the environment of service composition, it is difficult to write static interface matching rules for any couple of services because of their autonomic and dynamic nature. As a result, there is a need to a novel and practical method to address this transformation problem.

In this paper, we focus on studying the information exchange among different services in our D3D-serv (the Dynamic, Demand-Driven Web Services Engine) framework. D3D-serv [17] is a composite services engine that supports dynamic implementation of users' request and business logics by web service composition. The main contributions of this paper are as follows:

1. Formally address the interface matching and information exchanging problem in web service composition. It is formulated as an XML data element matching and extraction problem.
2. Present a feasible and practical grammar based methods to solve this problem. This approach processes XML data with limited amount of memory and high performance. This method can also handle the large XML data stream among the services.
3. Implement our proposed method in D3D_Serv, which is a service composition engine. Preliminary experimental comparison between the benchmark of our method and that of the others showed that our approach is effective and efficient.

This paper is organized as follows. Section 2 presents related work. Section 3 describes D3D_Serv architecture, and addresses the interface matching problem formally. An algorithm is presented in Section 4. Experiments and evaluations are given in Section 5. Section 6 concludes our work and discusses future work.

2   Related Work

Web services have recently received much attention from academia and industries. A lot of standards and protocols have been drafted by international organization or industrial community. Notable ones include WSDL [21] (Web Service Description Language), UDDI [20] (Universal Description, Discovery, Integration), and SOAP (Simple Object Access Protocol). Both industrial and academic, the information exchange among services is an important factor for service composition and communication. However, former work did not address this problem explicitly and formally.

Boualem et al. described SELF-SERV system [15, 16], which takes the advantage of P2P computing and presented Coordinator, Wrapper and Routing-table concepts for message communicating. However, SELF-SERV neither supports service interface change nor allows dynamic self-adjustment.

BPEL4WS (Business Process Execution Language for Web Services) [7] is an industrial protocol for composite services. BPEL4WS describes how a business process is created by composing the basic activities that it performs into those structures that express control patterns, data flow, the handling of faults and external events, and the coordination of message exchanges between the processes involved in a business protocol. However, it did not propose any strategy about information exchange between different services in view of changing data schema. Our method can therefore provide a useful complement to BPEL4WS.

There is much related work on XML information integration [4, 5, 8, 12], such as Biztalk (http://biztalk.com), Oasis (http://www.oasisopen.org), XSLT and AXML [18]. None of them provides a feasible method for interface matching and information exchange in the composition of Web services. Although various studies on XML streaming processing have been made to provide solutions for information extraction and publishing, but they mostly depend on static queries or requests. These solutions are also impractical for dynamic Web services, where input and output interfaces are susceptible to change [2, 6].

3 Preliminaries

In this section, we illustrate the D3D_Serv architecture, and then address formally the service interface matching problem.

3.1 Architecture of D3D_Serv

D3D_Serv adopts three Web services standards, SOAP, WSDL and XTM (XML Topic Map). SOAP defines an XML messaging protocol for basic service interoperability. WSDL provides a common language for describing services. XTM provides the infrastructure for organizing semantic information in XML. These specifications together allow applications to describe and discover Web services, and to interact with each other by following a loosely coupled, platform independent model.

The architecture of D3D_Serv is shown in Fig. 1. The emphasis of D3D_Serv is on reusing existing services and composing new services or business processes from existing ones rather than building a complex new service. A Web application system involves a number of computers connected together by the Internet. Each computer can play a different role, such as service provider and service requester, in different applications. Service composition allows the realization of a complex business task by feeding the output data of a service to the input of another service.

We have implemented this framework, including the functions for service execution, service publishing, and service querying using the P2P technology. In this paper, we focus on how to implement interface matching and information exchange among different services in the "Services Selection and Execution Engine".

3.2 Interface Processing Problem Among Services

We model a Web service as <WSDL, XTM>, where WSDL and XTM describe the interface information and the semantics information, respectively, as shown in Fig. 2. Due to space limitation, this paper focuses on the underlying technology for interface matching among different services. The problem of service discovery based on semantic information in XTM will be studied in another work.

To exchange information, the output of one service must match the input of the immediate service in the service workflow. For example in Fig. 3, the output of "Order Management Service" must match the input of "Shipping Service", and similarly the output of "Shipping Service" must match the input of "Customer Service". "Interface Processing" module is to deal with the information exchange among different services.

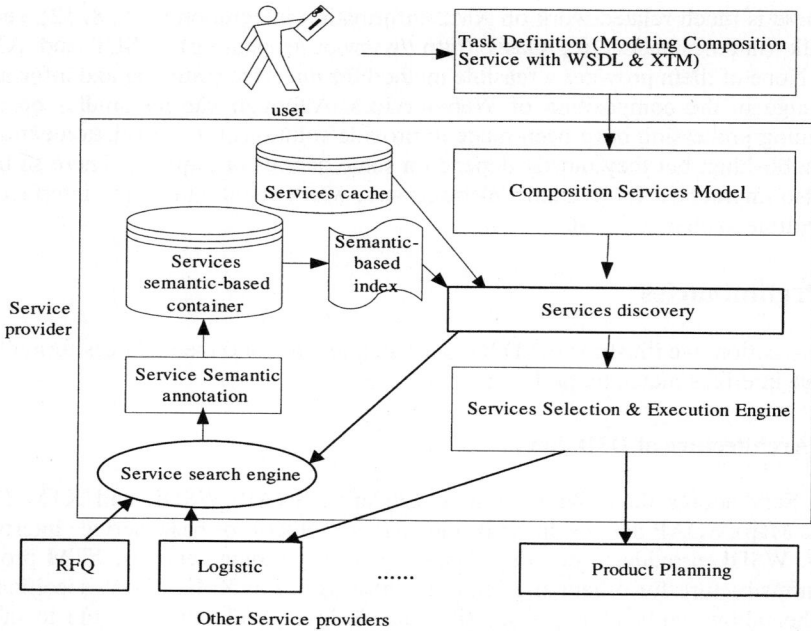

Fig. 1. The architecture of D3D-Serv

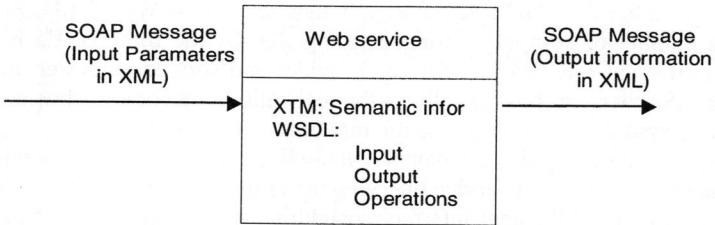

Fig. 2. Web services description

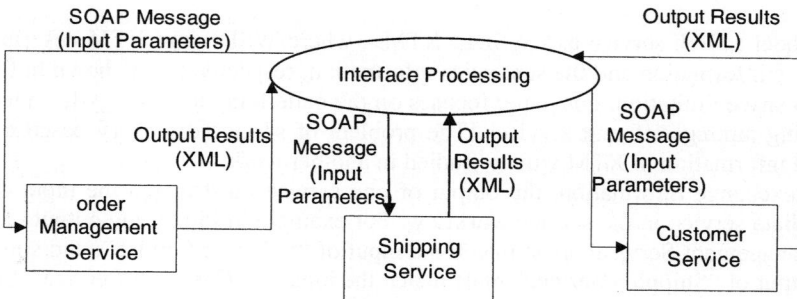

Fig. 3. Interface Processing in Web Service composition

In Web service composition, input data for a service are obtained from the output data of another service. Let us consider the input and output as data sets and their elements the XML data elements. Take the Order Results in Fig. 3 as an example. Order results are outputted by the "Order Management Service". Since the "Shipping Service" needs the input parameters of "Product", "Date" and "Destination", we extract related XML data before passing the information to "Shipping Service". Suppose we have an "Order" XML document as follows.

```
<Order>                      somenotes
<Product>                    </Notes>
   Chair                     <Detail>
</ Product >                    <Price>
<Date>                            100
  2004-6-1                    </Price>
</Date>                       <Quantity>
<Destination>                    8000
  HongKong                    </Quantity>
</Destination>               </Detail>
<Notes>                      </Order>
```

We can use an XML data graph to represent the "Order" XML document. In the XML graph of Fig. 4, each node represents the data element of an XML datum and the edge represents its tag. For example, nodes 1 and 2 represent "Order" XML element and "Product" XML element, respectively. We can formulate this interface matching problem in terms of XML data element extraction and matching.

There are three relationships between input XML data elements and output XML data elements. Let B denote the output data set of a service, and A denote the input data set of the immediate service in a workflow.

- $A = B$: Two data sets A and B are equal, that is, B's output is A's input exactly.
- $A \cap B = \varnothing$: Two sets are disjoint if $A \cap B = \varnothing$ (the empty set), that is B's output cannot be the input of A.
- $A \cap B \neq \varnothing$: Assuming the common subset is C. Then some input data of A, that is C, come from B. In order to implement extracting data C, we define *transform(XML doc from B, required Schema of A)* to transfer data from B into some requirement of A.

The naïve method to implement the *transform* operation is to parse output XML document into a DOM tree in memory, and then extract the associated XML elements. However, this method is not effective for service communication because it requires hard coding. In addition, it is too expensive to prepare the transforming codes for all service pairs. An alternative is to use SAX to parse XML documents and extract XML segments. This method needs to guarantee the completeness and correctness of the extracted segments. In the next section, we will introduce XML grammar and propose a solution which guarantees the completeness and correctness of extracted XML segments.

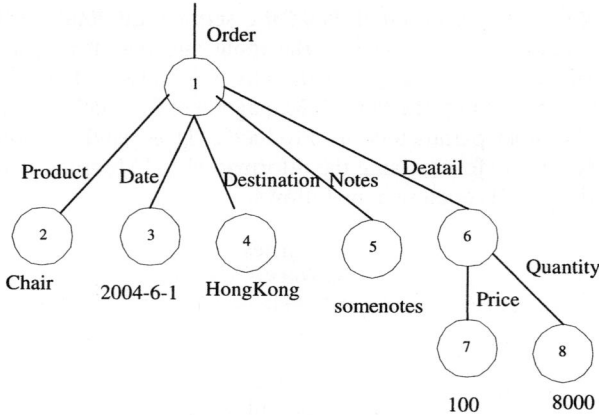

Fig. 4. "Order" XML data graph

4   Grammar Based Method for Transform

The problem of service interface transformation is equivalent to the problem of recognizing input stream elements subject to a given language or schema definition. In this section, we introduce the XML and its grammar characteristics, and then present a grammar based approach for the interface information exchange problem of dynamic Web service composition.

4.1   XML and Its Schema

XML is self-descriptive. The XML DTD (Document Type Declaration) and XML Schema [1] are two standards of XML schema information. A DTD defines a class of XML data using a language that is essentially a context-free grammar with several restrictions. The corresponding DTD in Fig 4 is given as follows.

```
<!ELEMENT Order     (Product, Date, Destination, Notes, Detail*)>
<!ELEMENT Product   (#PCDATA)>
<!ELEMENT Date      (#PCDATA)>
<!ELEMENT Destination (#PCDATA)>
<!ELEMENT Notes  (#PCDATA)>
<!ELEMENT Detail     (Price, Quantity)>
<!ELEMENT Price (#PCDATA)>
<!ELEMENT Quantity (#PCDATA)>
```

Note that the XML DTD follows notational convention, such as ?, * and + denoting respectively zero or one, zero or more, and one or more occurrences of the preceding construct.

4.2   XML Grammar

Before introducing the XML grammar, let us review some fundamental concepts of the conventional grammar theory. Chomsky founded the Formal Language Theory

[11] in 1956, which is to describe some language logically by grammar and languages. A grammar $G = (V_N, V_T, P, S)$ is a 4-tuple, where $S \in V_N$ is a distinguished element of V_N called the start symbol, V_N is a finite set of symbols or variables, V_T is a finite set of terminal symbols, P is a finite set of productions or rules, which are production functions mapping elements of V_N to the finite sets of strings over $(V_N \bigcup V_T)$.

In the grammatical point of view, XML DTD is a context-free grammar. The set of element in DTD is regarded as V_N, and V_T includes PCDATA, and the content of every element or attribute. Now, we introduce our XML Grammar.

Definition 1 A grammar $G = (V_N, V_T, P, S)$ is an XML grammar if all production functions in P satisfy the following rules:

- P has the form $A \rightarrow < A > \alpha < / A >$, where both $< A >$ and $< / A >$ are symbols of V_T, α is a list of variable symbol or PCDATA.
- There is no production functions like $A \rightarrow XY$, where A, X, Y are V_N.

In above definition, let $< A >$ denote Begin-V_T, and $< / A >$ End-V_T. That is, the left-hand side must consist of a single variable, and the right-hand side consists of two terminals and variables. Then, we can represent one DTD element <!ELEMENT Product (#PCDATA)> with   Product $\rightarrow$ < Product > #PCDATA < / Product >. Then, each DTD constrains can be transformed to XML grammar rules equally. Now we show some interesting results from XML grammar.

Theorem 1 Begin-V_T and End-V_T always appear simultaneously in the derivation string ω_n.

Proof: According to Definition 1, each rule is formally expressed as $A \rightarrow < A > \alpha < / A >$. As a result, Begin-$V_T$ and End-V_T are always coupled with each other. □

This theorem is to prove that neither Begin-V_T nor End-V_T appear separately.

Theorem 2 If there is End-V_T $< / A >$ in a derivation string ω_n, there must apply a rule $A \rightarrow < A > \alpha < / A >$ in the derivation process.

Proof: According to Definition 1, each production function is like $A \rightarrow < A > \alpha < / A >$; so End-$V_T$ $</A>$ always appears after Begin-V_T $<A>$ and End-V_T always appears at the right hand side of the rules. Consequently, if there is $</A>$ in the sentential form, some rules like $A \rightarrow <A> \alpha </A>$ must have been applied in the derivation process. □

Theorem 3 For each sentential form $S \xrightarrow{\ *\ } \alpha (\alpha \neq S)$ in the derivation process, the first End-V_T in a is $</A>$ IFF the leftmost variables are in the form of $<A> \gamma </A>$.

Proof: Firstly, we prove "the first End-V_T in a is $</A> \Rightarrow$ the leftmost variables are in the form of $<A> \gamma </A>$."

According to Theorem 1 "Begin-V_T and End-V_T always appear at the same time", there must exist $<A>$, which matches $</A>$. So, $S \xrightarrow{\ *\ } \alpha_1 </A> a_2$ can be rewritten as $S \xrightarrow{\ *\ } \alpha_3 <A> a_4 </A> a_2$, where $\alpha_3 <A> a_4 = \alpha_1$. As we have assumed, $</A>$ is the most left End-V_T, so there is no End-V_T in α_3 and a_4. According to Theorem 1, there is no any Begin-V_T and End-V_T in a_4. As a result, α_4 is composed of variables and strings. According to Theorem 2, $A \longrightarrow <A> \alpha_4 </A>$ holds.

Here, let us verify that $<A> \alpha_4 </A>$ is the leftmost variable.

Suppose the leftmost variable in a_3 is $<B> \delta </B>$, rather than $<A> \alpha_4 </A>$. Then, in $S \xrightarrow{\ *\ } \alpha_3 <A> a_4 </A> a_2$, $</B>$ appears before $</A>$. This inference contradicts our initial assumption that "$</A>$ is the leftmost End-V_T". So, $B \longrightarrow <B> \delta </B>$ does not exist in a_3 and $<A> \alpha_4 </A>$ is the leftmost variable.

Secondly, to prove "if $<A> \gamma </A>$ is leftmost $\Rightarrow$ the leftmost End-V_T in a is $</A>$."

Supposing the leftmost End-V_T in a is $</B>$, rather than $</A>$. So, there exists a leftmost sentential form $<B> \delta </B>$. Because $</B>$ appears before $</A>$, $<B> \delta </B>$ is leftmost. This result contradicts the condition in the proposition that "$<A> \gamma </A>$ is the leftmost variable". So, this hypothesis does not hold and the most left End-V_T in a is $</A>$.

Thus, this proposition holds. □

The conventional grammar theory has proved that the leftmost variable is the induction unit. When the reading header reads End-V_T $</A>$, the structure of A in the stack from <A> to </A> can be inducted correctly. Theorem 3 guarantees that we can perform induction when we find the End tag during the parsing phase and we can ignore data element that are not required by the target DTD and keep the required ones. Based on this theorem, we will present our Grammar based Transform (GBT) algorithm in the next section.

4.3   Grammar Based Transform (GBT) Approach

This section will describe how to take advantage of the feature of XML Grammar to implement information extraction correctly.

Theorem 3 guarantees the correctness of extracting the requirement information from XML data and skip no-needed data elements during scanning XML data. We construct a pushdown automata machine in Fig. 5 to implement this function.

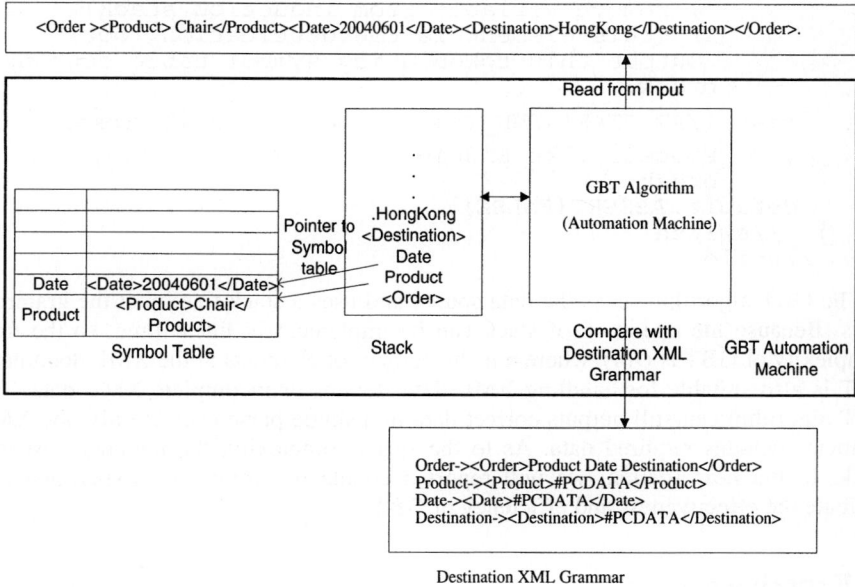

Fig. 5. Grammar based automata machine Model

Now we describe the GBT algorithm in detail.

```
Input: XML data Doc, Required destination XML schema S
Output: XML document according to desired XML schema
while (1)   //run this process until no data
{ bRetGetToken = GetToken(&InputStream);
  if (!bRetGetToken)  // no token from input, exit
    break;
  switch(Token. Type)
  {case TYPE_START_TAG:   //if token is Begin-$V_T$
        Stack[SP++] = Token; //Put it into work stack
        Break;
    case TYPE_DATA:   //if token is the content
        Stack[SP++] = content value;
        Break;

    case TYPE_END_TAG: //If token is End-$V_T$
      //search the corresponding begin tag in the stack
        PopToBeginTag(Token);
```

```
          aRule=induction(Token);
          stack[iStackTop]=inductionVariable;
          CheckRuleInDestRule(aRule);
          If(Desired) //is a rule in desired schema
            { Put the induction variable into Symbol
              table, replace all variables in right part
              of rule with symbol table content, remove
              the unused symbols}
          Else //not the desired schema
            { remove variable from induction stack}
          If (this element is the destination root)
            Output this token using symbol table content.
          break;
      case TYPE_START_END_TAG:  //Begin-End-V_T Token
          Process like Endtag;
          break;
      default:ASSERT(FALSE);
    }  //switch
  }  //while
```

The GBT algorithm scans the data source and uses a stack to extract the grammar rules. Because the operation of stack can be implement in linear time, so the time complexity of GBT is $O(n)$, where n is the number of elements in the XML document. GBT is also suitable for handling XML data streams or incomplete XML data. The GBT algorithm can still outputs correct data in a single parse even if only one XML segment includes required data. As to the space complexity, the memory cost is a stack. In the next section, we will report a couple of preliminary experiments to evaluate the effectiveness and efficiency of GBT.

5 Experiments

We have implemented the GBT algorithm and conducted some experiments to evaluate the effectiveness and efficiency.

1. We studied the robustness of our algorithm with workloads represented by different XML benchmark, and the files sizes are from few KB to several MB.
2. In order to examine the performance of our method, we compared our methods with a typical DOM based method.

The experiments were conducted on a P4 machine with a 933MHz CPU and 256MB memory. The test data and workload used in our tests included the Shakespeare XML data sets [3], the SIGMOD XML data [9] and the XMark [10]. We tested GBT against different sizes of XML data. Our test file size covered a large range, from KB to MB. Table 1 show that our algorithm can be effective for both small data size and large data size.

Table 1. GBT Performance on different data sets

Data sets	Sigmod xml	Data generated by XMark
XML data size(KB)	482	115775
GBT time(second)	0.18	24.875

The next test is to compare GBT with DOM based method. We tested the first 20 files from Shakespeare XML data sets. The result is shown in Fig 6. It verifies the efficiency of GBT. The X-axis indicates the size of XML documents, and the Y-axis gives the running time to parse and extract related elements from the original XML documents. Since the DOM-based method needs to construct a tree in memory and then extract related parts from the tree, it takes more time. In addition the GBT approach can obtain desired elements at the scan and induction process. The performance is better than that of the DOM-based method.

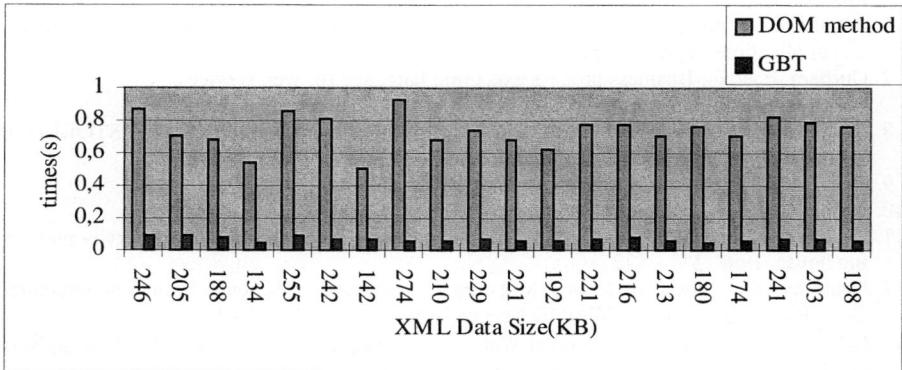

Fig. 6. GBT algorithm vs. DOM-based method

6 Conclusions

This paper studies the interface information processing among different services. Because of the dynamic and autonomic nature of these services, they can have different input and output data for the same function. We presented a grammar based method GBT to guarantee the correctness of XML element extraction. Preliminary experiments show that the proposed approach is promising and can offer better performance than that of conventional DOM based methods. There are several potential advantages of deploying grammar based approaches, such as not requiring special codes between any couple of services, robust enough to handle different sizes of XML data, and requiring little spaces. We are studying the problem of semantic-based interface exchanging in service composition and the support of more complex schema, such as composition conditions.

Acknowledgment

We would like to thank Professors Lionel Ni and Qiong Luo for their valuable comments.

References

1. http://www.w3.org/XML/
2. McIlraith, S. and Martin, D.: Bringing Semantics to Web Services. IEEE Intelligent Systems, 18(1):90--93, January/February, 2003.
3. Shakespeare XML data sets. ftp://sunsite.unc.edu/pub/sun-info/standards/xml/eg/
4. Dolin, R., Agrawal, D. and Abbadi, A.: Scalable collection summarization and selection. In Proc. of ACM Conference on Digital Libraries, pages 49-58, 1999.
5. Florescu, D., Koller, D., Levy A. Y.: Using probabilistic information in data integration. In Proc. of VLDB97. 216 - 225.
6. Alonso, G., Casati, F., Kno, H. and Machiraju, V.: Web services: concepts, architectures and applications. Springer ISBN 3-540-44008-9. 2004
7. Curbera, F. et al.: Business process execution language for web services. http://www-106.ibm.com/developerworks/webservices/library/ws-bpel/.
8. Deutsch, A., Fernandez, M. and Suciu, D.: Storing semistructured data with STORED. In Proc. of ACM SIGMOD, Philadelphia, PN, 1999.
9. Sigmod XML example. http://www.acm.org/sigs/sigmod/record/xml/
10. Xmark XML example. http://www.xml-benchmark.org
11. Yingzhi, L., SuQin, Z. and WeiDu, J.: Principle of compile. Tshinghua university publishing house. 1998. 117 - 122
12. Nestorov, S., Abiteboul, S. and Motwani, R.: Extracting Schema from Semistructured Data. Proc SIGMOD'98, SIGMOD Record 27, 2, June 1998. 295 - 306.
13. Oded, S.: Architectures for Internal Web Services Deployment. VLDB 2001, Roma, September, 2001
14. Tsur, S., Abiteboul, S. and Agraval, R.: Are Web Services the Next Revolution in E-Commerce? VLDB 2001, Roma, September, 2001
15. Benetallah, B., Dumas, M.: Declarative Composition and Peer-to-Peer Provisioning of Dynamic Services. ICDE 2002, San Jose, February 2002
16. Sheng, Q. Z. and Benatallah, B.:SELF-SERV: A Platform for Rapid Composition of Web Services in a Peer-to-Peer Environment. VLDB 2002, Hong Kong, August, 2002
17. Wang, X., Yue, K. and Zhou, A.: Service Selection in Dynamic Demand-Driven Web Services. ICWS 2004, CA, July 2004
18. Abiteboul, S. and Benjelloun, O.: Active XML: Peer-to-Peer Data and Web Services Integration. VLDB 2001, Roma, September, 2001
19. Hull, R., Benedikt, M., Christophides, V. and Su, J.: E-Services: A Look Behind the Curtain. PODS 2003, San Diego, CA, June 2003.
20. UDDI Home Page. http://www.uddi.org.
21. Web Services Description Language (WSDL) 1.1. http://www.w3.org/TR/wsdl

Query Rewriting for Extracting Data Behind HTML Forms

Xueqi (Helen) Chen[1], David W. Embley[1], and Stephen W. Liddle[2]

[1]Department of Computer Science
[2]School of Accountancy and Information Systems
Brigham Young University
{chen,embley}@cs.byu.edu, liddle@byu.edu

Abstract. Much of the information on the Web is stored in specialized search-able databases and can only be accessed by interacting with a form or a series of forms. As a result, enabling automated agents and Web crawlers to interact with form-based interfaces designed primarily for humans is of great value. This paper describes a system that can fill out Web forms automatically according to a given user query against an ontological description of an application domain and, to the extent possible, can extract just the relevant data behind these Web forms. Experimental results on two application domains show that the approach can work well.

1 Introduction

With the enormous amount of information on the Internet that can be accessed only by interacting with a form or a series of forms, it becomes useful to enable automated agents and Web crawlers to interact with form-based interfaces designed primarily for humans. Several researchers, besides ourselves, have recognized this challenge. The Hidden Web Exposer (HiWE) [17] extends crawlers by giving them the capability to fill out Web forms automatically. HiWE, however, tends to retrieve all the information behind the sources, and human assistance is critical to ensure that the Exposer issues queries that are relevant to the particular task. Microsoft's Passport and Wallet system [11] encrypts a user's personal information and then automatically fills out Web forms with the user-provided information whenever it is applicable, but the system makes no attempt to retrieve information behind those forms. The commercial system ShopBot [3] is a general-purpose mechanism for comparison shopping. Its form filling process is automatic, but simplistic. In earlier work within our research group [10], a fully automated system aims to extract all the information from one Web site (behind one Web form), regardless of what a user wants. In our current work, we aim to provide only the information a user wants.

There are significant technical challenges in automating the form filling process. First, an automated agent must understand a user's needs by interpreting a user's query. Second, an automated agent must understand Web forms, which provide for site queries, and map the user's query to a site query. This is challenging because different Web forms, even for the same application, provide different ways to query their databases. Figure 1 shows three different Web forms for the same application, car advertisements, from three different information providers.

S. Wang et al. (Eds.): ER Workshops 2004, LNCS 3289, pp. 335–345, 2004.

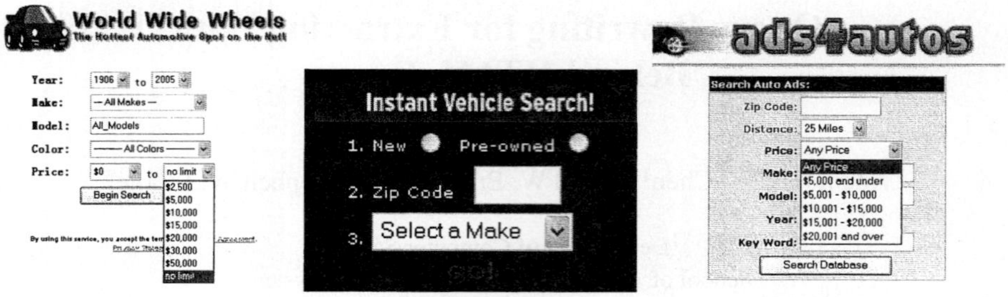

| (a) wwwheels.com | (b) dealernet.com | (c) ads4autos.com |

Fig. 1. Web Forms for Car Advertisement Search

Since Web forms are designed in a variety of ways, handling all kinds of Web forms according to user queries by one automated agent is challenging. Although seemingly simple, direct matches between user-query fields and form fields can be challenging because synonymy and polysemy may make the matching nontrivial. Moreover, problems arise when user queries do not match with form fields. Consider, for example, the three forms in Figure 1, the user query, "Find green cars that cost no more than \$9,000," and the returned information in Figure 2 for the form in Figure 1(b) filled in for one of the makes, "Ford". Mismatches occur as follows. (1) Fields specified in a user query are not contained in a Web form, but are in the returned information (e.g. color in Figures 1(b) and 2). (2) Fields specified in a user query are not contained in a Web form, and are not in the returned information (e.g. color, which is not in the returned information for the form in Figure 1(c)). (3) Fields required by a Web form are not provided in a user query, but a general default value, such as "All" or "Any", is provided by the Web form (e.g. make and model in Figure 1(a)). (4) Fields required by a Web form are not provided in a user query, and the default value provided by the Web form is specific, not "All" or "Any", which leads to the need for multiple form submissions to satisfy the query (e.g. for Figure 1(b) a make and either new or pre-owned must be selected). (5) Values specified in a user query do not match with values provided in a Web form, which leads to the problem that the desired information cannot be retrieved using a single form query (e.g. no \$9,000 in either Figures 1(a) or (c)).

To solve the problems described above, we have built a prototype system designed to fill out Web forms automatically according to a given user query against an ontological description of an application domain. To the extent possible, the system extracts just the relevant data behind Web forms. We have implemented the system using Java [8], Java Servlets [1], and MySQL [14]. Our prototype system has two central parts, the input analyzer (described in Section 2) and the output analyzer (described in Section 3). Our input analyzer interacts with the user to obtain a query, fills in a site form according to the user query, and retrieves relevant Web pages. The output analyzer retrieves Web data contained in multiple pages using "next" or "more" links, extracts data from all retrieved Web pages, populates our database, and displays the final results to the user. In Section 4, we analyze experimental results and discuss the advantages and disadvantages of our system. We conclude with summary remarks in Section 5.

☑	N/U	Year	Vehicle	Price	Miles	Color	Location	Distance ⬇
☐	U	1995	Ford Explorer 📷	$8,995	110,155	Black	Chantilly, VA	3
☐	U	1998	Ford Econoline 📷	$9,995	51,978	White	Chantilly, VA	3
☐	U	2000	Ford F-250 Series 📷	$22,995	117,425	Red	Chantilly, VA	3
☐	U	2001	Ford Explorer Spo... 📷	$16,995	69,978	Black	Chantilly, VA	3
☐	U	2000	Ford Explorer 📷	$16,995	67,049	Burgund...	Chantilly, VA	3
☐	U	1998	Ford Escort 📷	$10,995	132,389	Green	Chantilly, VA	3
☐	U	2000	Ford Mustang 📷	$7,995	65,742	White	Chantilly, VA	3
☐	U	2003	Ford Escape 📷		18,558	Blue	Herndon, VA	4
☐	U	1993	Ford Escort	$4,895	67,917	Burgand...	Fairfax, VA	6
☐	U	1998	Ford Windstar	$16,795	12,844	Green	Fairfax, VA	6
☐	U	1996	Ford Taurus	$9,995	47,165	White	Fairfax, VA	6
☐	U	1996	Ford Windstar	$13,695	47,030	Green	Fairfax, VA	6

Fig. 2. Partial Retrieved Data from *dealernet.com*, February 2004

2   Input Analyzer

Our system starts by allowing a user to choose an application from a list of applications for which we have ontological descriptions encoded as domain extraction ontologies [4]. The input analyzer then parses the extraction ontology, collects data from the user query, matches the fields in the user query to the fields in a given site form, generates a set of one or more queries, and submits the set for processing at the form's site.

An extraction ontology is a conceptual-model instance that serves as a wrapper for a narrow domain of interest such as car ads. The conceptual-model instance consists of two components: (1) an object/relationship-model instance that describes sets of objects, sets of relationships among objects, and constraints over object and relationship sets, and (2) for each object set, a definition of the lexical appearance of constant objects for the object set and keywords that are likely to appear in a document when objects in the object set are mentioned.

Our prototype system provides a user-friendly interface for users to enter their queries. In order to make the interface user-friendly and make query specification easy to understand for our system, we construct an intermediate application-specific form from the object sets of the chosen application ontology. Given a list of fields, which are the object sets in the extraction ontology designated as potential fields, the user chooses which fields are of interest. For fields for which the user can specify ranges our system allows the user to select the type of input—*exact value*, *range values*, *minimum value*, or *maximum value*. After the user selects the desired fields and range types, our system provides a final, ontology-specific search form to the user. The user fills out this ontology-specific form to create a query.

We then process the user-specified query for an HTML page that contains a form that applies to our chosen application. To do so, we first parse the content of the HTML page into a DOM tree and find the site form. For site form fields created with an input tag, we are interested in the type, name, and value attributes. After parsing the input tag, we store the field name, field type, and field value for fields with type text, hidden, checkbox, radio, and submit. For site form fields created with a textarea tag, we store the field name with field type textarea. For site form fields created with a select tag, we obtain the field name, the option values (values inside the option tags), and the displayed values (values displayed in the selection list on the Web page). Our system fills in the HTML site form by generating a query or a set of queries that correspond, as closely as possible, to a given user query. The form filling process consists of three parts: (1) field name recognition, (2) field value matching, and (3) query generation.

2.1 Form Field Recognition

Because site forms vary from site to site, even for the same application domain, site-form field recognition is difficult. Because of the way we allow a user to specify a query, field-name recognition is essentially a problem of matching the fields in a site form to object sets in our extraction ontology.

We first group all radio fields and checkbox fields that have the same value for the name attribute and consider each group as one field. Then, for fields with values provided (i.e., select fields, grouped radio fields, and checkbox fields), we apply our constant/keyword matching rules to determine the field names. If more than half of the values in a field belong to the same object set, we conclude that the field corresponds to that object set.

For all input fields of type text and all textarea fields in the site form, we compare the field-tag names to the object-set names using similarity measures from 0 (least similar) to 1 (most similar), and we choose the object set with the highest similarity as long as the similarity is above an empirically-determined match threshold. When the field tag names and the object set names are exactly the same, we assign 1 to the similarity measure and conclude that there is a match between the two fields. Otherwise, we calculate the similarity between the two strings that represent the names of the object set and the field using heuristics based on WordNet [6, 12]. WordNet is a readily available lexical reference system that organizes English nouns, verbs, adjectives, and adverbs into synonym sets, each representing one underlying lexical concept. We use the C4.5 decision tree learning algorithm [16] to generate a set of rules based on features we believe would contribute to a human's decision to declare a potential attribute match from WordNet, namely (f0) same word (1 if A = B and 0 otherwise), (f1) synonym (1 if "yes" and 0 if "no"), (f2) sum of the distances of A and B to a common hypernym ("is kind of") root, (f3) the number of different common hypernym roots of A and B, and (f4) the sum of the number of senses of A and B. We calculate the similarity between an object-set name and a field name based on a set of rules generated by the C4.5 decision tree. If the similarity between the object-set name and the field name reaches a certain threshold, we match the two. If there is still no match, we calculate the similarity between an object-set name and a field name by a combination of character-based string matching techniques. First, we apply standard information-retrieval-style stemming to get a root for each name [15]. Then, we combine variations of the Levenshtein edit distance [9], soundex [7], and longest common subsequence algorithms to generate a similarity value.

2.2 Form Field Recognition

Section 1 describes five ways that mismatches can occur between site-form and user-query fields. We treat these various cases in different ways. In the field matching process, we offer solutions to Case 0, which is the direct match between user-query fields and site-form fields, and two of the five aforementioned mismatch types (Cases 3 and 4). We also offer a partial solution to Case 5, but leave Cases 1, 2, and the remainder of 5 to the output analyzer.

Case 0: Fields specified in the user query have a direct match in a site form, both by field name and by field value(s). For example, a user searches for cars around a certain *Zip Code*, and *Zip Code* is a field of type text in the site query.

Solution: We simply pair and store the user-provided value with the attribute.

Case 3: Fields required by a site form are not provided in the user query, but a general default value, such as "All", "Any", "Don't care", etc. is provided by the site form. For example, a user does not specify any particular *Make* for cars of interest, and *Make* is a field with a select tag in a site form with a list of option values including a general default value as a selected default value.

Solution: We find the general default value and pair and store the corresponding option value with the field name.

Case 4: Fields that appear in a site form are not provided in the user query, and the default value provided by the site form is specific, not "All", "Any", "Don't care", etc. For example, a user does not specify any particular *Make* for cars of interest, and *Make* is a field with a select tag in a site form with a list of option values. Unfortunately, no general default value is provided in the option list.

Solution: We pair and store the field name with each of the option values provided for the fields by the site form. Later in the submission process, we submit the form once for each name/value pair.

Case 5: Values specified in the user query do not match with values provided in a site form. For example, a user searches the form in Figure 1(c) for "cars that cost no more than $9,000."

Solution: This case happens only for range fields. As human beings, we know that we should find the least number of ranges that cover the user's request. For our example, we should submit the form for "$5,000 and under" and "$5,001 - $10,000". Our system converts range fields into a standard form and selects the appropriate ranges for the query.

2.3 Form Query Generation

Once our system has Web form fields paired with values, it can "fill out" the form, i.e. generate queries for the site form. Our system selects one name/value pair from each form field, concatenates all selected pairs together, and appends them to the site URL constructed from the meta information of the HTML page and the action attribute of the form tag. Using the form in Figure 1(a) for our sample query, "Find green cars that cost no more than $9000," our output analyzer generates the following string:

```
http://wwwheels.com/cfapps/autosearchresults.cfm?
ExteriorColor=GREEN&PriceHigh=10000&SqlStatement=&
carLowPrice_range="MIN=1 MAX=500000"&
carHighPrice_range="MIN=1 MAX=500000"&
carLowMileage_range="MIN=1 MAX=500000"&
carHighMileage_range="MIN=1 MAX=500000"&
MinYear=1926&MaxYear=2005&cfr=cfr&YearBegin=1926&
YearEnd=2005&Make=&Model=All_Models&PriceLow=0
```

Observe that the system specifies Exterior Color=GREEN and PriceHigh=10000 as required by the query. It also specifies default values for all remaining fields as required by the form.

3 Output Analyzer

Our system stores the Web result pages the input analyzer collects and then sends them to the output analyzer. The output analyzer examines each page and extracts the information relevant for the user query. The output analyzer then filters the extracted information with respect to the user query and displays the final results to the user in HTML format. At this point, the records displayed to the user are, to the extent possible, just the data relevant to the user's original query.

Sometimes, the results for one query come in a series of pages, but by submitting the query, we can only retrieve the first page of the series. To obtain all results, our system iteratively retrieves consecutive *next* pages. It may need to follow the value of the href attribute of an anchor node with the keyword "next" or "more" or a sequence of consecutive numbers that appear as a text child node of the anchor node. Or it may need to submit a form with the keyword "next" or "more" appearing in the value attribute of an input node with type submit in the form node.

In the next step, the output analyzer takes one page at a time, runs it through a record separator [5] and then through BYU Ontos [2, 4], a data extraction system, to populate a database. To extract information from Web pages using Ontos requires recognition and delimitation of records (i.e. a group of information relevant to some entity, such as an individual car ad in a car-ads application). Our record separator captures the structure of a page as a DOM tree, locates the node containing the records of interest, identifies candidate separator tags within the node, selects a consensus separator tag, removes all other HTML tags, and writes the modified page to a record file. Given the output from the record separator, the next step is to invoke Ontos. For each document, Ontos produces a data-record table containing a set of descriptor/string/position tuples for the constants and keywords recognized in the record, resolves all conflicts found in the data-record table, and constructs SQL insert statements for database tuples from the modified data-record table. After the system processes all pages obtained from the Web site, it generates a fully populated database. Once the database is populated, it is simple for the system to execute the SQL query generated from the user's original filled-in form.

This generated SQL query performs the post-processing necessary to return the data and just the data requested by the user. It therefore resolves **Case 1** (as described in Section 1) by obtaining the information requested in a user query that is returned but not specifiable in a site form. For example, it obtains the records for just green cars for the form in Figure 1(b), the sample returned results in Figure 2, and our sample query, "Find green cars that cost no more than $9,000." It also resolves the remaining part of **Case 5** by eliminating extraneous values returned by the site form but not requested in the user query. For our sample query and the returned results in Figure 2, for example, it eliminates cars costing more than $9,000. Since **Case 2** is irresolvable in the sense that the information is simply nowhere available, the system reports to the end user that the site does not supply some requested information.

4   Experimental Results and Analysis

In this project, we experimented on seven Web sites for each of two applications: car ads and digital camera ads. The approach, however, is not limited to the two applications on which we experimented. It can work with other applications as long as those applications have Web sites with forms, and we have ontologies for those applications. The process of rewriting queries in terms of site forms is the same.

4.1   Experimental Results

In this study of automatic form-filling, we are interested in three kinds of measurements: field-matching efficiency, query-submission efficiency, and post-processing efficiency.[1]

To know if we properly matched the fields in a user query with the fields in a site query, we measure the ratio of the number of correctly matched fields to the total number of fields that could have been matched (a recall ratio R_{fm} for field matching, *fm*), and we measure the ratio of the number of correctly matched fields to the number of correctly matched fields plus the number of incorrectly matched fields (a precision ratio P_{fm}). To know if we submitted the query effectively, we measure the ratio of the number of correct queries submitted to the number of queries that should have been submitted (a recall ratio R_{qs} for query submission, *qs*), and we measure the ratio of the number of correct system queries submitted to the number of correct queries submitted plus the number of incorrectly submitted queries (a precision ratio P_{qs}). We also compute an overall efficiency measurement which we obtain by multiplying the recall measurements together and the precision measurements together. Because the two kinds of metrics measure two stages of one single process, we use products to calculate the overall performance of the process with respect to our extraction ontology.

Car Advertisements. We experimented on seven Web sites containing search forms for car ads. We issued five queries to each of the sites and obtained the following

---

[1] Because we do not want to measure the effectiveness of the existing application extraction ontologies and plug-in programs, such as the record separator and Ontos, which are outside the scope of our work, we do not measure their effectiveness (see [4] for measures of their effectiveness). Finally, we do not measure the effectiveness of the post-processing part of our system because, by itself, it cannot fail.

results. We found 31 fields in the seven forms. Among them, there were 21 fields that are recognizable with respect to our application extraction ontology. The system correctly matched all 21 of them. There were no false positives. If we ignore nonapplicable fields, the system should have submitted 249 original queries generated from the five user-provided queries and 1847 queries for next links. For just the applicable fields, the system actually submitted 301 original queries and 1858 queries for next links. Table 1 shows the precision and recall ratios calculated with respect to recognizable fields for the measurements we made, and it also shows the overall efficiency.

Digital Camera Advertisements. We experimented on seven Web sites containing search forms for digital camera ads. We issued four queries to each of the sites and obtained the following results. We found 41 fields in the seven forms. Among them, 23 fields were applicable to our application extraction ontology. The system correctly matched 21 of them. There were no false positives. According to the four queries and the 21 matched fields, the system should have submitted 31 original queries and 85 queries for retrieving all next links. It actually submitted 31 original queries and 85 queries for retrieving next links. Table 2 shows the precision and recall ratios for the measurements we made, and it also shows the overall efficiency.

Table 1. Experimental Results for Used-Cars Search

Number of Forms: 7 Number of Fields in Forms: 31 Number of Fields Applicable to the Ontology: 21 (67.7%)			
	Field Matching	**Query Submission**	**Overall**
Recall	100% (21/21)	100% (249/249)	100%
Precision	100% (21/21)	82.7% (249/301)	82.7%
		97.1% (249+1847)/(301+1858)[*]	97.1%[*]

*These numbers are calculated including queries submitted for retrieving next links.

Table 2. Experimental Results for Digital-Cameras Search

Number of Forms: 7 Number of Fields in Forms: 41 Number of Fields Applicable to the Ontology: 23 (56.1%)			
	Field Matching	**Query Submission**	**Overall**
Recall	91.3% (21/23)	100% (31+85)/(31+85)	91.3%
Precision	100% (21/21)	100% (31+85)/(31+85)	100%

4.2 Results Analysis and Discussion

Field-matching efficiency is a measurement for field-name matching. This matching is affected by the value for the name attribute and the type attribute the site form designer chooses for each field. For fields in the site form with no values provided or fields having less than half their values recognized, our system depends only on the values of name attributes. If the site form designer assigns meaningful names to each tag, our field-matching efficiency is high. In our experiment, we found respectively 7 and 6 such fields from the two domains, and our system recognized 95.7% of the fields for the two domains. We found respectively 14 and 17 fields from the two domains with values provided, and the result was 100% for both precision and recall for the two domains tested. We have no way of recognizing fields that are not described in our extraction ontology, so we did not consider those fields when calculating name matching efficiency.

Query-submission efficiency is a measurement of field-value matching. When calculating query-submission efficiency, we consider only the fields where names matched correctly. This efficiency is greatly affected when fields are designed in a way that our system cannot handle. For the form in Figure 3, we found two fields for *Price*; together, they form a range. Range fields formed by two independent fields normally are of the same type, e.g. both fields are text fields or both fields are selection lists. The range in Figure 3, however, is formed by a text field and a selection list. Our system does not recognize these two fields as a range. So, when it fills out the *Price* fields, it puts both the lower value and the upper value a user specifies in the first *Price* field. For the second *Price* field, which is a selection list, our system chooses all three values in the selection list. This generates six queries instead of one query—properly chosen, one query would be sufficient. For the particular form in Figure 3, our system always generates 6 queries if either *Price* or *Year* is specified by the user, among which 5 of the queries are not necessary. When the user specifies both fields, our system submits 36 queries, among which 35 of the queries are not necessary. This result significantly affects the precision of query submission. The recall, however, is 100% because all queries that could have been submitted are submitted correctly.

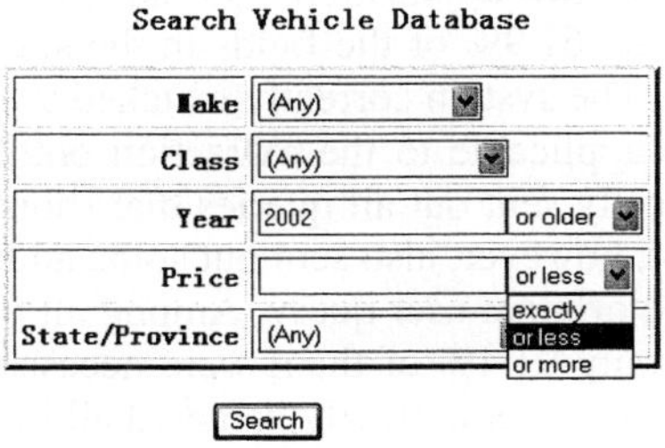

Search Vehicle Database

Fig. 3. Web form for Car-Ads Search at *http://www.autointerface.com/vfs.htm*, February, 2004

Even though our experimental results turned out well, the results can be adversely affected if text fields come with poor internal names or if extraction ontologies are poorly written. Poor internal names can make the name-matching process for text fields impossible, and poorly written extraction ontologies can make the name-matching process for fields with values inaccurate. Both cases would decrease the field-matching efficiency dramatically. With considerably more work, which we did not do because others have already solved this problem [13, 17], it would be possible to resolve the first problem by locating the displayed names, which should be human readable, rather than the internal names, which need not be human readable. The solution for the second problem is to improve the quality of a poorly written extraction ontology.

In addition, our system is not designed to handle all kinds of forms on the Web. It does not try to handle form chains (one form leads to another), dynamic forms (the value filled in for a field generates the next form field to fill in, or other such unusual behaviors controlled by script code), forms with required fields when the user does not provide values for the fields in the user-query form, combined fields (e.g. *Make* and *Model* together in one field), and forms whose actions are coded inside scripts. As future work, we could program our system to handle most of these problems. For scripts, however, we note that Web application developers can write dynamic form-processing instructions that may be difficult to reverse-engineer automatically (for example the submit action of a form may be encoded within JavaScript rather than using the standard `action` attribute of a form). Since code analysis is computationally complex in general, there may not be a complete automated solution for all dynamic forms.

5 Conclusion

In this research, we designed and implemented a system that can fill out and submit Web forms automatically according to a given user query against a corresponding application extraction ontology. From the returned results, the system extracts information from the pages, puts the extracted records in a database, and queries the database with the original user query to get, to the extent possible, just the relevant data behind these Web forms.

We tested our system on two applications: car advertisements and digital camera advertisements. On average, 61.9% of the fields in the site forms were applicable to the extraction ontologies. The system correctly matched 95.7% of them. Considering only the fields that were applicable to the extraction ontologies and were correctly matched, the system correctly sent out all queries that should have been submitted to the Web sites we tested. It, however, also sent out some additional queries that are not necessary according to the original user query. Among all queries our system submitted for our experiments, only 91.4% of them were necessary. Further, for the Web sites we tested, our output analyzer correctly gathered all linked pages. Finally, of the records correctly extracted by Ontos, our system always correctly returned just those records that satisfied the user-specified search criteria.

Acknowledgements

This work is supported in part by the National Science Foundation under grant IIS-0083127 and by the Kevin and Debra Rollins Center for eBusiness at Brigham Young University.

References

1. The Apache Jakarta Project. http://jakarta.apache.org/tomcat/index.html.
2. Brigham Young University Data Extraction Group home page. http://www.deg.byu.edu/

3. R.B. Doorenbos, O. Etzioni, and D.S. Weld. "A Scalable Comparison-Shopping Agent for the World-Wide Web," In *Proceedings of the First International Conference on Autonomous Agents*, Marina del Rey, California, USA, February 5-8, 1997, pp. 39–48.
4. D.W. Embley, D.M. Campbell, Y.S. Jiang, S.W. Liddle, Y.-K. Ng, D. Quass, and R.D. Smith. "Conceptual-Model-Based Data Extraction from Multiple-Record Web Pages," *Data & Knowledge Engineering*, 31(3):227–251, November 1999.
5. D.W. Embley, Y.S. Jiang, and Y.-K. Ng. "Record-Boundary Discovery in Web Documents," In *Proceedings of the 1999 ACM SIGMOD International Conference on Management of Data*, Philadelphia, Pennsylvania, May 31 – June 3, 1999, pp. 467–478.
6. C. Fellbaum. "WordNet: An Electronic Lexical Database." MIT Press, Cambridge, MA, 1998.
7. P.A. Hall and G.R. Dowling. "Approximate String Matching." ACM Computing Surveys, 12(4):381–402, 1980.
8. Java Technology. http://java.sun.com/.
9. V.I. Levenshtein. "Binary codes capable of correcting spurious insertions and deletions of ones." Problems of Information Transmission, 1:8–17, 1965.
10. S.W. Liddle, D.W. Embley, D.T. Scott, and S.H. Yau. "Extracting Data Behind Web Forms," In *Proceedings of the Workshop on Conceptual Modeling Approaches for e-Business*, Tampere, Finland, October 7-11, 2002, pp. 402–413.
11. Microsoft Passport and Wallet Services. http://memberservices.passport.com
12. G.A. Miller. "WordNet: A Lexical Database for English." Communications of the ACM, 38(11):39–41, November 1995.
13. G. Modica, A. Gal, and H. Jamil. "The Use of Machine-Generated Ontologies in Dynamic Information Seeking, " In Proceedings of the 9th International Conference on Cooperative Information Systems (CoopIS 2001), Trento, Italy, September 5-7, 2001, pp. 433–448.
14. MySQL. http://www.mysql.com/.
15. M. Porter. "An Algorithm for Suffix Stripping." Program, 14(3):130–137, 1980.
16. J.R. Quinlan. "C4.5: Programs for Machine Learning." San Mateo, CA: Morgan Kaufmann, 1993.
17. S. Raghavan and H. Garcia-Molina. "Crawling the Hidden Web," In *Proceedings of the 27th Very Large Data Bases (VLDB) Conference*, Rome, Italy, September 11-14, 2001, pp. 129–138.

Third International Workshop on Evolution and Change in Data Management (ECDM 2004)

at the 23rd International Conference
on Conceptual Modeling (ER 2004)

Shanghai, China, November 9, 2004

Organized by

Fabio Grandi

Preface to ECDM 2004

Change is a fundamental but sometimes neglected aspect of information and database systems. The management of evolution and change and the ability for database, information and knowledge-based systems to deal with change is an essential component in developing and maintaining truly useful systems. Many approaches to handling evolution and change have been proposed in various areas of data management and this forum seeks to bring together researchers and practitioners from both more established areas and those from emerging areas to look at this issue. The third ECDM workshop (the first ECDM workshop was held with ER'99 in Paris and the second one with ER'02 in Tampere) deals with the manner in which change can be handled, and the semantics of evolving data and data structure in computer based systems, including databases, data warehouses and Web-based information systems.

With respect to the main ER conference, the ECDM workshop series aim at stressing the evolutionary aspects involved in conceptual modelling and in development and implementation of systems, ranging from the modelling of information dynamics to the dynamics of the modelling process itself. Another explicit aim of ECDM workshops is to bring together scientists and practitioners interested in evolution and change aspects in different research fields and, thus, often belonging to completely separate communities. As a result, such an interaction could become tighter and cross-fertilization more useful in the context of a collaborative workshop like ECDM than in the context of the main conference sessions. Moreover, since the emphasis is on the evolutionary dimension, a special insight is sought upon this specific aspect, that hardly could find an appropriately broad coverage in the scope of the main ER conference.

Following the acceptance of the workshop proposal by the ER 2004 organizing committee, an international and highly qualified program committee was assembled from research centers worldwide. As a result of the call for papers, the program committee received 21 submissions from 14 countries and after rigorous refereeing 8 high quality papers were eventually chosen for presentation at the workshop and publication in these proceedings. The workshop program was organized into three sessions, Systems Evolution Support in Conceptual Modeling, Temporal and Evolution Aspects in Internet-based Information Systems, Schema Evolution and Versioning in Data Management.

I would like to express my thanks to the program committee members and the additional external referees for their timely expertise in reviewing, the authors for submitting their papers, the ER 2004 organizing committee and, in particular, the Workshop Chairs and Publication Chair for their support.

November 2004   Fabio Grandi

Automatic Integrity Constraint Evolution
due to Model Subtract Operations

Jordi Cabot[1,2] and Jordi Conesa[2]

[1] Estudis d'Informàtica i Multimèdia, Universitat Oberta de Catalunya
jcabot@uoc.edu
[2] Dept. Llenguatges i Sistemas Informàtics, Universitat Politècnica de Catalunya
jconesa@lsi.upc.es

Abstract. When evolving Conceptual Schemas (CS) one of the most common operations is the removal of some model elements. This removal affects the set of integrity constraints (IC) defined over the CS. Most times they must be modified to remain consistent with the evolved CS. The aim of this paper is to define an automatic evolutionary method to delete only the minimum set of constraints (or some of their parts) needed to keep the consistency with the CS after subtract operations. We consider that a set of constraints is consistent with an evolved CS when: 1) none of them refer to an element removed from the original CS and 2) the set of constraints is equal or less restrictive than the original one. In this paper we present our method assuming CS defined in UML with ICs specified in OCL, but it can be applied to other languages with similar results.

1 Introduction

Evolution is critical in the life span of an information system (IS) [16]. Until now, the issues that arise when evolving an IS have been studied from an implementation point of view, specially in the field of schema evolution in relational and object-oriented databases [1, 2, 8, 15]. However, nowadays conceptual schemas (CS) are gaining a lot of relevance in the life cycle of the IS; for instance, in the context of the MDA [12], they are taken as a basis for the code generation of the IS. In such cases, the evolution must take place at the conceptual schema level.

The evolution of conceptual schemas has been poorly studied. Most of the evolution operations are included in the concepts of refactorings and model transformations [13, 3, 14, 10, 5].

A complete conceptual schema must include the definition of all relevant integrity constraints [7]. When evolving conceptual schemas one of the most common operations is the removal of some model elements. This removal affects the set of integrity constraints (IC) defined over the CS. Obviously these constraints must be modified to remain consistent with the evolved CS when they reference any of the removed model elements. As far as we know, none of the existing approaches deal with the evolution of integrity constraints in such a case.

S. Wang et al. (Eds.): ER Workshops 2004, LNCS 3289, pp. 350–362, 2004.
© Springer-Verlag Berlin Heidelberg 2004

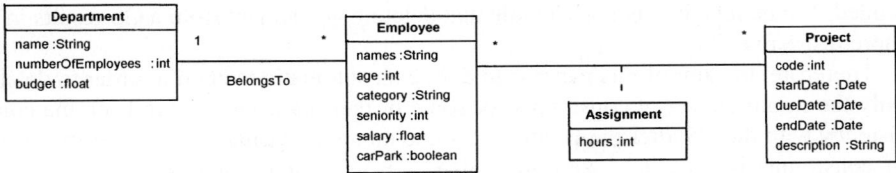

Fig. 1. Example of a conceptual schema

In this paper we define a method to automatically evolve ICs assuming CS defined in UML [9] with ICs specified in OCL [11]. However, our method could be adapted to other conceptual modeling languages that provide a formal (or semi-formal) language to write ICs, such as EER [4] or ORM [6].

A naïve approach to solve this problem would consist of removing all the IC where a removed element appears. However, this approach removes constraints that still need to be enforced in the evolved CS. As an example, assume that we have the conceptual schema of Figure 1, devoted to represent the relationship between employees, their departments and the projects where they work, with the following constraints:

```
-- Every employee must be older than 16 and he/she cannot work for the
company during more than 40 years
    context Employee inv IC1: self.age>16 and self.seniority<40

-- Bosses with seniority >10 must earn more than 30000 euros
    context Employee inv IC2:
    (self.category='boss' and seniority>10) implies salary>30000

-- Bosses or employees with seniority > 5 have a car park
    context Employee inv IC3:
        (self.category='boss' or seniority>5) implies carPark

-- Employees under 18 must be scholarship holders
    context Employee inv IC4:
        self.age>18 or self.category='scholarship holder'

-- The assignment for each employee must be <= 40
    context Employee inv IC5: self.assignment.hours->sum()<=40
```

If we delete the *seniority* attribute this naïve approach will remove the first three constraints. We consider this is an incorrect solution, since then, for instance, we can insert a new *employee* younger than 16 years old or a *boss* without *carPark*, which was not allowed in the original CS.

We would also like to point out that another simple solution, consisting in deleting only the part of the OCL constraint where the deleted elements appear, does not work either. For example, this approach would leave IC2 as *"category='boss' implies salary>30000"*. Note that this constraint is more restrictive than the original one, since before the deletion of *seniority*, a boss could earn a salary lower than 30000 (when his/her seniority was under 10) but now all bosses are forced to grow their salary up to 30000 or more. Thus, employees satisfying the constraint before the removal of the attribute now may violate it, and of course, this is not what we intended. We believe it is not acceptable that deleting an element from a CS results in a more restrictive CS.

Therefore, the aim of this paper is to define an automatic method in order to delete only the minimum set of constraints (or some of their parts) needed to keep the consistency with the CS after the deletion of some model element. We consider that a set of constraints is consistent with an evolved CS when none of them refers to a removed element of the original CS and the set of constraints is equal or less restrictive than the original one. Note that this also ensures that if the original CS was satisfiable (i.e. there exists an instantiation of the model for which no IC is violated) the resulting CS will also be satisfiable.

The rationale of our method is to translate each OCL constraint into a sort of conjunctive normal form, and then, apply an algorithm that, taking into account the specific structure of the constraint, decides whether the constraint (or a part of it) is still relevant for the evolved CS.

The rest of the paper is structured as follows. Next section describes the different removal operations that can be executed over a CS. Section 3 describes the transformation of an OCL constraint to its equivalent conjunctive normal form. Then, in Section 4, we introduce the OCL metamodel, needed in Section 5 to present our method to evolve the constraints. Afterwards, we present some optimizations in Section 6. Finally, we give our conclusions and point out future work in Section 7.

2   Schema Evolution Operations

Conceptual schema changes can be seen as additive, subtractive or both [8]. In additive changes new elements are created without altering the existing ones. On the other hand, a subtractive change involves the removal of some model elements.

Several subtractive operations are defined in the field of schema evolution; much of these operations have been created to be applied to software or databases [1, 2, 8, 15], but not to conceptual schemas. Therefore, we extend their set of subtractive operations to allow the user to delete any model element of a conceptual schema. We deal with the following list of operations:

- Removal of an attribute.
- Removal of a specialization/generalization link. We admit CS with multiple inheritance.
- Removal of a class: a deletion of a class implies the deletion of all its attributes, associations and generalizations links in which it takes part.
- Removal of an association: the deletion of an association implies the deletion of all its association ends and generalization relationships.
- Removal of an association end: To delete an association end its owner association must be a n-ary association with n>2.
- Removal of an association class: this deletion is treated as an elimination of the association followed by the deletion of its class

3   Transforming OCL Constraints to Conjunctive Normal Form

The transformation of an OCL constraint to a conjunctive normal form (CNF) is necessary to reduce the complexity of its treatment in the next sections. We could also have chosen the disjunctive normal form instead. In this section we describe the rules for transforming a logical formula into a conjunctive normal form and we adapt them to the case of OCL constraints.

A logical formula is in conjunctive normal form if it is a conjunction (sequence of ANDs) consisting of one or more clauses, each of which is a disjunction (OR) of one or more literals (or negated literals).

OCL expressions that form the body of OCL constraints can be regarded as a kind of logical formula since they can be evaluated to a boolean value. Therefore, we can define a conjunctive normal form for OCL expressions exactly in the same way as that of the logical formulas. The only difference is the definition of a literal. We define a literal as any subset of the OCL constraint that can be evaluated to a boolean value and that does not include a boolean operator (*or*, *xor*, *and*, *not* and *implies* [11]). We say that an OCL constraint is in conjunctive normal form when the OCL expression that appears in its body is in conjunctive normal form. For instance, constraints IC1, IC4 and IC5 are already in CNF but not IC2 and IC3.

Any logical formula can be translated into a conjunctive normal form by applying a well-known set of rules. We use the same rules in the transformation of OCL expressions with the addition of a new rule to deal with the *if-then-else* construct. The rules are the following:

1. Eliminate non-basic boolean operators (*if-then-else,implies, xor*), using the rules:
    a.   A *implies* B == *not* A *or* B
    b.   *if* A *then* B *else* C == (A *implies* B) and (*not* A *implies* C) ==
         (*not* A *or* B) *and* (A *or* C)
    c.   A *xor* B == (A *or* B) *and* (*not* A *or not* B)
2. Move *not* inwards until negations be immediately before literals by repeatedly using the laws:
    a.   *not* (*not* A) == A
    b.   DeMorgan's laws: *not* (A *or* B) == *not* A *and not* B
         not (A *and* B) == *not* A *or not* B
3. Repeatedly distribute the operator *or* over *and* by means of:
    a.   A *or* (B *and* C) == (A *or* B) *and* (A *or* C)
    Once these rules have been applied, constraints IC2 and IC3 are transformed into:

IC2: not self.category='boss' or not seniority>10 or salary>30000
IC3: (not self.category='boss' or carPark) and (not seniority>5 or carPark)

It is important to note that in order to apply these transformations we do not need to use any Skolemization process to get rid of existential quantifiers, since all variables that appear in OCL expressions are assumed to be universally quantified.

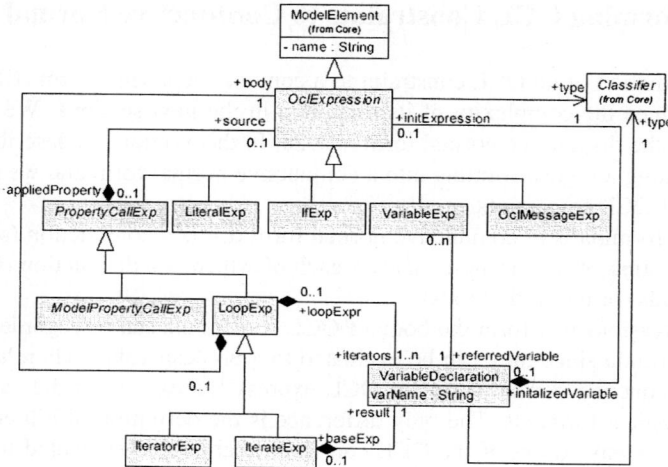

Fig. 2. OCL Metamodel

4   The OCL Metamodel

Our method assumes that OCL constraints are represented as instances of the OCL metamodel [11, ch. 8]. Working at the metamodel level offers us two main advantages. First, we avoid problems due to different syntactic possibilities of the OCL. Second, we can use the taxonomy of the metamodel in order to simplify the evolutionary algorithms.

In this section we briefly describe the OCL metamodel. Its basic structure (Figure 2) consists of the metaclasses *OCLExpression* (abstract superclass of all possible OCL expressions), *VariableExp* (a reference to a variable, as, for example, the variable *self*), *IfExp* (an if-then-else expresion) , *LiteralExp* (constant literals like the integer '1') and *PropertyCallExp* which is an expression that refers to a property and is the superclass of *ModelPropertyCallExp* and *LoopExp*.

ModelPropertyCallExp can be splitted in *AtributteCallExp* (a reference to an attribute), *NavigationCallExp* (a navigation through an association end or an association class) and *OperationCallExp*. This later class is of particular importance in our approach, because its instances are calls to operations defined in any class or metaclass. This includes all the predefined operations of the types defined in the OCL Standard Library [11, ch.11], such as the add operator ('+') and the 'and' operator. These operations can present a list of arguments if they have parameters.

LoopExp represents a loop construct over a collection of elements. OCL offers some predefined iterators to loop over the elements. The most important ones are *forAll*, *exists*, *one*, *select*, *any* and *reject*, which are fully explained in the section 5.4.

As an example, Figure 3 shows the constraint IC3 as an instance of the metamodel.

context Employee inv IC3:
(not self.category="boss" or carPark) and (not seniority>5 or carPark)

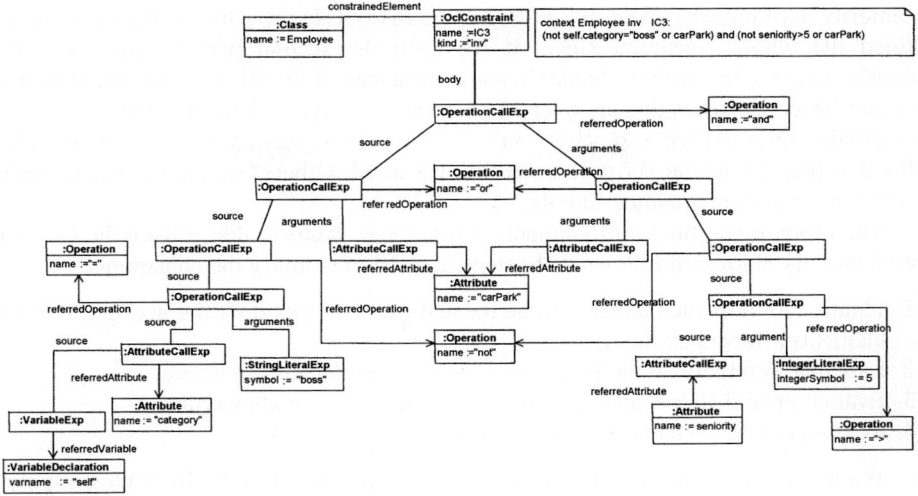

Fig. 3. OCL Constraint expressed as an instance of OCL Metamodel

5 Evolutionary Algorithm

So far, we have transformed all the ICs into CNF. Therefore, their literals (boolean expressions) can only be connected by *and*, *or* and *not* operators.

Our algorithm begins with marking as undefined every literal that references to an element about to be removed, since then it cannot be evaluated. Next, to compute which parts of the OCL expression need to be deleted we apply the following rules for the *and*, *or* and *not* operators to the set of literals that form the OCL constraint:

1. literal AND undefined → literal
2. literal OR undefined → undefined
3. NOT undefined → undefined

The above rules are deduced from the well-known implication: *A and B* → *A*. To justify rule number one, assume that we have a constraint *c* in the CS with the body "*A and B*" (where *A* and *B* represent two literals, according to our definition). This constraint forces all the instances of the Information Base (IB) to verify both of them. The implication states that no matter what happens with *B* (for example, *B* can become undefined), *A* must still hold in the IB. On the other hand, the rule *A or B* does not imply *A*, so when *B* is undefined we cannot force the IB to verify *A*.

Obviously, if a literal is undefined its negation is undefined as well.

As an example, assume we want to delete the attribute *age* from the CS. In this case, the integrity constraints IC1 (*self.age>16 and self.seniority<40*) and IC4 (*self.age>18 or self.category='scholarship holder'*) will be affected and need to be reviewed. In the former, we keep the second part of the constraint (*self.seniority<40*) since all the instances of the IB that held the constraint (and thus, the value of their seniority attribute was under 40) will continue to do it. In the latter, we mark as undefined the whole constraint since, if we keep the second part of the constraint (*self.category='scholarship holder'*), some instances of the IB could violate the constraint (those instances that are over 18 years old not being scholarship holders).

Finally, after applying the above rules, if the expression consists only of an undefined value, the whole constraint will be discarded. Otherwise, the constraint body will consist of the remaining literals.

Therefore, given the set of elements that the user wants to delete from the CS, our evolutionary algorithm follows these steps in order to simplify the constraints:

1. Identification of the elements to be deleted (those selected by the user plus all the elements affected by them).
2. Labeling as undefined the literals that refer to the elements to delete
3. Evaluation and simplification of the constraint using the above rules
4. Removal of the elements of step number one from the CS.

We would also like to note that these algorithms are efficient. Both have a polynomial order of complexity. In particular, $O(N)$, where N represents the number of the nodes of the tree that represents a given constraint, the same cost as the preorder and postorder algorithms over which our algorithms are based as we will see below.

Next subsections explain deeply the first three steps. The fourth does not present any further complexity.

5.1 Identification

Given a set S_0 containing the model elements the user wants to remove from the CS, we obtain a set S where $S = S_0 \cup X$, and X is the set of model elements that need to be deleted due to the removal of classes and associations of S_0.

It is easy to see the removal of attributes or association ends does not add any other element to the set X. The removal of an association class can be seen as the removal of its class followed by the removal of its association. The consequences of the deletion of a generalization link are addressed in section 5.2.

More concretely, $X = indClass(S_0) \cup indAssoc(S_0)$. The function *indClass* returns, for each class, the set of associations and generalizations where the class participates and all their features (attributes and association ends). *indAssoc* returns, for each association, the set of generalizations where the association participates and all its association ends.

The formalization in OCL of these operations using the UML 2.0 metamodel is:

```
context System::indClass(S₀ : Set(Element)) : Set(Element)
body:let S_class:Set(Class) = S₀-> select(s| s.oclIsKindOf(Class))
    in
    S_class.feature->union(Sclass.feature.association)->
    union(S_class.generalization->union(Generalization.allInstances()->
    select(g| S_class->includes(g.general))
```

```
context System::indAssoc(S₀ : Set(Element)) : Set(Element)
body:let S_assoc:Set(Association) = S₀-> select(s|
    s.oclIsKindOf(Association))
  in
  S_assoc.memberEnd->union(S_assoc.generalization)->
  union(Generalization.allInstances()->
  select(g| S_assoc->includes(g.general))
```

In order to compute the elements of the set X we can use the following algorithm:

```
S₀=Elements selected by the user
X₀=S₀
Repeat
      X_{i+1}=X_i ∪indClass(X_i) ∪ indAssoc(X_i)
Until X_{i+1}=X_i
S=X_i
```

As an example, assume we want to delete the attribute *seniority* and the association class *Assignment* from the CS of Figure 1. In this case, S_0 will be composed by *seniority* and *Assignment* elements. In the first iteration of the algorithm, the deletion of *Assignment* causes the selection of the attribute *hours* (*Assignment* regarded as a class), and its two association ends (*Assignment* regarded as an association) by the execution of the *indClass* and *indAssoc* operations. Thus, at the end of the iteration, X_0 will contain *Assignment, hours, seniority,* and the association ends *Assignment-Employee* and *Assignment-Project*. In the second iteration no more elements will be selected, so the algorithm finishes with the following result set:

$S = \{Assignment, hours, Seniority, Assignment-Employee, Assignment-Project\}$.

5.2 Marking as Undefined

An OCL expression expressed as an instance of the OCL metamodel can be regarded as a binary tree. Each node represents an atomic subset of the OCL expression (an instance of any metaclass of the OCL metamodel: an operation, an access to an attribute or an association ...). The left child is the source of the node (the part of the OCL expression previous to the node). When the node represents an operation the right child is the argument of the operation, if any. When the node represents a loop expression, the right child is the body (the expression that is evaluated for each element in the collection) of the iterator.

To mark as undefined the parts of the constraint that refer to some of the elements of S (the set of elements to be deleted), we traverse the tree of the constraint in preorder. For each node we check if its referred element appears in S; if it does, we mark the node as undefined. Otherwise, we recursively repeat the process for its children.

Besides, we also mark a node as undefined when it refers to a feature (attribute, association end or operation) that now cannot be accessed since its owner is a superclass which cannot be reached because of the deletion of a generalization link.

Figure 4 shows the labeling of constraint IC3 with the set of deleted elements computed in the previous example. In this case, the algorithm marks as undefined the access to the attribute *seniority* (see the grey circle in the figure).

5.3   Constraint Evolution

Once we have marked as undefined all the affected nodes of the OCL expression we are ready to evolve the constraint.

Now, we traverse the tree in postorder. For each node, if the node or any of its children are undefined we pull up the undefined value up to its parents. This applies always except for nodes representing the boolean operator *AND*, where we pull up the defined child, if any.

After the execution of this second algorithm, the constraint will consist only of those parts that are consistent with the evolved CS. If the whole constraint is undefined it will be removed from the CS. Figure 4 shows the execution of the algorithm over the constraint IC3 marked in the previous section.

Following our example (*S = {Assignment, hours, Seniority, Assignment-Employee, Assignment-Project}*), after applying this last step the results will be the deletion of two integrity constraints (IC2 and IC5) and the simplification of another two (IC1 and IC3) while the other (IC4) will remain untouched. The final set of constraints is:

IC1:   *self.age>16*
IC3:   *not self.category='boss' or carPark*
IC4:   *self.age>18 or self.category='scholarship holder'*

5.4   Loop Expressions

Literals containing loop expressions deserve a special treatment in order to try to keep them in the evolved CS, even when they refer to some of the elements of S.

In an iterator expression, an element of S can appear either in the *source* of the iterator expression or in its *body*. When it appears in the source, the whole iterator expression must be marked as undefined by following the steps explained in the previous sections. On the contrary, if it appears inside the iterator body (and not in the source), depending on the concrete kind of iterator, we will be able to preserve the iterator expression evolving its body.

The candidate iterators for this special treatment are those presenting a boolean expression in their body. We apply the process explained in sections 3 and 5 over that boolean expression. As a result, the original boolean expression may have been left untouched (the iterator did not contain references to elements of S), completely deleted (its whole body expression is undefined) or simplified. When simplified it means that although the body referred to some element of S, by applying the rules defined before, there is still a part of the original boolean expression that could be applied over the new CS. Then, we have two different alternatives depending on the iterator semantics:

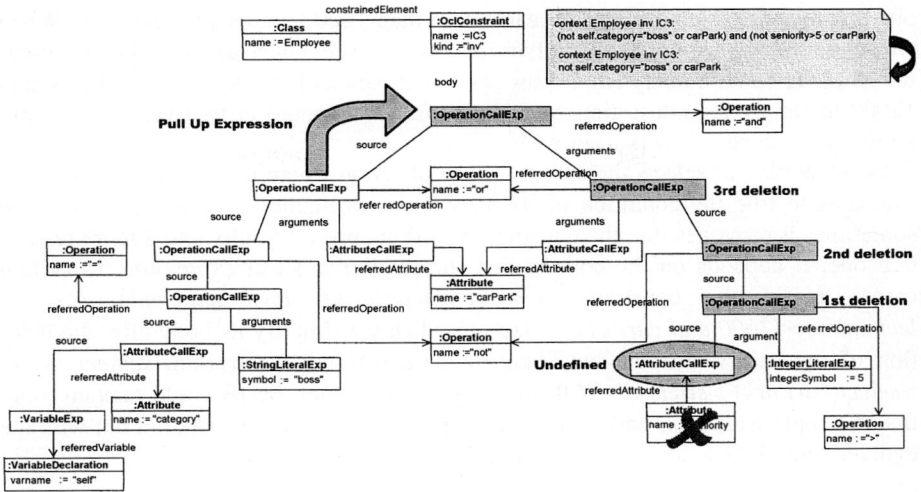

Fig. 4. Evolution of the constraint IC3

- To keep the iterator replacing the original body with the simplified one. This is the case of the iterators *forAll, exists* and *any*
- To mark the whole iterator expression as undefined, which is the case of the rest of the iterators (*select, reject* and *one*). This happens when the simplified body may result in a more restrictive constraint, which is not the desired effect.

In the following we justify the possible options for each iterator type.

A *forAll* iterator defines a condition that must be accomplished by all the instances of the set over which we apply the iterator. Thus, if every element in the set satisfied the original body it implies that also every element must satisfy a subset of the original body, since it is less restrictive. Therefore, we can keep it on the evolved CS with the defined parts of its original body. As an example, the constraint: *context Department inv: self.employee->forAll(salary>12000 and carPark)* would result in *self.employee->forAll(carPark)*, after the removal of the attribute *salary*. Note that if all employees assigned to a department earned more than 12000 and had a car park, when we delete seniority it is still true that all of them have a car park.

The *exists* iterator defines a condition that must be true for, at least, one element of the set where we apply the iterator. In a similar way as before, if one element satisfied the original boolean expression, it can be followed that at least that element will satisfy a subset of that condition. Again, we can keep expressions containing this iterator since it does not imply a more restrictive constraint over the CS. A similar reasoning can be used for the *any* iterator.

The *one* iterator defines a condition that must be true for, exactly, a single element in the set. When this kind of iterators refers to an element of S in its body, we must delete the whole iterator, since the resulting expression is more restrictive than the original one. Using the previous example, but replacing *forAll* with *one* we obtain the constraint: *self.employee->one(salary>12000 and carPark)* which states that only one employee per department can earn more than 12000 and have a car park. When we delete *salary* the constraint evolves to *self.employee->one(carPark)*. In the former the IB could contain many employees with a car park as long as they earned less than 12000 to satisfy the constraint. Now the constraint is more restrictive, since it only allows a single person with a car park.

A *select* iterator selects the subset of the collection where the select is applied that evaluates to true the condition of its body. We can neither keep this iterator since sometimes it provokes that the constraint where it appears results into a more restrictive one. It depends on the context that surrounds the select expression. In certain cases, as in the constraint: *context Department inv: self.employee-> select(salary>12000 and carPark)->*size()<5 when we simplify its body after the deletion of the *salary* attribute we obtain a more restrictive constraint *self.employee->select(carPark)->size()<5*. With the original constraint the IB could contain more than 5 people with a car park as long as they earned less than 12000, but not with the evolved one. We can apply a similar reasoning to the *reject* iterator.

6 Optimizations

We would like to finish sketching some optimizations to our process. These optimizations could reduce the amount of OCL expressions we are forced to remove due to the removal of elements of the conceptual schema.

So far, all the binary OCL operations except for the AND operation, like *union, intersection*, '+', '-' and so forth, are marked as undefined when any of the two arguments is undefined. However, sometimes we could avoid it if we had enough information about the context where the operation is placed. This is the same problem we have found related to the *select* iterator.

For instance, the expression *self.a + self.b < self.c*, where *a*, *b* and *c* are assumed to be three attributes with a natural value, is marked as undefined if the designer deletes *a*, *b* or *c* from the CS. However, when *a* or *b* are deleted, we could still preserve the expression by transforming it into *b < c* (or *a<c*, depending of the attribute that is deleted). The reason is that if *a* plus *b* returns a value lower than *c*, it implies that *a* or *b* alone must be also lower than *c*.

More optimizations may be obtained taking into account the relations between different parts of the constraint. For instance, *a<b and b<c* could be simplified to *a<c* after the deletion of b.

To implement these optimization mechanisms we would need to propagate through the tree some kind of information to decide in each case which are the operations or iterators (for *select* and *reject* iterators) that must be really marked as undefined. Note that it is not enough to examine the immediate nodes around the operation since, in some cases, the information needed is placed several levels up or down in the tree.

Another possible optimization is to find and remove those constraints that become tautologies after the process due to the removal of some of their parts. For instance, constraints of the form '*(A and B) or (not A)*' become a tautology when the attribute

B is deleted (after converting the constraint into a CNF, it results in: *(A or not A) and (B or not A)*, and once the attribute is deleted it turns into *(A or not A)*).

7   Conclusions and Further Work

In this article we present a method to automatically evolve integrity constraints after the removal of some model elements of a conceptual schema (CS) in order to keep them consistent with the evolved schema. To the best of our knowledge, ours is the first approach to address this issue at the conceptual level. Moreover, we have developed a tool[1] that implements our approach to prove the feasibility of our method.

We focus our explanation in the evolution of OCL expressions appearing in OCL constraints, but we could generalize the method to evolve OCL expressions appearing in other model elements such as derived elements or operation pre and postconditions.

We apply our method to the particular case of conceptual schemas specified in UML with integrity constraints defined using OCL. However, we could adapt our method to other conceptual modeling languages.

We plan to continue our work in (at least) two directions. First, we would like to extend the method with the optimizations sketched in section 6 and , second, we plan to study the evolution of OCL expressions after any kind of schema evolution operations over a conceptual schema.

Acknowledgments

We would like to thank Vera Kiessling for implementing a prototype of this work and the GMC group for many useful comments to previous drafts of this paper. We would also like to thank the reviewers for their suggestions.

This work has been partially supported by the Ministerio de Ciencia y Tecnologia and FEDER under project TIC2002-00744.

References

1. J. Banerjee. Data Model Issues for Object-Oriented Applications. ACM Transactions on Office Information Systems, 5(1):3-26, January 1987.
2. P. Brèche. Advanced Primitives for Changing Schemas of Object Databases. 8th Int. Conf. on Advances Information Systems Engineering (CAiSE'96), LNCS 1080, pp. 476-495.
3. M. Fowler. Refactoring: Improving the Design of Existing Code. Addison-Wesley, 1999.
4. M. Gogolla, U. Hohenstein, Towards a Semantic View of an Extended Entity-Relationship Model. ACM Transactions on Database Systems, 1991. 16(3): p. 369-416.
5. C. Gomez, A. Olivé, Evolving Partitions in Conceptual Schemas in the UML. 12th Int. Conf. on Advances Information Systems Engineering (CAiSE'02), LNCS 2348. pp. 467-483.
6. T.A. Halpin, Information Modeling and Relational Databases, Morgan Kaufmann, 2001.

---

[1] It can be downloaded from http://www.lsi.upc.es/~gmc/Downloads/ICEvolution.html.

7. ISO/TC97/SC5/WG3. Concepts and Terminology for the Conceptual Schema and Information Base, J.J. van Griethuysen (ed.), 1982.
8. S. Monk. A Model for Schema Evolution in ObjectOriented Database Systems. PhD thesis, Lancaster University, 1993.
9. OMG, OMG Adopted Specification. UML 2.0 Superstructure Specification, 2002.
10. OMG. Request for proposal: MOF 2.0 Query/Views/Transformations. OMG, 2002
11. OMG, OMG Adopted Specification, UML 2.0 OCL, 2003.
12. OMG. Model Driven Architecture (MDA). OMG, 2003.
13. W. F. Opdyke. Refactoring Object-Oriented Frameworks. PhD thesis, University of Illinois, 1992.
14. Ivan Porres. Model refactorings as rule-based update transformations. UML 2003 – The Unified Modeling Language, 6th Int. Conf. LNCS 2863, pp. 2-17.
15. John F. Roddick, Noel G. Craske, Thomas J. Richards. A Taxonomy for Schema Versioning Based on the Relational and Entity Relationship Models. 12th Int. Conf. on the the Entity Relationship Approach (ER'93). LNCS 823, pp. 137-148.
16. Dag Sjø, Quantifying Schema Evolution. Information and Software Technology, Vol. 35, No. 1, pp. 35-44, January 1993.

An Ontological Framework
for Evolutionary Model Management

Atsushi Matsuo[1], Kazuhisa Seta[1],
Yusuke Hayashi[2], and Mitsuru Ikeda[2]

[1] Osaka Prefecture University, Graduate School of Science
Mathematics and Information Science, 1-1, Gakuen-cho, Sakai, Osaka, 599-8531, Japan
atsushi@kbs.cias.osakafu-u.ac.jp, seta@mi.cias.osakafu-u.ac.jp
[2] JAIST, School of Knowledge Science
1-1, Asahidai, Tatsunokuchi, Nomi, Ishikawa, 923-1292, Japan
{yusuke,ikeda}@jaist.ac.jp

Abstract. Ontology provides a system of basic primitives, terms, concepts and axioms that constitute the model. We present an ontology-based evolutionary model management framework that authors can build their models without concern for the complicated consistency and evolutionary management among multiple view models based on the ontology. Our ontological approach to evolutionary management is that we deeply analyze and clarify generic principles of knowledge construction and modeling premises. In this paper, we first consider the fundamental issues of conceptual relations and identity. And we suggest our philosophy that can deal explicitly with authors' viewpoint and their viewpoint change/movement. Then, we show our evolutionary model management framework taking simple examples.

1 Introduction

Ontological Engineering. Ontology is a specification of conceptualization [4]: it specifies a principle system of vocabulary of a target world. Therefore, by developing an information system based on the ontology, we can improve its quality, sharability and reusability [6, 7]. The underlying philosophy of our research is to carefully elucidate humans' perception of existence and develop a high fidelity modeling framework that is compatible with engineering requirements. We have continued to develop an Ontology-based Model Management System called OntoMMS to realize that philosophy.

Roles of Ontology. An ontology provides a system of basic primitives, terms, concepts and axioms that constitute the model. Roughly speaking, there are two types of ontologies: (I) task ontology, which captures problem-solving processes and (II) domain ontology that targets worlds. Building ontology libraries for various tasks and domains and implementing an integration mechanism for them, modeling environment (authoring environment) can provide users with appropriate support for their requirements. Ontology serves basic roles to realize useful support functions for modeling: (1) it provides users with a guideline for modeling; (2) guarantees the consistency of models; and (3) provides users with simulation results of problem-solving

S. Wang et al. (Eds.): ER Workshops 2004, LNCS 3289, pp. 363–376, 2004.
© Springer-Verlag Berlin Heidelberg 2004

processes at the conceptual level. The concepts of model "change" and "evolution" strongly relate to (2) and (3). For (2), we must clarify the principle knowledge to manage the propagation of effects triggered by change in a model.

Motivating Scenario. Figure 1 shows an abstract image of our motivating scenario. Two authors, A and B, share information of target modeling concept X. In general, despite modeling the same target X, the conceptual information that author A focuses on and that which author B focuses on are not similar if they have different modeling objectives. For example, author A's attempts to characterize a subject (e.g. an employee) from the viewpoint of training may model the "learner." On the other hand, author B's attempts to characterize the same subject from the viewpoint of human resource

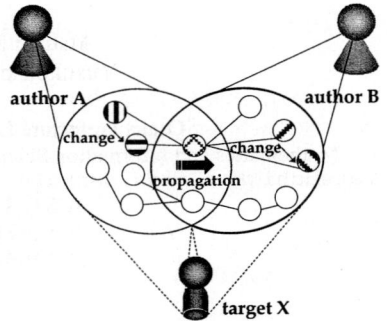

Fig. 1. Motivating Scenario

management may model the "manager." The learner modeled by A and the manager modeled by B share some kinds of common information, such as name, age, and skills. They also have their specific information, such as training history in the learner and on-going projects in the manager. In such a case, if the shared common information is modified at one viewpoint, then its effects must be propagated appropriately to another viewpoint. For example, if the state of the supervisory skills of the learner is modified by training(author A), then the effect must be propagated to the manager(author B).

Our Approach to Ontology-Based Evolutionary Model Management. It is difficult to realize the evolutionary model management for the reason that the modeling premises didn't be made explicit; how to model the entity and attributes and how to set the identity in modeling the same target from different viewpoints are not explicitly specified. Thus the system can't adequately manage the models based on respective implicit modeling premises. Consequently, our approach to build a methodology for modeling the world from various viewpoints and evolutionary model management is to clarify generic principles of knowledge construction based on our identical perception and to make modeling premises explicit. In this paper, we first consider the fundamental issues of conceptual relations (such as is-a, part-of) and identity (version). Thereby, a way is suggested of (i) handling each author's viewpoint and (ii) managing the propagation effects triggered at one viewpoint.

2   Identity

2.1   Identity

The "is-a" is widely used as a basic primitive to represent a generalization-specification relation between two concepts. For instance, "human is-a mammal" relation holds. Its fundamental meaning is represented in a logical formula as

$$\text{Human (X)} \Rightarrow \text{Mammal (X)}. \tag{a}$$

In object-oriented modeling and a knowledge representation paradigm, the meaning of the formula is used by way of the concept of "inheritance." In this case, the facts attributable to the mammal are also attributable to the human. For instance, if formulas (a) and (b) hold, then (c) is inferred.

$$\text{Mammal (X)} \Rightarrow \text{Has (X, navel)} \tag{b}$$

$$\text{Human (X)} \Rightarrow \text{Has (X, navel)} \tag{c}$$

In most object-oriented and knowledge processing systems, the abstract level is that level at which one can produce instances that are fixed to most specific levels of concepts in an is-a hierarchy: one can make an instance of human in the above is-a hierarchy and bind it to the variable X. This mechanism can be viewed as an approximation based on the idea that (1) conceptual definitions are systemized prior to the instance generation and instances are made based on the conceptual definitions and, (2) inheritance of the conceptual definition is important from a practical viewpoint.

However, on the other hand, there exists another viewpoint of rigorous uses of models based on the ontology. "Entities" exist prior to the conceptual definitions and their "conceptual entities" are conceptualized and generated based on conceptual definitions. Based on the idea, we can infer a framework that can deal explicitly with the so-called "viewpoint change/movement." For instance, we can deal with an entity X that was recognized as a human (mammal) at first as a mammal (human) by generalization (specification). To realize such a system that can follow this kind of viewpoint movement, we must advance a modeling framework that can allow both X as a mammal and the X as a human to have the same identity even as conceptualized entities differ from each other.

In OntoMMS, we distinguish an un-conceptualized entity from a conceptualized one and call the former "perceptual entity (Pe)" and the latter "conceptual entity (Ce)." Furthermore, we call identities with Pe and Ce "perceptual identity" and "conceptual identity", respectively.

Figure 2 shows basic conceptual structures in OntoMMS. Figure 2(a) represents a perceptual entity $Pe(P_i)$ with perceptual identity P_i. We do NOT directly relate it to the conceptual definitions; it simply represents an "existence" perceived. In (b), the conceptual entity $Ce(C_k)$ is an entity recognized based on the conceptual definition that concept name is C_{name}. $Ce(C_k)$ has existential perceptual identity P_i and target perceptual identity P_j. Intuitively, the existential perceptual identity represents the existence and target perceptual identity does the oneness. Furthermore, the conceptual entity also has a conceptual identity C_k that represents the identity as the conceptual entity. Figure 2(c) shows relations among conceptual entities and conceptual definitions. In that figure, conceptual entities $Ce(C_{12})$ and $Ce(C_{13})$ conceptualize an identical target perceptual entity $Pe(P_1)$. The perceptions of existences as different concepts are represented by the existential perceptual identity $Pe(P_2)$ and $Pe(P_3)$. They are related to each perceptual entity based on the concept definitions of C_{lower} and C_{upper}. Consequently, the model employed in our framework explicitly represents one's perceptual changes of identity according to the viewpoint movement. Figure 2(d) shows UML representation of (c). Broadly speaking, the conceptual definition, conceptual relation "is-a" and conceptual entities in (c) correspond to the class definition, generalization relationship and instances in (d), respectively. In addition to these, all of existential perceptual identity, target perceptual identity and conceptual identity in (c) are represented as attributes in (d). The important point to note is that the identities and the constraints held among them are defined in our framework at the meta-level as a specification of conceptualization. But, one can NOT explicitly represent its specification in UML. Thus, it is not suitable to employ UML diagrams to represent our principles of knowledge as a specification of conceptualization while it is useful to use UML diagrams to represent the models of the target domain. Therefore, we use our notation represented in (c) in the following.

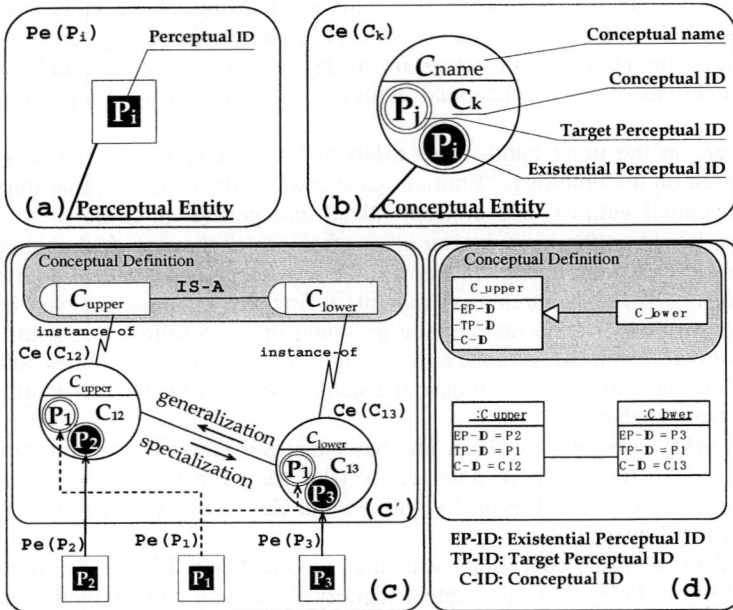

Fig. 2. Perceptual and Conceptual Identity

2.2 Basic Concept and Augmentation Concept

Integration of the task concept and domain concept, one subject of this paper, can be realized by the inheritance mechanism via "is-a" links. Nevertheless, we require a new aspect of conceptual definitions and is-a links when we examine the use of conceptualization and identity. For instance, consider the relation between "a president" in the business context and "a human" in the generic context. In this case, we can recognize it not only as a simple relation of "a president is-a human", but also as an integration of "a human in a generic context and a role of president in business task" like multiple inheritances. The difference between these two aspects is clarified by considering the "identity of a president." The former can be viewed as different conceptualizations of a same target perceptual entity at different abstract levels. Therefore, two different conceptual entities must share the same target perceptual entities.

For the latter case, on the other hand, we must model three conceptual entities, i.e., "a human", "role of president", and "a president." The targets of conceptualization and their existence are unified when conceptualizing humans. In contrast, when conceptualizing the "role of president" and "human as a president", the targets of conceptualization and their existence are not unified because the conceptualization target is human.

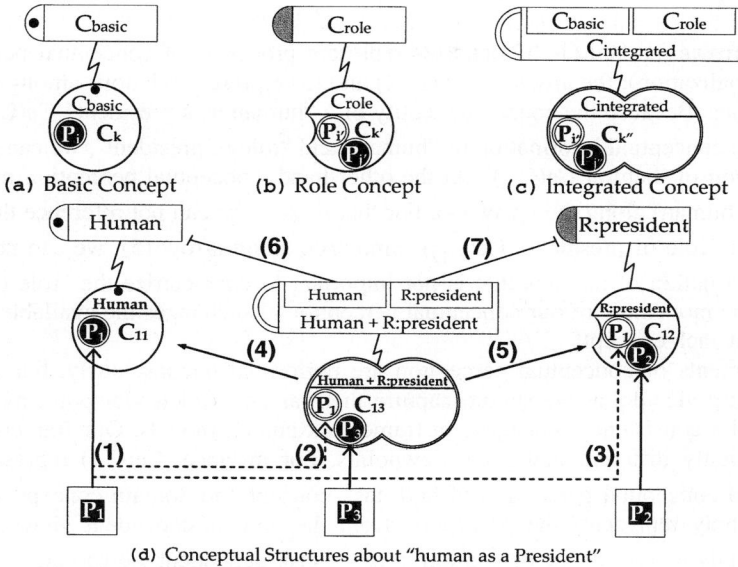

(d) Conceptual Structures about "human as a President"

Fig. 3. Basic Concept, Role Concept, Integrated Concept

In OntoMMS, the conceptual structures mentioned above are modeled as follows. "Human in a generic context" simply has an existential perceptual identity, whereas each entity of the "role of president" and "human as a president" has both an existential perceptual identity and a target perceptual identity. Each target perceptual identity of the two concepts is set with the existential perceptual identity of "human", which is the conceptualization target. Consequently, the "role of president" is modeled as a role concept which accompanies "human." "Human as a president" is modeled as an integrated concept that combines both "human" and "role of president." We call a concept that has only an existential perceptual identity (human) a basic concept and one that has both an existential perceptual identity and a target perceptual identity (role of president and human as a president) an augmentation concept.

Figures 3(a), 3(b) and 3(c) show the notations of basic concept, role concept and integrated concept, respectively. Their upper portions represent concept definitions; the lower ones show their entities. In 3(a), the small black circle in the left semi-circle represents that the concept is a basic concept. Similarly, in 3(b) and 3(c), the gray semi-circle and double line of semi-circle represent that the concepts are a role concept and an integrated concept, respectively.

Figure 3(d) represents conceptual structures of the "human as a president." $Pe(P_1)$ simply represents that the perception of the existence differs from other existences. $Ce(C_{11})$ and $Ce(C_{12})$ represent conceptualization of $Pe(P_1)$. $Ce(C_{11})$ is a conceptual perception of $Pe(P_1)$ as a human and $Ce(C_{12})$ is a conceptual perception of that human in the role of president. Furthermore, $Ce(C_{13})$ represents conceptual perception as a "human as a president" that integrates the conceptual perception of $Ce(C_{11})$ and $Ce(C_{12})$.

The arrows marked (1), (2) and (3) represent processes of conceptual perception (conceptualization); the arrows marked (4) and (5) represent relations among conceptual entities. Because the conceptual entity of a "human as a president" $Ce(C_{13})$ integrates the conceptual information of "human" and "role of president", we can refer all information of them at $Ce(C_{13})$. On the other hand, conceptual perception moves to $Ce(C_{11})$ (human) along the arrow (4). For that reason, we can not reference the information of "role of president" $Ce(C_{12})$. Similarly, along arrow (5), we can not reference information of humans that is not important to characterize the "role of president." The movement of our conceptual perception also changes the available axioms defined at each concept.

Movements of conceptual perception are performed unconsciously, but from an engineering viewpoint, we should capture them as an explicit viewpoint movement and build a consistency management framework among models. Our framework offers explicitly different modeling viewpoints. For instance, $Ce(C_{13})$ represents the integrated conceptual perception of both task concept and domain concept, whereas $Ce(C_{11})$ only represents the perception from the domain-dependent viewpoint and $Ce(C_{12})$ only represents the perception from the task-dependent viewpoint.

3   Axioms on Changes

Roughly speaking, there are two types of our perception of changes: identical changes and version changes. The former shows or hides the conceptual identity, whereas the latter has the same conceptual identity and still allows change of the conceptual version. The concepts related to the former are the role concepts; those related to the latter are state concepts. Clarification of the principle knowledge to realize adequate evolutionary management captures the conceptual perceptions mentioned above because the management processes are clear and acceptable to users. Hereafter, we explain our basic ontological principle of conceptual relations and then introduce ontological axioms of changes.

3.1   Conceptual Relations: Part-of, B-IS-A, R-IS-A

Figure 4 shows conceptual relations among part-of, B-IS-A, and R-IS-A in our Ontology-based Model Management System (OntoMMS).

In that figure, "C_{whole}" and "Role" represent the whole concept and role concept that are recognized from the viewpoint of the whole concept, respectively. "C_{part}"

represents a concept that can play a role as the role concept. The right semicircle – the one at C_{part} in 4(a) – refers its conceptual definition defined at another place: the one at C_{part} in 4(b).

In that figure, it is notable that the box under "Role" is interpreted as the conceptual augmentation entity "C_{13}", which is an instance of the "Role." The "Role" supports and accompanies the "C_{part}", which specifies a class membership constraint on the conceptual entity filled in the box. The $Ce(C_{12})$ represents a conceptual entity that is recognized independent of the existence of the whole entity $Ce(C_{11})$ with P_1, whereas $Ce(C_{13})$ represents a conceptual augmentation entity of $Ce(C_{12})$ recognized dependent on its whole concept $Ce(C_{11})$ and its target concept $Ce(C_{12})$. "$Ce(C_{14})$" is the conceptual integrated entity of $Ce(C_{12})$ and $Ce(C_{13})$.

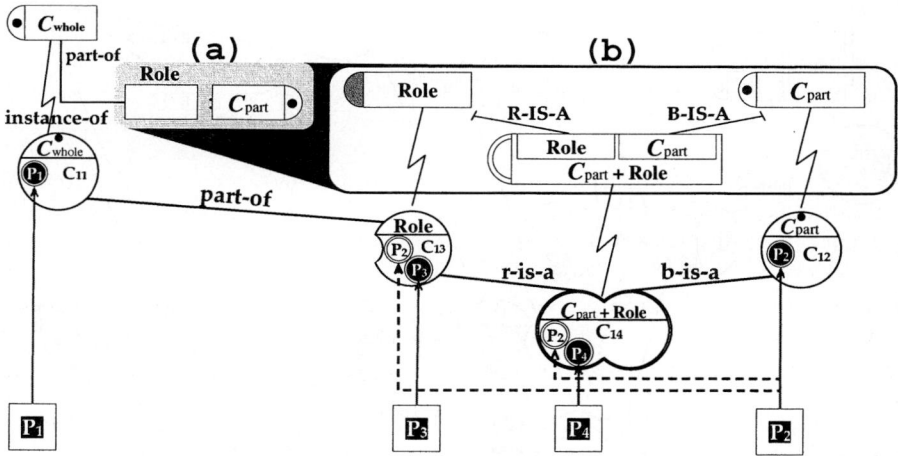

Fig. 4. Conceptual relations: part-of, B-IS-A, R-IS-A

In this figure, conceptual entities are connected by conceptual relations of part-of, B-IS-A and R-IS-A. A "part-of" relation connects an augmentation concept with a whole concept. On the other hand, "B-IS-A" and "R-IS-A" are kinds of "IS-A" relations. B-IS-A plays a role to inherit characteristics of a basic concept to an integrated concept, whereas R-IS-A plays the role of inheriting characteristics of a role concept to an integrated concept. Consequently, an integrated concept has characteristics of both basic and role concepts; it manages correspondences among properties of the basic concept and the role concept. Furthermore, the axioms of "B-IS-A" and "R-IS-A" in meta-ontology specify characteristics of the correspondences among identities and the appearance and disappearance of these conceptual entities. These axioms are defined intensionally at the meta level (domain in-dependent level) while the identical relations in the target world are provided (modeled) extensionally at the instance level (domain dependent modeling level).

Constructing a modeling framework like that depicted in Fig. 4, allows discrimination of two viewpoints regarding an entity: the viewpoint as a part of a whole concept and the one as an entity that is independent of the whole concept.

3.2 Role Concept and Its Changes

The role concept is an augmentation concept that is recognized from the viewpoint of the whole concept. The existential perceptual identity of a role concept depends on the existence of its whole concept, whereas the target perceptual identity depends on the existence of the basic concept that one modifies.

Hereafter, we use Fig. 5 to explain role-concept characteristics and changes of its identity. In that figure, "W", "B", "R", and "B+R" represent definitions of the whole, basic, role, and integrated concepts, respectively. Furthermore, they represent the role changes of the conceptual entity. For instance, they show a role change by which R_1 that modifies B changes to R_2 as changes of the conceptual information of P_0.

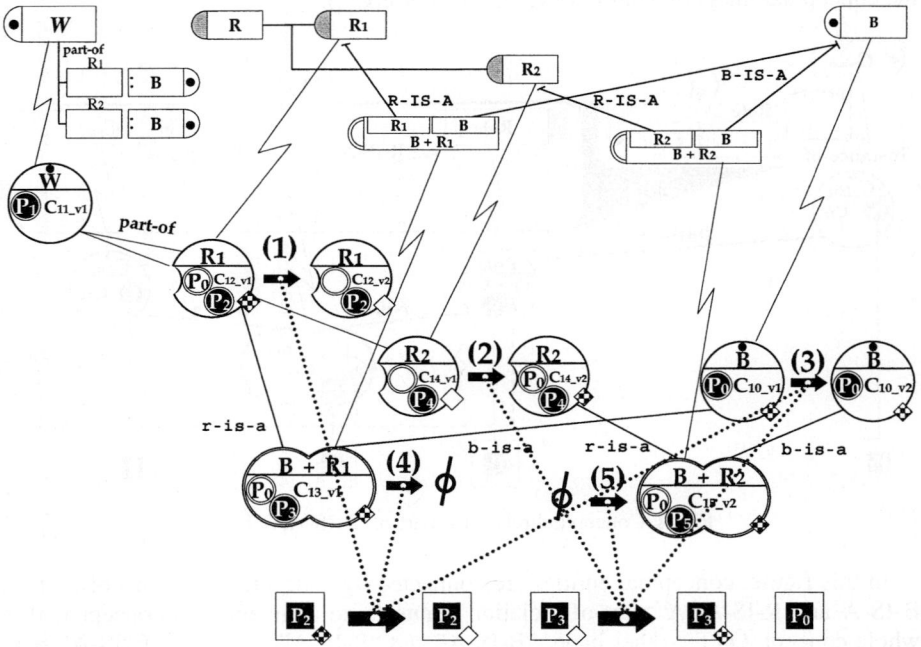

Fig. 5. Role concept and changes

That figure shows six conceptual entities: $Ce(C_{10})$, $Ce(C_{11})$, $Ce(C_{12})$, $Ce(C_{13})$, $Ce(C_{14})$, and $Ce(C_{15})$. Arrow (1) shows the conceptual role entity $Ce(C_{12})$, which is an instance of the role concept "R_1" (e.g. post of assistant manager), updates its version as the "r-is-a" relation disappears. By updating the version, it released its target perceptual identity P_0 (e.g. as the promotion of a conceptual basic entity $Ce(C_{10})$), whereas it kept its existential perceptual identity. This update captures the perception of the disappearance of R_1 with $Ce(C_{10})$ as well as the changeless existence of R_1 as the part of the whole concept $Ce(C_{11})$.

On the other hand, arrow (2) shows the conceptual role entity $Ce(C_{14})$, which is an instance of the role concept "R_2" (e.g. post of manager) updates its version as the "r-is-a" relation appears. By updating the version, it obtains the target perceptual identity P_0 (e.g. as the promotion of a conceptual basic entity $Ce(C_{10})$). Furthermore, arrow (3) shows the conceptual basic entity $Ce(C_{10})$, which is an instance of the basic concept "B" (e.g. employee), updates its version as the disappearance of the integrated concept $Ce(C_{13})$. The "b-is-a" relation between $Ce(C_{10})$ and $Ce(C_{13})$ marked (4) and the appearance of the integrated concept $Ce(C_{15})$ and "b-is-a" relation between $Ce(C_{10})$ and $Ce(C_{15})$ marked (5). By updating the version, existential perceptual identity P_0 does not changes. Instead, they only update their version of ($Ce(C_{10\_v1})$ to $Ce(C_{10\_v2})$).

Consequently, for instance, the perception that an employee $Ce(C_{10})$ has released the post of assistant manager $Ce(C_{12})$ and assumes a new post of manager $Ce(C_{14})$, and that the post of assistant manager $Ce(C_{12})$ is not filled and the post of manager is filled are explicitly represented in the model.

The axioms among "part-of", "B-IS-A" and "R-IS-A" in meta-ontology specify characteristics of the (dis)appearance of the identity of conceptual role entity.

It is notable that "role concept" in our framework specifies the identical relations and characteristics of changes while "role-restriction/property" in UML and most description-logic style modeling languages simply do class membership constraints.

3.3   State Concept and Its Changes

The state concept is an augmentation concept that adds conceptual information of state (e.g. hot, cold) to basic concept. Both the existential perceptual identity and the target perceptual identity of a state concept depend on the existence of the basic concept that one modifies.

Next, we explain characteristics of the state concept and changes of its version shown in Fig. 6. In that figure, "B", "S" and "B+S" represent definitions of the basic, state and integrated concept, respectively. Furthermore, the figure represents the state changes of a conceptual entity.

It depicts a state change by which S_1 changes to S_2 as the changes of the conceptual information of P_1. That figure shows three conceptual entities: $Ce(C_{11})$, $Ce(C_{12})$, and $Ce(C_{13})$. Arrow (2) shows that the conceptual state entity $Ce(C_{12})$, which is an instance of the state concept "S_1" (e.g. supervisory skill level 3), updates its version.

In the case of state change, it is noteworthy that the state changes its membership. In this case, it does so from an instance of S_1 to an instance of S_2 without changing its identities. Characteristics of the identity of the state concept are specified as axioms of the "s-is-a" relation. Figure 6(1) shows that the conceptual basic entity $Ce(C_{11})$ updates its version as its state $Ce(C_{12})$ changes its version.

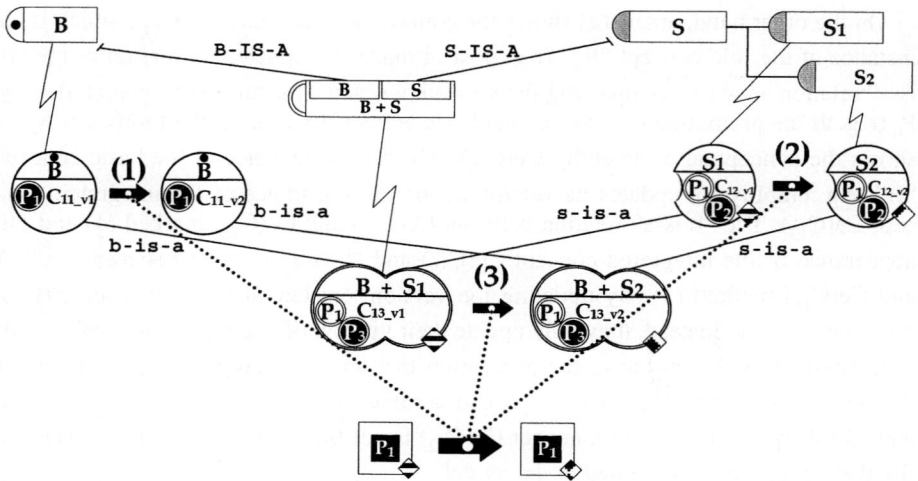

Fig. 6. State concept and state changes

4 Ontology-Based Model Management

4.1 Roles of Integrated Concept

Integrated concepts play an important role to enable sophisticated evolutionary model management. Basic functions of the integrated concept are to manage correspondence among concepts, their (dis)appearances, and their propagations of changes in the model. In other words, the system can capture important conceptual changes and propagations based on the integrated concept. We will elucidate basic functions of the integrated concept.

In the figure, upper portions marked (i), (iii), and (ii) represent the task ontology of human resource development (TO-1), which involves education for employees, and organizational structure management (TO-2), which entails staffing and job assignment, and domain ontology of organizational resource (DO), respectively. The lower portion represents a model that comprises three view models: the view model of human resource development (VM-1), organizational structure management (VM-2), and organizational resource (DM) that is shared by both VM-1 and VM-2. In other words, we can capture VM-1 and VM-2 as models of DM from the viewpoint of human resource development (VP-1) and organizational structure management (VP-2). Furthermore, DM(t) represents the states of DM at time t, similarly, VM-1(t) and VM-2(t) represent the states of VM-1 and VM-2 at time t.

We define a viewpoint as a set of concepts that an author recognizes. For example, we define the human resource development viewpoint (VP-1) for human resource developers as a set of conceptual entities that are instances of concepts defined in TO-1, whereas the organizational structure management viewpoint (VP-2) for organizational structure manager is a set of conceptual entities that are instances of the concepts defined in TO-2.

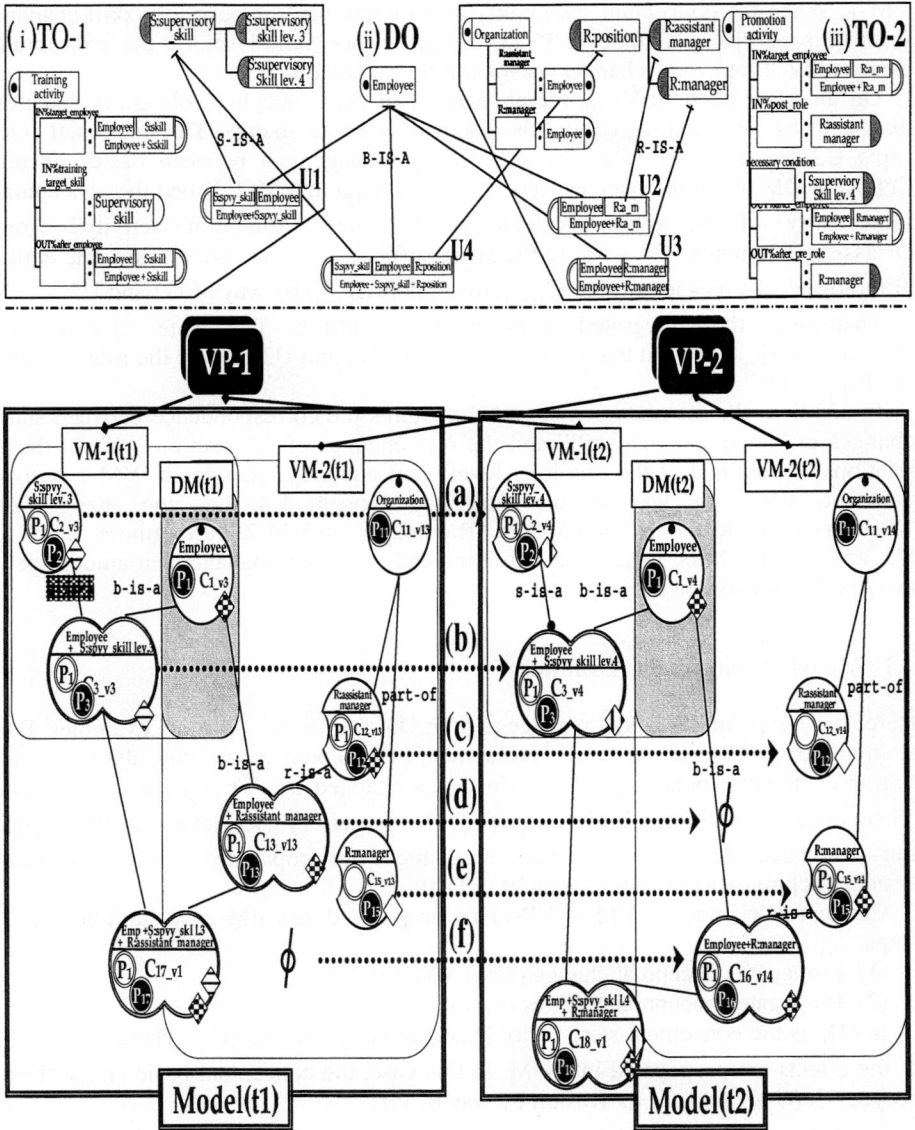

Fig. 7. Conceptual identity and the part-of relation

Figure 7 shows the evolutionary processes of the model, showing the growth in an employee's supervisory skills by training in VM-1 propagated to the employee's promotion VM-2.

At VP-1, the change of an employee's supervisory skill by the "Training activity" is modeled as a state change. The integrated concept of U1 plays a role in consistency management between the basic concept $Ce(C_1)$ in DM and state concept $Ce(C_2)$ in

VM-1. When the conceptual state entity of an employee's supervisory skill changes from "S:supervisory_skill level.3" to "S:supervisory_skill level.4", the conceptual basic entity "Employee" changes its version by way of U1.

On the other hand, at VP-2, the appearance of the conceptual role entity "post of manager" by "Promotion activity" is modeled as a role change. The integrated concepts U2 and U3 play roles for consistency management between basic concept $Ce(C_1)$ in DM and role concepts $Ce(C_{12})$ and $Ce(C_{15})$ in VM-2. When the conceptual role entity of employees is changed by the "Promotion activity" from "R:assistant_manager(C_{12-v13})" to "R:manager(C_{13-v14})", the conceptual basic entity "Employee" changes its version(C_{1-v3}) to version(C_{1-v4}) by way of U2 and U3.

Therefore, the integrated concept U1 captures the state changes of "S:supervisory_skill" and the integrated concepts U2 and U3 capture the role changes of "R:position."

In addition, the integrated concept U4 captures the correspondence between state changes of "S:supervisory_skill" and the role changes of "R:position." All of these relations are defined at the ontology level. Consequently, changes in VM-1 propagated to VM-2 by way of the integrated concept U4, i.e., the changes of "S:supervisory_skill lev.4" in VM-1 to "R:manager" in VM-2. Furthermore, U1, U2, U3, and U4 play basic roles for consistency and viewpoint management among view models and domain models.

4.2 Model Management: Example

Herein, we explain the model management in OntoMMS, taking some examples. For example, from a human resource development (VP-1) perspective, consider the situation that the conceptual state entity $Ce(C_2)$ is changed by "Training activity" from "S:supervisory_skill level.3 ($Ce(C_{2-v3})$)" to "S:supervisory_skill level.4 ($Ce(C_{2-v4})$)." Let us consider the manner in which the changes are propagated into the organizational structure management viewpoint (VP-2).

Conceptual changes in VM-1 (VP-1) are propagated into VM-2 by the following 2 steps.

(1) Propagate conceptual changes from VM-1 to DM
(2) Propagate conceptual changes from DM to VM-2

In (1), as the conceptual state entity "S:supervisory_skill $Ce(C_2)$" changes in VM-1, the effects are propagated into DM. In this case, the conceptual basic entity "Employee ($Ce(C_1)$)" updates its version by way of U1.

In (2), as the conceptual basic entity "Employee($Ce(C_1)$)" updates its version in DM, the effects are propagated into VM-2. In this case, it is encouraged to perform the promotion activity to promote the employee from assistant manager to manager. Because the necessary condition to perform the promotion activity that manager needs level_4 supervisory skill described in its definition is satisfied.

By performing the activity, the conceptual role entity "R:assistant_manager ($Ce(C_{12})$)" releases its target perceptual identity P_1 (Fig. 7(c)) and the conceptual role entity "R:manager ($Ce(C_{15})$)" attains its target perceptual identity P_1 (Fig. 7(e)). Consequently, the conceptual integrated entity "Employee + R:assistant_manager $(Ce(C_{13}))$" disappears and the conceptual integrated entity "Employee + R:manager($Ce(C_{16})$)" appears. These changes are propagated by way of integrated concept U4.

Therefore, a state change of the conceptual entity in VP-1 is propagated as a meaningful conceptual change in VP-2. For that reason, OntoMMS helps the author to manage human resources adequately according to the changes.

The propagation mechanism is realized based on the axioms of identity which characterizes each concept (basic concept, role concept, state concept and integrated concept) and conceptual relations (part-of, B-IS-A, S-IS-A and R-IS-A) among them. Those mechanisms are implemented as background models in OntoMMS. Thereby, authors do not need to be concerned with the complicated propagation management.

5 Concluding Remarks

Many researchers have studied the art of view maintenance especially in the area of relational databases [1, 2, 3, 5, 8, 9]. However, we believe that specifications and premises for data modeling are not explicit. Our ontological approach provides explicit specifications and premises for data management as an ontology. We lay a foundation to evolutionary model management based on the characteristics of target-world concepts.

This paper presented an ontology-based evolutionary model management framework. We first clarified our underlying philosophy to systemize an ontology based on the analysis of "identity." Using that philosophy, the system can properly manage the consistency of models that combine multiple viewpoints. Then, we showed our evolutionary model management framework taking simple examples.

Consequently, an advantage of our approach is that we can advance beyond *ad hoc* modeling and implementation of a system based on ontology. Our approach provides model guidelines and explicit specifications for evolutionary management of models.

We believe that we can build a foundation of modeling framework, as a first step, whereby authors can build their models without concern for the complicated consistency and evolutionary management among multiple view models based on the ontology.

On the other hand, it is also important to consider ontology versioning issues [10]: we also need a methodology with methods to update ontologies adequately according to the evolutions of the real world and to maintain the consistency between models and ontologies by a sophisticated manner. We will discuss this matter further in the future. Furthermore, we will implement a system whereby various agents can share and use models effectively from different viewpoints as a future work.

References

1. Blakeley, J.A., Larson, P., Tompa, F.W.: Efficiently updating materialized views. In SIGMOD (1986) pages 61–71

2. Dimitrova, K., El-Sayed, M., Rundensteiner, E.A.: Order-Sensitive View Maintenance of Materialized XQuery Views. Proc. of the Conceptual Modeling (ER 2003) (2003) pages 144–157

3. Griffin, T., Libkin, L.: Incremental maintenance of materialized views with duplicates. In SIGMOD (1995) pages 328–339

4. Gruber, T. R.. A translation approach to portable ontologies. Knowledge Acquisition(1993) pages 5:199-220

5. Gupta, A., Mumick, I.S., Subrahmanian, V.S.: Maintaining views incrementally. In SIGMOD (1993) pages 158–166

6. Van Heijst, G., Schreiber, A.Th., Wielinga, B.J.: "Using Explicit Ontologies in KBS Development", Int. J. Human-Computer Studies (1997) pp.183–292

7. Ikeda, M., Seta, K., Mizoguchi, R.: "Task Ontology Makes It Easier To Use Authoring Tools", Proc of IJCAI-97 (1997) pp. 342–347

8. Quan, L.P., Chen, L., Rundensteiner, E.A.: Argos: Efficient refresh in an XQL-based web caching system. In WebDB (2000) pages 23–28

9. Quass, D.: Maintenance expressions for views with aggregation. In SIGMOD (1996) pages 110–118

10. Klein, M., Fensel, D., Kiryakov, A., Ognyanov, D.: Ontology versioning and change detection on the web. In 13th International Conference on Knowledge Engineering and Knowledge Management (EKAW02) (2002).

Elementary Translations:
The Seesaws for Achieving Traceability
Between Database Schemata*

Eladio Domínguez[1], Jorge Lloret[1], Ángel L. Rubio[2], and María A. Zapata[1]

[1] Dpto. de Informática e Ingeniería de Sistemas
Facultad de Ciencias, Edificio de Matemáticas
Universidad de Zaragoza, 50009 Zaragoza, Spain
ccia@posta.unizar.es
[2] Dpto. de Matemáticas y Computación, Edificio Vives
Universidad de La Rioja, 26004 Logroño, Spain
arubio@dmc.unirioja.es

Abstract. There exist several recent approaches that leverages the use of model transformations during software development. The existence of different kinds of models, at different levels of abstraction, involves the necessity of transferring knowledge from one model to another. This framework can also be applied in the context of metadata management for database evolution, in which transformations are needed both to translate schemata from one level to another and to modify existing schemata. In this paper we introduce the notions of translation rule and elementary translation which are used within a forward database maintenance strategy.

1 Introduction

Several recent research efforts focus on automating the generation and management of transformations between models or schemata representing the same system at different levels of abstraction [3, 6, 13, 16, 22]. For example, the importance of transformation management has been recently acknowledged by the UML community with the advent of the Model Driven Architecture (MDA) [16] in which transformations between platform independent models (PIM) and platform specific models (PSM) must be managed. The transformation modeling framework proposed in [3] within the 'model management' approach [2] is another example. In general, the management of model transformations is recognized as a necessary goal for achieving the ambitious goal of automating system implementation. This issue is particularly difficult when models or schemata evolve over time and consistency between them must be kept.

* This work has been partially supported by DGES, project TIC2002-01626, by Ibercaja-University of Zaragoza, project IB 2002-TEC-03, by the Government of La Rioja, project ACPI2002/06, by the Government of Aragon and by the European Social Fund.

S. Wang et al. (Eds.): ER Workshops 2004, LNCS 3289, pp. 377–389, 2004.

In the database field, the evolution issue is related to the existence of changes within the different phases of the life cycle, from early design stages to exploitation and maintenance. In particular, a recent paper highlighted the existence of "a lack of support (methods and tools) in the database maintenance and evolution domain" [13]. Among the several problems related to evolution activities (see [11]), one of the most important is the 'forward database maintenance problem' (or 'redesign problem', according to [20]). This problem is how to reflect in the logical and extensional schemata the changes that have occurred in the conceptual schema of a database. One way of tackling this problem is to ensure the traceability of the translation process between levels. But there is no agreement about which artifacts and mechanisms are needed for assuring traceability [19, 22].

As a contribution towards achieving a satisfactory solution to this problem, in this paper we propose a metadata approach to ensure the traceability of the translation between the conceptual and logical levels. Some ideas about our approach are presented in [7], and its relationship with a model–driven framework is in [5]. The way in which we propose to achieve traceability is making use of a specific translation component in which information related with the translation process is stored by means of elementary translations. When an evolution process is carried out in a conceptual model, it is propagated to the logical model applying a propagation algorithm which makes use of the information stored in the translation component.

The remainder of the paper is organized as follows. Section 2 is devoted to the presentation of both an outline of our proposed architecture and an example that we use throughout the paper. In Section 3 we present concepts we use for creating the component which ensures traceability, whereas in Section 4 we detail how the elementary translations of this component are used in the evolution process. The paper is completed with some related work and conclusions.

2 Preliminaries

2.1 Evolution Architecture Overview

Many current research efforts aim at contributing to the solution of the several problems related to database evolution activities [4, 13, 10, 15]. As a contribution towards achieving a satisfactory solution to one of these problems (that of 'forward database maintenance problem'), some of the authors of the present paper have presented in [7] an architecture and a prototype tool for managing database evolution whose graphical representation is shown in Figure 1. The way of working of this architecture is as follows: the conceptual component reacts to database evolution related external events and changes the conceptual database schema according to the semantics of the received event. The information of the fact that these changes have taken place is propagated through the rest of the components and they are appropriately changed in order to reflect the changes at the conceptual level.

It is worth noting the two main characteristics of this approach that make it different from other database evolution proposals. On the one hand, it includes an explicit translation component that stores information about the way in which a concrete conceptual database schema is translated into a logical schema. This component ensures the traceability of the translation process. On the other hand, a metamodeling approach [8] has been followed for the definition of the architecture. Within this architecture, three meta–models are considered which capture, respectively, the conceptual, logical and translation modeling knowledge.

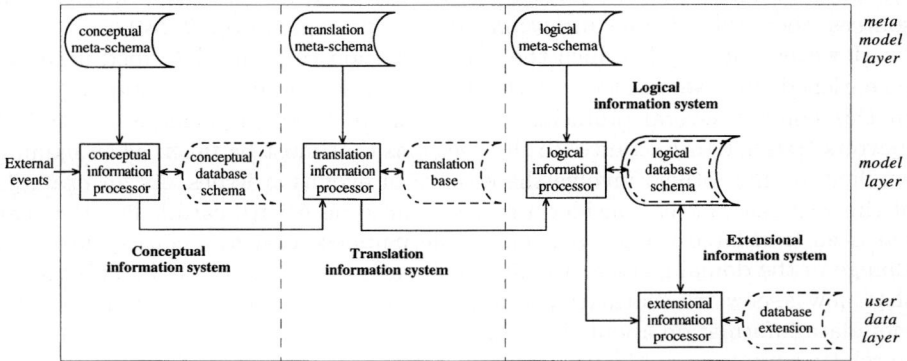

Fig. 1. Architecture for Database Evolution

2.2   Running Example

The architecture described in [7] and outlined in the previous subsection has been developed with the property of being independent of any particular modeling technique, by means of the adoption of a metamodeling approach. However, in order to clarify the way the architecture works we will be considering a particular example, and for this reason some concrete techniques must be established. For this example, we have chosen the most common techniques in the context of the database field: the E/R technique as the conceptual modeling technique and the relational model as the logical modeling technique.

The conceptual evolution example we are going to use during the paper is the following. The initial conceptual schema consists of an isa hierachy H with supertype E and subtype S. The entity type E has two attributes, e1 and e2, and the primary key is e1. We assume the existence, within our architecture, of the external event type 'change of primary key in the root of an ISA hierarchy'. This event is issued against the initial conceptual schema, resulting in a modified conceptual schema where the attribute e2 is the new primary key of E. A comprehensive, detailed study about this event and its processing within the architecture is included in [6]. This archetypal example has been chosen with the aim of covering two aspects that, to our knowledge, have not been fully explored in the evolution context. These aspects (that are being incorporated into our current implementation) are the evolution of ISA relationships and the evolution of integrity constraints.

3   Creation of Elementary Translations

Traditional database engineering processes include, among others, the transformation of conceptual schemata into logical schemata. Usually this transformation process consists of applying a set of transformation rules to a conceptual schema in order to obtain a logical one. In the literature, there is a multitude of sources where sets of transformation rules are presented (see [9, 18]).

However, the way in which the transformation rules are described needs to be enriched when we want to use them in a forward database evolution context. In this context several problems arise. Some of these problems are technical matters (related, for instance, to the decisions that must be taken when naming conflicts occur); others involve the essence of evolution issues. As an example of this last case, let us consider a relationship type A with cardinality 1–n that has been translated using a particular rule. Suppose that as a consequence of a change in the domain, the cardinality of A must be changed to n–n. This means that now a new transformation rule must be applied. However, if there is no metadata information about the transformation rule that was originally applied to relationship type A, the new transformation can not benefit from the original one and the complete translation process must be redone from scratch.

Within our architecture, we have enriched the notion of transformation rule giving rise to a notion we call *translation rule* (see Section 3.1). These translations rules are used for building the *translation algorithm* (see Section 3.2). When this algorithm is applied to a conceptual schema, it produces not only a logical one (as in the traditional transformation process) but also a set of *elementary translations* stored in the translation base. An elementary translation is the smallest piece of information reflecting the correspondence between the conceptual elements and the logical ones.

3.1   Translation Rules

As architect designers, we specify the set of translation rules which can be applied to each kind of conceptual building block. Basically, a translation rule defines how the conceptual elements of an instance of a conceptual building block (of the conceptual schema) are translated into logical elements of instances of logical building blocks (of the logical schema).

Table 1 shows a sketch of the syntax of a translation rule. A translation rule determines the procedures that have to be applied to instances of the specified conceptual building block B and each procedure calculates the value of the different logical elements (table name, column name, column type, constraint...). In order to determine the procedure to be applied, exactly one of the conditions from 1 to m must be *true*. For instance, a particular translation rule applicable to relationship types must establish the procedure for determining the name of the

Table 1. Sketch of a translation rule

<rule_name> (b instance of conceptual building block <B>)
  when <condition 1> then
    $a_1 \leftarrow procedure_{11}$;
    $a_2 \leftarrow procedure_{12}$;
    . . .
    $a_n \leftarrow procedure_{1n}$;
  endwhen
    $\vdots$
  when <condition m> then
    $a_1 \leftarrow procedure_{m1}$;
    $a_2 \leftarrow procedure_{m2}$;
    . . .
    $a_n \leftarrow procedure_{mn}$;
  endwhen
endrule

table into which the relationship type is buried. The conditions serve to decide which procedure is selected according to the cardinality of the relationship.

3.2 Translation Algorithm

This algorithm takes as input conceptual building block instances of the conceptual schema and creates 1) the elementary translations that relate the conceptual elements and the logical ones and 2) the logical elements of the logical schema. For the modeling techniques we have chosen, the conceptual building blocks are entity type, relationship type and ISA hierarchy, and the only logical building block is table.

In general, for each conceptual building block there are different translation rules that can be applied. In our approach, the translation rule to be applied is chosen by the user. This process can be modeled by the following algorithm sketch:

```
For each set C of instances of building block B
 For each instance b of C
   (a) election=get_building_block_translation_rule(B)
   (b) apply_elementary_translation_creation_procedure(b, election)
   (c) apply_logical_element_creation_procedure(b, election)
```

In the first iteration of the outer loop of this sketch, C represents the set of entity types not involved in any ISA hierarchy. In the second iteration, C represents the set of ISA hierarchies and, in the third, C represents the set of relationship types. As a consequence, each building block instance is translated by exactly one translation rule: the entity types that belong to an ISA hierarchy are translated by the translation rule for the ISA hierarchy chosen by the DBA and the rest of building blocks instances are translated by the corresponding translation rules chosen by the DBA.

At the time of this writing, this algorithm is completely automated for the chosen metamodels and we think it is easily extensible to other metamodels like UML [17] and object–relational [21].

Let us describe the steps of the algorithm.

Step (a). The user selects the translation rule to be applied to the corresponding building block instance of the initial conceptual schema. The metadata stored in the conceptual information base of the architecture representing the initial conceptual schema example are shown in Figure 2(a). Let us suppose that the designer has chosen to apply a translation rule called *isa1_tr* to the ISA hierarchy number 7 in Figure 2(a). This rule is based on Elmasri's rule on chapter 7 of [9] and transforms each entity type of the hierarchy into a table.

Step (b). Taking into account the translation rule selected in step (a), the appropriate elementary translations which model the transformation process of each building block instance are stored in the translation base. To achieve this, each translation rule has an elementary translation creation procedure associated to it. For example, we have a translation rule for entity types (namely *entity_type1_tr* rule) and a translation rule for ISA hierarchies (namely *isa1_tr* rule) whose elementary translation creation procedures are shown below. In these procedures several auxiliary functions are used (shown in Table 2) and the created elementary translations can be of different types (some of them are shown in Table 3).

Creation procedure for the translation rule entity_type1_tr

1. For the entity type E add to the translation base
   $new\_elementary\_translation('ETT01', get\_id(E), new\_id(E), get\_id(E))$
2. For each attribute $a_i, i = 1..n$, add to the translation base
   $new\_elementary\_translation('ETT20', get\_id(a_i), new\_id(a_i), get\_id(E))$
3. For the primary key pkE of E add to the translation base
   $new\_elementary\_translation('ETT60', get\_id(pkE), new\_id(pkE),$
   $get\_id(E))$

Creation procedure for the translation rule isa1_tr

1. For the supertype E of the ISA hierarchy H, apply the creation procedure for the translation rule *entity_type1_tr*, but *get_id(H)* is used as the final parameter instead of *get_id(E)*
2. For each subtype $S_i, i = 1..m$ of the ISA hierarchy H, add to the translation base
   $new\_elementary\_translation('ETT01', get\_id(S_i), new\_id(S_i), get\_id(H))$
3. For each attribute b_{ki} of $S_i, k = 1..r_i, i = 1..m$, add to the translation base
   $new\_elementary\_translation('ETT20', get\_id(b_{ki}), new\_id(b_{ki}), get\_id(H))$
4. For each attribute $a_i, i = 1..j$, of the primary key of E and each subtype $S_i, i = 1..m$, add to the translation base
   $new\_elementary\_translation('ETT22', get\_id(a_i), new\_id(a_i), get\_id(H))$

(a) Conceptual database schema

entity_type		attribute			
entity_type_id	name	attribute_id	name	datatype	entity_type_id
1	E	3	e1	integer	1
2	S	4	e2	integer	1

isa_hierarchy			conceptual_constraint				
isa_hier_id	supertype	subtypes	conc_constr_id	type	name	source	target
7	1	2	5	pk	pk1	1	3
			6	isa	isa1	2	1

(b) Translation base

elementary_translation				
elem_transl_id	type	conceptual_element	logical_element	belongs_to
16	ETT20	E.e1	E.e1	7
17	ETT20	E.e2	E.e2	7
18	ETT01	E	E	7
19	ETT60	pk1	pk1	7
20	ETT22	E.e1	S.e1	7
21	ETT62	pk1	pk2	7
22	ETT01	S	S	7
23	ETT65	isa1	fk1	7

(c) Logical database schema

table		column				logical_constraint				
table_id	name	column_id	name	datatype	table_id	logic_constr_id	type	name	source	target
8	E	10	e1	integer	8	13	pk	pk1	8	10
9	S	11	e2	integer	8	14	pk	pk2	9	12
		12	e1	integer	9	15	fk	fk1	12	10

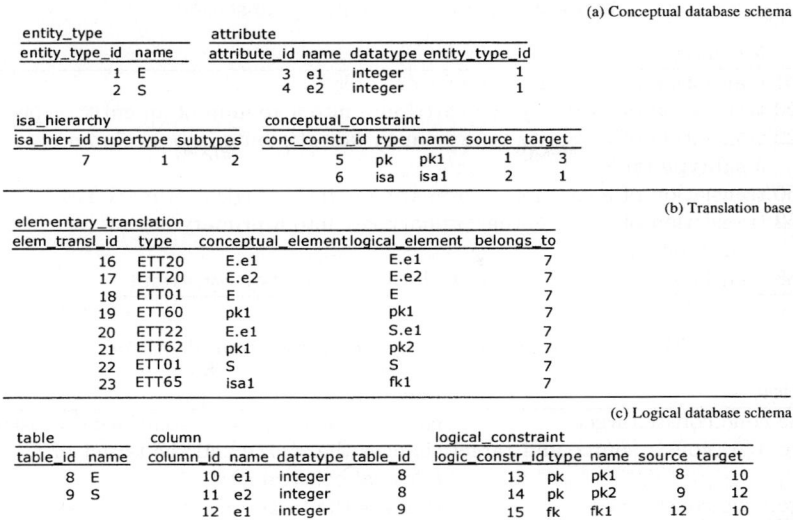

Fig. 2. Information bases

Table 2. Functions used in the translation algorithm

Function	Meaning
new_elementary_translation	creates a new elementary translation in the translation base
new_id	provides a unique identifier for a table, column or logical constraint
get_id	queries the conceptual base in order to find the identifier of an attribute, entity type, isa hierarchy or conceptual constraint

5. For the primary key of the entity type E and for each subtype $S_i, i = 1..m$, add to the translation base
   $new\_elementary\_translation('ETT62', get\_id(pkE), new\_id(pkE), get\_id(H))$

6. For each isa constraint with source S_i and target $E, i = 1..m$, add to the translation base
   $new\_elementary\_translation('ETT65', get\_id(isa_i), new\_id(isa_i), get\_id(H))$

When we apply step (b) to the ISA hierarchy numbered 7 in Figure 2(a), the elementary translation creation procedure generates the elementary translations of Figure 2(b).

Step (c). The appropriate logical elements are stored in the logical database schema. We omit for space reasons details about this step. The metadata representing the logical elements that are obtained in step (c) for our example are shown in Figure 2(c).

Table 3. Types of elementary translations

Type	Meaning
ETT01	translation of an entity type into a table
ETT20	translation of an entity type attribute into a column of an entity type table
ETT22	translation of an entity type primary key attribute into a column of a subtype table
ETT60	translation of a conceptual primary key into a logical primary key
ETT62	translation of a conceptual primary key into a primary key of a subtype table
ETT65	translation of an isa constraint into a foreign key constraint

Table 4. Functions used in the propagation algorithm

Function	Meaning
change_conc_constr_target	changes the target of a conceptual constraint
change_logic_constr_source_and_target	changes the source and the target of a logical constraint
change_logic_constr_target	changes the target of a logical constraint

4   The Elementary Translations Act as Seesaws

In this section we explain how the traceability achieved by means of the translation component is used to propagate the changes between the conceptual and the logical levels. The changes in the conceptual database schema are specified by means of external events. We show the *propagation subalgorithm for the translation information system* and how the changes made in the conceptual database schema are used by this subalgorithm in order to change the translation base appropriately. Then we show how the changes in the translation base are used by the *propagation subalgorithm for the logical information system*, which makes the appropriate changes to the logical database schema. Some functions used in these subalgorithms are shown in Table 4.

External Events. When the DBA wants to change the database, (s)he issues an external event. An external event is composed of a set of conceptual modification primitives. For example, according to our running example, we suppose that the DBA wants to make the attribute E.e2 the new primary key of entity type E. Then, (s)he issues the external event change_of_primary_key_in_ISA('E','e2').

We, as architect designers, have designed the external event type change_of_primary_key_in_ISA so that it performs only one conceptual modification for evolution that in our running example is:

$$change\_conc\_constr\_target('pk', 'E', 'E.e1', 'E.e2') \qquad (1)$$

where the target of the primary key of the entity type E is modified so that the attribute E.e2 is the new target.

For this external event type, only one conceptual modification is performed, but, in general, an external event type can perform more than one modification. For example, we, as architect designers, could have designed the external event type change_of_ primary_key_in_ISA so that it also retains an uniqueness constraint for the old attributes of the primary key. In this case, another conceptual modification for evolution, apart from (1), must be performed. In our particular example this modification is new_conceptual_constraint('uniq','E','E.e1'), which declares a new unique constraint on E.e1.

Propagation Subalgorithm for the Translation Information System. This subalgorithm takes as input the conceptual changes produced by an external event and updates the translation base so as to reflect these changes. Its input is the set of changes (specified as a set of conceptual modifications) which has been made in the conceptual schema. The output is the added, deleted and affected elementary translations. An *affected elementary translation* is an elementary translation that must be modified or an elementary translation that is not modified but that involves a logical element which will have to undergo some change. Here we sketch this subalgorithm:

```
For each change in the conceptual database schema
  (a) Detect new elementary translations to be added to the
      translation base and add them.
  (b) Detect those elementary translations that are affected by
      the conceptual change and modify them when needed.
  (c) Detect elementary translations to be deleted from the
      translation base and delete them.
```

Let us show how this subalgorithm works for our external event example. The subalgorithm is applied to the conceptual modification (1) mentioned above. In step (a) for (1), since the new primary key attribute of the entity type E must be translated into a column of the primary key of the subtype table S, the following elementary translation is added to the translation base:

$$new\_elementary\_translation('ETT22', get\_id('E.e2'), new\_id('S.e2'), 7) \quad (2)$$

In step (b) the elementary translations involving the primary key of E and the ISA between S and E (translations numbers 19, 21 and 23 in Figure 2(b)) are identified as affected. In our example, the elementary translations 19 and 21 are affected because their logical element (the primary key constraint of E) will change their target. The elementary translation 23 is affected because its logical element(the foreign key fk1) will change its source and its target.

In step (c), the elementary translation that informs the attribute e1 has been added to the table of S (translation 20) is deleted because the conceptual element E.e1 is no longer an attribute of the primary key of the entity type E.

Here we explain the seesaw metaphor. A seesaw is an artifact by means of which the actions performed on the left side (for example, putting a weight on it) provoke a reaction on the right side (the right side goes up). In our framework, each elementary translation acts like a seesaw. The changes made in the conceptual part (the left side of the seesaw) provoke modifications in the logical part (the right side). This reaction is captured by the propagation subalgorithm for the logical information system, explained in next subsection, which updates the logical database schema.

Propagation Subalgorithm for the Logical Information System. The information about the added, deleted and/or affected elementary translations is sent to the logical information system, which applies the propagation subalgorithm to the logical information system. We do not include this subalgorithm here and only show how it works in our example. Thus, for the new elementary translation added in (2), the subalgorithm adds to the logical information schema the following logical element:

$$new\_column(`e2', `integer', `S') \tag{3}$$

For the affected elementary translations 19 and 21, their logical elements must change their target. For this reason, the subalgorithm performs the following changes in the logical database schema:

$$change\_logic\_constr\_target(`pk', `E', `E.e1', `E.e2') \tag{4}$$

$$change\_logic\_constr\_target(`pk', `S', `S.e1', `S.e2') \tag{5}$$

where the parameters are the type of constraint, the table on which the constraint is specified and the old and new targets.

For the affected elementary translation 23, its logical element must change its source and its target. For this reason, the subalgorithm performs the following change in the logical database schema:

$$change\_logic\_constr\_source\_and\_target(`fk', `S', `S.e1', `S.e2', `E.e1', `E.e2') \tag{6}$$

where the parameters are the type of constraint, the table on which the constraint is specified, the old and new sources and the old and new targets.

Finally, the deletion of the elementary translation 20 produces the following deletion in the logical information system:

$$delete\_column(`S.e1') \tag{7}$$

By means of the two algorithms explained above, the translation base now reflects the correspondence between the new conceptual and logical schemas.

5   Related Work

The necessity of capturing information about translations performed between models has been recognized in the literature and very varied approaches have been proposed. In the MDA approach [16], mapping techniques between PIM's or between PSM's are considered as association classes so that a mapping will be represented as an object (relating models) and not only as a link. With respect to PIM to PSM mappings, the MDA approach proposes the 'incremental consistency' as a desirable feature [14]. Within our architecture this property is achieved by means of the translation component.

The consideration of our architecture has lead us to propose in [5] several modifications in the MDA architecture. For example, it seems surprising to us that in the original MDA Metamodel Description there are explicit association classes to represent 'PSM Mapping Techniques' (from PSM to PSM) and 'PIM Mapping Techniques' (from PIM to PIM), but there are no analogous association classes to represent 'PIM to PSM Mapping Techniques' or 'PSM to PIM Refactoring Techniques'. According to the proposal we present in this paper, the class 'PIM to PSM Mapping Techniques' would also be included in the MDA architecture. In our opinion, the addition of this association class is suitable for ensuring the traceability needed for propagating the changes in a PIM model to its corresponding PSM model.

The innovative trend of model management [2, 3] also advocates the representation of mappings by means of objects. It is argued that this reification is often needed for satisfactory expressiveness [2]. This proposal aims to be generic in the sense that it can be applied to different kinds of models while our work is specific to database evolution issues.

Within the specific database evolution field, several papers deal with the evolution of object-oriented databases and relational databases [1] but, in general, they lack the consideration of a conceptual level which allows the designer to work at a higher level of abstraction [15]. A proposal that also considers the conceptual level is presented in [13]. In this case, the traceability is achieved storing all the sequence of operations (called history) performed during the translation of the conceptual schema into a logical one. In order to use this history for propagating the conceptual schema modifications, a process of history cleaning (eliminating redundant operations) must be performed [12]. The main difference between this approach and ours is the type of information stored for assuring traceability. Whereas in [13] the idea is to store the history of the process performed (probably with redundancies), in our case the goal of the elementary translations is to reflect the correspondence between the conceptual elements and the logical ones (in this case, there is no room for redundancies). Another difference is that we follow a meta–modelling approach. An in depth explanation of the use of meta–models within our proposed architecture appears in [7].

6   Conclusions and Future Work

The main contribution of this work is the presentation of a method for ensuring traceability in the context of database evolution. We explain how to use this traceability for propagating changes between database design levels. More specifically, we have explained how to create a component that reflects the correspondence between conceptual and logical database schemas. For this purpose, we have used elementary translations that reflect the relations between the conceptual elements and the logical ones and that facilitate evolution tasks.

There are several possible directions for future work. One is how to apply our ideas to other metadata management problems, in particular to the problems of schema integration or round–trip engineering [2]. Another direction is how to apply the architecture when more modeling constructs are defined in the conceptual and logical meta–schemata. For example, we could consider models with richer constructs, such as UML [17] and object-relational models [21].

References

1. L. Al-Jadir, M. Léonard, Multiobjects to Ease Schema Evolution in an OODBMS, in T. W. Ling, S. Ram, M. L. Lee (eds.), *Conceptual modeling, ER-98*, LNCS 1507, Springer, 1998, 316–333.
2. P. A. Bernstein, Applying Model Management to Classical Meta Data Problems, *First Biennial Conference on Innovative Data Systems Research- CIDR 2003*, Online Proceedings, 2003.
3. K. T. Claypool, E. A. Rundensteiner, Gangam: A Transformation Modeling Framework. *International Conference on Database Systems for Advanced Applications-DASFAA 2003*, IEEE Computer Society, 2003, 47–54.
4. K. T. Claypool, E. A. Rundensteiner, G. T. Heineman, ROVER: flexible yet consistent evolution of relationships, *Data Knowl. Eng.* 39(1), 2001, 27–50.
5. E. Domínguez, J. Lloret, A. L. Rubio, M. A. Zapata, An MDA–Based Approach to Managing Database Evolution (position paper), in A. Rensink, (Editor), *Proceedings of MDAFA 2003. Model–Driven Architecture: Foundations and Aplications*, CTIT Technical Report Series, No. 03-27, 2003, 97–102.
6. E. Domínguez, J. Lloret, A. L. Rubio, M. A. Zapata, Evolving the implementation of ISA Relationships, submitted for publication.
7. E. Domínguez, J. Lloret, M. A. Zapata, An architecture for Managing Database Evolution, in A. Olivé et al. (eds) *Advanced conceptual modeling techniques- ER 2002 Workshops*, LNCS 2784 , 2002, 63–74.
8. E. Domínguez, M. A. Zapata, J. J. Rubio, A Conceptual Approach to Meta–Modelling, in A. Olivé, J. A. Pastor (Eds.), *Advanced Information Systems Eng.-CAiSE'97*, LNCS 1250, 1997, 319–332.
9. R. A. Elmasri, S. B. Navathe, *Fundamentals of Database Systems (4th ed.)*, Addison-Wesley, 2003.
10. F. Ferrandina, T. Meyer, R. Zicari, G. Ferran, J. Madec, Schema and Database Evolution in the O2 Object Database System, *Very Large Data Bases- VLDB'95*, 1995, 170–181.
11. J. L. Hainaut, V. Englebert, J. Henrard, J. M. Hick, D. Roland, Database Evolution: the DB-MAIN approach, in P. Loucopoulos (ed.), *Entity-Relationship approach- ER'94*, LNCS 881, 1994, 112–131.
12. J.M. Hick, *Evolution of relational database applications: Methods and tools*, PhD Thesis, University of Namur, 2001 [in French].
13. J.M. Hick, J.L. Hainaut, Strategy for Database Application Evolution: The DB-MAIN Approach, in I.-Y. Song et al. (eds.) *Conceptual Modeling- ER 2003*, LNCS 2813, 2003, 291–306.
14. Anneke Kleppe, Jos Warmer, Wim Bast, *MDA explained. The Model Driven Architecture: Practice and Promise*, Addison–Wesley, 2003.
15. J. R. López, A. Olivé, A Framework for the Evolution of Temporal Conceptual Schemas of Information Systems, in B. Wangler, L. Bergman (eds.), *Advanced Information Systems Eng.- CAiSE 2000*, LNCS 1789, 2000, 369–386.

16. J. Miller, J. Mukerji (eds.), MDA Guide Version 1.0, Object Management Group, Document number omg/2003-05-01, May 1, 2003.

17. OMG, *UML Specification version 1.5 formal/2003-03-01*, available at http://www.omg.org, March, 2003.

18. H. A. Proper, Data Schema Design as a Schema Evolution Process. *Data Knowl. Eng* 22(2), 1997, 159-189

19. B. Ramesh, Factors influencing requirements traceability practice, *Communications of the ACM*, 41 (12), December 1998, 37-44.

20. A. S. da Silva, A. H. F. Laender, M. A. Casanova, An Approach to Maintaining Optimized Relational Representations of Entity-Relationship Schemas, in B. Thalheim (ed.), *Conceptual Modeling- ER'96*, LNCS 1157, 1996, 292–308.

21. M. Stonebraker, D. Moore, P. Brown, *Object Relational DBMSs: Tracking the next great wave (2nd ed.)*, Morgan Kaufmann Publishers, 1999.

22. W. M. N. Wan-Kadir, P. Loucopoulos, Relating evolving business rules to software design, *Journal of Systems Architecture*, article in press, 2003.

Towards the Managment of Time
in Data-Intensive Web Sites

Paolo Atzeni and Pierluigi Del Nostro

Dipartimento di Informatica e Automazione – Università Roma Tre
{atzeni,pdn}@dia.uniroma3.it

Abstract. The adoption of a logical model for temporal, data-intensive Web sites is proposed together with a methodology for the development. The model allows the definition of page-schemes with temporal aspects (which could be related to the page as a whole or to individual components of it). The design process follows a development that starts with a traditional E-R scheme; the various steps lead to a temporal E-R scheme, to a navigation scheme and finally to a T-ADM scheme. A tool associated with the methodology has been developed: it automatically generates both the relational database (with the temporal features needed) supporting the site and the actual Web pages, which can be dynamic (JSP) or static (plain HTML), or a combination thereof.

1 Introduction

The systematic development of Web sites has attracted the interest of the database community as soon as it was realized that the Web could be used as a suitable means for the publication of useful information of interest for community of users (Atzeni et al. [1], Ceri et al. [2], Fernández et al. [3]). Specific attention has been devoted to *data-intensive* sites, where the information of interest has both a somehow regular structure and a possibly significant volume; here the information can be profitably stored as data in a database and the sites can be generated (statically or dynamically) by means of suitable expressions (that is, queries) over them (Merialdo et al. [4]). In this setting, the usefulness of high-level models for the intensional description of Web sites has been advocated by various authors, including Atzeni et al [1,4] and Ceri et al. [2], which both propose logical models in a sort of traditional database sense and a *model-based development* for data intensive Web sites.

When accessing a Web site, users would often get significant benefit from the availability of time-related information, in various forms: from the history of data in pages to the date of last update of a page (or the date the content of a page was last validated), from the access to previous versions of a page to the navigation over a site at a specific past date (with links coherent with respect to this date). As common experience tells, various aspects of a Web site often change over time: (i) the actual content of data (for example, in a University Web site, the instructor for a course); (ii) the types of data offered (at some point we could decide to publish not only the instructor, but also the teaching assistants,

S. Wang et al. (Eds.): ER Workshops 2004, LNCS 3289, pp. 390–401, 2004.

TAs, for a course); (iii) the hypertext structure (we could have the instructors in a list for all courses and the TAs only in separate detail pages, and then change, in order to have also the TAs in the summary page); (iv) the presentation. Indeed, most current sites do handle very little time-related information, with past versions not available and histories difficult to reconstruct, even when there is past data. Clearly, these issues correspond to cases that occur often, with similar needs, and that could be properly handled by specific techniques for the support to time-related features. Therefore, we have here requirements that are analogous to those that led to the development of techniques for the effective support to the management of time in databases by means of *temporal database* (see Jensen and Snodgrass [5] for a recent survey and Snodgrass [6] for a textbook discussion).

It is well known that in temporal databases there are various dimensions along which time can be considered. Beside *user-defined time* (the semantics of which is "known only to the user", and therefore is not explicitly handled), there are *valid time* ("the time a fact was true in reality") and *transaction time* ("the time the fact was stored in the database"). In order to highlight the specific aspects of interest for Web sites, let us concentrate on valid time, even if transaction time could have some specific, additional facets.

In a Web site, the motivation for valid time is similar to the one in temporal databases: we are often interested in describing not only snapshots of the world, but also histories about its facts. However, there is a difference: in temporal databases the interest is in storing histories and in being able to answer queries about both snapshots and histories, whereas in Web sites the challenge is on how histories are offered to site visitors, who browse and do not query. Therefore, this is a design issue, to be dealt with by referring to the requirements we have for the site. The natural (and not expensive) redundancy common in Web sites could even suggest to have a coexistence of snapshots and histories.

This paper is aimed at giving a contribution to the claim that the management of time in Web sites can be effectively supported by leveraging on the experiences made in the database field, and precisely by the combination of the two areas we have briefly mentioned: temporal databases on the one hand and model-based development of Web sites on the other. In particular, attention is devoted to models and design: models in order to have a means to describe temporal features and design methods to support the developer in his/her decisions on which are the temporal features of interest to the Web site user.

The paper extends the experiences in the Araneus project [1, 4, 7, 8] where models, methods and a tool for the development of data-intensive Web sites were developed. Indeed, we propose a logical model for temporal Web sites, a design methodology for them and a tool to support the process (whose features have been recently demonstrated and sketched in a short paper, Atzeni and Del Nostro [9]).

The rest of the paper is organized as follows. Section 2 is devoted to a brief review of the aspects of the Araneus approach that are needed as a background. Then, Section 3 illustrates the temporal extensions for the models we use in our process and Section 4 the methodology with the associated tool, with the help of an example. Finally, in Section 5 we briefly sketch possible future developments.

2   The Araneus Models and Methodology

The Araneus approach (Merialdo et al. [4]) focuses on data-intensive Web sites and proposes a design process (with an associated tool) that leads to a completely automatic generation of the site extracting data from a database. The design process is composed of several steps each of which identifies a specific aspect in the design of a Web site. Models are used to represent the intensional features of the sites from various points of view:

1. the Entity Relationship (ER) model is used to describe the data of interest at the conceptual level (then, a translation to a logical model can be performed in a standard way, and is indeed handled in a transparent way by the associated tool);
2. a "navigational" variant of the ER model (initially called NCM and then N-ER) is used to describe a conceptual scheme for the site. The main constructs in this model are the major nodes, called *macroentities*, representing atomic units of information, which often consolidate concepts from the ER model (one or more entities/relationships), and navigation paths, expressed as *directed relationships*;
3. a logical scheme for the site is defined using the Araneus Data Model (ADM), in terms of *page schemes*, which represent common features of pages of the same "type" with possibly nested attributes, whose values can come from usual domains (text, numbers, images) or be links to other pages.

The design methodology (sketched in Figure 1, see Atzeni et al. [7]), supported by a tool called Homer (Merialdo et al. [4]), starts with conceptual data design, which results in the definition of an ER scheme, and then proceeds with the specification of the navigation features, macroentities and directed relationships (that is, a N-ER scheme). The third step is the description of the actual structure of pages (and links) in terms of our logical model, ADM.

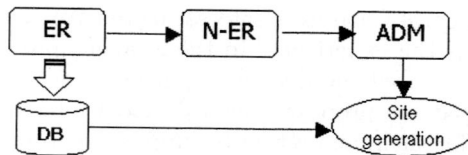

Fig. 1. The Araneus design process

Three simple schemes for the Web site of a University department, to be used in the sequel for comments, are shown in Figures 2, 3, and 4, respectively.

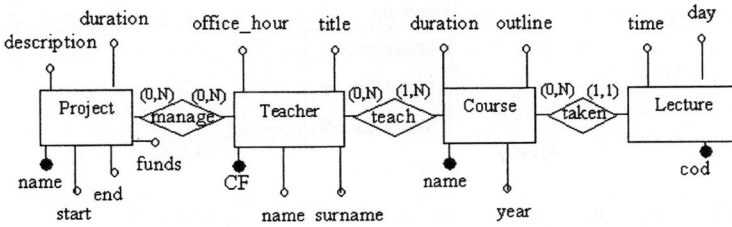

Fig. 2. The example ER schema

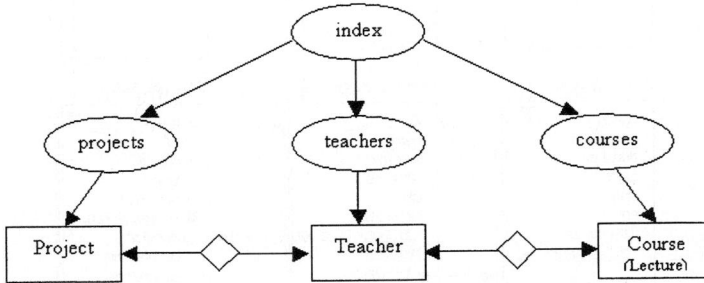

Fig. 3. The example N-ER schema

A fourth step is the specification of the presentation aspects, which are not relevant here. In the end, since all the descriptions are handled by the tool and the various steps from one model to the other can be seen as algebraic transformations, the tool is able to generate, in an automatic way, the actual code for pages, for example in JSP or in plain HTML, with access to a relational database built in a natural way from the ER scheme.

3   Models for the Management of Temporal Aspects of Web Sites

We believe that there is a need for the representation of temporal aspects at various levels during the design process, and therefore in each of our models, by means of features that are coherent with the focus of the phase of the development process the model is used in.

We propose a development process that follows the same path as we discussed in Section 2, with some extensions. and the tool we are implementing supports all phases as well. We start with brief comments on the models we use for describing our data, which follow known extensions from the temporal database literature, and then illustrate the conceptual and logical hypertext models.

3.1   Models for the Representation of Data

The temporal extension for the conceptual data model refers to standard proposals in the literature for temporal E-R models (see Gregersen and Jensen [10] for

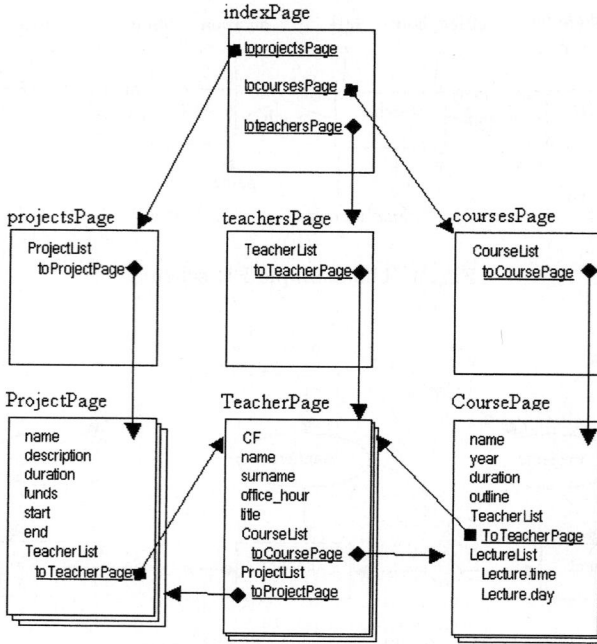

Fig. 4. The example ADM schema

a survey). In the data conceptual design phase, temporal features are added to the ER scheme, by indicating which are the entities, relationships and attributes for which the temporal evolution is of interest. The temporality can be specified either for an entity (or a relationship) as a whole or for single attributes; if an entity is temporal, then it is not allowed (as not needed) to assign the temporality property to its individual attributes. By specifying an entity as temporal, the designer indicates interest in maintaining a history about its life-cycle, and especially about its creation instant and its possible deletion instant, as well as about changes for the values of all and each its attributes.

The database used to handle the data for our Web sites is relational, as in the Araneus approach, with temporal features added to it (if using our tool the temporal features are generated automatically and the developer need not have any specific competence). If an entity is defined as temporal, when a change is applied to one or more of its attributes (at the same time), the system stores the tuple as a whole. If one single attribute is marked as temporal, its history is maintained independently from the other attributes.

3.2 Temporal Aspects of Web Sites at a Conceptual Level

The two models we use to describe the structure of a Web site refer to different levels: the N-ER model considers concepts whereas ADM refers to the actual structure of pages. The same disticntion applies to their temporal extensions.

The temporal N-ER model allows the specification of whether versions have to be managed for the concepts (macroentities and directed relationships) of interest for the site, and how.

More precisely, it allows the definition of the temporality feature for each of its concepts (macroentities, direct relationships, and attributes). A concept can be defined as temporal if its origin in the ER scheme has at least a temporal component, but not necessarily vice versa. For example, a macroentity can be defined as temporal if it involves temporal elements from the T-ER model; however, we could have macroentities that are not defined as temporal even if they involve temporal elements, for example because the temporality is not relevant within the macroentity itself (indeed, Web sites often have redundancy, so an attribute or an entity of the ER scheme could contribute to various macroentities, and, even if temporal, it need not be temporal in all those macroentities).

For each temporal element, a major facet is relevant here: which version(s) are of interest from the conceptual point of view? Currently, we consider this as a choice from a set of alternatives, such as (i) the last version with a timestamp; (ii) versions at a given granularity (to be specified by the designer); (iii) all versions.

3.3   Temporal Aspects of Web Sites at a Logical Level

The logical design of a temporal Web site has the goal of refining the description specified by a temporal N-ER scheme, by introducing all the details needed at the page level: how concepts are organized in pages and how the versioning of temporal elements is actually implemented. The temporal extension of ADM (hereinafter *T-ADM*) includes all the features of ADM (and so allows for the specification of the actual organization of attributes in pages and the links between them), and those of the higher level models (the possibility of distinguishing between temporal and non-temporal page schemes, and for each page, the distinction between temporal and non-temporal attributes; since the model is nested, this distinction is allowed at various levels in nesting, with some technical limitations), and some additional details, on which we concentrate. A major choice here is the implementation of the versioning requirement specified at the conceptual level. Out of the three cases (i)-(iii) mentioned at the end of the previous section, the non-trivial ones are the second and the third, which offer the same alternatives, as follows:

- A first possibility is that the various versions are included together in the same page, each annotated with the respective validity interval.
- A second alternative is to separate the "current value" from the previous versions, correlated by means of links. The current value could be associated with the date of the last change, whereas the versions could have validity intervals.
- A completely different organization is the "time-based selective navigation," where, for a page, the user selects the instant of interest and sees the corresponding version (and then can navigate over pages as of that date).

Additional features allow for the emphasization of recent changes (on a page or on pages reachable via a link).

The above features are expressed in T-ADM by means of a set of constructs, which we now briefly illustrate.

LAST MODIFIED This is a special, predefined attribute used to represent the date/time (at the granularity of interest) of the last change applied to a temporal element. This is a rather obvious, and widely used technique, but here we want to have it as a first class construct offered by the model (and managed automatically by the support tool) and also we think it should be left to the site designer to decide which are the pages and/or attributes it should be actually used for, in order to be properly informative but to avoid overloading.

VALIDITY INTERVAL This is another standard attribute that can be associated with any temporal element.

TIME POINT SELECTOR This is a major feature of the model, as it is the basis for the time-based selective navigation. It can be associated with pages and with links within them, in such a way that navigation can proceed with reference to the same time instant; essentially, in this way the user is offered the site with the information valid at the selected instant.

TARGET CHANGED This feature is used to highlight a link when the destination is a page that includes temporal information which has recently (according to a suitable metric: one day, one week, or whatever the designer chooses) changed. This property can be used in association with LAST MODIFIED to add the time the modification has been applied. The TARGET CHANGED feature is illustrated in Figure 5: a Department page (source) has a list of links to teacher pages (target). In a teacher page the office hours have been modified. When the user visits the department page he is informed which teacher pages have been modified (and when) so he can follows the link to check what is new. The example refers to just one source and one target page, but things may become more interesting when we consider non-trivial hypertextual structures: this gives the opportunity to propagate this kind of information through a path that leads to the modified data (see Figure 6). When a new lecture is introduced, then both the teacher and the department page are informed (and highlight the change) so the user can easily know which are the site portions with modified data.

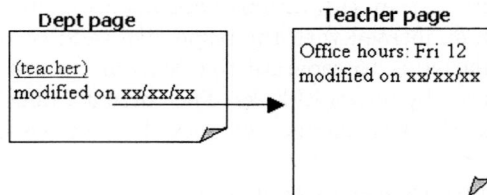

Fig. 5. The TARGET CHANGED feature

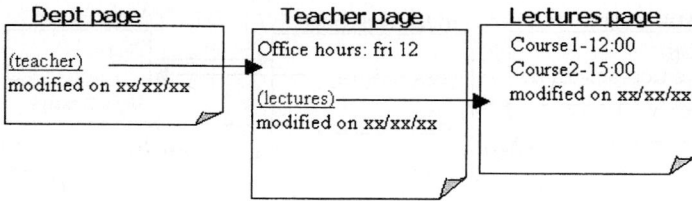

Fig. 6. The TARGET CHANGED feature along a path

REVISION LIST This feature allows for the specification that all versions of a temporal element are shown in the same page as a list of revisions.

LINK TO VERSIONS This is a special type of link that has as a target a VERSION STRUCTURE (to be illustrated shortly), handling the versions of a temporal element called. It is used when the designer chooses to have just the last version in the main page and the others held in other pages.

VERSION STRUCTURE These are "patterns" for pages and page schemes, used to organize the different versions of a temporal element and referred to by the LINK TO VERSIONS attribute. There are various forms for this construct involving one or more pages:

— SIMPLE VERSION STRUCTURE: a single page presenting all the versions for the temporal element with timestamps.

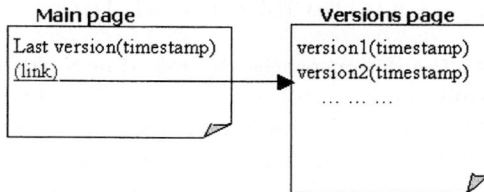

Fig. 7. The SIMPLE VERSION STRUCTURE pattern

— LIST VERSION STRUCTURE: an "index" page with a list of links labelled with the validity intervals that point to pages showing the particular versions and include links back to the index.
— CHAIN VERSION STRUCTURE: this is a list of pages each of which refers to a specific version. It is possible to scan versions in chronological order, by means of the "previous" and "next" links available in each page.
— SUMMARY VERSION STRUCTURE: similar to the previous case but the navigation between versions is not chronological. Each version page has a list of links that works as an index to all versions.

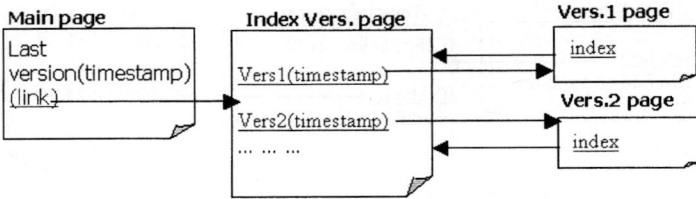

Fig. 8. The LIST VERSION STRUCTURE pattern

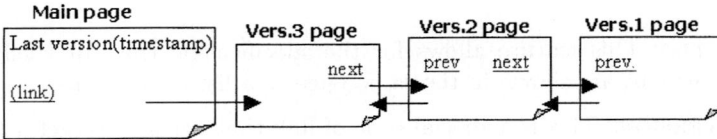

Fig. 9. The CHAIN VERSION STRUCTURE pattern

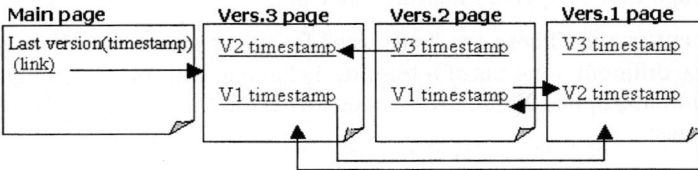

Fig. 10. The SUMMARY VERSION STRUCTURE pattern

4   The T-Araneus Methodology and Tool

Let us now exemplify the design process by referring to the example shown in Section 2 which, despite being small, allows us to describe the main issues in the methodology. We also sketch how the tool we are implementing supports the process itself. Rather than showing a complete example, where it would be heavy to include all the temporal features, we refer to a non-temporal example, and comment on some of its temporal extensions.

The first step is to add temporal features to the snapshot ER schema. Let us assume that the requirements specify that we need: (i) to know the state (with all attribute values) of the entity *Project* when a change is applied to one or more attribute values; and (ii) to keep track of the modifications on the *office hour* and *title* attributes for the *Teacher* entity. The first point means that the whole *Project* entity needs to be temporal, whereas for the second point, indeed, the designer has to set the temporality only for the attributes *office hour* and *title* in the *Teacher* entity.

In Figure 11 we illustrate the portion of interest of the resulting T-ER schema: the elements tagged with **T** are those choosen as temporal.

With respect to the conceptual design of the navigation, we have already shown in Section 2 the overall N-ER scheme. Let us concentrate here on the

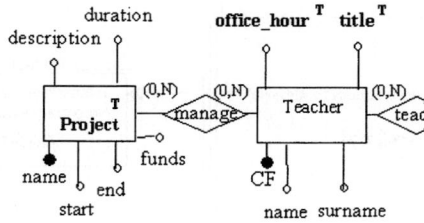

Fig. 11. The example T-ER schema

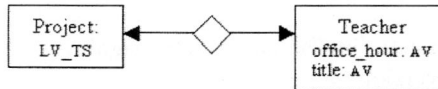

Fig. 12. Temporal features in the N-ER model

temporal features: at this point it is possible to choose how to manage versions for each macroentity and attribute. We have defined the *Project* entity as temporal in the T-ER model so the temporal database will handle the modifications but we don't want the site to show all versions so we choose here to have only the last version with a timestamp in the *Project* macroentity. It will be possible in the future to change this choice and add the projects version handling in the site simply changing this property. For the temporal attributes in *Teacher* we want all the versions to be managed. In Figure 12 the temporal N-ER scheme is shown with explicit indication of the version management choices (with the following codes: AV: all versions; LV: last value; LV_TS: last value with timestamp; VG: versions at a given granularity).

Let us then consider the logical design. As we have specified in Section 2, a standard ADM scheme can be automatically generated as an algebraic transformation based on the conceptual models (it can then be restructured if needed). During this automatic generation, for each temporal element in a page scheme, it is possible to specify how to present versions starting from the choices made in the N-ER scheme. On the basis of the requirements, we could decide to handle versions for *CoursePage* by means of a CHAIN VERSION STRUCTURE. For the *TeacherPage* page scheme we could choose to handle versions in different ways for the various attributes: all versions in the main page for the *title* attribute and just the last version in the main page for the *office hour* attribute with an associated SIMPLE VERSION STRUCTURE page presenting all versions.

The T-ADM page scheme for the *TeacherPage* is illustrated in Figure 13. The *office hour* attribute is associated with the LAST MODIFIED information and a LINK TO VERSION that point to the SIMPLE VERSION STRUCTURE page scheme presenting all the versions each with a VALIDITY INTERVAL. The simple attribute *title* has instead been transformed into a REVISION LIST due to the designer

TeacherPage

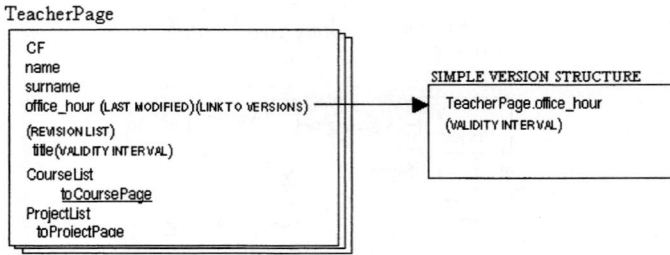

Fig. 13. A T-ADM page scheme

choice to have all versions in the same page; again, a VALIDITY INTERVAL is associated with each version value.

At the end of the design process, the tool can be used to generate the actual site, which can be static (that is, plain HTML) or dynamic (JSP); actually some of the features (such as the time point selector) are allowed only in the dynamic environment.

Currently, there is an implementation of the tool that is complete with respect to the development process even some of the features of the models are not available.

5   Future Works

In this paper we have focused on one of the aspects of interest for the management of temporal evolution: the content. Other dimensions are obviously of interest for real-world, complex Web sites, and we plan to consider them in the near future. They include: (i) the presentation; (ii) the hypertext structure; (iii) the database structure. Among them, the most challenging is probably the last one: as we consider data intensive Web sites, the hypertext structure is obviously strongly related to the database structure so it could be very important to keep track of the schema evolution. If you change the ER schema (and, as a consequence, the underlying database schema), for example deleting an entity and a relationship, it can result in a change in the hypertext structure and/or the presentation. Essentially, this would be a variation of a maintenance problem, with the need to keep track of versions.

Another issue we are considering is the relationship between a temporal Web site and an associated Content Management System (CMS). Indeed, the management of the temporal database could be delegated to the CMS: updates to data would generate histories if the involved data is defined as temporal. This would completely support the management of the temporal data of interest if updates are allowed only via the CMS. We are working on a prototype for a "data-intensive" CMS to be used in conjunction with the T-Araneus tool.

References

1. Atzeni, P., Mecca, G., Merialdo, P.: To weave the Web. In: VLDB'97, Proceedings of 23rd International Conference on Very Large Data Bases, August 25-29, 1997, Athens, Greece, Morgan Kauffman, Los Altos (1997) 206–215
2. Ceri, S., Fraternali, P., Bongio, A., Brambilla, M., Comai, S., Matera, M.: Designing Data-Intensive Web Applications. Morgan Kauffman, Los Altos (2002)
3. Fernandez, M., Florescu, D., Levy, A., Suciu, D.: Declarative specification of Web sites with Strudel. VLDB Journal **9** (2000) 38–55
4. Merialdo, P., Atzeni, P., Mecca, G.: Design and development of data-intensive web sites: The araneus approach. ACM Trans. Inter. Tech. **3** (2003) 49–92
5. Jensen, C., Snodgrass, R.: Temporal data management. IEEE Transactions on Knowledge and Data Engineering **11** (1999) 36–44
6. Snodgrass, R.: Developing Time-Oriented Database Applications in SQL. Morgan Kauffman, Los Altos (1999)
7. Atzeni, P., Merialdo, P., Mecca, G.: Data-intensive web sites: Design and maintenance. World Wide Web **4** (2001) 21–47
8. Merialdo, P., Atzeni, P., Magnante, M., Mecca, G., Pecorone, M.: Homer: a model-based case tool for data-intensive web sites. In: Proceedings of the 2000 ACM SIGMOD International Conference on Management of Data, May 16-18, 2000, Dallas, Texas, USA, ACM (2000) 586
9. Atzeni, P., Del Nostro, P.: T-Araneus: Management of temporal data-intensive Web sites. In: Int. Conf. on Extending Database Technology (EDBT 2004), Crete, Lecture Notes in Computer Science 2992. Springer-Verlag (2004) 862–864
10. Gregersen, H., Jensen, C.: Temporal entity-relationship models – a survey. IEEE Transactions on Knowledge and Data Engineering **11** (1999) 464–497

Evolution of XML-Based Mediation Queries
in a Data Integration System

Bernadette Farias Lóscio[1] and Ana Carolina Salgado[2]

[1] Faculdade 7 de Setembro
Av. Maximiniano da Fonseca, 1395, Edson Queiroz, 60811-020 Fortaleza - CE, Brasil
{berna@fa7.edu.br}
[2] Centro de Informática - Universidade Federal de Pernambuco
Av. Professor Luis Freire s/n, Cidade Universitária, 50740-540 Recife - PE, Brasil
{acs@cin.ufpe.br}

Abstract. One of the main challenges in data integration systems is the maintenance of the mappings between the mediation schema and the source schemas.
In a dynamic environment, such mappings must be flexible enough in order to
accommodate new data sources and new users' requirements. In this context,
we address a novel and complex problem that consists in propagating a change
event occurring at the source level or at the user level into the mediation level.
We propose an incremental approach to develop the mediation schema and the
mediation queries based on the evolution of the data source schemas and the
evolution of the users' requirements. The proposed approach allows the mediation level to evolve incrementally and modifications can be handled easier increasing the system flexibility and scalability.

1 Introduction

In recent years, the problem of integrating data from heterogeneous and autonomous
data sources has received a great deal of attention from the database research community. This problem consists in providing a uniform view of these data sources, called
mediation schema, and defining a set of mediation queries which compute each object
in the mediation schema.

Previous work on data integration can be classified according to the approach used
to define the mediation mappings between the data sources and the global schema [5,
12]. The first approach is called global-as-view (GAV) and requires that each object
of the global schema should be expressed as a view (i.e. a query) on the data sources.
In the other approach, called local-as-view (LAV), mediation mappings are defined in
an opposite way; each object in a given source is defined as a view on the global
schema.

One of the main problems in the context of the GAV approach is the maintenance
of the mappings between the mediation and the source schemas. Each change at the
source schema level may lead to the reconsideration and possibly the change of all
mediation queries. In [2] we present a solution to the problem of mediation queries
maintenance for data integration systems which adopt the relational model as the
common data model. In this paper, we extended such approach in order to manage the
evolution of XML-based mediation queries.

S. Wang et al. (Eds.): ER Workshops 2004, LNCS 3289, pp. 402–414, 2004.

The statement of this evolution problem is as follows: given a change event occurring at the source level or at the user level, how to propagate this change into the mediation queries. In this context, mainly two kinds of evolution have to be dealt within a mediation-based system: i) *the evolution of the user needs* - it may consist in adding, removing or modifying an user requirement. These changes impact the mediation schema by adding, modifying or deleting an element from the mediation schema. If these changes can be reflected in the mediation queries, the modifications on the mediation schema are committed; otherwise the user is informed that his or her new requirements cannot be satisfied, and ii) *the evolution of the data source schemas* - if a change occurs in a source schema, it has to be propagated to the mediation queries. The later are modified if the source elements on which they were defined are modified or when a source element is added or deleted. In this paper we deal both with the evolution of the user needs and the evolution of the data source schemas instead of just the evolution of data source schemas as proposed in [2].

The mediation queries evolution process was developed as part of a data integration system [6], which adopts the GAV approach. Due to its flexibility to represent both structured and semi-structured information, XML [3] is used by the system as the common language to data exchange and integration.

This paper is organized as follows. Section 2 describes the problem of propagating users' requirements and data source schemas changes to the mediation schema and the mediation queries. Section 3 introduces the X-Entity model. Section 4 presents some definitions. Section 5 briefly describes the process of generation of mediation queries. Section 6 presents some rules proposed for propagating the data source schemas changes and the users' requirements changes into the mediation level. Section 7 discusses some related works and Section 8 presents our conclusions.

2   Propagation of Schema Changes to the Mediation Queries

In this work, we address the problem of how to maintain the consistency of the mediation schema and the mediation queries when users' requirements or data sources' schemas change. We deal with this problem by considering a specific context of data integration, where a mediation schema (Fig. 1) represents the reconciliation between users' requirements and the data sources' capabilities. Thus, users pose queries in terms of the mediation schema, rather than directly in terms of the source schemas. Each element in the mediation schema is virtual, i.e. it is not actually stored anywhere, and it is computed from the integration of the distributed data.

In our approach, the mediation schema and the data source schemas are defined in the XML Schema language [4]. The XML schema is used to validate the local data returned by the data sources as well as the integrated data returned by the mediator in response of a user query. Although being very useful for these tasks, an XML schema is not suitable for tasks requiring knowledge about the semantics of the represented data. For such tasks, as generation and maintenance of mediation queries, the system needs a high level description. To provide a high-level abstraction for information described in an XML schema we proposed a conceptual data model, called X-Entity model [7] (detailed in section 3), an extension of the Entity-Relationship (ER) model. The main concept of the X-Entity model is the entity type, which represents the structure of XML elements composed by other elements and attributes.

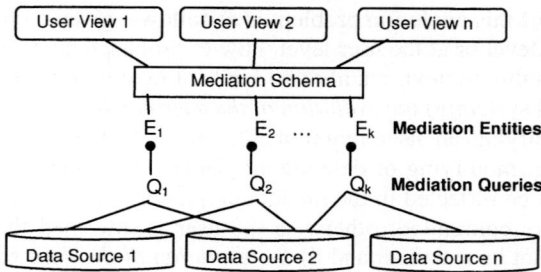

Fig. 1. Mediation schema definition

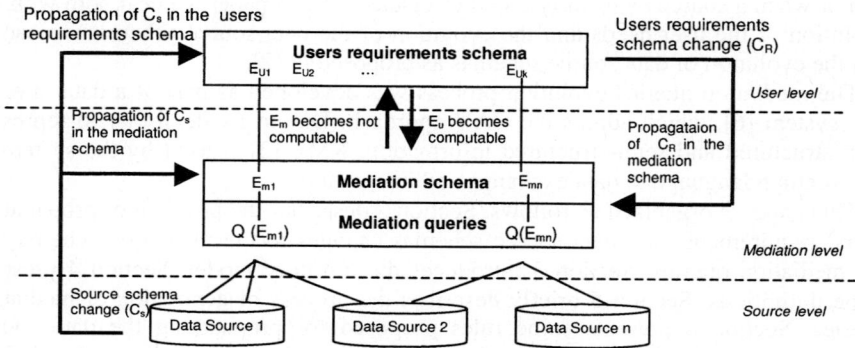

Fig. 2. Propagation of data source schemas changes and users' requirements changes

The mediation schema contains all the entities needed to answer the user queries, which can be computed from the data sources. Each entity E_i in the mediation schema is associated with a mediation query Q_i that computes the entity E_i over the set of data sources. Based on the mediation queries, previously defined, the system computes at run-time the answer for a user query by simply executing the necessary mediation queries and combining their results.

In this paper, we propose a solution to the problem of managing the evolution of mediation queries in dynamic environments. Fig. 2 describes the impact of schema changes in the mediation level. As we may observe, the addition of new data sources and the modifications in the data source schemas are changes that must be propagated both to the user level and the mediation level. In the first case, the propagation involves an analysis of existing users' requirements to identify relevant source entities to compute user entities (E_u), i.e., entities participating in the users' requirements schema. The propagation of data source schemas changes to the mediation level consists mainly in changing the mediation queries when the changes raised at the source level still allow the computation of the mediation entities (E_m). However, some mediation entities may become no longer computable concerning to the changes raised at the source level. Therefore, in these cases, the mediation entities must be removed from the mediation schema.

Concurrently with the evolution of the data source schemas, the users' requirements may continue to change. The evolution of the users' requirements schema originates directly change operations in the mediation schema. If these changes can be reflected in the mediation queries, the modifications on the mediation schema are committed; otherwise, the user is informed that his or her new requirements cannot be satisfied. This occurs when the requirement is not computable from the data available in the data sources. Each change raised in the mediation schema may lead to the re-definition of some mediation queries or the generation of new ones.

3 X-Entity Model

In this section, we describe the X-Entity model [7], which provides a high-level abstraction for information described in an XML schema. In [6] we present in details how to generate an X-Entity schema from a schema defined in the XML Schema language. In the following, we introduce the X-Entity model constructs and the X-Entity schema change operations.

3.1 X-Entity Concepts

An X-Entity schema S is denoted by $S = (E, R)$, where E is a set of entity types and R is a set of relationship types.

Entity Type: an entity type E, denoted by $E(\{A_1,...,A_n\},\{R_1,...,R_m\})$, is made up of an entity type name E, a set of attributes $A_1,...,A_n$ and a set of relationship types $R_1,...,R_m$. An entity type represents a set of elements with a complex structure, composed by attributes and other elements (called subelements). An instance of an entity type represents a particular element in the source XML document. Each entity type has attributes $\{A_1,...,A_n\}$ that describe it. An attribute A_i represents either an attribute or a subelement, which is not composed by other elements or attributes. Each attribute A_i is associated with a domain, denoted $Dom(A_i)$, which specifies its value set. A_i is also associated with a cardinality, denoted $Card(A_i)=(min, max)$, which specifies the minimum and the maximum number of instances of A_i that can be related to an instance of E.

Containment Relationship Type: a containment relationship type between two entity types E_1 and E_2, specifies that each instance of E_1 contains instances of E_2. It is denoted by $R(E_1, E_2,(min,max))$, where R is the relationship name and (min, max) defines the minimum and the maximum number of instances of E_2 that can be associated with an instance of E_1. Each entity type E may be associated with one or more containment relationships, which describes the element-subelement relationship between E and other entity types.

Reference Relationship Type: a reference relationship between two entity types E_1 and E_2, denoted by $R(E_1, E_2,\{A_{11},...,A_{1n}\}, \{A_{21},...,A_{2n}\})$, specifies that the entity type E_1 references the entity type E_2. $\{A_{11},...,A_{1n}\}$ and $\{A_{21},...,A_{2n}\}$ represent the referencing attributes between entities of E_1 and E_2 such that the value of A_{1i}, $1\leq i \leq n$, in any entity of E_1 must match a value of A_{2i}, $1\leq i \leq n$, in some entity of E_2.

Fig. 3 presents an example of an X-Entity schema. The entity type *movie* has two attributes: *title* and *actor*. The attribute *title* is a key attribute of the entity type *movie*. The attribute *actor* is multivalued, which means that an element *movie* may have multiple occurrences of the subelement *actor*. In this example, there is one containment relationship, which specifies that an instance of *movie* has at least one subelement *director* and may have unlimited occurrences of the subelement *director*.

Fig. 3. Example of an X-Entity schema

Since XML may be used to represent both structured and semi-structured data, X-Entity schemas are flexible, i.e., instances of the same entity type may have different structures. To represent this, both attributes and containment relationships are associated with participation constraints, which represent the occurrence constraints of elements and attributes in an XML document.

3.2   X-Entity Schema Change Operations

To specify modifications performed in the mediation schema or in the local schemas we defined a set of X-Entity schema change operations, which are summarized in Table 1. In the following consider: i) *E.A* is the set of attributes of the entity type *E* and ii) *E.R* is the set of relationships of the entity type *E*. The removal or the addition of a data source can be represented in terms of a sequence of change operations, for example, the addition of a source can be viewed as a set of *add_entity* operations.

Table 1. X-Entity schema change operations

Change Operation	Definition
add_entity(*E*, *S*)	Adds a new entity type *E* into the schema *S*
remove_entity(*E*, *S*)	Removes an existing entity type *E* from the schema *S*
add_attribute(*E*, *A*)	Adds the attribute *A* into *E.A*
remove_attribute(*E*, *A*)	Removes the attribute *A* from *E.A*
add_contains_rel(*E*, *R*)	Adds the containment relationship *R* into *E.R*
remove_contains_rel(*E*, *R*)	Removes the containment relationship *R* from *E.R*
add_reference_rel(*E*, *R*)	Adds the reference relationship *R* into *E.R*
remove_reference_rel(*E*, *R*)	Removes the reference relationship *R* from *E.R*
add_relationship(*R*,*S*)	Adds a new relationship type *R* into the schema *S*
remove_relationship(*R*,*S*)	Removes an existing relationship type *R* from the schema *S*

4   Some Definitions

Our approach is widely based on the existence of metadata describing individual sources and the mediation schema, and on correspondence assertions [11] specifying relationships between the mediation schema and the source schemas. In [6] we identify various types of correspondence assertions to formally describe the relationships between the concepts of different X-Entity schemas. In the following, we describe just the correspondence assertions relevant to this work.

☐ *Entity correspondence assertions*: specify the relationship between entity types of distinct schemas.

Definition 1 (Entity correspondence assertion): Let E_1 and E_2 be entity types. The correspondence assertion $E_1 \cong E_2$ specifies that E_1 and E_2 are semantically equivalent, i.e, they describe the same real world concept and they have the same semantics.

For each pair of semantically equivalent entity types we specify the correspondences between their subentities and attributes using subentity correspondence assertions and attribute correspondence assertions. A subentity of an entity type E is an entity type E' which is associated with E through a containment relationship $R(E, E', (min, max))$.

☐ *Subentity correspondence assertions*: specify the correspondence among subentities of semantically equivalent entity types.

Definition 2 (Subentity correspondence assertion): Let $R_1(E_1, E', (min, max))$ and $R_2(E_2, E'', (min, max))$ be two containment relationships, such that $E_1 \cong E_2$ and $E' \cong E''$. The correspondence assertion $E_1.R_1.E' \cong E_2.R_2.E''$ specifies that the subentity E' of the entity type E_1 is semantically equivalent to the subentity E'' of the entity type E_2.

☐ *Attribute correspondence assertions*: specify the correspondence among common attributes of semantically equivalent entities.

Definition 3 (Attribute correspondence assertion): Let A_1 be an attribute of the entity type E_1 and A_2 an attribute of the entity type E_2, such that $E_1 \cong E_2$. The correspondence assertion $E_1.A_1 \cong E_2.A_2$ specifies that the attributes A_1 and A_2 are semantically equivalent (correspond to the same concept in the real world).

☐ *Path correspondence assertions*: specify special types of correspondences between attributes and subentities of semantically equivalent entity types having different structures. To define a path assertion, we have to define the concepts: link and path.

Definition 4 (Link): Let X_1 and X_2 be elements of an X-Entity schema (an element can be an entity type, a containment relationship type or an attribute), $X_1.X_2$ is a link if:

i) X_2 is an attribute of the entity type X_1, or

ii) X_1 is an entity type participating in the relationship type X_2 (or vice-versa).

Definition 5 (Path): If $X_1.....X_n$ are elements of a schema, such that $\forall X_i, 1 \leq i \leq n$, $X_i.X_{i+1}$ is a link, then $X_1.X_2. ... .X_n$ is a path from X_1 (the path is directed from X_1 to X_n).

Definition 6 (Inverse path): If there is a path from X_1 denoted by $X_1.X_2. ... .X_n$, such that X_n is an entity type, then there is a path denoted by $(X_1.X_2. ... .X_n)^{-1}$, which is called the inverse path of $X_1.X_2. ... .X_n.$

Definition 7 (Path correspondence assertion): Let P_1 and P_2 be two paths:

Case 1: $P_1 = X_1.X_2.....X_n$ and $P_2 = Y_1.Y_2.....Y_m$, where $X_1 \cong Y_1$. The correspondence assertion $P_1 \cong P_2$ specifies that the entity types X_n and Y_m are semantically equivalent.

Case 2: $P_1 = X_1.X_2.....X_n.A$ and $P_2 = Y_1.Y_2.....Y_m.A'$, where $X_1 \cong Y_1$. The correspondence assertion $P_1 \cong P_2$ specifies that the attribute A of the entity type X_n and the attribute A' of the entity type Y_m are semantically equivalent.

Case 3: $P_1 = X_1.X_2.....X_n$ and $P_2 = (Y_1.Y_2.....Y_n)^{-1}$, where $X_1 \cong Y_n$. The correspondence assertion $P_1 \cong P_2$ specifies that the types X_n and Y_1 are semantically equivalent.

Case 4: $P_1 = X_1.X_2.....X_n.A_k$ and $P_2 = (Y_1.Y_2.....Y_n)^{-1}.A_k'$, where $X_1 \cong Y_n$. The correspondence assertion $P_1 \cong P_2$ specifies that the attribute A_k of the entity type X_n and the attribute A_k' of the entity type Y_1 are semantically equivalent.

Definition 8 (Atributte derivation path): A path assertion $E_i.A_k \cong P.A_k'$, where A_k and A_k' are attributes and $P = Y_1.Y_2.....Y_n$ or $P = (Y_1.Y_2.....Y_n)^{-1}$, specifies that $P.A_k'$ is a derivation path of the attribute A_k.

Definition 9 (Entity derivation path): A path assertion $E_i.R_j.E_i' \cong P$, where R_j is a relationship type, E_i and E_i' are entity types and $P = Y_1.Y_2.....E_i''$ or $P = (E_i''.Y_2.....Y_n)^{-1}$, specifies that P is a derivation path of the entity E_i'.

5   Mediation Queries Definition

In [8] we presented the process for generating XML-based mediation queries in the context of the GAV approach. As the mediation schema is represented by an X-Entity schema, the process of mediation queries generation consists in discovering a computing expression for each entity in the mediation schema.

More formally, we can say that defining a mediation query for a mediation entity E_m consists in decomposing E_m into n entity types $E_{p1},...,E_{pn}$ such that $E_m = E_{p1}\theta_1 E_{p2}\theta_2...\theta_n E_{pn}$, where θ_i is a binary operator and each entity type E_{pi} is derived using an expression $Exp(E_i)$ over a single source entity type E_i. An entity type E_i is called a relevant source entity to compute E_m and an entity type E_{pi} is called a mapping view (m-view).

At the end of the mediation queries generation process, each mediation entity will be associated with a mediation query, which is represented by an operation graph. The operation graph describes all information that is relevant to compute a given integrated view. Another important issue to be considered is that an operation graph can be incrementally created and it can be easily modified.

The use of operation graphs to represent mediation queries facilitates the identification of the mediation entities that are affected by a data source schema change and that, consequently, should be rewritten. A mediation entity will be affected by an entity source change if one of its mapping views will be affected by the data source change. Consequently, the problem of propagating the source changes or users' requirements changes into the mediation queries consists, first in propagating these changes into the mapping views, and second in modifying the mediation queries in order to take into account the modifications in the set of mapping views.

To illustrate the mediation queries definition consider the mediation schema and the source schemas presented below.

Mediation Schema S_{med} =
$(\{movie_m(\{title_m, director_m\},$
$\{movie_m\_actor_m\}),$
$actor_m(\{name_m, nationality_m\}, \{\}))$,
$\{movie_m\_actor_m(movie_m, actor_m, (1,N))\})$

Schema of data source S_1 =
$(\{movie_1(\{title_1, duration_1\},$
$\{movie_1\_actor_1, movie_1\_director_1\}),$
$actor_1(\{name_1, nationality_1\}, \{\})$
$director_1(\{name_1, nationality_1\}, \{\}))\},$
$\{movie_1\_actor_1(movie_1, actor_1, (1,N)),$
$movie_1\_director_1(movie_1, director_1, (1,N))\})$

Schema of data source S_2 =
$(\{movie_2(\{title_2, actor_2\}, \{movie_2\_director_2\}),$
$director_2(\{name_2, nationality_2\}, \{\}))\},$
$\{movie_2\_director_2(movie_2, director_2, (1,N))\})$

Schema of data source S_3 =
$(\{director_3(\{name_3, nationality_3\},$
$\{director_3\_movie_3\}),$
$movie_3(\{title_3, year_3\}, \{\}),$
$\{director_3\_movie_3(director_3, movie_3, (1,N))\})$

Fig. 4 presents an example of an operation graph that describes the possible operators to combine the mapping views associated with the mediation entity $movie_m$. The nodes of the operation graph represent the mapping views (V_{Movie1}, V_{Movie2} and V_{Movie3}) and the edges between these mapping views represent the mapping operators. At the end of the mediation queries generation process, the mediation query $Q(movie_m) = (E_{pMovie1} \cup E_{pMovie3}) \cup E_{pMovie2}$ is obatined from the operation graph G_{moviem}. More details about the mediation queries generation process can be found in [6].

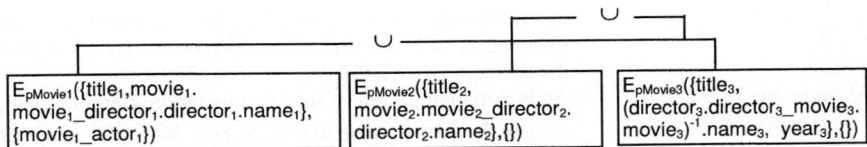

Fig. 4. Operation graph G_{moviem}

6   Mediation Queries Evolution

Data source schemas or users' requirements changes are propagated to the mediation queries through a set of event-condition-action (ECA) rules, which are triggered according to the different schema changes. The propagation process consists of two main tasks: first, the triggering, evaluation and execution of the rules in order to update the mapping views and the operations among them, and secondly, new mediation queries are generated using the modified operation graphs.

6.1 Propagation Primitives

We consider that each mediation entity E_m is associated with an operation graph $G_{Em}(M_{Em}, O_{Em})$ corresponding to the mediation query defining E_m, where M_{Em} is the set of nodes of G_{Em}, representing the set of m-views associated with E_m, and O_{Em} is the set of edges of G_{Em}, labeled with one of the mapping operators. If a change occurs in the data source schemas or in the mediation schema, some checking operations have to be performed on this graph to test if the mediation query associated with E_m are still valid. If not, a new mediation query has to be defined. Each entity E_m in the mediation schema is associated with two attributes, *MAPSET_STATUS* and *OPSET_STATUS*. These attributes have boolean values and represent the status of the set of mapping views associated with $E_m(M_{Em})$ and the set of candidate operations to combine these entities (O_{Em}). They determine if the set of mapping views associated with E_m (M_{Em}) and the set of candidates operations to combine these views (O_{Em}) were modified during the propagation of the schema changes. These two attributes are set to *False* at the beginning of the propagation process, and they will be set to *True* if a change occurs in the set of mapping views or the set of operations respectively.

The set of propagation primitives is presented in Table 2. These primitives are used in the mapping view evolution rules.

Table 2. Mediation queries propagation primitives

Mediation Level Propagation Primitive	Definition
search_operation(G_{Em})	Searches new operations for combining pairs of mapping views in the operation graph G_{Em}. If new operations are generated, then the attribute *OPSET_STATUS* is set to *TRUE*.
remove_operations(G_{Em}, V, A)	Removes all edges in the operation graph G_{Em} that become invalid because of the removal of the attribute *A* from the mapping view *V*. If at least one operation is removed, then the attribute *OPSET_STATUS* is set to *TRUE*.
add_mapping(V, G_{Em})	Adds a mapping view *V* into the operation graph G_{Em} and assigns the *TRUE* value to the attribute *MAPSET_STATUS*. If G_{Em} does not exist then this primitive creates G_{Em} from *V*.
remove_mapping(V, G_{Em})	Removes the mapping view *V* from the operation graph G_{Em} and assigns the *TRUE* value to the attribute *MAPSET_STATUS*.

6.2 Mapping Views Evolution Rules

Given a change represented by one of the schema change operations described in section 3.2, we will first propagate these changes in the set of mapping views associated with each entity of the mediation schema. To specify this propagation, we use event-condition-action (ECA) rules. Due to space limitations, in this section, we present just some rules to illustrate the propagation of schema changes into mapping views. In the following, consider:

- E_m: *mediation entity*
- *V: mapping view*
- E_i: *source entity*
- *ADP: attribute derivation path*
- *EDP: entity derivation path*
- G_{Em}: *operation graph of the mediation entity E_m*

- M_{Em}: *is the set of mapping views corresponding to the mediation entity E_m*
- $X(E_i)$: *is the set of mapping attributes between the source entity E_i and the other source entities which originated mapping views belonging to M_{Em}*
- $V = Exp(E_i)$: *specifies that the mapping view V is derived from the source entity E_i*

Propagation of data source schemas changes

We propose a set of rules to propagate data source schemas changes into mapping views. As described in section 4, a mapping view specifies how to compute attributes and subentities of a mediation entity E_m from a source entity E_i. A mapping view $V(\{X_1,...,X_n\},\{Y_1,...,Y_m\})$ is a special entity type where X_i is an attribute or an attribute derivation path and Y_i is a relationship or an entity derivation path. So, the propagation of a data source schema change into a mapping view V may result in the addition or removal of an attribute, relationship or derivation path from V. Each rule has a name and a parameter denoted E_m, which represents a mediation entity. To illustrate the propagation of a data source schema change consider the following rules (Rule 1 and Rule 2), which update the set of mapping views associated with the entity E_m in the mediation schema after the deletion of a source entity E_i. It is important to observe that may exist more than one rule for each one of the source schema change operations.

Rule 1(E_m)

> **When** remove_entity(E_i, S)
> **If** $\exists V \in M_{Em} \mid V = Exp(E_i)$
> **Then** remove_mapping(V, G_{Em})

Rule 1: the condition part of this rule checks if there is a mapping view V associated with E_m over E_i. To reflect the deletion of the local entity E_i, the corresponding mapping view V must be removed from the operation graph G_{Em}, along with all the operations involving the mapping view V. It is important to observe that just the mapping view corresponding to the source entity E_i is removed from the operation graph G_{Em}.

Rule 2(E_m)

> **When** remove_entity(E_i, S)
> **If** $\exists E_j \cong E_m \mid \exists V \in M_{Em} \wedge V = Exp(E_j)$
> **Then**
>> **If** $\exists \{ADP_1, ...,ADP_n\}$, where $ADP_t = (E'.....E_i.R_k. ... E_j)^{-1}.A_k \mid$
>>> $\forall t = 1,...,n, \exists E_m.A_m \cong ADP_t$ **or** $\exists \{EDP_1, ...,EDP_p\}$, where $EDP_t = (E'.....E_i.R_k. ... .E_j)^{-1} \mid$
>>> $\forall t = 1,...,p, \exists E_m.R_m.E_m' \cong EDP_t$
>> **Then** $\underline{V.A} := \underline{V.A} - \{ADP_1,...,ADP_n\}$, $\underline{V.R} := \underline{V.R} - \{EDP_1,...,EDP_p\}$
>>> remove_operations(G_{Em}, V, $\{ADP_1,...,ADP_n\}$)

Rule 2: the condition part of this rule checks if there is a source entity E_j that is semantically equivalent to E_m and if there is a mapping view V associated with E_m over E_j. In this case, the condition part of the rule also verifies if there are derivation paths from the entity type E_j that cannot be computed after the deletion of E_i. In this case, these derivation paths (attribute derivation path or entity derivation path) must be removed from the mapping view V. Then, it is necessary to verify if there are some mapping operators between the mapping view V and other mapping views V' which depend on one of the removed attribute derivation paths. In this case, the mapping operator must be removed.

Propagation of users' requirements changes

The propagation of users' requirements changes into mapping views is done through a set of mapping views evolution rules, which identifies information from the source entities relevant to the computation of the new requirements and performs the necessary modifications into the set of mapping views. Depending on the operation, the rules are evaluated over the whole set of source entities or over the mapping views associated with the mediation entity that is being modified. Each rule has a name and a parameter denoted V, which represents a mapping view or a parameter E_i which represents a source entity. To illustrate the propagation of a user requirement change consider the following rule (Rule 3), which identifies the mapping views relevant to compute a new mediation entity E_m. Rule 3 is evaluated for each source entity E_i. If E_i is semantically equivalent to E_m then the attributes, containment relationships and derivation paths of E_i, which are considered relevant to compute E_m, are identified. Such elements will compose the content of the new mapping view V, which will be inserted into the operation graph G_{Em}.

Rule 3(E_i)

> **When** add_entity($\underline{E_m}$, S_m)
>
> **If** $E_i \in \underline{S.E} \cong E_m$
>
> **Then** if $\exists \{A_1,...,A_n\} \in \underline{E_i.A} \mid \forall t = 1,...,n, \exists E_m.A_m \cong E_i.A_t$ **or**
>
> $\qquad \exists \{ADP_1, ...,ADP_n\}$,where $ADP_t = E_i. \ ... .A_k \vee ADP_t = (E'. \ ... .E_i)^{-1}.A_k \mid$
>
> $\qquad \forall t = 1,...,n, \exists E_m.A_m \cong ADP_t$ **or**
>
> $\qquad \exists \{R_1,...,R_k\} \in \underline{E_i.R} \mid \forall t = 1,...,n, \exists E_m.R_m.E_m' \cong E_i.R_t.E'$ **or**
>
> $\qquad \exists \{EDP_1, ...,EDP_p\}$, where $EDP_t = E_i. \ ... .E' \vee EDP_t = (E'. \ ... .E_i)^{-1} \mid$
>
> $\qquad \forall t = 1,...,n, \exists E_m.R_m.E_m' \cong EDP_t$
>
> **Then** $\underline{V_{E_i}.A}:=\{A_1,...,A_n\} \cup \{ADP_1,..., ADP_n\} \cup X(E_i)$, $\underline{V_{E_i}.R} :=\{R_1,...,R_p\} \cup \{EDP_1, ...,EDP_p\}$
>
> $\qquad$ add_mapping($V_{\underline{E_i}}$, G_{Em}), search_operation(G_{Em})

As the structure of XML data is more flexible than the structure of conventional data, the process of propagating a schema change into a set of XML-based mediation queries is more complex. Such propagation process consists mainly in updating the attributes, containment relationships and reference relationships of relevant mapping views rather than just updating attributes. One advantage of our approach is that we use path correspondence assertions to capture the correspondences between elements with different structures.

7 Related Works

The problem of mediation queries evolution is also discussed in other works [1, 9,10]. The work presented in [1] adopts an approach similar to ours for defining mediation queries. The algorithm to discover integration axioms is incremental, which means that when new sources are added, the system can efficiently update the axioms, but no details on how this could be achieved nor examples are given. In the case of deleting a source the algorithm must start from scratch. They use the LAV approach to define the mappings between the global model and the local sources. In [9] an approach is presented to handle both schema integration and schema evolution in heterogeneous database architectures. This approach is different from ours, because instead of defining mediation queries as our approach does, they use primitive transformations to automatically translate queries posed to the global schema to queries over the local schemas. In [10] modifications are directly executed in the mediation query definition rather than in the metadata that describes the mediation query. Moreover, a view must evolve just when a source schema change makes the view definition obsolete; i.e., just the cases of removal of relations or attributes are dealt with.

8 Conclusion

In this paper, we have presented the process of managing the evolution of XML-based mediation queries. Changes to mediation queries may be due to changes in the users' requirements or in data source schemas. The proposed solution was developed as part of a data integration system which adopts the GAV approach. A prototype is currently being implemented and some tests have been done using two real medical databases. The current version of the prototype includes the implementation of: data source schemas extraction, data source schema changes identification, X-Entity schemas generation, mediator schema generation (partially) and user queries decomposition. One problem to be investigated is the optimization of the synchronization between the update of the mapping views and the generation of mediation queries. As the propagation process is always running in background, other aspects concerning concurrency and failure recovery must also be investigated.

References

1. Ambite, J., Knoblock, C., Muslea, I., Philpot. A.: Compiling Source Description for Efficient and Flexible Information Integration, Journal of Intelligent Information Systems, Vol. 16, N. 2, (2001) 149-187
2. Bouzeghoub, M., Lóscio, B. F., Kedad, Z., Salgado, A.C.: Managing the Evolution of Mediation Queries, In Proceedings of the International Conference on Cooperative Information Systems (COOPIS 2003),Catane Italy, (2003) 22-37
3. Bray, T., Paoli, J., Sperberg-McQueen, C. M.: Extensible Markup Language (XML) 1.0, World Wide Web Consortium. Available at: http://www.w3.org/TR/REC-xml, October 2000
4. Fallside, D., C.: XML Schema Part 0: Primer, World Wide Web Consortium. Available at: http://www.w3.org/TR/xmlschema-0/, May 2001

 5. Halevy, Y.: Theory of answering queries using views, SIGMOD Record, Vol. 29, N. 4, (2000) 40-47
 6. Lóscio, B. F.: Managing the Evolution of XML-based Mediation Queries. PhD Thesis. UFPE, Cin – Centro de Informática, Recife Pernambuco Brazil, (2003)
 7. Lóscio, B. F., Salgado, A. C., Galvão, L. R.: Conceptual Modeling of XML Schemas, In Proceedings of Fifth International Workshop on Web Information and Data Management (WIDM 2003), New Orleans Louisiana USA, (2003) 102-105
 8. Lóscio, B. F., Salgado, A. C.: Generating Mediation Queries for XML-based Data Integration Systems, In Proceedings of XVIII Brazilian Symposium on Databases (SBBD 2003), Manaus Amazonas Brazil, (2003) 99-113
 9. Mcbrien, P., Poulovassilis, A.: Schema Evolution in Heterogeneous Database Architectures, A Schema Transformation Approach, In Proceedings of Conference on Advanced Information Systems Engineering (CaiSE 2002), Toronto Canada, (2002) 484-499
10. Nica, A., Rundensteiner, E. A.: View maintenance after view synchronization, In Proceedings of International Database Engineering and Application Symposium (IDEAS'99), (1999) 215-213
11. Spaccapietra, S. and Parent, C.: View integration: a step forward in solving structural conflicts, IEEE Transactions on Knowledge and Data Engineering, Vol. 6, N. 2, (1994)
12. Ullman, J. D.: Information integration using logical views, In Proceedings of 6th International Conference on Database Theory (ICDT'97), (1997) 19-40

Schema Versioning in Data Warehouses

Matteo Golfarelli[1], Jens Lechtenbörger[2], Stefano Rizzi[1], and Gottfried Vossen[2]

[1] DEIS, University of Bologna, Italy
[2] Dept. of Information Systems, University of Muenster, Germany

Abstract. As several mature implementations of data warehousing systems are fully operational, a crucial role in preserving their up-to-dateness is played by the ability to manage the changes that the data warehouse (DW) schema undergoes over time in response to evolving business requirements. In this paper we propose an approach to schema versioning in DWs, where the designer may decide to undertake some actions on old data aimed at increasing the flexibility in formulating cross-version queries, i.e., queries spanning multiple schema versions. After introducing an algebra of DW schema operations, we define a history of versions for data warehouse schemata and discuss the relationship between the temporal horizon spanned by a query and the schema on which it can consistently be formulated.

1 Introduction

Data Warehouses (DWs) are databases specialized for business intelligence applications, and can be seen as collections of *multidimensional cubes* centered on facts of interest for decisional processes. A cube models a set of *events*, each identified by a set of *dimensions* and described by a set of numerical *measures*. Typically, for each dimension a hierarchy of *properties* expressing interesting aggregation levels is defined. A distinctive feature of DWs is that of storing historical data, hence, a temporal dimension is always present.

Data warehousing systems have been rapidly spreading within the industrial world over the last decade, due to their undeniable contribution to increase the effectiveness and efficiency of decision making processes within business and scientific domains. Today, as several mature implementations of data warehousing systems are fully operational within medium to large contexts, the continuous evolution of the application domains is bringing to the forefront the dynamic aspects related to describing how the information stored in the DW changes over time from two points of view:

- *At the extensional level*: Though historical values for measures are easily stored due to the presence of temporal dimensions that timestamp the events, the multidimensional model implicitly assumes that the dimensions and the related properties are entirely static, which is clearly unrealistic.
- *At the intensional level*: The DW schema may change in response to the evolving business requirements: new properties and measures may become necessary, while others may become obsolete.

S. Wang et al. (Eds.): ER Workshops 2004, LNCS 3289, pp. 415–428, 2004.

Note that, in comparison with operational databases, temporal issues are more pressing in DWs since queries frequently span long periods of time; thus, it is very common that they are required to cross the boundaries of different versions of data and/or schema. Besides, the criticality of the problem is obviously higher for DWs that have been established for a long time, since unhandled evolutions will determine a stronger gap between the reality and its representation within the database, that will soon become obsolete and useless.

So far, research and DW vendors have mainly addressed changes at the extensional level (see [1, 2] for instance); schema versioning has only partially been explored and no dedicated commercial tools or restructuring methodologies are available to the designer. Thus, both an extension of tools and a support to designers are urgently needed. In this paper we propose an approach to schema versioning in DWs, specifically oriented to support the formulation of *cross-version queries*, i.e., queries spanning multiple schema versions. The main contributions are:

1. *Schema graphs* are introduced in order to univocally represent DW schemata as graphs (Section 2), and an algebra of graphs operations to determine new versions of a DW schema is defined (Section 3.1).
2. *Augmented schemata* are introduced in order to increase flexibility in cross-version querying (Section 3.2). The augmented schema associated with a version is the most general schema describing the data that are actually recorded for that version and thus are available for querying purposes.
3. The sequencing of versions to form *schema histories* in presence of augmented schemata is discussed (Section 3.3), and the relationship between the temporal horizon spanned by a query and the schema on which it can consistently be formulated is analyzed (Section 4).

1.1   Approach Overview and Motivating Example

In this section we introduce our approach based on a working example. Consider a schema S_0 modeling the shipments of parts to customers all over the world. A conceptual schema for the shipment fact is depicted in Fig. 1(a) using the DFM formalism [3]. The fact has two measures, QtyShipped and ShippingCostsDM, and five dimensions, namely Date, Part, Customer, Deal, and ShipMode. A hierarchy of properties is attached to each dimension; the meaning of each arc is that of a many-to-one association, i.e., a functional dependency.

Suppose now that, at $t_1 = 1/1/2003$, S_0 undergoes a major revision. In the new version S_1: (a) The temporal granularity has changed from Date to Month; (b) A classification into subcategories has been inserted into the part hierarchy; (c) A new constraint has been modeled in the customer hierarchy, stating that sale districts belong to nations; (d) The incentive has become independent of the shipment terms. Then, at $t_2 = 1/1/2004$, another version S_2 is created as follows: (a) Two new measures ShippingCostsEU and ShippingCostsLIT are added; (b) The ShipMode dimension is eliminated; (c) A ShipFrom dimension is added; (d) A descriptive attribute PartDescr is added to Part. The conceptual schemata for S_2 is depicted in Fig. 1(b).

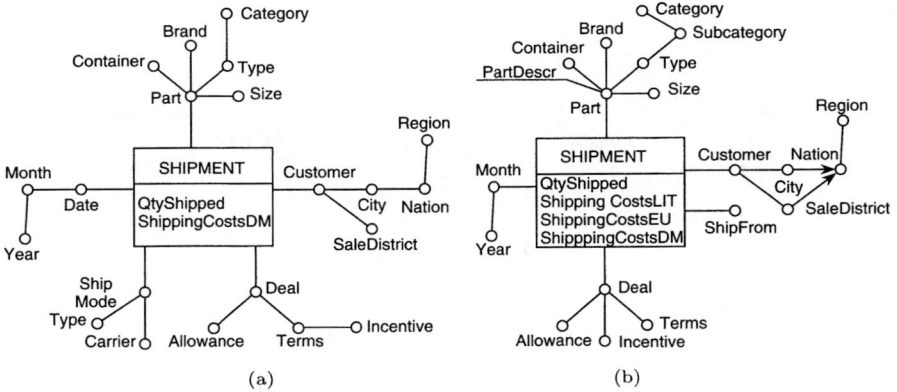

Fig. 1. Conceptual schemata for two versions of the shipment fact: S_0 (a) and S_2 (b)

In our approach, the schema modifications occurring in the course of DW operation lead to the creation of a *history of schema versions*. All versions are available for querying purposes, and those relevant for a particular analysis scenario may either be chosen explicitly by the user or implicitly by the query subsystem. In order to improve cross-version querying, when creating a new schema version the designer may choose to create *augmented schemata* that extend previous schema versions to reflect the current schema extension, both at the schema and the instance level. To be more precise, let S be the current schema version and S' be the new version. Given the differences between S and S', a set of possible *actions* on past data are proposed to the designer; they may entail checking past data for additional constraints or inserting new data based on user feedback. The set of actions the designer decides to undertake leads to defining and populating an augmented schema S^{AUG}, associated with S, that will be used, instead of S, to determine if a given query q spanning the validity interval of S is correct, i.e., if the data required by q can be consistently queried under the schema specified by q. Importantly, S^{AUG} is always an extension of S, i.e. the instance of S can be computed as a projection of the instance of S^{AUG}.

Consider for instance the schema modification operation that introduces attribute Subcategory, performed at time $t_1 = 1/1/2003$ to produce version S_1. For all parts shipped after t_1 (including both parts introduced after t_1 and parts already existing before t_1), a subcategory will clearly have to be defined, so that queries involving Subcategory can be answered for all shipments from t_1 on. However, if the user is interested in achieving cross-version querying on years 2002 and 2003, i.e. if she asks to query even old data (shipments of parts no more existing at t_1) on Subcategory, it is necessary to: (1) define an augmented schema for S_0, denoted S_0^{AUG}, that contains the new attribute Subcategory; (2) migrate old data entries from S_0 to S_0^{AUG}; and (3) assign the appropriate values for Subcategory to old data entries in S_0^{AUG}. This process will allow queries involving Subcategory to be answered on old data via the instance of S_0^{AUG}. Note

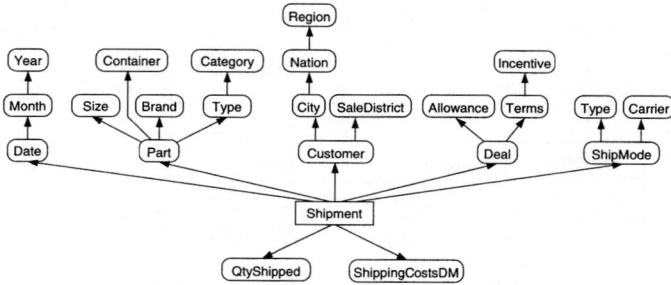

Fig. 2. Schema graph S_0

that, while the first two steps will entirely be managed by the versioning system, the last one requires further input from the designer.

2   Schema Graphs

Following standard notation for relational databases, capital letters from the beginning (ending) of the alphabet denote single (sets of) attributes. We restrict our attention to simple FDs, i.e., FDs of the form $A \to B$, and in this setting we use F^* to denote the usual graph-theoretic transitive closure of F (where sets of simple FDs are interpreted as directed graphs). We also recall that a set F of FDs is *canonical* if: (a) $(X \to Y \in F) \Longrightarrow (|Y| = 1)$; (b) $(X \to A \in F) \wedge (Y \subsetneq X) \Longrightarrow (Y \not\to A)$; and (c) $(F' \subsetneq F) \Longrightarrow (F' \not\equiv F)$. For every set F of FDs there is at least one *canonical cover*, i.e., a canonical set of FDs equivalent to F [4].

In order to talk about versioning, we first have to fix a representation for DW schemata. In this section, we introduce a graph-based representation for schemata called *schema graph*, which captures the core of multidimensional models such as the DFM and allows to define schema modifications by means of four elementary graph manipulations, namely adding and deleting nodes and arcs, and to analyze the schema versioning problem in a simple and intuitive setting.

Definition 1 (Schema Graph). *A schema graph is a directed graph* $S = (\hat{U}, F)$ *with nodes* $\hat{U} = \{E\} \cup U$ *and arcs* F, *where*

1. *E is called* fact node *and represents a placeholder for the fact itself;*
2. *U is a set of attributes (including properties and measures);*
3. *F is a canonical set of simple FDs defined over $\{E\} \cup U$;*
4. *E has only outgoing arcs, and there is a path from E to each attribute in U.*

The schema graph for the shipment fact in Fig. 1(a) is shown in Fig. 2.

Adopting a canonical form for representing the FDs in schema graphs has the important advantage of providing a non-redundant, i.e., *compact* representation. On the other hand, in order to obtain unique results for schema modification operations, we also need to make sure that we are dealing with a *uniquely determined* representation. In the following we will discuss how a set F of simple

FDs can be put in a uniquely determined canonical form, denoted with F^- and called *reduced form*.

We begin by observing that, for *acyclic* sets of simple FDs, canonical covers are uniquely determined [5] and can be computed via transitive reduction [6]. Thus, if F is acyclic, F^- can be defined as the transitive reduction of F. On the other hand, in real-world DW schemata, cycles in F may occur since either (a) a property is associated with one or more univocal descriptions – e.g., the technical staff may refer to products by their codes while sale agents may use product names; or (b) two measures may be derivable from each other – e.g., Euros can be transformed to Italian Liras by applying a constant conversion factor. The approach we follow to *uniquely* select a canonical form for cyclic sets of FDs is formalized in [5]; for space reasons, here we report only an intuitive description.

Consider a schema graph $S = (\hat{U}, F)$, where F is neither necessarily canonical nor acyclic. First, we define an equivalence relation $\equiv_F$ on $\hat{U}$ as: $A \equiv_F B$ iff $(A \to B \in F^*) \wedge (B \to A \in F^*)$ for $A, B \in \hat{U}$. Then, we consider the acyclic directed graph where each node is one equivalence class X induced by $\equiv_F$ and an arc goes from X to Y if there are attributes $A \in X$ and $B \in Y$ such that $A \to B \in F$. The transitive reduction of this graph, denoted S', is acyclic and uniquely determined (as transitive reduction is unique for acyclic graphs). Now, let a total order on $\hat{U}$ be given (e.g., user-specified or system-generated based on some sorting criterion such as attribute name). The reduced form for F, F^-, is composed as follows: (1) for each arc (X, Y) in S', one FD from the minimum attribute in X to the minimum attribute in Y; (2) for each equivalence class $X = \{A_1 \ldots A_n\}$ where $A_1 < \ldots < A_n$, a set of FDs $\{A_1 \to A_2, \ldots, A_{n-1} \to A_n, A_n \to A_1\}$. It is possible to prove that F^- is a transitive reduction of F and a canonical cover of F [5]. In the remainder of the paper, we will always consider schema graphs where the set of FDs is in reduced form, and we will use the terms *schema* and *schema graph* interchangeably.

Example 1. Consider the shipment fact in Fig. 1(b), where Part has an equivalent property PartDescr and shipping costs are expressed in EU, LIT, and DM. The schema graph based on the total order induced by attribute names is shown in Figure 3.

Given a schema graph $S = (\hat{U}, F)$ and a set $X \subseteq \hat{U}$, the *projection of F to X* is defined as $\pi_X(F) := \{A \to B \in F^* \mid AB \subseteq X\}$. Based on [7] and [5] it is possible to show that, if $E \in X$, projection is closed on schema graphs.

3   Versioning

In Section 3.1 we define four modification operations that manipulate schema $S = (\hat{U}, F)$, namely $\mathtt{Add_A}()$ to add a new attribute, $\mathtt{Del_A}()$ to delete an existing attribute, $\mathtt{Add_F}()$ to add an FD involving existing attributes, and $\mathtt{Del_F}()$ to remove an existing FD. Then, in Section 3.2 we show how a new version is created by applying a sequence of schema modification operations, and we explain

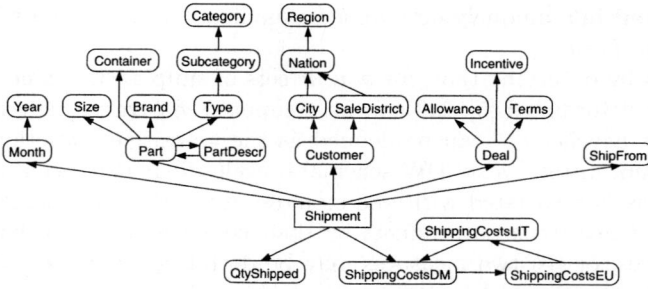

Fig. 3. Schema graph S_2

how augmented schemata are derived based on actions undertaken by the designer. Finally, in Section 3.3 we consider schema versioning based on sequences of versions, which influence the history of schemata and augmented schemata.

3.1 Schema Modification Operations

For each operation $M(Z)$ (where M is $\mathtt{Add_A}$ or $\mathtt{Del_A}$ and Z is an attribute, or M is $\mathtt{Add_F}$ or $\mathtt{Del_F}$ and Z is an FD), we define the new schema $\mathtt{New}(S, M(Z))$ obtained when applying M on current schema S.

Definition 2. *Let* $S = (\hat{U}, F)$ *be a schema graph and* A *be a fresh attribute[1]. Then* $\mathtt{New}(S, \mathtt{Add_A}(A)) := (\hat{U} \cup \{A\}, F \cup \{E \to A\})$.

Definition 3. *Let* $S = (\hat{U}, F)$ *be a schema graph and* $A \in U$ *be an attribute. Then* $\mathtt{New}(S, \mathtt{Del_A}(A)) := (\hat{U} \setminus \{A\}, \pi_{\hat{U} \setminus \{A\}}(F)^-)$.

Definition 4. *Let* $S = (\hat{U}, F)$ *be a schema graph and* $f = A_1 \to A_2$ *be an FD involving attributes in* U. *Then* $\mathtt{New}(S, \mathtt{Add_F}(f)) := (\hat{U}, (F \cup \{f\})^-)$

Definition 5. *Let* $S = (\hat{U}, F)$ *be a schema graph and* $f = A_1 \to A_2$ *be an existing FD in* F, *where* $A_1 \neq E$. *Then* $\mathtt{New}(S, \mathtt{Del_F}(f)) := (\hat{U}, (F \setminus \{f\} \cup \{E \to A_2\} \cup \{A_1 \to A_3 \mid (\exists A_3 \in U) \, A_2 \to A_3 \in F\})^-)$.

It is possible to prove that the four operators above are closed, i.e., they yield schema graphs.

Example 2. The sequence of operations applied to add Subcategory to parts is $\mathtt{Add_A}(\mathsf{Subcategory}), \mathtt{Add_F}(\mathsf{Type} \to \mathsf{Subcategory}), \mathtt{Add_F}(\mathsf{Subcategory} \to \mathsf{Category})$.

---

[1] A *fresh* attribute is an attribute that does not occur in S and never occurred in the past.

3.2   Augmented Schemata

We call a *version* a schema that reflects the business requirements during a given time interval. Thus, a version is populated with the events occurring during that interval and can be queried by the user. A new version is the result of a sequence of modification operations, which we call *schema modification transaction*, or simply *transaction*. In analogy to the usual transaction concept, intermediate results obtained after applying single schema modifications are invisible for querying purposes. Moreover, intermediate schemata are neither populated with events, nor are they associated with augmented schemata.

In temporal databases, versioning is generally associated with the definition of some *data migration strategies* used to consistently move data from the old version to the new one. Data migration is not described in this paper for space reasons; we only briefly note that:

- Events occur at a particular moment in time, so an event occurring at time t conforms to the version that is valid at t and no data migration is necessary.
- Instances of hierarchies are generally active during time intervals (e.g., a part is active from the time it is first shipped to the time it is declared obsolete), so their lifetime may span several versions. Thus, if a new version S is created at time t, for each hierarchy instance that is active both before and after t it may be necessary to migrate it to S. For instance, if S adds a new FD f between already existing attributes, migration requires to check if f holds for all active instances and possibly to enforce it by modifying, under the user guidance, the values of one or both attributes involved in f.

A transaction produces (1) a new version and (2) an augmented schema for each previous version, all of which are (either physically or virtually) populated with data and are visible to the querying process. We emphasize that usage of augmented schemata in the course of the querying process is handled transparently to the DW users, i.e., users are only aware of regular versions but not of augmented schemata. Concerning (1), given a version S, let the sequence of operations $M_1(Z_1), \ldots, M_h(Z_h)$ be the executed transaction. Then, the new version S' is defined by executing the modification operations one after another, i.e., $S' = \text{New}(S_h, M_h(Z_h))$, where $S_1 = S$ and $S_i = \text{New}(S_{i-1}, M_{i-1}(Z_{i-1}))$ for $i = 2, \ldots, h$. In the remainder of this section we address (2) by illustrating how augmented schemata are created at the end of transactions to increase flexibility in cross-version querying.

We anticipate that we only consider augmentations for previous versions and in response to operations that *add* attributes or FDs: in fact, the utility of augmenting the current version with deleted attributes/FDs seems highly questionable[2]. As motivated in Section 1.1, such schema extensions can be reflected

---

[2] Designers can, of course, delete attributes and FDs from schema versions: the point is that such deletions do not lead to augmentations. Besides, while in a "pure" versioning approach attributes would be deleted only logically, for practical purposes designers might want to delete data physically to free disk space.

Table 1. Actions associated to each new attribute/FD

Element	Condition		Action
$A \in \text{Diff}_\text{A}(S, S')$	$(E \to A) \in F'$	A is measure	estimate values for A
		A is dimension	disaggregate measure values
	$(E \to A) \notin F'$	A is derived measure	compute values for A
		A is property	consistently add values for A
$f \in \text{Diff}_\text{F}(S, S')$	-		check if f holds

on previous versions by undertaking appropriate actions, which enable flexible cross-version querying.

Let $S' = (\hat{U}', F')$ be the new version obtained by applying a transaction to version $S = (\hat{U}, F)$, and let $\text{Diff}_\text{A}(S, S') := U' \setminus U$ (set of new attributes) and $\text{Diff}_\text{F}(S, S') := F' \setminus F^*$ (set of new FDs). The set of *potential* actions that may be undertaken by the designer on past data in order to increase querying flexibility only depends on $\text{Diff}_\text{A}(S, S')$ and $\text{Diff}_\text{F}(S, S')$. The possible actions are reported in Table 1 and defined as follows:

1. *Estimate values for A:* A new measure A has been added. The designer may provide values for A for past events, typically by deriving an estimate based on the values of the other measures. For instance, if a new measure Discount is added to the shipment fact, and the discount applied depends on the shipped quantity bracket, its values for past events may be easily estimated from measure QtyShipped.

2. *Disaggregate measure values:* A new dimension A has been added. The designer may disaggregate past events by A according to some business rule or by adopting a statistical interpolation approach that exploits multiple summary tables to deduce the correlation between measures [8]. For instance, a likely reason for adding dimension ShipFrom is that, while in the past all shipments were made from the same warehouse w, now they are occasionally made from other warehouses: in this case, all past events can be easily related to w.

3. *Compute values for A:* A derived measure A has been added (e.g. Shipping-CostsEU); by definition, the values of A for past events can be computed by applying some known computation to another measure.

4. *Consistently add values for A:* A new property A has been added. The designer may provide values for A in such a way that all specified FDs are satisfied. For instance, when Subcategory is added to the part hierarchy, then each part type must be associated with exactly one subcategory and all types included in each subcategory must belong to the same category.

5. *Check if f holds:* A new FD f has been added. In order to augment f on old version S the designer needs to check whether f also holds for S. To support the designer, the system might perform the necessary (but not sufficient) check that f really holds by inspecting the augmented instance. For instance, when SaleDistrict $\to$ Nation is added, the system may check

that no sale district including customers from different nations exists. If the check performed by the system fails, f cannot be augmented.

Among the potential actions according to Table 1, the designer may choose to perform some of them but to ignore others: in particular, she will decide to undertake a given action only if, considering the business requirements, she believes that the space/time overhead implied by the action is counterbalanced by the increased querying flexibility. Since *only chosen actions contribute to augmentation*, the overall data volume increase is under the designer's control.

Now, to formalize the notion of schema augmentation, consider a schema modification transaction that starts from version $S = (\hat{U}, F)$ and creates the new version $S' = (\hat{U}', F')$. As explained in the next section, in the presence of *histories* of versions, augmentations may be back-propagated even to older versions. Thus, we assume that for every schema S_i there is already an augmented schema S_i^{AUG}, which is initially identical to S_i but may be augmented in response to every transaction. To this end, we next define the augmentation operation $\mathtt{Aug}(S_i^{AUG}, S, S')$ that (further) augments the augmented schema $S_i^{AUG} = (\hat{U}_i, F_i)$ based on the actions chosen by the designer in response to the schema change from S to S'.

Let $\widehat{\mathtt{Diff}_A}(S, S')$ and $\widehat{\mathtt{Diff}_F}(S, S')$ be the subsets of $\mathtt{Diff}_A(S, S')$ and $\mathtt{Diff}_F(S, S')$, respectively, including only the attributes and FDs whose related actions have been performed by the designer. We note that all attributes occurring in FDs of $\widehat{\mathtt{Diff}_F}(S, S')$ must be contained in $\hat{U}_i \cup \widehat{\mathtt{Diff}_A}(S, S')$, as only those FDs can be augmented whose attributes occur in the augmented schema. Then the *new augmented version for* S_i is defined as follows:

$$\mathtt{Aug}(S_i^{AUG}, S, S') := (\hat{U}_i \cup \widehat{\mathtt{Diff}_A}(S, S'), (F_i \cup \pi_{\hat{U}_i \cup \widehat{\mathtt{Diff}_A}(S,S')}(\widehat{\mathtt{Diff}_F}(S, S')))^-)$$

Informally, the new augmentation for S_i is obtained by adding (1) all the new attributes for which the related actions have been undertaken and (2) all the new valid FDs.

Example 3. In the shipment example, initially we start from version S_0 where $S_0^{AUG} = S_0$. When new version $S_1 = (\hat{U}_1, F_1)$ is created from S_0, we have:

$$\mathtt{Diff}_A(S_0, S_1) = \{\mathsf{Subcategory}\}$$
$$\mathtt{Diff}_F(S_0, S_1) = \{\mathsf{SaleDistrict} \to \mathsf{Nation}, \mathsf{Type} \to \mathsf{Subcategory},$$
$$\mathsf{Subcategory} \to \mathsf{Category}\}$$

Thus, the actions the designer can undertake to augment S_0^{AUG} are: (1) provide values for Subcategory and (2) check if SaleDistrict $\to$ Nation holds on S_0. Assuming the designer decides to undertake both actions, that the FD SaleDistrict $\to$ Nation actually holds on S_0 and that subcategory values are added consistently with F_1, we have $\widehat{\mathtt{Diff}_A}(S, S') = \mathtt{Diff}_A(S, S')$ and $\widehat{\mathtt{Diff}_F}(S, S') = \mathtt{Diff}_F(S, S')$. Thus, the new augmented schema for S_0 is $S_0^{AUG} = \mathtt{Aug}(S_0, S_0, S_1) = S_1$.

3.3   Version Histories

A *history* is a sequence H of one or more triples representing versions of the form (S, S^{AUG}, t), where S is a version, S^{AUG} is the related augmented schema, and t is the start of the validity interval of S [3]:

$$H = (\ (S_0, S_0^{AUG}, t_0), \ldots, (S_n, S_n^{AUG}, t_n)\ ),$$

where $n \geq 0$ and $t_{i-1} < t_i$ for $1 \leq i \leq n$. Note that, in every history, for the last triple (S_n, S_n^{AUG}, t_n) we have $S_n^{AUG} = S_n$ as augmentation only enriches *previous* versions using knowledge of the current modifications.

Given version S_0 created at time t_0, the initial history is $H = ((S_0, S_0^{AUG}, t_0))$, where $S_0^{AUG} = S_0$. Schema modifications then change histories as follows. Let $H = ((S_0, S_0^{AUG}, t_0), \ldots, (S_{n-1}, S_{n-1}^{AUG}, t_{n-1}), (S_n, S_n^{AUG}, t_n))$ be a history, and let S_{n+1} be the new version at time $t_{n+1} > t_n$; then the resulting history H' is

$$H' = (\ (S_0, \texttt{Aug}(S_0^{AUG}, S_n, S_{n+1}), t_0), \ldots,$$
$$(S_n, \texttt{Aug}(S_n^{AUG}, S_n, S_{n+1}), t_n), (S_{n+1}, S_{n+1}^{AUG}, t_{n+1})\ ).$$

where $S_{n+1}^{AUG} := S_{n+1}$.

We point out that a schema modification might potentially change any or all augmented schemata contained in the history; e.g., adding a new FD at time $n + 1$, which has been valid but unknown throughout the history, may lead to a "back propagation" of this FD into every augmented schema in the history. Moreover, note that new augmentations of previous schemata are based on the *augmented* schemata as recorded in the history, not on the schemata themselves. Thus, augmentations resulting from different modifications are accumulated over time, resulting in augmented schemata whose information content – hence, potential for answering queries – is growing monotonically with every modification.

Example 4. Consider again the two schema restructurings described in Section 1.1, assuming that all actions are undertaken. The initial history is $((S_0, S_0, t_0))$. At time t_1, when S_1 is created, the history becomes $((S_0, S_1, t_0), (S_1, S_1, t_1))$. Then, at time t_2, when S_2 is created, the history becomes $((S_0, S_2, t_0), (S_1, S_2, t_1), (S_2, S_2, t_1))$.

4   Querying

In this section we discuss how our approach to versioning supports cross-version queries, i.e., queries whose temporal horizon spans multiple versions.

---

[3] In accordance with [9] we argue that there is no need to distinguish valid time from transaction time in the context of schema versioning. Thus, if a new version S' is created from S at (transaction) time t', the valid time of S' is $[t', +\infty]$, while the valid time of S ends at t'. In other words, we assume that the valid time is defined as an interval that starts upon schema creation time and extends until the next version is created.

Preliminarily, we remark that OLAP sessions in DWs are aimed at effectively supporting decisional processes, thus they are characterized by high dynamics and interactivity. A session consists of a sequence of queries, where each query q is transformed into the next one q' by applying an OLAP operator. For instance, starting from a query asking for the total quantity of parts of each type shipped on each month, the user could be interested in analyzing in more detail a specific type: thus, she could apply a drill-down operator to retrieve the total quantity of each part of that type shipped on each month. Then, she could apply the roll-up operator to measure how many items of each part were shipped on the different years in order to catch a glimpse of the trend. Hence, since OLAP operators mainly navigate the FDs expressed by the hierarchies in the multidimensional schema, specifying the version for query formulation in the OLAP context does not only mean declaring which attributes are available for formulating the next query q', but also representing the FDs among attributes in order to determine how q' can be obtained from the previous query q.

In this sense, the *formulation context* for an OLAP query is well represented by a schema graph. If the OLAP session spans a single version, the schema graph is the associated one. Conversely, when multiple versions are involved, a schema under which *all* data involved can be queried uniformly must be determined. In our approach, such a schema is univocally determined by the temporal interval T covered by the data to be analyzed, as the largest schema that retains its validity throughout T. In particular, since T may span different versions, we define an *intersection* operator, denoted by $\otimes$, for determining the common schema between two different versions.

Definition 6. *Let* $S = (\{E\} \cup U, F)$ *and* $S' = (\{E\} \cup U', F')$. *Then the* intersection *of* S *and* S', *denoted by* $S \otimes S'$, *is the schema defined as* $S \otimes S' = (\{E\} \cup (U \cap U'), (F^* \cap F'^*)^-)$.

Intuitively, the intersection between two versions S and S' is the schema under which data recorded under S or S' can be queried uniformly: in fact, it includes only the attributes belonging to both S and S', as well as their common FDs.

Given a history H and a (not necessarily connected) temporal interval T, we call the *span* of T on H the set $\mathtt{Span}(H, T) = \{S_i^{AUG} \mid (S_i, S_i^{AUG}, t_i) \in H \wedge [t_i, t_{i+1}[\cap T \neq \emptyset\}$ (conventionally assuming $t_{n+1} = +\infty$).

Definition 7. *Given a history* H *and a temporal interval* T, *the* common schema *on* H *along* T *is defined as* $\mathtt{Com}(H, T) = \bigotimes_{\mathtt{Span}(H,T)} S_i^{AUG}$.

Let q be the last query formulated, and T be the interval determined by the predicates in q on the temporal dimension (if no predicate is present then $T =]-\infty, +\infty[$). The formulation context for the next query q' is expressed by the schema graph $\mathtt{Com}(H, T)$. Note that the OLAP operator applied to transform q into q' may entail changing T into a new interval T'; in this case, the formulation context for getting a new query q'' from q' will be defined by $\mathtt{Com}(H, T')$.

Example 5. Let $H = ((S_0, S_0^{AUG}, t_0), (S_1, S_1^{AUG}, t_1), (S_2, S_2^{AUG}, t_2))$ be the history for the shipment fact, recall that we have $t_1 = 1/1/2003$ and $t_2 = 1/1/2004$,

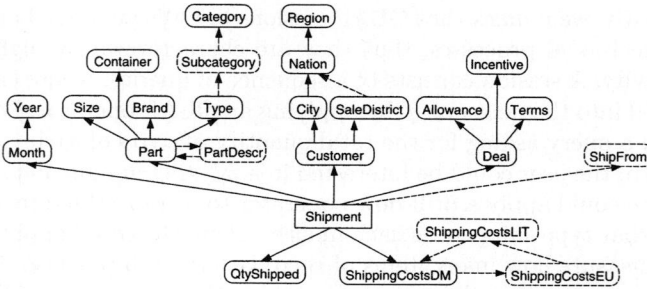

Fig. 4. Formulation contexts for the query in Example 5 without augmentation (in plain lines) and with augmentation (in plain and dashed lines)

and let $q =$ "*Compute the total quantity of each part category shipped from each warehouse to each customer nation since July 2002*". The temporal interval of q is $T = [7/1/2002, +\infty[$, hence $\mathtt{Span}(H, T) = \{S_0, S_1, S_2\}$. Fig. 4 shows the formulation context, defined by $S_0^{AUG} \otimes S_1^{AUG} \otimes S_2^{AUG}$, in two situations: when no augmentation has been made, and when all possible augmentations have been made. First of all, we observe that q is well-formulated only if ShipFrom has been augmented, since otherwise one of the required attributes does not belong to the formulation context. Then we observe that, for instance, (1) drilling down from Category to Subcategory will be possible only if subcategories and their relationship with categories have been established also for 2002 data; (2) drilling down from Nation to SaleDistrict will be possible only if the FD from sale districts to nations has been verified to hold also before 2003.

As to the querying interface, we argue that two approaches for identifying the version for querying, namely *implicit* and *explicit*, should be supported [10]. In this section we considered the implicit approach: given the time interval T of a query, the system computes the widest common schema associated to it. Conversely, in the explicit approach the user chooses a specific version for querying, and the system calculates the widest time interval T that preserves that schema. Obviously, the second approach is best suited for OLAP sessions that analyze data under a specific configuration of attributes and FDs (e.g., "*Compute the total quantity shipped for each month and each subcategory since subcategories have been introduced*").

5   Conclusions and Related Work

In this paper we have presented an approach towards DW schema versioning. Based on the standard graph operations of transitive closure and reduction, we have defined four intuitively appealing schema modification operations in the context of graphical DW schemata. We have shown how single schema modifications lead to a history of versions that contain augmented schemata in addition to "ordinary" schemata, and we have defined an intersection operator that allows to determine whether a given query, possibly spanning several versions, can be answered based on the information contained in augmented schemata. With reference to the terminology introduced in [11] our approach is framed as *schema versioning* since past schema definitions are retained so that all data may be accessed both retrospectively and prospectively through user-definable version interfaces; additionally, with reference to [12] we are dealing with *partial schema versioning* as no retrospective update is allowed to final users.

In the DW field, mainly four approaches to evolution/versioning can be found in the literature. In [13], the impact of evolution on the quality of the warehousing process is discussed in general terms. In [14] a prototype supporting dimension updates at both the extensional and intensional levels is presented. In [15], an algebra of basic operators to support evolution of the conceptual schema of a DW is proposed. In all these approaches, versioning is not supported and the problem of querying multiple schema versions is not mentioned. Finally, [16] proposes the COMET model to support schema evolution: though the problem of queries spanning multiple schema versions is mentioned, the discussion of how to map instances from one version to another is only outlined.

On the commercial side, the versioning problem has only marginally been addressed, for instance in the *Oracle Change Management Pack* [17] and in KALIDO [18]. In both cases, the possibility of formulating a single query on multiple databases with different schemata is not even mentioned.

References

1. Eder, J., Koncilia, C.: Changes of dimension data in temporal data warehouses. In: Proc. DaWaK. (2001) 284–293
2. Yang, J.: Temporal data warehousing. PhD thesis, Stanford University (2001)
3. Golfarelli, M., Maio, D., Rizzi, S.: The dimensional fact model: a conceptual model for data warehouses. IJCIS **7** (1998) 215–247
4. Maier, D.: The theory of relational databases. Computer Science Press (1983)
5. Lechtenbörger, J.: Computing Unique Canonical Covers for Simple FDs via Transitive Reduction. Technical report, Angewandte Mathematik und Informatik, University of Muenster, Germany. To appear on Information Processing Letters (2004)
6. Aho, A.V., Garey, M.R., Ullman, J.D.: The transitive reduction of a directed graph. SIAM Journal on Computing **1** (1972) 131–137
7. Lechtenbörger, J., Vossen, G.: Multidimensional normal forms for data warehouse design. Information Systems **28** (2003) 415–434
8. Pourabbas, E., Shoshani, A.: Answering Joint Queries from Multiple Aggregate OLAP Databases. In: Proc. 5th DaWaK, Prague (2003)
9. McKenzie, E., Snodgrass, R.: Schema evolution and the relational algebra. Information Systems **15** (1990) 207–232
10. Roddick, J., Snodgrass, R.: Schema versioning. In: The TSQL2 Temporal Query Language. Kluwer Academic Publishers (1995) 425–446
11. Jensen, C.S., Clifford, J., Elmasri, R., Gadia, S.K., Hayes, P.J., Jajodia, S.: A consensus glossary of temporal database concepts. ACM SIGMOD Record **23** (1994) 52–64

12. Roddick, J.: A survey of schema versioning issues for database systems. Information and Software Technology **37** (1995) 383–393
13. Quix, C.: Repository Support for Data Warehouse Evolution. In: Proc. DMDW. (1999)
14. Vaisman, A., Mendelzon, A., Ruaro, W., Cymerman, S.: Supporting dimension updates in an OLAP server. In: Proc. CAiSE. (2002) 67–82
15. Blaschka, M.: FIESTA - A framework for schema evolution in multidimensional databases. PhD thesis, Technische Universitat Munchen, Germany (2000)
16. Eder, J., Koncilia, C., Morzy, T.: The COMET Metamodel for temporal data warehouses. In: Proc. CAiSE. (2002) 83–99
17. Oracle: Oracle change management pack. Oracle Technical White Paper (2000)
18. Kalido: Kalido dynamic information warehouse - a technical overview. KALIDO White Paper (2004)

Facilitating Database Attribute Domain Evolution Using Mesodata

Denise de Vries and John F. Roddick

School of Informatics and Engineering
Flinders University of South Australia
PO Box 2100, Adelaide, South Australia 5001
{Denise.deVries,roddick}@infoeng.flinders.edu.au

Abstract. Database evolution can be considered a combination of schema evolution, in which the structure evolves with the addition and deletion of attributes and relations, together with domain evolution in which an attribute's specification, semantics and/or range of allowable values changes. We present a model in which mesodata – an additional domain definition layer containing domain structure and intelligence – is used to alleviate and in some cases obviate the need for data conversion or coercion. We present the nature and use of mesodata as it affects domain evolution, such as when a domain changes, when the semantics of a domain alter and when the attribute's specification is modified.

1   Introduction

The way we view and deal with information evolves. In paper-based manual systems this evolution did not present a great problem – we turned the page and ruled it up differently, renamed columns, used different terminology and proceeded to store our information. We could always review what had been stored historically by viewing the information exactly as it had been recorded. The static nature of this method means that notations that were recorded retained their semantics in context, that is the headings and layout of the form/paper imparted the structures and conventions as well as the values themselves. We, the human, translated and transformed the information when we retrieved it. It was simple. It was also so time consuming that much of what we now consider to be basic tasks, such as sorting, aggregating, summarising and reporting was infeasible.

The development of RDBMS and automated systems put the layout and *form* into the unchanging metadata and gave us *record once* systems. Database technology has provided the power to store and manipulate information in a variety of ways, however we still cannot reproduce the simplicity of dealing with information evolution as we used to. A major introduced problem that has not yet been completely solved is that of attribute domain evolution. For example, if one were searching for a particular value, time consuming though it was, subtle differences in data values were captured because the searcher understood the

S. Wang et al. (Eds.): ER Workshops 2004, LNCS 3289, pp. 429–440, 2004.

domain and therefore included or excluded records based on their own knowledge of the domain. For instance, a database query searching through historical medical records for an illness matching 'rubella' generally uses a string comparison only, thus the string 'german measles' would not be retrieved even though semantically it matched.

There have been many techniques developed to deal with database evolution but none can currently deal with all aspects of evolution and few of them deal specifically with the problem of attribute domain evolution. Middleware, using various approaches, has been used to alleviate evolution problems by translating, transforming or coercing data and metadata. In this work we use a new approach that introduces complex data structures with embedded intelligence that lie between metadata and data – *mesodata* [1]. Mesodata allows domains to be engineered so that attributes can be defined to possess additional intelligence and structure and thus reflect more accurately ontological considerations, including changes in the domain itself.

Attribute domain evolution refers to the evolution of the valid range of values that a database attribute (field) may store and the semantics they infer. For example, an integer field of 4 bytes can store values in the range of -2,147,483,648 to 2,147,483,647 whereas a float field of 4 bytes has a range of negative values from -3.402823E+38 to -1.401298E-45 and positive values from 1.401298E-45 to 3.402823E+38. The domain has changed even though the storage requirement has not altered. Domain evolution can be broadly categorised into three types;

Attribute Representation Change: expansion or contraction of field, for example, CHAR(15) to CHAR(20) or vice versa, change of base type: integer to float, numeric to character, character to enumerated list.

Domain Constraint Change: the possible range of values that may be recorded has changed without the metadata changing or the currently stored data changing, for example, the minima and/or maxima change. The new constraints may, or may not, be applied retrospectively.

Perception (Meaning) Change: the semantics of the data change, for example, Reference 116Q15 no longer is interpreted as 'Burbridge Road' and is now 'Sir Donald Bradman Drive', however, both interpretations are required for historical purposes.

Currently when a schema changes two events typically occur - the application is modified and recompiled to deal with the changes and the data is converted to the new format, either by *strict*, *lazy* or *no* conversion [2]. Lazy conversion performs data conversion only when data are accessed and they are still recorded with superseded formats (or values), no conversion is done if the data are not accessed. The advantage of this approach is that only the data that are used are converted and the whole database does not need to be locked or taken off-line to perform the conversion. The disadvantages are that a record of schema changes must be recorded and accessible and that every time data is accessed it must be checked to see if it conforms to the current schema. Until all data have been accessed there exist some that are invalid, incomplete or uncertain.

Strict conversion requires that as soon as there is a modification to the schema all data are converted to conform with the current definition. The advantage of this approach are that all data are consistent with the new schema. The disadvantage is that all applications interacting with the database must be stopped and the database locked while the conversion takes place. Depending upon the nature of the modifications this can take a long time. In addition, information is lost and changes cannot be reversed.

By adding a mesodata layer to the structure, some of the data would not need to be converted and the application itself may not need to change. Mesodata can store semantics as well as operators and operations in data structures other than base types.

2   Related Work

The primary goal for schema evolution in databases is to preserve the integrity of the data. Sjøberg [3] observed several reasons for schema change. These included that:

- people do not know in advance, or are not able to express, all the desired functionality of a large-scale application system. Only experience from using the system will enable the needs and requirements to be properly formulated.
- the application world is continually changing. A viable application system must be enhanced to accommodate these changes.
- often the scale of the task requires incremental design, construction and commissioning. This results in requirements to change the installed subsystems.

Sjøberg's case study, a health management system, revealed that schema changes were significant both during the six months of development and the twelve months after the system was operational. In the study, the changes covered the gamut of possibilities including each relation being changed, 139% increase in the number of relations, 274% increase in the number of fields, and 35% more additions than deletions. During the development phase (5 months) there were 65 changes (additions and deletions) to relations and 470 changes (additions and deletions) to attributes. During the operational phase (11 months) the corresponding numbers were 299 and 2324 respectively. In Sjøberg's study changes to the type/domain of an attribute was not captured in the statistics, however, in the last month of the study the changes to fields were 18 renamings, 4 changes of unique/non-nulls, 23 changes of length and 4 changes of representation. That is 31 changes at attribute level.

A consensus glossary [4] provides the definition that a database supports *schema evolution* if it allows modification to the schema without loss of extant data and that no support for previous schemas is required, whereas it supports *schema versioning* if it allows the querying of all data, both retrospectively and prospectively, through user-definable version interfaces. Roddick *et al.* [5] present a taxonomy of schema versioning issues with respect to the Entity-Relationship Model and the effects on the relational database model. The work discussed in this paper does not deal with all issues of schema evolution and versioning, but concentrates on the specifics of domain/attribute evolution. The pertinent evolutionary operations are:

- Expanding an attribute domain
- Restricting an attribute domain
- Changing the domain of an attribute

There has not been a great deal of research done in the field of evolving relational databases over the past few years, however work done in object-oriented database evolution, data warehousing, data integration and schema integration research has areas of relevance to the evolution of domains.

2.1   Information Capacity

Hull [6] and later Qian [7] provided formal approaches to evaluate the *information capacity* of schemata. The four relative information capacity measures between database structures as defined by Hull are, in progressively less restrictive order, calculus dominance, generic dominance, internal dominance and absolute dominance. These measures are used to evaluate the information capacity of two or more schemata by mathematically mapping between the schemata. An important point to note is that even when two schemas can be proved to have the same information capacity, it does not then follow that they are equivalent semantically. Qian's formalisation of Abstract Data Types (ADT) for schema transformations presents a slightly different notion of 'information preservation', which is strictly less restrictive than calculus dominance, strictly more restrictive than absolute dominance and incomparable to generic and internal dominance. These formal approaches are the foundation of later work into *schema equivalence* and *schema integration*.

Miller [8] describes *Equivalence* as the requirement that all data stored in one schema (S_1) can be accessed and updated through another schema (S_2),

- for queries the transformation function (f) must be total: $q(i_2) = q(f(i_1))$;
- to access all data f must be injective: i_1 must correspond to a unique i_2, a 1-1 cardinality;
- for updates f must be onto: $I(S_2)$;
- for equivalence $(S_1 \equiv S_2)$ there exists a bijective function: $f : I(S_1) \rightarrow I(S_2)$;

and *Dominance* as $S_1 \preceq S_2$ allows all data stored under schema S_1 to be queried through S_2,

- to access all data f must be injective: 1-1 cardinality;
- every instance of S_1 can be transformed to an instance of S_2 without loss of information;
- S_2 may hold more information.

Miller *et al.* [9, 8, 10] point out that whilst equivalent information capacity is a required condition, it is not sufficient to guarantee a natural correspondence between schemas and in practice database administrators rely on their own intuition when defining *transformations* between schemas. Schemas, in practice, contain constraints that define which instances of a schema are meaningful in a certain context. Their research in this area shows that deciding information capacity equivalence and dominance of schemas is an undecidable problem. As a result, they developed tests to evaluate equivalence and dominance more restrictively. These tests utilise a set of *schema transformations* that declare that Schema S_1 is dominated by schema S_2 if and only if there is a sequence of transformations that converts S_1 to S_2. These transformations use Schema Intension Graphs (SIG) data models to aid in understanding the relative information capacity of schemas containing constraints. The authors have developed algorithms for deciding equivalence of schemas with constraints. The SIG model must be data-centric rather than type-centric in order to reason about constraints on collections of entities rather than the internal structure of a single entity. (This approach ignores the problem of data type changes and the conflicts type changes present.) They define the Schema Translation problem as follows:

> Given two schemas one needs to know with respect to information capacity if each instance of the first schema can be represented as an instance in the second schema and whether the translation can be reversed?

Table 1 identifies possible conflicts that can occur between semantically equivalent schemas with regard to attributes and values. Capacity and equivalence, therefore, is not sufficient, data integration must considered to take into account query and view requirements.

2.2   Schema Integration

Xu and Poulovassilis [11], when addressing the integration of deductive databases, considered both the extensional and intensional parts of the component databases for integration. The Common Data Model (CDM) uses a binary relational Entity-Relationship model with subtyping to integrate the extensional parts. The authors proposed a semi-automatic method which requires only the declaration of the relationships between schema constructs to perform the integration. For the purposes of their model, they defined a database as the quintuple:

```
< Schema, Extensional Database, Intensional Database,
    Constraint Database, Procedural Database >
```

Each of these sets is integrated in turn and in that order. The schemas and the extensional databases are represented by directed graphs and from those graphs correspondences between nodes are declared. These mappings are then used to perform the integration into a CDM. The intensional database is integrated by integrating the rules from the component databases according to their denotational semantics employing a method of comparing the semantics of the rules.

Table 1. Schematic Conflicts between Attributes and Values

	Value	Attribute
Value	Domain conflicts between schema S_1 and schema S_2, such as expression conflicts, data unit conflicts, precision conflicts	The values in S_1 are used as attributes in S_2
Attribute		Different definitions for semantically equivalent attributes — 1-1 one attribute used to model the same information in each schema, such as naming conflict, integrity constraint conflict, data representation conflict. — 1-N and N-N different numbers of attributes used to model the same information in each schema. — The N-N conflict is a generalisation of the 1-N conflict

2.3   Schema Transformation

Transformation consists of the tasks schema conforming, schema merging and schema restructuring. McBrien *et al.* [12] present a formal framework again using the CDM for ER schema transformation in which they have defined a set of primitive transformations based on schema equivalence. This is achieved by formalising a database instance as a set of sets containing entity type names, subtypes, attribute names and associations. These are also integrated in order, however, in this work there is a distinction made between transformations which require knowledge of the instances in the database and those that do not.

Continuing this work [13] and combining schema integration and schema evolution activities, the authors propose using a Hypergraph Data Model (HDM) to build a global schema from heterogeneous source schemata and from this transformations may be used to translate queries between the global and source schemas. Schema transformations defined on the HDM are reversible. Every *add* transformation step is reversed by a *del* transformation with the same parameters, *renaming* (ren) transformations from $S_1 \rightarrow S_2$ are the reverse of $S_2 \rightarrow S_1$. Contract transformations map to void, queries and sub-queries over such constructs then translate to void. *Extend* transformations requires domain knowledge, either from a human expert or a domain ontology and cannot be automated. Higher level modelling only works with names, tables, relations but not at attribute data type level.

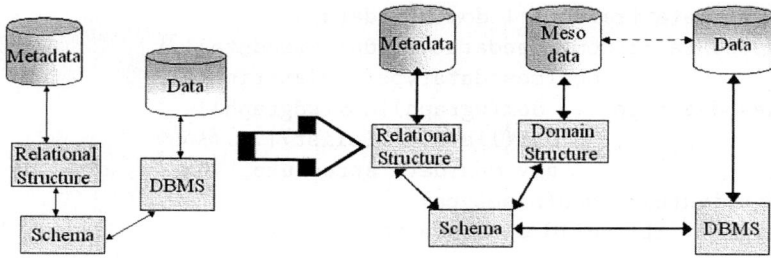

Fig. 1. Mesodata is an additional layer in RDBMS

Davidson *et al.* [14] recognising that information capacity preserving trans-formations do not necessarily preserve the semantics of databases developed the declarative language *WOL* (Well-founded Object Language) for expressing database transformations and constraints. They argue that approaches which allow a fixed set of well defined transformations to be applied in series (for most methodologies the outcome is dependent on the order in which the schemas are integrated - they are not associative) are inherently limited in the class of trans-formations that can be expressed and that while using a high-level language for transformations is necessary for general transformations, it is difficult to reason about, and prove, properties of transformations. This work tackles the difficulty of correctly transforming complex data structures (sets, records and variants) and recursive structures. Constraints on the source and target databases are crucial to notions of information preservation, but typically are not, or cannot, be expressed in the models of the underlying databases.

3   Model

In this research, we use the term 'intelligent domains' for our enhanced meso-data generated domains as they provide increased semantic content over the domains. Mesodata is an additional domain definition layer containing structure and operators.

A traditional relational database can be viewed as consisting of relations that are a subset of the Cartesian product of their attributes' domains [15].

$$R \subseteq (dom(A_i) \times dom(A_j) \times \ldots dom(A_n)) \tag{1}$$

where R is the relation
        A is an attribute
        dom is the domain of the attribute A.

The mesodata layer extracts the domain to a separate level such that the Mesodata Domain (Mdom) is the domain of the mesodatatype of the basetype, for example a weighted graph of strings or a list of graphs of strings. Mdom is defined as:

```
Mdom :: dom(attribute) | dom(mesodata)
dom(mesodata):: dom(mesodatatype(dom(mesodata)) |
                dom(mesodatatype(dom(attribute)))
dom(mesodatatype) :: dom(wgraph)|dom(wdgraph)|
                dom(list)|dom(clist)|...
                any mesodata structure
dom(attribute):: dom(basetype)
basetype::  all valid database base types.
```

The redefinition of the relation R thus becomes

$$R \subseteq (Mdom_i \times Mdom_j \times \ldots \times Mdom_n) \qquad (2)$$

that is, the cartesian product of the mesodata defined domains.

Table 2. Examples of domain structures suitable for adoption as mesodata types

Domain Structure	Operations (Extended SQL Op.)	Source Relation(s)
Unweighted Graph (GRAPH)	Adjacency (NEXTTO)	Binary relation (FROM, TO)
Weighted Graph (WGRAPH)	Adjacency, Proximity (CLOSETO)	Ternary relation (FROM, TO, WEIGHT)
Directed Graph (DGRAPH)	Adjacency	Binary relation (FROM, TO)
Directed Weighted Graph (DWGRAPH)	Adjacency, Proximity	Ternary relation (FROM, TO, WEIGHT)
Tree (TREE)	In subtree (DESCENDENT), Parent(PARENT), Ancestor(ANCESTOR), Child(CHILD), Sibling(SIBLING)	Binary Relation(PARENT, CHILD)
Weighted Tree (WTREE)	In subtree, Parent, Ancestor, Sibling, Proximity	Ternary Relation(PARENT, CHILD, WEIGHT)
List (LIST)	Next (NEXT), Previous (PREV), First(FIRST), Last(LAST), Between(BETWEEN)	Binary Relation(SEQUENCE, ITEM)
Circular List (CLIST)	Next, Previous, Between	Binary Relation(SEQUENCE, ITEM)
Set (SET)	In Set(INSET)	Unary Relation(ITEM)
Tri-State Logical	Maybe Equal (MAYBE)	None

Table 2 of mesodata types presents a few examples of domain structures with their intrinsic operations. It is important to differentiate between *mesodata* structures and Abstract Data Types (ADT). For instance, the specification of a graph as a mesodata type and the specification of a graph as an abstract data type (ADT). In the former, the attribute would take as its value an instance of a base type that exists *within a graph* while in the latter the type is a graph for which code must be included in the application. The mesodata type is not directly accessible through the attribute. The semantics of information (S) held in a database can be considered as a function of the data value. That is,

$$S = F(v) \qquad (3)$$

where F is a mapping external to the database which maps the data value (such as 116Q15) to an understood concept (such as Burbridge Road). The introduction of a mesodata layer allows regularly used mappings to be accommodated in the database, ie.

$$S = F(M(v)) \qquad (4)$$

where M is the mesodata layer mapping. Conceptually, it is then possible to have mappings of mappings

$$S = F(M_1(M_2(\ldots M_k(v) \ldots))) \qquad (5)$$

where M_i are mesodata layer mappings. Therefore, $S = F(M_1(M_2(116Q15)))$ where $F(M_2(116Q15)) = $ 'Burbridge Road' and $F(M_1(\text{'Burbridge Road'})) = $ 'Sir Donald Bradman Drive'. For more information about using mesodata types in DBMS refer to [1].

4   Domain Evolution

Domain evolution can be broadly categorised into three types - *Attribute Representation Change, Domain Constraints Change* and *Domain Perception (meaning) Change* - that to date require a manual solution to their successful incorporation into a DBMS. Using a mesodata layer in the database can reduce these problems. We illustrate these through examples. The mesodata types selected for the examples are not proscriptive, just as the DBA judges which attribute data type to use, so too must the decision of which mesodata type to employ lie with the DBA.

4.1   Attribute Representation Change

Example: A character code is replaced by a number code. The specification CHAR(20) is altered to an INTEGER.

Current Typical Solution: Add new attribute of type INTEGER to relation, convert old data values to new values and store in new attribute, delete old attribute, rename attribute to old name, update application to handle different type.

Mesodata Solution: Use the mesodata type, LIST, that maps the existing CHAR(20) values to the new INTEGER values. The attribute in the relation remains unchanged as does the application, as the operators to access the changed attribute type are built into the mesodata type.

For example:

old AppCode = 'widgetA'
new AppCode = 2131

using the mesodata domain layer, we have, (see Eq 4),

AppCode = Mdom('widgetA') = 2131

Both code values **2131** and **widgetA** are accessible and valid. Information capacity holds as both *equivalence* and *dominance* requirements are met. It is recognised that not all attribute type changes can be handled using mesodata, for example from BLOB to INT, however there are many instances where the evolutionary process can be alleviated.

4.2   Domain Constraints Change

Example: 'Country of birth' is an attribute contained in a number of databases, the allowable values of which have changed significantly during the twentieth century. When a country name changes it may be a one-to-one change, such as *Rhodesia* to *Zimbabwe*, or one-to-many change, for example *Yugoslavia* to {*Bosnia Herzegovina, Croatia, Macedonia, Serbia, FYR Montenegro, Slovenia.*}

Current Typical Solution: Convert all old values and replace them with new values. This could be an ongoing task and it results in loss of information.

Mesodata Solution: Utilise the mesodata types WGRAPH or TREE to map old values to the new values. The domain of 'countries' includes *All* country names, current and superseded, which are then accessible to the DBMS with the extended SQL operators and original values are not lost.

4.3   Domain Perception (Meaning) Change

Example: A perception change may entail an absolute change where there is new interpretation or it may be the addition of synonyms. The days of the week stored numerically from 1 to 7 inclusive may interpret the value '1' as 'Monday', equally valid are the interpretations 'lunes', 'lundi', 'maandag', 'Montag', 'segunda-feira' and so forth.

Current Typical Solution: The application may be parameter driven to select a single preferred interpretation (such as language setting) or the users must learn the dominant term.

Mesodata Solution: A mesodata layer allows regularly used mappings to be accommodated in a database (see Eq. 5). Therefore we have
1 = 'Monday' = 'lundi' etc.

Mesodata helps to reduce potential systems changes to one of two simpler solutions

1. A change to the schema definition that requires no change to either the application or data.
2. A change to the mesodata reference relation with or without a change to the schema but again, without the need to change either the application or the data.

Schema integration and transformation is not required as the mesodata type has the operators and 'intelligence' to replace these tasks. Information capacity is not only maintained, as both requirements of equivalence and dominance are met, but also in many cases expanded as the cartesian product of the mesodata domains is greater than the original domain of the relation.

Our implementation of this model uses MySQL [16] software with wrappers to transform our extended SQL, as described in [1]. Space precludes a more detailed report of the implementation which will be the subject of a future paper.

5   Conclusion and Further Research

Though an attribute change in itself may not be a complex process it is not a trivial task. Database evolution and maintenance consists of many such simple steps as shown in [3] most of which also necessitate changes to application code and system down time. The mesodata layer, an additional domain definition layer containing domain structure and intelligence, provides the means to manage some aspects of attribute domain evolution. We have shown that its use when a domain changes, when the semantics of a domain alter or when the attribute's specification is modified can reduce or remove the necessity of schema conversion, schema integration, data conversion and application change as well as maintain or expand the schema's information capacity.

Work in this area is progressing, particularly in the use of ontology frameworks to describe and incorporate evolving domains into the mesodata layer.

References

1. De Vries, D., Rice, S., Roddick, J.F.: In support of mesodata in database management systems. In: 15th International Conference on Database and Expert Systems Applications DEXA 2004. Lecture Notes in Computer Science, Zaragoza, Spain (2004)
2. Ferrandina, F., Meyer, T., Zicari, R.: Implementing lazy database updates for an object database system. In: Twentieth International Conference on Very Large Databases, Santiago, Chile (1994) 261–272
3. Sjøberg, D.: Quantifying schema evolution. Information and Technology Software **35** (1993) 35 – 44
4. Jensen, C.S., Clifford, J., Elmasri, R., Gadia, S.K., Hayes, P., Jajodia, S., Dyreson, C., Grandi, F., Kafer, W., Kline, N., Lorentzos, N., Mitsopoulos, Y., Montanari, A., Nonen, D., Peressi, E., Pernici, B., Roddick, J.F., Sarda, N.L., Scalas, M.R., Segev, A., Snodgrass, R.T., Soo, M.D., Tansel, A., Tiberio, P., Wiederhold, G.: A consensus glossary of temporal database concepts - february 1998 version. In Etzion, O., Jajodia, S., Sripada, S., eds.: Temporal Databases - Research and Practice LNCS. Volume 1399. Springer-Verlag (1998) 367–405
5. Roddick, J.F., Craske, N.G., Richards, T.J.: A taxonomy for schema versioning based on the relational and entity relational models. In: Proc. Twelfth International Conference on Entity-Relationship Approach. (1993) 143–154
6. Hull, R.: Relative information capacity of simple relational database schemata. Society for Industrial and Applied Mathematics **15 No 3** (1986) 856 – 886
7. Qian, X.: Correct schema transformations. In Apers, P.M.G., Bouzeghoub, M., Gardarin, G., eds.: Advances in Database Technology - EDBT'96, 5th International Conference on Extending Database Technology. Volume 1057 of Lecture Notes in Computer Science., Avignon, France, Springer (1996) 114–128
8. Miller, R.J., Ioannidis, Y.E., Ramakrishnan, R.: Schema equivalence in heterogeneous systems: Bridging theory and practice. Information Systems **19** (1994) 3 31
9. Miller, R.J., Ioannidis, Y.E., Ramakrishnan, R.: The use of information capacity in schema integration and translation. In Agrawal, R., Baker, S., Bell, D., eds.: Nineteenth International Conference on Very Large Data Bases, VLDB'93, Dublin, Ireland, Morgan Kaufmann, Palo Alto, CA (1993) 120–133
10. Miller, R.J., Ioannidis, Y.E., Ramakrishnan, R.: Schema intension graphs: A formal model for the study of schema equivalence. Technical report, University of Wisconsin-Madison (1994)

11. Xu, L., Poulovassilis, A.: A method for integrating deductive databases. In: British National Conference on Databases. (1997) 215–231
12. McBrien, P., Poulovassilis, A.: A formal framework for er schema transformation. In: International Conference on Conceptual Modeling / the Entity Relationship Approach. (1997) 408–421
13. McBrien, P., Poulovassilis, A.: Schema evolution in heterogeneous database architectures, a schema transformation approach. In: CAiSE'02, Birkbeck College and Imperial College (2002)
14. Davidson, S., Buneman, P., Kosky, A.: Semantics of Database Transformations. Volume 1358. Springer-Verlag, Berlin (1998)
15. Elmasri, R., Navathe, S.B.: Fundamentals of database systems. 3rd edn. Addison-Wesley, Reading, Mass ; Menlo Park, Calif. (2000)
16. MySQL: Sql shareware software: documentation and source code (2003)

Multitemporal Conditional Schema Evolution

Ole G. Jensen[1] and Michael H. Böhlen[2]

[1] Department of Computer Science, Aalborg University
Fredrik Bajers Vej 7E, DK-9220 Aalborg Øst, Denmark
guttorm@cs.aau.dk
[2] Faculty of Computer Science, Free University of Bozen-Bolzano
Dominikanerplatz 3, I-39100 Bolzano, Italy
boehlen@inf.unibz.it

Abstract. Schema evolution is the ability of the database to respond to changes in the real world by allowing the schema to evolve. The *multidimensional conditionally evolving schema*(MD-CES) is a conceptual model for conditional schema changes, which modify the schema of those tuples that satisfy the change condition. The MD-CES is lossless and preserves schemas, but has an exponential space complexity. In this paper we restrict conditional schema changes to timestamp attributes. Specifically, we develop 1D-CES for schema versioning over one time dimension, and 2D-CES for schema versioning over two time dimensions. We show that the space complexity of these new evolution models is linear or polynomial. 1D-CES and 2D-CES are compared to temporal schema versioning, and we show that, unlike valid time versioning, they are lossless and achieve the same space complexity as temporal versioning if the schema changes are ordered.

1 Introduction

Conditional schema changes modify the schema of those tuples that satisfy the change condition [5], and they properly subsume unconditional and temporal schema changes. The semantics of conditional schema evolution is defined in terms of the *multi-dimensional conditionally evolving schema* (MD-CES) [4]. The MD-CES has many desirable properties that are essential for relations that contain tuples with different intended schemas. First, the MD-CES is *lossless*, i.e., the intended schema of each tuple is consistent with the initial schema definition and the subsequent conditional schema changes. Thus, at any point during the evolution process the correct intended schema of a tuple is known[1]. Second, the MD-CES is *schema preserving*, i.e., all schemas are preserved and tuples never have to be migrated to a schema with more or less attributes. On the downside, however, the space complexity of the MD-CES grows exponentially with the number of schema changes, and, consequently, the MD-CES is not a practical model.

In this paper, we specialize conditional schema changes to conditions over timestamp attributes. This leads to the 1D-CES and 2D-CES with linear or polynomial space

---

[1] Models with an ad-hoc approach to schema evolution often violate this property. For example if NULL values are used to "flag" attributes that are not part of the schema and with multiple schema changes the correct identification of current and past intended schemas is often a problem.

S. Wang et al. (Eds.): ER Workshops 2004, LNCS 3289, pp. 441–454, 2004.
© Springer-Verlag Berlin Heidelberg 2004

complexity for monoconditional and biconditional schema changes, respectively. We prove that 1D-CES and 2D-CES are still lossless and, if the history of conditional schema changes is ordered, ensure schema preservation. We also provide the cost of ensuring schema preservation for 1D-CES and 2D-CES with unordered histories. Thus, by restricting the MD-CES to mono- and biconditional schema changes, we get a practical model for schema evolution that is particularly relevant to temporal schema versioning.

An important aspect of our work is that it offers an alternative approach to investigate the semantics of temporal evolution models. As pointed out before the semantics of temporal schema versioning, in particular for bitemporal databases, is non-trivial. We compare 1D-CES and 2D-CES to transaction-time, valid-time, and bitemporal schema versioning. For transaction-time the two models are equivalent. In valid-time and bitemporal schema versioning, a schema change is applied to a single schema version specified by the user irrespective of the validity of the change and the version. In contrast, with MD-CES a schema change modifies the schema of all segments with a validity that overlaps the validity of the change. We show that valid-time and bitemporal schema versioning are not lossless, and that valid-time schema versioning does not ensure schema preservation.

2   Preliminaries

2.1   Multi-dimensional Conditionally Evolving Schemas

A *multi-dimensional conditionally evolving schema* (MD-CES), $E = \{S_1, \ldots, S_n\}$, generalizes a relation schema and is defined as a set of schema segments. A *schema segment*, $S = (\mathcal{A}, P)$, consists of a schema $\mathcal{A}$ and a qualifier P. Throughout, we write $\mathcal{A}_S$ and P_S to directly refer to the schema and qualifier of segment S, respectively. As usual, a *schema*, $\mathcal{A} = \{A_1, \ldots, A_n\}$, is defined as a set of attributes. No distinction is made between schemas and sets of attributes. A *qualifier* P is either TRUE, FALSE, or a conjunction/disjunction of attribute constraints. An *attribute constraint* is a predicate of the form $A\theta c$ or $\neg(A\theta c)$, where A is an attribute, $\theta \in \{<, \leq, =, \neq, \geq, >\}$ is a comparison predicate, and c is a constant. A MD-CES may have segments with different schemas. Consequently, some tuples may be missing attributes that appear in other segments. In order to evaluate attribute constraints on such tuples, $A\theta c$ is an abbreviation for $\exists v(A/v \in t \wedge v\theta c)$ where t is a tuple and A/v is an attribute/value pair. Likewise, $\neg(A\theta c)$ is an abbreviation for $\neg\exists(A/v \in t \wedge v\theta c)$. Note that this implies that the constraints $\neg(A = c)$ and $A \neq c$ are not equivalent.

A tuple t is a set of attribute values where each attribute value is an attribute/value pair: $\{A_1/v_1, \ldots, A_n/v_n\}$. The value must be an element of the domain of the attribute, i.e., if $dom(A)$ denotes the domain of attribute A, then $\forall A, v, t(A/v \in t \Rightarrow v \in dom(A))$. A tuple t *qualifies* for a segment S, $qual(t, S)$, iff t satisfies the qualifier P_S. A tuple satisfies a qualifier, $P(t)$, iff the qualifier is TRUE or the tuple makes the qualifier true under the standard interpretation. If a tuple t qualifies for a segment S in a MD-CES E, then $\mathcal{A}_S$ is the *intended schema* of t, i.e., $\forall t, S, E(S \in E \wedge qual(t, S)) \Rightarrow is(t, E) = \mathcal{A}_S$). A tuple t *matches* a segment S iff the schema of S and t are identical: $match(t, S)$ iff $\forall A(A \in \mathcal{A}_S \Leftrightarrow \exists v(A/v \in t))$. If a tuple t matches a segment S in the MD-CES E, then $\mathcal{A}_S$ is the *recorded schema* of t, i.e., $\forall t, S, E(S \in E \wedge match(t, S) \Rightarrow rs(t, E) = \mathcal{A}_S)$.

2.2 Conditional Schema Changes

A *conditional schema change* is an operation that changes the set of segments of a MD-CES. The condition determines the tuples that are affected by the schema change. A *condition* C is either TRUE, FALSE, an attribute constraint, or a conjunction of attribute constraints. For the purpose of this section we consider two conditional schema changes: adding an attribute, $\alpha(A, E, C)$, and deleting an attribute, $\beta(A, E, C)$. An extended set of schema changes that includes mappings between attributes and a discussion of their completeness can be found elsewhere [4].

$\alpha(A, E, C)$: An attribute A is added to the schemas of all segments that do not already include the attribute. For each such segment two new segments are generated: a segment with a schema that does not include the new attribute and a segment with a schema that includes the new attribute. Segments with a schema that already includes A are not changed.

$\beta(A, E, C)$: The attribute A is deleted from the schemas of all segments that include the attribute. For each such segment two new segments are generated: a segment with a schema that still includes the attribute and a segment with a schema that does not include the attribute. Segments with a schema that does not include A are not changed.

The precise formal definitions of conditional attribute additions and deletions are given in Figure 1. Note that conditional schema changes properly subsume regular (i.e., unconditional) schema changes. It is possible to have the condition (TRUE) select the entire extent of a relation.

$$\alpha(A, \emptyset, C) = \emptyset$$

$$\alpha(A, \{(\mathcal{A}, P)\} \cup E, C) = \begin{cases} \{(\mathcal{A}, P)\} \cup \alpha(A, E, C) & \text{iff } A \in \mathcal{A} \\ \{(\mathcal{A} \cup \{A\}, P \wedge C), (\mathcal{A}, P \wedge \neg C)\} \\ \cup \alpha(A, E, C) & \text{iff } A \notin \mathcal{A} \end{cases}$$

$$\beta(A, \emptyset, C) = \emptyset$$

$$\beta(A, \{(\mathcal{A}, P)\} \cup E, C) = \begin{cases} \{(\mathcal{A}, P)\} \cup \beta(A, E, C) & \text{iff } A \notin \mathcal{A} \\ \{(\mathcal{A} \setminus \{A\}, P \wedge C), (\mathcal{A}, P \wedge \neg C)\} \\ \cup \beta(A, E, C) & \text{iff } A \in \mathcal{A} \end{cases}$$

Fig. 1. Adding ($\alpha(A, E, C)$) and Deleting ($\beta(A, E, C)$) Attribute A on Condition C

A *history* $H = [\gamma_1(A_1, E, C_1), \ldots, \gamma_n(A_n, E, C_n)]$ where $\gamma_i \in \{\alpha, \beta\}$ is a sequence of conditional schema changes. We say that (E_0, H) is the *evolution history* of E where $E_0 = \{S\}$ consists of a single segment, iff $E = \gamma_n(A_n, \ldots \gamma_1(A_1, E_0, C_1) \ldots, C_n)$.

2.3 Lossless and Schema Preserving Properties

The MD-CES is both *lossless* [6] and *schema preserving* [5].

The lossless property ensures that the intended schema defined by a MD-CES for each tuple is consistent with the initial schema definition and the conditional schema changes. Intuitively, the intended schema of a tuple t can be determined from the initial schema by applying only those schema changes with a condition satisfied by t in the sequence given by the history.

The schema preserving property guarantees that all schemas defined by segments in a MD-CES also appear in segments after a conditional schema change. Schema preservation is a requirement if the segments of the MD-CES are to be used as relations to store the (heterogeneous) tuples of evolving relations.

3   Schema Change Conditions on a Single Attribute

Conditional schema evolution allows schema changes to be conditioned by any attribute in the schema. Since each conditional schema change potentially splits every segment in the MD-CES into two new segments, conditional schema evolution leads to MD-CESs where the number of segments is exponential in the size of the history [5]. Therefore, solutions that rely on the segments of a MD-CES are not tractable, including using the segments as relations to record tuples or querying individual segments.

In this section we restrict the conditions of conditional schema changes to a single attribute. This leads to *one-dimensional conditionally evolving schemas* (1D-CES). We show that the number of segments in a 1D-CES becomes proportional to the size of the history. This is achieved by the elimination of segments with false qualifiers. We show that this optimization preserves the lossless property of the MD-CES but can violate schema preservation. To achieve schema preservation additional segments must be kept in the 1D-CES increasing its space complexity to be polynomial in the size of the history. We show that if the history is ordered, then schema preservation is achieved without recording additional segments. First, however, we present a running example to be used throughout.

3.1   Running Example

We use a relation storing information about employees as a running example. The employee relation records a name, an address, and a phone number for each employee. Additionally, a timestamp attribute records when the fact was recorded in the database (a detailed discussion about different notions of time is deferred until Section 5). The schema $E_{\text{mployee}} = (N_{\text{ame}}, A_{\text{ddress}}, P_{\text{hone}}, T_{\text{ime}})$ defines the (initial) employee relation schema.

A conditional schema change $\gamma_1 = \alpha(S_{\text{sn}}, E_{\text{mployee}}, T_{\text{ime}} \geq 2003\text{-}01\text{-}01)$ adds a social security number to employees recorded on and after January 1st 2003. The schema change splits the employee schema into two segments: $S_1 = (\{N, A, P, T\}, T < 20030101\})$ and $S_2 = (\{N, A, P, S, T\}, T \geq 20030101\})$.

The schema change only adds a social security number to the schema of employees recorded with a timestamp value equal to or greater than 2003-01-01. Employees recorded with a timestamp before 2003-01-01 are not affected.

We require that all tuples have a timestamp value and that schema changes cannot drop the timestamp attribute from the schema. The evaluation of qualifiers is affected by this requirement. Recall that qualifiers are existentially quantified, so $\neg(T\theta c) \equiv \neg \exists T(T\theta c)$. This implies that conditions $\neg(T = c)$ and $T \neq c$ are not equivalent. E.g.

tuples without a T attribute value evalues to true for the former condition and false for the latter condition. However, since we require that all tuples *have* a timestamp value, the equivalence holds, and the negation can be pushed down to the comparison predicate as is the case for S_1.

A second schema change drops the address from the schema of employees recorded from the 1st of March 2003 and replaces it with a C_{ity} attribute. In response, two conditional schema changes are applied to the $E_{mployee}$ MD-CES:

$$\beta(A_{ddress}, E_{mployee}, T_{ime} \geq 2003\text{-}03\text{-}01) \qquad \alpha(C_{ity}, E_{mployee}, T_{ime} \geq 2003\text{-}03\text{-}01)$$

The conditional schema change on the left drops the A_{ddress} attribute from the schema of employees with a timestamp value of at least 2003-03-01. The conditional schema change on the right adds the C_{ity} attribute on the exact same condition. Both conditional schema changes are applied in sequence as seen below. Note that the order in which the two schema changes are applied does not change the result, because the schema changes affect different attributes (A_{ddress} and C_{ity} respectively).

$$E_2 = \alpha(C, \beta(A, E_1, T \geq 2003\text{-}03\text{-}01), T \geq 2003\text{-}03\text{-}01) =$$

$\{((\{N, C, P, S, T\}, T \geq 2003\text{-}01\text{-}01 \wedge T \geq 2003\text{-}03\text{-}01 \wedge T \geq 2003\text{-}03\text{-}01)^1,$
$(\{N, P, S, T\}, T \geq 2003\text{-}01\text{-}01 \wedge T \geq 2003\text{-}03\text{-}01 \wedge T < 2003\text{-}03\text{-}01)^2,$
$(\{N, A, C, P, S, T\}, T \geq 2003\text{-}01\text{-}01 \wedge T < 2003\text{-}03\text{-}01 \wedge T \geq 2003\text{-}03\text{-}01)^3,$
$(\{N, A, P, S, T\}, T \geq 2003\text{-}01\text{-}01 \wedge T < 2003\text{-}03\text{-}01 \wedge T < 2003\text{-}03\text{-}01)^4,$
$(\{N, C, P, T\}, T < 2003\text{-}01\text{-}01 \wedge T \geq 2003\text{-}03\text{-}01 \wedge T \geq 2003\text{-}03\text{-}01)^5,$
$(\{N, P, T\}, T < 2003\text{-}01\text{-}01 \wedge T \geq 2003\text{-}03\text{-}01 \wedge T < 2003\text{-}03\text{-}01)^6,$
$(\{N, A, C, P, T\}, T < 2003\text{-}01\text{-}01 \wedge T < 2003\text{-}03\text{-}01 \wedge T \geq 2003\text{-}03\text{-}01)^7,$
$(\{N, A, P, T\}, T < 2003\text{-}01\text{-}01 \wedge T < 2003\text{-}03\text{-}01 \wedge T < 2003\text{-}03\text{-}01)^8\}$

The employee MD-CES, E_2, now consists of eight segments (marked with small numbers in the above expression). Clearly, the number of segments in E_2 increases exponentially with the application of conditional schema changes.

3.2 Elimination of Segments with False Qualifiers

Consider the qualifier of segment 2 in E_2:

$$T \geq 2003\text{-}01\text{-}01 \wedge T \geq 2003\text{-}03\text{-}01 \wedge T < 2003\text{-}03\text{-}01 \Leftrightarrow$$

$$T \geq 2003\text{-}03\text{-}01 \wedge T < 2003\text{-}03\text{-}01 \Leftrightarrow$$

FALSE

Clearly, the qualifier is false. No tuple can qualify for such segments. Future conditional schema changes applied to a segment S with a false qualifier P_S cannot result in segments with non-false qualifiers, because those qualifiers are either identical to P_S or in conjunctive form with P_S (cf. Section 2) and, clearly, FALSE $\wedge p \equiv$ FALSE.

A MD-CES defines the intended schema of all tuples for a given initial schema and history. The association of tuples and their intended schemas is done through qualification of tuples with individual segments. Since tuples cannot qualify for segments with false qualifiers, such segments define no intended schemas. We can therefore omit segments with false qualifiers from a MD-CES without loss to the set of intended schemas defined by it, i.e., without losing the lossless property. This is stated by Lemma 1. All proofs have been omitted due to space considerations, but can be found in [3].

Lemma 1. *Let E and E' be MD-CESs and let γ be a conditional schema change. If E' consists of exactly the non-false segments of E, then $\gamma(A, E, C)$ and $\gamma(A, E', C)$ define the exact same intended schema for any tuple, i.e., $\forall \gamma, t(is(t, \gamma(A, E, C)) = is(t, \gamma(A, E', C)))$.*

3.3   MD-CES with Linear Space Complexity

Schema changes conditioned by a single attribute result in a MD-CES, where the qualifier of each segment is a conjunction of predicates over that attribute only. We say that such MD-CESs are one-dimensional (denoted as 1D-CES). The intended schema of a tuple depends exclusively on its recorded attribute value for the attribute used to condition the schema changes. Moreover, for every value of that attribute, the 1D-CES defines exactly one intended schema, because the qualifiers of two segments part of the same MD-CES never overlap (a property of the MD-CES [5]).

Note that the number of segments in E_{mployee} scales with the number of conditional schema changes applied. In the general case conditional schema changes cause an exponential increase of segments, as each schema change potentially splits every segment into two new segments. However, because all the conditional schema changes applied to E_{mployee} are conditioned by the same attribute, at most one segment is split into two non-false segments after each schema change (every other segment yields only one non-false segment).

Unbounded Conditions. We shall assume that all conditional schema changes have *unbounded* conditions, i.e. conditions of the form $A\theta c$ where $\theta \in \{\geq, >, <, \leq\}$. Bounded conditions can be specified easily. E.g. $\alpha(A, E, T \geq t_1 \wedge T \leq t_2)$ has a bounded condition affecting only tuples with a T value in the interval from t_1 to t_2. Schema changes with bounded conditions can split up to two different segments (one per predicate in the condition). However, any conditional schema change with a non-false bounded condition has an equivalent pair of unbounded conditions. Table 1 shows the equivalent conditional schema changes for attribute addition and deletion with bounded condition, respectively. Conditional schema changes applying to a single point, i.e. with a condition of the form $A = c$, is a specialization of a bounded condition (since $(A = c) \equiv (A \geq c \wedge A \leq c)$).

We can assume unbounded conditions without loss of generality. Unbounded conditions result in segments where the qualifier is satisfied for a contiguous set of attribute values (an interval). This facilitates the comparison between segments of a 1D-CES (and 2D-CES) and schema versions in Section 5. Next, we give the main result of this section.

Lemma 2. *Let both E and E' be a MD-CES. Let $H = [\gamma_1(A_1, E, C_1), \ldots, \gamma_n(A_n, E, C_n)]$ be a history where $\gamma_i \in \{\alpha, \beta\}$ and each condition C_i is unbounded and over the same attribute T. If $E' = \gamma_n(A_n, \ldots \gamma_1(A_1, E, C_1) \ldots, C_n)$ then E' has a number of additional non-false segments that is proportional to the size of the history, i.e. $|E'| - |E| \leq |H|$.*

Table 1. Conditional Schema Change Equivalences between Bounded and Unbounded Conditions

Bounded Condition	Unbounded Conditions
$\alpha(A, E, T \geq t_1 \wedge T \leq t_2)$	$\alpha(A, E, T \geq t_1), \beta(A, E, T > t_2)$
$\beta(A, E, T \geq t_1 \wedge T \leq t_2)$	$\beta(A, E, T \geq t_1), \alpha(A, E, T > t_2)$

Lemma 2 shows that solutions to conditional schema evolution based on the segments of the MD-CES become tractable when schema changes are conditioned by a single attribute.

3.4 Ordering of Histories

While Lemma 2 guarantees the scalability of the 1D-CES, a conditional schema change can potentially change the schema of every segment in the 1D-CES. The 1D-CES is no longer guaranteed to ensure schema preservation once segments with false qualifiers have been eliminated. Using the segments of the 1D-CES as relations to record the (heterogeneous) tuples of an evolving relation, requires that existing schemas are preserved for tuples already recorded in the database with those schemas. In the worst case, a conditional schema change results in a 1D-CES where all the schemas of the original segments appear in new segments with false qualifiers. Therefore, they would all need to be kept in addition to the new segments created by the schema change. An upper bound of $\frac{n^2+n}{2}$ segments, where n is the size of the history, would have to be kept to ensure schema preservation. In this section we show that by ordering the history it is possible to keep the number of segments linear.

Example 1. Consider a 1D-CES with segments: $(\{A, C, T\}, T \geq 3)$ and $(\{B, C, T\},$ $T < 3)$. Assume a conditional schema change that drops the C attribute on the condition $T \geq 1$. The result is a 1D-CES with the following segments:

1. $(\{A, T\}, T \geq 3 \wedge T \geq 1) \equiv (\{A, T\}, T \geq 3)$
2. $(\{A, C, T\}, T \geq 3 \wedge T < 1) \equiv (\{A, C, T\}, \text{FALSE})$
3. $(\{B, T\}, T < 3 \wedge T \geq 1)$
4. $(\{B, C, T\}, T < 3 \wedge T < 1) \equiv (\{B, C, T\}, T < 1)$

Note that the 1D-CES no longer contains a segment with the schema $\{A, C, T\}$ (since it was dropped due to a false qualifier).

Unbounded conditions over a single attribute leads to segments with qualifiers that are satisfied for a contiguous interval of attribute values. The problem illustrated by Example 1 arises when the condition of a schema change can be satisfied by attribute values within the interval qualified by more than one segment. From Lemma 2 we have that at most one segment is split by the conditional schema change resulting in two non-false segments. For all other segments, the conditional schema change creates only one non-false segment. The interval of attribute values qualified by each of these segments either all satisfy the condition or none satisfy the condition. In the former case, the segment with the schema of the original segment, will have a false qualifier.

To ensure schema preservation while still eliminating segments with false qualifiers, we impose an ordering on the conditional schema changes. Definition 1 defines ordered histories of conditional schema changes.

Definition 1. (ordered history) *Let* $H = [\gamma_1(A_1, E, C_1), \ldots, \gamma_n(A_n, E, C_n)]$ *be a history. If each conditional schema change applies to a proper subset of the tuples which the previous conditional schema change applied to, i.e., if* $\{t | t \in dom(T) \wedge C_{i+1}(t)\} \subset \{t | t \in dom(T) \wedge C_i(t)\}$ *where* $dom(T)$ *is the domain of* T *attribute values, then* H *is an* ordered history.

In general, all schemas are preserved by a conditional schema change if it splits exactly one segment into two non-false segments and no attribute value in the interval defined by any other segment satisfy the condition. This will ensure that all previously defined schemas appear in segments with non-false qualifiers.

Lemma 3 states that a 1D-CES with an ordered history preserves all previously defined schemas.

Lemma 3. *Let* $H = [\gamma_1(A_1, E, C_1), \ldots, \gamma_n(A_n, E, C_n)]$ *be a history where* $\gamma_i \in \{\alpha, \beta\}$ *and each condition* C_i *is unbounded over a single attribute* T. *Let* E *be a MD-CES with a single segment, and let each* $E_i = \gamma_i(A_i, \ldots \gamma_1(A_1, E, C_1) \ldots, C_i)$ *be a MD-CES. If* H *is an ordered history then* E_n *preserves all schemas defined by segments in* E *and* E_1 *to* E_{n-1}, *i.e.,* H *is ordered implies that* $\forall S(S \in E \cup E_1 \cup \ldots \cup E_{n-1} \wedge \exists S'(S' \in E_n \wedge \mathcal{A}_S = \mathcal{A}_{S'}))$.

Lemma 2 and 3 ensure that a 1D-CES with an ordered history of conditional schema changes using unbounded conditions over a single attribute is both lossless and schema preserving and has a space complexity, which is linear in the size of the history.

4   Schema Change Conditions on Different Attributes

So far conditional schema changes with a single attribute constraint have been considered. We refer to these schema changes as monoconditional schema changes. This section investigates histories where monoconditional schema changes are over different attributes. We show that as the number of different attributes used in the conditions increases, so does the space complexity of the MD-CES. We also consider biconditional schema changes, i.e., schema changes where the condition is a conjunction of two attribute constraints over different attributes. Biconditional schema changes lead to *two-dimensional conditionally evolving schemas* (2D-CES). A 2D-CES with a history of biconditional schema changes has a polynomial space complexity. We show that an ordered history of biconditional schema changes ensures a 2D-CES, which is lossless and schema preserving and has a linear space complexity.

Monoconditional schema changes over different attributes are orthogonal. These conditional schema changes split every segment in a MD-CES into two non-false segments. Since the qualifier P of the segment S and the condition C of the change are over different attributes, the logical conjunction $P \wedge C$ cannot be equivalent to false (unless either P or C is already false), so both segments resulting from applying the conditional schema change to S have non-false qualifiers.

Lemma 4 establishes the upper bound on the number of segments with non-false qualifiers in a MD-CES with a history of monoconditional schema changes over different attributes.

Lemma 4. *Let H be a history of monoconditional schema changes with unbounded conditions. Let $\mathcal{A}_H = \{A_1, \ldots, A_n\}$ be the set of attributes appearing in conditions in H, and let $num(A, H)$ be the number of schema changes in H conditioned by A. Let E and E' be a MD-CES and let E consist of a single segment. If (E, H) is the evolution history of E' then the upper bound on the number of segments with non-false qualifiers in E' is $(1 + num(A_1, H)) \times \ldots \times (1 + num(A_n, H))$.*

4.1 Ordering of Biconditional Schema Changes

In the general case, biconditional schema changes lead to a polynomial number of segments in a 2D-CES. This occurs when the condition of a schema change overlaps the qualifiers of all segments. However, if the condition is contained by the qualifier of a single segment then the number of segments increase by at most one.

This is the case for ordered histories of conditional schema changes. Recall that a history is ordered iff each conditional schema change applies to a proper subset of the tuples which the previous conditional schema change applied to. For biconditional schema changes this occurs when the conditional of each schema change is contained by the qualifier of the latest segment.

Lemma 5 states that the number of segments in a MD-CES defined by a history of biconditional schema changes is proportional to the size of the history, if there exists a sequence of those schema changes such that the sequence is an ordered history.

Lemma 5. *Let H be a history of biconditional schema changes with conditions C_i of the form $T \geq t_i \wedge V \geq v_i$. Let E and E' be a MD-CES and let E consist of a single segment. Let (E, H) be the evolution history of E'. If there exists a sequence H' of the schema changes in H such that H' is an ordered history, then E' has a number of additional non-false segments that is proportional to the size of the history, i.e., $|E'| - |E| \leq |H|$.*

Lemma 5 does not guarantee that all schemas are preserved by segments in the 2D-CES after a biconditional schema change has been applied. Schema preservation is achieved by ordered histories as stated in Lemma 6.

Lemma 6. *Let $H = [\gamma_1(A_1, E, C_1), \ldots, \gamma_n(A_n, E, C_n)]$ be a history of biconditional schema changes where $\gamma_i \in \{\alpha, \beta\}$. Let E be a MD-CES with a single segment, and let each $E_i = \gamma_i(A_i, \ldots \gamma_1(A_1, E, C_1) \ldots, C_i)$ be a MD-CES. If H is ordered then E_n preserves all schemas defined by segments in E and E_1 to E_{n-1}, i.e., H is ordered implies that $\forall S(S \in E \cup E_1 \cup \ldots \cup E_{n-1} \wedge \exists S'(S' \in E_n \wedge \mathcal{A}_S = \mathcal{A}_{S'}))$.*

5 Related Work

A versioning approach to schema evolution has been proposed within the context of both OODBs and temporal databases. In OODBs, a new version of the object instances is constructed along with a new version of the schema. The *Orion* [1] schema versioning mechanism keeps versions of the whole schema hierarchy instead of the individual classes or types. Every object instance of an old schema can be copied or converted to become an instance of the new schema. The class versioning approach CLOSQL [8] provides update/backdate functions for each attribute in a class to convert instances from the format in which the instance is recorded to the format required by the application. The *Encore* [11] system provides exception handlers for old types to deal with new attributes that are missing from the instances. This allows new applications to access undefined fields of legacy instances.

Schema changes relating to time have been investigated in the context of temporal schema versioning, where proposals have been made for the maintenance of schema versions along one [7, 9, 10] or more time dimensions [2]. Two time dimensions are usually considered: *transaction time*, which tells when facts are logically present and events occur in the database, and *valid time*, which tells when facts are true and events occur in the reality [12].

In schema versioning each version associates a schema with its time pertinence specified as an interval of time stamp values. The symbol "0" denotes the special values *initiation* in transaction time (i.e. the time when the system was started) and *beginning* in valid time (i.e. the minimum value of valid time). The symbol "∞" denotes the special values *until_changed* in transaction time (which is used to timestamp a still current fact) and *forever* in valid time (i.e. the maximum value of valid time).

In this section we use 1D-CES and 2D-CES with mono- and biconditional schema changes over timestamp attributes as a yard stick to investigate temporal schema versioning. There is a strong similarity between segments in a MD-CES and schema versions. This facilitates a comparison between the effects of schema changes in both frameworks. Only changes at the intensional (schema) level are considered. The interaction between intensional and extensional versioning is considered elsewhere [5].

5.1   Transaction-Time Schema Versioning

In transaction-time schema versioning, one of the versions is the *current* schema version. Only the current version can be affected by a schema change. When this occurs, the current version is archieved and replaced by a new current version obtained by applying the schema change to the old schema. The implicit transaction-time pertinence of a schema change is always $[now; \infty]$, i.e., the schema change takes effect when recorded in the database and remains in effect until changed by another schema change [2].

Example 2. Consider the schema version with schema (N, A, P, T) and time pertinence $[0; \infty[$. Two schema changes are applied: 1) on the $2003-01-01$ an S attribute is added to the schema, and 2) on the $2003-03-01$ the A attribute is dropped.

The first schema change results in a new current schema version with schema (N, A, P, S, T) and the time pertinence of the schema change. The old schema version is archieved and its time pertinence is restricted to $2003-01-01$:

$$V_1 : (N, A, P, T) \quad [0; 2003-01-01[$$
$$V_2 : (N, A, P, S, T) \quad [2003-01-01; \infty[$$

The second schema change is then applied to the new current schema version V_2. V_1 is not considered. The result is a new current schema version V_3 and the restriction of the time pertinence of V_2:

$$V_1 : (N, A, P, T) \quad [0; 2003-01-01[$$
$$V_2 : (N, A, P, S, T) \quad [2003-01-01; 2003-03-01[$$
$$V_3 : (N, P, S, T) \quad [2003-03-01; \infty[$$

Note the similarity between the time pertinence of schema versions in Example 2 and the qualifiers of segments in 1D-CES. Assuming that T records transaction-time, then both schema versions and segments define the intended schema for an interval a timestamp values.

The implicit time pertinence of schema changes in transaction-time schema versioning corresponds to unbounded monoconditional schema changes of the form $T \geq c_{now}$, where c_{now} is the timestamp value of *now* when the schema change is recorded.

Due to the nature of transaction-time, c_{now} will increase with each additional schema change. This implies that the history of schema changes in transaction-time schema evolution is always ordered.

Lemma 2 and 3 ensure that all schemas are preserved (so schema changes do not affect archieved schema versions) and only one schema (the current schema version) is affected by a schema change splitting it into a version with the old schema and a restricted time pertinence (corresponding to a conjunction with the negated condition for the schema change) and the new current schema version. Although, transaction-time schema versioning considers only the current version when a schema change is applied, the resulting schema versions are still equivalent to the segments in 1D-CES due to transaction-time providing a natural ordering of the schema changes.

5.2   Valid-Time Schema Versioning

In valid-time schema versioning, a schema change creates a new schema version by applying the changes to a specified schema version. The new version is assigned the validity of the schema change. Previous schema versions completely overlapped by the validity of the new schema segment are deleted and schema versions which are only partially overlapped have their validity restricted accordingly [2].

The sequence of schema changes as well as the versions picked in each evolution step determine which schema versions are created.

Example 3. Consider the schema version with schema (N, A, P, V) and validity $[0; \infty[$. Two schema changes are applied: 1) attribute S is added with validity $[c_2; \infty[$, and 2) attribute C is added with validity $[c_1; \infty[$, where $c_1 = 2003-01-01$ and $c_2 = 2003-03-01$.

The table below shows the possible outcomes of applying the two schema changes. The sequence is the order in which the two attribute additions are applied. The second column indicates the result of applying the second schema change to the initial schema version, and the third column contains the schema versions resulting from applying the second schema change to the new version created by the first schema change.

Sequence	1st version		2nd version	
α_S, α_C	(N, A, P, V)	$[0; c_1[$	(N, A, P, V)	$[0; c_1[$
	(N, A, P, C, V)	$[c_1; \infty[$	(N, A, P, S, C, V)	$[c_1; \infty[$
α_C, α_S	(N, A, P, V)	$[0; c_1[$	(N, A, P, V)	$[0; c_1[$
	(N, A, P, C, V)	$[c_1; c_2[$	(N, A, P, C, V)	$[c_1; c_2[$
	(N, A, P, S, V)	$[c_2; \infty[$	(N, A, P, C, S, V)	$[c_2; \infty[$

Consider the schema changes in Example 3 isolated. According to the lossless property, we should expect that the S and C attributes both appear in the schema of tuples valid after 2003–03–01, and that none of them are in the schema prior to 2003–01–01. Additionally, between $2003 - 01 - 01$ and $2003 - 03 - 01$, only the C attribute should appear in the schema. This is achieved by three segments in 1D-CES regardless of the order in which the schema changes are applied.

In valid-time schema versioning, this is achieved by first adding the C attribute and then apply the addition of S to the new schema version resulting from applying the attribute addition of C. Note that in this case, the schema changes are ordered, and in each evolution step the schema change is applied to the schema version with a validity that extends forever (∞). The situation corresponds exactly to transaction-time schema versioning.

Valid-time schema versioning and monoconditional schema evolution differ on two points, when the conditions for transaction-time schema versioning are not met. First, there is no relation between the schema version chosen as the target for the schema change and the validity of the schema change[2]. Second, the schema change does not apply to any of the schema versions with a validity that overlaps the validity of the schema change. Instead, these schema versions are either deleted (if their validity is completely overlapped by the validity of the schema change) or restricted accordingly.

Valid-time schema versioning is not schema preservering. While it is possible for the number of schema versions to shrink as a result of a schema change (if the validity of the change overlaps several existing schema versions), the upper bound on the number of schema versions in valid-time schema versioning is proportional to the number of schema changes applied. The space complexity of 1D-CES and valid-time schema versioning are therefore the same.

5.3 Bitemporal Schema Versioning

In bitemporal schema versioning, schema versions are maintained along both transaction-time and valid-time. A schema change creates a new version by applying the changes to a specified schema version which is current. The new current version is assigned the validity of the schema change, and previous schema versions completely overlapped by the change validity are archieved, rather than being deleted as happens in valid-time schema versioning.

---

[2] There is, however, one case, where the schema version and the validity of the schema change are related. If no validity is specified for the schema change, then the schema change assumes the validity of the schema version.

Table 2. Properties of Different Evolving Schema Models

	Lossless	Schema Preserving	Space Complexity
Transaction-time Schema Versioning	Yes	Yes	$O(n)$
Ordered 1D-CES	Yes	Yes	$O(n)$
Valid-time Schema Versioning	No	No	$O(n)$
Unordered 1D-CES	Yes	No	$O(n)$
Unordered 1D-CES with schema preservation	Yes	Yes	$O(n^2)$
Bitemporal Schema Versioning	No	Yes	$O(n)$
Unordered 2D-CES	Yes	No	$O(n^2)$
Unordered 2D-CES with ordered reordering	Yes	No	$O(n)$
Ordered 2D-CES	Yes	Yes	$O(n)$
MD-CES	Yes	Yes	$O(2^n)$

Example 4. Consider the schema version with the schema (N, A, P, T, V), time pertinence $[0; \infty[_t$, and validity $[0; \infty[_v$. First, an S attribute is added on $2003-02-01$ with
validity $[2003-03-01; \infty[$. We are left with two current schema versions.

$$V_1 : (N, A, P, T, V) \quad [0; \infty[_t \qquad\qquad [0; 2003-03-01[_v$$
$$V_2 : (N, A, P, S, T, V) \; [2003-02-01; \infty[_t \; [2003-03-01; \infty[_v$$

Next, a C attribute is added on $2003-04-01$ with validity $[2003-01-01; \infty[$.
The schema change is applied to a current version. Since both V_1 and V_2 are current,
we choose one of them (V_1).

$$V_1 : (N, A, P, T, V) \quad [0; \infty[_t \qquad\qquad\qquad [0; 2003-01-01[_v$$
$$V_2 : (N, A, P, S, T, V) \; [2003-02-01; 2003-04-01[_t \; [2003-03-01; \infty[_v$$
$$V_3 : (N, A, P, C, T, V) \; [2003-04-01; \infty[_t \qquad\qquad [2003-01-01; \infty[_v$$

The validity of V_2 is completely overlapped by the validity of the schema change,
so V_2 has to be archieved by updating its time pertinence.

Note that for bitemporal schema versioning "holes" can appear in the bitemporal
domain, where no schemas are defined. E.g., the intended schema for a tuple recorded
before $2003-02-01$ and valid after $2003-03-01$ cannot be determined for bitemporal
schema versioning. Therefore, bitemporal schema versioning is not lossless. However,
it is schema preserving, because it archieves the versions with a validity that is completely overlapped by the validity of the schema change rather than deleting them as in
valid-time schema versioning. Only one new version is created in response to a schema
change, so the space complexity of bitemporal schema versioning is linear.

6 Summary

The MD-CES is a conceptual model for conditional schema changes with many desirable properties that are essential for relations that tuples with different intended
schemas. The MD-CES is both lossless and schema preserving, but has an exponential space complexity.

The paper investigates MD-CES where conditional schema changes are restricted to conditions over one or two timestamp attributes. This leads to 1D-CES and 2D-CES with a linear or polynomial space complexity. Both 1D-CES and 2D-CES are lossless and if the conditional schema changes in their histories are ordered, then they are also schema preserving.

Figure 2 compares 1D-, 2D-, and MD-CSE with transaction-time and valid-time. Ordered and unordered refers to whether the history of conditional schema changes is ordered or not, and ordered reordering refers to unordered 2D-CES, where there exists a different ordered sequence of the same conditional schema changes.

References

1. J. Banerjee, W. Kim, H.-J. Kim, and H.F. Korth. Semantics and Implementation of Schema Evolution in Object-Oriented Databases. In *ACM SIGMOD International Conference on Management of Data*, pages 311–322. ACM Press, 1987.
2. C.D. Castro, F. Grandi, and R.R. Scalas. Schema Versioning for Multitemporal Relational Databases. *Information Systems*, 22(5):249–290, 1997.
3. O.G. Jensen. *Multi-Dimensional Conditional Schema Evolution in Relational Databases*. PhD thesis, Aalborg University, 2004.
4. O.G. Jensen and M.H. Böhlen. Evolving Relations. In *Database Schema Evolution and Meta-Modeling*, volume 9th International Workshop on Foundations of Models and Languages for Data and Objects of *Springer LNCS 2065*, page 115 ff., 2001.
5. O.G. Jensen and M.H. Böhlen. Current, Legacy, and Invalid Tuples in Conditionally Evolving Databases. In *ADVIS*, volume Second International Conference, ADVIS 2002, Izmir, Turkey, October 23-25, 2002, Proceedings of *Springer LNCS 2457*, pages 65–82, 2002.
6. O.G. Jensen and M.H. Böhlen. Lossless Conditional Schema Evolution. In *ER*, volume 22nd International Conference on Conceptual Modeling, ER 2004, Shanghai, China, November 8-12, 2004, Proceedings, 14 pages, 2004.
7. L.E. McKenzie and R.T. Snodgrass. Schema Evolution and the Relational Algebra. *Information Systems*, 15(2):207–232, 1990.
8. Simon R. Monk and Ian Sommerville. Schema Evolution in OODBs using Class Versioning. *SIGMOD Record*, 22(3):16–22, 1993.
9. J.F. Roddick. SQL/SE - A Query Language Extension for Databases Supporting Schema Evolution. *ACM SIGMOD Record*, 21(3):10–16, 1992.
10. J.F. Roddick and R.T. Snodgrass. *Schema Versioning. In: The TSQL92 Temporal Query Language*. Noewell-MA: Kluwer Academic Publishers, 1995.
11. Andrea H. Skarra and Stanley B. Zdonik. The Management of Changing Types in an Object-Oriented Database. In *OOPSLA, 1986, Portland, Oregon, Proceedings*, pages 483–495, 1986.
12. R.T. Snodgrass et al. TSQL2 Language Specification. *ACM SIGMOD Record*, 23(1), 1994.

First International Workshop on Conceptual Modeling for Agents (CoMoA 2004)

at the 23rd International Conference
on Conceptual Modeling (ER 2004)

Shanghai, China, November 9, 2004

Organized by

Eleni Mangina

Preface to CoMoA 2004

Computational intelligence research originally focused on complicated, centralized intelligent systems with expertise in certain domains. However, the increasing demand for distributed problem solving led to the development of multi-agent systems. The latter are formed from a collection of independent software entities whose collective skills can be applied in complex and real-time domains. The target of such systems is to demonstrate how goal directed, robust and optimal behavior can arise from interactions between individual autonomous intelligent software agents. Their functionality and effectiveness has proven to be highly depended on the way they are represented in conceptual models. The way intelligent agents work, how the software seems to act in intelligent ways on its own, whether over computer networks or not, can be difficult to be perceived from people.

The workshop on Conceptual Modelling for Agents (CoMoA 2004) focus on the growing field of Intelligent Agents, and in particular on the way they can be modelled, designed and visualized. The target of the workshop was to bring together researchers, agent-based software developers, users and practitioners involved in the area of multi-agent systems in order to discuss the different fundamental principles for construction and design of agents, how they co-operate and communicate, what tasks can be set to accomplish and how properties can be modelled. Existing perspectives of agents' conceptual modelling within different application domains were welcome for presentation during the workshop, as well as the different conceptual model languages that can be used within different agent building environments for various application domains. The workshop provided a forum for research contributions and experiences of agents' conceptual models with the following main questions to be answered:

- Which are the currently used conceptual modelling languages for agents?
- What properties of agents within the conceptual models should be emphasized?
- What level of abstraction should an agent conceptual model have?
- Could we standardize the procedure for agents' conceptual model design?

Based on these questions, the submissions from the audience targeted included both theoretical and practical issues on conceptual modelling for agents. The organizer also wanted to give the opportunity to Ph.D. students to present their results and discuss with the experts in the field for valuable feedback. I would like to thank all the reviewers from the program committee for their help and feedback to all the papers submitted to CoMoA 2004.

November 2004   Eleni Mangina

A Sociological Framework
for Multi-agent Systems Validation and Verification

Rubén Fuentes, Jorge J. Gómez-Sanz, and Juan Pavón

Universidad Complutense Madrid, Dep. Sistemas Informáticos y Programación*
28040 Madrid, Spain
{ruben,jjgomez,jpavon}@sip.ucm.es
http://grasia.fdi.ucm.es

Abstract. Social and intentional behaviours appear as two main components of the agent paradigm. Methods of conventional software engineering do not seem to be appropriate to gain a full knowledge of these behavioural aspects, as they are not traditional software components. Their study involves new concepts and techniques, belonging to social sciences, to be integrated into software development. In this paper we show how to use the Activity Theory as a support for current Multi-Agent System methodologies. The application of the same social concepts at every stage of the development cycle allows requirements traceability, and provides methods to describe social properties and decide whether the specification satisfies them or not. The use of the approach is shown with a case study.

Keywords: Multi-Agent Systems Development, Activity Theory, Activity Checklist, Requirements Elicitation

1 Introduction

Under the agent-oriented paradigm, systems are conceived as collections of autonomous entities that collaborate in order to achieve their common goals. This conceptualization demands the study of the organizational, cognitive, developmental, and motivational components present in many Multi-Agent Systems (MAS). We find that most of existing MAS methodologies do not fully address these *social properties*. Usually, they consider these as a problem of defining roles, responsibilities, and power relationships (as it is the case for INGENIAS [16], KAOS [5], or DESIRE [2]). However, social features comprehend a wider range of information that demands additional theoretical background, abstractions, and techniques.

We think that the most evident source for the required knowledge to work with these features is to be found in social sciences, such as Philosophy, Sociology, or Psychology. The basis for exporting the knowledge of social disciplines into the field of Agent Oriented Software Engineering is found in the existence of certain features which are shared by agents and humans. Despite obvious similarities from a conceptual point of view, their respective meanings and applicability are not quite compara-

---

* This work has been funded by Spanish Ministry of Science and Technology under grant TIC2002-04516-C03-03.

S. Wang et al. (Eds.): ER Workshops 2004, LNCS 3289, pp. 458–469, 2004.

ble. Software Engineering makes use of "design" languages whilst Psychology or Sociology make use of natural language; furthermore, the algorithmic methods applied during systems development are very different from the ones applied to the study of populations. Thus, a thorough review of theories and the development of adapting tools are required to allow transferring results from one field to another.

Our previous work [6], [7] establishes the suitability of one of these social theories, the Activity Theory (AT) [19], as a model for certain social features in MAS. AT is a cross-disciplinary framework for the study of human doings that considers all labour as simultaneously individual and social. The social component includes the actual society in which a subject carries out the activity, but also the historical development of those activities. AT cannot be used directly in a development process since its concepts are quite abstract. To deal with this limitation, we have proposed to describe AT concepts using a UML notation. This approach to modelling social aspects was presented in two works. In [6] we used the AT as a basis to detect contradictions in the MAS design process. In [8], a similar AT based schema was the tool to identify requirements. This research suggests that *contradictions* and *requirements* are particular cases of a more generic framework that we associate with problems addressed in the definition of *social properties*. Therefore, here we generalize those results to apply AT like a social framework for MAS validation and verification.

So far, we have identified three types of useful social properties: For any given specification, these properties can represent configurations of the system that developers have to preserve (e.g. requirements), or to avoid (e.g. contradictions in the information), or knowledge about the MAS (e.g. the existing organization). In all these cases, developers may need to adapt the information about the system along the development process. For this purpose, social properties are described by match patterns, to detect them, and solution patterns, to advise changes in the specifications.

The application of these patterns to a concrete MAS methodology is possible once concepts from AT are mapped to those in the MAS methodology. The use of mappings allows the applicability of this approach in different MAS methodologies and a semi-automated method to work with social properties.

The present paper is organized in six additional sections. The following one discusses about the need of considering new tools to deal with the social and intentional knowledge involved in the specifications of MAS. Section 3 shows the way of describing social properties in terms of the language used in our approach to express the AT abstractions. Some examples of these properties in the three identified categories are provided in section 4. Then, section 5 describes a method, which can be automated, to use the social properties in the validation and verification of specifications. A case study on a real specification follows in section 6. Finally, in the conclusions section, a discussion of the results obtained in the validation and verification of social and intentional properties of MAS with this approach is provided.

2   The Need of Tools for Social Features in MAS

Software Engineering involves a continuous research about new concepts and methodologies that make possible building more complex software systems. One of the main advantages of the agent paradigm is that it constitutes a natural metaphor for systems with purposeful interacting agents, and this abstraction is close to the human way of thinking about our own activities [20]. This foundation has lead to an increasing interest in social sciences (like in the works described in [14], [17], and [20]) as a source for new concepts and methods to build MAS. However, this interest has hardly covered some interesting social features of the MAS design. Here, the term *social features* encompasses organization culture, politics, leadership, motivation, morale, trust, learning, or change management, in the line of [4]. There are two main reasons in MAS research to give a novel and special attention to these features: the human context and the own essence of the agent abstractions.

In the first place, the environment of a software system, defined as the real world outside it, is usually a human activity system [15]. Its study must consider then the *social features* of humans and their societies. Software Engineering research in branches like Requirements Engineering [9], CSCW [10], or HCI [3], already makes an extensive use of sciences like Ethnography, Sociolinguistics, Organization Theory, or Cognitive Psychology, to grasp the relevant information about the human context.

Besides the human context, MAS are modelled in a very alike fashion to human organizations, as societies of collaborating intentional entities [18]. It gives a possibility of describing some properties of the system and its context, the social and intentional ones, in a quite uniform way, taking advantage of the knowledge extracted from human sciences.

These two reasons take us to consider the need of generic mechanisms to model these social properties of the MAS and to validate the specifications against them. These properties do not appear in traditional software methodologies. If MAS design has to be a new level of abstraction for system design, it needs to develop innovative tools that consider these social concepts. Previous experience with the description and checking of social properties about contradictions [6] and requirements [8], take us to propose the use of the AT [19] for this purpose. This work in [6], [7], and [8] shows that AT can provide developers a language to describe social properties, methods to verify them, and a rich library of case studies.

3   Describing *Social Properties* with AT

From a development point of view, a social property can represent several types of information; it can correspond to a configuration to keep, i.e. a pattern property, or to avoid, i.e. an anti-pattern property, or a feature to discover, i.e. a descriptive property. In the first case, a problem with the property occurs when there is no match with the pattern or just a partial one. For configurations to avoid, problems arise when there is a complete match. The descriptive properties help developers to discover new information about a given system.

In our approach, the representation of social properties has to fulfil three aims. Firstly, the properties have to support the development. Their description should be in a language understandable by developers and suitable for automated processing. Secondly, the representation should serve as a tool for communication between customers and developers. Social abstractions are close to both to customers and developers and should, therefore, facilitate their mutual understanding [1]. Finally, the representation of properties should help to solve the problems related with themselves, such as the non-accomplishment of requirements or the appearance of contradictions.

To satisfy these requirements, social properties are represented with two components: a set of match patterns and other of solution patterns. Each pattern of these components has two possible representations, a textual form and another one based on the use of UML stereotypes to represent AT concepts [6]. The UML form is the basis for the automated process of pattern detection and problem solving. Only the UML form of a match pattern is mandatory for the checking process. In this process, the stereotypes and names of the UML form can be variables or fixed values. This allows fixing some values of the properties before the detection procedure and combining patterns through shared values. On the other hand, the textual form is intended to give further information about the patterns. It helps customers to understand the meaning of the used UML notation and enables both customers and developers to know the social interpretation of the pattern.

According to our description, a social property can include two different sets of patterns: one for detection and other for solution. A match pattern describes a set of entities and their relationships, which represents the property. It acts as a frame that has to be instantiated with information from the specification. If a set of artefacts in the specification fits into the pattern, the property is satisfied. Both in the case of properties to keep and properties to avoid, the solution pattern is a rearrangement of its corresponding match pattern, maybe with additional elements. Solution patterns can correspond to partial or full matches. For the total absence of match, there is no point to define a solution.

It is remarkable to note that patterns for detection and solution are not tied in fixed pairs; moreover, they can be reused or combined through shared variables to describe new situations. An example of this possibility can be found in the case study described in section 6, which combines match patterns to describe situations that are more complex than originally.

The following section provides several examples for properties definition according to these guides.

4   Social Properties for MAS

Following the previous introduction, this section presents examples of properties used in MAS development for every identified type and gives their representation. The textual form of these properties is included as part of the explanation of the diagrams. So, we will show a pattern property, an anti-pattern property, and a descriptive property. The first one corresponds to a requirement describing a social setting in the system environment, and it is adapted from the Activity Checklist [12]. The second one describes a contradiction according to the AT, the Social Objective contradiction extracted from [13]. Finally, there is a description of a hierarchical organization as described in [18].

The Activity Checklist [12] is a guideline to elicit contextual knowledge about an *activity*. It is composed by a set of aspects that have questions expressed in terms of natural language to grasp their information. One of these aspects is the influence that the access to resources or devices has over *activities*. A related question with this aspect is "Has every user access to the device? If not, who has it? Who does not have it?". The representation of this question can be seen in Fig. 1. The diagram shows that the user's access to the device has to be considered in the context of an *activity* in

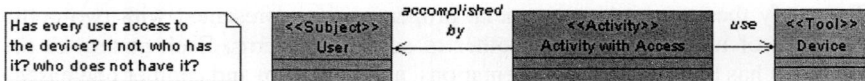

Fig. 1. A question about the system context.

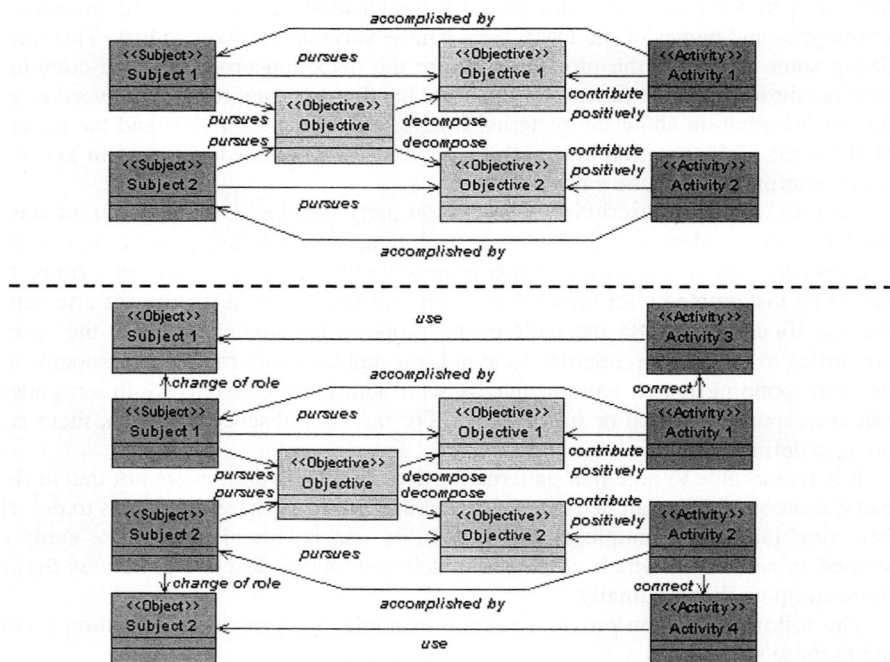

Fig. 2. The Social Objective contradiction above and a possible solution below.

which he participates. The study of these activities helps developers and customers to establish relationships among users according to their dependencies.

The second example is an adaptation of the Social Objective contradiction described by [13]. This contradiction emerges in a MAS when a goal that is individually pursued by the members of the *community* is satisfied through partial goals. Every agent only accomplished tasks related to these partial goals and have no information about the current state of the other goals. If there is no communication between the agents, they cannot know whether the global goal that they pursue is accomplished or not. The upper middle of Fig. 2 shows this situation.

One possible solution to a Social Objective situation is making explicit the need that agents have of sharing information about the results of their own *activities* with the other members of the *community*. This can be done including new *activities* to inform other agents about the status of the partial goals. The model could be changed to include this information as it is shown in the lower middle of Fig. 2. The relation "change of role" for *subjects* illustrates that they are the target of communication *activities*.

The final example is an identification of the social structures in MAS. In the overview about MAS by [18], several types of organizations for the community of agents were identified. One of them is the *hierarchy*, which is as an organization where "The authority for decision making and control is concentrated in a single problem solver [...]. Superior agents exercise control over resources and decision making.". This situation can be described with Fig. 3. The *Superior Agents* can give orders that *Subordinated Agents* have to accomplish, that is, *Superior Agents* are able to generate new *objectives* for their *Subordinated Agents*. The relationships labelled "*change of role*" represents that the *outcome* of the *activity* that generates the orders, i.e. the *Objective*, becomes an *objective* for the *Subordinated Agent*.

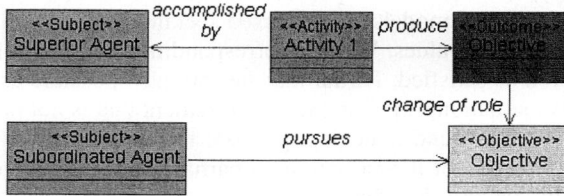

Fig. 3. A hierarchical organization in a MAS.

Detecting this kind of descriptive-properties about the MAS gives developers information about how is their system. This acquired knowledge can help them to decide the best design solution to challenges in their systems. Examples of these are ways to solve negotiation or communication problems in the MAS according to its organizational structure.

5   Checking a Specification with Social Properties

This section introduces a method to check social properties against a specification. The method is a generalization of the one used for AT contradictions [7]. The process begins with the mappings between the AT vocabulary and that of the agent oriented methodology, a set of social properties to check, and the MAS specification to verify.

This method intends to be generic in the sense that it could be applied to any agent oriented methodology, without demanding a vocabulary based on AT concepts. One of its parameters is the mappings between the concepts of AT and the given agent oriented methodology. These mappings allow the translation between both groups of abstractions. A more detailed description on how to build these mappings and an example with the INGENIAS methodology [16] can be found in [7].

The properties to be verified are social properties as described in preceding sections. They can be predefined, e.g. requirements [8] or contradiction patterns [6] of the AT, or defined by users. The process itself can then be regarded as "validation", i.e. when its patterns are requirements, and "verification", i.e. when its patterns represent other kind of properties.

The third parameter is a set of views that describe the MAS to check with the social properties. Since the process uses mappings, there are no prerequisites about the language of the specification, as long as it based in the agent paradigm.

The checking process includes the following steps:

1. Translate the MAS specifications to the AT language. The mappings describe how structures defined in one language can be written in the other. Translation is not a trivial task since the correspondences between structures and their translations are not unique. In general, mappings are relationships "many to many" between structures of two languages. So, the translated structure needs to keep reference to the original one.
2. For every property, look for correspondences of its match patterns in the specification. The process of detecting properties is one of pattern matching. A pattern is a set of entities and relations between them. Some of the slots in the pattern, such as names, stereotypes, or specific properties of entities, can be variables or fixed values. According to them, models are traversed seeking groups of elements with the same structure and slot values. When a corresponding structure is found, the property is considered as satisfied. Partial matches are also possible as they can represent a conflictive situation, for example a requirement that is not preserved.
3. For a property with a found matching, propose its customized solution pattern. A match pattern, or a part of it representing a partial match, can have a related solution pattern. This pattern describes a change in the models to solve a problem or improve some aspect of the specifications. Typically, these solutions are reconfigurations of the elements involved in the match pattern where additional elements can be involved.

This method allows developers and users checking their models for the satisfaction of social properties in a semi-automated way. User interaction is needed to determine values for the variables in patterns, decide when a match makes real sense, or the best manner to modify models.

The main advantages of the overall approach over others applied to the use of methods from social sciences in Software Engineering (like those in [1], [9], [10] and [14]) are that this proposal uses UML, which is well known for developers and more adequate for users than formal languages, and it provides a structured method to work with its social properties.

6   Case Study

In order to show how to apply the social properties and the previous checking process, this paper presents a case study based on agent teams in *Robocode* [11]. *Robocode* simulates tank battles through robots that actuate with predefined primitives (like ahead, turn left, and fire) and perceive the situation with position and radar sensors and the detection of some events. Developers have to program the behaviour of those tanks to destroy their enemies and survive the battle. The winner of the battle is the team with the last surviving tanks. The proposed case study considers collaboration between tanks in this frame. It is modelled with the INGENIAS methodology [16] and its full specification can be found at *http://ingenias.sourceforge.net*.

The case study regards armies of collaborative tanks (i.e. *DynamicArmies*) composed by squadrons (i.e. *DynamicSquadrons*). In every squadron, agents can play one of two roles:

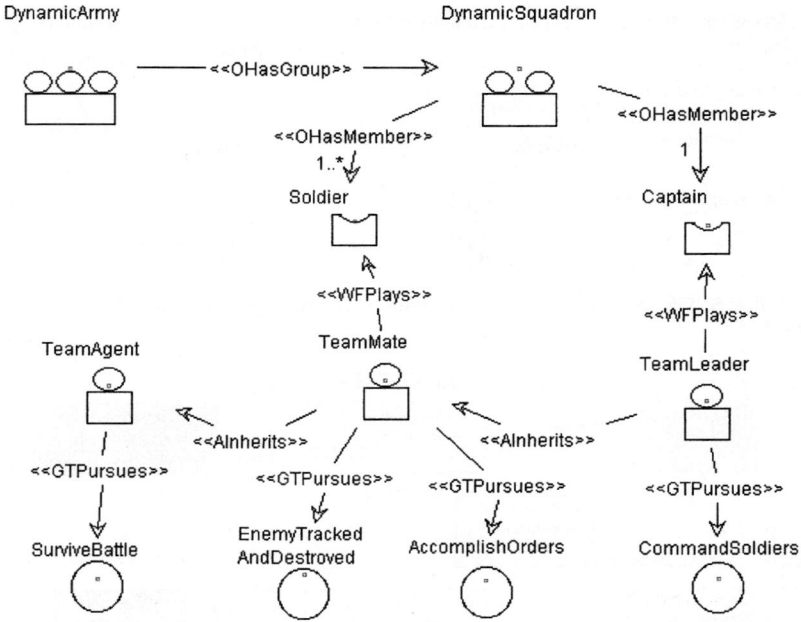

Fig. 4. Specification of a *Robocode* army.

- *Soldiers*. These are the basic members of the squadron. They try to survive the battle while destroying their enemies and accomplishing the orders of their *captain*.
- *Captains*. They are the leaders of the squadrons. Showing the basic performance of a soldier, they also have to command their troops. A captain plans a strategy for its soldiers, communicates it to them, tracks the course of the battle, and makes modifications to the planning if needed.

A squadron can have just one captain but several soldiers. The previous situation is summarized in Fig. 4 and Fig. 5.

Fig. 4 shows the structure of the army. The elements involved are: *DynamicArmy*, which is an organization; *DynamicGroup*, which is a group; *TeamAgent*, *TeamMate* and *TeamLeader*, which are agents; *Soldier* and *Captain*, which are roles; circles represent goals. A *TeamLeader* has the goal of *CommandSoldiers*. When it executes the task to satisfy this objective, it generates and communicates orders to its troops. A *TeamMate* of these troops tries to satisfy the goal *AccomplishOrders*. Therefore, it tries to obey the orders of its *TeamLeader*. A common goal for all the agents, which is inherited from the *TeamAgent*, is *SurviveBattle*. This goal forces the agents to preserve their life, whatever the situation can be.

Fig. 5 shows how the chain of command works. The *TeamLeader* studies the battle situation and *GeneratePlan* according to it. Then, the *TeamLeader* communicates its plan to their troops as *SoldierOrders*. *TeamAgents* follow these orders to satisfy their goal of *AccomplishOrders*.

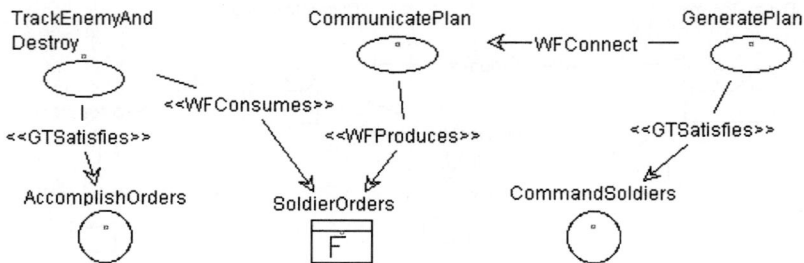

Fig. 5. The implementation of the chain of command in a *DynamicSquadron*.

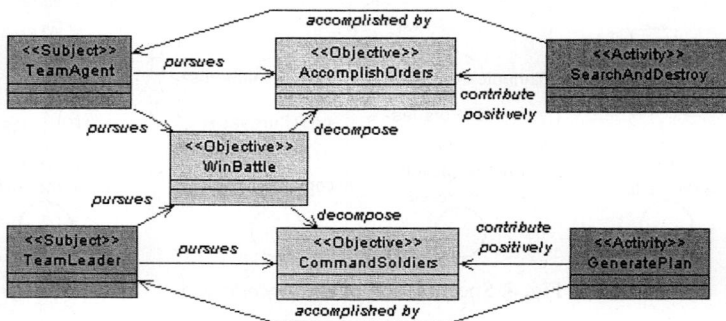

Fig. 6. Social Objective contradiction for the chain of command.

Since the final goal of a *DynamicSquadron* is to win the battle, this is not an optimal strategy. To win the battle, the squadron has to destroy its enemies and keep alive their own tanks. With this implementation, the *TeamLeader* has to plan, track down enemies, and control their troops, what can involve a work overload. So, reports of *TeamAgents* about their own state or the battle situation can help the *TeamLeader* to focus about conceiving the best strategy. This kind of information is not provided now. This situation corresponds to a variant of the Social Objective contradiction, as shown in Fig. 6. In this case, the *TeamLeader* provides feedback to the *TeamAgents* with the *SoldierOrders*, but the *TeamAgents* to the *TeamLeader* do not.

Given the Social Objective contradiction, the solution previously introduced would suggest the addition of a new *activity* to be carried out by the *TeamAgent*, which would allow him to inform its *TeamLeader*. This solution appears in Fig. 7. Of course, the solution pattern just provides a hint about how the contradiction can be solved. Developers have to decide, for instance, what the reported information is or when the *TeamAgent* should inform.

Independently of the solution adopted, the final decisions about the adequacy of the match, the proposed solution, and other patterns always require human judgment. For this reason, the use of additional analysis techniques combined with those of the AT will certainly provide new insights into the field of MAS.

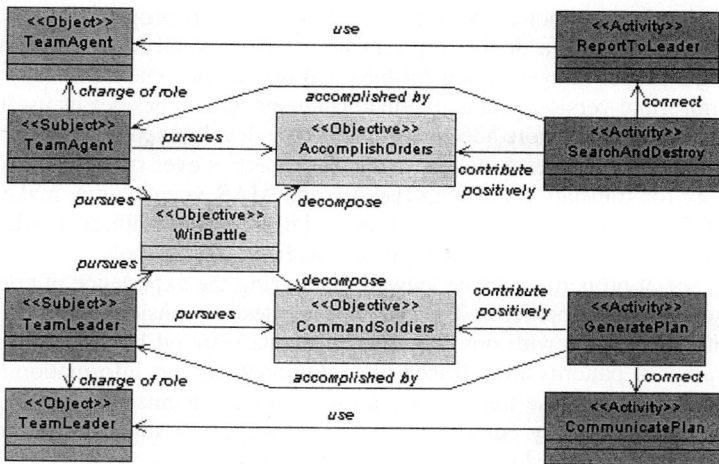

Fig. 7. Solution for the Social Objective contradiction in the chain of command.

7 Conclusions

This paper introduces a new approach to model and check social properties in MAS specifications. Social sciences provide an adequate framework for the explanation of many crucial features of the agent paradigm. More research is still needed for the generic modelling of social properties and their use in the development process. In this sense, the Agent Oriented Software Engineering community could obtain a better understanding of the human environment and its mutual ways of interaction with the system, and an improvement in the conceptual and methodological tools in use.

Based on earlier work [6], [7], [8], this proposal uses the Activity Theory as the foundation to study these social properties:

- Its theoretical framework related to social sciences gives a rich groundwork to understand the meaning of these properties and their effects in the society. It encompasses aspects such as organization, motivation, contradictions, or adaptability, which are major concerns for MAS.
- Its vocabulary gives a comprehensive and limited set of concepts to model properties. Given that this vocabulary involves common concepts for human life, it proves to be useful for communication between customers and developers.
- Its methodology inspires a guide for software development that comprises the description of social patterns, their detection in the current organization, and suggestions for the possible development according to previous experiences.
- Its case studies constitute a source of properties that can be relevant in MAS.

This knowledge of the AT has contributed to MAS development by providing a vocabulary for the definition of social properties, a method to verify them in specifications, and a wide range of properties and solutions.

From our point of view, customers, who live in the real environment, know the social properties of the system, but properties also need to be understood by the developers who implement them. The representation of social properties is intended to cover both aspects with a double representation. The textual form explains the patterns, being a proper representation for informal discussion. The UML representation gives a semiformal version with a graphical language which, besides its usefulness for discussion, proves to be more adequate for software development.

This paper also proposes a method to check properties over specifications. The use of mappings for translation between the AT and MAS vocabularies makes it independent of the methodology. Representing social properties with an UML language gives the possibility of supporting the process with automated tools.

Finally, social properties can be viewed as carrying the experience of concrete AT studies and MAS developments. The social properties are considered as pairs where a given context is related with possible improvements inspired by earlier experiences. Thus, libraries of patterns are released like additional support information for development. Examples of these libraries are those for requirements based in the Activity Checklist, which currently comprises twenty properties, or those for contradictions according to AT studies, which includes ten properties at the moment.

We have identified three open issues in the current status of our work. The first one is how to get a higher degree of automation. Social properties are a tool for the interactive study of specifications. That is to say that customers and developers customise properties of their interest and have to judge results due to the inherent uncertainty and ambiguity of mutual understandings and specifications in evolution. Probably, more detailed patterns and increased reasoning capabilities in the checking method are in way to reduce users work. The second issue, highly connected with the first one, is the richness of the patterns vocabulary. AT allows concepts to express the social properties but they still need to be used by practitioners who work with natural language. Since we try to express that shades of meaning in a semiformal language, proper primitives have yet to be defined. Finally, the study and enrichment of a most suitable set of properties is pending work too.

References

1. [Bødker & Grønbœk 1991] Bødker, S., Grønbœk, K.: *Cooperative prototyping: users and designers in mutual activity*. International Journal of Man-Machine Studies 34 (3). 1981.
2. [Brazier *et al.* 1997] Brazier, F. M. T., Dunin-Keplicz, B. M., Jennings, N. R., and Treur, J.: *DESIRE: Modelling Multi-Agent Systems in a Compositional Formal Framework*. International Journal of Cooperative Information Systems, special issue on Formal Methods in Cooperative Information Systems: Multi-Agent Systems. 1997.
3. [Button & Dourish 1996] G. Button, P. Dourish: *Technomethodology: Paradoxes and Possibilities*. In ACM Proceedings of CHI'96, pp. 19-26, Vancouver. 1996.
4. [Constantine 1995] L. L. Constantine: *Constantine on Peopleware*. Englewood Cliffs, NJ: Yourdon Press. 1995.
5. [Dardenne *et al.* 1993] A. Dardenne, A. van Lamsweerde, S. Fickas: *Goal-directed Requirements Acquisition*. Science of Computer Programming, Vol. 20 pp. 3-50. 1993.
6. [Fuentes *et al.* 2003a] R. Fuentes, J.J. Gómez-Sanz, J. Pavón: *Activity Theory for the Analysis and Design of Multi-Agent Systems*. In Proceedings of the 4th International Workshop on Agent Oriented Software Engineering (AOSE 2003), Melbourne, Australia, July 2003. Vol. 2935 of LNCS, pp. 110–122. Springer Verlag. 2003.

7. [Fuentes *et al.* 2003b] R. Fuentes, J.J. Gómez-Sanz, J. Pavón.: *Social Analysis of Multi-Agent Systems with Activity Theory.* In Proceedings of CAEPIA 2003, San Sebastian, Spain, November 2003. Vol. 3040 of LNAI. Springer Verlag. 2004.
8. [Fuentes *et al.* 2004] R. Fuentes, J.J. Gómez-Sanz, J. Pavón.: *Towards Requirements Elicitation in Multi-Agent Systems.* In Proceedings of the 4th International Symposium From Agent Theory to Agent Implementation (AT2AI 2004), Vienna, Austria, April 2004.
9. [Goguen 1992] Goguen J. A.: *The Dry and the Wet. Information System Concepts*, pp. 1-17. Elsevier North-Holland, 1992.
10. [Hughes *et al.* 1994] Hughes, J., King, V., Rodden, T., Andersen, H.: *Moving out from the control room: ethnography in system design.* In Proceedings of the ACM 1994 Conference on Computer Supported Cooperative Work (CSCW'94). ACM Press, pp. 429-439. 1994.
11. [IBM 2002] IBM alphaWorks: *Robocode.* 2002. http://robocode.alphaworks.ibm.com
12. [Kaptelinin *et al.* 1999] V. Kaptelinin, B. A. Nardi, C. Macaulay: *The Activity Checklist: A tool for representing the "space" of context.* Interactions, 6 (4), pp. 27-39. 1999.
13. [Leontiev 1981] Aleksie Nikolaevich Leontiev: *Problems of the development of the mind.* Moscow, Progress. 1981.
14. [Malsch 2001] Thomas Malsch: *Naming the Unnamable: Socionics or the Sociological Turn of/to Distributed Artificial Intelligence.* Autonomous Agents and Multi-Agent Systems 4 (3): 155-186. 2001.
15. [Nuseibeh & Easterbrook 2000] B. Nuseibeh and S. Easterbrook: *Requirements Engineering: A Roadmap.* In Proceedings of the 22nd International Conference on Software Engineering (ICSE-2000), Limerick, Ireland, June 2000. ACM Press.
16. [Pavón & Gómez-Sanz 2003] Pavón J., Gómez-Sanz, J.: *Agent Oriented Software Engineering with INGENIAS.* In V. Marík, J. Müller, and M. Pechoucek, eds, Multi-Agent Systems and Applications III, Vol. 2691 of LNCS, pp. 394-403. Springer Verlag. 2003.
17. [Sichman & Demazeau 2001] J. Sichman, Y. Demazeau: *On Social Reasoning in Multi-Agent Systems.* Inteligencia Artificial, Revista Iberoamericana de Inteligencia Artificial, Special Issue on Development of Multi-Agent Systems, n°13, pp. 68-84, AEPIA. 2001.
18. [Sykara 1998] Katia P. Sykara: *Multiagent systems.* AI Magazine 19(2). 1998.
19. [Vygotsky 1978] Vygotsky: *Mind and Society.* Cambridge MA, Harvard University. 1978.
20. [Wooldridge & Ciancarini 2001] Wooldridge M., Ciancarini P.: *Agent-Oriented Software Engineering: The State of the Art.* In P. Ciancarini and M. Wooldridge, editors, Agent-Oriented Software Engineering. Springer-Verlag LNAI Vol. 1957. January 2001.

Spatio-temporal Relevant Logic as the Logical Basis for Specifying, Verifying, and Reasoning About Mobile Multi-agent Systems

Jingde Cheng

Department of Information and Computer Sciences, Saitama University
Saitama, 338-8570, Japan
cheng@aise.ics.saitama-u.ac.jp

Abstract. To specify, verify, and reason about mobile multi-agent systems, we need a fundamental logic system to provide us with a criterion of logical validity as well as a representation and specification language. Because design and development of mobile multi-agent systems primarily concern that what decisions and how the decisions can be made by mobile agents with incomplete or even inconsistent knowledge acting concurrently in spatial regions changing over time, the fundamental logic must be able to underlie truth-preserving and relevant reasoning in the sense of conditional, ampliative reasoning, paracomplete reasoning, paraconsistent reasoning, spatial reasoning, and temporal reasoning. Since no existing logic system can satisfy the requirements, this paper proposes a new family of logic, named "spatio-temporal relevant logic," and shows that it is a hopeful candidate for the fundamental logic we need.

1 Introduction

Mobile multi-agent computing systems, originally studied in the area of computer science, are a general and natural computational model for understanding, representing, simulating, analyzing, and solving various problems in not only computer science but also other scientific disciplines, and therefore, have many applications in natural science, human science, and social science.

A *multi-agent computing system* is one that consists of a number of computing entities running concurrently on a computer or computer network, called *agents*, which are capable of independent and autonomous action in some degree and capable of interacting (cooperating, coordinating, negotiating) with one another in order to successfully carry out their own tasks [25-27]. In general, the agents in a multi-agent system is representing or acting on behalf of their users or owners with different motivations and/or goals. To build agents that are capable of independent and autonomous action and interacting with one another, we have to program each agent such that it knows its own task, reasons about the knowledge and behavior of other agents based on its own knowledge, and makes correct decisions to carry out its own task. A *mobile multi-agent computing system* is a multi-agent system where all agents are capable of moving or of being moved readily from place to place in a computer net-

S. Wang et al. (Eds.): ER Workshops 2004, LNCS 3289, pp. 470–483, 2004.

work. In general, the tasks, knowledge, and behavior of mobile agents are concerned with spatial and/or temporal objects.

To specify, verify, and reason about mobile multi-agent systems, we need a fundamental logic system to provide us with a criterion of logical validity as well as a representation and specification language. Because design and development of a mobile multi-agent system primarily concern that what decisions and how the decisions can be made by mobile agents with incomplete or even inconsistent knowledge acting concurrently in spatial regions changing over time, the fundamental logic must be able to underlie truth-preserving and relevant reasoning in the sense of conditional, ampliative reasoning, paracomplete reasoning, paraconsistent reasoning, spatial reasoning, and temporal reasoning.

There are many classical, modal, epistemic logic systems proposed for mobile systems, agent systems, and multi-agent systems [4, 14, 17, 18, 21, 26]. These logics are some how based on classical logic, and therefore, they cannot underlie truth-preserving and relevant reasoning in the sense of conditional, ampliative reasoning, paracomplete reasoning, and paraconsistent reasoning. Relevant logics cannot underlie spatial reasoning and temporal reasoning, even though them can underlie relevant reasoning, truth-preserving reasoning, ampliative reasoning, paracomplete reasoning, and paraconsistent reasoning well [1, 2, 5, 7, 8, 13, 16, 20]. Therefore, no existing logic system can satisfy the above requirements.

This paper proposes a new family of logic, named "*spatio-temporal relevant logic*," and shows that it is a hopeful candidate for the fundamental logic to underlie specifying, verifying, and reasoning about mobile multi-agent systems.

2 Basic Notions

Reasoning is the *process* of drawing *new conclusions* from given premises, which are already known facts or previously assumed hypotheses to provide some *evidence* for the conclusions (Note that how to define the notion of 'new' formally and satisfactorily is still a difficult open problem until now). Therefore, reasoning is intrinsically ampliative, i.e., it has the function of enlarging or extending some things, or adding to what is already known or assumed. In general, a reasoning consists of a number of arguments *in some order*. An *argument* is a set of *statements* (or *declarative sentences*) of which one statement is intended as the *conclusion*, and one or more statements, called "*premises*," are intended to provide some evidence for the conclusion. An argument is a conclusion standing in relation to its supporting evidence. In an argument, a claim is being made that there is some sort of *evidential relation* between its premises and its conclusion: the conclusion is supposed to *follow from* the premises, or equivalently, the premises are supposed to *entail* the conclusion. Therefore, the correctness of an argument is a matter of the *connection* between its premises and its conclusion, and concerns the *strength* of the relation between them (Note that the correctness of an argument depends neither on whether the premises are really true or not, nor on whether the conclusion is really true or not). Thus, there are some fundamental questions: What is the criterion by which one can decide whether the conclusion of an argument or a reasoning really does follow from its premises or not? Is there the only one criterion, or are there many criteria? If there are many criteria, what are the intrinsic differences between them? It is logic that deals with the validity of argument and reasoning in a general theory.

A *logically valid reasoning* is a reasoning such that its arguments are justified based on some *logical validity criterion* provided by a logic system in order to obtain correct conclusions (Note that here the term 'correct' does not necessarily mean 'true'). Today, there are so many different logic systems motivated by various philosophical considerations. As a result, a reasoning may be valid on one logical validity criterion but invalid on another. For example, the *classical account of validity*, which is one of fundamental principles and assumptions underlying classical mathematical logic and its various conservative extensions, is defined in terms of *truth-preservation* (in some certain sense of truth) as: an argument is valid if and only if it is impossible for all its premises to be true while its conclusion is false. Therefore, a classically valid reasoning must be *truth-preserving*. On the other hand, for any correct argument in scientific reasoning as well as our everyday reasoning, its premises must somehow be *relevant* to its conclusion, and vice versa. The *relevant account of validity* is defined in terms of *relevance* as: for an argument to be valid there must be some connection of meaning, i.e., some relevance, between its premises and its conclusion. Obviously, the relevance between the premises and conclusion of an argument is not accounted for by the classical logical validity criterion, and therefore, a classically valid reasoning is not necessarily relevant.

Proving is the process of finding a justification for an explicitly *specified statement* from given *premises*, which are already known facts or previously assumed hypotheses to provide some *evidence* for the specified statement. A *proof* is a description of a found justification. A *logically valid proving* is a proving such that it is justified based on some logical validity criterion provided by a logic system in order to obtain a correct proof. The most intrinsic difference between reasoning and proving is that the former is intrinsically prescriptive and predictive while the latter is intrinsically descriptive and non-predictive. The purpose of reasoning is to find some new conclusion previously unknown or unrecognized, while the purpose of proving is to find a justification for some specified statement previously given. Proving has an explicitly given target as its goal while reasoning does not. Unfortunately, until now, many studies in Computer Science and Artificial Intelligence disciplines still confuse proving with reasoning.

Logic deals with *what entails what* or *what follows from what*, and aims at determining which are the correct conclusions of a given set of premises, i.e., to determine which arguments are valid. Therefore, the most essential and central concept in logic is the *logical consequence relation* that relates a given set of premises to those conclusions, which validly follow from the premises. To define a logical consequence relation is nothing else but to provide a logical validity criterion by which one can decide whether the conclusion of an argument or a reasoning really does follow from its premises or not. Moreover, to answer the question what is the correct conclusion of given premises, we have to answer the question: correct for what? Based on different philosophical motivations, one can define various logical consequence relations and therefore establish various logic systems.

In logic, a sentence in the form of 'if ... then ...' is usually called a **conditional proposition** or simply **conditional** which states that there exists a relation of sufficient condition between the 'if' part and the 'then' part of the sentence. In general, a conditional must concern two parts which are connected by the connective 'if ... then ...' and called the **antecedent** and the **consequent** of that conditional, respectively. The truth of a conditional depends not only on the truth of its antecedent and consequent but also, and more essentially, on a necessarily relevant and conditional relation between them. The notion of conditional plays the most essential role in reasoning because any reasoning form must invoke it, and therefore, it is historically always the most important subject studied in logic and is regarded as the heart of logic [1].

When we study and use logic, the notion of conditional may appear in both the **object logic** (i.e., the logic we are studying) and the **meta-logic** (i.e., the logic we are using to study the object logic). In the object logic, there usually is a connective in its formal language to represent the notion of conditional, and the notion of conditional, usually represented by a meta-linguistic symbol, is also used for representing a logical consequence relation in its proof theory or model theory. On the other hand, in the meta-logic, the notion of conditional, usually in the form of natural language, is used for defining various meta-notions and describing various meta-theorems about the object logic.

From the viewpoint of object logic, there are two classes of conditionals. One class is empirical conditionals and the other class is logical conditionals. For a logic, a conditional is called an **empirical conditional** of the logic if its truth-value, in the sense of that logic, depends on the contents of its antecedent and consequent and therefore cannot be determined only by its abstract form (i.e., from the viewpoint of that logic, the relevant relation between the antecedent and the consequent of that conditional is regarded to be empirical); a conditional is called a **logical conditional** of the logic if its truth-value, in the sense of that logic, depends only on its abstract form but not on the contents of its antecedent and consequent, and therefore, it is considered to be universally true or false (i.e., from the viewpoint of that logic, the relevant relation between the antecedent and the consequent of that conditional is regarded to be logical). A logical conditional that is considered to be universally true, in the sense of that logic, is also called an **entailment** of that logic. Indeed, the most intrinsic difference between various different logic systems is to regard what class of conditionals as entailments, as Diaz pointed out: "The problem in modern logic can best be put as follows: can we give an explanation of those conditionals that represent an entailment relation?" [12]

3 The Logical Basis for Specifying, Verifying, and Reasoning About Mobile Multi-agent Systems

The question, "Which is the right logic?" invites the immediate counter-question "Right for what?" Only if we certainly know what we need, we can make a good choice. The fundamental logic system to underlie specifying, verifying, and reasoning about mobile multi-agent systems must satisfy the following four essential requirements.

First, as a general logical criterion for the validity of reasoning as well as proving, the logic must be able to underlie relevant reasoning as well as truth-preserving reasoning in the sense of conditional, i.e., for any reasoning based on the logic to be valid, if its premises are true in the sense of conditional, then its conclusion must be relevant to the premises and true in the sense of conditional.

Second, the logic must be able to underlie ampliative reasoning in the sense that the truth of conclusion of the reasoning should be recognized after the completion of the reasoning process but not be invoked in deciding the truth of premises of the reasoning. From the viewpoint to regard reasoning as the process of drawing new conclusions from given premises, any meaningful reasoning must be ampliative but not circular and/or tautological.

Third, the logic must be able to underlie paracomplete reasoning and paraconsistent reasoning. In particular, the so-called principle of Explosion that everything follows from a contradiction cannot be accepted by the logic as a valid principle. In general, our knowledge about a domain as well as a scientific discipline may be incomplete and/or inconsistent in many ways, i.e., it gives us no evidence for deciding the truth of either a proposition or its negation, and/or it directly or indirectly includes some contradictions. Therefore, reasoning with incomplete and/or inconsistent information and/or knowledge is the rule rather than the exception in our everyday lives and almost all scientific disciplines. In a rational and practical agent system, every agent should have the ability of paracomplete reasoning and paraconsistent reasoning. Thus, this requires that the fundamental logic system logic must be able to underlie paracomplete reasoning and paraconsistent reasoning.

Finally, the logic must be able to underlie spatial reasoning and temporal reasoning. In a mobile multi-agent system, agents act concurrently in spatial regions changing over time. Therefore, the tasks, knowledge, and behavior of mobile agents are primarily concerned with spatial and/or temporal objects. This naturally requires that the fundamental logic system logic must be able to underlie spatial reasoning and temporal reasoning.

Classical mathematical logic (**CML** for short) was established in order to provide formal languages for describing the structures with which mathematicians work, and the methods of proof available to them; its principal aim is a precise and adequate understanding of the notion of mathematical proof. **CML** was established based on a number of fundamental assumptions. Among them, the most characteristic one is the classical account of validity that is the logical validity criterion of **CML** by which one can decide whether the conclusion of an argument or a reasoning really does follow from its premises or not in the framework of **CML**. However, since the relevance between the premises and conclusion of an argument is not accounted for by the classical validity criterion, a reasoning based on **CML** is not necessarily relevant. On the other hand, in **CML** the notion of conditional, which is intrinsically intensional but not truth-functional, is represented by the notion of material implication, which is intrinsically an extensional truth-function. This leads to the problem of 'implicational paradoxes' [1, 2, 13, 16, 20].

CML cannot satisfy any of the above four essential requirements because of the following facts: a reasoning based on **CML** is not necessarily relevant; the classical truth-preserving property of a reasoning based on **CML** is meaningless in the sense of conditional; a reasoning based on **CML** must be circular and/or tautological but not ampliative; reasoning under inconsistency is impossible within the framework of **CML** [5, 7, 8]. These facts are also true to those classical conservative extensions or non-classical alternatives of **CML** where the classical account of validity is adopted as the logical validity criterion and the notion of conditional is directly or indirectly represented by the material implication.

Temporal (classical) logics was established in order to represent and reason about notions, relations, and properties of time-related entities within a logical framework, and therefore to underlie temporal reasoning, i.e., reasoning about those propositions and formulas whose truth-values may depend on time. As a conservative extension of **CML**, they have remarkably expanded the uses of logic to reasoning about human (and hence computer) time-related activities [3, 23, 24]. However, since any of temporal (classical) logics is a classical conservative extension of **CML** in the sense that it is based on the classical account of validity and it represents the notion of conditional directly or indirectly by the material implication, no temporal (classical) logic cannot satisfy the first three of the above four essential requirements for the fundamental logic system.

Spatial logics was proposed in order to deal with geometric and/or topological entities, notions, relations, and properties, and therefore to underlie spatial reasoning, i.e., reasoning about those propositions and formulas whose truth-values may depend on a location [9-11, 17, 19, 22]. However, these existing spatial logics are classical conservative extensions of **CML** in the sense that they are based on the classical account of validity and they represent the notion of conditional directly or indirectly by the material implication. Therefore, these spatial logics cannot satisfy the first three of the four essential requirements for the fundamental logic system.

Some other modal and/or epistemic logic systems have been proposed for reasoning about behavior of agent and/or multi-agent systems [14, 26]. These logics are also some how based on **CML**, and therefore, they cannot satisfy the four essential requirements for the fundamental logic system.

Traditional relevant (or relevance) logics ware constructed during the 1950s in order to find a mathematically satisfactory way of grasping the elusive notion of relevance of antecedent to consequent in conditionals, and to obtain a notion of implication which is free from the so-called 'paradoxes' of material and strict implication [1, 2, 13, 16, 20]. Some major traditional relevant logic systems are 'system **E** of entailment', 'system **R** of relevant implication', and 'system **T** of ticket entailment'. A major characteristic of the relevant logics is that they have a primitive intensional connective to represent the notion of (relevant) conditional and their logical theorems include no implicational paradoxes. The underlying principle of the relevant logics is the relevance principle, i.e., for any entailment provable in **E**, **R**, or **T**, its antecedent and consequent must share a propositional variable. Variable-sharing is a formal notion designed to reflect the idea that there be a meaning-connection between the antecedent and consequent of an entailment. It is this relevance principle that excludes those implicational paradoxes from logical axioms or theorems of relevant logics. Also, since the notion of entailment is represented in the relevant logics by a primitive intensional connective but not an extensional truth-function, a reasoning based on the relevant logics is ampliative but not circular and/or tautological. Moreover, because the relevant logics reject the principle of Explosion, they can certainly underlie paraconsistent reasoning.

In order to establish a satisfactory logic calculus of conditional to underlie relevant reasoning, the present author has proposed some strong relevant (or relevance) logics, named **Rc**, **Ec**, and **Tc** [5, 7, 8]. The logics require that the premises of an argument represented by a conditional include no unnecessary and needless conjuncts and the conclusion of that argument includes no unnecessary and needless disjuncts. As a modification of traditional relevant logics **R**, **E**, and **T**, strong relevant logics **Rc**, **Ec**, and **Tc** rejects all conjunction-implicational paradoxes and disjunction-implicational paradoxes in **R**, **E**, and **T**, respectively. What underlies the strong relevant logics is the strong relevance principle: If A is a theorem of **Rc**, **Ec**, or **Tc**, then every propositional variable in A occurs at least once as an antecedent part and at least once as a consequent part. Since the strong relevant logics are free of not only implicational paradoxes but also conjunction-implicational and disjunction-implicational paradoxes, in the framework of strong relevant logics, if a reasoning is valid, then both the relevance between its premises and its conclusion and the validity of its conclusion in the sense of conditional can be guaranteed in a certain sense of strong relevance.

Although strong relevant logics can satisfy the first three of the essential requirements for the fundamental logic system, they cannot satisfy the fourth requirement.

From the above discussions, we can see that what we need is a suitable combination of strong relevant logics, temporal logics, and spatial logics such that it can satisfy the all four requirements for the fundamental logic system.

4   Spatio-temporal Relevant Logics

We now propose a new family of relevant logic systems, named *spatio-temporal relevant logic*, which can satisfy the all four essential requirements for the fundamental logic system to underlie specifying, verifying, and reasoning about mobile multi-agent systems. The logics are obtained by introducing region connection predicates and axiom schemata of RCC [9-11, 19], point position predicates and axiom schemata, and point adjacency predicates and axiom schemata into temporal relevant logics [8]. Therefore, they are conservative extensions of temporal relevant logics as well as strong relevant logics.

Let $\{r_1, r_2, r_3, \ldots\}$ be a countably infinite set of individual variables, called *region variables*. Atomic formulas of the form $C(r_1, r_2)$ are read as 'region r_1 connects with region r_2.' Let $\{p_1, p_2, p_3, \ldots\}$ be a countably infinite set of individual variables, called *point variables*. Atomic formulas of the form $I(p_1, r_1)$ are read as 'point p_1 is in region r_1.' Atomic formulas of the form $Arc(p_1, p_2)$ are read as 'points p_1, p_2 are adjacent such that there is an arc from point p_1 to point p_2, or more simply, points p_1 is adjacent to point p_2.' Note that an arc has a direction. Atomic formulas of the form $Path(p_1, p_2)$ are read as 'there is a directed path from point p_1 to point p_2.' Let $\{a_1, a_2, a_3, \ldots\}$ be a countably infinite set of individual variables, called *agent variables*. Atomic formulas of the form $A(a_1, p_1)$ are read as 'agent a_1 arrives at point p_1.' Note that here we use a many-sorted language.

The logical connectives, temporal operators, region connection predicates, point position predicates, point adjacency predicates , axiom schemata, and inference rules are as follows:

Primitive Logical Connectives: $\{ \Rightarrow$ (entailment), $\neg$ (negation), $\wedge$ (extensional conjunction) $\}$

Defined Logical Connectives: $\{ \otimes$ (intensional conjunction, $A \otimes B =_{df} \neg(A \Rightarrow \neg B))$, $\oplus$ (intensional disjunction, $A \oplus B =_{df} \neg A \Rightarrow B)$, $\Leftrightarrow$ (intensional equivalence, $A \Leftrightarrow B =_{df}$ $(A \Rightarrow B) \otimes (B \Rightarrow A))$, $\vee$ (extensional disjunction, $A \vee B =_{df} \neg(\neg A \wedge \neg B))$, $\rightarrow$ (material implication, $A \rightarrow B =_{df} \neg(A \wedge \neg B)$ or $A \rightarrow B =_{df} \neg A \vee B)$, $\leftrightarrow$ (extensional equivalence, $A \leftrightarrow B =_{df} (A \rightarrow B) \wedge (B \rightarrow A)) \}$

Temporal Operators: $\{ G$ (future-tense always or henceforth operator, GA means 'it will always be the case in the future from now that A'), H (past-tense always operator, HA means 'it has always been the case in the past up to now that A'), F (future-tense sometime or eventually operator, FA means 'it will be the case at least once in the future from now that A'), P (past-tense sometime operator, PA means 'it has been the case at least once in the past up to now that A') $\}$ Note that these temporal operators are not independent and can be defined as follows: $GA =_{df} \neg F \neg A$, $HA =_{df} \neg P \neg A$, $FA =_{df} \neg G \neg A$, $PA =_{df} \neg H \neg A$.

Primitive Binary Predicate: $\{ C$ (connection, $C(x, y)$ means 'x connects with y'), I (be in, $I(p_1, r_1)$ means 'p_1 is in r_1'), Arc (arc, $Arc(p_1, p_2)$ means 'p_1 is adjacent to p_2'), $Path$ (path, $Path(p_1, p_2)$ means 'there is a directed path from p_1 to p_2'), A (arrive at, $A(a_1, p_1)$ means 'a_1 arrives at p_1') $\}$

Defined Binary Predicates:

$DC(r_1, r_2) =_{df} \neg C(r_1, r_2)$ 　　　　　($DC(r_1, r_2)$ means 'r_1 is disconnected from r_2')

$Pa(r_1, r_2) =_{df} \forall r_3(C(r_3, r_1) \Rightarrow C(r_3, r_2))$ 　($Pa(r_1, r_2)$ means 'r_1 is a part of r_2')

$PrPa(r_1, r_2) =_{df} Pa(r_1, r_2) \wedge (\neg Pa(r_2, r_1))$ ($PrPa(r_1, r_2)$ means 'r_1 is a proper part of r_2')

$EQ(r_1, r_2) =_{df} Pa(r_1, r_2) \wedge Pa(r_2, r_1)$ 　($EQ(r_1, r_2)$ means 'r_1 is identical with r_2')

$O(r_1, r_2) =_{df} \exists r_3(Pa(r_3, r_1) \wedge Pa(r_3, r_2))$ ($O(r_1, r_2)$ means 'r_1 overlaps r_2')

$DR(r_1, r_2) =_{df} \neg O(r_1, r_2)$ 　　　　　($DR(r_1, r_2)$ means 'r_1 is discrete from r_2')

$PaO(r_1, r_2) =_{df} O(r_1, r_2) \wedge (\neg Pa(r_1, r_2)) \wedge (\neg Pa(r_2, r_1))$

($PaO(r_1, r_2)$ means 'r_1 partially overlaps r_2')

$EC(r_1, r_2) =_{df} C(r_1, r_2) \wedge (\neg O(r_1, r_2))$

($EC(r_1, r_2)$ means 'r_1 is externally connected to r_2')

$TPrPa(r_1, r_2) =_{df} PrPa(r_1, r_2) \wedge \exists r_3(EC(r_3, r_1) \wedge EC(r_3, r_2))$

($TPrPa(r_1, r_2)$ means 'r_1 is a tangential proper part of r_2')

$NTPrPa(r_1, r_2) =_{df} PrPa(r_1, r_2) \wedge (\neg \exists r_3(EC(r_3, r_1) \wedge EC(r_3, r_2)))$

($NTPrPa(r_1, r_2)$ means 'r_1 is a nontangential proper part of r_2')

Axiom Schemata:

E1	$A \Rightarrow A$
E2	$(A \Rightarrow B) \Rightarrow ((C \Rightarrow A) \Rightarrow (C \Rightarrow B))$
E2′	$(A \Rightarrow B) \Rightarrow ((B \Rightarrow C) \Rightarrow (A \Rightarrow C))$
E3	$(A \Rightarrow (A \Rightarrow B)) \Rightarrow (A \Rightarrow B)$
E3′	$(A \Rightarrow (B \Rightarrow C)) \Rightarrow ((A \Rightarrow B) \Rightarrow (A \Rightarrow C))$
E3″	$(A \Rightarrow B) \Rightarrow ((A \Rightarrow (B \Rightarrow C)) \Rightarrow (A \Rightarrow C))$
E4	$(A \Rightarrow ((B \Rightarrow C) \Rightarrow D)) \Rightarrow ((B \Rightarrow C) \Rightarrow (A \Rightarrow D))$
E4′	$(A \Rightarrow B) \Rightarrow (((A \Rightarrow B) \Rightarrow C) \Rightarrow C)$
E4″	$((A \Rightarrow A) \Rightarrow B) \Rightarrow B$
E4‴	$(A \Rightarrow B) \Rightarrow ((B \Rightarrow C) \Rightarrow (((A \Rightarrow C) \Rightarrow D) \Rightarrow D))$
E5	$(A \Rightarrow (B \Rightarrow C)) \Rightarrow (B \Rightarrow (A \Rightarrow C))$
E5′	$A \Rightarrow ((A \Rightarrow B) \Rightarrow B)$
N1	$(A \Rightarrow (\neg A)) \Rightarrow (\neg A)$
N2	$(A \Rightarrow (\neg B)) \Rightarrow (B \Rightarrow (\neg A))$
N3	$(\neg (\neg A)) \Rightarrow A$
C1	$(A \wedge B) \Rightarrow A$
C2	$(A \wedge B) \Rightarrow B$
C3	$((A \Rightarrow B) \wedge (A \Rightarrow C)) \Rightarrow (A \Rightarrow (B \wedge C))$
C4	$(LA \wedge LB) \Rightarrow L(A \wedge B)$, where $LA =_{df} (A \Rightarrow A) \Rightarrow A$
D1	$A \Rightarrow (A \vee B)$
D2	$B \Rightarrow (A \vee B)$
D3	$((A \Rightarrow C) \wedge (B \Rightarrow C)) \Rightarrow ((A \vee B) \Rightarrow C)$
DCD	$(A \wedge (B \vee C)) \Rightarrow ((A \wedge B) \vee C)$
C5	$(A \wedge A) \Rightarrow A$
C6	$(A \wedge B) \Rightarrow (B \wedge A)$
C7	$((A \Rightarrow B) \wedge (B \Rightarrow C)) \Rightarrow (A \Rightarrow C)$
C8	$(A \wedge (A \Rightarrow B)) \Rightarrow B$
C9	$\neg (A \wedge \neg A)$
C10	$A \Rightarrow (B \Rightarrow (A \wedge B))$
T1	$G(A \Rightarrow B) \Rightarrow (GA \Rightarrow GB)$
T2	$H(A \Rightarrow B) \Rightarrow (HA \Rightarrow HB)$

T3	$A{\Rightarrow}G(PA)$
T4	$A{\Rightarrow}H(FA)$
T5	$GA{\Rightarrow}G(GA)$
T6	$(FA{\wedge}FB){\Rightarrow}F(A{\wedge}FB){\vee}F(A{\wedge}B){\vee}F(FA{\wedge}B)$
T7	$(PA{\wedge}PB){\Rightarrow}P(A{\wedge}PB){\vee}P(A{\wedge}B){\vee}P(PA{\wedge}B)$
T8	$GA{\Rightarrow}FA$
T9	$HA{\Rightarrow}PA$
T10	$FA{\Rightarrow}F(FA)$
T11	$(A{\wedge}HA){\Rightarrow}F(HA)$
T12	$(A{\wedge}GA){\Rightarrow}P(GA)$
IQ1	$\forall x(A{\Rightarrow}B){\Rightarrow}(\forall xA{\Rightarrow}\forall xB)$
IQ2	$(\forall xA{\wedge}\forall xB){\Rightarrow}\forall x(A{\wedge}B)$
IQ3	$\forall xA{\Rightarrow}A[t/x]$ (if x may appear free in A and t is free for x in A, i.e., free variables of t do not occur bound in A)
IQ4	$\forall x(A{\Rightarrow}B){\Rightarrow}(A{\Rightarrow}\forall xB)$ (if x does not occur free in A)
IQ5	$\forall x_1 \ldots \forall x_n\,(((A{\Rightarrow}A){\Rightarrow}B){\Rightarrow}B)$ (n$\geq$0)
RCC1	$\forall r_1\forall r_2(C(r_1, r_2){\Rightarrow}C(r_2, r_1))$
RCC2	$\forall r_1(C(r_1, r_1))$
PRCC1	$\forall p_1\forall r_1\forall r_2((I(p_1, r_1){\wedge}DC(r_1, r_2)){\Rightarrow}{\neg}I(p_1, r_2))$
PRCC2	$\forall p_1\forall r_1\forall r_2((I(p_1, r_1){\wedge}Pa(r_1, r_2)){\Rightarrow}I(p_1, r_2))$
PRCC3	$\forall p_1\forall r_1\forall r_2((I(p_1, r_1){\wedge}PrPa(r_1, r_2)){\Rightarrow}I(p_1, r_2))$
PRCC4	$\forall p_1\forall r_1\forall r_2((I(p_1, r_1){\wedge}EQ(r_1, r_2)){\Rightarrow}I(p_1, r_2))$
PRCC5	$\forall r_1\forall r_2(O(r_1, r_2){\Rightarrow}\exists p_1(I(p_1, r_1){\wedge}I(p_1, r_2)))$
PRCC6	$\forall p_1\forall r_1\forall r_2((I(p_1, r_1){\wedge}DR(r_1, r_2)){\Rightarrow}{\neg}I(p_1, r_2)$
PRCC7	$\forall r_1\forall r_2(PaO(r_1, r_2){\Rightarrow}$
	$\exists p_1(I(p_1, r_1){\wedge}I(p_1, r_2)){\wedge}\exists p_2(I(p_2, r_1){\wedge}{\neg}I(p_2, r_2)){\wedge}\exists p_3({\neg}I(p_3, r_1){\wedge}I(p_3, r_2)))$
PRCC8	$\forall p_1\forall r_1\forall r_2((I(p_1, r_1){\wedge}EC(r_1, r_2)){\Rightarrow}{\neg}I(p_1, r_2))$
PRCC9	$\forall p_1\forall r_1\forall r_2((I(p_1, r_1){\wedge}TPrPa(r_1, r_2)){\Rightarrow}I(p_1, r_2))$
PRCC10	$\forall p_1\forall r_1\forall r_2((I(p_1, r_1){\wedge}NTPrPa(r_1, r_2)){\Rightarrow}I(p_1, r_2))$
PAC1	$\forall p_1\forall p_2(Arc(p_1, p_2){\Rightarrow}Path(p_1, p_2))$
PAC2	$\forall p_1\forall p_2\forall p_3((Path(p_1, p_2){\wedge}Path(p_2, p_3)){\Rightarrow}Path(p_1, p_3))$
APRCC1	$\forall a_1\forall p_1\forall r_1\forall r_2((A(a_1, p_1){\wedge}I(p_1, r_1){\wedge}DC(r_1, r_2)){\Rightarrow}{\neg}A(a_1, r_2))$

Inference Rules:

$\Rightarrow$E : from A and $A{\Rightarrow}B$ to infer B (Modus Ponens)

$\wedge$I : from A and B to infer $A{\wedge}B$ (Adjunction)

TG : from A to infer GA and HA (Temporal Generalization)

$\forall$I : if A is an axiom, so is $\forall xA$ (Generalization of axioms)

Thus, various relevant logic systems may now defined as follows, where we use 'A | B' to denote any choice of one from two axiom schemata A and B.

$\mathbf{T}_{\Rightarrow} =_{df} \{E1, E2, E2', E3 \mid E3''\} + \Rightarrow E$

$\mathbf{E}_{\Rightarrow} =_{df} \{E1, E2 \mid E2', E3 \mid E3', E4 \mid E4'\} + \Rightarrow E$

$\mathbf{E}_{\Rightarrow} =_{df} \{E2', E3, E4''\} + \Rightarrow E$

$\mathbf{E}_{\Rightarrow} =_{df} \{E1, E3, E4'''\} + \Rightarrow E$

$\mathbf{R}_{\Rightarrow} =_{df} \{E1, E2 \mid E2', E3 \mid E3', E5 \mid E5'\} + \Rightarrow E$

$\mathbf{T}_{\Rightarrow,\neg} =_{df} \mathbf{T}_{\Rightarrow} + \{N1, N2, N3\}$

$\mathbf{E}_{\Rightarrow,\neg} =_{df} \mathbf{E}_{\Rightarrow} + \{N1, N2, N3\}$

$\mathbf{R}_{\Rightarrow,\neg} =_{df} \mathbf{R}_{\Rightarrow} + \{N2, N3\}$

$\mathbf{T} =_{df} \mathbf{T}_{\Rightarrow,\neg} + \{C1{\sim}C3, D1{\sim}D3, DCD\} + \wedge I$

$\mathbf{E} =_{df} \mathbf{E}_{\Rightarrow,\neg} + \{C1{\sim}C4, D1{\sim}D3, DCD\} + \wedge I$

$\mathbf{R} =_{df} \mathbf{R}_{\Rightarrow,\neg} + \{C1{\sim}C3, D1{\sim}D3, DCD\} + \wedge I$

$\mathbf{Tc} =_{df} \mathbf{T}_{\Rightarrow,\neg} + \{C3, C5{\sim}C10\}$

$\mathbf{Ec} =_{df} \mathbf{E}_{\Rightarrow,\neg} + \{C3{\sim}C10\}$

$\mathbf{Rc} =_{df} \mathbf{R}_{\Rightarrow,\neg} + \{C3, C5{\sim}C10\}$

$\mathbf{TQ} =_{df} \mathbf{T} + \{IQ1{\sim}IQ5\} + \forall I$

$\mathbf{EQ} =_{df} \mathbf{E} + \{IQ1{\sim}IQ5\} + \forall I$

$\mathbf{RQ} =_{df} \mathbf{R} + \{IQ1{\sim}IQ5\} + \forall I$

$\mathbf{TcQ} =_{df} \mathbf{Tc} + \{IQ1{\sim}IQ5\} + \forall I$

$\mathbf{EcQ} =_{df} \mathbf{Ec} + \{IQ1{\sim}IQ5\} + \forall I$

$\mathbf{RcQ} =_{df} \mathbf{Rc} + \{IQ1{\sim}IQ5\} + \forall I$

Here, $\mathbf{T}_{\Rightarrow}$, $\mathbf{E}_{\Rightarrow}$, and $\mathbf{R}_{\Rightarrow}$ are the purely implicational fragments of $\mathbf{T}$, $\mathbf{E}$, and $\mathbf{R}$, respectively, and the relationship between $\mathbf{E}_{\Rightarrow}$ and $\mathbf{R}_{\Rightarrow}$ is known as $\mathbf{R}_{\Rightarrow} = \mathbf{E}_{\Rightarrow} + A{\Rightarrow}LA$; $\mathbf{T}_{\Rightarrow,\neg}$, $\mathbf{E}_{\Rightarrow,\neg}$, and $\mathbf{R}_{\Rightarrow,\neg}$ are the implication-negation fragments of $\mathbf{T}$, $\mathbf{E}$, and $\mathbf{R}$, respectively; $\mathbf{Tc}$, $\mathbf{Ec}$, $\mathbf{Rc}$, $\mathbf{TcQ}$, $\mathbf{EcQ}$, and $\mathbf{RcQ}$ are strong relevant (relevance) logics proposed by the present author.

We can obtain minimal or weakest propositional temporal relevant logics as follows:

$\mathbf{T_0Tc} = \mathbf{Tc} + \{T1{\sim}T4\} + TG$

$\mathbf{T_0Ec} = \mathbf{Ec} + \{T1{\sim}T4\} + TG$

$\mathbf{T_0Rc} = \mathbf{Rc} + \{T1{\sim}T4\} + TG$

Note that the minimal or weakest temporal classical logic $\mathbf{K_t}$ = all axiom schemata for $\mathbf{CML} + {\rightarrow}E + \{T1{\sim}T4\} + TG$. Other characteristic axioms such as T5~T12 that correspond to various assumptions about time can be added to $\mathbf{T_0Tc}$, $\mathbf{T_0Ec}$, and $\mathbf{T_0Rc}$ respectively to obtain various propositional temporal relevant logics. Various predicate temporal relevant logics then can be obtained by adding axiom schemata IQ1~IQ5 and inference rule $\forall I$ into the propositional temporal relevant logics. For examples, minimal or weakest predicate temporal relevant logics are as follows:

$$\mathbf{T_0TcQ} = \mathbf{T_0Tc} + \{IQ1{\sim}IQ5\} + \forall I$$
$$\mathbf{T_0EcQ} = \mathbf{T_0Ec} + \{IQ1{\sim}IQ5\} + \forall I$$
$$\mathbf{T_0RcQ} = \mathbf{T_0Rc} + \{IQ1{\sim}IQ5\} + \forall I$$

We can now obtain various spatio-temporal relevant logics by adding region connection, point position, and point adjacency axiom schemata into the various predicate temporal relevant logics. For examples:

$$\mathbf{ST_0TcQ} = \mathbf{T_0TcQ} + \{RCC1, RCC2, PRCC1{\sim}PRCC10, PAC1, PAC2, APRCC1\}$$
$$\mathbf{ST_0EcQ} = \mathbf{T_0EcQ} + \{RCC1, RCC2, PRCC1{\sim}PRCC10, PAC1, PAC2, APRCC1\}$$
$$\mathbf{ST_0RcQ} = \mathbf{T_0RcQ} + \{RCC1, RCC2, PRCC1{\sim}PRCC10, PAC1, PAC2, APRCC1\}$$

The above spatio-temporal relevant logics have the following characteristics. First, as conservative extensions of strong relevant logics satisfying the strong relevance principle, the logics underlie relevant reasoning as well as truth-preserving reasoning in the sense of conditional, ampliative reasoning, paracomplete reasoning, and paraconsistent reasoning. Second, the logics are a family of logics but not a single logic system. We can select any one of them according our purpose in an application from various aspects of relevance, temporality, and spatiality. Third, in order to investigate spatio-temporal properties of the fundamental logic system to underlie specifying, verifying, and reasoning about mobile multi-agent systems in a way independent of epistemic properties of agents, we did not add epistemic operators and related axiom schemata into the logics. To represent, specify, verify, and reason about epistemic states of agents, those epistemic operators and related axiom schemata should be added into the logics.

5   Concluding Remarks

The spatio-temporal relevant logics proposed in this paper have the following possible applications: First, because the logics can underlie relevant, truth-preserving, ampliative, paracomplete, paraconsistent, spatial, and temporal reasoning, they provide us with criteria of logical validity for reasoning about behavior of mobile agents with incomplete or even inconsistent knowledge acting concurrently in spatial regions changing over time. This is in fact the major purpose to propose and develop the logics. Second, once we modeled behavior of mobile agents in a mobile multi-agent system and specified desirable properties with the formal language of the spatio-temporal relevant logics, we can verify the properties based on the logics. For the first and second applications, an automated reasoning and verifying engine based on the spatio-temporal relevant logics is indispensable. Based on EnCal, an automated forward deduction system for general-purpose entailment calculus [6], we are developing an automated reasoning and verifying engine for the spatio-temporal relevant logics. Third, the spatio-temporal relevant logics provide us with a foundation for constructing more powerful logic systems to deal with those interaction and security issues in mobile multi-agent systems. For examples, we can add epistemic operators and related axiom schemata into the logics in order to reason about interaction among agents as well as epistemic states of agents; we can also add deontic operators and related axiom schemata into the logics in order to reason about information security and assurance in mobile multi-agent systems.

The work presented in this paper is our first step for establishing a fundamental logic system to underlie specifying, verifying, and reasoning about mobile multi-agent systems. There are many challenging theoretical and technical problems that have to be solved in order to apply the spatio-temporal relevant logics to practices in the real world.

References

1. Anderson, A.R., Belnap Jr., N.D.: Entailment: The Logic of Relevance and Necessity, Vol. I. Princeton University Press, Princeton (1975)
2. Anderson, A.R., Belnap Jr., N.D., Dunn, J. M.: Entailment: The Logic of Relevance and Necessity, Vol. II. Princeton University Press, Princeton (1992)
3. Burgess, J. P.: Basic Tense Logic. In: Gabbay, D., Guenthner F. (eds.): Handbook of Philosophical Logic, 2nd edition, Vol. 7. Kluwer Academic, Dordrecht (2002) 1-42
4. Cardelli, L., Gordon, A.D.: Anytime, Anywhere: Modal Logics for Mobile Ambients. In: Proc. 27th ACM Symposium on Principles of Programming Languages. ACM Press, New York (2000) 365-377
5. Cheng, J.: The Fundamental Role of Entailment in Knowledge Representation and Reasoning. Journal of Computing and Information, Vol. 2, No. 1, Special Issue: Proceedings of the 8th International Conference of Computing and Information. Waterloo (1996) 853-873
6. Cheng, J.: EnCal: An Automated Forward Deduction System for General-Purpose Entailment Calculus. In: Terashima, N., Altman, E. (eds.): Advanced IT Tools, Proc. IFIP World Conference on IT Tools, IFIP 96 – 14th World Computer Congress. Chapman & Hall, London Weinheim New York Tokyo Melbourne Madras (1996) 507-514
7. Cheng, J.: A Strong Relevant Logic Model of Epistemic Processes in Scientific Discovery. In: Kawaguchi, E., Kangassalo, H., Jaakkola, H., Hamid, I.A. (eds.): Information Modelling and Knowledge Bases XI. IOS Press, Amsterdam Berlin Oxford Tokyo Washington DC (2000) 136-159
8. Cheng, J.: Temporal Relevant Logic as the Logical Basis of Anticipatory Reasoning-Reacting Systems. In: Daniel M. Dubois (ed.): COMPUTING ANTICIPATORY SYSTEMS: CASYS 2003 - Sixth International Conference. AIP Conference Proceedings, Vol. 718. American Institute of Physics, Melville (2004)
9. Cohn, A.G., Bennett, B., Gooday, J., Gotts, N.M.: RCC: A Calculus for Region based Qualitative Spatial Reasoning. GeoInformatica, Vol. 1 (1997) 275-316
10. Cohn, A.G., Bennett, B., Gooday, J., Gotts, N.M.: Representing and Reasoning with Qualitative Spatial Relations About Regions. In: Stock, O. (ed.): Spatial and Temporal Reasoning. Kluwer Academic, Dordrecht Boston London (1997) 97-134
11. Cohn, A.G., Hazarika, S.M.: Qualitative Spatial Representation and Reasoning: An Overview. Fundamenta Informaticae, Vol. 45 (2001) 1-29
12. Diaz, M.R.: Topics in the Logic of Relevance. Philosophia Verlag, Munchen (1981)
13. Dunn, J.M., Restall, G.: Relevance Logic. In: Gabbay, D., Guenthner, F. (eds.): Handbook of Philosophical Logic, 2nd Edition, Vol. 6. Kluwer Academic, Dordrecht Boston London (2002) 1-128

14. Fagin, R., Halpern, J.Y., Moses, Y., Vardi, M.Y.: Reasoning About Knowledge. The MIT Press, Cambridge (1999)
15. Egenhofer, M.J., Golledge, R.G. (eds.): Spatial and Temporal Reasoning in Geographic Information Systems. Oxford University Press, New York Oxford (1998)
16. Mares, E.D., Meyer, R.K.: Relevant Logics. In: Goble, L. (ed.): The Blackwell Guide to Philosophical Logic. Blackwell, Oxford (2001) 280-308
17. Merz, S., Wirsing, M., Zappe, J.: A Spatio-Temporal Logic for the Specification and Refinement of Mobile Systems. In: Pezze, M. (ed.): Fundamental Approaches to Software Engineering. Lecture Notes in Computer Science, Vol. 2621. Springer-Verlag, Berlin Heidelberg New York (2003) 87-101
18. Milner, R.: Communicating and Mobile Systems: the π-Calculus. Cambridge University Press, Cambridge (1999)
19. Randell, D., Cui, Z., Cohn, A.,: A Spatial Logic Based on Regions and Connection. In: Proc. 3rd International Conference on Knowledge Representation and Reasoning (1992) 165-176
20. Read, S.: Relevant Logic: A Philosophical Examination of Inference. Basil Blackwell, Oxford (1988)
21. Sangiorgi, D., Walker, D.: The Pi-calculus: A Theory of Mobile Processes. Cambridge University Press, Cambridge (2001)
22. Stock, O. (ed.): Spatial and Temporal Reasoning. Kluwer Academic, Dordrecht Boston London (1997)
23. van Benthem, J.: Temporal Logic. In: Gabbay, D.M., Hogger, C.J., Robinson, J.A. (eds.): Handbook of Logic in Artificial Intelligence and Logic Programming, Vol. 4. Oxford University Press, Oxford (1995) 241-350
24. Venema, Y.: Temporal Logic. In: Goble, L. (ed.): The Blackwell Guide to Philosophical Logic, Blackwell, Oxford (2001) 203-223
25. Weiss, G. (ed.): Multiagent Systems: A Modern Approach to Distributed Artificial Intelligence. The MIT Press, Cambridge (1999)
26. Wooldridge, M.: Reasoning about Rational Agents. The MIT Press, Cambridge (2000)
27. Wooldridge, M.: An Introduction to Multiagent Systems. John Wiley & Sons, (2002)

Towards an Ontology for Agent Mobility

Rosanne Price[1], Shonali Krishnaswamy[2], Seng W. Loke[2], and Mohan B. Chhetri[2]

[1] School of Business Systems, Monash University, Australia
Rosanne.Price@infotech.monash.edu.au
[2] School of Computer Science and Software Engineering, Monash University, Australia
{Shonali.Krishnaswamy,Seng.Loke,
Mohan.Chhetri}@infotech.monash.edu.au

Abstract. The benefits of agent mobility have been recognized for a variety of applications such as wireless ad hoc environments and network management. Mobile agent toolkits have been developed in response; however, they do not address the need for conceptual modeling techniques required for the analysis and design phases of mobile agent applications. On the other hand, conceptual level modeling techniques developed as part of Agent Oriented Software Engineering methodologies do not specifically address the requirement for modeling mobility. This paper represents a first step towards developing support for conceptual modeling of mobile agents, in terms of an ontology describing the abstract concepts and inter-relationships required to model agent mobility. The specific aim of this paper is to present an ontology that is widely applicable, not constrained by the peculiarities of any one mobile agent toolkit or application, and specified using an easily understood informal vocabulary.

1 Introduction

Mobile agents are a class of software agents that have the ability to move from one host to another, yielding enormous flexibility in the placing of computations. Agent mobility has been recognized as beneficial in a number of applications [13], including information-centric applications [10], wireless ad hoc environments [17], and network management [7]. In this context, a sizable number of mobile agent toolkits are currently available[1] to facilitate the implementation of mobile agent applications. The European company Tryllian[2] has attracted enormous support for applying mobile agent technology to enable adaptive enterprises.

While there has been considerable focus on the development of toolkits and environments to facilitate implementations of mobile agent applications, conceptual modeling of mobile agent applications is an area that has received little attention to date [11]. The increasing focus on mobile agents as an important technology for developing mobile and distributed applications and the inherent disadvantages of confining mobility of agents to the implementation phase necessitate techniques for modeling such systems prior to embarking on implementation. Agent Oriented Software Engineering (AOSE) is defined as software engineering for agent based computing [22].

---

[1] http://mole.infomatik.unituttgart.de/mal/preview/preview.html
[2] http://www.tryllian.com/

S. Wang et al. (Eds.): ER Workshops 2004, LNCS 3289, pp. 484–495, 2004.

The major objective of AOSE is to establish methodologies for re-using and maintaining agent systems as well as to facilitate development of structured agent-based software [20]. AOSE methodologies aim to provide tools and techniques for modeling, analyzing, and designing agent systems prior to implementation. Several methodologies have been proposed to model multiagent systems (MAS) [22]. However, the focus of these methodologies has been on multiagent systems at a generic level: they do not address specific modeling issues that pertain to mobile agent systems. For those multi-agent applications that require mobility, we contend that it can be disadvantageous if agent mobility is a concept only considered in the implementation phase or as an after-thought in engineering systems using generic multiagent tools or techniques, as is now typically the case. Possible consequences of such an approach are that the full advantages of agent mobility might remain unexploited. For example, the ad-hoc addition of mobility during the implementation phase would result in insufficient consideration of required mobility concepts and inconsistent system analysis and design. There is the real danger that the current implementation would influence and unnecessarily constrain the possibilities considered for representing mobility (specific examples of this problem are discussed in Sections 2 and 3). Furthermore, the implementation resulting from an ad-hoc addition of mobile agents deviates from the original multi-agent system design because mobility was not considered in the earlier phases of software engineering process. The 'why' and 'how' of agent mobility should be considered and its rationale documented at analysis and design time. On the other hand, developing an application using a specialized implementation-level mobile agent toolkit for code generation that does not fall under the auspices of an integrated agent-oriented analysis and design is clearly less than ideal. Therefore, there is a need for AOSE methodologies that take into account agent mobility, a still largely unexplored area.

Our preliminary investigation has yielded mGAIA [18], an extension of the GAIA methodology [24] to model mobile agent systems. This work identified several issues that need to be considered in developing multiagent systems with agent mobility and revealed the need for a more comprehensive approach that is decoupled from any specific AOSE methodology. Furthermore, there is a need for a conceptual model that supports both analysis and design phases of mobile agent applications. From this preliminary work, we propose that agent mobility and its associated aspects be modeled in their own right, independent of toolkit and methodology. An aim is that such a model of agent mobility can be independently conceived and constructed, and then systematically incorporated into an AOSE methodology, thereby extending the methodology to be used for mobile multiagent systems.

We see our work as complementary to other work that has attempted to identify key concepts of software agents for teaching purposes or simply to aid developers new to the idea of agents. For example, the work in [23] identified the concepts of action, percept, decision, goal, event, plan and belief as key for intelligent agents (inspired by the Belief-Desire-Intention model). Our work can be viewed as complementary to such work where we identify concepts such as location, task, visit, resource, itinerary, environment and others as described below as key for mobile agents. A broader outlook on our approach is that it might be possible to conceptually model aspects (e.g. properties) of agents separately and then establish links between such models, towards a *comprehensive* but *modular* ontology of agent (and agent-related) concepts. Then, regardless of judgments about the relative importance of the different aspects or properties of agents, which would tend to depend on the application at hand, parts of such a modular ontology might be selected as needed for application modeling purposes and as a basis for (perhaps aspect-specific) development and comparison of methodologies and toolkits.

This paper describes our initial work in conceptual modeling of agent mobility in order to facilitate analysis and design of mobile agent systems and enable integration of the model with existing AOSE methodologies. Our work is informed not only by our own work on mGAIA but also by the abundance of mobile agent toolkits each with its own conception of mobility (not necessarily orthogonal) and associated concepts, and the several methodologies where agent mobility have indeed been considered [6] [18] [16] (discussion in Section 2), even if not emphasized. An initial result which we are aiming towards is an ontology of key concepts and considerations one is faced with, and would need to take into account, when thinking about mobility for agents in building applications. The goals of this work are that the ontology be:

1. general and comprehensive enough to encompass the needs of the full range of mobile agent applications,
2. focusing on the essential concepts required to model mobility rather that features that are specific to a given toolkit, methodology, or application (i.e. not artificially constrained by implementation-level artifacts), and
3. specified using an informal vocabulary that is easily understood and not biased by terminology peculiar to a specific toolkit, methodology, or application.

Such an ontology is intended to serve as a basis for:

1. discussing high-level modeling of mobility and stimulating debate as to the essential concepts required for such modeling,
2. comparing mobile agent toolkits with respect to conceptual coverage, and
3. developing conceptual modeling techniques, languages, or language extensions to support analysis and design of agent mobility.

The rest of this paper is organized as follows. Section 2 reviews existing work that addresses issues of mobility beyond those relating to implementation (i.e. relevant to the analysis or design development phases). Section 3 presents our proposed ontology and illustrates the concepts by means of two example scenarios. Finally, we conclude and discuss future research work in Section 4.

2   Related Work

In this section, we first give an overview of Agent Oriented Software Engineering (AOSE) and the approaches used to develop AOSE methodologies and then discuss in detail the extent to which agent mobility issues are considered in current AOSE work.

Current AOSE methodologies have been developed based either on one of three different approaches [22] or by integrating these three approaches.

- The first approach is based on agents and multi-agent technology. This approach focuses on abstractions, which are above the object level and instead focus on capturing social-level abstractions such as agent, group and organization. Examples include GAIA [24], SODA [16], mGAIA [18], and EXPAND [3].

- The second approach is characterized by an attempt to extend existing object oriented technologies so that they capture the notion of an agency. Examples include Agent UML [1], MaSE [6], MASSIVE [12] among others.
- The third approach is based on Knowledge Engineering. It is characterized by the emphasis on identifying, acquiring and modeling knowledge to be used by the agents in the MAS. Examples of this approach include CoMoMAS [8] and MAS-CommonKADS [9].
- Certain AOSE methodologies have been developed by combining the best features of the above three approaches. MESSAGE [5] is an agent-oriented methodology, which combines the best features of GAIA, Agent UML and MAS-CommonKADS.

There are a few methodologies that are beginning to provide some support for the modeling of mobile multi-agent systems. The principal ones are Agent UML, mGAIA, SODA, Petri Nets and MaSE. However, none of them address the concept of mobility satisfactorily, in particular specific mobility issues such as agent mobility, agent itineraries, migration constraints, locations, environment, etc. Most AOSE methodologies do not consider mobility at all, and even those that do, fail to consider it as a separate concept, which is independent of and co-exists with the concept of an agent. Instead they consider it as an attribute of an agent and model it during runtime. As discussed in Section 1, this can result in implementation artifacts artificially constraining design. Specific examples are given below and in Section 3 with respect to individual AOSE methodologies. In general, such an ad-hoc approach is naturally less effective in fully describing agent mobility than research intended to explicitly consider mobile agent concepts in a rigorous and comprehensive manner independent of implementation concerns. Furthermore, there is no breakdown or clear definition of mobility into general mobility concepts, core mobility concepts and meta-level concepts in any of the current AOSE methodologies. Another major limitation of current AOSE methodologies is that they are designed primarily to design and support closed domain agent-systems as opposed to the modeling, design and support of open systems, thus limiting the range of agent supported applications.

At a more fundamental level, even the term mobility has not been used consistently in the context of AOSE. The different uses of this term are described and categorized in [4] as virtual mobility, actual agent mobility and physical agent mobility. Virtual mobility is defined in terms of transfer of roles between agents. Actual agent mobility refers to the capability of agents to migrate from one host/node to another across a distributed network along with its code data and state. Physical agent mobility refers to the physical movement of the device actually hosting the agent. Using the terminology described in [4], in this paper, we are concerned primarily with actual agent mobility. However, we additionally need to consider the possibility that such mobile agents could themselves be located on mobile hosts. Thus physical agent mobility must also be addressed to the extent needed to support modeling issues relating to mobile agent location, e.g. tracking or constraining mobile agent locations. However, virtual mobility is outside the scope of this paper and, in fact, is more closely related to (and thus more accurately referred to as) role transfer (implying a shift in responsibilities or tasks) rather than mobility (implying movement).

Given this overview of mobility in AOSE, we next review specific methodologies that include some consideration of mobility and describe, in general terms, their approach to and limitations with respect to modelling mobility. These issues are revisited in more detail in Section 3, in the context of evaluating these methodologies in terms of the specific concepts identified as key in our proposed ontology.

MaSE [6]: Although MaSE acknowledges the possible requirement for an agent to move in order to complete a task, it does not model the concept of mobility explicitly or at a conceptual level. Consideration is restricted to the implementation requirements for a single movement of an agent as part of executing a task. MaSE is also restricted to modelling closed domain systems. A further limitation of MaSE is that it allows only one-to-one agent interactions and no multicasting [20].

PetriNets [14]: Petri Nets have been applied for modelling mobile agent systems and mobile agent itineraries. While the control structures in Petri Nets map well to those in itineraries, the concepts concerning agent mobility have not been clearly articulated or described.

mGAIA [18]: mGAIA does not explicitly handle analysis of mobility but instead models it in the design phase. Although mGAIA considers mobility at a level higher than that of implementation, mobility is considered only in conjunction with a specific agent rather than as an orthogonal concept that can be associated as required with different components of agent systems (e.g. roles, tasks). Furthermore, mGAIA is not suited for the open and unpredictable domain of Internet applications because it requires that inter-agent relationships (organization) and agent abilities be specified before run-time.

SODA [16]: SODA is an extension of the GAIA methodology which tries to address the shortcomings of GAIA. It also considers mobility in the design phase. In the agent model, each agent is characterized by its location as being either fixed for static agents, or variable, for mobile agents. However, apart from this, SODA does not address any other mobility related issue.

Agent UML [1]: Agent UML is another methodology, which is an important player in the field of AOSE. It is an extension of the object-oriented modelling language, the Unified Modeling Language (UML) to support modelling of agents. Agent UML does not have the concept of agent at its centre and inherits all the limitations of UML, since it is an extension of an object-oriented methodology. Along with deployment, mobility has been identified as one of the principle modelling areas for Agent UML. In terms of mobility, Agent UML considers mobility to be a characteristic of individual agents [2] and thus—as with mGAIA—does not consider mobility as a separate and orthogonal concept. Another limitation of Agent UML is that it considers the mobility of agents only between physical hosts (a computational resource with memory and processing power), which it calls nodes. Thus Agent UML considers only movement between nodes and does not consider other types of movement, e.g. between two places within the same node. These limitations result from bias evidenced towards the implementation view in their exposition of mobility concepts. A further example of this bias its consequences in terms of limiting representation of mobility, is the explicit representation of return paths whereas mobile agent travel could be circular in the most general case.

3 Key Concepts for Modeling Mobile Agent Systems

This section presents the key concepts that are essential for modelling mobile agent systems, followed by an example scenario illustrating the use of such concepts in modelling a mobile agent application. We note that the identification of key concepts and their inter-relationships is typically a pre-requisite stage in any conceptual modelling exercise and is the precursor to the formal specification of concepts and their inter-relationships. In fact, these different stages can essentially be considered to be specification of an ontology for mobile agent applications at increasing levels of formality. The term ontology is defined in [21, p. 96] as "some sort of world view with respect to a given domain...often conceived as a set of concepts..., their definitions and their inter-relationships" whose semantic definition ranges from the informal (in natural language) to the formal and rigorous. An informal specification of an ontology for modelling mobile applications is presented in this section.

We distinguish the core concepts that are specific to modelling mobility of agents from orthogonal concepts that are applicable to agent systems in general. Orthogonal concepts are typically addressed in current AOSE methodologies that do not focus on mobility of software agents. These include concepts such as: Permissions, Responsibilities, Cloning, Communication and Coordination, Billing, Internal Reasoning Models and Services. Core concepts identified are Roles, Agents, Home, Itineraries, Visits, Tasks, Locations, Migration Constraints, and Resources. Of these, itineraries represent the single most important concept for modelling mobility. Furthermore, we will see that it is critical that itineraries, and thus mobility, be treated as an orthogonal concept that can be freely associated with any other core concept requiring the attribute of mobility in order to ensure that the ontology be generally applicable and at a truly conceptual level.

3.1 Core Concepts

Role: A role by definition is a position or purpose that any object/entity/agent has in a system, situation, organization, society or relationship. The notion of roles is widely prevalent in AOSE methodologies (e.g., [16], [24], [5], [1]), particularly in the analysis phase, where it is essential to identify the requisite purpose of agents that need to be created. Typically, an agent in a system needs to acquire or take at least one role and may fulfill several roles as deemed necessary.

In the context of modeling mobile agent systems, we perceive roles as a core concept because a role may require the attribute of mobility in order to fulfill its responsibilities, commitments and obligations. Thus, an agent that embodies a role that has the attribute of mobility would need to be a mobile agent. This view of mobility of roles explicates the addressing of mobility at the analysis phase of developing mobile agent systems. We note that, while often mobility of agents is focused on in the design phase [1][18][5], it is largely ignored at the analysis phase, much to the detriment of the conceptual modeling process of such systems. Furthermore, we note that in [19], the notion of role mobility is specified as roles that can be transferred between agents. We see this as an orthogonal concept and one that is more aptly termed a role transfer as it does not imply migration or mobility required by a role, which is the primary concern in terms of mobile agent systems.

Agent: In general, an agent can be defined as a proactive, reactive, autonomous and communicative software entity situated in an environment. Other attributes such as learning are possible [15].

Mobility of agents is obviously a core concept that needs to be modeled. The mobility of agents can be:

1. Role Driven: The role that an agent is required to fulfill requires it to be mobile.
2. Resource Driven: The unavailability (or lack) of specific resources requires it to be mobile.
3. Functionality Driven: The functional requirements of a specific task require the agent to be mobile.

In conceptually modeling the mobility of agents, it is essential to capture the rationale that necessitates the mobility. Such a conceptual model enforces planned/ engineered mobility of agents (e.g., necessitated by functionality and/or improved performance) rather than ad-hoc migration in the implementation.

In mGAIA, agents that possess the capability to migrate are specified in the design phase, while AUML models agent mobility in the Deployment diagram and the Activity Diagram. In MASE, the mobility of agents is modeled in terms of the saving of state and transfer, which is not so much a conceptual model of mobility as it is an implementation model of a "move" command. We note that none of these methodologies, actually, capture the rationality for having mobility in an agent in terms of both functional (roles and tasks) and extra-functional (resources) needs.

Home: The notion of home is explicitly catered for in several mobile agent implementation environments. Home refers to the origin or place of creation of a mobile agent. Any agent – mobile or stationary - does indeed have a "home", but this concept becomes particularly important when an agent possess the ability to move and visit several locations. This integral concept is not modeled in mGAIA or AUML.

Itinerary: Mobile agents or roles that require mobility implicitly have itineraries that determine a migration or travel path. This is the concept most critical to the conceptualization and modeling of mobility in agents. An itinerary typically specifies an ordering of visits to various locations to accomplish specific tasks. Furthermore, an itinerary is associated with migration constraints that determine the sequence of visits within an itinerary, and why and when a particular itinerary is chosen and undertaken. In order to ensure that the conceptualization of mobility is at a truly abstract level and thus supports the full range of possible mobile agent applications, it is essential that the concept of itineraries be treated as an orthogonal modeling concept. This means that itineraries, and hence mobility, can be associated freely with any other component of agent systems that might require the attribute (i.e. characteristic) of mobility, including agents, roles, tasks, agent locations (e.g. mobile hosts/nodes such as PDAs). The importance of associating mobility with agents and roles has been discussed previously. An itinerary could additionally be directly associated with a high-level task or goal (e.g., one that requires the agent to migrate to certain locations in a specific ordered fashion), which can then be further specified in terms of the specific sub-tasks to be performed during individual visits in the itinerary. The need to associate locations with mobility is discussed under the section on locations.

Itineraries have been modeled using Petri Nets in [14], as travel paths in mGAIA and as mobility paths in Agent UML [2]. Each of these approaches recognizes the need to model itineraries and they do focus on some components of it, however, they fail to capture and specify in entirety the why, what, where and when of itineraries. Furthermore, and most critically, they fail to treat itineraries, and thus mobility itself, as an orthogonal concept that could be associated directly with different components (i.e. types of components, e.g. roles, agents) of agent systems as required.

Visit: As discussed in the context of itineraries, a visit refers to one stop within an itinerary. Like an itinerary itself, a visit too involves specific locations and tasks that need to be performed at those locations. Furthermore, the migration from one location to another is also often associated with constraints that determine why a migration occurs, when a migration occurs and to where a migration occurs. The individual visits within an itinerary are not modeled in current approaches [1] [18].

Task: Tasks are activities that an agent is required to complete and are modeled in AOSE methodologies. However, in the case of mobile agents, tasks are typically associated with different locations. In mGAIA, the activity that a mobile agent is engaged in is modeled as part of the Role Schema, but this is not linked with the mobility model's travel schema. In AUML the deployment diagram models the origin, destination, and the mobility path of mobile agents [2]. A simple note is used to indicate the purpose of the movement. Thus the relation between the task and the location is not clearly defined. However, MaSE does associate tasks with mobility. Thus, in the context of mobile agents, it is essential to model a task in terms of both the activity performed and associate that activity with the location(s) where it needs to be performed. Furthermore, a task could also be directly associated with an itinerary, as discussed in the section on itineraries.

Location: The concept of location is integral to mobile agent systems. Most mobile agent modeling methodologies such as mGAIA and AUML include constructs/notations for modeling location. However, they limit their scope to the agency (also known as place or context) when modeling location. An agency is the run-time environment that allows hosting of mobile agents and supports their interaction with external computer systems. We take a holistic view of location wherein it is essential to model geographic/administrative domains and hosts/devices in addition to agencies. The modeling of location in these terms is to facilitate specification of access/permissions that constrain the locations that mobile agents can visit. It also allows specification of certain groupings of locations where a specific task may be performed. By incorporating the notions of geographic/administrative region and host/node in the modeling of location, we facilitate more general modeling. In modeling location, it is also important to specify whether the location itself possesses the attribute of mobility. Increasingly mobile agent applications are being viewed as being particularly suitable for pervasive environments which involve mobile and resource constrained devices. Thus, many agencies as well as hosts/nodes could be mobile devices such as PDAs or agencies running on PDAs (i.e. essentially an example of the concept of physical mobility discussed in Section 2).

Migration Constraint: In specifying mobility through visits and itineraries, a concept that is often ignored is the constraints that determine the migration. To the best of our knowledge, we are not aware of this concept being modeled in current mobile agent modeling methodologies. We note that these constraints can be associated with an individual visit as well as with itineraries as a whole. Migration constraints specify priorities that determine when an agent migrates, where an agent migrates to and why an agent migrates. There are three types of migration constraints.

1. Spatial constraints specify where an agent should (or should not) migrate.
2. Temporal constraints specify when an agent migrates (including time bounds).
3. Non Spatial Temporal constraints include all other conditions that are associated with migration and can include:
   - Agent Population - If the number of agents at a particular location exceeds a specified number, the agent under consideration should migrate to another location where it would face less competition for resources.
   - Resource Unavailability – If a particular resource is not available or accessible, the agent under consideration should migrate.
   - Task Requirement – A task requires the agent to be in a particular location, which necessitates the agent under consideration to migrate.
   - Forced Migration – In addition to voluntary migration (i.e. based on the previously discussed constraints), involuntary migration may also occur (i.e. when an agent is forced to move, e.g. due to host/node failure).

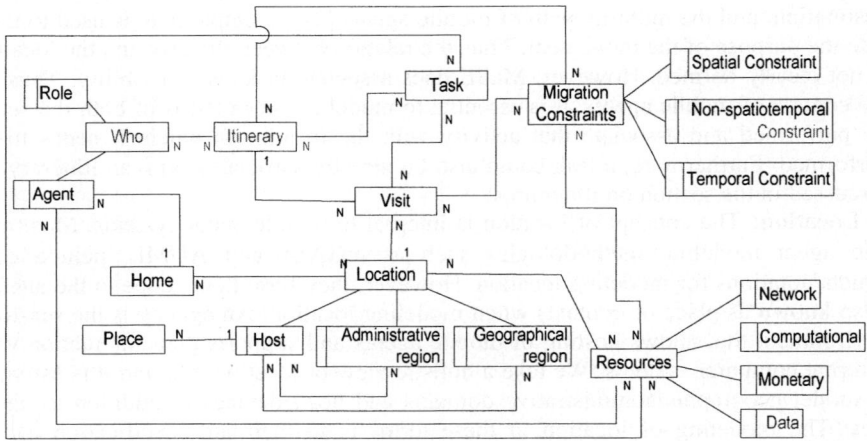

Fig. 1. A diagram showing the concepts and the links between them, including the cardinalities of the links: many-to-many, many-to-one or one-to-many, and subclassing.

Resource: This concept is associated with migration constraints, tasks and agents. An agent can obviously require resources or carry/have resources. A migration can be determined due to the lack of resources. Tasks can require specific resources (e.g. the availability of a network connection in order for the agent to book a movie for a mobile user). However, this concept has not been integrated into existing mobile agent conceptual modeling approaches. A resource includes:

1. Computational resources
2. Network resources
3. Data resources
4. Monetary resources (An agent might be given what is akin to monetary units in order to pay for computational resources they use.)

The described concepts and the links between them are depicted in Figure 1. Subclasses are shown by entities having an extra small rectangle on the left.

3.2 Two Scenarios for Illustration

Finally, we give two examples of how such concepts might be used to describe a mobile agent application for the purpose of specification or design.

Consider a system which launches (from a **home** running on some machine) mobile **agents** to remote **locations** to retrieve information. Each agent has the **role** of searching a number of sites for a specified kind of information and has an **itinerary** of **places** to **visit** and **tasks** to perform at these places. Each place is maintained within its own **administrative region** and is realized as a server running within a host and these hosts are distributed over several **geographical regions**. The agents are given **monetary units** to pay for **computational resources** used in these remote locations. Host and application-specific **data** are important for the functioning of the agents. The agents are not permitted to visit certain sites (i.e., the agents are **spatially constrained**) and must come back within a certain user-specified time and, for fairness and load control, are only allowed to run within a certain host for a limited time (i.e., the agents are **temporally constrained**).

Consider another scenario of sending teams of mobile agents out to a computational grid in order to perform tasks in parallel. As in the other application, **agents**, each with the **role** of getting a job (e.g., a unit of parallel computation) done in a cost-effective manner (where costs are measured in **monetary units** both in terms of response times and resources utilized) and are launched from a **home** place. Each agent is given an **itinerary** of **visits**, each associated with a computational **task**. The agents have to negotiate permissions for each **host** they visit since the hosts might be under different **administrative regions**. The hosts are spread over different **geographical regions** so that network migration delays cannot be ignored. The resources involved are **computational** and **network bandwidth**, required application-specific **data** for the computation, as well as some means of accounting for resources used (in the form of monetary units). The agents might reason in order to select suitable **locations** so that user-specified **temporal constraints** are satisfied.

The point is that a number of concerns about mobility are elicited and recorded when thinking is guided by our (albeit only a core) ontology given in Section 3.1. Moreover, the concepts involved are clearly evident in both scenarios, highlighting the underlying "similarity" between two different applications; with possibly quite different uses of agents and implementations.

4   Conclusion and Future Work

This paper presents an ontology for mobile agents that is intended to have general applicability and intelligibility independent of a particular AOSE methodology or toolkit and to focus on the essential concepts required to model mobility independent of implementation level concerns. Thus the main contribution of our work, differentiating it from compared previous mobile agent research, is the treatment of mobility as an independent concept based on these stated goals and, consequently, as orthogonal to other concepts in the agent domain.

The ontology is intended to serve as a basis for further discussion of the key concepts required to model agent mobility (e.g. by considering scenarios from different application contexts) and for evaluation of the conceptual coverage of methodologies and toolkits with respect to mobility. Future work includes the formalization of the ontology and its integration with existing AOSE methodologies/languages.

Another possible area of future research relates to tracking agent movement. The presentation of the ontology in this paper is from the perspective of specification of mobile agents; however, such an ontology could also potentially be used to specify historical information relating to agent movements that should be recorded. Such information could be useful, for example, for monitoring system utilization, for billing, or for improving agent effectiveness (e.g. by re-designing agents or allowing agents to learn from historical information). For example, system utilization could be studied if information was recorded regarding the details of actual agent movements with respect to the location of, time duration of, resources used in, tasks accomplished during, and actual migration trigger responsible for each agent visit.

References

1. http://www.auml.org
2. FIPA Modeling Area: Deployment and Mobility (Working Document), 2003, http://www.auml.org/auml/documents/DeploymentMobility.zip
3. Brauer, W., Nickles, M., Rovatsos, M., Weiß, G. and Lorentzen, K.F. Expectation-oriented analysis and design, in (eds.) M. Wooldridge, G. Weiß, and P. Ciancarini, Agent-Oriented Software Engineering II, Second International Workshop, AOSE 2001, Vol. 2222 of Lecture Notes in Computer Science, Montreal, Canada, Springer-Verlag, 2001, pp. 226-244.
4. Cabri, G. Leonardi, L. Mamei, M., and Zambonelli, F. Dipartimento di Scienze dell'Ingegneria, Mobile Agent Organizations. Poster presentation at The 5th IEEE International Conference on Mobile Agents, Atlanta, Georgia, USA, 2001.
5. Caire, G., Leal, F., Chainho, P., Evans, R., Garijo, F., Gomez, J., Pavon, J., Kearney, P., Stark, J., and Massonet, P. Agent oriented analysis using MESSAGE/UML, In (eds.) M. Wooldridge, P. Ciancarini, and G. Weiss, Proceedings of the 2nd International Workshop on Agent-Oriented Software Engineering (AOSE-2001), 2001, pp. 101-108.
6. DeLoach, S.A., Wood, M.F., Clint, H., and Sparkman, H. Multiagent Systems Engineering, in International Journal of Software Engineering and Knowledge Engineering, 11(3), 2001, pp. 231-258.
7. T.C. Du, E.Y. Li, and A. Chang. Mobile Agents in Distributed Network Management. Communications of the ACM, Vol. 6, No. 7, 2003, pp. 127-132.
8. Glaser, N. Contribution to Knowledge Modelling in a Multi-agent Framework, PhD Thesis, Universite Henry Poincare, Nancy, France, 1996.
9. Iglesias, C.A., Garijo, M., Gonzalez, J.C., and Velasco, J.R. Analysis and design of multi-agent systems using MAS-CommonKADS, In AAAI'97 Workshop on Agent Theories, Architectures and Languages, Providence, RI, 1997. ATAL. (An extended version of this paper has been published in INTELLIGENT AGENTS IV: Agent Theories, Architectures, and Languages, Springer Verlag, 1998).
10. Klusch, M., Zambonelli, F. (ed): Cooperative Information Agents – Best Papers of CIA 2001. International Journal of Cooperative Information Systems, Vol. 11, No. 3 and 4, 2002.
11. Krishnaswamy, S., and Loke, S, W. On Modelling Agent Mobility in Multiagent Methodologies, Position Paper, Workshop on Agent Oriented Information Systems (AOIS 2003), Held in conjunction with the Second International Joint Conference on Autonomous Agents and Multiagent Systems (AAMAS 2003), Melbourne, Australia, July.

12. Lind, J. Iterative Software Engineering for Multiagent Systems: The MASSIVE Method, Lecture Notes in Computer Science, Vol. 1994. Springer-Verlag, 2001.

13. Loke, S.W. An Overview of Mobile Agent Technology for Distributed Applications: Possibilities for Future Enterprise Systems. Informatica: An International Journal of Computing and Informatics, 2001, Vol. 25, No. 2, pp. 247-260.

14. Loke, S.W. and Ling, S. Mobile Agent Itineraries and Workflow Nets for Analysis and Enactment of Distributed Business Processes, in Proceedings of the International Symposium on Multi-Agents and Mobile Agents in Virtual Organizations and E-Commerce (MAMA'2000) at the International ICSC Congress on Intelligent Systems and Applications (ISA'2000), Wollongong Australia, (ed.) H. Tianfield, 2000, ICSE Academic Press, pp. 459-466.

15. Nwana, H.S. Software Agents: An Overview. Knowledge Engineering Review: Intelligent Systems Research, AA&T, BT Laboratories, UK, 1996, pp. 206-244.

16. Omicini, A. SODA: Societies and Infrastructures in the Analysis and Design of Agent-based Systems. Proceedings of the 1st international workshop, AOSE 2000 on Agent-oriented software engineering, Limerick, Ireland, 2001, pp. 185-193.

17. Qi, H. and Wang, F. Optimal Itinerary Analysis for Mobile Agents in Ad Hoc Wireless Sensor Networks. Proceedings of the 13th International Conference on Wireless Communications, vol. 1, Calgary, Canada, July, 2001, pp. 147-153.

18. Sutandiyo, W., Chhetri, M.B., Krishnaswamy, S., and Loke, S.W. Experiences with Software Engineering of Mobile Agent Applications, Proceedings of the Australian Software Engineering Conference (ASWEC 2004), Melbourne, Australia, April 13-16, IEEE Press.

19. Tiropanis, T. Offering Role Mobility in a TINA Environment, in the Proceedings of the 5th International Conference on Intelligence in Services and Networks, IS&N '98, Antwerp, Belgium, May 1998.

20. Tveit, A. A Survey of Agent-Oriented Software Engineering, NTNU Computer Science Graduate Student Conference, Norwegian University of Science and technology, 2001.

21. Uschold, M. and Gruninger, M. Ontologies: Principles, Methods, and Applications, Knowledge Engineering Review, Vol. 11, No. 12, 1996, pp. 93-136.

22. Weiß, G. Agent Orientation in Software Engineering, Knowledge Engineering Review, Vol. 16, No. 4, 2002, pp. 349-373.

23. Winikoff, M., Padgham, L., and Harland, J. Simplifying the Development of Intelligent Agents. In Proceedings of the 14th Australian Joint Conference on Artificial Intelligence (AI'01), Adelaide, 2001.

24. Wooldridge, M., Jennings, N.R. and Kinny, D. The GAIA Methodology for Agent-Oriented Analysis and Design, Journal of Autonomous Agents and Multi-Agent Systems, Vol.3, No. 3, 2000, pp. 285-312.

Subjective Trust Inferred by Kalman Filtering vs. a Fuzzy Reputation

Javier Carbo, Jesus Garcia, and Jose M. Molina

Group of Applied Artificial Intelligence, Computer Science Dept.
Universidad Carlos III de Madrid, Leganes Madrid 28911, Spain

Abstract. Inferring Trust in dynamic and subjective environments is an interesting issue in the way to obtain a complete delegation of human-like decisions in autonomous agents. Complex mathematical algorithms have been proposed by academic researchers to be applied when there are no universally accepted evaluation criteria. In this paper we propose a method based on fuzzy logic that has been successfully compared with several of these alternatives. Our innovative approach is compared with a classic estimation method known as kalman. The convergence of both methods has been tested in a multiagent scenario where the subjectivity of evaluations adopt prefixed variability.

1 Introduction

From the popularity of Internet remote interactions have become very frequent. It is out of discussion the importance of trusting the partners in such kind of interactions. Therefore the study of computational trust models has acquired great relevance in the last years. Most of them consists of a central entity that certifies the satisfaction of some given evaluation criteria. This is the approach of Information Security discipline. It is a kind of solution that relays on the assumption of objective evaluation criteria that are universally accepted. Unfortunately, this requirement can not be satisfied in many domains, and distributed solutions should be considered. For instance, humans often evaluate sites, people, products and services according to personal subjective preferences. And the emulation of such subjective evaluation through computerized social metaphors is strongly linked with Distributed Artificial Intelligence Foundations: autonomous computer programs with human-like reasoning that act in behalf of humans (these programs are the so called agents) [6].

In DAI-inspired models, each agent computes reputation values, that represent particular opinions of such agent about each target of the evaluation according to subjective criteria. Then, trust is inferred from reputation, and agent-based trust models can be classified according to two sources of reputation [1]: direct experiences and witness information. The first of them is, without doubt, the most reliable source of reputation for a trust model since it is the experience based on the direct interaction with the other side. The second one (also called word-of-mouth), comes from recommendations of third parties. Since

S. Wang et al. (Eds.): ER Workshops 2004, LNCS 3289, pp. 496–505, 2004.

these third parties may have different subjective evaluation criteria, or they even lie intentionally, this source of reputation is surrounded of some level of uncertainty and can not be considered as completely transitive. Although different particular views of the world may coexist (as many as members of the system), and malicious agents may lie, a certain level of cooperation is required in low-density populations in order to improve the filtering of trustworthy people, sites, services and products [9].

In this paper we will focus on how this filtering has been implemented. On the commercial side, trust models have been applied to several online marketplaces. The problem of filtering partners and products has been solved with reputation based on direct experiences computed in simple ways: item-to-item correlation (Amazon), sum of ratings (eBay), and average sum of ratings (Onsale Exchange).

On the other hand, academic community has analyzed alternatives involving more complex computations that include how to weight recommendations from third parties. Furthermore, the academic proposals that come from the DAI discipline are inspired in the factors that humans are supposed to apply in real-life filtering. From these automated filtering mechanisms should emerge the intended intelligent behaviour of agents. In the next subsections we will describe some of the main academic DAI-inspired trust models, including our proposed model (noted as AFRAS), in order to compare them with a classic estimation method known as kalman.

1.1   SPORAS and HISTOS, from M.I.T.

SPORAS [11] is based on ELO rating system – a method to evaluate player's relative strength in pairwise games such as chess. The main point of this trust model is that trusted agents with very high reputation experience much smaller changes in reputation than agents with low reputation. SPORAS also computes the reliability of agents' reputation using the standard deviation of such measure.

HISTOS [11] is an evolved version of SPORAS that include witness information as source of reputation through a recursive computation of weighted means of ratings.

Based on these principles, the reputation value of a given agent at iteration i, R_i, is obtained recursively from the previous one R_{i-1} and from the subjective evaluation of the direct experience DE_i:

$$R_i = R_{i-1} + \frac{1}{\theta} \cdot \Phi(R_{i-1}) \cdot (DE_i - R_{i-1}) \tag{1}$$

Let θ be the effective number of ratings taken into account in an evaluation ($\theta > 1$). The bigger the number of considered ratings, the smaller the change in reputation is.

Furthermore, Φ stands for a damping function that slows down the changes for very reputable users:

$$\Phi(R_{i-1}) = 1 - \frac{1}{1 + e^{\frac{-(R_{i-1} - Max)}{\sigma}}} \tag{2}$$

Where dominion D is the maximum possible reputation value and σ is chosen in a way that the resulting Φ would remain above 0.9 when reputations values were below $\frac{3}{4}$ of D.

1.2   REGRET, from the Spanish Artificial Intelligence Research Institute

REGRET [8] takes into account direct experiences and witness information. It also considers the role that social relationships may play. It provides a degree of reliability for the reputation values, and it adapts them through the inclusion of a temporal dependent function in computations. The time dependent function ρ gives higher relevance to direct experiences produced at times closer to current time. The reputation held by any part at a iteration i is computed from a weighted mean of the corresponding last θ direct experiences. The general equation is of the form:

$$R_i = \sum_{j=i-\theta}^{j=i} \rho(i,j) \cdot W_j \tag{3}$$

Where $\rho(i,j)$ is a normalized value calculated from the next weight function:

$$\rho(i,j) = \frac{f(j,i)}{\sum_{k=i-\theta}^{k=i} f(k,i)} \tag{4}$$

Where $i \geq j$. Both of them represent the time or number of iteration of a direct experience. For instance, a simple example of a time dependent function f is:

$$f(j,i) = \frac{j}{i} \tag{5}$$

REGRET also computes reliability with the standard deviation of reputation values computed from:

$$STD - DVT_i = 1 - \sum_{j=i-\theta}^{j=i} \rho(i,j) \cdot \mid W_j - R_i \mid \tag{6}$$

But REGRET defines reliability as a convex combination of this deviation with a measure, $0 < NI < 1$, of whether the number of impressions, i, obtained is enough or not. REGRET establishes an intimate level of interactions, itm, to represent a minimum threshold of experiences to obtain close relationships. More interactions will not increase reliability. The next function models the level of intimate with a given agent:

$$if(i \in [0, itm]) \rightarrow NI = \sin(\frac{\pi}{2 \cdot itm} \cdot i), Otherwise \rightarrow NI = 1 \tag{7}$$

1.3   Unnamed, from the University of South Caroline

This trust model [10] uses Dempster-Shafer theory of evidence to aggregate recommendations from different witnesses. The main characteristic of this model is the relative importance of fails over success. It assumes that deceptions (noted as β, and valued negatively) causes stronger impressions than satisfactions (noted as α, and valued positively). So mathematically: $|\beta| > \alpha \geq 0$.

It then applies different gradients to the curves of gaining/losing reputation in order to lose easily reputation, while it is hard to acquire it. The authors of this trust model define different equations to the sign (positive/negative) of the received direct experience (satisfaction/deception) and the sign of the previous reputation corresponding to the given agent.

So in the case that both of them had the same sign, then the new reputation would take the form of the next equation:

$$R_j = \frac{DE_i + R_{i-1}}{1 - min|DE_i|, |R_{i-1}|} \tag{8}$$

In the case that those variables had different sign, then the corresponding opinion would be computed from equation 7:

$$R_j = reputation_{i-1} + DE_i \cdot (1 + R_{i-1}) \tag{9}$$

2   A Fuzzy Reputation Agent System (A.F.R.A.S.)

The main characteristic of this trust model [4] is the use of fuzzy sets to represent reputation values. This formalism makes sense since human opinions about others are vague, subjective and uncertain (in other words, reputation is a fuzzy concept, valued in fuzzy terms).

In this trust model, direct experiences and witness information are considered, both of them are aggregated through a weighted mean of fuzzy sets. Aggregation of fuzzy sets is computed with Mass Assignment assumptions based on Baldwin's theory of evidence [2] . In the case of direct experiences, weights depend on a single attribute that represents the *memory* of the agent ($0 < memory < 1$). We associated such meaning to this value because it determines the importance of past direct experiences (R_{i-1}) over a new one (DE).

$$R_i = R_{i-1} + \frac{(DE_i - R_{i-1}) \cdot (1 - memory)}{2} \tag{10}$$

It is computed as a function of the overlapping between two fuzzy sets that represents the level of success of last estimation. If the satisfaction provided by a partner was similar to the reputation estimation assigned to such partner, the relevance of past experiences (memory) would be increased. On the other hand, if they were different, is the relevance of the last experience what would be increased (the corresponding agent is 'forgetting' past experiences, 'losing' memory).

Once similarity is computed in this way, an average sum of previous memory $memory_{i-1}$ with similarity $SIM(R_{i-1}, DE_i)$ is applied to obtain $memory_i$:

$$memory_i = \frac{memory_{i-1} + SIM(R_{i-1}, DE_i)}{2} \qquad (11)$$

This equation follows the next simple principles: If the prediction fitted well the rating ($SIM \approx 1$) then memory (the importance given to the past reputation over the last direct experience) would increase in $1/2 + memory/2$. On the other hand, when they were not similar at all ($SIM \approx 0$), memory would become useless, and its relevance in the next estimations of reputation would be halved.

These properties avoid *memory* being below zero and above one. The initial value of *memory* associated to any agent joining the system should be minimum (zero), although it would be soon increased when there was any success in the estimations of reputation.

Reliability of reputation values is modeled through the fuzzy sets themselves. It is implicit in them, graphically we can interpret the gradient of the sides of a trapezium representing a fuzzy reputation as its reliability. A wide fuzzy set representing a given reputation represents a high degree of uncertainty over that reputation estimation, while a narrow fuzzy set implies a reliable reputation.

Recommendations are aggregated directly with direct experiences in a similar way (weighted sum of fuzzy sets). But in this case, the weight given to each part (recommendation vs. previous opinion) is dependent on the reputation of the recommender. A recommendation would (at most) count as much as a direct experience if the recommender had the highest reputation.

Finally, to update the reputation of recommenders, an agent computes similarity (level of overlapping between the corresponding fuzzy sets) with a posteriori results of the direct experience with the recommended partner. Then, reputation of recommender would be increased or decreased accordingly.

3   Estimation with Kalman Filter

The Kalman filter [7] is a linear recursive algorithm to estimate an unknown state variable from noisy observations. In our application, the state variable would be the reputation, while observations would be the results from direct experiences. The Kalman filter assumes certain linear stochastic models for the state dynamics and observation processes, so that it would achieve the optimum estimator (in the Bayesian sense of Minimum Squared Error) under those conditions. It has been extensively applied to different fields, outstanding the tracking systems based on sensor data [3].

As indicated in the figure 1, it sequentially processes the observations, z[k], combining them with predictions computed accordingly to the state dynamic model, to recursively update the state estimator and associated covariance matrix, P[k]. Z-1 denotes a delay between sequential observations, so it does not require a uniform updating sequence.

The models assumed by the algorithm can be summarized in the following equations:

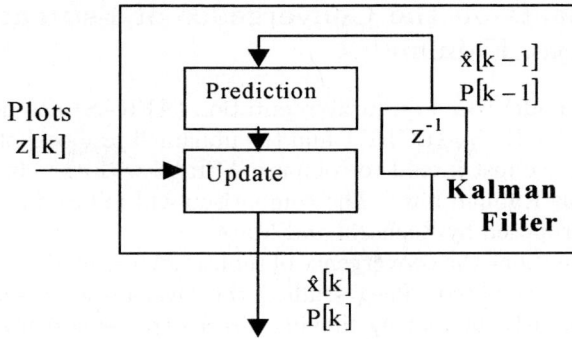

Fig. 1. Kalman filter outline

– The state variable has a linear behaviour, with a certain uncertainty characterized by a parameterized random variable (plant-noise model):

$$x(t + Dt) = F(Dt)x(t) + q(Dt) \tag{12}$$

where Dt is the time delay between last update and current observation, tk-tk-1, F(Dt) is the transition matrix and q(t) is characterized by its covariance matrix, Q(Dt).

– Observations result from a linear operation on state variable corrupted by additional noise:

$$z(t) = Hx(t) + n(t) \tag{13}$$

being n(t) a random variable with covariance given by matrix R.

Under these models, the equations for Kalman filter to compute the best estimation for x(t) are the following:

– Prediction

$$\widehat{x}[k|k - 1] = F(\Delta t) \cdot \widehat{x}[k - 1] \tag{14}$$

$$P[k|k - 1] = F(\Delta t) \cdot P[k - 1] \cdot (F(\Delta t))^t + Q(\Delta t) \tag{15}$$

– Updating

$$W[k] = P[k] \cdot H^t \cdot (R[k] + H \cdot P[k] \cdot H^t)^{-1} \tag{16}$$

$$\widehat{x}[k|k] = \widehat{x}[k|k - 1] + W[k] \cdot (z[k] - H \cdot \widehat{x}[k|k - 1]) \tag{17}$$

$$P[k|k] = P[k - 1] \cdot (I - H^t \cdot W[k]) \tag{18}$$

Usually, the exact models for dynamics and observation are not known, so the design for a certain application is a trade-off to adjust the parameters. Matrix R is usually adjusted from observed data variability, while matrix Q is tuned to achieve satisfactory balance between noise filtering (when the prediction model is much better than observation noise) and reactions to maneuvers (intervals while the model fails to accurately predict the state variable).

4   Experiments on the Convergence of Estimations: AFRAS vs. Kalman

Since our trust model based on fuzzy reputation (AFRAS) has been previously compared with SPORAS, REGRET and the unnamed proposal of Singh and Yu [5], in this paper we just intend to compare AFRAS with the classic estimation method known as Kalman filter. The comparison will follow an evolved version of the testbed proposed by Zacharia and Maes.

In those simulations the convergence of reputation estimations towards direct experiences were evaluated. They studied the level of success obtained from an agent continuously interacting (having direct experiences) from a particular agent of reputation chosen in each iteration randomly among all the possible partners.

Our experiments have tested the results of one agent randomly having direct experiences with 10 different agents in 200 iterations. The same order of partners was selected along the 200 iterations and the same response from those agents was applied to AFRAS and Kalman in order to obtain a consistent comparison. Due to this reason, we have also adapted the range of possible values of reputation to be from 0 to 100.

Kalman has been tested with two different R parameters: one of them constant, an another one that is iteratively adapted to the average error produced. Both alternatives will be reflected in the corresponding figures.

In order to evaluate the convergence of reputation estimations, the simulation has satisfied the next properties:

- Each of the partners has assigned a prefixed subjective evaluation (uniformly distributed). And all of them use such prefixed behavior along the 200 iterations.
- The satisfaction provided at any time is drawn from a normal distribution. The mean of that distribution is equal to the prefixed evaluation. Three different standard deviations will be tested in the experiments: 3, 10 and 33.
- Initially the reputation of all agents is 10 (over 100) with a reliability of 1 (over 100).

The error produced in each purchase has been quantified in AFRAS through the difference between the center of gravity defuzzification of previous reputation and the result of the direct experience. Next, we will show the average error committed in the estimations computed by both alternatives in order to compare the speed of convergence. We assumed that the standard deviation of direct experiences from partners was prefixed to be 3, 10 and 33.33 (over 100).

These values represent three different scenarios. The first of them is a situation where each partner is evaluated all the time with very similar values (Standard deviation is a 3%). On the other hand, the third of them represents a situation where subjective evaluation may change suddenly ($StdDvt = 33\%$). The second one is in the middle of these two extremes ($StdDvt = 10\%$).

Figures will show the evolution of the average error produced by AFRAS and the two studied alternatives of kalman. We will also see how both algorithms

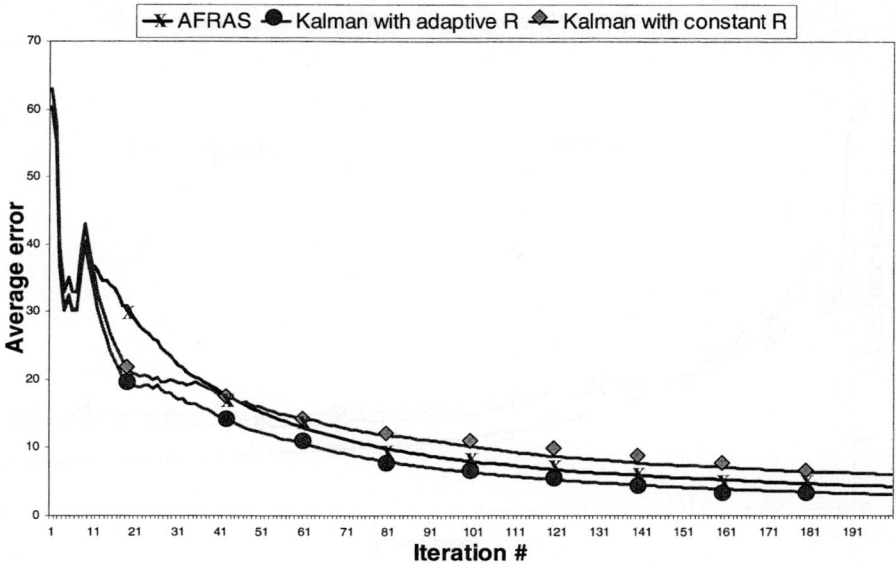

Fig. 2. Convergence of AFRAS vs. Kalman with StdDvt=3

converge to the prefixed subjective evaluation of each agent since the average error decreases continuously. The velocity of convergence will be then implicit in the gradient of the curves.

The curves of figure 2 show: how both kalman alternatives have soon a faster convergence than AFRAS. However after that initial 40 iterations, AFRAS show a slightly better average error than kalman with constant R. It also can be noted that in kalman, adaptive R causes a light improvement of constant R as it was obviously expected. Anyway the three of them show small differences. Next, we tried with other variability of partners. They may change more the results of direct experiences between one interaction and the next one. The standard deviation due to the subjectivity of evaluations considered in figure 3 was of 10 over 100 (10%). We can observe there that the convergence is now much more similar between AFRAS and any of both kalmans in the first 40 iterations. After that phase, the error of AFRAS got a clear improvement respect to the error of both kalmans. Finally an extreme situation is also analyzed in figure 4. It shows an scenario with sudden changes of subjective evaluations of direct experiences (the standard deviation is 33%). In such situation AFRAS behaves better than both kalmans since its convergence is much faster. We can also note that in these circumstances kalman with constant R becomes very similar to kalman with adaptive R.

5   Conclusion

We have presented a comparison between kalman estimation method and our fuzzy reputation based trust model. The chosen testbed is an evolved version

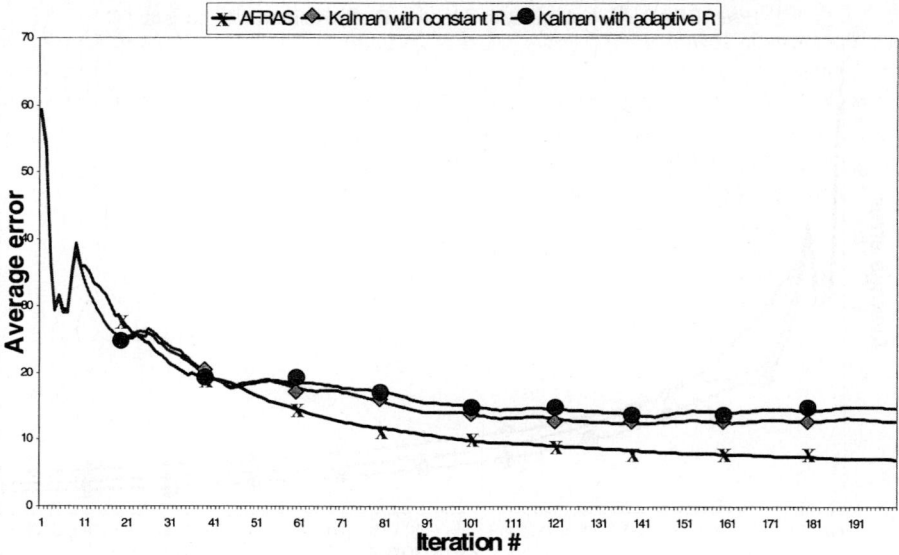

Fig. 3. Convergence of AFRAS vs. Kalman with StdDvt=10

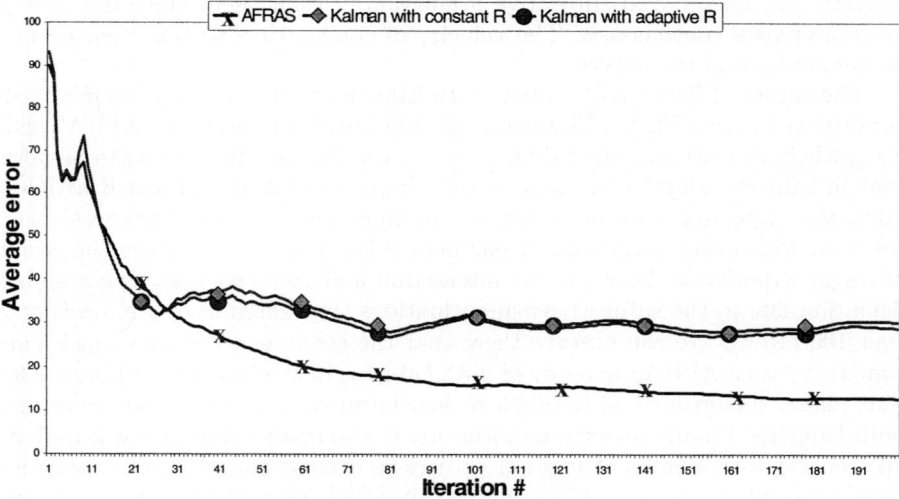

Fig. 4. Convergence of AFRAS vs. Kalman with StdDvt=33

of a classic one proposed by Zacharia and Maes to test the speed of convergence of the estimations from SPORAS. Analogous comparisons were previously implemented with other subjective trust models.

From the tested simulation we can conclude that AFRAS behaves better than kalman in the worst circumstances (the extreme situation of highly variable evaluations). This was also the conclusion and the main point for AFRAS when it was compared with REGRET, Singh and Yu and SPORAS. However, its results under normal circumstances are not able to improve the estimations obtained from kalman. In the next future we will also try to achieve further comparisons with kalman such as the reactivity facing sudden changes, the influence of cooperation (witness information). We have also the intention of comparing AFRAS with other classic estimation and filtering methods such as alfa-beta.

References

1. Abdul-Rahman, A.,Hailes, S.: Supporting trust in virtual communities. Procs. of 33th IEEE Int. Conf. on Systems Sciences (Maui, 2000).
2. Baldwin, J.F.: A calculus for mass assignments in evidential reasoning. Advances in Dempster-Shafer Theory of Evidence, eds. M. Fedrizzi, J. Kacprzyk and R.R. Yager (John Wiley, 1992).
3. Bar-Shalom, Y., Xiao-Rong, L.: Estimation and tracking principles, techniques and software. (Artech House, 1991).
4. Carbo, J., Molina, J.M., Davila, J.: A fuzzy model of reputation in multi-agent systems. Procs. of the 5th Int. Conf. on Autonomous Agents (Montreal, 2001).
5. Carbo, J., Molina, J.M., Davila, J.: Trust management through fuzzy reputation. Int. Journal of Cooperative Information Systems $12(1)$ (2003) 135–155.
6. Castellfranchi, C., Falcone, R.: Principles of trust for multiagent systems: Cognitive anatomy, social importance and quantification. Procs. of the 3rd Int. Conf. on Multi-Agent Systems (1998) 72–79.
7. Gelb, A.: Applied Optimal Estimation. (MIT press, Cambridge Mass, 1974).
8. Sabater, J., Sierra, C.: Reputation and social network analysis in multiagent systems. Procs. of the 1st Int. Joint Conf. on Autonomous Agents and Multiagent Systems (Bologna, 2002) 475–482.
9. Shardanand, U., Maes, P.: Social information filtering: algorithms for automatic word of mouth. Procs of the ACM Conf. on Human Factors in Computing Systems (1995) 210–217.
10. Yu B., Singh, M.P.: A social mechanism for reputation management in electronic communities. Lecture Notes in Computer Science **1860** (2000) 154–165.
11. Zacharia, G., Maes, P.: Trust Management through Reputation Mechanisms, Applied Artificial Intelligence **14** (2000) 881–907.

Conceptual Model of BDI Agents within an ITS

Frances Mowlds and Bernard Joseph Roche

University College Dublin, Belfield, Dublin 4
frances.mowlds@ucd.ie

Abstract. The purpose of this paper is to illustrate the conceptual modelling of Belief, Desire and Intention (BDI) Agents within an Intelligent
Tutoring System (ITS). A two stage modelling process is presented during which a combination of existing modelling techniques are employed.
Initially an overall analysis and functional decomposition of the complex
ITS is performed using recognized modelling procedures from the area
of object oriented software engineering. Finally, the BDI agents' state
(belief, desire, intention, goals and plans) are modelled using techniques
particular to the area of agent oriented software engineering. Epistemic
and modal logic are employed as a modelling language. Specific attention
is paid to the modelling and observations of social interactions among
Intelligent Agents (IA) in the ITS.

1 Introduction

The Intelligent Tutoring System (ITS) discussed in this paper, is under construction in University College Dublin (UCD). We will present a model of an
ITS whose core reasoning ability is realized by adapting the standard Belief
Desire Intention (BDI) model of agency[1]. BDI agents are a popular and well
established model for Multi Agent Systems (MAS)[2][1][3]. The BDI approach
is valuable for use within a dynamic environment such as an ITS. They operate
flexibly and appropriately to changing circumstances, despite incomplete information about the state of the world and other Intelligent Agents (IAs) in it[2][3].
The IAs' mentalistic attitudes of belief, desire, goals and intention guide their
behavior.

A complex application such as an ITS is composed of many types of agents
and objects. Recognized branches of software engineering such as Object-Oriented (OO) and Agent-Oriented (AO) software engineering have well established
tools for the construction of conceptual models for a given application. From OO
software engineering we have many modelling languages such as UML[4]. IAs
and objects share many similarities[3][5][6][7], but OO techniques fail to meet
all the requirements for the modelling of IAs. The state of an IA is composed
of beliefs, goals and plans and their behaviour is driven by properties such as
autonomy, adaptation, interaction, learning, mobility and collaboration[7]. A
formal framework must provide a conceptual infrastructure for the analysis and
modelling of agents and multi-agent systems, while also enabling implemented
and deployed systems to be evaluated and compared[8]. Much research has been

S. Wang et al. (Eds.): ER Workshops 2004, LNCS 3289, pp. 506–517, 2004.
© Springer-Verlag Berlin Heidelberg 2004

carried out in the area of AO software engineering and the formal modelling of agents (e.g.[9], [7], [8], [10], [11] and [12]) The approach followed in this paper for the construction of a conceptual model of an ITS considers aspects from both arenas. The modelling process is completed in two principal stages. Each stage examines the system at differing levels of detail. First a functional analysis of the structure of the ITS is performed. This stage is concerned primarily with how the overall system will operate. What the IAs should do, what are its goals, responsibilities, tasks and expertise. After that, a formal model of the agents reasoning behaviour is constructed. A mental model is built to observe how the agents will function as a group.

2   ITS

The ITS is being integrated into an existing lecturing system, it does not operate as a separate entity. Figure 1 illustrates the basic components of the ITS: The Intelligent User Interface (IUI), the agent engine consisting of Course Agents(CA), Lecturer Agents (LA), Student Agents (SA) and Tutor Agents (TA), resources and finally subordinate agents.

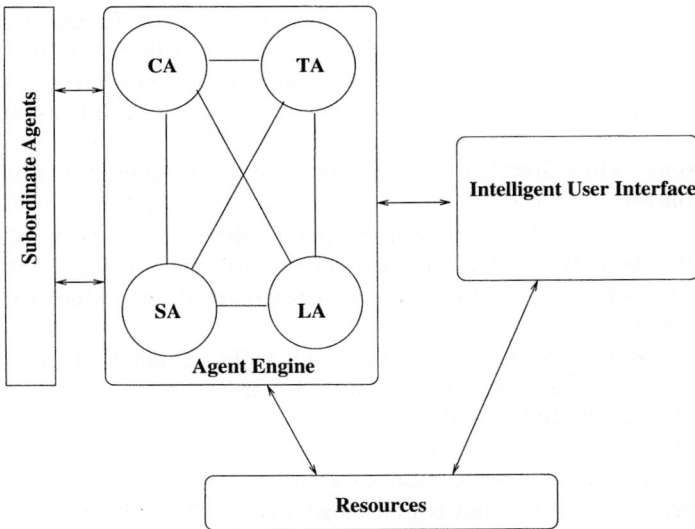

Fig. 1. Architecture of Intelligent Tutoring System

The IUI consists of various modules: a user modelling system, a page generation engine, a collaboration agent and a web browser. The agent engine consists of a society of interacting IAs, which together deliver much of the system functionality. Resources consist of an initial data set, which is used to initialize the IAs' belief sets. The belief sets will be supplemented with extra data throughout the life cycle of the ITS. Subordinate agents are agents of lesser importance that carry out tasks on behalf of IAs existing within the agent engine. They have a simple architecture, they may be purely reactive or carry out simple reasoning or data processing tasks. An example subordinate agents is an information agent, which helps users manage information[7]. In the next section, the core components of the agent engine are explored in greater detail through the use of agent role-modelling[5].

3   Agent Roles

The first stage of our conceptual model of the ITS consists of role modelling. This form of modelling originates in the area of OO software engineering[5], an emphasis is put on how agents interact with one another. This level of abstraction enables users to gain a deeper understanding of the overall system functionality. Roles are abstract characterizations of the behavior of an actor (physical agent such as student and lecturer or a software agent), within some specialized context[13]. This is what makes them particularly suitable for the modelling of IAs within an ITS. The Gaia methodology[9] for AO analysis and design defines roles as having four attributes by which it will be defined. These attributes are: responsibilities, permissions, activities and protocols. Figure 2 shows the role schema for a student agent. The student agent will represent a single student within the ITS. One agent will exist per student. Its core functionality is concerned with ensuring that the student is not deficient in any aspect of their course. Through continuously monitoring the students attendance, tutorial scores, online quiz scores, etc. the agent may identify students who are in difficulty. Consequently, these problem students will be promptly identified. The student agent will deliver these findings to the tutor agent who will act to rectify the situation.

The protocol AlertTA (figure 2) illustrates how the student agent will interact with a tutor agent. When the safety property is not met (i.e when a students tutorial or quiz score, for example is below a pre-defined low value) the tutor agent will be alerted.

When undertaking a role the IA will carry out tasks on behalf of the relevant figure that they are representing. These core roles are:

Student, represents a student within the ITS

Lecturer, represents the lecturer of a given course

Tutor,   represents the tutors associated with a course

Course,   represents the different courses available to the student

Admin,   represents the administrative departments of the university

Additional tasks that the student IA undertakes are: to present all information regarding a course to the student, alert the student of assignments that are due and provide the course material to the student using an appropriate pedagogical model[1]. The main tasks to be carried out by an IA assuming the

---

[1] The pedagogical model, is the teaching system used to present the course material to the student. One common system used is that of rote learning, where the student learns by doing, focuses on memorizing the material so that it can be recalled by the learner exactly the way it was written. The IA will then hold beliefs regarding the students knowledge of the material, and can test them to ensure they understood.

Role Schema: Student Agent (SA)

Description:
Identifies students in need of attention on a topic basis and also detects class trends.
Signals information to Lecturer Agent

Protocols and Activities:
AlertTA, MonitorAttendance, MonitorTutorialScore, MonitorQuizScore

Permissions:

reads	supplied		
		studentDetails	*// student information, e.g. student number*
		Studentattendance	*//student attendance*
		tutorialScore	*// Tutorial grades*
		onlineQuizScore	*//Quiz grades*
	generate	*studentUnderstanding*	*//indicates students progress*

Responsibilities:
Liveness:
StudentAgent = (MontiorTutorialScore | |MonitorAttendance| |MonitorQuizScore)ᵂ

Safety:
tutorialScore >= low
onlineQuizScore >= low
StudentAttendance >= low

Fig. 2. Gaia Role Schema for Student Agent SA within the ITS

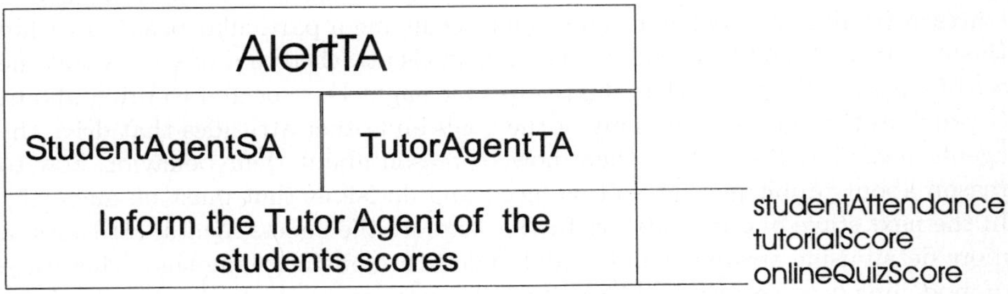

AlertTA		
StudentAgentSA	**TutorAgentTA**	
Inform the Tutor Agent of the students scores		studentAttendance tutorialScore onlineQuizScore

Fig. 3. Gaia protocol definition: AlertTA

lecturer role, is to consolidate information regarding any course that a lecturer is teaching with information on students attending these courses etc. The lecturer IA will engage in social activity with the student IA, the tutor IA, the course IA and the admin IA.

Each role can be further subdivided into smaller roles, resulting in a hierarchical structure (figure 4), which illustrates the decomposition of a student role.

A top-level agent fulfilling a particular role may rely on subordinate agents to carry out some tasks. These subordinate agents, such as a knowledge agents, will respond to different triggers provoking it to complete its objective. When goals related to the sub-roles are complete or achieved, a higher-level agent will resume its course of action. The highest role in the hierarchy of sub-roles in figure 4, will model the desired trajectory of a student's progress during its lifetime within the system.

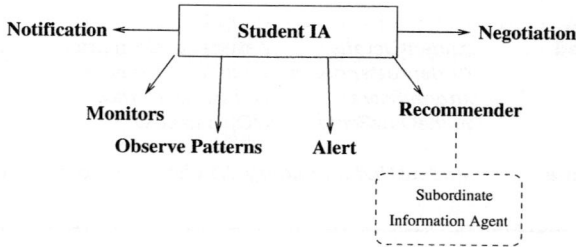

Fig. 4. Hierarchical Structure of Student Role

A further benefit of the analysis of the ITS into roles is to help identify key characteristics and associated data regarding the actors represented by an individual role. Taking the student agent as an example, inferences can be made about its expected levels and actual levels of knowledge. The student's learning style can be observed and modelled; this data can be used to help the system adapt, helping fine tune the teaching process to benefit future students. Finally, a role as described by[14] can be viewed as a set of mental attitudes or mental states governing the behavior of an agent occupying a particular position within the structure of a multi-agent system. The goals attached to a role represent the set of states of the world that the role-player might be expected to bring about. It provides the agents with many of the goals and other attitudes that drive the agents behavior. Instructing them how to reason about their behavior, how to reason about problems that will be faced and decisions that must be made[15]. In the next stage of our model we further explore the agents mental attitudes in more detail using epistemic and modal logic as a modelling language. This stage of modelling is to ensure the ITS will exhibit predictable and reliable behaviour. At this level the emphasis shifts from viewing IAs as role-players to studying the interaction of the IAs as they work together to deliver the overall system functionality.

4   Social Interaction of BDI Agents Within the ITS

Coordination among agents is considered a form of interaction devoted to goal attainment and to task completion[13]. The interaction of an IA with other IAs takes place through the role it plays within a scenario. Scenarios are defined in terms of roles, rather than in terms of specific IAs[16]. In this stage of the modelling process interaction modal logic, as defined by Alvarado and Sheremetov[16] is used. Interaction modal logic defines an agent as a 3-tuple of beliefs, goals (intentions) and actions. Everything else is expressed in terms of the latter. In this logical model, the following system is presented: let 1,2, ..., n be a set of indexes to represent agents. The information being believed by IAs corresponds to propositions or first order formulas, $\varphi_1, \varphi_2, \ldots, \varphi_n$. Similarly goals denoted, $g_1, g_1, \ldots, g_n$ and actions denoted $\alpha_1, \alpha_1, \ldots, \alpha_n$. Each set of formulas are combined in the classical manner using logical connectives $\neg, \wedge, \vee, \rightarrow, \leftarrow, \leftrightarrow$. The logic closures of the these sets are joined to obtain the object language Λ. Modal operators to model beliefs, goals and actions are also introduced.

$B_i\varphi$ Agent i believes φ.
$G_i g$ Agent i has the goal g.
$[do_i\alpha]$ g when agent i does action α the result is g.

A model language is the result of the closure of the modal formula with the logical connectives. Finally, the formal framework consists of Kripke structure $M = \langle W, \prod, I, D \rangle$. W consists of a non-empty set of worlds, $\prod$, I, D correspond to the interpretation of semantic (true or false), information (beliefs or desires) and dynamic (do) modal operators respectively. Let $r^*(i, \alpha)(M, w_0)$ be a set of worlds being accessed by the execution of an action α over the worlds (M, w_0) operators. Satisfaction of the modal formulas of belief, goals and actions of agent i in world w_0 is as follows:

- $w_0 \models B_i\varphi$ iff w $\models \varphi$, for all w with $w_0 R^{Bi} w$
- $w_0 \models G_i g$ iff w $\models$ g, for all w with $w_0 R^{Gi} w$,
- $w_0 \models [do_i\alpha]$ g iff w $\models$ g, for all w with $w^o r^*(i, \alpha)(M, w_o)$.

Relations R^{Bi}, R^{Gi} and $r^*(i, \alpha)(M, w_o)$ define a set of possible worlds, where belief φ of agent i is satisfied, goal g of agent i is satisfied or action α of agent i achieves g satisfaction, respectively. Action α should be performed under the fulfillment of w_o preconditions and the result is a set of worlds satisfying postconditions. An agent maintains commitment to the goal of the role until it leaves that role. This only happens upon achieving the relevant goals, or when the goals are believed unachievable or there is an unexpected event that inhibits goal realization or finally when the goal is no longer a scenario goal.

In the ITS the role will be instantiated with an initial default set of beliefs, goals and actions. It will then develop personality throughout the role that it plays inside different scenarios[16]. As an example we will provide a conceptual model of the student IA using interaction modal logic[16]. Initial beliefs will be related to factual data about the student. A student model[17], provides information about the object being modelled, a model is created at runtime as the student inputs to the system. Initial beliefs will include information such as student number, name, age, date of birth, date of registration and courses they are enrolled in. This knowledge will form part of the permanent beliefs of the student IA, as this data remains the same for the duration of the students life time within the ITS. For example the following initial beliefs could be present in the student agent SA_i's belief set.

B_{SA_i} (studentNumber, 10000000)
B_{SA_i} (studentName, SamJones)
B_{SA_i} (studentDoB, 11-11-1979)
B_{SA_i} (coursesReg, (Comp2001, Comp2001, Comp2013, Comp1003))

The agent will also have beliefs that do not need to be held for the duration of a students life span within the system. Such short term beliefs will be kept for a period of time while they remain relevant, upon their expiration they will be deleted. In our example, the student agent will hold beliefs about course participation. These beliefs are necessary in order to monitor the students progress, so that it is possible to positively guide the students learning. Once a student has completed a course these beliefs become irrelevant and therefore should be deleted, so that the agent's belief set will not be cluttered with irrelevant beliefs. These short term beliefs will have a time stamp associated with them, which is used to establish when they expire, for example at the end of a term. Examples of short term beliefs relate to deadlines that the students must meet, exams they must sit and classes they should attend.

B_{SA_i} (assign-due(DateD))
B_{SA_i} (meeting, Lecturer-2001,(TimeT))
B_{SA_i} (exam,Comp2001, (TimeT))

A scenario, discussed in more detail in[16] is a place where roles may interact. It consists of a set of roles, plans, goals, structures and a type of interaction. This is expressed formally as a tuple Sc: (R, Pl, G, St, Int) where

R $= R_1, R_2, \ldots, R_s$ is a set of roles in the scenario,

Pl $= g_1, g_2, \ldots, g_f$ is a sequence of goals that should be achieved by actions executed by agents concerned with it, and

G $= \mathbf{g}_f$ is a plans final goal

St is a scenario structure, it iterates the different types of relationships between the roles that the agents are engaged in. These are subordination, authority, peer and acquaintance etc.

Int is a subset of $\{Col, Cord, Comp, Inch\}$ that correspond to the type of scenario interaction in order to achieve the scenario purpose. Agents in a scenario interact to collaborate, coordinate, compete or interchange.

In the ITS, an example scenario is that of the student IA who must cooperate with other representative IAs so that they may gather information and carry out their respective tasks. It will negotiate with the tutor IA and the lecturer IA in order to arrange meetings with the student etc. Finally, it will make a request to the course IA requesting extra material for the student. The scenarios that exist in the ITS contain a set of roles, goal, structure and type of interaction as previously discussed. Expressed more formally, where i ϵ A represents an IA in the set of agents A = 1, ... ,n, SA_i is a student IA, LA_i is a lecturer IA and TA_i is a tutor IA. j ϵ S represents a student in the set of students S = 1,..., n

The tuple Sc: (R, Pl, G, St, Int) for the various interaction roles in the ITS is as follows:

R = {Student, Lecturer, Course, Tutor, Admin}
      these are the upper level roles

Pl = {monitorAttendance(X_j), monitorProgress(X_j),
      (alert (LA, X_j)), (alert (TA, X_j)), help(X_j)}

G $_f$ is the final goal in a plan, such as (alert (LA,X_j)

St   Authority (LA$_i$, SA$_i$), Authority (LA$_i$, TA$_i$),
      Peer(SA$_i$, SA$_i$), Peer(LA$_i$, LA$_i$),
      Peer(TA$_i$, TA$_i$), Peer(SA$_i$, TA$_i$)

Int   {Col, Cord, Comp}

A scenario plan Pl, may consist of the student IA identifying students in need of attention on a topic basis. This information is then signalled to the lecturer IA. Social mental shaping[14] occurs here as the lecturer IA will examine this new information and other information provided by the collaborating student IA and will modify its mental state, by adopting new beliefs and goals. This results in a change of the IA's behavior. The social relationship the two agents are engaged in is a cooperative one. The two agents have a common implicit goal to help the student. The lecturer IA may contact the student IA to arrange a meeting with the actual student, or contact the tutor IA informing it of the students problem so that the tutor may take the appropriate action. A further option would be to supply the course IA with supplementary reading material. This material will also be made available to all students registered for the course, including the student who is in difficulty. For example, a class trend may be detected where a number of students are having difficulty with some portion of the course. The multiple options available to the agent in a particular situation and the possible outcomes of an action are represented as a time tree[10] (see figure 5).

The student IA consequently will hold beliefs regarding the state of the student it is modelling, about the actions it has performed and about the mental state of the lecturer IA. Using the belief modal operator B for student agent SA, the following subset of beliefs, indicates the student j from the set X of students is in difficulty, as the student has both a poor score for tutorial number 1 and a poor score for an onlineQuiz.

B_{SA_i} (Student(X_j))
B_{SA_i} (PoorScore (Tutorial1, X_j))
B_{SA_i} (PoorScore (OnlineQuiz, X_j))

Upon carrying out action *alertLecturer()* SA$_i$ will hold a belief regarding the mental state of the lecturer IA, i.e. it believes that the lecturer IA knows the student needs help and also has a goal to help a student in need of help.

B_{SA_i} (B_{LA_i}, NeedsHelp, (Student(X_j)))
B_{SA_i} (G_{LA_i}, HelpStudent, (Student(X_j)))

In order for the student IA to carry out the course of action just specified it must first want or desire to notify the Lecturer IA under these circumstances.

Past   Present   Future   **Arrange with SA$_i$ a**
                                                                       suitable time for
                                                                       the lecturer to meet
                                                                       the student

Problem   **Notify TA$_i$**
Student   **Acknowledge**   **about the**
reported   **Report**   **problem**
received   **student**
from SA$_i$

                                                                       Supply extra
                                                                       teaching
                                                                       material to
                                                                       CA$_i$

                                                                       Take no action
                                                                       for now

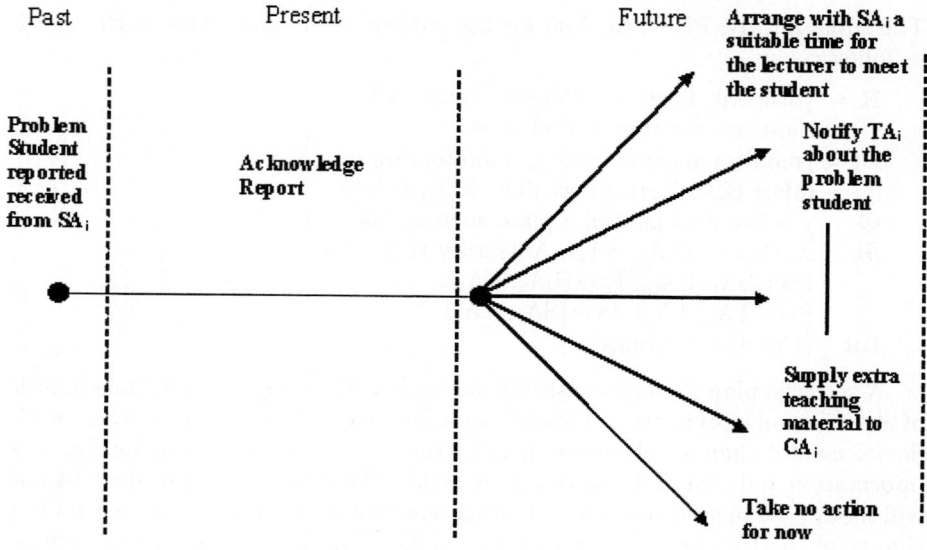

Fig. 5. Time Tree for Lecture Agent LA$_i$

The goals in Pl illustrate this. Finally, the subset Int, contains the interactions:
Collaboration for example, two agents may have the same goal to help the student such as the lecturer and the tutor IA.
Coordination IAs may need to schedule their activities in order to accomplish their goals, this could be between a student and a lecturer IA.
Competition may occur between two agents of the same type for example two student IAs, who both require the attention of the lecturer or tutor IA.

The life cycle of the agent is defined in the global (universal) scenario. This universal scenario contains a number of different scenarios forming the hierarchical structure as discussed previously. The life cycle of an agent is determined by the scenario in which it participates, assuming the corresponding roles. Achieving one particular goal the agent abandons the sub-role in the corresponding scenario and continues in the scenario of the upper level of the hierarchy[16].

4.1   Modelling of an IA's Life Cycle Within the ITS

The final stage of our conceptual model deals with mapping the duration of the IA's life cycle. During its life span, various sub-roles within the ITS will be fulfilled. It is important for the IA to be able to make inferences about past, current and future events in order to simulate the learning process of the student and extract general models relating to them, while ensuring that the students progress is satisfactory.

As the student progresses through a course, it is expected that they should complete specific requirements. The corresponding student agent SA$_i$ will update its beliefs as these requirements are met. These beliefs have a time stamp appended to them. This time stamp will help identify how far they have progressed with regards to their expected levels of progress. The following includes examples of beliefs that will be true for SA_i at some point and also some requirements that the student should meet.

Beliefs:
- $B_{SA_i}(\text{AssigmentSubmitted}(A))(T_1)$. SA_i believes that student A hands in an assignment at time T_1.
- $B_{SA_i}(\text{CourseRegristration}(C))(T_1)$. SA_i believes the relevant student is registered for course C at time T_1
- Hand in assignments $A_1, \ldots, A_n$ where n is the total number of assignments the student must complete to finish the course. SA_i will communicate this to the course IA CA_i and the lecturer IA LA_i.

Requirements:
- Achieve minimum grade ϕ in each assignment.
- Achieve an overall pass grade α in assignments
- Complete the final exam and achieve a minimum pass grade β
- Student passes course if $\alpha + \beta \geqslant \varphi$ where φ is some arbitrary value assigned by the lecturer of a course to be a minimum pass grade.

Grades will be calculated by CA_i who will share the values with both SA_i and LA_i and any relevant parties respectively. These are ideal steps that the students will follow through. Some possible variations of the ideal stages the student will follow through and the actions that the student IA SA_i must undertake are:

1. A particular student suffers an illness and misses lecturers. Resulting from this they fail the assignment part of the course. SA_i will have to take into consideration the students circumstances. Some possible courses of action are: SA_i could request that the tutor IA TA_i deals with the case, TA_i might arrange for extra lesson to be given to the student at an appropriate time. Otherwise SAi could contact a lecturer IA LA_i requesting supplementary exercises for the student to complete so that he may have an opportunity to achieve a pass grade in the assignment part of the course.
2. A student for no apparent reason fails to hand in an assignment for a course. In this case SA_i will notify LA_i and TA_i for that particular course. These agents will then decide what action to take. SA_i will hold the belief that LA_i and TA_i are capable of dealing with the situation.

5   Discussion

In this paper, a formal model of an ITS founded on the BDI agent model was introduced. BDI agents are a useful agent paradigm and are particularly suited for modelling human thinking and reasoning behaviour. The conceptual model presented, consists of two primary stages. The first of which presents a view of the overall architecture of the ITS and how it will be conceived. This helps identify key aspects of the system to be implemented. These aspects are examined through the use of a role-modelling technique. The second stage of modelling looks at how the agents will interact to deliver system functionality. An interaction model using modal logic[16] is used to provide a formal description.

For future work it is conceivable that weights will be incorporated into the agents belief and goal sets. This weighting of the IAs' beliefs and goals will further help to guide its behaviour in the desired direction. The motivation for this is that there are many conceivable possible worlds available to an agent, but not all these worlds are sought-after. The most desirable worlds, for example would consist of each student completing their course and also that they achieve the best grade they are capable of. Guided by past experiences, a weighting function can be applied to the belief and goal set in order to guide the IAs' behaviour.

References

1. A. S. Rao and M. P. Georgeff. BDI-Agents: From Theory to Practice. In Victor Lesser and Les Gasser, editors, *Proceedings of the First International Conference on Multiagent Systems, ICMAS 95*, pages 312–319, California, 1995. AAAI Press.
2. F. Dignum, D. Moreley, E Sonenberg, and L. Cavendon. Towards Socially Sophisticated BDI Agents. In *Proceedings of the ICMAS 2000*, pages 111–118, 2000.
3. I. Rahwan, R. Kowalczyk, and Y. Yang. Virtual Enterprise Design – BDI Agents vs. Objects. In R. Kowalczyk and M. Lee, editors, *Recent Advances in Artificial Intelligence in e-Commerce, Lecture Notes in Artificial Intelligence*, volume 2112, pages 147–157. Springer-Verlag, 2001.
4. P. Stephens and R. Pooley. *Using UML: Software Engineering with Objects and Components*. Pearson Education Limited, Essex, England, 2000.
5. E. A. Kendall. Role Modelling for Agent System Analysis, Design, and Implementation. In *1st International Symposium on Agent Systems and Applications*. IEEE CS Press, October 1999.
6. M. Wooldridge and N. R. Jennings. Intelligent Agents: Theory and Practice. *Knowledge Engineering Review*, 10(2):115–152, 1995.
7. A. F. Garcia, C. J. P. Lucena, and D. D. Cowan. Agents in Object-Oriented Software Engineering. *Software Practice and Experience*, 34(5):489–521, April 2004.
8. M. d'Inverno and M.Luck. Formal Agent Development:Framework to System. In *Rash, J. L., Rouff, C. A., Truszkowski, W., Gordon, D. and Hinchey, M. G., Eds. Formal Approaches to Agent-Based Systems:First International Workshop, FAABS 2000*, pages 133–147, Bologna, Italy, 2001. Springer.
9. M. Wooldridge, N. R. Jennings, and D. Kinny. The Gaia Methodology for Agent-Oriented Analysis and Design. *Autonomous Agents and Multi-Agent Systems*, 3(3):285–312, 2000.
10. A. S. Rao and M. P. Georgeff. Modelling Rational Agents within a BDI Architecture. In Allen, J. Fikes, and E. Sandwell, editors, *Proceedings of the Second International Conference on Principles of Knowledge Representation and Reasoning*, pages 473–484, San Mateo, CA, 1991. Morgan Kaufmann.
11. A. Perini, P. Bresciani, P. Giorgini, and F. Giunchiglia. Towards an Agent Oriented Approach to Software Engineering. In *Proceedings of the Workshop Dagli oggetti agli agenti: tendenze evolutive dei sistemi software*, Bologna, Italy, September 2001. Pitagora Editrice.

12. C. Castelfranchi. Modelling Social Action for AI Agents. *Artificial Intelligence*, 103:157–182, 1998.
13. A. Perini, A. Susi, and F. Giunchiglia. Designing Coordination among Human and Software Agents. Technical Report DIT-02-0060, University of Trento, 38050 Povo, Trento, Italy, Via Sommarive 14, 2002.
14. P. Panzarasa, N. Jennings, and T. Norman. Social Mental Shaping: Modelling the Impact of Sociality on the Mental States of Autonomous Agents. *Computational Intelligence*, 17(4):738–782, 2001.
15. P. Panzarasa, T. J. Norman, and N. Jennings. Modeling Sociality in the BDI Framework. In J. Liu and N. Zhong, editors, *Intelligent Agent Technology: Systems, Methodologies, and Tools. Proceedings of the First Asia-Pacific Conference on Intelligent Agent Technology*, pages 202–206. World Scientific Publishing, 1999.
16. M. Alvarado and L. Sheremetov. Interaction Modal Logic for Multiagent Systems based on BDI Architecture. In *Proc of the 3er. Encuentro Internacional de Ciencias de la Computación (ENC' 01)*, pages 389–396, Aguascalientes, Mexico, September 2001.
17. J. Self. *Student Modelling: The Key to Individualized Knowledge-Based Instruction (NATO Asi Series. Series F, Computer and Systems Sciences, Vol. 125)*. Springer Verlag, Berlin, Germany, 1994.

Model-Driven Agent-Based Web Services IDE

Yinsheng Li[1], Weiming Shen[2], Hamada Ghenniwa[3], and Xiaohua Lu[1]

[1] Software School, Fudan University, Shanghai, P.R. China
{liys,0014010}@fudan.edu.cn
http://www.software.fudan.edu.cn
[2] Integrated Manufacturing Technologies Institute, National Research Council Canada
London, Ontario, Canada
Weiming.shen@nrc.ca
[3] Department of Electrical and Computer Engineering, University of Western Ontario
London, Ontario, Canada
hghenniwa@eng.uwo.ca

Abstract. Web services have not exploited sufficient semantics and approaches to dynamic service-oriented operations. Software agents have been developed for these operations. Current efforts on integrating Web services and agents are predominantly concerned with enabling the agents in existing systems to request, provide or broker Web services. This paper presents the notion of agent-based Web services (AWS) and describes an integration framework that utilizes software agents' model to construct semantic and dynamic Web services. Essential concerns about the integration have been addressed. Basic AWS meta model and model-driven integrated development environment (IDE) have been developed. The IDE has been evaluated by a case study of an e-Marketplace service, reverse auction.

1 Introduction

Web services are featured with application, platform and provider independence. They provide an appropriate paradigm for open large-scale application environments, such as e-Marketplaces. These environments can be viewed as a collection of services. Services collaborate to achieve their business goals. An essential aspect for Web services is the semantic model by which the service is recognized and able to interoperate with the external world. However, numerous efforts have been focused on Web services choreography and orchestration through delegation agents or proxies [1,2,3]. Web services are not treated as autonomous entities in themselves.

On the other hand, software agents are endowed with autonomy, responsiveness, pro-activeness and social ability. They can interact with other agents and behave on behalf of their own interests. A few researches have been ongoing with agent-based e-Marketplaces [4,5,6,7]. Software agent paradigm has achieved technical advantages in software construction, legacy systems integration, transaction-oriented composition and semantics-based interaction [8,9,10,11,12,13].

This paper is a further development of the concept, i.e., agent-based Web services (AWS) that we proposed in [14]. AWS is to integrate software agents and Web services technologies into a cohesive body that attempts to avoid the weaknesses of each

S. Wang et al. (Eds.): ER Workshops 2004, LNCS 3289, pp. 518–528, 2004.

individual technology, while capitalizing on their individual strengths. There are a number of other studies on the integration of software agents with Web services. However, some were on software agents with semantic Web rather than Web services technologies, e.g., specifying software agents with standard mark-up languages like RDF/Schema and RuleML [15] or DAML-S [16,17]. Some are predominantly concerned with enabling the agents in existing systems to request, provide or broker Web services [18,19]. Some only concern one specific aspect of agent-based Web services, e.g., integrating DAML-S-based Web services and an agent communications language [20].

In our agent-based Web services model, software agents are not used for services communication front ends or as proxies. Rather, they are basic entities that encapsulate Web services. We believe that agent-based services implementation, i.e., AWS, can embody essential features and functions of Web services. Agent orientation for software construction and semantics-based interaction can be effectively applied for Web services applications.

2 Agent-Based Web Services

The notion of agent-based Web services (AWS) is to capture, model and implement service functionalities with autonomous and dynamic interactions [14]. The idea is to exploit agents' capabilities of dynamic interactions to enhance Web services' behaviors. We strongly believe that AWS could improve Web services at functional and interaction in themselves, instead of using external interaction delegations or communication proxies.

Technically the principle for AWS is to exploit agent-oriented software construction, knowledge representation [10, 17] and interaction mechanisms as integral aspects of service-oriented business transactions and application environments. The major efforts towards this end are devoted to determining and integrating strengths of both Web services and software agents. As observed, Web services have proved to be prosperous in e-Business applications while still undergoing fundamental setbacks in semantic integration, interaction behaviors and implementation frameworks. Software agents could naturally offer solutions for overcoming these setbacks. We have developed an integration framework for AWS.

The framework addresses several aspects of AWS, i.e., software construction, technical stacks and business processes. Figure 1 illustrates the framework in terms of six layers: business applications, Web services operation protocols, knowledge representation, functional entity, communication, and transportation layers. At the business applications layer, such as e-Marketplace or supply chain, service-oriented and business-centric protocols for business transactions are applied by AWS to collaborate with each other. At the Web services operation layer, Web service-oriented protocols, such as UDDI [21], WSDL [22] and BPEL4WS [3], are applied to support business processes and transactions. At the knowledge representation layer, ontology-level languages, such as OWL-S [23], are applied to support AWS' semantic description and knowledge models. At the functional entity layer, agent-oriented modeling and construction, such as CIR-Agent [11] and AUML [13], are applied to build the AWS problem solving entities. With such entity structure, AWS could capture service capabilities and interaction patterns. The interaction components handle the conversation among AWS entities. The problem-solving component provides functionality of the service. And the communication component is responsible for the service-oriented messaging mechanism. At the communication layer, a set of protocols are applied to implement application coordination using agent patterns. The application-oriented protocols include e-Auction and Contract-Net. These protocols are supported by ACL [9, 24] based communication. ACL messages and protocols are incorporated into service-oriented protocols, such as SOAP [25]. Finally at the transportation layer, basic Web services applicable Internet protocols, such as HTTP, FTP and HTTPS, are applied to get through the data.

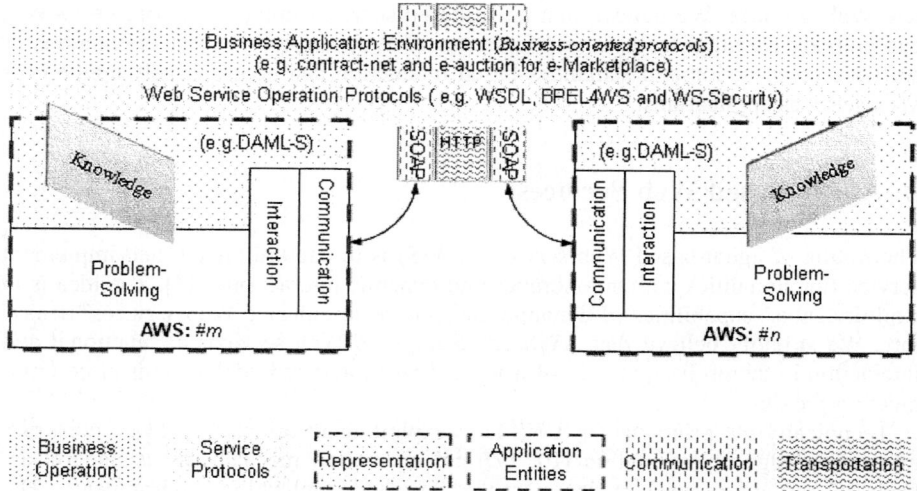

Fig. 1. Integrated technical framework for agent-based Web services

3 AWS Modeling and Knowledge Representation

The concept of an AWS is viewed as a computational model that enables a designer to capture and represent complex applications in open environments, such as e-Marketplaces. AWS applications are modeled with open technologies. The OMG MDA approach offers a specialized and concerns-focused framework for application designers in order to express their applications independently of a particular technology [26]. A meta-model is a good start point and essential tool for AWS based application development. To facilitate systems design and integration, the AWS meta-model expresses and relates the fundamental AWS classes and operations concepts. The meta-model is treated at two levels, (i) UML-based model for service capabilities and processes, and (ii) agent-oriented model for service interactions. Figure 2 illustrates the links of its top packages. The involved classes are based on AWS' profile, processes, resources, and derived concepts. Figure 3 illustrates package relationship among three most important classes 'profileclasses', 'processclasses' and 'groundclasses'.

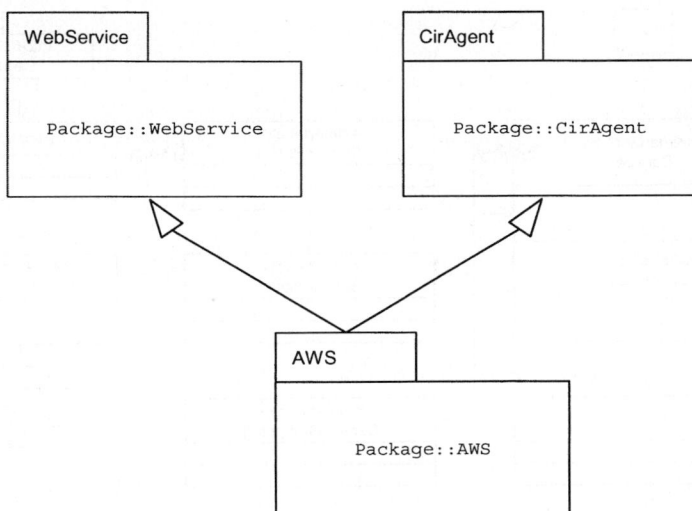

Fig. 2. Applied meta-model for AWS

AWS use ontology-level languages to describe the designated services and to support service composition and execution. For AWS, the semantic descriptions of the related services are captured using the meta-model. These descriptions are further arranged to build AWS knowledge. AWS knowledge is equipped with a rule base, which incorporates execution logics, business logics and policy constraints into decision-supporting and task-oriented rules. The agent-oriented information sensing and handling mechanisms enforce timely update of the descriptions and knowledge of AWS. A consistent analyzer runs over the rule base to support the other AWS modules. AWS therefore apply reasoning principles, and accomplish automatic interaction patterns in terms of business-specific commitments and service-oriented process orchestration.

There is a concern that representation languages should account for efficient exchanges between AWS knowledge and service-oriented protocols. This issue has been addressed by using ontology-level language with our proposed integrated description method [14]. This method integrates knowledge representation with Web services protocols through semantic mapping. AWS description is technically independent and can be supported by different languages with rich semantics, e.g., XML/Schema, RDF/RDFS, DAML+OIL/OWL and DAML-S (OWL-S). Also it supports several protocols, e.g., UDDI, SOAP and WSDL, related to Web services operations. As illustrated in Figure 4, the semantic mapping between the description and protocols is machine-translatable since both sides are based on XML/Schema. The mappings are handled at the meta-model level.

4   Model-Driven AWS Implementation and IDE

With the proposed framework, meta-model and knowledge representation, AWS can be implemented as unified software entities that integrate the strength of software

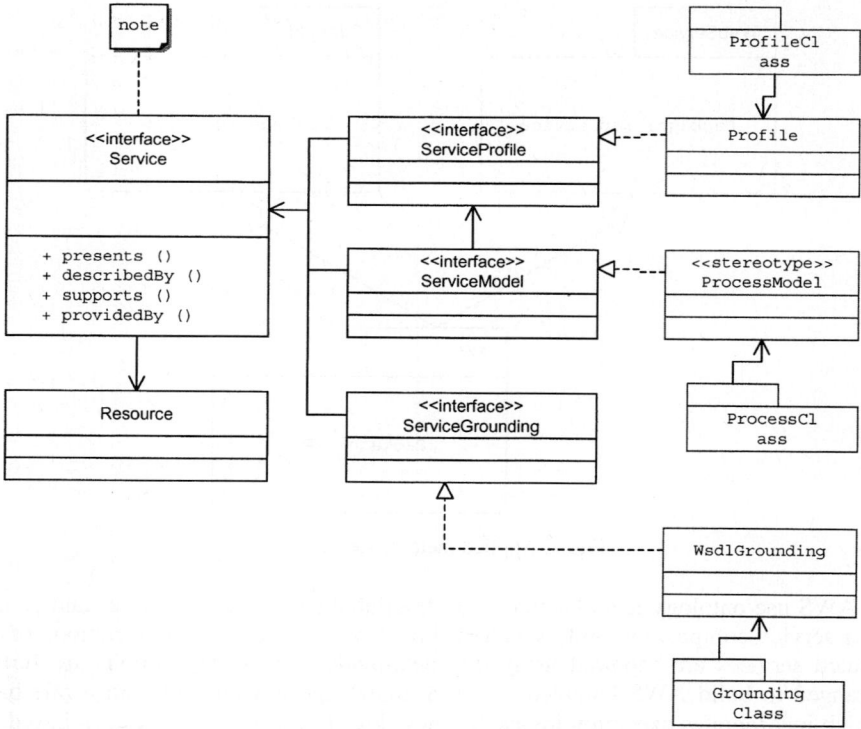

Fig. 3. Package relationship among 'profileclasses', 'processclasses' and 'groundclasses'

agents and Web services. They have the 'S' caps as Web services and 'A' bodies as agents in Figure 5. They recognize Web services protocols with the integrated description based knowledge, while behaving as software agents since they incorporate the CIR-Agent model. The service-oriented operations are handled through agent interaction patterns. The ontological descriptions of them are organized into AWS' agent-oriented knowledge modules. AWS can therefore recognize and deal with Web service-oriented protocols. For example, through them are WSDL interfaces (boxes with the letter 'I' in Figure 4) generated and exposed, published and looked up through UDDI-based registrar, and interact through SOAP and HTTP. Being built as Web applications, AWS physically resides on Web servers to take advantages of Web technologies.

An IDE called 'SOAStudio' has been developed for the mentioned AWS implementation framework and related concepts. It is an integrated development environment, and includes development workspaces and mechanisms as well as basic application programming interfaces. It can be used to develop AWS, and further build business systems like e-Marketplaces. Agent-based Web Services are the entities to be developed. They are based on the proposed meta-model, and have agent constructs in software entity and DAML-S description to build ontologies. The extracted WSDL interfaces are published through UDDI-compliant registrar, which is also an agent-based Web service.

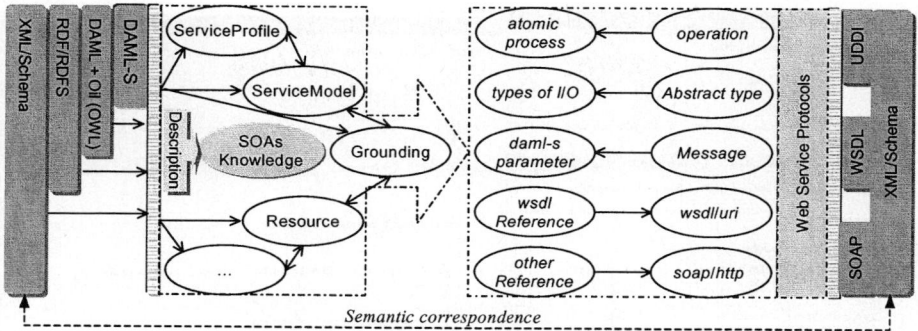

Fig. 4. Meta-model based AWS description and mapping

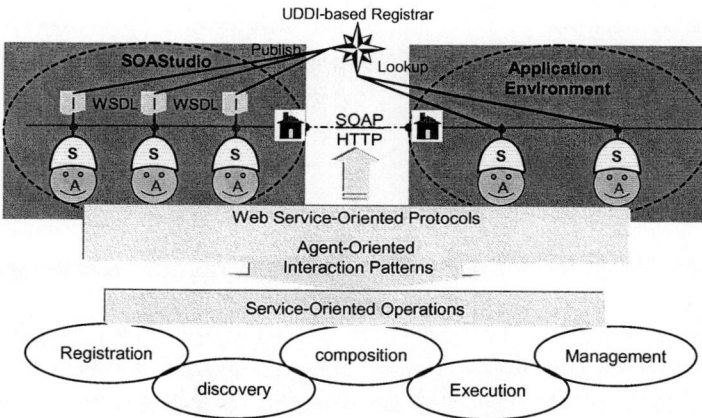

Fig. 5. AWS implementation framework

As shown in Figure 6, SOAStudio is designed with a unified graphical user interface for users to interactively access system functionalities. The unified interface provides users with four functional workspaces: 'AWS design workspace', 'AWS development workspace', 'AWS Debugging workspace', and 'AWS runtime environment'. The 'design workspace' is for AWS definition and project management. The definition function collects information about the ongoing projects and AWS. This workspace covers complete service description items by DAML-S (also applicable for OWL-S) in terms of profile, process and grounding. The development workspace is supported by functions to create descriptions and agent files based on the predefined meta-model and user-defined service properties. The description files include DAML-based semantic description files and associated WSDL-based files. Basic agent-oriented class files are generated based on JADE [24] agent classes which can be defined according to the CIR-Agent model. The debugging workspace is composed of a building tool, an operation and output window and a view tool. The building tool is provided to generate and deploy executable byte-stream classes on the generated and customized service (agent) files. The operation and output window

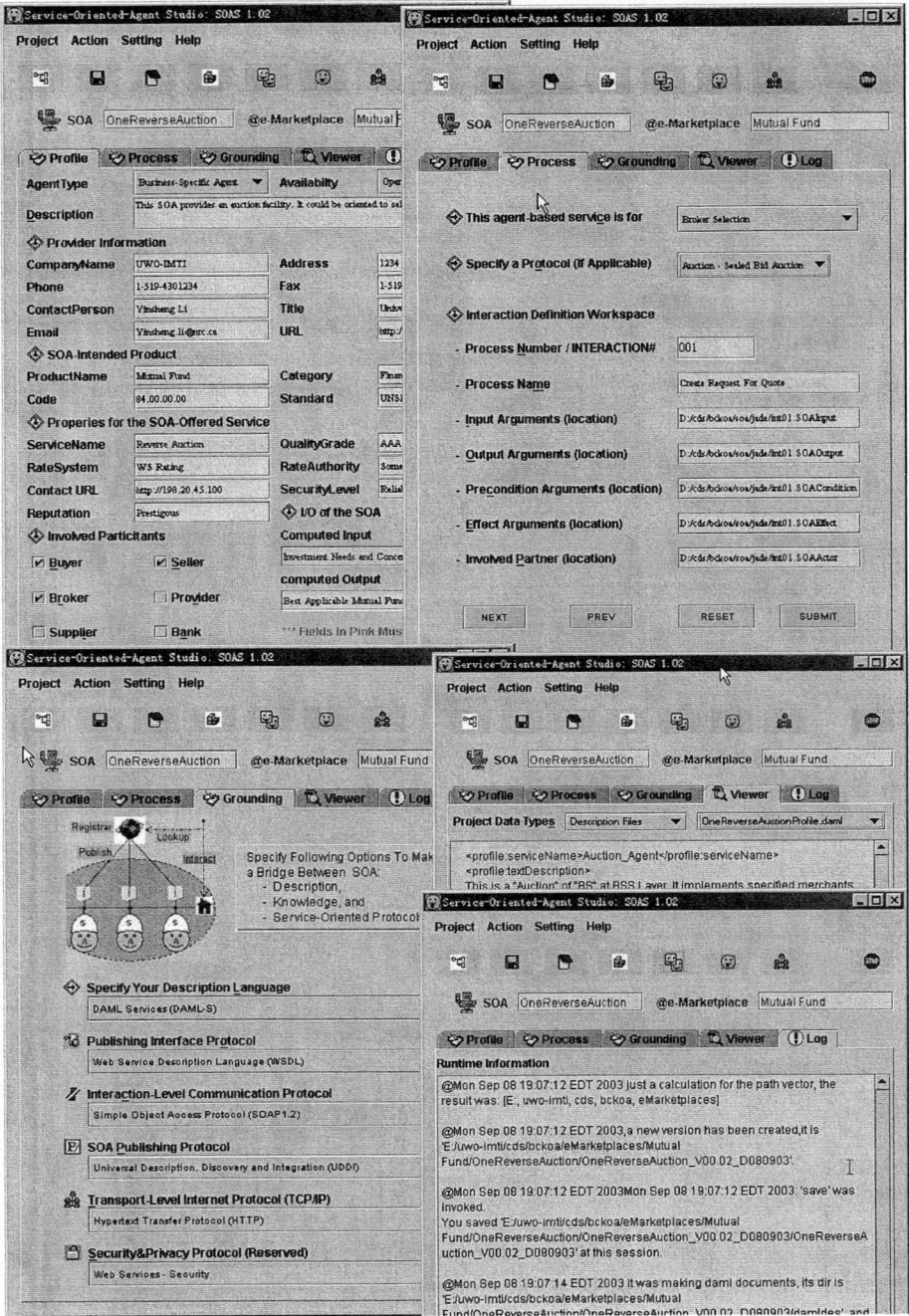

Fig. 6. Agent-based Web services Studio (SOAStudio)

monitors the performance of the studio and integrated platforms through operation logs, outputs and error reports. The view tool is used to check the generated packages, descriptions, code and project structure and data. Developers can complement pertinent resources to the integrated platforms like JADE and Tomcat, and make the generated AWS functional and refined. The 'runtime environment' is achieved by the JADE platform and Tomcat application server [27]. They work together to provide execution, monitoring and management environment for the AWS.

5   Case Study

An e-Marketplace reverse auction service is processed as follows [28]. A buyer submits a request for bids to a marketplace auctioneer through its user interface AWS. The auctioneer starts with an auction session and sends requests for bids to all matching sellers. The sellers decide if they want to participate and submit bids through their interface AWS. Both the buyer and sellers may spell out some attached conditions with their bids. For example, the buyer may set criteria for bid evaluation, and sellers may ask for a minimum price to be paid for the good. The auctioneer collects sellers' bids, calculates their credits based on the buyer's preferences, and recommends a winner to the buyer.

The auction goal is mutual fund in the evaluated case. Figure 7 shows some snapshots for the main items of this process. The execution platform 'JADE', the auctioneer AWS 'SOAService', the UDDI AWS 'UDDIFacility', the buyer AWS 'buyer_with_SOAService' and three seller AWSs 'bider*_with_SOAService' are present at the top of the Figure 8. The bidding actions are recorded and can be viewed by SOAStudio's 'Debugging Workspace'. At the bottom of Figure 8 there are six dialogs during the reverse auction case. The 'Set CFB Options' is for the buyer to specify requirements. The 'Set Weighted Facts for Requirement' is for the buyer to specify weights of the required items. In this case, the buyer is to buy 'Mutual Fund' of $100,000. The required 'Interest' is '10%'and 'Risk' is '2%%'. The weight for 'Interest', 'Risk' and 'Provider Reputation' is 10:2:100. The buyer AWS 'buyer_with_SOAService' sends this message to the auctioneer AWS 'SOAService'. The auctioneer sends this call-for-bid message to three seller AWSs 'bider1_with_SOAService', 'bider2_with_SOAService' and 'bider3_with_SOAService'. Three sellers receive the message and reply by setting on their provided 'Set Offer Options' dialogs. They input their offers with 'Interest', 'Risk' and 'Reputation' and send back to the auctioneer. The auctioneer collects these three offers, calculates their credits with buyer-defined item weights, and determines a winner. It recommends it to the buyer by sending back a message. This message is displayed through the buyer's AWS in a dialog 'The Result of This session of Reverse Auction'.

6   Conclusion

The integration of Web services and software agents has been envisioned as appropriate computational paradigm for service-oriented dynamic and open environments. We

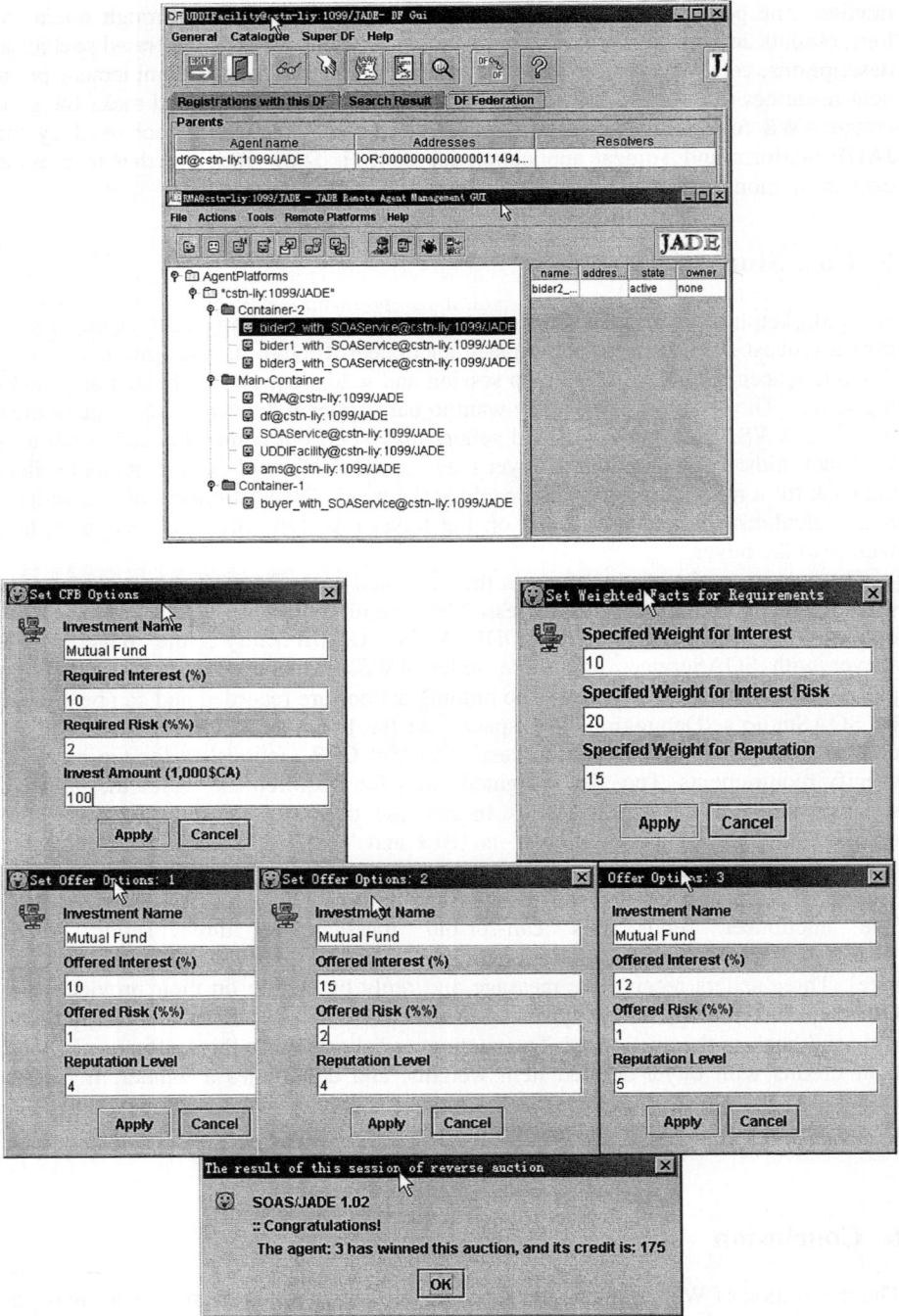

Fig. 7. A reverse auction with SOAStudio

believe it destined to combine them and integrate related technologies into a cohesive body that attempts to avoid the weaknesses of each individual technology, while capitalizing on their individual strengths. The current Web services technologies are market-driven and business-oriented. They, however, have not exploited sufficient semantics and approaches to automatic and dynamic services operations. In this paper we attempt to address these issues through sophisticated agent-oriented knowledge representation and interaction approaches. This attempt is intended both for Web services and for software agents.

Current efforts on integrating Web services and agents are predominantly concerned with enabling the agents in existing systems to request, provide or broker Web services. This paper described the notion of agent-based Web Services (AWS), to integrate the concepts and the technologies from Web services and software agents. Several critical issues have been addressed, including components structure and meta-model framework. A model-driven IDE called SOAStudio for the proposed AWS concepts, knowledge representation and implementation framework has been implemented. The prototype has been tested through an e-Marketplace-oriented service, reverse auction. Note that new frameworks and tools are emerging from Web services and software agents' communities to support the proposed AWS approach. We believe that AWS is a promising methodology for service-oriented applications. It provides an appropriate implementation framework for service-oriented technologies and applications.

References

1. W3C: Web Services Architecture. [Online]. Available: http://www.w3.org/2002/ws/arch/ (2003)
2. IBM, Microsoft, and BEA: Web Services Coordination (WS-Coordination). [Online]. Available: http://www-106.ibm.com/developerworks/webservices/library/ws-coor (2003)
3. McIlraith, S., Mandell, D: Comparison of DAML-S and BPEL4WS. Stanford University, Knowledge Systems Lab (2002)
4. Chavez, A., Maes, P. Kasbah: An Agent Marketplace for Buying and Selling Goods. In: Proceedings of the First International Conference on the Practical Application of Intelligent Agents and Multi-Agent Technology, London (1996)75-90
5. Bjornsson, H., et al: FX-Agents Description - White Paper. [Online]. Available: http://fxagents.stanford.edu/vision.php (2002)
6. The Intelligent Software Agents Lab: RETSINA. [Online]. Available: http://www-2.cs.cmu.edu/~softagents/retsina.html (2001)
7. Ghenniwa, H.H.: eMarketplace: Cooperative Distributed Systems Architecture. In: Proceedings of the 4th International Conference on Electronic Commerce Research, Southern Methodist University, Texas (2001)
8. Emorphia Limited: Toolkit Overview. [Online]. Available: http://www.emorphia.com/research/features.htm (2003)
9. FIPA: FIPA ACL Message Structure Specification. [Online]. Available: http://www.fipa.org/specs/fipa00061/SC00061G.html (2002)
10. Genesereth, M.R., Fikes, R.E. (eds.): Knowledge Interchange Format, Version 3.0 Reference Manual. CA: Stanford University (1992)
11. Ghenniwa, H.H., Kamel, M.: Interaction Devices for Coordinating Cooperative Distributed Systems. Automation and Soft Computing, 6 (2) (2000)173-184

12. IntelliOne Technologies: Agent Construction Tools: Academic and Research Projects. [Online]. Available: http://www.agentbuilder.com/AgentTools/index.html (2001)
13. FIPA Modeling Technical Committee. 2003. Working Documents. [Online]. Available: http://www.auml.org
14. Li, Y., Ghenniwa, H.H., Shen, W.: Integrated description for Agent-based Web Services in eMarketplaces. In: Proceedings of the Business Agents and the Semantic Web Workshop, Halifax, Nova Scotia, Canada (2003)11-17
15. Eberhart, A.: OntoAgent: A Platform for the Declarative Specification of Agents. In: Proceedings of the international Workshop on Rule Markup Languages for Business Rules on the Semantic Web (2002)58–71
16. Dale, J., Ceccaroni, L., Zou, Y., Agam, A.: Implementing Agent-Based Web Services, AAMAS 2003, Workshop on Challenges in open agent environments, Melbourne Australia (2003)
17. The DAML Services Coalition: DAML-S: Semantic Markup for Web Services. [Online]. Available: http://www.daml.org/services/daml-s/0.7/daml-s.html (2002)
18. Lyell, M., Rosen, L., Casigni-Simkins, M., Norris, D.: On software agents and Web services: Usage and design concepts and issues. In: Proceedings of the AAMAS2003 Workshop on Web Services and Agent-based Engineering (2003)
19. Maamar, Z., Sheng, Q.Z., Benatallah, B.: Interleaving web services composition and execution using software agents and delegation. In: Proceedings of the AAMAS2003 Workshop on Web Services and Agent-based Engineering (2003)
20. Gibbins, N., Harris, S., Shadbolt, Nigel: Agent-based Semantic Web Services. In: The Twelfth International World Wide Web Conference, Budapest, HUNGARY (2003)
21. OASIS: UDDI Version 2.04 API Specification. [Online]. Available: http://uddi.org/pubs/ProgrammersAPI-V2.04-Published-20020719.htm (2002)
22. W3C: Web Services Description Language (WSDL) 1.1. [Online]. Available: http://www.w3.org/TR/2001/NOTE-wsdl-20010315 (2001)
23. W3C: OWL Web Ontology Language Overview. [Online]. Available: http://www.w3.org/TR/owl-features/ (2003)
24. TILAB: Java Agent DEvelopment Framework . [Online]. Available: http://sharon.cselt.it/projects/jade/ (2003)
25. W3C: SOAP Version 1.2 Part 1: Messaging Framework. [Online]. Available: http://www.w3.org/TR/soap12-part1 (2003)
26. Object Management Group: MetaObjectFacility(MOF) Specification (Version 1.4). [Online]. Available: http://www.omg.org (2002)
27. Apache Software Foundation: The Tomcat 5 Servlet/JSP Container. [Online]. Available: http://jakarta.apache.org/tomcat/tomcat-5.0-doc/index.html.
28. Laskey, B., Parker, J.: Microsoft BizTalk Server 2000: Building a Reverse Auction with BizTalk Orchestration. [Online]. Available: http://msdn.microsoft.com/library/default.asp?url=/library/en-us/dnbiz/html/bizorchestr.asp (2000)

AFDM: A UML-Based Methodology for Engineering Intelligent Agents

Yanjun Tong, Gregory M.P. O'Hare, and Rem Collier

Department of Computer Science, University College Dublin, Ireland

Abstract. Agents are a potential technology with many applications. It is urgent to develop appropriate methodologies for the development and deployment of agent-oriented applications. This paper presents Agent Factory Development Methodology (AFDM), a cohesive methodology, to support the development of agent-oriented applications. On the other hand agent cloning is considered one of the possible approaches to avoid system overloading. In later sections of this paper, agent cloning is designed and implemented as an example through the AFDM.

1 Introduction

With the continuing emergence of the Agent-Oriented paradigm, it is urgent to develop appropriate methodologies for the development and deployment of agent-oriented applications. This paper motivates and presents the Agent Factory Development Methodology (AFDM) for the design and development of intelligent agent systems. This methodology has been employed in the development of a number of large scale agent-based applications including the WAY System[13], Gulliver's Genie[7] and the award-winning ACCESS Architecture[3].

The Agent Factory System is a cohesive framework that delivers structured support for the development and deployment of agent-oriented applications. It is comprised of four distinct layers, namely: Agent Factory Programming Language (AF-APL), Agent Factory Run Time Environment (RTE), the Agent Factory Development Environment and the Agent Factory Development Methodology. The RTE and AF-APL, which provide supports for agent migration, agent management and resource management and reside at the core of the system. Figure1 presents a demonstration of the AF framework.

The AF-APL represents a dedicated Agent Oriented Programming language with logical, deliberative and imperative features. Agent within Agent Factory [4] can be considered as an amalgam of three sets of components: the mental state, actuators, and perceptors.

The mental state contains the agent's current model of itself and the environment it is situated. This knowledge is stored as 'Beliefs'. Each 'Belief' represents a piece of knowledge an agent thinks it knows about the environment. For instance Belief(like(rice)) can be interpreted as meaning 'the agent likes rice'.

Perceptors are those components that convert raw sensory data into beliefs which are subsequently adopted and augmented into the existing into the agents'

S. Wang et al. (Eds.): ER Workshops 2004, LNCS 3289, pp. 529–538, 2004.

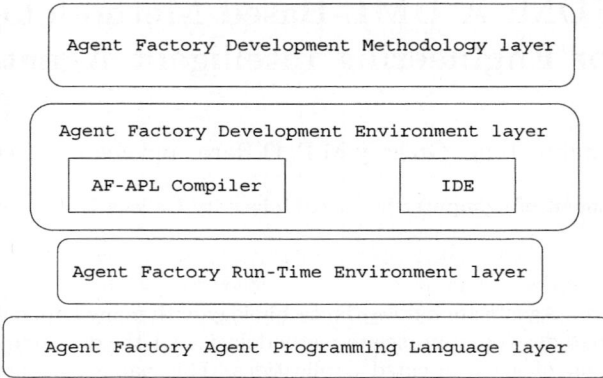

Fig. 1. AF Framework

belief set. For example an agent may have a perceptor which obtains the current time and generates a belief like: Belief(curretTime(17:00 pm)).

Within AF-APL, actions are realized through the triggering of an associated actuator unit. As with perceptor units, actuator units are associated with specific agents as part of the agent program through the ACTUATOR keyword. Actions can be combined into plans that form more complex behaviors by using one or more plan operators - currently there are four plan operators: sequence (SEQ), parallel (PAR), unordered choice (OR), and ordered choice (XOR). For example an actuator which clones an agent may be triggered by the action 'Clone'. Details about BELIEF, ACTUATOR and PERCEPTOR are presented in [10].

One of the primary objectives behind AF is the development of a cohesive software engineering methodology that delivers structured support for the design, implementation, and deployment of multi-agent systems. In designing this methodology, we sought to address a number of objectives: (1) to employ, where possible, pre-existing industry recognisee design notations; (2) to focus upon the definition of visual notations; (3) to use models that promote design reuse; and (4) to maintain a strong link between design and implementation, opening the way for automated code generation.

Within Multi-Agent Systems (MAS), it is imperative that task distribution strategies take cognizance of load-balancing issues. Load-balancing mechanisms thus need to be formulated and delivered which prevent individual agents and/or processors becoming over-loaded. Agent migration is one key technology which may be commissioned in this regard. Another mechanism is that of imbuing individual agents with the autonomy to invoke and negotiate load-balancing regimes. Agent cloning is considered as one instrument and possible responses to the above approach. Embedded with cloning ability, when over-loaded, an agent may generate several concurrent agent instances with the same mental states and behaviors, and discharge the computing tasks to appropriate clone instances.

The work presented in this paper is concerned with the design and implementation of an agent cloning capability. In particular, we adopt the Agent Factory

Development Methodology in the design of the agent cloning model (section 2). Section 3 highlights the use of this methodology in agent cloning as an exemplar case study. Finally, section 4 introduces some related work and gives some concluding marks.

2 The Design Phase of Agent Cloning

The design phase is concerned with the translation of system requirements into a well defined model of the target system that may easily be implemented using agent technologies. Five key models have been developed within the Agent Factory Development Methodology layer. These models include the System Behavior Model, the Interaction Model, the Activity Model, the Protocol Model and the Agent Model [5]. Figure 2 presents a diagrammatic overview of this methodology.

Consequently in the later sections we consider each of the system behavior, activity, protocol and agent models in turn.

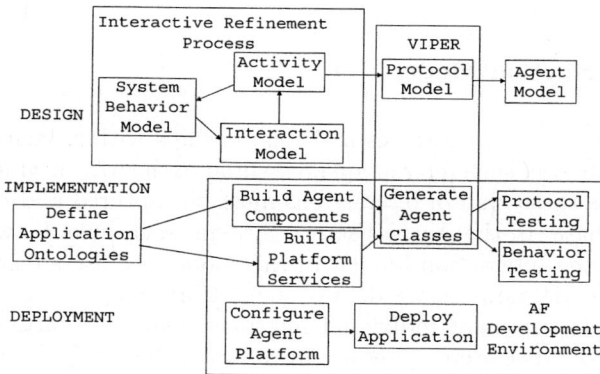

Fig. 2. The AF Development Methodology

2.1 System Behavior Model

The first model we introduce in this paper is the System Behavior Model (SBM). A system behavior is any distinct set of activities and / or interactions that take place during the operation of the system. For the cloning category, key system behaviors may include agent movement updates, agent registration, and agent de-registration etc.

The SBM is formalized by a customized form of the UML Use Case Diagram. UML Use Case Diagrams is a well-understood approach to describe the functionality of a system in a horizontal way. From an agent-oriented perspective, we adopt the actors as agents that are playing a specific role, and use cases as the behaviors that one associates with these roles [5]. Formally, we introduce two stereotypes to customize those UML Use Case Diagrams: the «role» stereotype

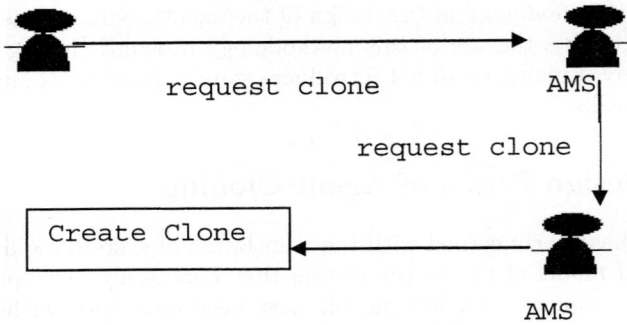

Fig. 3. System Behavior Model

identifies actors that represent roles that agents are playing, and the «role-use-case» stereotype identifies use cases that occur between agents that are engaged in specific roles.

2.2   The Interaction Model

There are inevitably dependencies and relationships within agents in a multi-agent system. Indeed, such interaction is essential to the way in which the system functions. Thus these kind of interactions must be captured and represented in our analysis phase. In the Agent Factory System, such links between roles are represented in the Interaction Model. In this model, we use a customization of UML Sequence Diagrams (more details about UML Sequence Diagram can be found in [12]) to represent individual interplay relations. Particularly, in our Interaction Model, two stereotypes are introduced: the «role» stereotype identifies which role does a specific agent play; and «fiap-acl» stereotype which constrains the message to be FIPA compliance.

2.3   The Protocol Model

The Protocol Model can be viewed as a formalization of the Interaction Model. It is used to refine the various interaction scenarios into a set of protocols that describe how the agents interact and it encapsulates each of the alternate scenarios associated with a given system behavior. Each protocol specified in this model is defined as an Agent UML Sequence Diagram [5]. In contrast to the Activity Model, this model concentrates on the interplays between agents.

Tool-based support for protocol creation is provided via the VIPER [11] visual protocol editor. VIPER performs two jobs within this methodology: (1) it supports visual editing of Agent UML Sequence Diagrams, and (2) it automatically generates agent code based upon these diagrams.

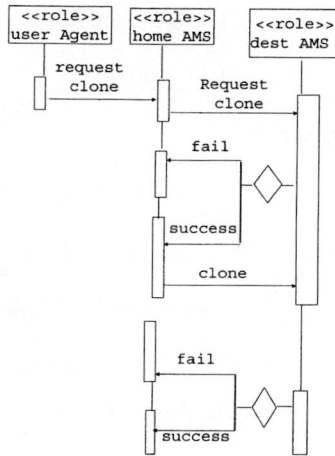

Fig. 4. Protocol Model

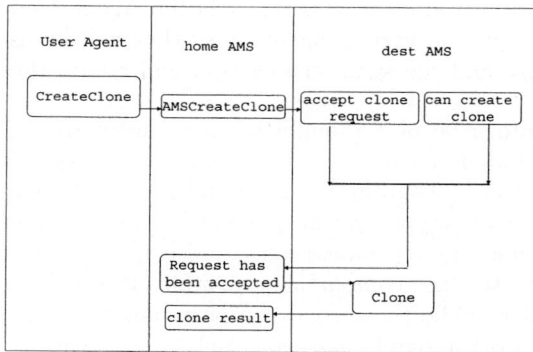

Fig. 5. Activity Model

2.4   The Activity Model

The agent Interaction Model can sometimes require specifications with very clear processing-thread semantics. The activity diagram expresses operations and the events that trigger them [8]. Compared with the Protocol Model (section 2.3), this model primarily concentrates on those activities that support the system behaviors. We customize the Activity Model through the customization UML Activity Diagrams [2]. Specifically, we customize this diagram through the introduction of a «role» stereotype, which associates swim lanes with roles [5].

Typically an individual agent can have various activities and similarly a given activity can be performed by more than one agent. On the other hand a role may be viewed as a collection of activities. Thus a *many-to-many* relationship exists between role, activity and agent class.

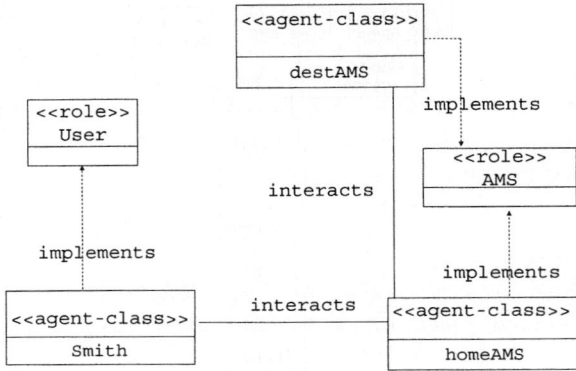

Fig. 6. Agent Model

2.5   The Agent Model

The final model we introduce in this paper is the Agent Model. This model is designed to document the various agent types that will be used in the system under development, and the agent classes that will realize these agent types at run-time.

An agent type can be best thought of as an agent role. The Agent Model within Agent Factory is formalized by way of UML Class Diagram (more details about UML Class Diagram can be found in [12]) that are customized to include: a «role» stereotype and an «agent-class» stereotype. According to [5] the «role» stereotype represents roles, and can be described as a box with two compartments: the first compartment contains the stereotype after which it is the role identifier; while the second compartment contains a list of protocol identifiers. On the other hand, the «agent-class» stereotype represents agent classes with the form of a box which includes three compartments: the first compartment contains the stereotype and the agent class identifier, the second compartment contains a list of protocols (not those in the associated roles), and the third compartment contains a set of activity identifiers.

Since an agent implemented by a specific agent class can play more than one role in its life circle and at the same time one role can be played by many agents, in the Agent Factory System we permit a *many-to-many* relationship between «role» and «agent-class». That is each agent class can be associated with many roles and each role can be associated with many agent classes.

3   Case Study

3.1   Cloning Design

To illustrate our development methodology, we now present the cloning case study. When analyzing this system, two key roles were identified: the User role,

which is responsible for requesting cloning, the AMS role, whose duty is to reason
for available resources. We formalism this analysis within the System Behavior
Model (Figure 3). By way of illustration, we expand upon the 'request clone'
and 'Create Clone' system behaviors. Our expansion of this behavior is through
the Activity Model. Figure 5 presents an example of the Activity Model. Here
an agent whose role is user believes that it wants to create a copy of itself. This
results in a *CreateClone* event (actually in Agent Factory it is a *CreateClone*
belief) that triggers another agent on the same platform with the role of an
Agent Management System (AMS) to deal with the request. The AMS agent
then forwards the cloning request to the destination AMS (an AMS agent that
is running on a different platform). The destination AMS will accept the request
if and only if it has the required capacity to execute the cloning. Once the request
is accepted, it will ask the home AMS for the clone materials. The clone result
will be sent back to the home AMS after cloning.

To help to clarify how the above interplay works, we further expand upon the
'request clone' and 'Create Clone' behaviors through the Protocol Model. Figure
4 presents an example of agent cloning interaction. The rectangle can express
individual or sets (i.e., roles or classes) of agents. For example, an individual
agent could be labelled '≪role≫ home AMS' which means the agent serves as
the AMS on the home platform. When it is necessary for the user agent to
initiate cloning, the user agent may send a *request_clone* message to its AMS
(home AMS). The home AMS thereafter forwards this request message to the
destination AMS (an AMS agent that may have sufficient resources to create
the clone and execute the distributed tasks) via a *fipa-inform* message. Upon
examining its capacity, the destination AMS will inform the home AMS whether
it is able to create the clones or not. Again this information is transformed
through a *fipa-inform* message. If the result is positive, the home AMS then can
send the entire clone material to the destination AMS. A result will be returned
to the home AMS through the *fipa-inform* message indicating the success or
failure of the operation.

As the System Behavior, Interaction, and Activity models become more stable (i.e. as a general agreement on how the system behavior emerges), work on
these three models stops, and the designer commences work on Agent Model.
Specifically, this model presents a view of the roles within the system, and the set
of agent classes that will implement those roles (section 2.5). Figure 6 presents
the Agent Model for our cloning mechanism. As can be seen in this model, two
roles are implemented via two agent classes. The Smith class implements the
User role while both *homeAMS* class and *destAMS* class implement the AMS
role.

3.2   Cloning Implementation

To implement the above cloning design, we now introduce the CloneExample
Role, an example role that contains two AF roles to (1)initiate the creation of
an agent (2) to pass some job to the agent.

```
BELIEF(CreateClone(?name,
?design, ?clonenumber,? task))&
BELIEF(agentID(?destinationAMS,
?addesses))=>COMMIT(Self,Now,BELI
EF(true),request(?destinationAMS,
cloneAgent(?name,?design,?clonenu
mber ,?task)));
```

Fig. 7. CloneExample Rule

```
BELIEF(ToughJob)=>COMMIT(Self,Now
,BELIEF(true),SEQ(ToughJob,HelloW
orld)));
```

Fig. 8. ToughJob Rule

The rule in Figure 7 states that if the agent believes that it wants to create a specified number of copies (by the parameter ?clonenumber) of itself with a specified design(by the parameter ?design). In order to do this, it sends a request to the related AMS to create the agent.

The second rule (Figure 8) delivers a tough job , let us say loop 10,000 times, and a 'prn' job (which is used to print a string *'HelloWord'*) to the agent. In order to do this it fires the *'ToughJobActuator'* actuator and the *'prnActuator'* actuator. Note here we use the SEQ operator [4] which means the 'prn' action will not be executed until the 'toughJob' action is completed. In this way, after cloning the *'prn'* task can be delivered to the cloned agent.

Here if an agent believes that the current commitment is a tough job (which will cause the system overloading) it will take another belief that it needs to create a specified number of copies of itself to release the system overloading (Figure 7). This belief leads to a *fipa-request* message which requires the AMS agent to clone (Figure 7). During the clone process the commitment with the SEQ key word (Figure 8) is separated into some sub-tasks which will be re-allocated to the cloned agents. In this example, the HelloWorld sub-task will be delivered to the cloned agent after cloning.

4   Discussion and Conclusion

The methodology presented in this paper represents one of a number of potential approaches to fabricating MAS. It is not possible in a paper to do justice to the large amount of work being done.

Gaia is a methodology that supports the analysis and design phases. It is applicable to a wide range of agent systems. Furthermore it can deal with both macro-level (societal) and micro-level (agent) aspects of systems [9]. Gaia takes the view that a system can be seen as a society or an organization of agents. The methodology is applied after the requirements are gathered and specified, and covers the analysis and design phases. However according to [14] the disadvantages of Gaia is that (1) Gaia does not directly deal with particular modelling techniques. (2) Gaia does not directly deal with implementation issues and (3) Gaia does not explicitly deal with the activities of the requirements capturing and modelling, and specifically of early requirements engineering.

Tropos [6] proposes both a software development methodology and a development framework which are founded on concepts used to model early requirements and complements proposals for agent-oriented programming platforms. Compared with AFDM, Tropos provides an early requirements phase, which the AF Methodology does not. However tool support for Tropos is currently only in the form of diagram editor [8], rather than the automatically generation of agent code that is part of VIPER.

Recently a cooperation has been established between the Foundation of Intelligent Physical Agents (FIPA) and the Object Management Group (OMG). As a result of this cooperation, the framework of Agent UML (AUML) was introduced [1]. The AUML approach introduces the Agent Interaction Protocol (AIP) for agent communication and constraints on messages.

In summary, this paper presents the key aspects of the Agent Factory Development Methodology that is founded upon the industry standard UML design notation. This methodology has been employed in the development of a number of large scale agent-based applications. Moreover the models underpinning this methodology perform two jobs i.e.design reuse and automated partial code generation. On the other hand agent cloning is considered as an approach to release the system bottlenecks. Finally, we illustrate our AFDM through a trial example namely design and implementation of agent cloning in the AF System.

References

1. B. Bauer. Extending uml for the specification of interaction protocols. Technical report, submission for the 6th Call for Proposal of FIPA and revised version part of FIPA 99, U. C. Berkeley, April 1999.
2. James Rumbaugh Booch, Grady and Ivar Jacobson. *The Unified Language User Guide*. Addison-Wesley, Reading, MA, 1999.
3. D.Phelan R.Strahan and R.Collier C.Muldoon, G.M.P.O'Hare. *Access:An agent architecture for ubiquitous service delivery*, volume LNAI 2782. Spinger Verlag, cooperative information agents vii edition, 2003.
4. Rem William Collier. Agent factory: A framework for the engineering of agent-oriented applications, 2001.
5. O'Hare G.M.P. Collier R., Rooney C. A uml-based software engineering methodology for agent factory. Banff, Alberta, Canada, June 2004. Proceedings of the 16th International Conference on Software Engineering and Knowledge Engineering (SEKE-2004).

6. F.Giunchiglia, J Mylopoulos, and A. Perini. The tropos software development methodology: Process, models and diagrams. Technical report, Technical Report DIT-02-008, Informatica e Telecomunicazioni, Universita degli Studi di Trento, 2001, 2001.
7. G.M.P.O'Hare and M.O'Grady. Gulliver's genie:a multi-agent system for uniquitous and intelligent content delivery. *Computer Communication*, 26(11):1178-1187, 2003.
8. H. V. D. Parunack J. Odell and B. Bauer. Extending uml for agents. In Proceedings of AOIS Workshop at AAAI 2000, 2000.
9. N. R. Jennings M. Wooldridge and D. Kinny. The gaia methodology for agent-oriented analysis and design. *Journal of Autonomous Agents and Multi-Agent Systems*, 1999.
10. Collier R., Rooney C.F.B., O'Donoghue R.P.S., and O'Hare G.M.P. Mobile bdi agents. *11th Irish Conference*, 2000.
11. C. F. B. Rooney, R. W. Collier, and G. M. P. O'Hare. Viper: Visual protocol editor. In Proceedings of COORDINATION 2004, Pisa, Italy, 2004.
12. Ivar Jacobson Rumbaugh, James and Grady Booch. *The Unified Modeling Language Reference Manual*. Addison-Wesley, Reading, MA, 1999.
13. G.O'Hare T.Lowen and P.O'Hare. Mobile agents point the way:context sensitive service delivery through mobile lightweight agents. Bologna,Italy, 2002. Proceedings of First International Joint Conference on Autonomous Agents and Multi-Agent Systens Conference(AAMAS2002), AAAI Publisher.
14. Franco Zambonelli, Nicholas R. Jenningsy, and Michael Wooldridgez. Developing multiagent systems: The gaia methodology, dipartimento di scienze e metodi dell's ingegneria. October 2003.

First International Workshop on Digital Government: Systems and Technologies (DGOV 2004)

at the 23rd International Conference
on Conceptual Modeling (ER 2004)

Shanghai, China, November 9, 2004

Organized by

Il-Yeol Song and Jin-Taek Jung

Preface to DGOV 2004

DGOV2004 (Digital Governments: Systems and Technologies) is the first international workshop held in conjunction with the 23rd International Conference on Conceptual Modeling, held in Shanghai, China, Nov. 8-12, 2004.

The notion of Digital Government or e-Government has recently begun to receive a lot of attention from governments, academia, and industry. For example, US National Science Foundation has been sponsoring an annual research program for Digital Government for several years. Some universities have created centers for digital governments. E-business vendors such as SAP and IBM have begun to provide digital government support in their products. Through DGOV2004 workshop, we want to provide an academic forum for the exchange of ideas in research and developments of digital governments with focus on systems and technologies.

The workshop received a total of 16 papers (and four late submissions which were returned to authors). The submitted papers were in the areas of e-voting, legal issues, m-governments, e-government services, e-community systems, security & trust, legal systems, and web-based applications. All the submitted papers were assessed by three members of the international program committee. The committee finally selected six full papers for the presentation and inclusion in the proceedings. The six papers were organized into two sessions, Digital Governments: Systems and Digital Governments: Technologies. The papers in Systems session cover e-community system, e-government web services, and a camera-detector system for traffic control. The papers in Technologies session cover ontology for legal knowledge, e-voting, and mobile government service.

We thank all authors who submitted their papers to this workshop and the program committee members as well as the external reviewers of the workshop who contributed to selecting quality papers. We also thank the ER2004 workshop chairs Professor Shan Wang and Katsumi Tanaka for excellent coordination and cooperation and the ER2004 organizing committee for their support and arrangement. A special thank goes to Prof. Shuigeng Zhou who invested a lot of energy in collecting camera-ready versions and copyright forms. Most of all, we thank all the participants for attending this workshop and stimulating the technical discussions. We hope all participants share recent advances and new research topics & developments in the areas of digital governments.

November 2004   Il-Yeol Song and Jin-Taek Jung

IECS: An Integrated E-Community System for Management, Decision and Service

Yu Bo, Yunhai Tong, Hongding Wang, Zhang Peng,
Shiwei Tang, and Dongqing Yang

School of Electronics Engineering and Computer Science
Peking University, Beijing, China
{yubo,yhtong,hdwang,zhp,tsw,dqyang}@db.pku.edu.cn

Abstract. The paper presents an Integrated E-Community System (IECS) for management, decision and service, designed for the e-government project of the Haishu District of Ningbo, Zhejiang, China. The project need is to promote the integration of management information and service information of all communities in the district, providing a unified platform on which different departments of the district government can share and exchange community information, government users can analyze information and make decisions, and outside users can access and request service information. To meet the project need, the IECS consists of five parts: 1) The Central Database (CDB) that stores all information related with management, decision and service of the communities; 2) The Information Extracting Subsystem (IES) that extracts data from many data sources, transforms and loads it into the CDB for system administrators; 3) The Information Management Subsystem (IMS) that provides information querying and sharing for government users, and information maintenance, rights management and log management for system administrators; 4) The Intelligent Analysis Subsystem (IAS) that extracts analysis-related data from the CDB and loads it into a data warehouse, and provides multi-dimensional analysis and decision-making function for government users; 5) The Information Service Website (ISW) by which the government users and system administrators can collect and release management and service information, and outside users can access required information. The IECS supports management, decision and service of a government based on a unified data platform – the CDB. It also ensures data security by providing different workplaces and rights for different users. In the real application, the system works well.

1 Introduction

A community is a basic unit of society in China, such as the Zhongguanyuan Community, Haidian District of Beijing, China. Because the development of a community may make a great influence on the management, the decision and the service of a city or district government, the e-community is an essential part of e-government.

E-government refers to government's use of technology, particularly web-based Internet applications to enhance the access to and delivery of government information and service to citizens, business partners, employees, other agencies, and government entities [1]. Nowadays, governments are going-on line and using the Internet to pro-

S. Wang et al. (Eds.): ER Workshops 2004, LNCS 3289, pp. 542–552, 2004.

vide public services to their citizens. Current e-government projects pay too much attention to Internet applications, but neglect what information and service citizens actually want e-government to provide [2] and what the value of e-government is [3]. How to find demands of citizens and provide more useful information and better services? How to exert the management function of a government in the e-government environment? Karen Layne and Jungwoo Leeb in [4] gave a development model, namely the four-stage model for full-functional e-government: 1) cataloguing, 2) transaction, 3) vertical integration, and 4) horizontal integration. Therefore, the future e-government could be an integrated information platform that covers all functions including not only the access to and delivery of information services on the Internet but also gathering, sharing and analyzing of information based on database technology or data warehousing. We develop an Integrated E-Community System (IECS) for management, decision and service. It has been applied to the e-government project of the Haishu District of Ningbo, Zhejiang, China, with encouraging results. The IECS has the following objectives:

- Building a central database that stores all information related with management, decision and service of all communities in a district, which is extracted from systems of different administrations (e.g. police station, health office, and family planning office) and external data sources.
- Developing a unified platform on which different departments of the district government can share and exchange information, government users can handle official businesses, and outside users can request information services.
- Supporting intelligent analysis and decision. By making use of data warehousing and OLAP technology, the system can find valuable information about the development of a community, such as the number of the poor or the handicapped and the current status of public security. Based on this information, government users can make more appropriate work planning and service goal.
- Enhancing the communication between governments and citizens. The system provides a website on which people can advance their requirements and the government can respond to them in time.

The remaining of the paper is organized as follows. Section 2 describes related works. Section 3 describes the design and implementation of the IECS, particularly the architecture of the IECS and functions of the four subsystems. Section 4 summarizes features of the system and suggests future research directions.

2   Related Work

Researches of e-government cover objectives, methods, technologies and policies. We describe work related to the IECS from two perspectives.

Challenges and Future Directions: A detailed analysis on the lessons, challenges and future directions of e-government is given in [3]. Ensuring access to useful information and services is viewed as one of important challenges influencing the development of e-government. The usefulness of information means that e-government should provide timely, right and comprehensive information to users. From the technology point of view, it is an information-discovering issue. In our project, we use data warehousing and OLAP technology to solve the issue. [4] describes different stages of e-government development and proposes a 'stages of growth' model for full-functional e-government. In terms of the model, the future direction of e-government is the horizontal integration across different functions. Following the direction, in the IECS, we implement the integration of management, decision and service based on a unified platform.

Functions/Services: Because of different requirements and objectives, e-government systems often provide different functions/services, most of them emphasis on how to deliver government services to citizens such as England, India and Canada [5]. But the functions/services of a government includes not only the service for citizens and public but also the processing of administrative work. Therefore, recent researches mainly focus on the development of full-functional e-government, especially the implementation of knowledge management. [5] proposes to consider the development of local e-government from four perspectives: stakeholders, business drivers, technology drivers, and methodology, and discusses some benefits of knowledge management for Thai local e-government. [6] analyses overall requirements of e-government systems including functional requirements (e.g. accesses to information, transaction services and participations) and non-functional requirements (e.g. efficiency, portability and reuse). [7,8] present a knowledge management system designed for the Italian Department of Technology and Innovation, which supports work and knowledge sharing for all actors involved. Similarly, the goal of the IECS is to implement a full-functional e-government, which not only provides information services for outside users, but also supports management and decision for government users.

3   Design and Implementation of the IECS

3.1   Architecture

The IECS adopts a three-tier architecture (Fig.1):

- Data tier: It consists of a central database (CDB) and a data warehouse (DW). The CDB is an information-sharing platform that stores all information related with the management, decision and service of communities, such as demographics, population, sanitation, culture, sports, and requests. Data from autonomous systems and external sources are extracted into the CDB using the special information extracting system (IES). The DW is an information-analyzing platform that stores analysis-related data extracted from the CDB according to the DW subject classification.
- Business logic tier: It consists of a management application server that implements information management logic, a service application server that implements information service logic, an OLAP server that implements information analysis logic, and a Web server that serves web pages to the representation tier across the Internet or an Intranet and handles HTTP requests.
- Representation tier: It contains data management tools, querying and browsing tools, analyzing tools and a call center.

From the function point of view, the IECS consists of three sub-systems and one website (Fig.1): 1) The Information Extracting Subsystem (IES) that provides functions of extracting data from data sources, transforming and loading it into the CDB for system administrators; 2) The Information Management Subsystem (IMS) that provides function of information querying and sharing for government users, and functions of information maintenance, rights management and log management for system administrators; 3) The Intelligent Analysis Subsystem (IAS) that extracts analysis-related data from the CDB and loads it into the DW, and provides the function of multi-dimensional analysis and decision-making based on data warehousing and OLAP technology for government users; 4) The Information Service Website (ISW) that provides the function of information releasing and collecting for government users and system administrators. It also allows outside users to browse, query and request information. Therefore, the IECS is an integrated e-community system that supports management, decision and service of a government based on a unified data platform – the CDB. Data security is also ensured by providing different workspaces and rights for different users.

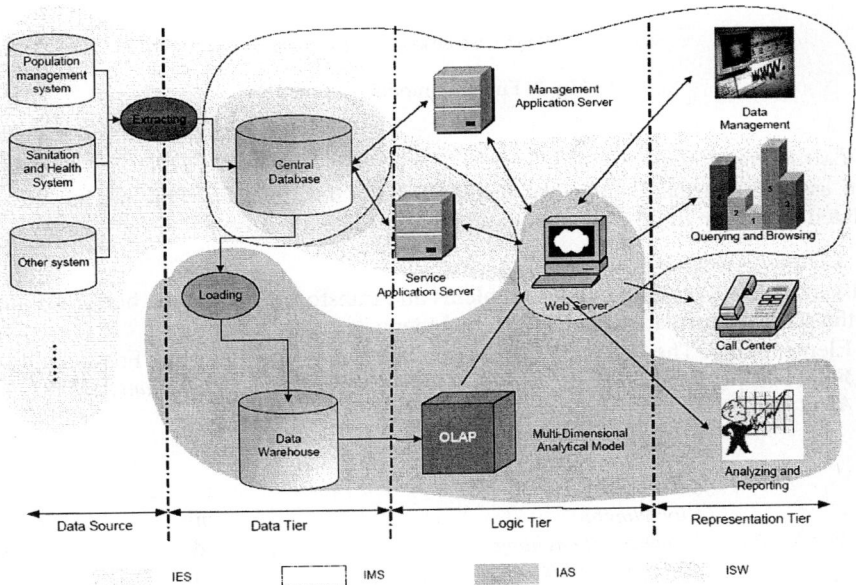

Fig. 1. Architecture of the IECS

3.2 Information Extracting Subsystem (IES)

The IES extracts data from many data sources, transforms and loads it into the CDB. It is the workspace of system administrators or data entry executives. For the governments that have not management information systems, the IES provides an on-line data-gathering tool with which data entry executives in different level governments can directly input related data into the CDB. For the governments that have some management information systems (e.g. population management system, culture & sports facility management system, Resident's Committee management system), the IES provides a data-extracting tool for system administrators to extract data from data sources and load it into the CDB. Because data sources usually are autonomous and heterogeneous, we have to solve data quality problems such as missing value, data conflicts, approximate duplicate records before loading. In the latter scenario, the IES has the following major modules (Fig.2):

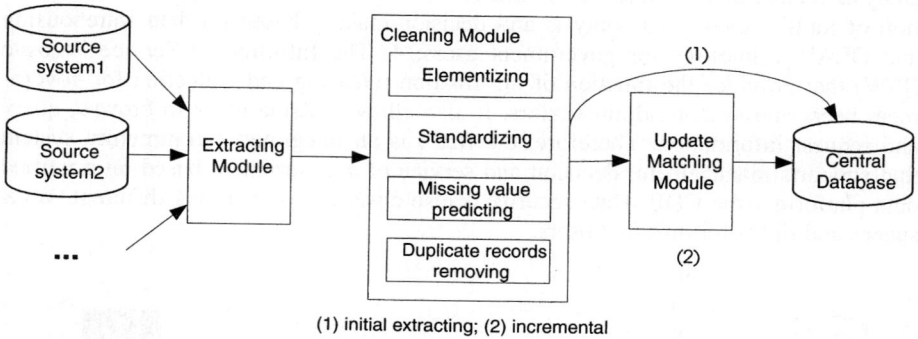

(1) initial extracting; (2) incremental

Fig. 2. Function modules of the IES

1) Extracting Module
We assume that data sources of the IECS have been chosen. This module extracts related data from these data sources and loads it into a buffer.

2) Cleaning Module
This module solves data quality problems and transforms data in the buffer according to the standard format of the CDB. It includes:

− Elementizing: The IES parses addresses and names into elements. For example,
  Before elementizing: *Jinfeng Road 13, Haishu District, Ningbo, Zhengjiang, 315002*
  After elementizing:
  > *Street Address Number: 13*
  > *Street Name: Jinfeng Road*
  > *District: Haishu*
  > *City: Ningbo*
  > *Province: Zhengjiang*
  > *Zipcode: 315002*
  > *Community: null*

− Standardizing: The IES maps attribute values from different data sources to canonical values. Attribute values are standardized by looking up the source value in a lookup table to directly obtain the standardized value or by applying some conversion functions. The lookup table is mainly used to address mapping from an abbreviation to a full name, such as "ZJU"→"Zhejiang University". Conversion functions mainly include: 1) Type conversion that converts one data type into another data type, such as STRING→DATE. 2) Domain conversion that converts values from different domains into canonical values in a same domain, such as the conversion of sex value: "0,1"→"M,F" or "Y,N" →"M,F". 3) Format conversion that mainly is used to convert phone number and ID number into standard format, such as the conversion of phone number: "8657482757756"→"86-574-8275-7756".

- Missing value predicting: The IES uses the most probable value to fill in the missing value. For different type missing values, there are three prediction methods: 1) Regression that is used to predict continuous or order values based on statistical models. 2) Lookup table that is used to fill missing address values. For example, by using the mapping rule "*Jinfeng Road 13 belongs to Wuhuang Community*" in the lookup table, the IES can predict the above null *community* value is *Wuhuang Commnunity*. 3) Inference-based methods (e.g. the C4.5 decision tree induction algorithm or the naïve Bayesian classification [9]) that is used to predict discrete or nominal values.
- Duplicate records removing: The IES uses the domain independent Priority-Queue algorithm [10] to detect duplicate records, and then, the IES submits detected records to domain experts or users who decide whether to remove duplications or not.

3) Update Matching Module
If it is the first time to extract data from source systems, namely initial extracting, the IES directly loads cleaned data in the buffer into the CDB. If it is an incremental update operation, the IES will call the Update Matching Module to find records in the CDB matching with the extracted and cleaned data in the buffer, then update them.

3.3 Information Management Subsystem (IMS)

The IMS is an information and office management platform (Fig.3) that provides different level rights of information querying and information sharing on the CDB for government users based on user roles, and functions of information maintenance, rights management and log management for system administrators. It is the workspace of government users and system administrators.

The IMS adopts the three-tier J2EE architecture (Fig.4): the representation tier, the business logic tier and the data tier. The representation tier accepts queries of government users and represents answers to the queries with HTML sheets and JSP. The business logic tier is the center of transaction processing which responds to queries of government users, processes data according to business logics, and returns answers to queries using EJB, JDBC and Servlet. The data tier, namely the CDB, is a core component of the IMS, which gathers, stores, and maintains the e-government information.

The IMS contains the following functions:

1) Information querying and browsing
Government users can query and browse the management information in the CDB based on user roles. They can set query conditions, browse detailed information satisfying query conditions, and print query results.

2) User right management
The IMS provides different-level access rights for different-level government users and ensures each user can only access the information within their right scopes. User rights are divided into four levels: the city government user, the district government user, the street government user, and the community user. Users in the same level may have different roles, such as the community official, the community worker, and

data entry executive. Different roles have different operation rights, such as querying, browsing, and modifying. Which operation right a user has and what kind of information this operation is exert on is decided by the user role. The system administrator sets user rights and user roles.

3) Log management

Logs are used to record the access and operation information of each user and help system administrators to maintain the system better. By utilizing the log function of DBMS, the log management of the IECS can record the time, contents and objects of user operations, monitor the running of the system and ensure the security of the system.

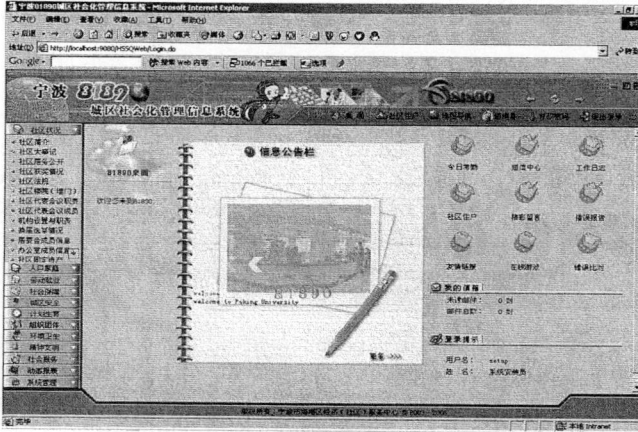

Fig. 3. Information management subsystem

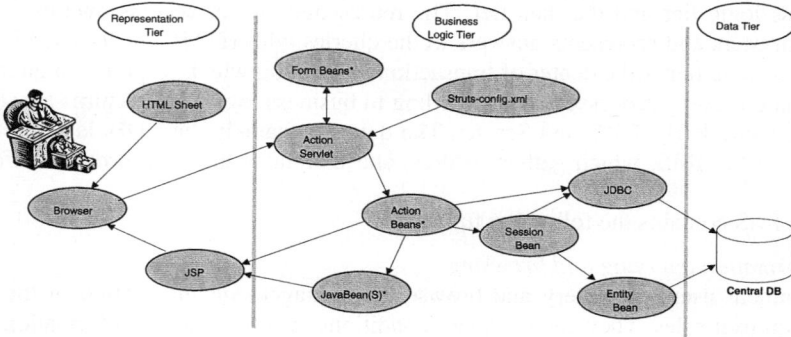

Fig. 4. Architecture of the IMS

3.4 Intelligent Analysis Subsystem (IAS)

The IAS is a decision-supporting platform based on data warehousing and OLAP technology, developed for multi-dimensional analysis and the generation of statistical reports for government users. It takes data from the CDB, transforms and loads it into the DW with an ETL tool, and then performs multi-dimensional on-line analytical processing based on user's analyzing requests. The IAS can help government officials make correct decisions about the development of a community and provide appropriate and timely services for citizens, enterprises or organizations. Therefore, it can improve the working efficiency of the government.

1) ETL

The ETL tool extracts data from the CDB termly, transforms and loads it into the DW in order to maintain data integrity and data consistency. An ETL task can be broken into the following steps:

– Extracting data from the CDB and loading it into the buffer of the DW. The ETL tool builds a materialized view log for tables from the CDB and extracts the changed data incrementally.
– Transforming data structure. The data in the CDB and the data in the DW are different in structure. Through building views over tables in the buffer, the ETL tool transforms the data into required structures in the DW.
– Updating the DW. The ETL tool only updates the changed data in the DW. For the unchanged data, it only modifies their time-stamps.
– Clearing the data in the CDB. Because the CDB only stores current data, when the data is out of date and extracted into the DW, it is cleared out from the CDB immediately.

2) Subject of DW and multi-dimensional analytical model

In the Haishu District's e-government project, we divided the community information into 10 subjects, as presented in Table 1.

Table 1. Subjects of the DW

Subject	Main information
Basic information	Brief, Memorabilia, Prize, Statute, Building, Meeting, Election, Finance, etc
Population and family	Permanent residents, temporary residents, children, etc
Employment statistics	Unemployment, re-employment, laid-off workers, re-employment of laid-off workers, hidden employment, etc
Social security	Poor, Handicapped, Old, Low income family, Donation, Funeral, Marriage, etc
Community security	Security guard, released man, arrested man, psychopath, house lease, etc
Organization and party	Political organization, women organization, sports party, etc
Environment and health	Health facility, Sanitation, infection prevention, etc
Civilization	Culture facility, School, Culture event, etc
Social service	Service item, Volunteer, Service facility (hotel, shop), etc
Family Planning	Child bearing age women, Married women, etc

Each subject has a number of information tables. Users can design one or several multi-dimensional analytical models based on these tables. The OLAP server provides the support of multi-dimensional analyzing. For example, if a user want to know the development information of culture facilities in a district, he/she can design the following model (Fig .5) in the *civilization* subject:

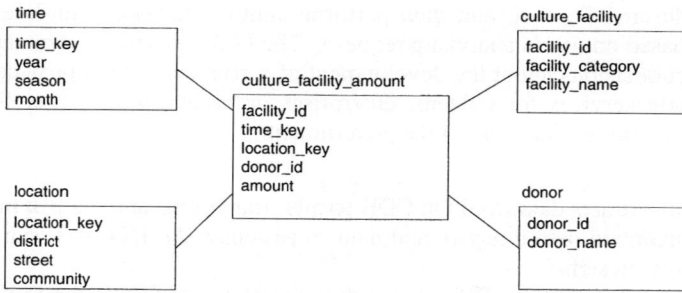

Fig. 5. A multi-dimensional analytical model on the *civilization* subject

The analysis results can be represented by crosstabs, reports, histograms, and pie charts (Fig.6). In the Haishu District's e-government project, we design 74 multi-dimensional analytical models in the 10 subjects in order to provide as many analyzing requests as possible for government users.

Just as the IMS does, the IAS also adopts the three-tier architecture and supports user right management.

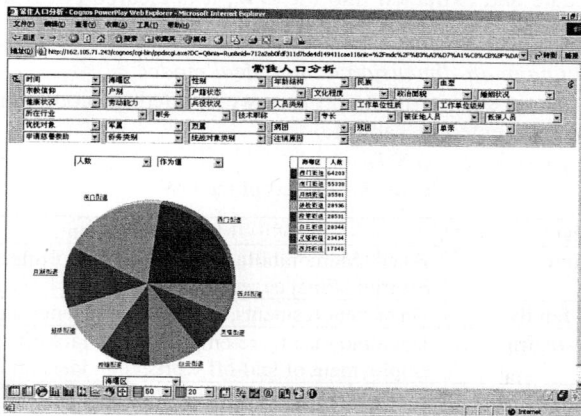

Fig. 6. Representation of analysis results in the IAS

3.5   Information Service Website (ISW)

The ISW is a citizen-friendly semantic network (Fig.7) that not only promulgates the management and service information but also supports resource discovery [11]. We employ RDF as the main metadata description language and have developed many metadata specifications to make the e-community more useful, particularly for the citizens in the district who use the e-community. The ISW offers 16-category service information (e.g. medical care, shopping, hotel, finance, and training), a number of management information (e.g. government notice and public policy) and some analytical reports (e.g. the public security reports and the environmental health reports).

The ISW is an information service platform on which governments can promulgate service information and respond to outside users' requests, and outside users can query service information and release new requests. Moreover, the ISW also loads valuable information into the CDB. The detailed information about the ISW can be seen in [12].

Fig. 7. Information service website – China 81890 Service Network

4 Summary and Future Work

We have introduced the design and implementation of the IECS. Different with other e-government systems, the IECS has three distinct features as follows:
– The IECS is an integrated e-community system, which not only supports work and knowledge sharing, but also stresses on knowledge management and the integration of management, decision and service of a government.
– The IECS is an analytical e-community system, which can help government officials to make correct decisions about the development of a community by making use of data warehousing and OLAP technology, and provide appropriate and timely information services for outside users.
– The IECS can facilitate the re-engineering of government process and service innovation. The integration of management, decision and service can effectively change the business process of a government, improve the effect of service, and finally accelerate process re-engineering and service innovation.

By taking the advantages of the IECS, the practical project designed for the Haishu District of Ningbo, China works well, and now, the district government is enhancing the efficiency of work and inhabitants have got more benefits from the e-community. Of course, to realize fully functional IECS, there is still much work to do, future work includes:
– How to add geographical information into the IECS and offer the query service of the geographical information.

- How to further integrate management information and service information and improve the efficiency of extracting, loading, querying and browsing.
- How to implement data mining and provide more valuable management information for governments and more appropriate service information for outside users.

References

1. McClure, D. L. Statement of David L. McClure, U.S. General Accounting Office, before the Subcommittee on Government Management, Information and Technology, Committee on Government Reform, House of Representatives. http://www.gao.gov. 2000.
2. Duncan Aldrich, John Carlo Bertot, and Charles R. McClure. E-Government: Initiatives, Developments, and Issues. Government Information Quarterly 19 (2002) 349–355.
3. Paul T. Jaeger, Kim M. Thompson. E-government around the World: Lessons, Challenges, and Future Directions. Government Information Quarterly 20 (2003) 389–394.
4. Karen Layne, Jungwoo Leeb. Developing Fully Functional E-government: a Four Stage Model. Government Information Quarterly 18 (2001) 122–136.
5. Wichian Chutimaskul and Vithida Chongsuphajaisiddhi. A Framework for Developing Local E-government. KMGov 2004, LNAI 3035, pp. 335–340, 2004.
6. V. Palade, R.J. Howlett, and L.C. Jain (Eds.). Object-Oriented Design of E-Government System: A Case Study. KES 2003, LNAI 2773, pp. 1039-1045, 2003.
7. Albolino S., Morici R., Schael T., and Sciarra G. A KM Solution for the Empowerment of Local Authorities, international Workshop "Knowledge Management in e-government-KMGov2002 "(2002).
8. Roberta Morici, Eugenio Nunziata, and Gloria Sciarra. A Knowledge Management System for E-government Projects and Actors. KMGov 2003, LNAI 2645, pp. 304–309, 2003.
9. Fu Qiang. The Designing and Implementing of Data Cleaning Tool Based on Classification Methods. Master Paper, 2004. (In Chinese)
10. Monge Alvaro E., Elkan Charles P. An Efficient Domain-independent Algorithm for Detecting Approximately Duplicate Database Records. Department of Computer Science and Engineering, University of California, 1997.
11. Youlin Fang, Weihua Zhang, Dongqing Yang, and Shiwei Tang. Towards a Semantic E-Community. EGOV 2003, LNCS 2739, pp. 476–479, 2003.
12. http://www.81890.gov.cn

WebDG – A Platform
for E-Government Web Services

Athman Bouguettaya[1], Brahim Medjahed[2], Abdelmounaam Rezgui[1],
Mourad Ouzzani[3], Xumin Liu[1], and Qi Yu[1]

[1] Department of Computer Science, Virginia Tech
{athman,rezgui,xuminl,qyu}@vt.edu
[2] Dept of Computer and Information Science, University of Michigan, Dearborn
brahim@umd.umich.edu
[3] Department of Computer Sciences, Purdue University
mourad@cs.purdue.edu

Abstract. Web services are deemed as the natural choice for deploying e-government applications. Their use enables e-government to fully get advantage of the envisioned *Semantic Web*. In this paper, we propose WebDG, a comprehensive *Web Service Management System* for e-government applications. It aims to improve government-citizen interactions through an infrastructure built around the "life experience" of citizens. WebDG provides a framework for automatically composing e-government services, optimized querying services, and preserving privacy.

1 Introduction

Digital government applications are becoming commonplace. A major propelling technology for e-government is the emerging concept of *Web services*. The nexus between the two is becoming very strong as Web services provide the platform of choice for deploying the different functionalities offered by governments and supporting interactions with both government and non-government applications. A *Web service* is a set of related functionalities that can be programmatically accessed and manipulated through the Web [16]. Examples of e-government Web services include electronic tax filing, department of motor vehicle driver's license service, and social services (e.g., health insurance for disadvantaged people). The powerful concept of Web service is taking root because of the convergence of government and business efforts to make the Web the place of choice for all types of human activities [9,15].

The ability to efficiently access and share e-government services is a critical step towards the full deployment of digital government. This requires the development of techniques to address various challenging issues. Required techniques include service description, discovery, querying, composition, monitoring, security, and privacy. All these techniques would be part of a comprehensive *middleware* for managing *autonomous* and *heterogeneous* Web services. For that purpose, we are investigating the architectural components of a *Web Services Management System* (*WSMS*). The aim of a *WSMS* is to do for Web services

S. Wang et al. (Eds.): ER Workshops 2004, LNCS 3289, pp. 553–565, 2004.

what DBMSs have done for data. Users will no longer need to think in terms of *data* but rather *services*. Web services will be treated as *first-class* objects that can be manipulated as if they were pieces of data.

In this paper, we present a comprehensive WSMS for e-government called *Web Digital Government* (WebDG). WebDG embraces emerging standards for describing (WSDL), discovering (UDDI), and invoking (SOAP) Web services [6]. Adopting Web services in WebDG enables: (i) *standardized* description, discovery, and invocation of welfare applications, (ii) *composition* of pre-existing services to provide *value added* services, and (iii) *uniform* handling of privacy. WebDG is built around a collection of features that include a framework for composing e-government services, optimized querying services, and preserving privacy.

The remainder of this paper is organized as follows. Section 2 gives a scenario of e-government applications. It shows the drawbacks of current system and what WebDG aims to do. Section 3 examines three features of WebDG, including composing e-government services, optimized querying services, and preserving privacy. Section 4 then describes the implementation of WebDG. Section 5 gives our concluding remarks.

2   A Scenario of E-Government Applications

To illustrate the drawbacks of the current system and how WebDG can help, we examine a typical scenario in this section. One of the major concerns of e-government is to improve government-citizen interactions using information and communication technologies [9]. In our e-government project, we have teamed up with *Indiana*'s FSSA. Collecting social benefits is currently a frustrating and cumbersome task in FSSA. Citizens must often visit different offices located within and outside their hometown. Additionally, case officers must delve into a wealth of proprietary applications to access welfare programs that best meet citizens' needs.

Let us consider the following scenario typical to FSSA application domain. A pregnant teen, let's call her Mary, goes to an FSSA office to collect social benefits. Mary needs a government-funded health insurance program. She would also like to receive nutritional advice. Because Mary will not be able to take care of the future newborn, she is also interested in finding a foster family.

The case officer, let's call him John, must access all relative services among various agencies to fulfill Mary's needs. He must first manually look up which social programs offer health insurance and food assistance for Mary. He then needs to find a foster family service. For each service, John needs to select an appropriate provider. The choice of the provider is mostly based on John's expertise and some information gathered through different means (e.g., Web sites, brochures). Since there might be a large number of candidate providers, choosing one that best fits Mary's requirement is by no means an easy task. Finally, John chooses Medicaid (a healthcare program for low-income citizens) and WIC (a federally funded food program for Women, Infants, and Children). Assuming

Medicaid is locally accessible, John can connect to the corresponding application and interact with it. Because WIC is a federal program, however, John has no direct access to the corresponding application. That means Mary must visit another agency, perhaps in a different town, to apply for the benefit.

Still more difficulties arise when John tries to find a foster family service. Using local resources, he finds no matching program, although Teen Outreach Pregnancy (TOP), an FSSA partner, does offer such services. To complicate things further, each time John connects to an application, he has to make sure that it abides by privacy rules related to the access to and use of sensitive information such as Mary's social security number (SSN).

This process of manually researching and accessing individual services is clearly time-consuming. It would be more efficient if John could specify Mary's needs once and address them all together. He could then seamless access all related services through a single access point. In addition, John manually selects a service or the combination of several services for Mary from a large number of candidate providers. This manual selection can hardly guarantee optimal outcomes. John needs a service query mechanism to help him efficiently select the best services. Moreover, during the process of applying services, Mary's privacy may be released because some providers may need such information to offer services. WebDG provides solutions for all above issues and offers comprehensive support for government-citizen interactions.

3   WebDG: A WSMS for Digital Government

WebDG is a WSMS for digital government. It provides a framework for efficiently accessing e-government services while preserving citizens' privacy. Its main contributions revolve around three features, including composing e-government services, optimized querying e-government services, and preserving privacy. We overview WebDG's approach for these features in this section.

3.1   Composing E-Government Services

We propose a new approach for the (*semi*) *automatic composition* of Web services. Automatic composition is expected to play a major role in enabling the envisioned *Semantic Web* [4]. WebDG's approach for service composition is particularly suitable for e-government applications. It focuses on a select type of users (citizens and case officers) to provide customized e-government services. Composing services in WebDG includes four phases: *specification, matchmaking, selection,* and *generation.* In the following, we focus on the matchmaking phase. The other phases are outlined for the sake of completeness.

Specification. Service composers define high level descriptions of the desired composition via an XML-based language called CSSL (*Composite Service Specification Language*). They simply provide abstract definitions of the operations to be performed without referring to existing Web services. CSSL uses a subset of WSDL service interface elements and extends it to allow the: (1) description of semantic features of Web services and (2) specification of the control flow between composite services operations. Defining a WSDL-like language makes the definition of composite services as simple as the definition of simple (i.e., non composite) services. It also allows the support of recursive composition.

Matchmaking. Once CSSL specifications are provided, the next step is to generate corresponding composition plans. The matchmaking phase includes two issues: how to choose e-government services to generate plans and how to ensure composabilities of these services. In the following, we propose our approach to address these two issues.

The matchmaking phase automatically generates *composition plans* that conform to users' specification. A *composition plan* refers to the list of outsourced services and the way they interact with each other (plugging operations, mapping messages, etc). To accelerate the discovery of component services, we organize WebDG services into *communities* [8, 5]. Communities provide means for an ontological organization of the available service space based on *categories*. All services that have similar category belong to the same community. We define an ontology for e-government service called *Category*. We assume that government social agencies would agree on the ontology ahead of time. The *Category* ontology contains four attributes: *name, synonyms, specialization*, and *operations*. The *name* gives the domain of interest of the current community (e.g., "healthcare"). The *synonyms* attribute contains a set of alternative names. For example, "medical" is a synonym of "healthcare". *Specialization* is a set of characteristics of the current category. For example, "insurance" and "children" are specializations of the "healthcare" category. The *Operations* attribute gives a list of generic operations provided by community services. Each operation has a set of *input* and/or *output* parameters. Each parameter has an XML Schema data type. An operation also has a *Type* element that belongs to an ontology *Type*. This ontology includes *name, synonyms*, and *specialization* attributes. The *name* gives the business functionality offered by the current operation (e.g., "eligibility", "counseling"). *Synonyms* and *specialization* attributes are defined as in *Category*.

Providers (e.g., FSSA bureaus) identify the community of interest and register their services with it. Services can leave and reenter a community at any time during their life-span. During the registration process, providers must define the mappings between generic operations defined in their community and those defined in their service. A service may offer all or some of the operations defined within a community. For each generic operation, it may use all operation's parameters, a subset of those parameters, and/or add new parameters.

A major issue addressed by WebDG's matchmaking algorithm is *composability* of the outsourced services [4]. We propose a set of rules to check composability for e-government services. These include *operation semantics* composability that compares the semantics of service operations and *composition soundness* that checks whether combining Web services in a specific way is worthwhile. We first give the definition of *e-government services* to formally describe the set of composability rules.

Definition 1 - *E-government Service.* An *e-government service* ES_i is defined by a tuple *(Description$_i$, OP_i, Bindings$_i$, Purpose$_i$, Category$_i$)* where:

- *Description$_i$* is a text summary about the service features.
- OP_i is a set of operations provided by ES_i.
- *Bindings$_i$* is the set of binding protocols supported ES_i.
- *Purpose$_i$* $= \{Purpose_{ik}(op_{ik}) \mid op_{ik} \in OP_i\}$ is a set of ES_i operations' purposes.
- *Category$_i$* $= \{Category_{ik}(op_{ik}) \mid op_{ik} \in OP_i\} \cup \{Category_i(ES_i)\}$ is a set of ES_i operations' categories. $\Diamond$

Operation semantic composability compares the categories or domains of interest (e.g., "healthcare", "adoption") of each pair of interacting operations. It also compares their purposes or functionalities (e.g., "eligibility", "counseling"). To define *compatibility* between operation categories, let us consider the two operations $op_{ik} = (D_{ik}, M_{ik}, In_{ik}, Out_{ik}, P_{ik}, C_{ik}, Q_{ik})$ and $op_{jl} = (D_{jl}, M_{jl}, In_{jl}, Out_{jl}, P_{jl}, C_{jl}, Q_{jl})$. We say that C_{ik} is *compatible with* C_{jl} if:

1. (C_{ik}.Domain $=$ C_{jl}.Domain) or (C_{ik}.Domain $\in$ C_{jl}.Synonyms) or (C_{jl}.Domain $\in C_{ik}$.Synonyms) or (C_{ik}.Synonyms $\cap$ C_{jl}.Synonyms $\neq \emptyset$); and
2. C_{ik}.Specialization $\subseteq$ C_{jl}.Specialization

Definition 2 - *Operation Semantics Composability.* We say that $op_{ik} = (D_{ik}, M_{ik}, In_{ik}, Out_{ik}, P_{ik}, C_{ik}, Q_{ik})$ is *operation semantics composable* with $op_{jl} = (D_{jl}, M_{jl}, In_{jl}, Out_{jl}, P_{jl}, C_{jl}, Q_{jl})$ if (i) P_{ik} is *compatible with* P_{jl} and (ii) C_{ik} is *compatible with* C_{jl}. $\Diamond$

Composition soundness checks whether combining a set of services in a specific way provides an added value. For that purpose, we introduce the notion of *composition template*. A *composition template* is built for each composition plan generated by WebDG. It gives the general structure of that plan. We also define a subclass of templates called *stored templates*. These are defined *a priori* by government agencies. Since stored templates inherently provide added values, they are used to test the *soundness* of composition plans.

Definition 3 - *Composition Soundness.* A composition of services is *sound* if its template is a subgraph of a *stored template*. $\Diamond$

Selection. At the end of the matchmaking phase, several composition plans may have been generated. To facilitate the *selection* of relevant plans, we define *Quality of Composition (QoC)* parameters. Examples of such parameters include time, cost, and plan's ranking. Composers define (as part of their profiles) thresholds corresponding to *QoC* parameters. Composition plans are returned only if the values of their *QoC* parameters are greater than their respective thresholds.

Generation. This phase aims at *generating* a detailed description of a composite service given a selected plan. This description includes the list of outsourced services, mappings between composite service and component service operations, mappings between messages and parameters, and flow of control and data between component services. Composite services are generated in emerging standards for service composition such as BPEL4WS [3], WSFL [7], and XLANG [2].

3.2   Optimized Querying of E-Government Services

WebDG provides a *query scheme* that offers *database-like* query facilities over
Web services [11]. Users submit queries that are answered through a combined
access to various Web services. The challenge is then to devise the "best" alternative of Web services combinations with respect to the delivered quality.
Fundamental premises of the querying scheme is that Web services are *a priori* unknown, their number is potentially very large, and they are usually autonomous, heterogeneous, and highly volatile [13]. We propose an *optimization
model* based on *Quality of Service* (*QoS*) that would capture users' requirements
for efficiency.

E-Government Service Querying. A fundamental challenge in enabling e-
government service queries is how to obtain the combination of actual operations
from the declarative expression of a query. For that purpose, WebDG contains
a three-level query paradigm where queries go through several transformations
that lead to the *service execution plan* [10, 12]. The query paradigm includes
query level, *virtual level*, and *concrete level*. Query level allows users to submit
database-like queries. Each relation defined at the query level is mapped to one
or a set of *virtual operations* at the virtual level. A virtual operation contains
elements of input variables, output variables, category, and function description.
The virtual operations are mapped to concrete e-government services at the
concrete level.

In the virtual operations matching phase, it is not always possible to find an
exact match for a given virtual operation. In addition, the same functionality
may be offered in various ways by different e-government services. Users may
be inclined to accept similar or close answers to their queries. This is especially
true in the context of social services where the objective is to get whatever social
benefits are available for a needy citizen. Consequently, we propose a multi-level
model for virtual operation matching.This model includes *exact match*, *overlap
match*, and *partial match*. These three levels reflect different *matching degree*
that quantifies how exact the matching is. This would help the citizen and the
case manager in assessing the results of their queries. The matching degree has
a direct impact on the quality of the query results and subsequently on the
optimization process.

Optimization Model. Given a query, an important challenge for the system
is to find the "best" query execution plan with respect to an objective function. Indeed, the resolution of any query may lead to various alternatives with
disparate qualities. Quality of service (*QoS*) is playing a crucial role in assessing the added-value of competing e-government services [17].In our approach,
e-government services are selected and combined based on the *QoS* they offer
adjusted through a *dynamic rating scheme* and *multilevel matching*. Each time
an e-government service is selected in solving a query, it is rated by comparing
its advertised *QoS* with its actual *QoS*. In addition, different *levels of matching*
have been considered in matching virtual and concrete operations. Each level
has a *matching degree* that is also used to adjust the objective function.

QoS is defined through a number of parameters supplied by the service providers. The objective of the optimization process is to maximize or minimize each value. In the proposed system, we consider the QoS parameters including *latency, availability*, and *security*. Latency represents the average time it takes for an operation to return results after its invocation. Availability defines whether the e-government service is present and ready to be invoked. It represents the probability that a service is available. Security reflects the ability to provide confidentiality and non-repudiation of exchanged information. This is crucial for digital government applications that indeed manipulates large amounts of sensitive information.It is clear that these are not the only parameters that may be used to assess the quality of e-government services. Other parameters include accessibility, reliability, etc.

Due to the different fluctuations that may occur with an e-government service, the QoS advertised by that e-government service may not be always fulfilled. Furthermore, the e-government service may change some of its QoS parameter values over time. To ensure that QoS parameters are used in a way that represents the *actual* quality of an e-government service, we propose to adjust the advertised values of those parameters. The idea is to *rate* e-government services by monitoring them and computing the *delivered QoS*. The *promised QoS* ($pQoS$) is the value of the QoS parameters advertised by the service provider. The *delivered QoS* ($dQoS$) is the value of the QoS parameters obtained by monitoring the Web service.

Any new e-government service receives initially the highest rating. Rates range over a [0, 100] scale where 100 is the highest value. E-government services with a negative QoS *distance* above a certain negative threshold will have their rating lowered. On the contrary, if the QoS *distance* has a positive value greater than a certain positive threshold, the Web service rating is increased if it does not have already the highest value. In subsequent queries, QoS parameters are weighted by the ratings of the corresponding e-government services.

3.3   Preserving Privacy

Privacy is a major issue that e-government needs to address [17]. Citizens generally must divulge sensitive information, such as their SSN or salary, to access e-government services. Two characteristics add to the complexity of the privacy problem in e-government: sharing of citizens information among government agencies and citizens differing privacy requirements. Privacy is generally misperceived as an issue whose natural solution consists of good security mechanisms. Security and privacy are tightly interrelated issues, but secure e-government infrastructures do not necessarily ensure privacy. Our system focuses on privacy enforcement; we assume that appropriate security mechanisms, such as secure communication channels, already exist within the e-government environment.

Privacy Model. A typical citizen-government interaction involves three participants: users, services, and databases. This naturally defines a three-layered model for privacy [14].

The first layer of the *privacy model* is *user privacy*. Users of an e-government service include persons (e.g., citizens and case officers), applications, and other e-government services. In many cases, users interacting with an e-government service are required to provide a significant amount of personal sensitive information (e.g., their SSN, credit card number and address). Users of e-government services, however, may expect or require different levels of privacy according to their perception of the *information sensitivity*. The user's perception of privacy also depends on the *information receiver* and the *information usage* [1]. The set of privacy preferences applicable to a user's information is called *user privacy profile*. A user privacy profile is typically defined by the user but can also be uniformly set for a group of individuals. Privacy profiles are *dynamic*: users can create, view, update, or delete their privacy profiles. To provide support for resolving legal disputes over privacy violation, the underlying Web service architecture must trace all of these operations. We also define a user's *privacy credentials* as a signature that is typically appended to any request that the user submits to the Web service. They determine the *privacy scope* for the corresponding user. A privacy scope for a given user defines the information that an e-government service can disclose to that user.

The second layer of the *privacy model* is *service privacy*. An e-government service generally has its own *privacy policy* that specifies a set of rules applicable to *all* users. Service privacy generally specifies three types of policy: *usage* policy, *storage* policy, and *disclosure* policy. The *usage* policy states the purposes for which the information collected can be used. For example, consider an e-government service Medicaid that provides healthcare coverage for low-income citizens. Medicaid may state that the information collected from citizens will not be used for purposes other than those directly related to providing health services to citizens. The *storage* policy specifies whether and until when the information collected can be stored by the service. For example, Medicaid may state that the information it collects from citizens will remain stored in the underlying databases one year after they leave the welfare program. The *disclosure* policy states if and to whom the information collected from a given user can be revealed. This information may relate to individual persons or to *groups* of individuals. For example, the privacy policy of the service Medicaid may state that external users cannot access statistical information that reveals general characteristics of the beneficiaries (e.g., average income, racial background distribution, etc).

The third layer of the *privacy model* is *data privacy*. A data object may be accessed by several e-government services. For example, consider the US National Database for New Hires (NDNH) that contains information about over 200 millions hired employees. A record in this database can be accessed (using an e-government service) by an IRS officer to check the accuracy of an employee's tax form. It may also be accessed (using another e-government service) by an officer at a child support agency to check whether a parent is compliant with his child support obligations. This shows that different e-government services may need different information from the same data object. Thus, data objects must be able to *expose* different views to different e-government services. For each data object, we define a *data privacy profile* that specifies the access views that it exposes to the different e-government services.

Privacy Enforcement. An important premise in our privacy enforcement approach is that users access databases through e-government services. When a service receives a request from a given user, it first checks that the user has the necessary credentials to access the requested operation according to its privacy policy. Lets look at the Medicaid service, which states that the only person who can update a citizens privacy profile is that citizen. If the request is valid, the service translates the users request into an equivalent data query and submits it to the appropriate government database management system.

Before the Medicaid service submits a query to the DBMS, it sends the query through a privacy preserving data filter (DFilter). This DFilter is composed of two modules: the credential checking module (CCM) and the query rewriting module (QRM). The CCM uses the credential received with the query to determine whether the service requester is authorized to access the requested information. If the credential authorizes access to only part of the requested information, the QRM redacts the query to enforce all the privacy constraints. For example, the QRM deletes the salary field from a service request that translates into the SQL query `select name, age, salary from Medicaid.enrollees` before submitting it to the DBMS if the credential of the user who made the request does not allow access to enrollees salaries.

The privacy profile manager (PPM) enforces privacy at a finer granularity than the CCM does. The local CCM might decide that an organization can access local information regarding health records for a group of citizens, but some of those citizens might explicitly request that parts of their records not be made available to third-party entities. In this case, the local PPM would discard those parts from the generated result. The PPM is a translation of the consent-based privacy model in that it implements individual citizens privacy preferences. It maintains a repository of privacy profiles that stores individual privacy preferences. The PPM also handles citizens requests for updating their privacy profiles.

4   Implementation

Figure 1 shows the WebDG system as implemented across a network of *Solaris* workstations. Citizens and case officers access the system via a graphical user interface implemented in HTML and Java servlets. WebDG currently includes seven FSSA applications implemented in Java (JDK 1.3). The Axis Java2WSDL utility in *IBMs Web Services Toolkit* automatically generates WSDL descriptions from Java class files, which WebDG publishes into a UDDI registry. WebDG uses the service management client within Apache SOAP 2.2 to deploy e-government services. Apache SOAP provides a server-side infrastructure for deploying and managing services, and a client-side API for invoking those services. Each service

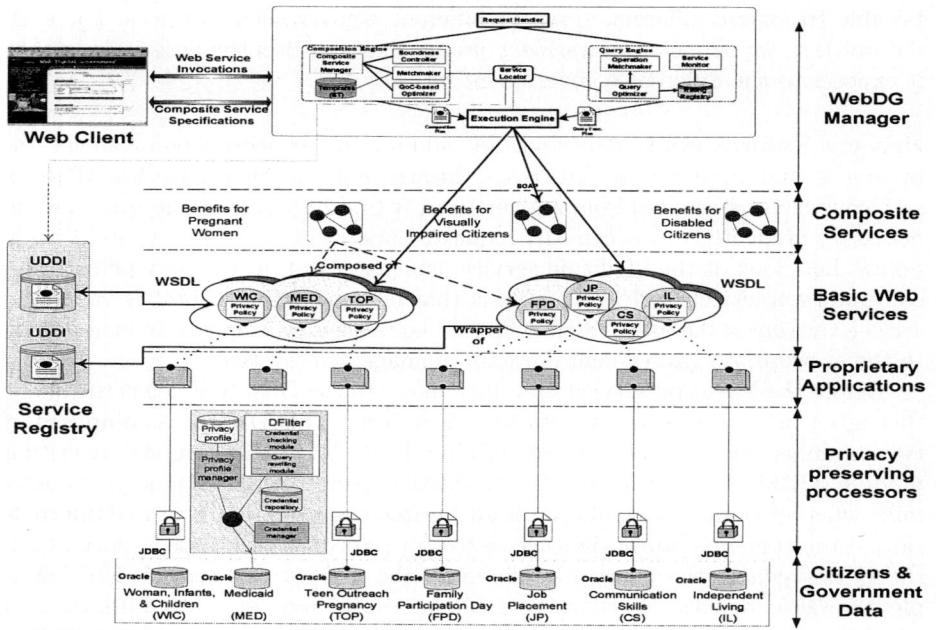

Fig. 1. WebDG Architecture

has a deployment descriptor that includes the unique identifier of the Java class to be invoked, the session scope of the class, and operations in the class available for the clients. WebDG deploys each service using its descriptor and the URL of the Apache SOAP servlet `rpcrouter` as input arguments.

The *WebDG manager* is at the system's core. The *service locator* (SL) looks up WSDL descriptions in the registry. Once the *execution engine* discovers a service, it invokes the services operations through a SOAP binding stub.

The *composite service manager* (CSM) uses the Java API for XML Processing (JAXP) to parse the XML-based composite service specifications. It then returns them to the *matchmaker*, which checks composability rules. The matchmaker then sends each composite service operations category to the SL. The SL parses each located services WSDL description and returns it to the matchmaker. The SL retrieves only services whose category is compatible with the operations category. After checking composability, the matchmaker generates composition plans and sends them to the *optimizer*, which selects plans based on quality of composition (QoC) parameters. Users define thresholds for such parameters as time, cost, and the plans relevance to the users specification. The optimizer returns plans to the matchmaker if the QoC parameter values are greater than the users thresholds. The matchmaker forwards the selected plans to the *soundness controller* (SC) to check that the way they are combined provides added value. The SC then returns plans, along with their compatible stored templates (if

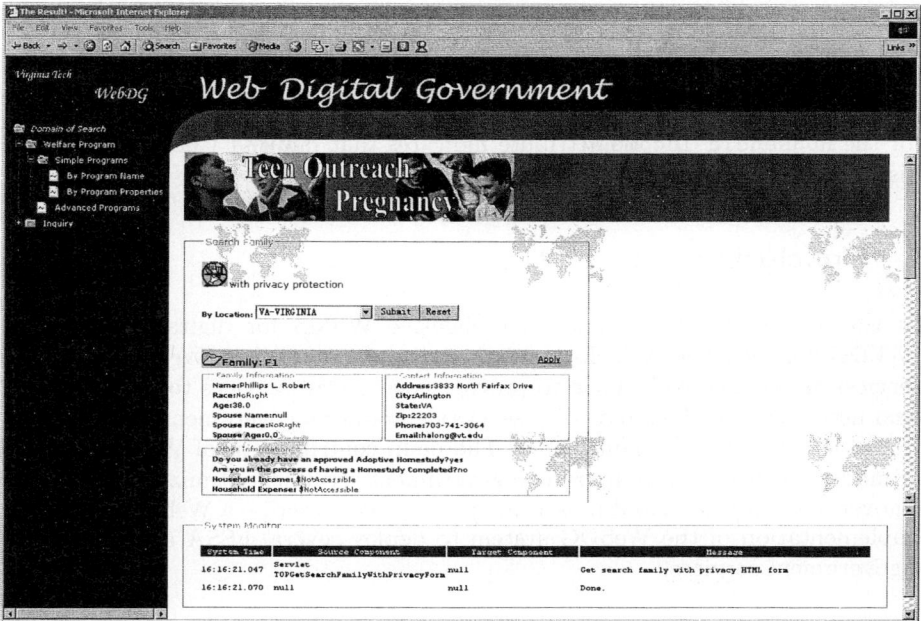

Fig. 2. WebDG Interface for Preserving Privacy

any), to the CSM. The CSM then sends the plans to the *execution engine*. The *execution engine* enacts the plans by actually invoking e-government services using SOAP. We use *SOAP Binding Stubs* which are implemented using Apache SOAP API for this purpose.

The *service query engine* (SQE) is responsible for the correct and optimal execution of e-government service queries in WebDG. The *operation Matchmaker* (OM) interacts with the SL to retrieve the services' descriptions in WSDL and determines the concrete operations to use in the service execution plan. WSDL descriptions (augmented with semantic attributes that we have defined) are parsed and concrete operations are matched to virtual operations using one of the matching modes. The *monitoring agent* (MA) monitors e-government service invocations. Its goal is to assess their behavior in terms of the delivered *QoWS*. The monitoring agent maintains a local repository for ratings and other information to compute those ratings. The *query optimizer* (QO) is the central component of the SQE. It determines the best service execution plan for a given query based on the optimization model that we presented in Section 3.2. After the optimizer generates an efficient service execution plan, the plan is handed over to the execution engine.

When service operations attempt to access the FSSA databases, a privacy-preserving processor intercepts the operation invocations and allows or disallows access based on privacy profiles, privacy credentials, and data filters. Figure 2 shows TOP's Search Family Adoption operation, which returns information about foster families in a given state, in this case Virginia. The value "no right" for the attribute Race means that the current user does not have system permission to access information about family F1's race. The value "not accessible" for the Household Income attribute indicates that family F1 does not want to disclose that sensitive information.

5   Conclusion

In this paper, we describe a comprehensive WSMS for digital government, WebDG, for the efficient deployment of e-government services. WebDG uses Web services as an effective vehicle to provide government services to those citizens who need the help the most. Web services empower e-government with features from the envisioned Semantic Web. The main contributions of WebDG revolve around three features: composing e-government services, optimized querying of e-government services, and preserving privacy. We describe a Web service based implementation of the WebDG system to deploy several FSSA applications as e-government services.

References

1. A. Adams. User's Perception of Privacy in Multimedia Environment. *PhD thesis, School of Psychology, University College London*, 2001.
2. S. Aissi, P. Malu, and K. Srinivasan. E-Business Process Modeling: The Next Big Step. *IEEE Computer*, 35(5), May 2002.
3. BEA, Microsoft, and IBM. *Business Process Execution Language for Web Services (BPEL4WS)*. http://xml.coverpages.org/bpel4ws.html, 2003.
4. T. Berners-Lee, J. Hendler, and O. Lassila. The Semantic Web. *Scientific American*, May 2001.
5. Bouguettaya, A. Elmagarmid, B. Medjahed, and M. Ouzzani. Ontology-based support for digital government. In *VLDB 2001*, Roma, Italy, September 2001.
6. F. Curbera, M. Duftler, R. Khalaf, W. Nagy, N. Mukhi, and S. Weerawarana. Unraveling the Web Services Web: An Introduction to SOAP, WSDL, and UDDI. *IEEE Internet Computing*, 6(2), February 2002.
7. IBM. *Web Services Flow Language (WSFL)*. http://xml.coverpages.org/wsfl.html, 2003.
8. B. Medjahed, A. Bouguettaya, and A. Elmagarmid. Composing Web Services on the Semantic Web. *The VLDB Journal, Special Issue on the Semantic Web*, September 2003 (to appear).
9. B. Medjahed, A. Rezgui, A. Bouguettaya, and M. Ouzzani. Infrastructure for E-Government Web Services. *IEEE Internet Computing*, 7(1), January 2003.
10. M. Ouzzani and A. Bouguettaya. A query paradigm for web services. In *1st International Conference on Web Services*, Las Vegas, NV, USA, June 2003.
11. M. Ouzzani, A. Bouguettaya, and B. Medjahed. Optimized querying of e-government services. In *The dg.o 2003 NSF Conference for Digital Government Research*, Boston, USA, May 2003.

12. M. Ouzzani and B. Bouguettaya. Efficient access to web services. *IEEE Internet Computing*, 37(3), March 2004.

13. M. Ouzzani and B. Bouguettaya. Query processing and optimization on the web. *Distributed and Parallel Databases, an International Journal*, 15(3), May 2004.

14. A. Rezgui, M. Ouzzani, A. Bouguettaya, and B. Medjahed. Preserving Privacy in Web Services. In *the 4th International ACM Workshop on Web Information and Data Management*, November 2002.

15. Stanley Y.W. Su, Herman Lam, Minsoo Lee, Sherman Bai, and Zuo-Jun Shen. An information infrastructure and e-services for supporting Internet-based scalable e-business enterprises. In *Enterprise Distributed Object Computing Conference, 2001. EDOC '01. Proceedings. Fifth IEEE International*, pages 2–13, Seattle, WA, USA, September 2001.

16. S. Tsur, S. Abiteboul, R. Agrawal, U. Dayal, J. Klein, and G. Weikum. Are Web Services the Next Revolution in e-Commerce? (Panel). In *Proceedings of 27th International Conference on Very Large Data Bases*, Rome, Italy, September 2001. Morgan Kaufmann.

17. S. Vinoski. Service discovery 101. *IEEE Internet Computing*, 7(1), Jan/Feb 2003.

A Study on Camera-Detector System
for Traffic Control in Korea

Chang-Hee Kwon[1] and Hong-Jin Park[2]

[1] Dept. of Computer Engineering, Div. of IT
Hansei University, 604-5, Dangjeond-dong, Gunpo-si, Kyunggi-do, Korea
kwonch@hansei.ac.kr
[2] School of Computer Information and Communication Engineering
SangJi University, Woosandong, Wonjusi, Kangwondo, Korea
hjpak1@sangji.ac.kr

Abstract. Actual detectors of traffic information in Korea adopt the loop detector system. The evaluation of alternative detector of the above system such as camera or microwave detector has been limited to the situation for uninterrupted flow instead of interrupted one. However, recently the installation of camera-detectors tends to be rapidly increased in freeways. In the case of camera detector, it appears to satisfy the requirement of measurement accuracy of traffic variables such as occupancy, traffic amount and speed in the interval unit of 5 or 15 min of uninterrupted flow situation. This study is to understand the traffic information data collection level required in the sophistication of future Korean traffic control in focusing on city map and freeways and to analyze and evaluate the loop detector as alternative for camera detector.

1 Introduction

1.1 Objectives

Installation of detectors have been largely increased with growth of intelligent traffic system and the relevant development by use of various in basic technology in Korea. But there have been confusions in selection of the most optimal detector with constant priority in traffic flow control and monitoring works. The currently-employed detectors for vehicles in traffic control system are overwhelmed by loop detector of magnetic induction system. This loop detector has the advantage in getting the exact history data and real-time data at lower cost. But it has difficulties to collect the delicate information of car classification. And when being adopted, it impedes traffic flow, decrease in performance according to pavement status and produces the large maintenance expenses due to road damages. This is recognized as considerable disadvantages in Korean road situation where daily traffic congestion are frequented.

Therefore, the data for the objective policy-decision-making about the alternative detector are required in order to overcome the problems of loop detector and ensure high quality and quantity of traffic information needed in the advanced traffic control. Thus, this study is to understand the traffic information data collection level required in the sophistication of future Korean traffic control in focusing on city map and freeways and to analyze and evaluate the loop detector as alternative for camera detector.

S. Wang et al. (Eds.): ER Workshops 2004, LNCS 3289, pp. 566–576, 2004.

1.2 Existing Study Case of Field Evaluation of Detectors

The current trend of Korean detection technology is explained on the result basis of field evaluation test implemented since 1996 since the ongoing research and development in the relevant industry is difficult to be examined. The most field evaluation tests were implemented in focusing the measurement accuracy of traffic variables. The characteristics and problems in this tests are summarized as follows.

- The most field evaluation tests were implemented in focusing the measurement accuracy of traffic variables.
- Most the data collection periods were adopted as under 24 hours (1 day) and then the factors of various weather status and other traffic conditions were not evaluated accordingly.
- This test focused on the measurement accuracy of traffic variables and then the integral evaluation for system's H/W, S/W, operation procedure, communication, interface, algorithm and noise are not lacked.
- The index of measurement accuracy adopted the equivalent coefficients and there are some restrictions of additional indexes such as deviation and correlation coefficient.
- Most field evaluation tests adopted one or more minute analysis unit instead of individual vehicle unit and then the constant and clear reliability of the result was not provided. And there were found some different results in the same detection system.
- The field evaluation tests were implemented under uninterrupted flow status of freeway or national roads but the city roads of signal-controllers were restrictedly evaluated.

1.3 Scope and Method

With Korean ITS vitalizations, the installation of numerous detectors is required in advanced traffic system. But there are no available objective evaluations for demand of traffic information and detector performance and then we have limitations on market expansion of alternative detectors. In this aspect, the study includes the analysis of market size of detector in the current traffic control system and field evaluation results in understanding the algorithm of detection technology, its level and availability of alternative detectors. This study consists of following 4 steps.

(Step 1)   Review of development technology and level of detector
(Step 2)   Investigation of construction cases of traffic control system in city road and freeways and analysis of demand of traffic information
(Step 3)   Provision of objective evaluation data for the selection and utilization of camera detector as alternative one in policy-decision of the public authorities, main demander of detector market
(Step 4)   Reliability evaluation of camera detector as an alternative one.

2   Performance Evaluation of Camera Detector

2.1   Evaluation Environment Setup

The field test of detectors were executed in approaching road of intersection of Yang-jae road, one of main trunk roads of Seoul. The camera detectors of three selected companies were installed in pole-type support of stop line as figure 1. And the arms in the height of 12 to 15 meters were installed in support of stop line for the vendors of the detectors' test.

Table 1. Detected data and transmission time unit

Data / Detector	Saturation	Saturation flow ratio	Saturated non-occu-pancycy	Speed	Occu-pancy	Normal status	Integral informa-tion	Traffic amount	Traffic amount by cars	Waiting distance
L O (straight) Stop line	O	O	O	×	O	O	O	O	×	×
O p Stop line (Left turn)	O	×	×	×	O	×	O	O	×	×
front-flow (waiting series)	×	×	×	O	O	×	O	O	×	×
front-flow(blocked)	×	×	×	O	O	O	O	O	×	×
C a m e r a Stop line(straight)	△	O	△	×	×	△	△	O	O	×
Stop line(Left turn)	△	O	×	×	×	△	△	O	O	×
front-flow (waiting series)	×	×	×	O	×	△	△	O	O	O
front-flow(blocked)	×	×	×	△	×	△	△	O	O	×
Transmission time unit	Interval	interval	interval	30s/interval	interval	in every event	interval	interval	In occurrence	30s/interval

※ O : information transmission, △ : available, × : unnecessary
※ Preventive control of blockage in front flow : transmission to R/C
※ 30-sec base transmission: 30-sec synchronous transmission
※ Saturated non-occupant time and saturation flow ratio : automatical update of saturated non-occupant time but the results are transmitted to R/C for intervention of monitor and operator in MMI.
※ In occurrence of event: realtime R/C communication by vehicle quantity base for determination of normal status (realization of second base communication)

Also, the saturation ratio is calculated for determination of interval distance in R/C. In general, occupancy/non-occupancy time is calculated in SCATS mode (Australia), but the green time in SCATS mode is replaced by the value of end-lag (about 1 sec) in yellow time added by 1. N meaning of number of vehicles adopts DS mode in using non-occupancy time calculated during green time (strictly speaking, including starting loss time). But in calculating DS by camera detector, DS value is converted as X expressing V/C. This X value is main interest of traffic flow control. This value cannot be gotten directly by loop detector. For alternative measure, it utilizes DS for traffic flow control. Thus, we get the X value by use of camera detector and utilize the concept of DS = F(X) and use the existing algorithm as non-corrected. DS calculation for straight drive and left turn are as follows:
    Straight drive saturation :

$$DS = \frac{EG - (\sum Space - N * t)}{EG}$$

Table 1. (Continued)

here,	DS = saturation ratio
	EG = effective green time (green time + End-Lag, sec)
	$\sum$ Space = Sum of non-occupant time (sec)
	N = number of non-occupant time (veh)
	t = average saturation non-occupancy time (sec)

and,

$$p\_occ(\%) = \frac{occ\_time(sec)}{30(sec)} * 100$$

$$speed(km\,hr) = \frac{volume(veh/30\,sec) * [r*l_1 + (1-r)*l_2]}{occ\_time(sec)}$$

$$= \frac{volume(veh/30) * [r*l_1 + (1-r)*l_2] * 100}{p\_occ * 30(sec)}$$

$$speed(km\,hr) = speed(m/sec) * 3.6$$

here,

> occ_time : occupant time of 30 Second
> p_occ : occupation ratio for 30 second
> r : mixing ratio by the large size Vehicle
> l_1: average length of large Vehicle(m)
> l_2: Length of small Vehicle(m)

The traffic variable collected in loop detector of signal control system were traffic amount, occupancy rate, non-occupancy numbers and non-occupancy time, which were needed in calculation of saturation flow rate in relevant direction (left turn and straight drive). But since the occupancy time was possibly converted to speed and occupancy time and non-occupancy time were available for the occupancy ratio, this field test evaluated the measurement accuracy of the measurement of the periodical traffic amount of vehicles in straight drive and occupancy/non-occupancy time of individual vehicles. Also the utilizable traffic variable in signal control system was the length of waiting queue.

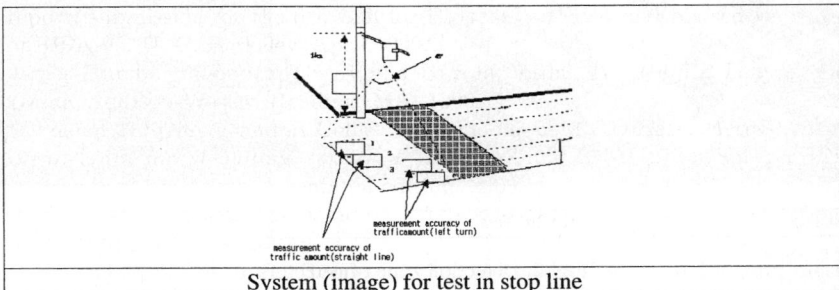

System (image) for test in stop line

Fig. 1. Environment structure of field test

The signal control system in Seoul estimated the length of waiting queue on the basis of occupancy time, occupancy ratio and traffic amount collected in upper part of loop detectors. But the camera detectors observed directly the length of waiting queue by the available images and then in the field test, the view of camera detector was directed toward the lower part for more measurement accuracy of length of waiting queue.

2.2 Test Description

The field tests were implemented for measurement accuracy of three companies' products in periodical traffic amount in straight line, occupancy/non-occupancy time of individual vehicles. The traffic conditions (peak time and non-peak time), day and night, transition time were applied. As shown in the following figure 1, the third test's modification of detection zone was made under the consideration of vicious impact of wires, traffic situation and geometric structure of test sites and then the the vendors were induced to modify sufficiently the detector's processing algorithm (S/W). The modified things in algorithm of camera detector were the relevant know-how of the vendors of the detectors but most field images were considered for the detection algorithm's parameters modification and selection of detection zone including image processing method.

○ Evaluation Result of Measurement Accuracy of Traffic Amount

To evaluate the measurement accuracy of traffic amount by camera detector, we analyzed those of vehicles of straight line and left turn. The analysis data of traffic amount was made in 342 periodicals and 8673 vehicles. The equivalent coefficient, difference of percentage and correlations in 17 tests of three sessions were evaluated in the measurement accuracy against the evaluation indexes as following Table 5. [Fig 2(a)] explained the comparison of measured value in a sampling test with standard value. [Fig 2(b)] expressed the variance of precision of occupancy/non-occupancy time by each test in respective the vendors of the detectors.

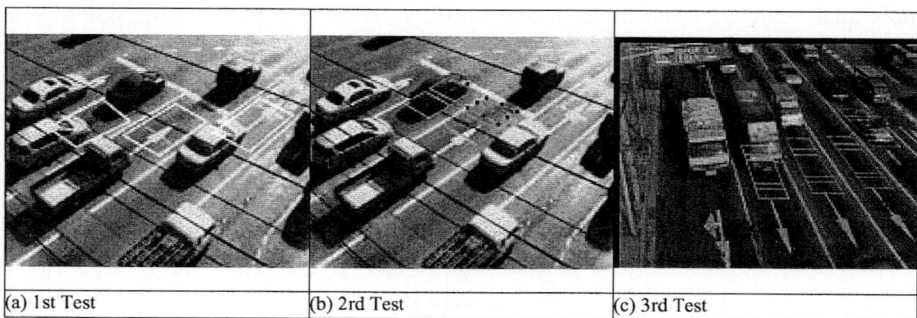

| (a) 1st Test | (b) 2rd Test | (c) 3rd Test |

Fig. 2. Detection zone of detectors

The measurement accuracy of traffic amount in straight line was best in A company with 9.90 with about 15% of percentage difference. And the following good results were B's and C's in order. The products of the vendors of the detectors were

Table 2. Field tests status

Division	Date	Test times	Data collection time	The vendors of the detectors			Remarks (weather)
				A	B	C	
1st test	21 Oct	ⓐ−1	non-peak time, pm (14:37 ~ 15:20)	○	○	×	Cloudy
		ⓐ−2	transition time, pm (17:43 ~ 18:24)	○	○	×	
		ⓐ−3	peak time, pm (18:35 ~ 19:39)	○	○	×	
	22 Oct	ⓐ−4	transition time, am(06:08 ~ 07:00)	○	○	×	Little foggy
		ⓐ−5	peak time, am (07:05 ~ 07:58)	○	○	×	Little foggy
		ⓐ−6	non-peak time, pm (13:25 ~ 14:25)	○	○	×	Clean
		ⓐ−7	transition time, pm (17:34 ~ 18:16)	○	○	×	
2nd test	1 Nov	ⓑ−1	transition time, am (06:10 ~ 07:10)	○	○	×	Cloudy
		ⓑ−2	peak time, am (07:25 ~ 08:25)	○	○	×	Rainy
		ⓑ−3	non-peak time, pm (13:45 ~ 15:10)	○	○	×	
		ⓑ−4	transition time, pm (17:05 ~ 18:07)	○	○	×	
		ⓑ−5	peak time, pm (18:08 ~ 19:00)	○	○	×	
3rd test	19 Nov	ⓒ−1	transition time, am (06:00 ~ 07:00)	○	×	○	Cloudy
		ⓒ−2	peak time, am (07:25 ~ 08:25)	○	×	○	
		ⓒ−3	non-peak time, pm (13:41 ~ 14:41)	○	×	○	Clean, little cloudy
		ⓒ−4	transition time, pm (16:40 ~ 17:40)	○	×	○	Little cloudy
		ⓒ−5	peak time, pm (17:40 ~ 18:40)	○	×	○	

different but there were no distinct performance differences in measurement accuracy in peak/non-peak time of camera detector under traffic conditions except the 1st test of A's product. The measurement accuracy of traffic amount in day/night and transition time was lower in night and transition time in the 1st test. The 3rd test with complementary detection algorithm showed somewhat lower performance of equivalent coefficient and % difference in nighttime than in daytime but there was no difference in measurement accuracy in day/night division.

○ **Evaluation Results of Measurement Accuracy in Occupancy/Non-occupancy Time**

The measurement accuracy for occupancy/non-occupancy time was evaluated under traffic conditions, day/night and transition time. The data were collected in the individual vehicles' corresponding periods on the basis of measurement accuracy of traffic amount. The difficult periods of which measured value didn't correspond to standard were excluded. Total 15 tests were implemented in 212 periods and 3687 vehicles and the analysis results for traffic condition, day/night and transition time's occupancy/non-occupancy time were summarized in following Table 3. Fig 3 was the comparison of measured value with standard in occupancy/non-occupancy time in a sampling test. The measurement accuracy in occupancy time was 0.72 as average equivalent coefficient of three companies and 39% in percentage difference. They were very lower than existing uninterrupted flow status. But in the results in each step during 3 test sessions, the measurement accuracy in equivalent coefficient and % difference in A company were gradually increased by 0.83 and 23% respectively. There were no differences in performance in peak/non-peak time of three companies' products but the performance in nighttime was weaker than that of daytime. The accuracy in transition time was somewhat lower but there was no big difference. The accuracy in non-occupancy time was similar to the evaluated results in occupancy time. In particular, the 3rd test expressed the equivalent coefficient as 0.92 and % difference

Table 3. Evaluation results of measurement accuracy of traffic amount by camera detector

Division	Traffic condition	measurement accuracy of traffic amount(straight line)								
		A			B			C		
		equivalent coeff.	% difference	correlation coeff.	equivalent coeff.	% difference	correlation coeff.	equivalent coeff.	% difference	correlation coeff.
1st	Transition time, am	0.87	20.86	0.90	0.78	36.93	0.62			
	Transition time, pm	0.80	27.87	0.10	0.83	29.71	0.37			
	Non-peak time, am									
	Non-peak time, pm	0.96	6.30	0.96	0.76	56.18	0.70			
	Peak time, am	0.65	42.62	0.32	0.87	23.74	0.73			
	Peak time, pm	0.77	35.71	0.47	0.85	25.07	0.66			
	Sub-total	0.81	26.67	0.55	0.82	34.33	0.62			
2nd	Transition time, am	0.93	10.26	0.93	0.89	13.67	0.87			
	Transition time, pm	0.95	7.19	0.95	0.91	16.13	0.86			
	Non-peak time, am									
	Non-peak time, pm	0.94	9.15	0.49	0.88	24.43	0.71			
	Peak time, am	0.95	8.53	0.32	0.92	14.27	0.54			
	Peak time, pm	0.93	11.96	0.93	0.85	33.15	0.87			
	Sub-total	0.94	9.42	0.72	0.89	20.33	0.77			
3rd	Transition time, am	0.95	5.25	0.97				0.94	29.37	0.28
	Transition time, pm	0.93	10.59	0.95				0.81	32.14	0.84
	Non-peak time, am									
	Non-peak time, pm							0.87	20.03	0.67
	Peak time, am	0.94	9.67	0.98				0.88	20.16	0.83
	Peak time, pm	0.95	9.12	0.99						
	Sub-total	0.94	8.66	0.97				0.88	25.43	0.66
	Total	0.74	0.90	14.91	0.75	0.85	27.33	0.69	Wrong Formula	W.F

Division	Traffic condition	measurement accuracy of trafficamount(left turn)								
		A			B			C		
		EQN.	%D	COC	EQN.	%D	COC	EQN.	%D	COC
1st	Transition time, am	0.5	52.54	0.24	0.77	39.24	0.79			
	Transition time, pm	0.79	28.69	0.48	0.77	66.92	0.37			
	Non-peak time, am									
	Non-peak time, pm	0.85	16.86	0.57	0.83	42.86	0.54			
	Peak time, am	0.60	30.90	0.58	0.86	27.79	0.75			
	Peak time, pm	0.68	50.84	0.48	0.78	65.65	0.21			
	Sub-total	0.68	35.96	0.47	0.80	48.49	0.53			
2nd	Transition time, am	0.67	48.21	0.34	0.82	12.04	0.97			
	Transition time, pm	0.73	41.23	0.56	0.76	27.64	0.91			
	Non-peak time, am									
	Non-peak time, pm	0.75	33.02	0.30	0.64	65.66	0.52			
	Peak time, am	0.81	29.29	0.54	0.79	44.96	0.45			
	Peak time, pm	0.56	63.58	0.36	0.78	30.27	0.86			
	Sub-total	0.70	43.06	0.42	0.75	36.11	0.74			
3rd	Transition time, am	0.78	29.43	0.86				0.94	17.46	0.99
	Transition time, pm	0.91	11.04	0.89				0.63	77.33	0.27
	Non-peak time, am									
	Non-peak time, pm							0.68	62.27	0.04
	Peak time, am	0.87	34.01	0.96				0.89	25.20	0.99
	Peak time, pm	0.79	26.87	0.75						
	Sub-total	0.83	25.33	0.86				0.78	45.56	0.57
	Total	0.74	34.78	0.58	0.78	42.30	0.64	0.78	45.56	0.57

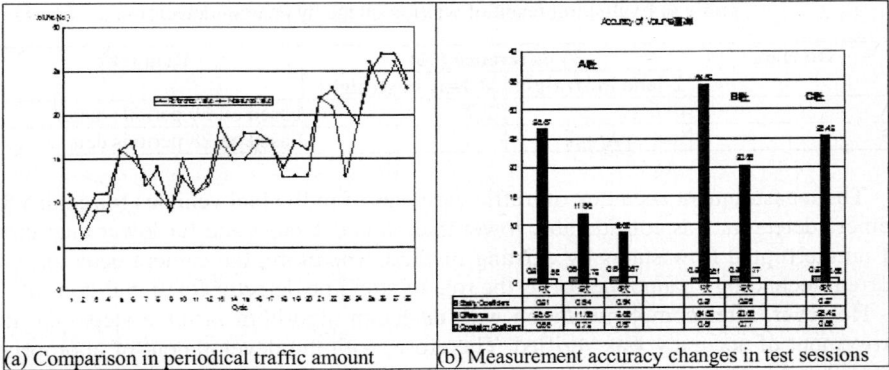

| (a) Comparison in periodical traffic amount | (b) Measurement accuracy changes in test sessions |

Fig. 3. Comparison of measured and accuracy valued in traffic amount and measurement accuracy changes in test sessions

as 18%, which appeared to approach the evaluation results of existing uninterrupted flow.

○ **Evaluation Results of Waiting Queue**

The field test of waiting queue collected total 50 periods (2 min 30 sec interval) data. The weather condition was cloudy and it rained after 30th period then the light was required. The evaluation index of waiting queue adopted % difference and the length of waiting queue was calculated on the basis of the length of the last vehicle in ending period. The evaluation result of waiting queue by camera detector were summarized in following Table 4.

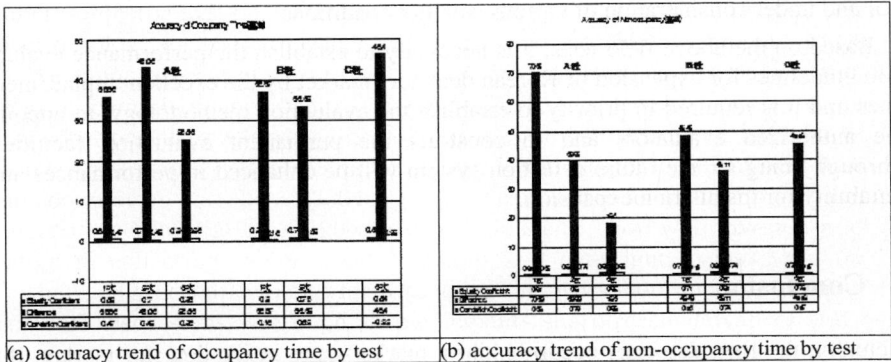

| (a) accuracy trend of occupancy time by test | (b) accuracy trend of non-occupancy time by test |

Fig. 4. Accuracy trends of occupancy/non-occupancy time by test

2.3 General Evaluation

The field evaluation tests were implemented in measurement accuracy of traffic amount, occupancy/non-occupancy time and in qualitative method in system installation and operation. Based on these tests, the evaluation results are summarized for alternative detectors as follows.

Table 4. Evaluation result of waiting queue by camera detector

Division	% difference (%)		Remarks
	2-lane in straight	3-lane in straight	
A	40.6	39.7	based on 50 periods data
B	36.1	-	based on 28 periods data

- The measurement accuracy of traffic variables of individual vehicles for control by camera detectors was considerably lower than loop detectors and far lower than those of uninterrupted flow status by existing method. Therefore, the camera detector with current technology cannot performs the role of stop line detector for signal-control.

- However, after complementation with detection algorithm in three steps, the improvement of accuracy was verified. Therefore, with the development of appropriate algorithm for traffic control, the camera detector can function its role as alternative detector for signal-control.

- After general evaluation of accuracy, there were considerable % differences of camera detector under traffic condition, day/night and transition time, but the performance of nighttime and transition time didn't show big differences from daytime even though the existing study showed its low performance. This is resulted from the main development of hardware and software based on uninterrupted flow status and then the improvement of detection algorithm, selection of detection zone and image processing will overcome the above-described difficulties.

- But, in these field tests, I could not lead the test under sufficient consideration of the lowered performances due to various weather conditions, it is difficult to judge the generalized possibility of replacement of detectors.

- The judgment of replacement by camera detectors can be made by constant field tests in coping with continuous development of detection system against traffic control and under consideration of various weather conditions.

Based on the above field tests, it is necessary to establish the performance evaluation guidelines for expansion of Korean detectors market by the excellent replacement ones and it is required in priority to establish the evaluation methodology, to appoint the authorized evaluators and to construct the permanent evaluation facilities. Through doing so, the future detection system will be enhanced in performances and reliability for installation expansion.

3 Conclusion

Actually, the loop detectors occupy highest market share in the detector market. But Japan depends exclusively on ultrasonic detectors and other countries replace them rapidly by camera and microwave detectors. Korea tend to adopt increasingly camera detectors around freeways. Until now, the evaluation of detectors realized in Korea were limited to uninterrupted flow instead of traffic control. In case of camera detector, it satisfies the requirement of detection system in measurement accuracy of traffic variables such as 15min or 5min-base occupancy ratio, traffic amount and speed. Different from existing field tests, this field test focused on the evaluation of the application of detection-based technology instead of detectors and evaluated the measurement accuracy of periodic data of traffic control. But, the evaluation result showed the high application of collection of 15min or 5min-based strategic data but lower performance in measurement accuracy of signal periods or data collected in short time than loop detector. It requires more study of the improvement of this aspect.

"This Study was supported by a grant from Hansei University, Korea."

References

1. Performance evaluation report of camera detector by Korea Highway Corporation, 1999. Sept.
2. 1st annual report of traffic signal control system development by Seoul Police Station, 1991. Oct.
3. Specification of design and installation of the Advanced traffic signal system by Seoul Police Station, 1996. July.
4. Performance study of non-laying detectors in street for traffic information collection by Seoul Municipal University, 1998. Nov.
5. Evaluation of field measurement accuracy of road detectors, by Traffic research center of Aju university, 1998. Nov.
6. MCD evaluation test for FTMS in freeways by Traffic research center of Aju university, 1998. July.
7. Inspection of camera detector by Oriental Electronic system, 1997. Nov.
8. Inspection for practical camera detection system and expansion measures by Road Traffic Safety Authority, 1996. Dec.
9. Study of basic plan of advanced road traffic system by Road Traffic Safety Authority, 1995. Dec.
10. General measure of safety control of city freeways by Road Traffic Safety Authority, 1997. Dec.
11. Study of construction of advanced road traffic system by Road Traffic Safety Authority, 1996. Dec.
12. Basic design construction and evaluation of traffic information system in Seoul area by Ministry of construction and transportation, 1997. Dec.
13. Investigation system and equipment of road traffic amount by Korea national land development institute, 1993. April.
14. Test of microwave detectors by LG Industrial , 1994. Nov.
15. Development of algorithm of vehicle recognition by magnetic detector by Kim, Soo-Hee (MA paper), 1999. Aug.
16. Study of improvement of new traffic signal control system by Road Traffic Safety Authority, 1994. Dec.
17. Joint development project of advanced traffic control and management system by Road Traffic Safety Authority, 1997. Dec.
18. Improvement of functions of new signal system by Seoul Police Station, 1999. Nov.
19. Study of developing promotion of ITS business by use of Kwachoen ITS exemplary cases. 1999. Dec.
20. Japan Traffic Management Technology Association, ITS Developed by Japanese Police, 1999.
21. Transport Technology Publishing, Advanced Traffic Detection, 1995.
22. FHWA, Detection Technology for IVHS(Executive Summary), 1995.
23. FHWA, Non-Intrusive Technologies(NIT) Phase I & II, 1998.

24. California Polytechnic State University, Evaluation of Video Image Technology Application in Highway Operation(Phase II), 1994. 12.
25. FHWA, Detection Technology(IVHS Volume 1), 1996. 6.
26. ITS International, Detection Systems Elemental Approach, 1999. 9.
27. City of Lincoln Traffic Studies, Traffic Studies Evaluation for Lincoln's Arterial Street System, 1999. 1.
28. TRB, Annual Meeting, Deployment of State-of-the-Art Technology for Incident Management(The Gowanus freeway Project), 1999
29. Traffic Technology International, Video Detection the Atlanta Experience, 1997. 1.
30. ASCE, Recent Advances in Implementation of Machine Vision Technology in Freeway, 1998. 1.
31. Traffic Technology International, Marriage Made in Michigan (Autoscope TM and SCATS together in FAST-TRAC), 1997. 9.
32. Cal Poly University, Machine Vision Technology, 1998.
33. California Polytechnic State University, Evaluation of Video Image Processing Technology(Application in Highway Operation, Phase II), 1998.
34. Image Sensing Systems, Inc., AUTOSCOPE TM Wide Area Video Vehicle Detection System, 1991.
35. Ljubisa Ristic, Artech House, Sensor Technology and Devices, 1994.

Ontology-Based Models of Legal Knowledge

Maria-Teresa Sagri and Daniela Tiscornia

ITTIG (Institute for Theory and Techniques for Legal Information)
Of the Italian Research Council, Florence, Italy
{Sagri,Tiscornia}@Ittig.cnr.it

Abstract. In this paper we describe an application of the lexical resource Jur-WordNet and of the Core Legal Ontology as a *descriptive vocabulary* for modeling legal domains. It can be viewed as the semantic component of a global standardisation framework for digital governments. A content description model provides a repository of structured knowledge aimed at supporting the semantic interoperability between sectors of Public Administration and the communication processes towards citizen. Specific conceptual models built from this base will act as a *cognitive interface* able to cope with specific digital government issues and to improve the interaction between citizen and Public Bodies. As a Case study, the representation of the click-on licences for re-using Public Sector Information is presented.

1 Introduction

The use of ontology-based methodologies has greatly expanded in recent years, and, as a consequence, the term 'ontology' has taken on a wide range of meanings. One of the distinctions that are most commonly accepted is that between *Semantic lexicons* (so called lightweight ontologies, Hirst G. 2003) and *Formal ontologies*. On the strength of our own experience, we have developed a legal semantic lexicon (Jur-WordNet) [Miller 1995, Sagri 2003] that is structured according to taxonomy and semantic relationships based on linguistic rules; the high level concepts of JurWord-Net have been framed and organised via a *Core Legal Ontology* (CLO) in order to remove terminological and conceptual ambiguities [Gangemi, Sagri and Tiscornia 2003]. At this second, more complex, level, *formal foundational ontologies* provide a powerful and logically sound base, because Core Ontology requires that cognitive assumptions underlying the meaning of concepts are made explicit and formally defined.

In this paper we describe an application of the lexical resource JurWordNet and of CLO as a *descriptive vocabulary* for modeling legal domains specific to the-Government issues. In the domain of the AI & LAW applications the two main streams of interest in the civil law countries are legal *advice* and *norm comparison* [Boer, Van Engers and Winkels R 2003, Breuker and, Winkels 2003]. We have elsewhere [Gangemi, Prisco et alii 2003] examined the use of ontology-based models in the light of norm comparison, and of normative conflicts handling. Here we will consider a third perspective, that is the creation of a *cognitive interface* [Borges et alii, 2001] for the description of legal knowledge, able to improve interaction between citizen and Public Bodies.

S. Wang et al. (Eds.): ER Workshops 2004, LNCS 3289, pp. 577–588, 2004.

The paper is structured as follows:

Sect. 1 introduces the basic theoretical framework underlying the ontology-based methodology and the components of this model. Some of the main structural aspects of the methodology will be described:

- on lexical level, the JurWordNet resource is a bridge between technical and common language and it allows multilingual access; moreover, at conceptual level, it lexicalizes the ontological entities;
- on ontological level, the Legal World interpretation according to the basic assumptions of the DOLCE+ foundational Ontology and the main classes of concepts in the Core Legal Ontology are described.

Sect. 2 describes an application of the ontology-based model for the building up of knowledge-based systems. In this context, the representation of licences for public sector information handling is sketched.

1.1 JurWordNet

JurWordNet is a formal ontology-based extension to the legal area of the Italian part of the *EuroWordNet* initiative[1]. As is the case for other WordNets[2] this is relevant to the class of computational lexicons that aim at making word content machine-understandable via the highly structured semantic representation of concepts. These are represented by *synsets,* a set of all the terms expressing the same conceptual area (*house, home, dwelling domicile...*) linked by a semantic relation of meaning equivalence. Semantic equivalences are limited (variants) in many terminology lexicons such as the legal one, which has a plethora of technical terms and where synonyms are rare. Conversely, it is important to create equivalence relations with normal language in order to make up for the imprecision of non-experts when searching for legal information, and to use common language terms instead of legal ones. Apart from having taxonomic vertical relations, the synsets of the law lexicon also have 17 associative horizontal relations based on the notions of meronymy, synonymy, and role.

One of the most interesting functions of the wordnet methodology is the distinction of meanings in polysemic terms, both within the domain and in relation with common language. Often, sense distinctions do not just concern language but also the differences in reality perception: for instance there is a need to separate within a concept the role played as opposed to the existence of a tangible physical entity. The entry *President of the Republic* indicates the physical person (referring to space and dimension), the constitutional body, and the holder of the state function.

The criteria followed to organise the concepts require, therefore, assumptions that are external to the language. These assumptions must be explicit so that the user is aware of the perspective according to which concepts are differentiated. This is the

---

[1] Currently, the Italian language coverage offered by IWN amounts to 50,000 terms (www.ilc.cnr.it); specialised sectors dealing with specific areas, e.g. EcoWordNet for economic/financial language; Euroterm is an extension of Eurowordnet with Public Sector Terminology funded by EC in the E-content Program. (www.ceid.upatras.gr/en/index.htm.)

[2] Since its initial release by Princeton University, WordNet has always been regarded as one of the most important resources in the NLP community (about 400 papers have been published on the subject).

role of ontology. This process also allows mapping terms between different languages. This is particularly effective in the legal field where corresponding terms are often absent in different languages but are present in concepts and legal systems. Shifting emphasis from the linguistic expression to content allows comparing concepts through properties and metaproperties, and to assess not only whether the concept itself occurs in different contexts, but also how the concept is processed in different regulatory structures. The project *LOIS,* funded by the European e-Content program[3], will extend the Italian legal network to five European languages (English, German, Portuguese, Czech, and Italian, linked by English).

The localisation methodology [4] is based on the automatic junction between already existing lexicons. The basic premise is that semantic connections between the concepts of a language can be mapped through the relationship between equivalent concepts in another language. This procedure serves to test what is covered by the lexicon with respect to the domain and provides an initial base of conceptual equivalents. From the first results of this intersection with the lexicon of EU laws (via the Eurodicautom[5] database) it was evident that out of the 2000 synsets of the Italian law lexicon 800 could be found in the German, 470 in the Dutch, 490 in the Portuguese and 580 in the English. The intersection with the Princeton WordNet showed 600 JurWordNet synsets in the English lexicon, and these were classified as legal terms.

2 Our Description Model: Types of Entities in the Legal World in DOLCE e D&S

The categories that bring together the top level of JurWordNet's taxonomical trees are the basic legal entities, which are held to be common to all the legal systems. We can give them a minimum set of properties shared by all the specific meanings of each system and/or language. They make up a *Core Ontology for law (CLO).* CLO is a specialisation of DOLCE (Descriptive Ontology for Linguistic and Cognitive Engineering) foundational Ontology [Gangemi et alii, 2002]. The four basic categories of DOLCE are *endurant* (including object- or substance-like entities, either physical or non-physical), *perdurant* (including state- or process-like entities), *quality* and *region* (including dimensional spaces of attributes such as time, geographical space, colour, etc.). DOLCE includes several primitive relations, such as *part, connection, constituency, and inherence* of qualities in entities, *participation* of endurants in perdurants, etc We refer to DOLCE documentation for a full description of DOLCE top categories.

In DOLCE extended to D&S a new top-category *situation* has been added. The Description and Situation Ontology is an extension of DOLCE aimed at providing a theory that supports a first order manipulation of theories and models. The basic assumption in D&S is that the cognitive structure emerging in cognitive processes refers

---

3 The project started on the 1st of March 2004
4 Amongst others, see the MultiWordNet project
   http://tcc.itc.it/projects/multiwordnet/multiwordnet.php
5 Eurodicautom is an aid for translators created by the European Commission
   http://europa.eu.int/eurodicautom/Controller

to high level descriptive structures: any State of Affairs becomes a Situation according to a possible Description of it "A Description is disjoint from situation. A description may be satisfied by a State of Affair. The satisfaction relation is reified in D&S as a first-order *referenced-by relation*. A description satisfied by a SOA is an s-description. A SOA satisfying a description is a situation". [Guarino and Mika, 2003]

According to this approach, the legal world is conceived as a representation, or a description of reality, an ideal view of the behaviour of a social group, according to a system of rules that is generally accepted and acknowledged. In the DOLCE+ D&S distinction, *descriptions* (in this domain legal descriptions, or conceptualisations) encompass laws, norms, regulations, crime types, etc., and *situations* (legal facts or cases) encompass legal states of affairs, non-legal states of affairs that are relevant to the right, and purely juridical states of affairs.

A norm is a legal *description* composed of *legal roles*, *legal courses* of events, and *legal parameters* on entities that result to be bound to *the setting* created by *a legal case*. The *satisfaction* relation holding between *legal descriptions* and *cases* is the reified counterpart of the semantic satisfiability relation: a legal description (the content of a norm, a regulation, a decision, etc.) is assumed to be the reification D_T of a (potentially formalized) theory T, while a legal case C_S is assumed to be the reification of a state of affairs S that can satisfy T.

According to the class they pertain to, norms may have *parts* and *components* that are the representation of:

– Legal functional roles (constitutive norms)
– Institutional agents (constitutive norms)
– Institutional powers (power-conferring norms)
– Behaviours (regulative norms)
– Incrimination acts as *legal courses*(incriminating norms)
– Cognitive states (presumptions).

Legal Roles are a sub-set of social roles, played by either physical or non-physical objects. Social Roles [Masolo et alii, 2004] have a relational nature, are anti-rigid[6] and are linked to *contexts* (in legal term, a *constitutive norm*, a *Description* created by intentional agents). Among legal roles, *legal subjects* and *legal assets* constitute the basic entities of the legal world. Legal-subject is an *agentive* legal role, while legal asset is *non-agentive*.

Institutional Agents, or Legally-constructed–Institutions pertain to Social Individuals; like Social Concepts, are defined by Descriptions, but, unlike roles, are rigid and agentive and can be classes (e.g. organisations, public bodies.) or instances; in many cases the same description defines both a class of Individuals (e.g. Ministry) and a Role (Minister) as a *representative* of the individual.

Modal Descriptions are *proper parts* of regulative norms that contain some *modality target* relation between legal roles (legal agents involved in the norm) and legal courses of events (descriptions of actions to be executed according to the norm).

Legal Information Objects *depend on* agents' cognitive objects and on mental process and can by *represented by legal* descriptions.

---

[6] Rigidity is a meta-property of particulars in Dolce [Guarino 2004].

Legal Cognitive Objects are internal descriptions, (e.g. *agreement* and *mistake*), which *descriptively-depend-on* information objects, are *participant-in* mental processes or *constituent-of* cognitive states. Cognitive objects have a *one-sided- specific-dependence* on agentive physical objects (e.g. a natural person).

Among **cognitive states** (that are *perdurants*), intentionality is subsumed by will, which is subsumed by consciousness:

Legal Facts (*cases*) are situations satisfying norms (only facts relevant for legal systems are legal facts). Subclasses or atomic components of legal facts can also be defined in terms of perdurant entities, such as:

- *Natural facts* (e.g. death) which are independent of human actions pertains to phenomenon
- *Human facts, depending on* consciousness (but not on will), are accomplishment; Among these:
- *Legal acts* (in a strict sense) *depending on* will
- *Legal transactions, depending on* intentionality.
- *Institutional facts:* are the functional counterparts of brute facts, legally constituted by (*satisfying*) constitutive rules.

Among **qualities** *inherent to* Legal transactions are temporal qualities as duration (quality region are deferment, expiration, term, etc) and the *validity-assessment quality* (valid, void, voidable).

3   The Regulative Domain: European Norms About the Re-use and Commercial Exploitation of PSI

We choose as a case study a specific topic in the domain of copyright Management, dealing with the re-use of information produced by Pubic Bodies. The reason of this choice is because Public sector information (PSI) has both a social and economic dimension and is one of the most important components of public services. In carrying out its tasks the public sector collects, collates, creates, stores and disseminates huge quantities of information: financial and business information, legal and administrative information, geographical, traffic, tourist information etc.

As part of the Action Plan for the Information Society, on the 24 October 2003 the European Commission has adopted a Directive on the exploitation of public sector information aimed at achieving a basic set of common rules in the European Community that at the same time do not or only minimally affect current public sector workloads and budgets[7].

The Directive aims at ensuring that in relation to the re-use of PSI the same basic conditions apply to all players in the European information market, that more transparency is achieved on the conditions for re-use and that unjustified market distortions are removed.

---

[7] Directive 2003/98/EC of the European Parliament and of the Council of 17 November 2003 on the re-use of public sector information, Official Journal L 345, 31/12/2003 P. 0090 – 0096. *http://europa.eu.int/eur-lex/pri/en/oj/dat/2003/l_345/l_34520031231en00900096.pdf*

According to the Directive[8], "Public sector bodies should be encouraged to make available for re-use any documents held by them The Directive should apply to documents that are made accessible for re-use when public sector bodies license, sell, disseminate, exchange or give out information, that is produced and charged for exclusively on a commercial basis and in competition with others in the market"[9].

Art. 1, sub-sects 2, 3,4 constraint the exercise of re-use, stating that it shall not affect the Access right of citizen[10], the personal data Protection[11] and the Intellectual Property Right[12].

The Directive sets some rules for the re-use of PSI: "In some cases the re-use of documents will take place without a license being agreed. In other cases a license will be issued imposing conditions on the re-use by the licensee dealing with issues such as liability, the proper use of documents, guaranteeing non-alteration and the acknowledgement of source. If public sector bodies license documents for re-use, the license conditions should be fair and transparent. Standard licenses that are available online may also play an important role in this respect."[13]. As a condition for licensing, Public sector bodies should respect competition rules when establishing the principles for re-use of documents avoiding exclusive agreements as far as possible.

3.1 Modeling a License

The UK Government has already started the process of implementing the EU rules. During 2003 a Consultation on a Partial Regulatory Impact Assessment (RIA[14]) on the Proposal for a Directive was held, considering the ways in which ensure that the public sector complies with the measures set out in the Directive. Moreover, the Her Majesty's Stationery Office (HMSO[15]), who is responsible of managing and licensing the re-use of Crown Copyright material, launched online *Click -Use Licence*. There are

---

[8] Art. 3, General prevision of the Directive 2003/98/EC: *"Member States shall ensure that, where the re-use of documents held by public sector bodies is allowed, these documents shall be re-usable for commercial or non-commercial purposes in accordance with the condition set out in Chapters III and IV".*

[9] Premises of the Directive.

[10] Art.1, sub-sect.3.: *"This Directive builds on and is without prejudice to the existing access regimes in the Member States.." .*

[11] Art., sub-sect.4.: *"This Directive leaves intact and in no way affects the level of protection of individuals with regard to the processing of personal data)".*

[12] Art.1, sub-sect.2.: *"This Directive does not apply to documents covered by industrial property rights, such as patents, registered designs and trademarks. The Directive does not affect the existence or ownership of intellectual property rights of public sector bodies, nor does it limit the exercise of these rights in any way beyond the boundaries set by this Directive. The obligations imposed by this Directive should apply only insofar as they are compatible with the provisions of international agreements on the protection of intellectual property rights".*

[13] Art. 8: **Licences** sub-sect.1 *"Public sector bodies may allow for re-use of documents without conditions or may impose conditions, where appropriate through a licence, dealing with relevant issues".*

[14] http://www.oqps.gov.uk/copyright/ria_consultation_03_archive.htm

[15] http://www.hmso.gov.uk/copyright/licences/click-use-home.htm

currently licences, which allow unrestricted use of *core* government information under licence. A new phase will extend the Click-Use approach to *value added licences* where fees are charged and collected online. Thus, we choose UK licences as a subject for testing the modeling framework[16].

Conceptualisations of IPR and of Digital Rights is a matter of great interest both from the ontological and the technical perspective [Dulong de Rosnay, 2003], [Delgado et alii, 2003]. Many international projects aim to define common models for the management of digital rights that involve both the substantial definition of normally accepted rules (I-Commons[17]) as well as the definition of languages and models for the digitalisation of the rules governing the use of digital information and web resources a (XrML[18], ODRL-Open Digital Rights Language[19]).

According to a common shared interpretation, a Licence is an Authorisation: it is a container of explicit permission to exercise rights. *Rights* set actions that an actor can exercise on a resource which can include *constraints* (limits), *conditions* (exceptions that expire permissions) and *requirements* (obligations that must be met before permissions can be exercised).

According to the D&S model, a licence is *composed by* a set of Descriptions stating permissions about use and re-use; *constraints, conditions* and *requirements* are expressed in term of S-Descriptions and M-Descriptions and are parts of Descriptions (Descriptions have only other descriptions as *parts).*

Descriptions are composed of:

- *Endurant*: activities set by the norms, as *see, copy, re-use, redistribute, republish;* such actions affect a non-physical entity (information*)* and are *sequenced by* a legal *course* of events (what is permitted or forbidden to do)
- *Perdurant* entities involved in actions are *Individual, Group* or *Organisation*:
- Agents play two categories of *roles: rights holders* and *users.* The role of *Rights holder* (who grants permissions) is played by Public Bodies as producers, by *Custodian* (who has the managing power over the resource, e.g. assignment and maintenance of access control markings.) and *by Representative.* Sub-classes of Users (who access electronic or digital versions of the Products) are *End-users* and *Re-user.*
- The role of *Author* is *a requirement for* [20] the role of *Rights holder. Author is defined by* a Description stating the existence of the right (Copyright Statement).

Legal parameters (qualities) are *requisites* for *roles* and *courses* (e.g. type and format of data can be a requisite for its delivery). Specific parameters for digital right management are *Accessibility* (whether particular users *will be able* to access or use the resource) and *Disclosability*: as a general rule, users are *allowed* to access data *disclosable* only. Material that is covered by security classification, legal or policy restrictions is excluded. More specifically, *Disclosability* is concurrently constrained

---

[16] Entities descriptions are based on the UK Metada Standard Framework.

[17] http://creativecommons.org/projects/international/

[18] http://www.xrml.org/

[19] http://odrl.net/

[20] Guarino et alii, 2004:"The requirement relation probably correspond to an often mentioned feature of reloes, coined as roles can play roles...this kind of double role-playng can be a consequence of the definition of one of the roles and therefore consitutes a case of requirement ".

by *DPA* Data Protection Act, Freedom of Information Act (FOIA), and Environmental Information Regulations. Declaration of Disclosability is subject to review. Therefore, time of the disclosability review is a requisite for stating it as *disclosable*.

– *Region* entities in a case setting must be values for some *legal parameter* (e.g. quantity of data required, The date of the formal decision regarding the disclosability review).
– *Legal roles* have a *modal target* in a *course* of events (e.g. citizens are allowed to access legal information and are forbidden to access *undisclosable* data).

The general rule states that re-use of disclosable information only is permitted. The requirement is expressed in our model, by a Modalized-Description where Endurant *play* roles of *re-user*, *accessing/handling PSI* is a course modalized as *permitted* and *undisclosable* is value for parameter *Disclosability,* as a *requisite for* accessing (fig.2):

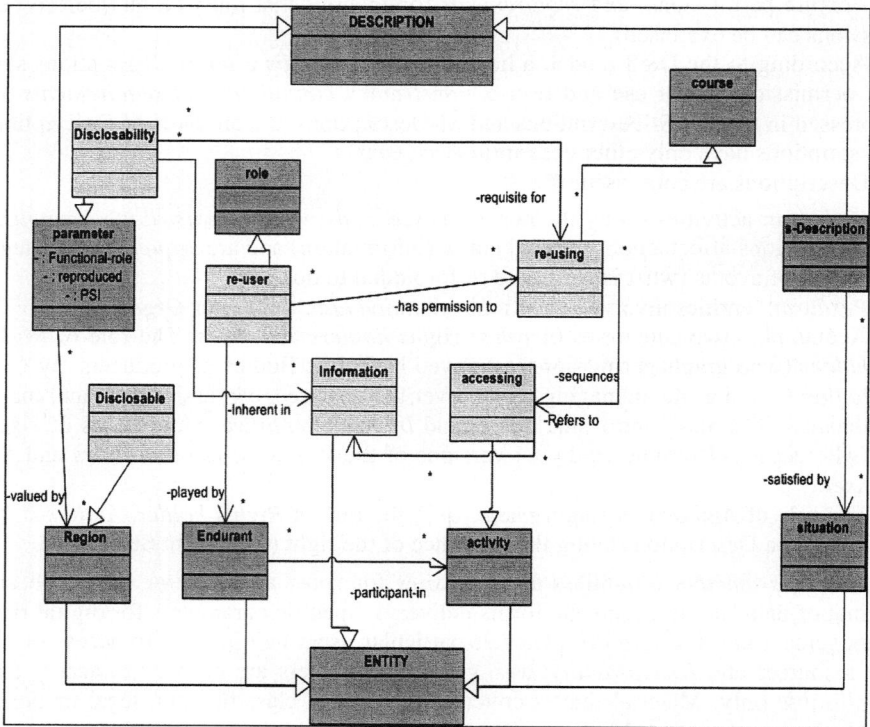

Fig. 1. Access to public information: general rule

As specific cases, three situations occur:
CASE 1) this covers material where Crown copyright is asserted[21,] but waived. Waiver material can be re-used free of charge without requiring a formal licence provided that it is:

---

[21] Public Sectore Information are subject in UK to Crown or Parlament Copyright .

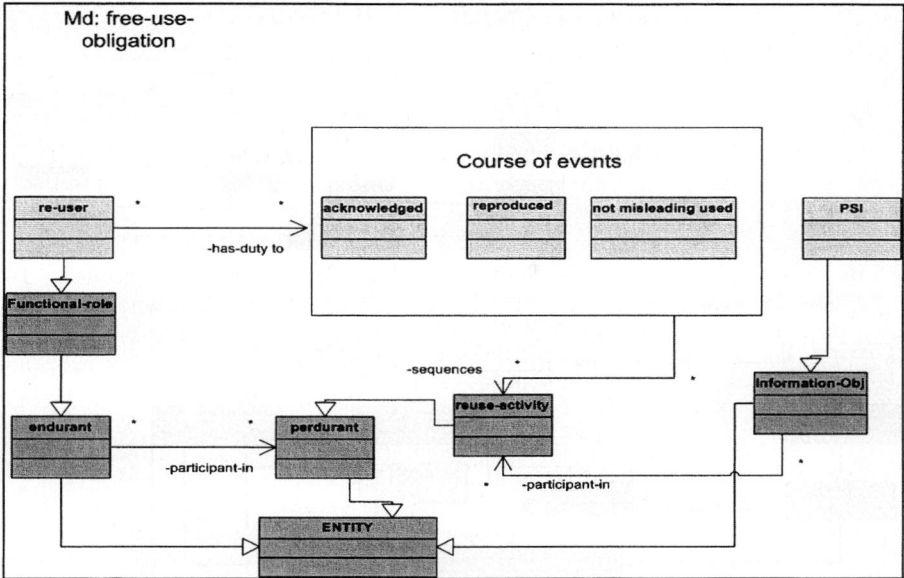

Fig. 2. Free-use-obligation

- acknowledged
- not used in a misleading way
- reproduced accurately and kept up to date

In *M-Description free-use-obligation* (fig. 3), re-user *has-duty to* acknowledge, reproduce accurately, up-date PSI (course).

The condition is part-of M-Description *md: free-use* :

CASE 2) A *Core Licence* covers core material, which is likely to satisfy the following conditions:
- It is essential to the business of government;
- It explains government policy;
- It is the only source of the information[22];

Among Obligations, re-user must:
- reproduce Material accurately from the current Official Source;
- identify the source of the Material and feature the following copyright statement if you publish the Material;
- Not use the Material for the principal purpose of advertising or promoting a particular product or service, etc.

Requirements are expressed by *parameters* which affect the quality inherent to public information, for instance the quality *exclusiveness* must have positive value; Conditions are expressed by M-Description (*Md: core-licence-condition*) constraining the *task* of reusing activity. It is *part-of* a *Md Core-licence.*

---

[22] Only part of the actual requirements and conditions has been listed.

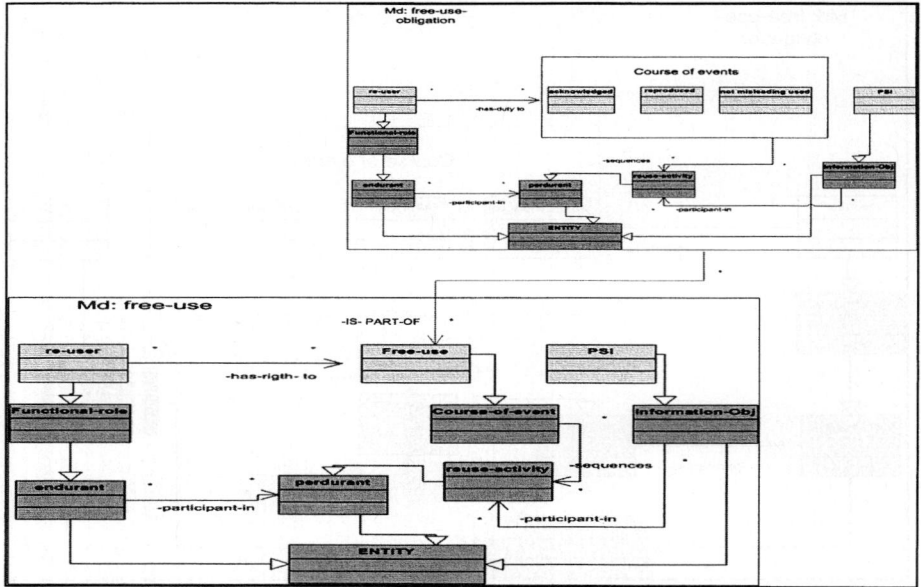

Fig. 3. Free-use

CASE 3) *Value added* material will usually satisfy the following conditions:
– It will bring together information from a variety of sources, Value will often be added by means of commentary, analysis, indexing, search facilities or text retrieval software
– There may be similar competing commercial services and products in the marketplace;
– Its creation is not vital to the workings of government. There will often be alternative suppliers of such information. Etc....

Among obligations re-user ought:
– to identify the source of the Material.... ecc.
and, in Specific to Royalty licences :
– to keep full and accurate records of the sales of your Product; ecc.

Here a specific condition: *fee or royalty payment* holds as a course which sets the reuse-activity, component of a M-description *Valued-added-condition*; requisite is a *parameter* valued by *kinds* and *quantity* of sales.

4 Conclusion and Future Work

This is a preliminary result of a project to develop domain ontology using entities defined in a Core Legal Ontology. The test is meant to assess the validity and coverage of ontology, the expressivity of the representation framework and the reliability of the model. We consider the work still in progress as further aspects need to be refined; among others:

1) the distinction of kinds of knowledge in term of the role the play within a system of norms; ontological features of constitutive and regulatory norms is clearly expressed in the norm structure, but the epistemologically different roles (the descriptive role of constitutive definitions and the prescriptive role of the norms imposing conditions and constraints) need a specialization of the *part-of relation*. Same ambiguity lays among properties that are *parameters* for roles and juridical effects, and conditions upon which effects are legally valid: the former meant as 'practical' requisites of an action (without which the action could not be carried out), and the latter as juridical requisites (without which the action is not validly carried out).

2) linking the domain ontology to the lexical level of JurWordNet: there is a two-way relation between semantic lexicon and ontology: "*it is possible that a lexicon with a semantic hierarchy might serve as the basis for a useful ontology, and an ontology may serve as a grounding for a lexicon. This may be so in particular domains, in which vocabulary and ontology are more closely tied than in more general domains*" [Hirst2004]. Up to now, we have tested the latter, while the former approach seems to offer interesting results in the field of cognitive interfaces [Borges et alii 2003].

3) testing the modelling framework in close domains, as those of digital rights and privacy regulations.

References

Boer A., Van Engers T. and Winkels R., *Using Ontologies for Comparing and Harmonizing Legislation*, in Proceedings of the 9 th ICAIL Conference, Edinburgh, 2003.

Borges F., Bourcier D., Andreewsky E. and Borges R.*, Conception of cognitive Interfaces for legal knowledge*, in Proceedings of ICAIL 2001, ACM Press.

Breuker J, and Winkels R, "*Use and reuse of legal ontologies in knowledge engineering and information management*", ICAIL03 Wks on Legal Ontologies and Web-based Information, Management, Edinburgh, Http://lri.jur.uva.nl/~winkels/legontICAIL2003.html, 2003.

Delgado J., Gallego I., Llorente S., García R., *IPROnto: An Ontology for Digital Rights Management*, in Proceedings of Jurix 2003, IOS Press.

Dulong de Rosnay M., *Cognitive Interface for Legal Expressions Description- Application to Copyrighted Works Online Sharing and Transactions*, in Proceedings of Jurix 2003, IOS Press.

Gangemi A. Prisco A, Sagri M-T., Steve G., Tiscornia D.,*Some ontological tools to support legal regulatory compliance, with a case study,* Proceedings of the second International Workshop on Regulatory Ontologies WORM 2004 -Part of the International Federated Conferences (OTM '04), Springer LNCS

Gangemi A., Mika P. 2003, *Understanding the Semantic Web through Descriptions and Situation,* Meersman R, et al. (eds.), Proceedings of ODBASE03, Springer, Berlin, 2003.

Gangemi A., Sagri M.T., Tiscornia D., *Metadata for Content Description in Legal Information*, Workshop Legal Ontologies, ICAIL 2003, Edinburgh. In press for Journal of Artificial Intelligence and Law, Kluwer.

Gangemi A., Guarino N., Masolo C., Oltramari, A., Schneider L. 2002. Sweetening Ontologies with DOLCE. in *Proceedings of EKAW 2002*, Siguenza, Spain, pp 166-178.

Hart H.L.A. 1961, *The Concept of Law*, Oxford(UK):Clarendon Press.

Hirst G., " Ontology and the lexicon " In Staab, Steffen and Studer, Rudi (editors) Handbook on Ontologies in Information Systems, Berlin: Springer, 2003, p.14.

Masolo C., Vieu L., Bottazzi E., Catenacci C., Ferrario R., Gangemi A., Guarino N., *Social Roles and their Descriptions, in Proceedings of* KR 2004

Masolo C., Borgo S., Gangemi A, Guarino N, Oltramari A, Schneider L, "The WonderWeb Library of Foundational Ontologies", IST Project 2001-33052 Wonder Web. http://wonderweb.semanticweb.org/deliverables/documents/D18.pdf, 2003

Miller G., *WordNet: A lexical database for English*, in Communications of the ACM 38(11), 1995, pp. 39-41.

Sagri M.T., 2003, *Progetto per lo sviluppo di una rete lessicale giuridica on line attraverso la specializzazione di ItalWornet*, in Informatica e Diritto, ESI, Napoli, 2003.

Valente A. 1995, *Legal Knowledge Engineering, A modelling Approach*, Amsterdam: IOS Press, 1995

Aspects of Regulatory and Legal Implications on e-Voting

Athanassios Kosmopoulos

Managing Authority fort the Operational Program "Information Society"
Ministry of National Economy, 105 57 Athens, Greece
kosm@aegean.gr
http://www.infosoc.gr

Abstract. This paper addresses the democracy-oriented regulatory and legal requirements that e-democracy impacts. The short term perspective of the questions put before the electorate obliterate the long term perspective in which many policy problems have to be seen. A well-designed e-voting system should produce an audit trail that is even stronger than that of conventional systems (including paper-based systems). Remote Internet voting systems pose significant risk to the integrity of the voting process, and should not be fielded for use in public elections until substantial technical and social science issues are addressed. Conclusively the paper focuses on the specific attributes an electronic voting (polling place) system should respect and ensure such as transparency, verifiability, accountability, security and accuracy in relation to the constitutional requirements such as General, Free, Equal, Secret, Direct and Democratic.

1 Introduction

The aim of this paper is to discuss whether an e-voting scheme could meet the constitutional and other legal requirements, as these are laid down in the international legal and regulatory framework. The significance of the issues addressed herein is clearly manifested by the volume of debate that lately has begun on them, in many countries over the globe. The most powerful and politically significant aspect of new technologies is for allowing people to collaborate and self-organize: not simply the ability to reorganize the relationships between governments and citizens, but to create new opportunities for citizens to organize them selves. Recent reports (i.e. CalTech-MIT Report, California Internet Voting Task Force, IPI National Workshop on Internet voting, European Union IST project, SERVE project in the USA etc.) describe the capabilities of e-voting systems, and at the same time identify their limitations, the risks and vulnerabilities they are exposed to, as well as the social concerns such systems give birth to. Despite the large volume of material published to support this debate, including several user requirements specifications, to the best of our knowledge no consolidated view on the requirements deriving from constitutional and legal consideration is available. These requirements and needs describe and identify the reflecting technical requirements that a voting system should comply with. This is the main and original contribution of the paper. The paper is structured as follows: In

S. Wang et al. (Eds.): ER Workshops 2004, LNCS 3289, pp. 589–600, 2004.
© Springer-Verlag Berlin Heidelberg 2004

section 2, we exhibit the definition of e-voting with respect to voting technologies and processes. Section 3 presents the main election principles. Section 4 discusses in depth details of the legal and constitutional requirements an e-voting system should respect, stemming from the democratic nature of the election process. Section 5 demonstrates the most important trends of direct democracy. Finally, section 6 summarizes our conclusions.

2   e-Voting Definitions

For the purpose of this paper we define e-voting as the use of a digital or analogue device, within a secure, authenticated environment, to cast a vote during an election process. We will consider both "voter present" and "voter non present e-voting" denoting whether the voter is physically present in the polling booth or remote from it. Electronic voting (e-voting) uses digital data to capture the voter selection. With Internet voting (I-voting) the voter casts his vote remotely via the Internet.

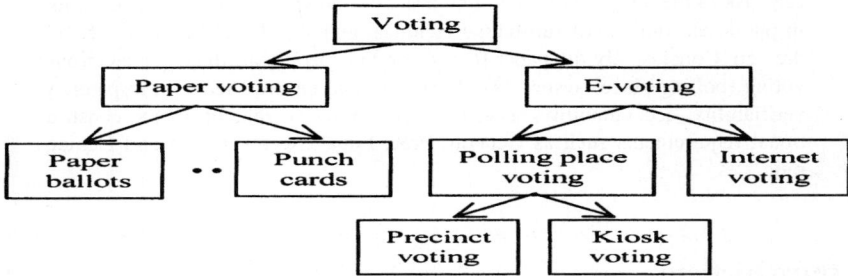

Fig. 1. Types of Voting

2.1   Voting Technologies and Processes

There are five broad classes of voting technologies in use today [1] throughout the world: Hand counted paper ballots, lever voting machines, punched card ballots, optical mark-sense ballots, direct-recording electronic voting machines

The most common and traditional voting process is voting at the poll site on Election Day. However, there are several alternative methods, including: Absentee ballots: Is the provision for the use of absentee ballots, which allow people to vote-by-mail before the election..Vote-by-mail: According to the Dutch law, only voters who reside outside the Netherlands, or live outside the country due to their professional activities, or because of their marital partners or their closest relatives are allowed to vote by mail. In Germany postal voting is allowed if only if the eligible voter applies for this option on important grounds as for illness, absence which prevents him from voting in his electoral district during voting day and hours. Oregon is the first, and so far only, state in the USA to adopt all mail voting [2]. Recently the significant report EVE [3] has shown that Internet voting is mostly being considered by countries that have already implemented changes regarding the polling methods, such as: placing electronic ballot boxes in polling stations, introducing postal voting, using the Internet as political campaign tool. Remote voting on election day is sometimes described as the vote anywhere model. In this model voters may use any polling place, however it was eliminated by its high susceptibility to fraud, nevertheless the combination of smart cards, biometrics and highly available on line voting systems may allow such a system to re-emerge. The arguments of the California Internet Voting Task Force against Internet voting on election day apply to most versions of the vote anywhere model so there is good reason to be skeptical about such systems [4]. In the context of our analysis we make a distinction between two types of e-voting: polling place voting and Internet voting and furthermore we distinguish two types of polling place voting, namely precinct voting and kiosk voting. In this paper we will focus mainly in polling place e-voting and not particularly in Internet voting. (see fig.1) In polling place voting, both the voting clients (voting machines) and the physical environment are supervised by authorized entities (closed regulated environment). Depending on the type of polling place (precinct or kiosk), validation may be either physical (e.g. by election officials) or electronic (with the use of digital identification). Casting and tallying of votes are electronic: the voting clients may be Direct Recording Electronic devices (DRE's) or they may send their tallies electronically to a central site (e.g. by using, a dedicated line or an ATM network).

3   Election Principles

Generally speaking, each election involves four distinct stages: Registration :Prior to the election, voters have to prove their identity and eligibility. An electoral roll is created. Validation: During the election, voters are authenticated before casting their vote. Only one vote per voter is authorized. Casting (Voting): Voters cast their vote. Tallying: At the end of the voting period, all votes are counted. Each of the above stages can take place by using *physical* or *electronic* procedures. It is generally accepted that parliamentary elections have to be free, equal and secret. At the same time, the election procedure has to be transparent and subject to public scrutiny. The constitutions of the main European Union member states demand that the parliamentary elections must be General, Free, Equal, Secret, and Direct. Adding up to the aforementioned elements the essential requirement of Democracy, and analyzing these requirements to the next level of detail we get hold of the first-level legal and regulatory e-voting requirements, which are summarized as follows [5] :

4   Analysis of Election Principles

4.1   General

Universal involvement is a basic principle. No one can be – directly or indirectly – excluded or discriminated. The main consequence deriving from the principle of

Table 1. Legal and regulatory e-voting requirements

1 General
1.1 Universal opportunity to participate
1.2 Eligibility (registration and identification)
2 Free
2.1 Uncoercibility
2.2 Eligibility (registration and identification)
3 Equal
3.1 Equality of candidates
3.2 Equality of voters
3.3 One voter – one vote
4 Secret
4.1 Secrecy
4.2 Balance security vs. transparency
5 Direct
5.1 Not monitored ballot recording and counting
6 Democratic
6.1 Trust and transparency
6.2 Verifiability and accountability

general elections is that every voter has the right to participate in an election process while the ability to participate to this process (eligibility) must be founded on the law and should be regulated according to the law. Furthermore, voting possibilities and technologies should be accessible by every voter, whilst considering the lack of necessary infrastructure and the digital divide, e-voting should be considered only as an alternative complementary way of exercising one's voting rights. The principle of universality requires that every eligible voter should be included in the election process, and therefore this principle results in the necessity for publicly available appropriate infrastructure (e.g. public internet kiosks, voting equipment in closed regulated areas such as government offices, etc.), in order to allow all citizens to exercise their voting rights. A critical question is whether the participation in the election through e-voting should be subject to the proof of special conditions as is the case with postal voting we described above. Adopting an e-vote capability as an exceptional one (i.e. on the ground of the proof of a special condition, which prevents the eligible voter from physically casting his/her vote), is - from the legal point of view - a legally and constitutionally "safe" choice. In opposition to this opinion, emanating from the historical and legal basis that voting in a physical voting station constitutes the rule, the following argument may be articulated: the evolution towards an information society has an improved impact on the ability of a citizen to exercise his/her rights and liberties. Having in mind the political decision to improve e-government and e-democracy, the introduction of an e-voting capability should be viewed as an institutionally equivalent and not as an exceptional and complementary option. Eligibility initially, can be guaranteed through the registration of voters, who meet the specific requirements of eligibility, and on a second phase through the identification of the citizens at the moment of registration. (Secure) Registration and authentication (identification) are the means to ensure that the principle of universal suffrage is being respected and that elections cannot be influenced. The purpose of voters' registers is to guarantee that only people eligible by law to vote can do so, and that no one can vote more than once.E-voting is, in a functional way, analogous to postal (absentee) voting. Where such a voting capability is introduced, a proper authorization or registration process is usually required. Such a procedure does not affect the principle of general elections for the following reasons: Supposing that there is no country-wide, online voter register, a pre-registration for e-voting is necessary in order to avoid vote fraud.. Such a registration supports the integrity of elections. For the same reason, an Internet-based voter registration system is not recommended because it could be vulnerable to large scale and automated vote fraud (i.e. the potential of undetected registration of large numbers of fraudulent voters) [6]. E-voting is considered as an alternative capability, which may facilitate the participation of the voters.A voter registration system must meet five standards [7]. First, registration information must be accurate and complete. The information on the voter registration rolls must cover all registered voters .Second, registration information must be immune from fraud. Third, registration information must be dynamic and up-to-date. Voter registration must be flexible to accommodate frequent moves made by previous voters, the addition of new voters, and late voter registrations A significant challenge is developing a fraud resistant system for last-minute registrations, including Election Day registration. Fourth, registration information must be usable by the election officials at the polling places. Fifth, it must be easy for voters to register. Registration should not be a burden to voters. Finally providing a secure identification and authentication scheme of eligible voters is a conditio sine qua non requirement for any public-election oriented e-voting system.

4.2 Free

The principle of free elections requires that the whole election process takes place without any violence, coercion, pressure, manipulative interference or other influences, exercised either by the state or by one or more individuals.

However, e-voting procedures in open non regulated environments (such as Internet) may indeed pose new threats to the freedom and integrity of voters' decision, beyond those that postal voting does. This becomes obvious in the workplace: even if the employer is not standing over the shoulder of the employee-voter intranets, system administrators may monitor or record the activity at each workstation and obtain a copy of the ballot. Moreover, the distributed nature of the Internet could facilitate large-scale vote selling or trading [8]. We suggest, as we will explain later, that the best solutions are voting systems exhibiting a "voter-verifiable audit trail," where a computerized voting system might print a paper ballot that can be read and verified by the voter [9]. However this paper "receipt" must be placed in a sealed box by the voter at the designated polling place, in order to prevent vote selling, intimidation or other coercion instances. These paper receipts may be used as well for additional recounts if such a need occurs.Uncoercibility and prevention of vote buying and extortion can be ensured by an e-voting system designed so that no voter can prove that he/she voted in a particular way (untraceability on the part of the voter). A possible solution is to develop a public accessible infrastructure, in closed regulated physical sites, thus allowing voters to exercise their rights free of the coercion of any third party.

The requirement of equality constitutes one of the political cornerstones of modern democracies.A blank vote is defined as a vote where the voter does not designate any candidate. During a paper vote, blank votes are usually counted as cancelled votes and cannot be distinguished from invalid ballot papers.The question of knowing whether blank votes should be officially taken into consideration is, what's more, a matter for debate. For political elections in France [10], a recent bill was aimed at enabling blank votes to be precisely defined and to be distinguished from invalid ballot papers.

The democratic legitimatization of e-voting relies on satisfying the generic voting criteria of a democratic election system. This includes the free expression of the preferences of the voter, even through casting a non-valid or a "white" paper ballot. In order to preserve the freedom of voters' decision, the possibility for casting a consciously invalid vote should be provided and guaranteed.

4.3 Equal

The requirement of equality constitutes one of the political cornerstones of modern democracies.The principle of equal suffrage is identified mainly in equality regarding the candidates and the political parties who participate in the public elections.

A requirement deriving from the principle of equality is that electronic ballots should be created and displayed in a way similar and equivalent to that used for the paper ballots. Electoral equality requires that there are no deviations between the printed ballot and its electronic equivalent. Furthermore, the placement of electronic ballots in the e-voting equipment (i.e. on the screen of the e-voting machine) should ensure equal accessibility. Thus, the structure and appearance of ballots should not favor or discriminate against any of the participating parties. Another aspect of equality among the parties to be elected is that a valid cast vote must not be altered or removed in the course of the voting process. Transparency should also be respected. All parties should have the opportunity for equal access to the elements of the voting procedure, in order to be able to establish its proper functioning. The other side of equality is the one regarding the voting rights of each voter. In view of the current technological and societal evolution, the right to "equal accessibility to the voting process" must be extended to the right of "equal accessibility to election technology". An adequate, non-discriminating procedure should be offered to the voters, in order to allow them to efficiently exercise their voting rights with no obstructions. As a result, universal access may become a constitutionally indispensable requirement. Digital and technological divide is a major issue in this context. An e-voting system should ensure that the "one voter, one vote" principle is respected. In other words, the system should ensure that only eligible voters vote. Every voter can vote only once for the specific election, either online or off-line. Therefore, an e-voting system should be designed in such a way as to prevent: firstly the "duplicability" of the vote (either by the voter her/himself or by someone else); secondly the "reusability" of the vote (either by voting more than once online or by voting both online and offline) and finally the "alteration" of the cast vote (after a voter has dispatched her/his vote).

4.4 Secret

If Secrecy and freedom are strictly related principles: Secrecy is the prerequisite of the voter's free political decision. In democratic elections the connection between the vote and the voter must be unachievable, in order to ensure that votes are cast freely. Secrecy and anonymity of the ballot also provide important checks against coercion, against a person being forced, lured or intimidated into voting one way or another by others. The following requirements are resulting from the principle of secrecy:

The secrecy of the vote has to be guaranteed during the casting, transfer, reception, storing and tabulation of votes. None of the actors involved in the voting process (organizers, election officials, trusted third parties, voters etc) should be able to link a vote to an identifiable voter. Ballot secrecy should be reconciled with transparency and auditability of the entire voting process. This is the main difficulty, that is to say the election system must be able to allow the verification of the authenticity of the ballot before the votes are viewed or counted. In order to protect secrecy, the voted ballots should be decrypted and counted after the authentication information is reviewed and "removed". The e-voting system should be designed in such a way as to make vote control and recount technically feasible, without re-identifying the voters. Atomic verifiability is a weaker version of universal verifiability in which voters can only check their own votes and correct mistakes without sacrificing privacy. The later is useful when the cost of achieving universal verifiability outweighs its benefit.

4.5 Direct

The principle of direct election states that there can be no mediators in the process of voting decision. This principle may be well adapted to match an e-voting procedure. The appropriate requirement is that each and every online ballot is directly recorded and counted. A problem may arise in the case where the voting period differs with the voting procedure (online or off-line) used to cast the vote. Online voting results may influence the outcome of the entire election process and limit the integrity and legitimacy of the whole process. A suggestion is to develop a system that allows the recording and maintaining of the cast vote, while prohibiting any counting before the end of the (off-line) conventional voting period.

4.6 Democratic

A democratically designed and deployed e-voting procedure should, at least, exhibit the requirements related to the preservation of attributes and characteristics, such as the transparency, accountability, security, accuracy, legitimacy and to the democratic legitimization of the election system. Voters should be able to understand how the elections are conducted. The traditional voting "technology" operates in a way that is transparent as well as understandable to the voters and to the other election actors, since in most countries votes are counted in the presence of the parties' representatives. Online voting procedures are not transparent, because the average voter does not have the knowledge necessary to understand how the system works. As a result, in the case of e-voting, much more trust in the technology used and the persons involved (election officials, technology providers etc) will be required by the voters. An additional requirement is the accountability of the system, meant as the logging and monitoring of all operations related to e-voting. A Provocative Scenario [11]: A programmer at SlickVotingMachines Corp. adds malicious code to a DRE (Direct Recording Electronic device) machine for the California 2004 Presidential election, so that every fiftieth vote for a Republican candidate is changed to a vote for the corresponding Democratic candidate. This only happens when the machine is in "real" mode as opposed to "test" mode, so the election officials never discover the fraud during their testing. The electronic audit trail made by the DRE machine is also affected, so "recounts" never discover anything amiss. Simplicity and accessibility of a system are not merely technical issues. They require additional educational procedures, as well as organizational measures (help desks, e-election officials, etc.), to be effectively resolved. Based on the above principles, the following, functionality-oriented, requirements are consequential: First of all there must be trusted certification procedures for hardware and software, while the entire infrastructure (including source code), as well as every system functionality, must be logged. On the other hand all operations (authentication, vote recording, vote tabulation etc) should be monitored, while secrecy is preserved. At the same time, the infrastructure should be open to inspection by authorized bodies, as voters, parties and candidates must be ensured that there has been no malpractice. Finally adequate system security must be ensured whilst the system must be simple and user-friendly. Reliability and security requirements are based on the democratic need to ensure that the result of the election reflects correctly the voters' will. A reliable system should ensure that the outcome of the voting process corresponds to the votes cast, i.e. that it guarantees eligibility, secrecy, equality and integrity. The ballot that is stored to the vote counting equipment must be an accurate and unmodifiable copy of the voter's real choice (integrity). Moreover, it should be impossible both to eliminate a valid vote from the tabulation, and to validate a non-valid one. Security is a multidimensional notion in the context of e-voting. As far as security is concerned, on the ground of this analysis, regarding a specific e-voting system, it must cover globally the following attributes [12] that we highlight as a set of overlapping characteristics: Accuracy, also referred to as correctness means that no one can change anyone else's vote (inalterability), all valid votes are included in the final tally (completeness) and no invalid vote is included in the final tally (soundness), Democracy, is safeguarded if only eligible voters are allowed to vote (eligibility) and if each eligible voter can only cast a single vote (unreusability), Privacy, states that nobody should be able to link a voter's identity to his vote after the latter has been cast Robustness, guarantees that no reasonably sized coalition of voters or authorities (either benign or malicious) may disrupt the election. (it

should also be provided against external threats and attacks eg denial of service attacks etc),Verifiability, implies that there are mechanisms for auditing the election in
order to ensure that it has been properly conducted, Uncoercibility, Fairness, this
property ensures that no one can learn the outcome of the election before the an-
nouncement of the tally,Verifiable participation, often referred to as declarability,
ensures that it is possible to find out whether a particular voter actually has participated in the election by casting a ballot or not. Unavailability of the system (or of one
of its components) may result to loss of the capability of a voter to exercise his/her
fundamental political rights. Traditional voting systems are relatively simple. On the
contrary, e-voting systems are inevitably complicated; furthermore, they usually involve more actors than traditional systems do. From the point of view of the voters,
the system should be easy to use and should require no particular skills. As a result,
an e-voting system should be developed in such a way as to facilitate its usability and
to preserve its controllability.

5 Direct Democracy vs Representative Democracy: Main Issues

The main weaknesses of existing democratic arrangements in most countries are that
members of the representative assemblies represent partisan interests under the guise
of the general interest. Often they tend to follow only their own partial understanding
of what is good for their constituencies, and they are more responsive to the requirements of the political party they belong to, than to the citizens whose mandate they
have received [13]. The growing popularity of referenda, co-production of policies
and interactive policy-making, underlines that people prefer direct democratic arrangements for the existing representative arrangements. Conversely, major disadvantages on the subject of direct democracy are observed. More specifically: Direct democracy would lead to a single issue approach. Successive majorities on single issues
would lead to incompatible policies within and between sectors. The complexities of
policies require intermittent and iterative decision cycles, which are not feasible
through referenda. Most political problems cannot be reasonably approached with a
simple "yes" or "no", as opinion polls and referenda do. Besides, the short term perspective of the questions put before the electorate obliterate the long term perspective
in which many policy problems have to be seen. In this light "push button" democracy is considered fragmentary as well as showing a deficit compared to representative democracy. On the other hand the transparency of public administration in the
information society, which results from the development of ICT applications, as analysed above, forces us to re-conceptualize the democracy theory [14]. The Internet
certainly isn't a panacea, but does have the potential to bring together large numbers
of people in a form of civic dialogue. It can also provide immense stores of information for people to access and interact with. Importantly, if universal access is
achieved, it allows those with few resources to have equal opportunities for political
debate and involvement. The fundamental challenge of e-democracy is to improve
and develop representative democracy towards processes based on the empowerment
of citizens [15]. The new civilization brought about by ICT cannot and should not

ignore the principles and values of democracy. The introduction of an e-voting system must also conform to this rule. Voting is undoubtedly one of the functions "e-citizens" would like to see performed online. On the other hand, two items must be considered: the digital divide and the intrinsic distrust in an e-voting procedure [16], considering that while computer scientists, for the most part, have been warning of the perils of such action, vendors have forged ahead with their products, claiming increased security and reliability. Relations between members of the public and holders of political authority are being transformed. New expectations and meanings of `citizenship' are being entertained and occasionally acted upon. People often expect to be heard and heeded on more occasions and matters than the ballot boxes of Polling Day can settle. Electronic voting (e-voting), as we already mentioned, uses digital data to capture the voter selection. With Internet voting (I-voting) we also get remote connectivity via the Internet. A few Internet-based elections have already taken place [17], while pilot elections are scheduled in several countries. The most famous project is the SERVE voting system (Secure Electronic Registration and Voting Experiment), an Internet-based voting system being built for the U.S. Department of Defense's FVAP (Federal Voting Assistance Program). A very important report [18] was published according to which:".. because SERVE is an Internet- and PC-based system, it has numerous other fundamental security problems that leave it vulnerable to a variety of well-known cyber attacks (insider attacks, denial of service attacks, spoofing, automated vote buying, viral attacks on voter PCs, etc.), any one of which could be catastrophic". The report finally recommends "*shutting down the development of SERVE immediately and not attempting anything like it in the future until both the Internet and the world's home computer infrastructure have been fundamentally redesigned, or some other unforeseen security breakthroughs appear.*" SERVE has eventually been cancelled by the Department of Defense.

6 Conclusions

This paper intends to limit the analysis to political voting and local or national elections or referendums. Technology should serve the goal to face the crisis of confidence that representative democracy is experiencing today. E-voting is not like a common electronic transaction. An e-voting procedure will only be acceptable under the condition that it safeguards the constitutional principles associated with the voting process, such as equality, freedom, secrecy, transparency and accountability. This must be open, accessible, interactive and secure, in order to enable citizens to participate in political life. Assuming that the relevant legal and the resulting "technical" requirements are met, e-voting systems will become a possibility for all citizens. The next step beyond poll site voting would be to deploy kiosk voting terminals in public places. This path toward greater convenience would enable technologists and social scientists to address the many issues that confront the voting process at each level of implementation. Many issues related to kiosk voting, such as setting standards for electronically authenticating voters, still need to be resolved.

Remote Internet voting systems pose significant risk to the integrity of the voting process, and should not be fielded for use in public elections until substantial technical (mainly security) and social science issues are addressed. Nevertheless, it is advisable to replace punch cards, lever machines, and older full-faced DREs (Direct Recording Electronic devices) with optical scanning systems that involve counting ballots in precincts, or with any electronic technology proven in field tests. The aim of this paper was to clarify the main legal concerns involved in electronic voting and to show that this system of voting could be introduced into the electoral process following a gradual and reasoned approach. The introduction of information and communication technologies (ICT) into voting operations does, in fact, considerably simplify the polling procedure, in particular, by making it faster and more functional. Electronic voting alone will not, however, change citizens' political attitudes. On the other hand, it would appear that ICT offers individuals new forms of expression and participation (discussion forums, debates, chat rooms, on-line public surveys etc.). Alternatively, security models such as the voter-verified audit trail allow for electronic voting systems that produce a paper trail that can be seen and verified by a voter. In such a system, the correctness burden on the voting terminal's code is significantly less as voters can see and verify a physical object that describes their vote. Even if, for whatever reason, the machines cannot name the winner of an election, then the paper ballots can be recounted, either mechanically or manually, to gain progressively more accurate election results. E-voting promises to improve accessibility for disabled voters. Furthermore, election results will be calculated quickly and efficiently, with less chance of human error, and long-term costs will be reduced by eliminating the expense of printing ballots. On the other hand e-data is likely to be more easily altered or destroyed than physical ballots. In addition, all kinds of e-voting systems are susceptible to a certain extent to insider attacks and Denial of Service (DOS) attacks. Our future work will be focused in examining the Kiosk voting methodology, from the technolegal point of view, as being the necessary phase between the polling place voting and the Internet voting. In info-Kiosk schemes, voting machines are located away from traditional polling places but under the control of election officials and also be appropriately monitored in order to meet security and privacy requirements as well as to prevent intervention (i.e. coercion). A well-designed e-voting system should produce an audit trail that is even stronger than that of conventional systems (including paper-based systems). Future of e-voting systems will exploit current technologies and tools including smart cards, biometrics (e.g. voice, fingerprint, retinal recognition - for identification), as well as mobile voting clients (e.g. hand-held organizers, cell phones, etc). Research is needed to determine to what extent such technologies are viable for e-voting.

References

1. Jones .D.W. Counting Mark- Sense Ballots – Relating technology, the Law and Common Sense, January 2002; http://www.cs.uiowa.edu/~jones/voting/optical/
2. Internet Policy Institute, Report of the National Workshop on Internet Voting, March 2001

3. («Evaluating practices &Validating technologies in E-democracy», www.eve.cnrs.fr/)
4. Final Report of the California Internet Voting Task Force , California Secretary of State, January 2000
5. Electronic Voting : Constitutional and legal Requirements and their technical Implications, Lilian Mitrou, D. Gritzalis, S. Katsikas, G. Quirchmayr , in "Secure Electronic Voting" 2003, Kluwer Academic Publishers.
6. "Internet-based voter registration poses significant risk to the integrity of the voting process, and should not be implemented until an adequate authentication infrastructure is available and adopted". Internet Policy Institute, Report of the National Workshop on Internet Voting: Issues and Research Agenda, March 2001.
7. California Institute of Technology – MIT, Voting Technology Project, Voting: What is, What could be, July 2001.
8. A Security Analysis of the Secure Electronic Registration and Voting Experiment (SERVE), January 21, 2004, Dr. David Jefferson, Dr. Aviel D. Rubin, Dr. Barbara Simons, Dr. David Wagner,
9. "Analysis of an Electronic Voting System" Tadayoshi Kohno, Adam Stubblefield, Aviel D. Rubin, Dan S. Wallach ,Johns Hopkins University Information Security Institute Technical Report TR-2003-19, July 23, 2003.
10. WHAT IS THE FUTURE OF ELECTRONIC VOTING IN FRANCE? The Internet rights forum, Published on 26 September 2003
11. Same as 6
12. Secure electronic voting: The current landscape, C. Lambrinoudakis, D. Gritzalis, V. Tsoumas, M. Karyda, S. Ikonomopoulos, in "Secure Electronic Voting" 2003, Kluwer Academic Publishers.
13. ICTS AND THE FUTURE OF DEMOCRACY by Ignace Snellen International Journal of Communications Law and Policy Issue , Winter 2000/2001
14. Realising Democracy Online: A Civic Commons in Cyberspace IPPR/Citizens Online Research, Publication No.2 - March 2001
15. European Commission, IST 2000 Programme, The Information Society for all, Final Re port, Brussels 2000
16. Analysis of an Electronic Voting System, Johns Hopkins Information Security Institute Technical Report TR-2003-19, July 23, 2003 by Tadayoshi Kohno, Adam Stubblefield, Aviel D. Rubin, and Dan S. Wallach.
17. Examples are: the Arizona Democratic party's election (legally binding), March 2000, Mohen, J., Gliden, J. " The case for Internet Voting". In Com. of the ACM, 44(1), 2001 the Military personnel Presidential election in the US and overseas (legally binding), 2000, Federal Voting Assistance Program. Voting over the Internet Project, www.fvap.gov/ the UK local and mayoral elections (non-binding), May 2002, DTLR News Release. "May Elections to Trial Online Voting", 2002,
http://www.press.dtlr.gov.uk/pns/DisplayPN.cgi?pb_id=2002_2003
Pilot schemes to test innovative voting and counting methods took place in 59 local authorities across England on 1 May 2003., The shape of elections to come, A strategic evaluation of the 2003 electoral pilot schemes, July 2003, The Electoral Commission UK In Switzerland, the first official ballot for which citizens can vote through Internet began on January the 7th 2003, in the municipality of Anières (Geneva)
http://www.geneve.ch/chancellerie/E-Government/e-voting.html
18. A group of experts in computerized election security was assembled by the FVAP to help evaluate SERVE., These four computer scientists published in January, 2004 a report entitled "A Security Analysis of the Secure Electronic Registration and Voting Experiment (SERVE)" (the "SERVE Security Report").

Architecture for Implementing
the Mobile Government Services in Korea

Yoojung Kim[1], Jongsoo Yoon[2], Seungbong Park[3], and Jaemin Han[3]

[1] IT Infrastructure Division, National Computerization Agency, Seoul, Korea
yjkim@nca.or.kr
[2] College of Business Administration, Kangnam University, Gyeonggi-Do, Korea
jongsoo@kangnam.ac.kr
[3] School of Business, Korea University, Seoul, Korea
{sbpark,jaemin}@korea.ac.kr

Abstract. Each nation's architecture for the m-Government including Korea's is being laid out somewhat differently according to its strategy, approach, and its own structural conditions and challenges. Despite such differences, all of them are making numerous efforts to establish the advanced service architecture for the m-Government. In the meantime, they are confused in securing and managing strategic resources which are needed to create constant mobile values, diversifying wired and wireless connection sections, and dealing with technological complexities and differences in various kinds of mobile technologies. Therefore, this study proposes the m-Government service architecture, which can effectively provide nationwide m-Governmental services. It also defines key components of the m-Government service architecture and specific characteristics of applied technologies.

1 Introduction

Mobile government (m-Government) has been designed to provide the public with more efficient mobile administrative services and more convenient access to public services anytime and anywhere through the wireless Internet. In an attempt to advance e-Government services and expand customer channels, this system is being put in place in some of the European nations, the United States, and Korea, all of which have succeeded in establishing advanced wireless Internet environments [3].

Some European countries such as Sweden have attempted m-Government in many different practical applications and Canada is proceeding with their m-Government project focusing on mobile (wireless) portal. In the United States, federal and local governments have taken the lead and emphasis has been placed on the wireless and mobile technology in order to develop applications for public safety and response, such as assistance for field staffs, information sharing, staff monitoring, and maintenance of network communications [3].

In the case of Korea, the number of wired Internet subscribers reached 29.22 million as of December 2003, and 33.093 million as of March 2004. As the first nation in

S. Wang et al. (Eds.): ER Workshops 2004, LNCS 3289, pp. 601–612, 2004.

the world that succeeded in commercializing the CDMA technology, Korea has now become fully equipped with the 3G mobile communications infrastructure such as cdma20001x EV-DO and W-CDMA with maximum transmitting speeds of 2.4 Mbps [6]. In addition, it has become more efficient and convenient to use the Internet and search and process information through mobile terminals, thanks to the faster spread of mobile virtual machines and further development of multimedia supportive technologies [9].

Based on these circumstantial advantages, the number of mobile businesses using the wireless Internet is growing and the informatization process is being rapidly adopted by the business sector. The m-Government is also being swiftly established in an effort to advance the wired e-Government service which was first launched in 2002.

Korea is applying mobile technologies to government agencies and each agency has a mobile page, similar to a web page, exclusively for the mobile terminals. A common gateway and platform are also in the process of being mapped out for the m-Government as well.

In the meantime, Korean local governments and governmental agencies have selectively adopted the mobile electronic signature systems and mobile field force automation service, by using the wireless Internet infrastructure and platforms of private mobile telecommunications service providers [5]. However, they have been facing various challenges and limits to applying these mobile technologies to the public sector. Specifically, they are having difficulties in securing and managing strategic resources which are needed to create constant mobile values, diversifying wired and wireless connection sections, and dealing with technological complexities and differences in various kinds of mobile technologies. Also, the waste of resources caused by overlapping development and vulnerability of the nation's information security are other urgent problems to address. To cope with these challenges, debates are heated in the Korean government to find a way to build public G/W and platforms of the m-Government service.

In short, each nation's architecture for the m-Government including Korea's is being laid out somewhat differently according to its strategy, its own structural conditions and challenges. Despite such differences, all of them are making numerous efforts to establish the advanced service architecture for the m-Government [3].

Therefore, this study takes a close look at current moves around the world to launch m-Government services and infrastructure, while comprehensively examining references related to this issue. By doing so, it come up with possible solutions to challenges that Korea is facing in implementing e-Government services, and thereby draft a blueprint best suited for Korea. The specific goal of the study is to propose the m-Government service architecture, which can effectively provide nationwide m-Governmental services. It also defines key components of the m-Government service architecture and specific characteristics of applied technologies.

2 Outlines of the m-Government

2.1 Definition and Characteristics of the m-Government

M-Government stands for the use of mobile wireless communication technology within the government administration and in its delivery of services and information to citizens and firms [11], [12]. By connecting a wireless section to a wired end section, the m-Government will create and guarantee mobility and portability for the public, business, and government. Furthermore, convenience in accessing information, real time access to information, and personalization of information access are guaranteed to maximize benefits of using information and, in turn, create further advanced e-Government services.

The merits and new services that m-Government can provide to e-Government can be summarized as the following: the enhanced convenience of the citizens and governmental employees that can get access to the e-Government anytime and anywhere; implementation of customized services and governmental CRM (Citizen Relationship Management) as mobile devices are more personalized; enhancement of productivity in the public sector by building mobile office; improvement of public safety through new mobile technologies such as LBS(Location-Based Services); government synergy effect through technological integration and convergence of government's wired and wireless IT infrastructure.

It seems that each country considers that the integration of wireless Internet infrastructure with the existing wired Internet infrastructure is timely and desirable.

Given these, the m-Government can be defined as a integrated service; through the wireless Internet, it provides not only governmental services for its people (G2C), such as information on civil affairs and locations, small financial transaction, and electronic identification, but also mobile administrative services for governmental bodies (G2G) that can enhance their administrative autonomies, cultural tourism, public health and welfare, and science technology. It also offers industry-based services for businesses (G2B) through the wireless Internet, such as information on procurement, distribution, payment, taxes and so on [9].

2.2 Types of the m-Government Service

As it is in e-Government services, the m-Government service may be classified into various types including mobile G2B, G2G, mobile G4C, and mobile G2E [2]. From the technological point of view, mobile G2B and G2G should belong to the mobile extranet, while mobile G4C and mobile G2E are included in the mobile Internet and mobile Intranet respectively.

However, in Korea, the m-Government service is often classified largely into two infrastructural services and six general services [5]. The former are services that can be used only with the support of general services, such as the authentication service designed to confirm a user's identification and location information service that uses information on a user's location. The latter is the services including electronic civil petitions, information search, and urgent notification. According to the service type offered, they are divided into six service categories such as information sarch, electronic civil petition, information collection, urgent notification, public commercial transaction and tax payment service.

2.3 Status of m-Government Services and Approaches to Building Infrastructure

M-Government Services in Foreign Countries. Canada's m-Government strategy and implementation is in the very early stages. Canada currently has a prototype on line that can be accessed by pointing a mobile device to www.gc.ca. However, Canada is still in the stages of identifying the services that should be made available via mobile channel. Canada is working with key stakeholders and representatives from other government departments to further define and evolve the m-Government strategy [3].

Canada's architecture of the m-Government service is based on G/W, infrastructure of mobile telecommunication service carriers and has launched m-Government portal in an attempt to integrate channels for m-Government services.

In Denmark, m-Government is seen in close relation to the implementation of e-Government. That is, m-Government is increasingly seen as a tool to make the e-Government vision encompass also those functions of government that are being carried out in the field, especially out of the traditional bureaucracies in town-halls, agencies and ministries [3].

The focus in Denmark is primarily on the promises of using wireless technologies in relation to public sector employees G2E (G2G). However, there are also a few examples of G2C solutions, for example in relation to parking. Most initiatives up to now have focused on improving work-processes and divisions of labor for public employees. As in every other area of e-Government, it is believed that the "e-Promises" of more efficiency and better quality in public service provision will only be fulfilled if digitization-initiatives (also in m-Government) are accompanied by changes in the working processes and organization.

So in conclusion, m-Government in Denmark is mainly focused on G2E. But that being the case, improving the working conditions of public employees will often result in better service for citizens and business. Denmark's m-Government service has been provided by each central and local government agency, which is using mobile telecommunications infrastructure.

In Sweden, Short Message Service (SMS) is often adopted in the public sector. One service is that it is possible to subscribe to information on job postings that match the profile of "the type of job I'm interested in". Hits will be emailed, and the email may be read using a mobile phone. Another service is that mobile phone numbers for SMS can be published in the job seeker's CV, so that job provider can get in touch via an SMS [3].

Another service is m-Parking. The city of Stockholm, Huddinge University Hospital, and airports operated by the Swedish Civil Aviation Authority, are examples of m-Parking providers at the three Swedish government levels.

Central and government agencies and local governments have been involved in planning and establishing the m-Government service in Sweden. In order to secure reliable interoperability between m-Government services, however, Government Agency Board has been the one that defined technical roaming, network roaming and information roaming [3].

In North Ireland, the OFMDFM e-Strategy work includes two examples of how staff might access information whilst on the move. CITU (NI) has led a pilot project for proofing the concept for the use of a variety of mobile devices that would enable the wider acceptance of truly mobile computing throughout the NICS [3].

North Ireland is currently developing policy, guidelines and standards to enable joined up services to the citizen and businesses in Northern Ireland by a variety of channels including mobile [3].

M-Government Services in Korea. In Korea, the m-Government service has been planned and introduced as one of the strategies to realize e-Korea. In particular, this has been driven with a goal to diversify existing e-Government channels and ultimately advance them into the ubiquitous e-Government [5].

Much of m-Government projects has been carried out by ministries and local governments, respectively. SMS is a killer application for mobile G4C and G2G services. Public agencies primarily send administrative and policy status information, events and notification through SMS. Also, the PDA-based services are provided in an interactive and informative way in some mobile G2G services such as m-Police and m-Tax Management.

Unfortunately, however, overlapping resource investment has been made into m-Government services, since different mobile systems have been adopted randomly to satisfy each government agency's demand and other specific needs at working unit levels. So, a consensus is being reached that these systems should be interconnected and a common foundation for information sharing should be established. On the other hand, current m-Government services are exposed to potential risks of serious security breaches because they are based on platforms of private service providers.

To straighten out these problems and successfully establish reliable m-Government service, the Korean government is laying out ISP/BPR plans and is aggressively seeking ways to establish more integrated m-Government service infrastructure based on ISP/BPR project.

Aproaches to Building m-Government Service Infrastructure. Different approaches to building m-Government service infrastructure can be adopted in accordance with the presence of an integrated channel for the m-Government service (for instance, the m-Government portal) and exclusive GW, and locations of exclusive G/W [5], [10]. Given this, approaches to building m-Government service infrastructure can be largely divided into seven categories and main characteristics of each approach are as follows (see Fig. 1).

Under the first approach 1, government agencies or local governments independently launch m-Government services by using G/W of existing mobile telecommunication service providers such as SKT, KTF, and LGT.

The second method 2 is basically the same as the first one. But, the m-Government service is provided through mobile portal channels of existing mobile telecommunications companies such as Nate, Magic-n, and ez-Web.

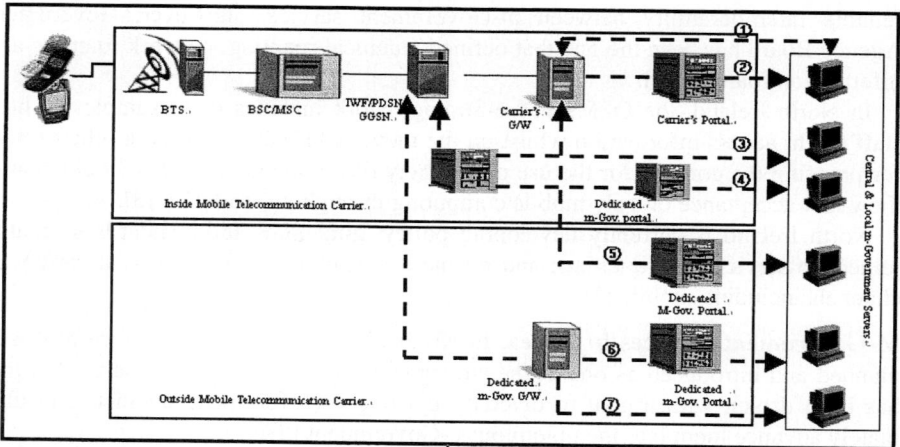

Fig. 1. Approaches to Building m-Government Service Infrastructure

Under the third approach 3, apart from G/W of mobile telecommunication companies, exclusive G/W for the m-Government service is established in the mobile IDC of mobile telecommunications carriers so that pubic infrastructure can be provided for government agencies and local governments. However, an integrated channel for the m-Government service is not offered.

The forth approach 4 is basically the same as the third one, but it provides the m-Government portal service as an integrated channel for mobile services.

The fifth method 5 to connect exclusive lines to G/W of existing mobile telecommunication service providers in order for the government to establish mobile portal servers in certain controlled areas. This eventually provides an integrated m-Government service channel for government agencies and local governments.

The sixth method 6 is to connect exclusive lines to IWF/PDSN/GGSN of mobile telecommunication operators to enable the government to build exclusive G/W for the m-Government services in certain controlled areas as m-Government infrastructure for public use.

The final approach 7 is basically the same as the sixth one. However, under this approach, the government launches an exclusive portal designed for the m-Government service, as an integrated service channel.

3 Architecture for the m-Government Service in Korea

3.1 Architecture and Components for the m-Government Service

From the beginning of 2004, the Korean government has been carrying out the ISP/BPR project to form m-Government service infrastructure that is key to implementing the m-Government. And, considering recent studies related to m-Government service architecture [10], the approach to establishing Dedicated m-Government

Gateway (see the sixth and seventh approaches in Fig. 2) may be considered as the best m-Government service architecture for Korea.

Under this approach, exclusive m-Government gateways can be built in certain area without using MIDC of mobile telecommunications service providers and provide foundations for central and local governments to establish m-Government infrastructure. There are several reasons for adopting this method.

First of all, by building public infrastructure as a strategic resource, every government agency can play a role in creating a reliable and constant mobile value and thereby maintain a comprehensive and positive view on service interoperability and information sharing [4].

Second, on the technological front, integration of wired and wireless access points is needed, in that mobile Internet services should be available in both wired and wireless spaces. Furthermore, this integration may eliminate technological complexities caused by using different terminal platforms, wireless Internet access protocols, and interworking modes [4].

Third, this approach is economically beneficial. By establishing a common platform rather than depending on the M:N Spaghetti connection method, overlapping development can be prevented and the waste of resources can be minimized. M-Government service infrastructure for central and local governments can be integrated as well, which may ultimately create effects of economies of scale, and reduce fees for using mobile networks of mobile telecommunications operators.

Fourth, making a collective contract with mobile telecommunications service providers can ensure enhanced convenience. Moreover, with only less than 10 staffs, more effective and systemic operation management can be achieved.

Finally, when it comes to security management, security breaches in the wireless Internet structure can be overcome by building gateways for the m-Government service in certain areas under the government control.

Along with overall directions and particular conditions of the mobile service in Korea, these advantages explains why Dedicated m-Government G/W should be selected as the architecture for Korea's m-Government service which is clearly shown in Fig. 2. In addition, given key components of the architecture suggested in Kalakota and Robinson [4], MB-net consortium [8], Lawrence and Littman [7], Balfanz et al. [1], and Yeoul [13], key components for the architecture should include mobile public services, wireless Internet platforms, Terminal access devices for users, and security devices.

Mobile Public Service. Mobile pubic service is the service designed for governmental organizations to not only boost mobility of internal works but also to make a mobile administrative work process more efficient and to deliver better public service. This service includes governmental agencies' back-end systems and mobile systems that provide mobile public services. Therefore, this service should be considered importantly in the technologies linked to wireless Internet platforms.

The mobile Internet here can serve as mobile Internet network infrastructure based on wireless networks of mobile telecommunications service providers (SKT, KTF, LGT), and wireless network infrastructure covers IS95A/B, 2G, 2.5G, and even 3G. Meanwhile, the mobile web-browser presentation and conversion service are the ones

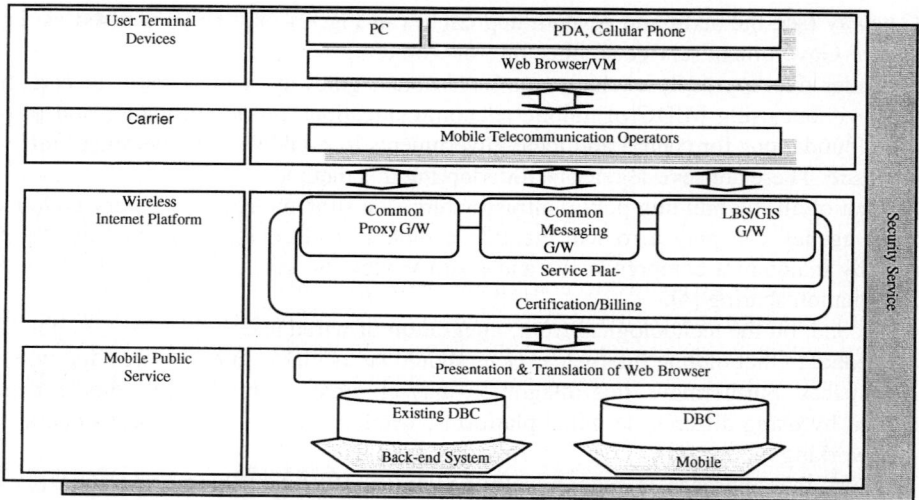

Fig. 2. Architecture for m-Government Service in Korea

that control interaction between mobile web-browsers and back-end systems/mobile systems. Specifically, they deliver user demands collected through mobile web-browsers to back-end or mobile systems. At the same time, they receive outcomes of user demands from back-end or mobile systems and then present those outcomes to mobile web-browsers.

Related technologies are web-page formats (WML, m-HTML, XML), client-side script languages (WML, scripts), server-side script languages (JSP, PHP, ASP) and web-browser built-in programs (Java Applet, Java Script).

Wireless Internet Platforms. Wireless Internet platforms include an integrated wireless Internet gateway, integrated messaging gateway, LBS gateway and service platform, which is designed to deliver the mobile public service more efficiently.

Among these, the integrated wireless Internet gateway is one of the most critical components in connecting wireless networks to wired Internet. WAP Proxy (SKT, LGT) and ME Proxy (KTF) are examples of this gateway. The integrated messaging gateway is part of the wireless data service and, along with SMSC of the three major mobile telecommunications operators, serves as a hub in the wireless data service adopted in the public sector.

In the meantime, the Location-Based System (LBS) gateway is the positioning service that, in effect, may realize real-time public service by positioning the current location through a variety of wireless terminals.

User Terminal Access Devices. A user terminal access device is the terminal device for mobile public service users and includes hand-held devices such as a mobile phone, PDA, and smart phone.

Finally, mobile web-browser is the standard user interface provided in user terminals. This include WAP browser and ME browser.

Security. The security service should be designed to protect each component of the m-Government service architecture. Communications security, user authentication, and encryption are included in this service.

3.2 Definition of Technology Applied to Key Components of m-Government Service Architecture

This study presents three major technological factors in the architecture for the m-Government service in Korea: technology for mobile Internet web browsers and middleware, technology applied to constructing exclusive gateways designed for the m-Government service as well as service platforms, and technology which are needed in linking mobile systems of governmental agencies to wireless Internet platforms. Table 1 clearly shows main characteristics of technologies needed in each architecture component.

Along with this, Table 2 presents m-Government service guidelines for each technological factor, profiles of commonly applied technology. This further specifies not only components of architecture for the Korean mobile service mentioned in Table 1 but also related technologies.

Table 1. Key Technologies Applied to Each Component of the m-Gov. Service Architecture

Components of m-Gov. service architecture			Technology	Outlines
Mobile Internet Web browser & middleware			Mobile web-browser	Browser that helps users get access to the Internet and web services through mobile terminals
			Virtual machine	Middleware that drives public mobile service in a terminal, after downloading it from a server
Wireless Internet platform	G/W		Wireless Internet G/W	Proxy that operates in IWP (PDSN, GGSN), with functions such as converting the mobile markup language, supporting security models of mobile telecommunications operators, and creating a real-time user access list
			Messaging G/W	Proxy that operates in the SMSC/SMS server, transmits a message, and confirms message arrival, with a function of Callback URL transmission
			LBS G/W	Interworking is possible with GMLC of mobile telecommunication companies and LBS application
	Service Platform		Contents conversion device	Program that manages XSLT according to each mobile Internet device, receives a required page in XML from the contents server, and converts it into the one to be suitable for the device that requests it
			Portal platform	Platform whose functions are subscriber management and authentication, contents branding, contents billing, and contents downloading and management
Mobile Public Service	Mobile web-browser representation		Mobile web page format	Formats and versions of mobile pages which are provided for diverse mobile terminals
			Client-side script language	Script language which belongs to the Mobile markup language and is run in the mobile web browser
	Mobile system integration		Data exchange format	Data exchange format between wireless Internet platforms and mobile systems Data exchange format mobile systems and back-end systems
			Communications linkage format	Communications protocol between wireless Internet platforms and mobile systems Protocol between mobile systems and back-end systems
Security			Telecommunications security	Protocol designed for end-to-End communications
			Encryption	Data encryption
			User authentication	Function of controlling a user's access and conforming the user's identification

Table 2. Profiles of Technologies Commonly Used in the m-Government Service

Mobile service architecture	Technology		Technology profile	Specifications of technology profile
Mobile web browser	Mobile web browser		User interface	-Browser
Mobile web browser representation and conversion	Mobile web page format			-Screen mgt(text, image), Data input format, Hyper text switching format, Style sheet
	Mobile terminal (client) script language			-Mobile web page script language
Middleware	Virtual machine			-Virtual machine program
Wireless Internet platform	G/W	Wireless Internet G/W	Proxy function	-Defining the proxy function for WAP 1.x, 2.0, ME, and VM -Defining the IWF (PDSN, GGSN) interworking format of mobile telecommunications companies -Defining the mutual conversion function for mobile markup language -Processing a security model which receives additional support from mobile telecommunications service providers
		Messaging G/W	Messaging linkage (Data and communications format)	-Defining a linkage format for SMSC/MMSC, and LMSC of mobile carriers -Defining the Call back URL transmission function -Defining the function of automatic conversion into the messaging format that can be supported in a terminal -Defining the interworking format with the terminal information system of mobile telecommunications service providers
		LBS G/W	LBS G/W Interworking	-Defining an interworking format between LBS G/W and GMLC of mobile telecommunications service providers -Defining common API for interworking between LBS G/W and LBS applications
		Converter	User interface	-Managing XSLT and converting it into a web page that can be represented in a terminal
	Service Platform	Portal platform	Service management	-Authenticating and managing a subscriber -Logging on to contents that a subscriber uses -Contents menu management function -Providing a menu for each subscriber or each subscriber group -Managing CP
		Information on terminals and subscribers		-Linkage to both subscribers of mobile telecommunications service providers and the terminal information system
		Contents downloading and management		- Menu function for various VM applications - Downloading VM applications into terminals
		Contents billing		- Managing information on each contents rate - Rating contents that a user utilizes -Ticketing function with issuing a notice -Diverse Pre-Paid function (ARS charge, credit card charge, mobile charge)
Mobile system linkage/ integration	Data switching format		Data switching	-Character set, Definition and verification of the data switching format structure, Data modelling, Switching format for various information modes
	Communications format		Communications linkage	-Messaging, Transport, Networking, Routing, Mail, File transmission, Hyper text transmission, Directory, Web service
Security	Communications security		Security	-Mobile web access security, Mail security, Wired and wireless network security
	Encryption			-Open Key algorithm
	User authentication			-Electronic signature, Electronic signature algorithm, Hash algorithm

4   Conclusions

This study not only takes look at foreign nations' moves to establish their own m-Government service and infrastructure, but also reviews references related to this matter. By doing so, it hopes to maps out the architecture for the m-Government service that is best suited to Korea and can effectively deliver nationwide m-Government services. Moreover, this study demonstrates key components of the proposed architecture and specific characteristics of technologies applied to them.

Major outcomes of this study are as follows.

First of all, the study comprehensively examines key factors in the establishment of the m-Government, such as the presence of exclusive G/W designed for the m-Government, location of exclusive G/W, and presence of the m-Government portal (integrated service channel). Based on this examination, it proposes Dedicated m-Government G/W as more suitable method for Korea to build m-Government service infrastructure than any other approaches used home and abroad.

It also presents components of the Dedicated m-Government G/W that are needed in building exclusive infrastructure for the m-Government service (G/W and Service platforms). Those components are Mobile Web Browser and Middleware (Virtual Machine, VM), Wireless Internet Platform (G/W, Translator, Service platform), Mobile Public Services (Presentation of mobile web browser, integration of Mobile system with legacy system), and security factors.

This study also delivers significant implications as follows.

First, it thoroughly reviews a variety of current domestic and overseas approaches to building infrastructure of m-Government services and suggests the most suitable and feasible architecture of the m-Government service for Korea.

Second, the proposed architecture, its components, technology applied to each component, and profiles of technologies commonly applied to the components, may serve as a guideline for Korea and other advanced nations in launching the integrated m-Government service and securing interoperability between diverse e-Government services.

However, there is a limit to this study. It presents the feasible and specific m-Government architecture and its components that fit well with nations like Korea, where, up to now, each government agency and local government has delivered m-Government services, using the infrastructure of private mobile telecommunications service providers.

Therefore, it might be hard for nations under different environments to apply outcomes of this study to their situations. Nevertheless, the outcomes of this study may be very useful for either nations that are going through similar conditions to Korea or nations that are less developed than Korea in terms of Internet infrastructure.

References

1. Balfanz, D., Schirmer, J., Grimm, M., Tazari, M.R.: Mobile Situation Awareness within the Project Map, Computers & Graphics Vol.27 (2003) 893–898

2. Dawes, S.S., Pardo, T.A., Cresswell, A.M.: Designing Electronic Government Information Access Programs - A Holistic Approach, Government Information Quarterly Vol.21 (2004) 3-23
3. GOL-IN: Implementation Strategy for Mobile Government, Mobile Government Project (2002)
4. KalaKota, R., Robinson, M.: M-Business: The Race to Mobility, McGraw-Hill (2001)
5. Kim, Y.: Understanding of Mobile Government, 2003 Fall Conference of Korea Society of Management Information Systems (2003) 63-86
6. Korea Network Information Center (KRNIC): Internal Statistics DB of Informatization, http://www.krnic.or.kr (2004)
7. Lawrence, P., Littman, L.: Preparing Wireless and Mobile Technologies in Government, http://www.businessofgovernment.org (2002)
8. MB-net Consortium: Final White Paper on m-Business Applications and Services Research Challenge (2001)
9. National Computerization Agency (NCA), White Paper Internet Korea 2004 (2004)
10. Oh, K., Byun, I.: The Effective Implementing Strategy for Mobile Government, 2003 Spring Conference of The Korean Society for Cyber Communications, Korean Society for Cyber Communications (2003) 75-87
11. Ostberg, O.: A Swedish View on Mobile Government, 2003 International Symposium on Digital Mobility and Mobile Government, KISDI (2003) 67-78
12. Ovum: Mobile E-commerce - Market Strategies, Ovum Press Release (2003)
13. Yeoul, H.: m-Government Initiatives for Korean e-Government, 2003 International Symposium on Digital Mobility and Mobile Government, KISDI (2003) 25-60

Fifth International Workshop on Conceptual Modeling Approaches for E-Business (eCOMO 2004)

at the 23rd International Conference
on Conceptual Modeling (ER 2004)

Shanghai, China, November 9, 2004

Organized by

Heinrich C. Mayr and Willem-Jan van den Heuvel

Preface to eCOMO 2004

The today's increasingly competitive and expanding global marketplace requires companies to cope effectively with rapidly changing market conditions. Conceptual business, enterprise and process models, either at the level of isolated or integrated enterprises, are heralded as an important mechanism for planning and managing these changes and transitions as well as for designing and constructing the necessary enterprise information systems. However, appropriate modeling methods and tools still are an issue under research and development. In addition, research issues in the area of business policy specification and change management of enterprise component based models are of paramount importance. Within that context, model-driven design of business processes is getting increasing interest from both industrial and academic communities to serve as the foundation for developing and wiring enterprise applications. At heart of this design philosophy lays the idea that conceptual (cross-) enterprise models can be constructed independently from any implementation context, but can still be mapped into platform specific models.

However, several problems are to be solved within that context, e.g., divorcing conceptual, platform independent models from the underlying implementation is problematic, as conceptual models may become siblings once they are mapped to the physical level. Moreover, it is not clear how mapping rules should be specified in such a way that it al-lows developers to smoothly roll forward and back between the conceptualized business processes that are captured by platform independent models, and supporting enterprise systems that are reflected using a platform specific models. In addition, the MDA philosophy is not geared towards facilitating collaborative modeling of shared business processes. Also, the automatic mapping and harmonization of the ontologies underlying enter-prise models is still a question under intensive research.

The eCOMO workshop is by now a well-respected event joining internal researchers to discuss the wide range of conceptual modeling issues to analyze and design e-Business collaboration. Thus, eCOMO'2004 is intended to continue and extend its four predecessor eCOMO workshops which were held during the ER conferences since 2000.

The program of eCOMO'2004 is the result of a careful review process in which each of the submitted papers was assessed by three experienced reviewers. More than the six papers printed here were rated worth to be published. However, the program committee had to make its final decision according to the rules of ER and LNCS. The selected six contributions mainly deal with

- e-Business Systems Requirements Engineering w.r.t. modeling business strategies, eliciting wide audience requirements for mobile systems, and developing a sectoral standard for EDI,
- e-Business Processes and Infrastructure w.r.t. classifying business process correspondences and associated integration operators, transforming behavi-

oural models in Web services, an model-driven planning and development of e-Commerce infrastructures.

Many persons deserve appreciation and recognition for their contribution to make eCOMO'2004 a success. First of all we thank both the authors for their valuable contributions and the members of the program committee for doing a great job in assessing submitted papers and participating in the iterated discussions on acceptance or rejection. Special appreciation is due to Christian Kop, who organized and co-coordinated the whole preparation process including the composition of these proceedings. Last but not least we thank the ER organizers and the ER workshop co-chairs Katsumi Tanaka and Shan Wang as well as the conference co-chairman Tok Wang Ling for their support in integrating eCOMO'2004 into ER'2004.

November 2004   Heinrich C. Mayr
                                              Willem-Jan van den Heuvel
                                                eCOMO2004 PC Co-Chairs

Modeling Business Strategy
in e-Business Systems Requirements Engineering

Steven J. Bleistein[1,2], Karl Cox[1,2], and June Verner[1]

[1] Empirical Software Engineering Research Program
National ICT Australia, Sydney 1430, Australia
{Steven.Bleistein,Karl.Cox,June.Verner}@NICTA.com.au
[2] School of Computer Science and Engineering,
University of New South Wales, Sydney 2052, Australia

Abstract. This paper proposes a conceptual modeling approach for requirements engineering for e-business systems that enables alignment of systems requirements with business strategy. Jackson problem diagrams and goal modeling are employed to capture business strategy. As a means of linking abstract, high-level business requirements to low-level system requirements, we leverage the paradigm of projection in both problem diagrams and goal models in parallel. We use Jackson context diagrams to describe the business model domain context. Goal modeling is used to both represent requirements and describe the objectives of business strategy. The feasibility of the approach is demonstrated by its application to a case study from the literature.

1 Introduction

Research shows that alignment of IT with business strategy leads to superior business performance [1, 2], and that CIOs have consistently considered alignment of IT with business strategy a top priority [3-5]. Requirements for e-business systems should therefore be in alignment with the objectives of the business strategy that stakeholders intend the e-business system to support. It is thus important that requirements analysis of e-business systems capture both the strategic objectives of the business, and the activities and processes by which those objectives are achieved within the context of a conceptual model.

However, current requirements engineering modeling approaches for e-business systems fail to capture business strategy and its objectives. Some highlight aspects of requirements such as dependency relationships of organizational actors or value analysis [6, 7], or instead focus on elicitation process without considering explicit and coherent statements of business strategy [8]. While these aspects are important, these modeling approaches make it difficult for requirements engineers to validate low-level requirements against the more abstract high-level requirements representing the business strategy that the system is intended to support.

Requirements represent "the effects in the real-world problem domain that your customer wants the system to guarantee" [9]. For organizations engaging in e-business, the "effects in the real-world problem domain" are the objectives of their business strategy that the e-business system is intended to ensure. It is therefore at the

S. Wang et al. (Eds.): ER Workshops 2004, LNCS 3289, pp. 617–628, 2004.

level of strategy that companies bound the requirements problem for their e-business systems.

We thus propose a requirements engineering approach for e-business systems that incorporates modeling of business strategy as part of a requirements model. Our approach integrates Jackson's problem diagrams [9] with requirements engineering goal modeling techniques. We employ Jackson's context diagrams to represent contextual properties of the business model, and goal modeling to capture all behavioral properties from high-level business strategy to low-level system requirements that the system must ensure. These include business goals, strategic objectives, activities and any other business or systems requirements. We leverage the paradigm of *projection* [9] in both approaches as a means of decomposing high-level requirements down to those of the system.

The rest of this paper is organized as follows: section 2 presents current modeling approaches for e-business systems requirements engineering; section 3 describes our approach and shows how Jackson problem diagrams and goal modeling can be integrated; section 4 presents a proof-of-concept case study from the literature describing Seven-Eleven Japan's e-business system; section 5 offers some conclusions.

This paper is based on previous work [10], part of which we present in section 3. In this paper however we introduce a refinement of our approach (based on lessons learned in [10]) in an updated case study, as well as additional discussion of the transition from goal models to traditional Jackson problem diagrams. This transition is important as it enables refinement of requirements down to machine specification while maintaining traceability to high-level, abstract, business requirements.

2   e-Business Systems Modeling in Requirements Engineering

An e-business system enables marketing, buying, selling, delivering, servicing, and paying for products, services, and information, primarily across non-proprietary networks. *Participants* include the business's current and target customers, agents, suppliers, and business partners [11]. Some requirements engineering research proposes modeling approaches for these systems.

The TROPOS project describes an industrial case study in the an approach using $i*$ goal-modeling notation to model e-business systems requirements [12]. However, the $i*$ notation is designed primarily to identify highly abstract requirements in terms of dependencies between organizational actors and goals. Objectives of business strategy are not explicitly represented.

The e^3-*Value* model is another requirements modeling approach tested in an industrial case study. It is a UML-like modeling notation for e-business systems that enables validation of requirements in terms of the system's potential to generate economic value [7]. However, e^3-*Value* has been recognized as ignoring key elements of business value analysis [13, 14]. Moreover, e^3-*Value* seems to miss the crucial point that while valuation is important for analysis of economic value, it is business strategy that creates it [15, 16].

While not intended for e-business systems specifically, a different approach is taken in *L'Ecritoire*, a CREWS project in which a tool is developed to support a methodology that links conceptual models to requirements of large and complex information systems [8, 17]. The methodology supports primarily a requirements elicitation process. While it has been shown to be effective in discovering high-level, abstract business requirements via interviews with multiple stakeholders, it does not support validation of those requirements against a coherent and explicitly described business strategy.

3 Strategy-Oriented Approach to Modeling e-Business Systems

In this section, we outline our proposed approach. Section 3.1 provides an overview of Jackson's Problem Frames approach. Section 3.2 justifies the application of a Problem Frames approach to business strategy. Section 3.3 discusses integration of problem diagrams with goal modeling and shows how goal modeling can represent the requirement set.

3.1 Overview of Jackson Problem Frames Approach

Jackson has proposed Problem Frames as a software requirements engineering approach to model, analyze, and decompose real-world problems in advance of modeling the software solution to solve those problems [9]. Jackson problem diagrams are a means of modeling and understanding real-world problem context to which software will provide a solution, as opposed to machine system context.

Jackson models the real world according to two moods (in the grammatical sense): *indicative* and *optative*. *Indicative* mood refers to the way the world is, or to problem context. *Optative* mood refers to the way in which we want the software to change the world, or the requirement [9]. A problem diagram consists of both the indicative (*Real World Problem Context*) and optative (*Requirement*) parts that the Machine is intended to solve.

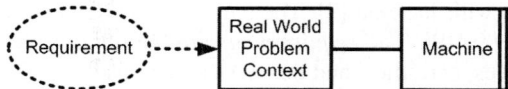

Fig. 1. Anatomy of a Problem Diagram

Fig. 1 illustrates the important elements of a problem diagram. *The Real World Problem Context* provides us with information about the structure, processes and tasks that are true of the problem domain. The *Requirement* addresses parts of the problem domain we wish to constrain and change according to our desires. The *Machine* is the software system solution we are trying to build. The connection between the *Real World Problem Context* and the *Machine* (what will eventually be the specification) is represented by the shared phenomena at the boundary between the problem and the solution.

While we do not provide a complete overview of Problem Frames here, we highly recommend the reader refer to Jackson's book [9].

3.2 Business Strategy as a Problem Diagram

Based on a broad survey of research literature, Oliver defines business strategy as "the understanding of an industry structure and dynamics, determining the organization's relative position in that industry and taking action either to change the industry's structure or the organization's position to improve organizational results" [18].

Oliver's definition of strategy is parallel to Jackson's concept of a problem diagram (see discussion of problem diagrams above). Oliver's "understanding of an industry structure and dynamics," and "determining the organization's relative position in that industry" refer to the *business model* [19, 20]. This is equivalent to Jackson's indicative mood or problem context. "Taking action either to change the industry's structure or the organization's position to improve organizational results," refers to activities and processes by which the organization achieves its *business objectives*. These are equivalent to Jackson's optative mood or the *requirement*. We consider all optative properties of a system to be requirements, including business goals, objectives, activities, business processes, policies, and any other business or systems requirements.

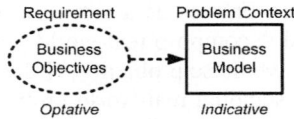

Fig. 2. Dividing Business Strategy into Optative and Indicative parts

We illustrate this division of strategy into optative and indicative parts in Fig. 2. We propose that an e-business strategy can be represented as a *problem diagram*, in which the e-business system is represented as the *machine* domain of interest. We recognize that an e-business system is in fact a collection of many machines working in concert, but at this level of abstraction, we represent the entire system as one *machine*, in accordance with Jackson [9]. The participants in an e-business system represent *domains of interest* [9]. The *requirements* are the *optative* part of business strategy; i.e., the objectives, activities, and business processes of the firm through which it attempts to succeed in its business.

3.3 A Progression of Problems and Goal Modeling: Linking High to Low Levels

E-business problems at the highest level of business strategy are very distant from the *machine*. To refine requirements from high-levels of abstraction down to the *machine*, the paradigm of a *progression of problems* is particularly useful (Fig. 3 below). The complexity of e-business systems as well as the need to align requirements with the highest levels of business strategy has pushed the requirements problem into what Jackson describes as "deep in the real world" [9]. The *domain* DA in Fig. 3 represents the *indicative* properties of the e-business problem context at the level of business strategy. *Requirement* RA represents the *optative* properties of strategy. Through analysis of DA and RA, it is possible to find a requirement RB that refers only to DB while satisfying RA [9]. DB represents the projection of DA, but at a lower level of abstraction. Through this process of analysis, problem projection, and refinement, ultimately the requirement refers just to the *machine*, yielding specification.

While the paradigm of a *progression of problems* serves as a powerful framework for decomposing e-business strategy down to *machine* requirements, the Problem Frames approach provides little explicit linkage between requirements at different levels of the progression. In the example above, requirement RB must satisfy requirement RA, and RC must satisfy RB, which satisfies RA, and so on. In order to ensure that system requirements are indeed in consistent with and provide support for business strategy, explicit traceability from lower level requirements to the highest level is necessary. However, while Jackson proposes analysis of DA and RA in order to find RB [9], Jackson describes no framework. Moreover, the Problem Frames approach provides no direct linkages between RA and RB. This is a potential "showstopper" in problem decomposition of complex systems, like e-business systems, where larger problems are projected into increasingly detailed smaller sub-problem diagrams. Thus while problem diagrams serve as powerful means of linking requirements to problem context and decomposing problem context, they are weak at decomposing requirements when projecting in a progression from high-level problem diagrams towards the *machine* level of abstraction.

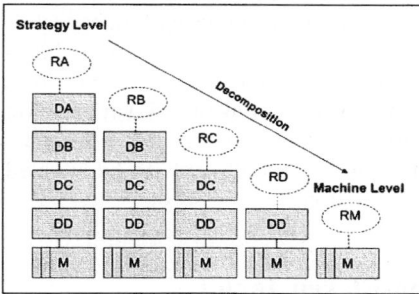

Fig. 3. Progression of Problems (adapted from [9] p. 103)

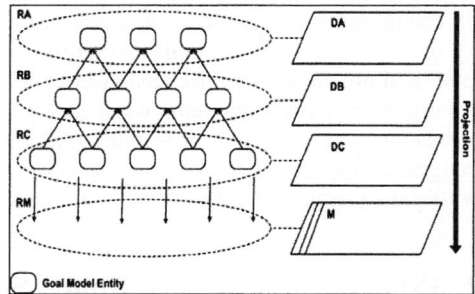

Fig. 4. Goal Model Integrated with Progression of Problems

In contrast, goal modeling is a useful technique to decompose higher-level requirements down to lower-level ones while maintaining traceability. Goals may be formulated at different levels of abstraction, from high-level strategic concerns to low-level technical ones [21]. Goals represent objectives that the system ought to achieve, and are thus requirements at a higher level of abstraction. Goals refer to behavioral properties that the system must ensure [22], so we treat goals as *optative*, as we would any requirement bounded by the problem domain [9]. Goal-oriented modeling techniques in requirements engineering thus provide a mechanism for requirements projection from high-level business requirements to low-level systems requirements through goal refinement.

Goal modeling techniques have been used for modeling business objectives and refining them toward system requirements [17, 23-26]. While these techniques treat business goals as discrete, independent entities, other approaches assemble business goals and their sub goals into structures representing complete business strategies, and then anchor requirements to the strategy model [27-29].

We therefore use goal-models to represent the *requirement* or *optative* part of the Jackson problem diagram in our approach, for it is the *requirement* that represents objectives of business strategy and system requirements. Goal models integrated with problem diagrams also serve as a means to trace requirements between problem diagrams in progression, and help ensure that the requirements at lower levels are in harmony with and provide support for objectives business strategy at higher levels.

The integration of a goal model with a progression of problems is illustrated in Fig. 4 above. The optative requirements at each level are described as a portion of a larger goal model. The goal model portions represent requirements at a level of abstraction equivalent to that of the *domain* to which they refer within the progression of problems. Each goal entity refers to specific *domains of interest* (though this is not indicated in the figure) within the referred domain, thus maintaining consistency between optative and indicative parts of the progression. The goal model enables explicit connections to requirements at adjacent levels in terms of super goals and sub goals. The sub goals are in fact projections of their super goals, and satisfaction of the sub goals guarantees satisfaction of the super goals in the same way that satisfaction of RB guarantees satisfaction of RA (Fig. 4). The context diagrams in the progression of problems (DA, DB, DC, ...) complement the goal model by providing problem context at various levels of abstraction with explicit linkage to requirements.

The integration of context diagrams with goal modeling also improves manageability of requirements of complex systems, in which many requirements may be high-level and abstract. The progression of problem diagrams toward the *machine* enables a partitioning of requirements while still maintaining explicit linkages. Each partition represents a smaller and more manageable portion of the goal model. Moreover, the integrated approach enables individual abstract business requirements to be situated within the problem context at explicit levels of abstraction.

4   Proof of Concept Case Study: Seven-Eleven Japan

We use a business case from a number of sources in the literature of Seven-Eleven Japan's e-business system [11, 30-32] to illustrate our approach.

Seven-Eleven Japan (SEJ), like its US progenitor, manages a national network of convenience stores. Unlike Seven-Eleven USA, SEJ generates value by leveraging and controlling ownership of information to optimize efficiency across a value chain with an unparalleled manner of sophistication. SEJ positions itself in the center of a value chain that includes suppliers, third-party logistics providers, and franchise shops, all of whom are independently-owned companies, yet all of whose objectives are maximizing throughput of products ultimately sold to franchise shop end-customers.

SEJ bases its strategy for competitive advantage on an extremely high level of competency at anticipating consumer purchases store-by-store, item-by-item, hour-by-hour, and then providing customers with products they want when they want them. SEJ's strategy leverages IT to accomplish its strategic objectives, and gain business advantage over its competitors. Its ownership of information enables sophisticated supply chain management to reduce inventories, lower costs, and increase sales. SEJ moves information between itself and its partner companies via an ISDN network. To better understand customer demand, SEJ actively gathers and analyzes purchasing information in real time, and correlates this with other social and environmental factors, including neighborhood demographics, planned local events like festivals, and the weather. SEJ then uses an acutely tuned just-in-time delivery system to meet that demand, generating remarkable value. It is these activities and their objectives that constitute the optative part of the SEJ e-business problem.

4.1 Progression of Problems of SEJ

Let us examine the progression of problems of SEJ's e-business system from the top, macro-level of business strategy down to the machine devices used in the franchise shops (see Fig. 5 below). Please note that for the purposes of describing the approach we only explore a particular aspect the requirements SEJ's large and complex e-business system. Also, at the lowest level we only explore one shop device in detail, the *Point-of-Sale (POS) Register*.

The progression of problems consists of an indicative part, which we describe as a progression of context diagrams, and an optative part, which we describe as a goal model. We represent the goal model using Goal-Oriented Requirements Language (GRL) notation [25, 33]. Note that we use only the *optative* entities of GRL, consisting of abstract goals (curved sponge shapes) and tasks (hexagons) in this model. We do not use *indicative*, contextual GRL entities of *resources* and *actors*, preferring instead to represent *indicative* information in Jackson context diagrams. Also note that the entities in the goal model are grouped by dashed-line ellipses (RA, RB, RC1 and RC2 in Fig. 5). The goal entities within the ellipses represent requirements referring to context diagrams in the progression at equivalent levels of abstraction (DA, DB, DC1 and DC2 in Fig. 5). The integration of the goal model and the context diagram at each level in the progression presents a *problem diagram* for that level of abstraction.

We now describe this progression in finer detail. The macro-level business strategy is the top-level problem, described by RA and DA. It is here that we bound the requirements engineering problem, because it is here that SEJ bounds the problem. To understand the optative part of the business strategy we explore the goal model at its highest level (RA).

SEJ's requirement is to *Stock products that customers want when they want them according to changing needs*. This meets the goals *Reduce lost opportunity/customer* and Minimize *unsold perishables* and is achievable by *Support just-in-time (JIT) delivery*, which in turn supports the goals of *Maximize use of limited floor space, Shorten inventory turns* and *Maintain constant freshness of perishable goods*. The scope of the requirement set RA can only be understood by an exploration of the context DA to which it refers. Context diagram DA shows the *machine* domain *SEJ System*. This retrieves the sales and customer data it needs from the *Franchise Store* domain (*interface a*). To know what to deliver *just in time* (a goal in RA), the needs of the *Shop Customer* must be understood (*interface b*). The *machine* domain provides the necessary information to the *Supplier* (*interface f*), which in turn uses a *Logistics Partner* to deliver the goods, supporting the goal *Support just-in-time (JIT) delivery*.

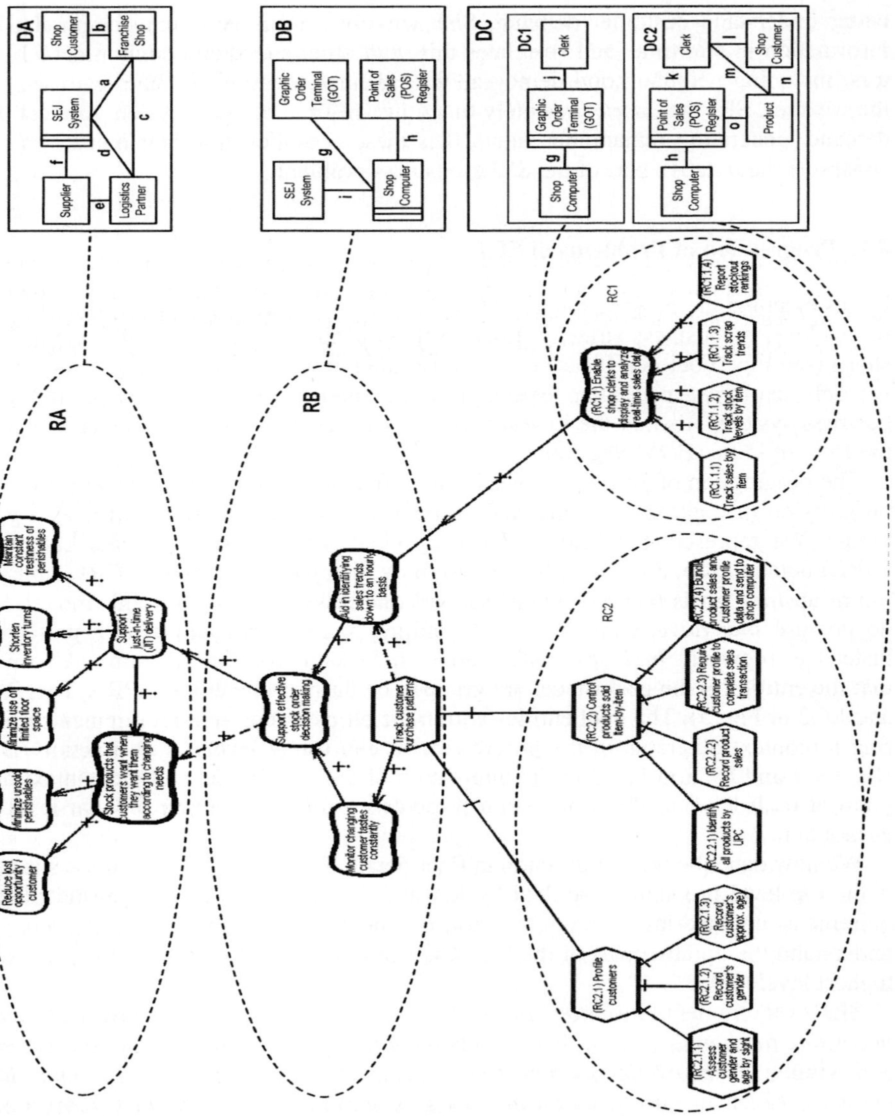

Fig. 5. SEJ Progression of Problems: Integrated Goal Model and Context Diagrams

The shared phenomena *e* represents the delivery schedule, the goods themselves and delivery address. The *Logistics Partner* must also provide its schedule details back to the *SEJ system* (*interface d*) about its delivery (*interface c*). The *Franchise Store* also provides details of the sales of perishable goods, how the store is stocked and how this affects the sale of goods.

The context diagram at DB is a progression from that of DA. The Requirement RB refers to context DB, which is the *Franchise Store* of DA with a focus on the *Shop*

Computer as the *machine*. The *Shop Computer* also aided by a two shop devices in DB, the *Point of Sales (POS) Register* and the *Graphic Order Terminal (GOT)*. RB focuses on how the *Franchise Store* can work effectively to meet SEJ's requirements with focus on the requirements for the *Shop Computer*.

The Requirement Set RB has three goals and a supporting task help achieve the requirements of RA. The RB requirement *Support effective stock order decision making* helps achievement of the RA goals of *Stock products that customers want when they want them according to changing needs* and *Support just-in-time (JIT) delivery*. Thus, in order to *Support effective stock order decision making* one must *Monitor changing customer tastes constantly* and *Identify sales trends down to an hourly basis*. The task of *Track customer purchase patterns* directly supports *Monitor changing customer tastes constantly,* and indirectly supports *Identify sales trends down to an hourly basis*.

The requirement set RC refers to domain DC, whose focus is on *the GOT and POS devices* represented in separate sub-context diagrams DC1 and DC2 referred to by requirement sets RC1 and RC2 respectively. Because of space constraints, we only explore RC2 in this paper (below). Note that in RC1 and RC2 we have labeled individual requirements such as *RC2.1*, or *RC2.2.1*, etc. as we explore part of the system in greater detail.

4.2   Transition from Goal Model to Standard Problem Diagram

At the RC*i*-DC*i* level of abstraction, we have decomposed both problem context and requirements to a relatively low level much closer to the *machine*. At this point it is useful to make a transition from RC2-DC2 as represented in Fig. 5 to a standard Jackson problem diagram shown in Fig. 6 below.

Goal models are useful for traceability and understanding linkages between requirements at different levels or abstraction, but they bloat and become unwieldy with low-level requirements. To avoid this, we transition to a standard problem diagram. Note that we have maintained a correspondence between requirements, interfaces, and domains of interest between Fig. 5 and Fig. 6 for traceability. The problem diagram Fig. 6 does not represent all aspects of RC2-DC2 in Fig. 5. We present only a partial example of RC2-DC2 below.

The *POS*, treated as the *machine* in DC2, has two tasks that have to be performed appearing in requirement set RC2 to satisfy *Tracking customer purchase patterns* (in RB). These are *Profile Customers (RC2.1)* and *Control products sold item-by-item (RC2.2)*. The *Clerk* also uses the *POS* (*interface k*) to process customer transactions. The *Shop Customer* takes his *Products* to the *Clerk* for checkout (*interfaces k and o*). The *Clerk* scans the *Product* information via the barcode to *Record all product sales (RC2.2.2,* interface *k* and *o*) into the *POS*. In order to achieve the task *Profile customer (RC2.1)*, the *POS* prompts the *Clerk* to *Record customer's gender (RC2.1.2, interface k)* and *Record customer's approximate age (RC2.1.3, interface k)* both of which the *Clerk* must *Assess...by sight (RC2.1.1, interface l)*. One of the requirements of the *POS* is that the cash drawer not open and sales cannot complete until the *Clerk* has entered this information satisfying *Require customer profile to complete sales transaction (RC2.2.3, interface k)*. As the clerk scans product barcodes, the *POS* records the sale enabling *Item-by-item control* of inventory as products are sold off the store's shelves (RC2.2). The POS must then *Bundle sales data customer profile data and send to shop computer (RC2.2.4, interface h)* for storage, processing, and transmission to the SEJ System (*interface k*). This sales data will be used for analysis, meeting the goal in RB of *Track customer purchasing patterns*.

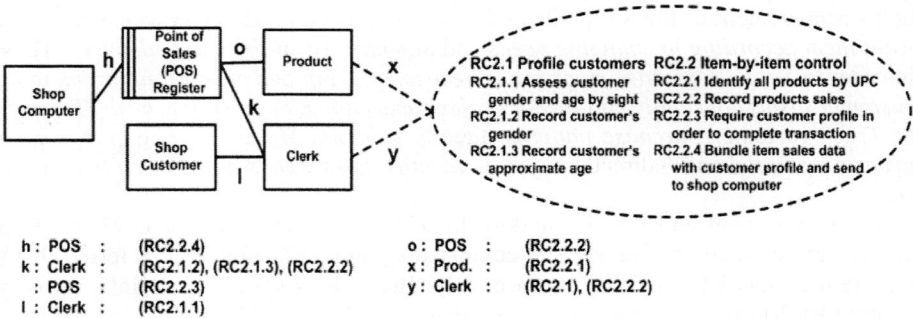

h : POS   :   (RC2.2.4)   o : POS   :   (RC2.2.2)
k : Clerk :   (RC2.1.2), (RC2.1.3), (RC2.2.2)   x : Prod. :   (RC2.2.1)
    : POS   :   (RC2.2.3)   y : Clerk :   (RC2.1), (RC2.2.2)
l : Clerk :   (RC2.1.1)

Fig. 6. Standard Jackson Problem Diagram RC2-DC2

It is possible to decompose the Jackson problem diagram in Fig. 6 to increasingly refined problem diagrams and ultimately down to a specification of the *machine*. We do not do this here, because this would be a standard exercise in Problem Frames, many examples of which are described in Jackson's book [9].

4.3   Discussion of the Integrated Approach

The indicative problem context diagrams in the progression of problems and the optative goal model mutually complement each other. Goal modeling provides explicit linkage between requirements in problem diagrams at different levels of abstraction as determined by the context diagrams. This integrated approach thus offers a means of helping ensure that requirements are consistent with and provide support for business strategy.

However, this is not enough. We recognize the importance of describing business processes in requirements for e-business systems, and that both GRL and Jackson problem diagrams are awkward notations for business process modeling (BPM). We have proposed that it could be beneficial to integrate recognized BPM notations into our approach to describe business processes in detail when needed [34].

Describing business processes of e-business systems without considering their business rationale is insufficient for requirements validation when modeling e-business systems [35]. As a business process is "a set of partially ordered activities intended to reach a goal" [36], we presumably have a means to integrate business process models into our larger requirements model by anchoring them to goals in the goal model. Anchoring business process models to the goal model is a possible means to justify how the business process contributes to the overall business strategy objectives of the system in context of the business model, enabling business process validation against the coherent objectives of business strategy.

5 Conclusion

In this paper, we present an integration of recognized requirements engineering approaches to model the business strategy requirements of e-business systems domain. Problem diagrams provide context for the indicative business problem and can be projected down to system requirements. Goal modeling captures the optative requirements that fit the problem context. Each projected sublevel of the goal hierarchy in itself represents the requirements set for its referred context.

While the approach we propose is based upon ongoing research that is still in its early stages, the integration of Jackson's Problem Frames approach and goal-oriented modeling techniques offer promise as a requirements engineering modeling tool for e-business systems.

References

1. Chan, Y., Huff, S., Barclay, D., Copeland, D.: Business Strategic Orientation: Information Systems Strategic Orientation and Strategic Alignment. Information Systems Research 8 (1997) 125-150
2. Croteau, A.-M.: Organizational and technological infrastructures alignment. In: Proc. Proceedings of the 34th Annual Hawaii International Conference on System Sciences (2001)
3. Brancheau, J. C., Janz, B. D., Wetherbe, J. C.: Key issues in information systems management: 1994-95 SIM delphi results. Mis Quarterly 20 (1996) 225-242
4. Watson, R. T., Kelly, G. G., Galliers, R. D., Brancheau, J.: Key Issues in Information Systems Management: an International Perspective. Journal of Management Information Systems 13 (1997) 91-115
5. Reich, B. H., Nelson, K. M.: In Their Own Words: CIO Visions About the Future of In-House IT Organizations. The DATA BASE for Advances in Information Systems 34 (2003) 28-44
6. Standing, C.: Methodologies for Developing Web Applications. Information Software and Technology 44 (2001) 151-59
7. Gordijn, J., Akkermans, J.: Value-based requirements engineering: exploring innovative ecommerce ideas. Requirements Engineering Journal 8 (2003) 114-135
8. Rolland, C., Prakash, N.: From conceptual modelling to requirements engineering. Annals of Software Engineering 10 (2000) 151-176
9. Jackson, M.: Problem Frames: Analyzing and Structuring Software Development Problem. 1st edn. Addison-Wesley Publishing Company (2001)
10. Bleistein, S., Cox, K., Verner, J.: Problem Frames Approach for e-Business Systems. In: Proc. The 1st International Workshop on Advances and Applications of Problem Frames (IWAAPF) at the International Conference on Software Engineering (ICSE'04) (2004) 7-15
11. Weill, P., Vitale, M.: Place to Space: Moving to eBusiness Models. Harvard Business School Publishing Corporation, Boston (2001)
12. Castro, J., Kolp, M., Mylopuolos, J.: Towards Requirements-Driven Information Systems Engineering: the Tropos Project. Information Systems Journal 27 (2002) 365-89
13. Ben Lagha, S., Osterwalder, A., Pigneur, Y.: Modeling e-Business with eBML. In: Proc. THE 5th INTERNATIONAL WORKSHOP ON THE MANAGEMENT OF FIRMS' NETWORKS (CIMRE'01) (2001)
14. Gordijn, J., Akkermans, J.: Does Business Modeling Really Help? In: Proc. The 36th Hawaii International Conference on Systems Sciences (HICSS-36) (2003)

15. Erdgogmus, H., Favaro, J., Strigel, W.: Return on Investment. IEEE Software (2004) 18-22
16. Porter, M.: Strategy and the Internet. Harvard Business Review 79 (2001) 62-78
17. Rolland, C., Souveyet, C., Ben Achour, C.: Guiding Goal Modeling Using Scenarios. IEEE TRANSACTIONS ON SOFTWARE ENGINEERING 24 (1998) 1055-71
18. Oliver, R. W.: What is Strategy, Anyway? Journal of Business Strategy (2001) 7-10
19. Chesbrough, H., Rosenbloom, R.: The Role of the Business Model in Capturing Value from Innovation: Evidence from Xerox Corporation's Technology Spin-Off Companies. Industrial and Corporate Change 11 (2002) 529-555
20. Magretta, J.: Why Business Models Matter. Harvard Business Review 80 (2002) 85-92
21. van Lamsweerde, A.: Goal-Oriented Requirements Engineering: A Guided Tour. In: Proc. 5th IEEE International Symposium on Requirements Engineering (2001) 249-63
22. Zave, P., Jackson, M.: Four Dark Corners of Requirements Engineering. ACM Transactions on Software Engineering and Methodology 6 (1997) 1-30
23. Gross, D., Yu, E.: From Non-Functional Requirements to Design Through Patterns. Requirements Engineering Journal 6 (2001) 18-36
24. Yu, E., Liu, L.: Modelling Strategic Actor Relationships to Support Intellectual Property Management. In: Proc. 20th International Conference on Conceptual Modelling, ER-2001 (2001)
25. Liu, L., Yu, E.: From Requirements to Architectural Design - Using Goals and Scenarios. In: Proc. ICSE-2001 (STRAW 2001) (2001) pp. 22-30
26. Anton, A. I., Potts, C.: The Use of Goals to Surface Requirements for Evolving Systems. In: Proc. ICSE-98: 20th International Conference on Software Engineering (1998) 157 - 66
27. Kolber, A. B., Estep, C., Hay, D., Struck, D., Lam, G., Healy, J., Hall, J., Zachman, J. A., Healy, K., Eulenberg, M., Fishman, N., Ross, R., Moriarty, T., Selkow, W.: Organizing Business Plans: The Standard Model for Business Rule Motivation. The Business Rule Group, http://www.businessrulesgroup.org/brghome.htm (2000)
28. Bleistein, S., Aurum, A., Cox, K., Ray, P.: Linking Requirements Goal Modeling Techniques to Strategic e-Business Patterns and Best Practice. In: Proc. Australian Workshop on Requirements Engineering (AWRE'03) (2003)
29. Bleistein, S., Aurum, A., Cox, K., Ray, P.: Strategy-Oriented Alignment in Requirements Engineering: Linking Business Strategy to Requirements of e-Business Systems using the SOARE Approach (to appear). Journal of Research and Practice in Information Technology (2004)
30. Bensaou, M.: Seven-Eleven Japan: Managing a Networked Organization. INSEAD Euro-Asia Centre, Case Study (1997)
31. Whang, S., Koshijima, C., Saito, H., Ueda, T., Horne, S. V.: Seven Eleven Japan (GS18). Stanford University Graduate School of Business, (1997)
32. Rapp, W. V.: Information Technology Strategies: How Leading Firms Use IT to Gain an Advantage. Oxford University Press, New York (2002)
33. University of Toronto. Goal-Oriented Requirements Language. University of Toronto [Online]. Available: http://www.cs.toronto.edu/km/GRL
34. Bleistein, S., Cox, K., Verner, J.: RE Approach for e-Business Advantage. In: Proc. Requirements Engineering Foundation for Software Quality (REFSQ) (2004)
35. Gordijn, J., Akkermans, H., Van Vliet, H.: Business modelling is not process modelling. Conceptual Modeling for E-Business and the Web, Proceedings 1921 (2000) 40-51
36. Hammer, M., Champy, J.: Reengineering the corporation : a manifesto for business revolution. 1st edn. HarperBusiness, New York, NY (1993)

A Method and Tool for Wide Audience Requirements Elicitation and Rapid Prototyping for Mobile Systems

Matti Rossi and Tuure Tuunanen

Helsinki School of Economics, Information Systems Science
Po Box 1210, FIN-00101, Helsinki, Finland
{mrossi,tuure.tuunanen}@hse.fi

Abstract. In recent years, consumer oriented information systems development has become increasingly important matter, as more and more complex information systems are targeted towards consumer markets. We argue that developing IS for non-organizational users creates new problems, which IS and requirement engineering (RE) community should attend to. First of all, the elicitation of requirements becomes more difficult as usually consumers do not explicitly know what they want, and it is difficult for them to express their ideas. To support different views of product development, project management and design, the method should present requirements in a 'rich enough' way to avoid overloading management, but in the same time giving designers the detailed information they need. Furthermore, to facilitate iterative requirements development the method should allow for rapid development of prototypes from designs. To support these goals we have constructed an enhanced requirements elicitation and mobile system construction method and its support environment within Metaedit+ Meta CASE tool. We based our method on Critical Success Chains (CSC) method, which supports top-down approach for planning, but also provides for wide participation of IS customers to get rich information. The high level results of CSC are turned into mobile applications running in Symbian platform by using a novel domain specific method that supports generation of executable environments from specifications.

1 Introduction

We argue that the traditional views of requirement engineering process and methods do not live up to the challenge that the information systems engineering community is facing when dealing with the development of software and embedded products for mass market audiences. Examples of these new types of software are applications for palm top devices, Java powered phones, and JAVA enabled Digital TV sets. The companies developing systems for these platforms have noticed that many novel innovations that look good on tests do not win the acceptance of customers.

The received view (e.g. [10, 21]) believes that requirements are out there to be gathered by the requirements engineers, and firms have used managers and engineers as proxies for end-users to develop applications without knowing what the customers want or are willing to pay for [18]. The problem has been that people have thought

S. Wang et al. (Eds.): ER Workshops 2004, LNCS 3289, pp. 629–640, 2004.
© Springer-Verlag Berlin Heidelberg 2004

they only have to find the right informants and use the right techniques to achieve the complete specification. Researchers [10, 21] have seen that by selecting and prioritizing the requirements into usable sets an agreement can be reached on the common goal.

In contrast, we present a view that often the end-users cannot express their needs [15], and as the end-users are scattered and outside the traditional information systems development (ISD) environment, a single organization, it is not a trivial task to reach them in the first place. Thus we argue that new methods are needed to elicit the hidden needs of users of information systems that are targeted at a very wide audience of end-users, such as consumers. As many of these systems do not exist currently, we should also consider how to extract requirements and relationships that are not historical. In addition, we argue that these goals and specifications should then be expressed in a semi-formal language that is accessible for both end-users and business representatives. We see this as critical for integrating the more sophisticated elicitation methods with the traditional development process. Furthermore we believe that rapid iterative prototyping using domain specific modeling languages supports refinement of the requirements engineering (RE) results. To support this, we have constructed an environment for generating new services for Symbian Series 60 mobile platform, which can be used to develop new service sketches based on user requirements.

In this paper, we seek to demonstrate how this kind of approach could work in practice by applying method engineering to develop a support environment for a new RE method. In the next section we present a method that supports a cognitive elicitation method, critical success chains (CSC). In the third section, we define method engineering and then create a Meta model of the CSC method. After this, we present a method construct based on the research done earlier on the CSC method. Finally, we discuss the possibilities of the constructed integrated environment and identify possibilities for further research.

2   Wide Audience Requirements Engineering

Within the Software Engineering (SE) literature, Requirements Engineering (RE) has been focusing on the issues surrounding the problems in eliciting and managing the changing requirements (e.g. [13]). Requirements are generally specified as something that the product must do, or something that it should achieve when considering quality [24]. However, if considered from the information systems viewpoint requirements can also be defined as descriptions of how the system should behave (i.e. application domain information, constraints on the system's operation, or specifications of a system property or attribute) [10]. In RE, the requirements elicitation has usually been done at the beginning of the SE process, or continuously during the process, as in spiral development approaches (for example [6]).

Most of the current RE approaches assume that the users are known, and therefore the requirements can be elicited from them, using some predefined semi-formal methods. However, in the case of new product development for wide audience end-users this is not the case [26]. Wide audience end-users have been defined as a group of end-users that are external to the organization developing the information systems (e.g. consumers or extranet users). This leads to problems in committing the end-users to participate in development and, more importantly, actually finding the end-users for the developed system. For systems like these, the traditional tools and techniques offered by SE and RE communities are not suitable, as in many cases we do not even have ways of getting at the users' opinions.

Researchers have realized that meeting consumers' demands is different than when developing an information system for an organization. They have pointed out that prioritization of requirements, continuous improvement of requirements, and a short period of time-to-market are vital [22]. However, RE has not been dealing heavily yet with consumer oriented products and eliciting information for these projects. These projects can easily involve something that the end-user has not even thought of being possible, like the ability to download JAVA applications to mobile phones and Digital TV sets. Some of the research on Commercial-of-the-Self (COTS) processes and market driven requirements determination deals with closely related areas, but it makes stringent assumptions about the availability of the end-users and the possibility of a relatively linear RE process. However, as Orlikowski [15] has pointed out, when the products are taken into use, they are reinterpreted and innovatively applied by end-users in ways not envisioned by the developers (an example of this is SMS messages, which were intended as operator messages for end users).

Pohl's [21] model of three dimensions of requirements engineering has been considered a good way of structuring the RE process, and we also use his approach as the main base of our method. The first dimension is specification. We prefer the term "elicitation" to "specification", to avoid the suggestion that requirements are out there to be collected simply by asking the right questions [7]. The elicitation methods used are mostly unstructured, such as brainstorming, open interviews and document inspection [11]. We suggest a structured cognitive method, i.e. critical success chains [17], for elicitation of the requirements of wide audience end-users [26]. The Representation dimension presents the gathered requirements, using some form of either diagrammatical notation or natural language prose. In this dimension, the important issues are such as the ease of understanding of the representation, its compactness etc. To these demands, we present an answer by means of rich presentation of end-users needs [20] and organizing them with a CASE tool in a structured way. The third dimension, the agreement, in turn deals with the issue of reaching a common vision, or agreement on the key requirements and the goals of the system. We suggest a preliminary solution for the agreement dimension, by using the concepts of CSC to communicate the requirements to different stakeholders, but at the same time managing the changing requirements with a CASE tool. In the next section, we develop an integrated support environment for the method within the MetaEdit+ tool.

3 Method Engineering

Method engineering provides methods and processes to specify, make explicit, codify, and communicate method knowledge as well as technical tools to enact such processes effectively. In order to model IS development methods we need a set of concepts during ME that can capture the content and form of any development method into a *meta-model*. In its simplest form, a meta-model is a conceptual model of a development method [2]. Consequently, meta-modeling can be defined as a modeling process, which takes place on one level of abstraction and logic higher than the primary modeling process [4].

3.1 Method Engineering Environment

MetaEdit+ is a customizable CASE environment that supports both CASE and meta-CASE functionality for multiple users within the same environment. It supports and integrates multiple methods and includes multiple editing tools for diagrams, matrices and tables. Architecture of MetaEdit+ is a client-server environment with the server containing a central meta-engine and object repository and various modeling tools (diagramming, matrix etc.) working as clients. The repository is implemented as a database running in a central server: clients communicate only through shared data and state in the server. All information in MetaEdit+ is stored in the Object Repository, including methods, diagrams, objects, properties, and even font selections.

The core conceptual types of a method are defined at the repository level and can be modified by the method developers. Method engineers can change components of a method specification even while system developers are working with older versions of the method. The method can be developed and simultaneously tested in method engineers' workstation much in the same way as described in [5]. The data continuity, (i.e. that specification data remains usable even after method schema changes), is confirmed by a number of checks and limitations to the method evolution possibilities. The idea is that the user can always be guaranteed data continuity while working with partial methods.

4   Construction of the Method

We begin the task of engineering the enhanced method by first choosing the specification language and process. Various methods have been used and are used for the elicitation and text books often mention interviews, scenario analysis, use-cases, Soft Systems Methods, observation and social analysis, ethnographic analysis, requirements reuse, and prototyping. The number of techniques and methods developed for this is almost unlimited and Lauesen [12] describes nineteen different ones. Nuseibeh and Easterbrook [14] have taken a more structured approach and they have developed a classification of methods, which divides them in six metagroups: 1) traditional techniques; 2) group elicitation; 3) prototyping, model-driven techniques; 4) cognitive techniques; 5) contextual techniques. For our paper, we have chosen a cognitive elicitation method.

The selected conceptual structure and process, critical success chains (CSC), originates from Information Systems Science and was developed by [16, 20]. The general process of using the CSC is described later in table 2 when we provide a step-by-step description of the method. However, before examining the process we must define a meta-model. Therefore, we did a simplified version of the Critical Success Chains model using GOPRR meta-modeling language [9] within MetaEdit+. The formal metamodel allows the analysis of the methods conceptual structure, as well as, immediate tool support for modeling using the method.

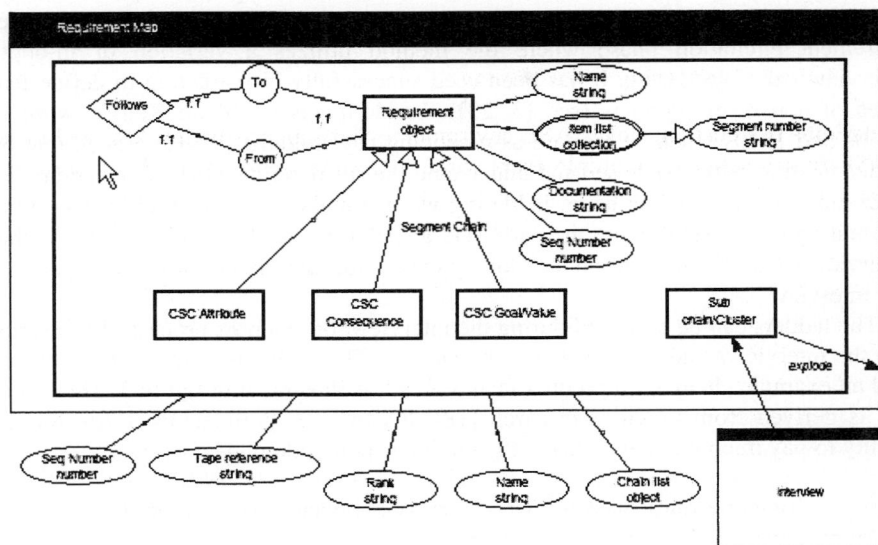

Fig. 1. Meta model of CSC

4.1   Conceptual Specification of the Method

Figure 1 above shows the conceptual definition of the CSC method according to GOPRR. The method definition is done by identifying the key objects of the method, connecting them by the CSC relationship types and adding properties to those. The CSC relationship types are further explained in section describing the CSC method in below. Roles define how different object types can participate in the relationships, and in our model it describes the interaction connection between the CSC meta-model and the data collect by an in-depth interviewing method stored in a spreadsheet. Objects can be connected by explosion/decomposition links between the diagram types. When these are defined with their graphical representations, we have a complete conceptual specification of the method, which can be used immediately as a template for new models (see [9] for details). In the end of the chapter, we tie the presented meta-model, figure 2, to the CSC method. This is offered in figure 5. We also present goals that can be archived with the engineered method.

4.2 Process Specification of the Method

The process of CSC starts by selection of participants for elicitation process. In case of wide audience IS [26] a lead-user concept has been used for finding potential users for the service/application who are adapting innovations at first [25]. After the selection process, the prospects are contacted and during, for example, a telephone conversation, stimuli for the system are collected informally. These are later used in requirement elicitation phase where the method utilizes a variation of in-depth interview called laddering. It has been used successfully in marketing to define features of consumer products (see [3, 23]). Laddering is based on Kelly's work in 1930s and 1940s when he worked as practicing psychologist [8]. He argued that by understanding how people see and understand the surrounding world, one can predict their behavior. Hence, by using laddering we can make implicit requirements more explicit and understand what the end-users actually want. The crucial benefit is that interviewees do not have to be technology experts or they do not have to prepare for the interview.

The ladders can be collected during the interview, or in a post-process of transcribing the interviews and reconstructing the ladders. The individual ladders form chains and an example chain, on a product feature level, is illustrated in figure 3. The example is derived from previous research [18]. It expresses participants desire for the ability to pay micropayments (0.1-2 €) with his/hers mobile phone.

Table 1. Critical Success Chains Process description, modified from [17]

CSC Process	Objectives
Prestudy Preparation Determine scope & participants. Collect project idea stimuli.	Determine scope to manage complexity. Select participants to represent views you want to understand. May be employees at various levels, suppliers, customers, and experts. Arrange for data collection. Collect interview stimuli.
Data Collection Elicit personal constructs from org. members.	Ask participant to rank-order stimuli on importance. Ask series of "why would this system be important..." questions to collect consequence and value data. Ask series of "what is it about this system that makes you think it would do that..." questions to collect attribute data. Record answers as linked chains. Collect several chains from each participant.
Analysis Aggregate personal constructs into CSC models.	Interpret individual statements and label consistently across participants. Cluster chains. Map clusters into network models.
Ideation Workshops Elicit feasible strategic IS from technical and business experts and customers.	Recruit workshop participants with technical and business skills. Evaluate CSC network models and develop 'back-of-envelope-level' ideas for IS projects that satisfy the relationships implicit in the models.

When considering the requirement explosion that usually happens in an RE project, as well as the described coherency and branching issues of the chains, an appealing way would be using a spreadsheet program, like Microsoft Excel, to structure the ladders. This would also enable researchers to use hyperlinks between the chains that branch out. In addition, it would be a simple task to add hyperlinks of the interviews themselves to a chain with a time index referring to a MP3 file for example.

In analysis phase of CSC, these are clustered using Ward's method [1] by researchers and the results are presented using a rich graphical presentation showing the aggregated features, the consequences resulting from them, and the values or the goals that explain why the end-users want these. An example of this is presented in Figure 4 that describes mobile wallet application for 3rd generation mobile phone reported in [19, 20]. It can be seen that 'access to account' and 'ability to pay small payments' are important systems features and one of the key reasons is avoiding fraud. These, in turn, result in more trust and economic security. The purpose of these 'CSC maps' are to show managers within few minutes what are the most beneficial features.

In the next phase, these graphical presentations, or CSC maps, are presented in an initial workshop in order to introduce the key features of the system to the client's R&D people who evaluate the maps and identify the feasible project ideas. The ideas are developed to a 'back-of-the-envelope' standard, so that participants can identify as many ideas as possible. For each system, the participants are supposed to label the system, briefly describe its nature, its likely architecture, the resources required to develop it, its cost, likely sourcing, the magnitude of the risk, and its expected impact on the organization. Following this, researchers have suggested that the results are then returned to the R&D function of the firm, and used in the system development process [17].

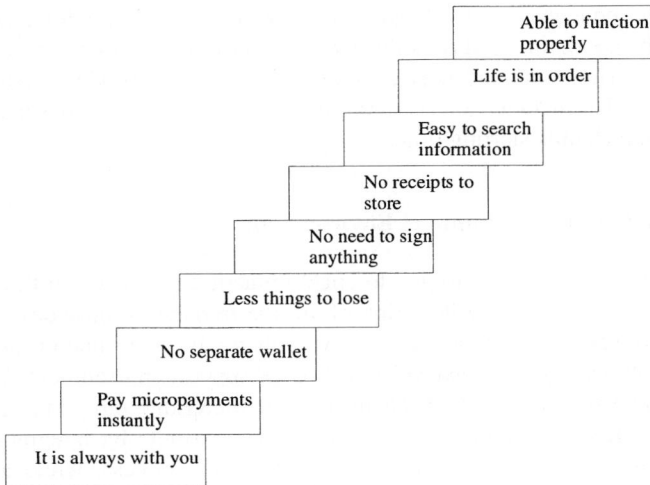

Fig. 2. Example chain collected from a participant, adapted from [18]

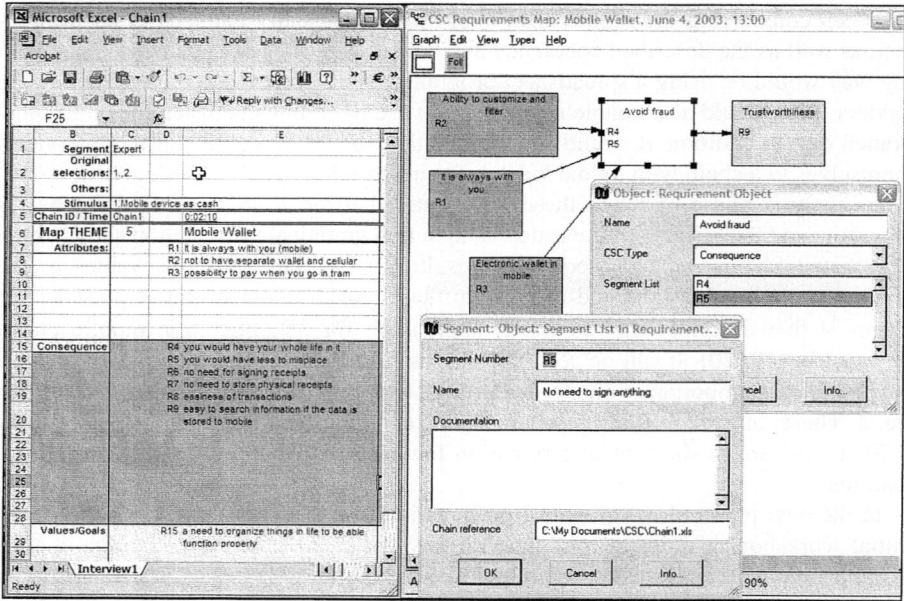

Fig. 3. The implemented CSC environment

We have integrated the CSC elicitation method and its philosophy of representation of the requirements into the Metaedit+ Method engineering environment. In figures 3 and 4 below, we demonstrate the engineered method and its support environment. We created an example model that constitutes a part of the CSC map, and the ladders are mapped back to this specific diagram. The model includes objects (attributes, consequences, and values) with references to the individual segments i.e. the ladders in the chain. Additionally, we show how it is possible to enable hyper linking to the original spreadsheet containing the actual interview data. The given example chain also includes a hyperlink into the interview recording that is indexed according to the chain's starting time.

4.3   Constructed Method Support Environment

Our theoretical framework is argued to enable practitioners to avoid three important issues faced daily basis in the RE. First of all, the framework provides an organized way of how to handle changing requirements coherently. The same problems that we were faced with a study that used MS Excel based support environment. To avoid this we developed a version of the CSC chains into the MetaEdit+ environment.

We believe that our method and tool together contribute by integrating complex requirement elicitation methods into the standard RE process, while avoiding the traceability dilemma. However, even more importantly the enhanced CSC method can be used to identify critical requirements for the project and needs of end-users, which are not explicit. On the practical side, the developed tools allow for immediate

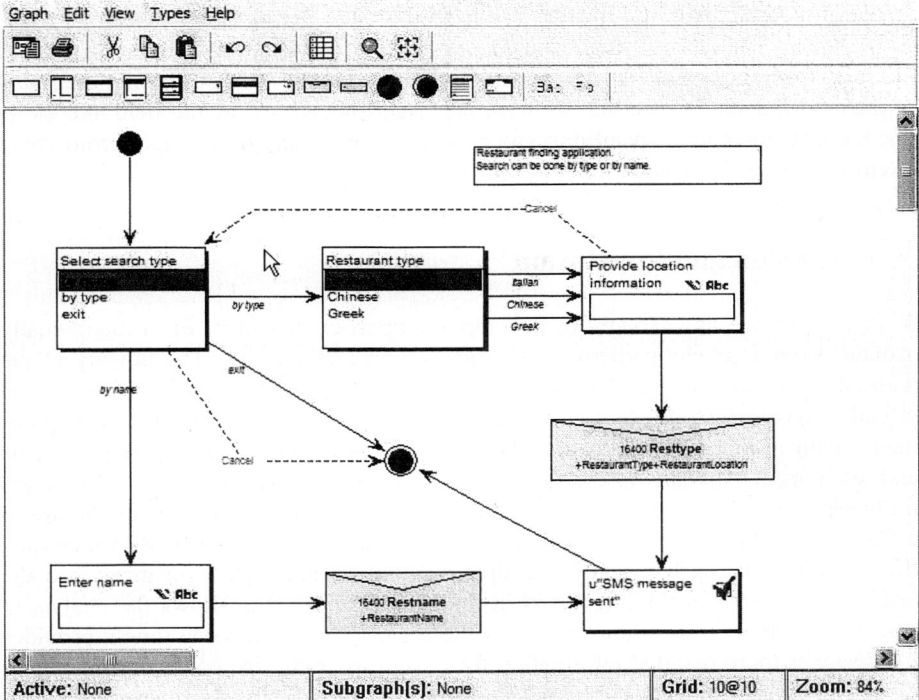

Fig. 4. Method for developing Symbian applications

tool support for the method and they allow us to change the method easily to accommodate for changed development needs. The developed environment is expected to go into field trials in several organizations during the fall of 2004.

The implemented environment, which can be seen in figure 4, supports the CSC method by linking together the original interviews of users, which are kept in MP3 files. After importing the requirements into the CASE tool, we can use the requirements information for product concept generation. The requirements are transformed into product prototypes or other form of presentation. We suggest that by using a domain specific modeling method within MetaEdit+ with proper domain knowledge we can produce prototypes from the requirements very rapidly.

There are several implementations available and we are currently working on an existing method that generates running prototypes for the Symbian[1] platform (see figure 4). As the platform has emulators, running on browser windows, that represent Symbian enabled phones, the prototypes can be produced and run seamlessly in either a workstation, over the internet, or in a Java enabled phone. This kind of environment enables us to provide prototypes or product concepts for validation almost immediately. It also makes the presentation of alternatives easier. Figure 4 describes an ex-

---

[1] About Symbian operating system, see http://www.symbian.com

ample of a domain specific method that can be used to define new services for Symbian phones. The defined method produces running software for Symbian Series 60 platform from these models without the need for coding by hand. Currently the status of the environment is such that we have the Excel sheets used by the field interviewers, the CSC method in MetaEdit+ and a preliminary version of the rapid prototyping environment available and running.

5 Discussion and Conclusion

In this paper, we have sought to develop a support environment for a theoretically grounded requirements gathering and organization method [17]. The novelty of the method is its integrated computer support environment for the whole gathering and organization process. We present a sophisticated way for elicitation of the requirements with a cognitive approach. The approach facilitates extracting requirements and relationships that are not historical i.e. do not have an existing system to compare to. In representation level, we enhanced the classic CSC model to facilitate the use of a CASE tool. In addition, we introduced a novel approach of using MP3 interview files enabling the system analyst to drill down from the high level maps into the original comments in the interviews in the source tapes. This allows the analyst to easily recall the critical requirements for the project, but in the same time it provides the designer more detailed information for design work. In this paper our main goal was to show how this could be used jointly with a prototyping tool to verify the requirements. We believe that this kind of environment will allow for very fast iteration cycles between the requirement gathering and user trials, which would make new service development far less risky.

There are several limitations in our current implementation. First of all, the selection of the participants needs more attention. We aim to create a technique that would produce easy tools for practitioners for prioritizing the requirements with a simple process, and additionally a little effort from the potential wide audience end-users. In addition, more research has to be done in many practical issues such as how to input information into the CASE tool in field research. The CASE tool may be too complex tool to use during laddering interviews and mitigating solutions probably is needed. However, the CASE tool provides extensive selection import filters, such XML. Thus more common tools can be used during data collection if needed and the data can be later imported to the CASE tool.

In the future, we seek to test the constructed support environment in real world settings. The method is currently in use in a few case organizations in a manual form [20]. In the next phase, we seek to deploy the developed support environment into a case to develop new innovative mobiles services with data gathering in three countries. This should lead into refinements in both the method and the support environment.

References

1. Aldenderfer, M.S. and R.K. Blashfield, *Cluster Analysis*. Sage University Paper series on Quantitative Applications in the Social sciences. Vol. series no. 44. 1984, Beverly Hills and London: Sage Pubns. 87.

2. Brinkkemper, S., *Formalisation of Information Systems Modelling*, in *Computer Science Department*. 1990, Univ. of Nijmegen: Amsterdam.

3. Gengler, C.E. and T.J. Reynolds, *Consumer understanding and advertising strategy: analysis and translation of laddering data*. Journal of Advertising Research, 1995. **35**(4): p. 19-33.

4. Gigch, J.v., *Systems design and modeling and metamodeling*. 1991, New York: Plenum Press.

5. Hedin, G., *Incremental Semantic Analysis*, in *Department of Computer Sciences*. 1992, Lund University: Lund.

6. Iivari, J., *Hierarchical spiral model for information system and software development Part 1: theoretical background*. Information and Software Technology, 1990. **32**(6): p. 386 - 399.

7. Jirotka, M. and J. Goguen, eds. *Requirements Engineering: Social and Technical Issues*. 1994, Academic Press.

8. Kelly, G.A., *Psychology of Personal Constructs*. 1955, New York, NY: W. W. Norton and Co.

9. Kelly, S., K. Lyytinen, and M. Rossi, *MetaEdit+: A Fully Configurable Multi-User and Multi-Tool CASE and CAME Environment*, in *Advanced Information Systems Engineering, proceedings of the 8th International Conference CAISE'96*, P. Constapoulos, J. Mylopoulos, and Y. Vassiliou, Editors. 1996, Springer-Verlag: Berlin. p. 1-21.

10. Kotonya, G. and I. Sommerville, *Requirements Engineering, Processes and Techniques*. 2002: John Wiley.

11. Kujala, S., *User involvement: a review of the benefits and challenges*. Behaviour & Information Technology, 2003. **22**(1): p. 1-16.

12. Lauesen, S., *Software Requirements, Styles and Techniques*. 2002: Addison-Wesley.

13. Neill, C.J. and P.A. Laplante, *Requirements Engineering: The State of the Practice*. IEEE Software, 2003. **20**(6): p. 40-45.

14. Nuseibeh, B. and S. Easterbrook, *Requirements engineering: a roadmap*, in *The Future of Software Engineering*, A. Finkelstein, Editor. 2000.

15. Orlikowski, W.J., *CASE Tools as Organizational Change: Investigating Incremental & Radical Changes in Systems Development*. MIS Quarterly, 1993. **17**(3): p. 309 - 340.

16. Peffers, K. and C. Gengler, *How to Identify New High-Payoff Information Systems for the Organization*. Communications of the ACM, 2003(1).

17. Peffers, K., C.E. Gengler, and T. Tuunanen, *Extending critical success factors methodology to facilitate broadly participative information systems planning*. Journal of Management Information Systems, 2003. **20**(1): p. 51-85.

18. Peffers, K. and T. Tuunanen. *Using Rich Information to Plan Mobile Financial Services Applications with Maximum Positive Impact: a Case Study*. in *Tokyo Mobile Round Table*. 2002. Tokyo, Japan.

19. Peffers, K. and T. Tuunanen. *Using Rich Information to Plan Mobile Financial Services Applications with Maximum Positive Impact: a Case Study*. in *Mobile Round Table*. 2002. Tokyo.

20. Peffers, K. and T. Tuunanen, *Planning for IS Applications: a Practical, Information Theoretical Method and Case Study In Mobile Financial Services.* Information & Management, 2004. **in press**.
21. Pohl, K., *Three Dimensions of Requirements Engineering: framework and its application.* Information Systems, 1994. **19**(3): p. 243-258.
22. Regnell, B., et al., *An Industrial Case Study on Distributed Prioritisation in Market-Driven Requirements Engineering for Packaged Software.* Requirements Engineering Journal, 2001(6): p. 51-62.
23. Reynolds, T.J. and J. Gutman, *Laddering theory, method, analysis, and interpretation.* Journal of Advertising Research, 1988. **28**(1): p. 11-31.
24. Robertson, S. and J. Robertson, *Mastering the requirements process.* 2002: Addison-Wesley.
25. Rogers, E.M., *New Product Adoption and Diffusion.* Journal of Consumer Research, 1976. **2**: p. 290-301.
26. Tuunanen, T., *A New Perspective on Requirements Elicitation Methods.* JITTA: Journal of Information Technology Theory & Application, 2003. **5**(3): p. 45-62.

Methodology for the Development of a Sectoral Standard for EDI

Reyes Grangel and Ricardo Chalmeta

Dep. de Llenguatges i Sistemes Informàtics, Universitat Jaume I
12071 Castelló, Spain
{grangel,rchalmet}@uji.es

Abstract. The introduction of an EDI (Electronic Data Interchange) system in a particular sector is a complex problem involving different organisational, technological and human aspects. Formal systematic methodologies must be used to resolve these problems and facilitate the implementation of new EDI systems. However, the literature offers no examples of this type of methodology.

Therefore, in this paper we present a methodology that describes the activities and techniques required to develop (1) a sectoral profile for EDI messages under internationally accepted standards and (2) a set of guidelines for implementation that helps EDI to be incorporated in a quick and simple manner. The methodology was developed from the experience acquired in the course of different EDI implementation projects carried out in different sectors. In order to illustrate how the process works, in this paper we have used a case study in which our methodology was put into practise to develop an EDI standard for enterprises in the tile sector.

1 Introduction

EDI can be defined as the computer-to-computer transfer of previously organised commercial information between business partners, generally following the North American ANSI X12 EDI or the international UN/EDIFACT (United Nations Electronic Data Interchange For Administration, Commerce and Transport) standards [4, 1]. These widely accepted standards consist of a set of formats, guidelines and steps that provide the basic controls needed to ensure the integrity of the transaction and that data be correct.

Therefore, EDI is a solution to the to-and-fro of paperwork between different organisations [9]. This, however, is not its only purpose. Enterprises that implement this system also gain important benefits, which can be direct (such as reductions in costs and more readily available higher quality data) and indirect (better customer service and improved relations with commercial partners, for example) [8, 6].

Nevertheless, in spite of its advantages there are also three decisive factors that have restricted its worldwide implementation: (1) the need to define a sectoral standard for messages which, while following one of the two international

S. Wang et al. (Eds.): ER Workshops 2004, LNCS 3289, pp. 641–652, 2004.

formats, can be adapted to meet the needs of each sector and is recognised and used by suppliers, manufacturers and customers; (2) the lack of methodologies that guide projects for the development and implementation of an EDI system in a sector; and (3) the difficulties involved in quantifying the benefits to be gained from implementing it [5].

To resolve these problems different research projects have been carried out worldwide. Some of the most interesting have been the MESSENGER (Manufacturing, Engineering and Service Status Electronically Negotiated for Greatly Enhanced Response) [10] or the EDIBOLD_SCS (EDI for Batch Operations Logistics & Design Supply Chain System) [3]. The results of these projects have taken the state of the art in EDI systems development a step forward, but they are very much tied to the sector they have worked on, like the EDIBOLD_SCS, or they are focus on the technological development as the MESSENGER. This means that there is a lack of a general purpose methodology that can help to develop and implement sectoral standards for EDI. In this context, we propose a general methodology that can be used as an aid to develop and implement a standard for EDI in any sector.

2 Methodology for the Development of a Sectoral Standard for EDI

The main aim of the methodology described below is to offer a series of activities and techniques that can be used (1) to quickly define a sectoral profile for EDI messages under a international standard (ANSI X12 EDI or UN/EDIFACT) and (2) to draw up a set of implementation guidelines that facilitates the integration of the EDI messages resulting in the information systems of the enterprises.

Among the factors taken into account when constructing the methodology was that as many enterprises from the sector as possible should be involved in the EDI standard development project, that they must offer the typical problems in the sector and that the number of people working directly on the project must be operational. Figure 1 shows the different phases and activities involved in the methodology.

2.1 Preliminary Phase

A project designed to develop a sectoral standard for EDI may arise (1) as an initiative by a group of enterprises with the need to implement EDI because they are required to do so by a supplier or for some other reason; or (2) as a project launched by a trade association, a college of professionals, a chamber of commerce, and so forth. In any case, the first thing that those in charge of the project or those promoting it must do is to define its goals and strategy. Following that, it is a good idea to hold an informative meeting for the enterprises in the sector and the scientific community linked with it, in order to collect the general information about sector enterprise. Once information on the enterprises interested in taking part in the project has been collected, the activities to be carried out in the first phase are as follows:

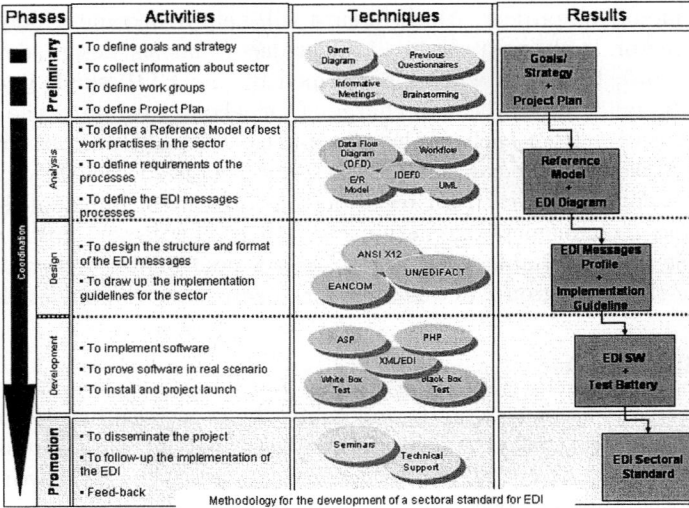

Phases	Activities	Techniques	Results
Preliminary	• To define goals and strategy • To collect information about sector • To define work groups • To define Project Plan	Gantt Diagram / Previous Questionnaires / Informative Meetings / Brainstorming	Goals/ Strategy + Project Plan
Analysis	• To define a Reference Model of best work practises in the sector • To define requirements of the processes • To define the EDI messages processes	Data Flow Diagram (DFD) / Workflow / E/R Model / IDEF0 / UML	Reference Model + EDI Diagram
Design	• To design the structure and format of the EDI messages • To draw up the implementation guidelines for the sector	ANSI X12 / UN/EDIFACT / EANCOM	EDI Messages Profile + Implementation Guideline
Development	• To implement software • To prove software in real scenario • To install and project launch	ASP / PHP / XML/EDI / White Box Test / Black Box Test	EDI SW + Test Battery
Promotion	• To disseminate the project • To follow-up the implementation of the EDI • Feed-back	Seminars / Technical Support	EDI Sectoral Standard

Methodology for the development of a sectoral standard for EDI

Fig. 1. Phases of the proposed methodology.

– Definition of the work groups and organisation the way the different members of the sector are to take part in the project.
– Determination of the scope of each of the phases and definition of the tasks to be performed by each of the participants in order to define the Project Plan.

The enterprises from the sector can take part by integrating them in one of the following four work groups, according to whether they want to be involved to a greater or lesser extent in the project:

– **Group 1:** enterprises from the sector that do not wish to be actively involved in the project but do wish to benefit from the results obtained from it: the EDI standard and the guidelines for the implementation of sectoral EDI. They can make their contributions to the project at any time.
– **Group 2:** enterprises interested in taking part in the project in a more active manner. They receive information about the progress made in the project as well as questionnaire about their work habits, which allows them to be taken into account in the different phases of the project.
– **Group 3:** enterprises interested in participating directly in the project. A representative from each of them will have to attend the work meetings in order to check how the project is progressing and to provide the information required by the developers.
– **Group 4:** from the enterprises in group 3 a smaller group will be selected as pilot enterprises for the development and promotion phase.

The idea of using this method of working in groups on different levels is to enable all the enterprises in the sector can reflect their basic business needs in the final outcome of the project, which in turn will help it become widely accepted within the sector. In the next phases of the project (analysis, design, development and promotion), the tasks will be performed by the **EDI work group**. This group will be made up of those in charge of, or who are running, the initiative, one representative from each of the enterprises that take part in **group 3**, a technological partner who can provide the technology needed in the development phase, and representatives from technological institutes, universities or other scientific and technological institutions related to the sector. From among all these people it is advisable to choose a project coordinator who will organise meetings of the work group, act as a moderator in them, document them, etc.

2.2   Analysis Phase

The aim of this phase is to identify and define the processes that are carried out in a typical enterprise from the sector in question. One of them-the most critical-must be chosen and first of all analysed as a pilot test in the **EDI work group**. Thus, once the pilot has been analysed, the rest are progressively examined in a parallel fashion so as to form work subgroups. To identify the most critical process a questionnaire can be conducted among the enterprises in **group 2**, and therefore it has the greatest need to be dealt with first by the whole **EDI work group**.

Before choosing the pilot process it is needed to carry out the analysis of the all processes of a typical enterprise of the sector. The result of this analysis must be a **Reference Model of best work practises in the sector**. The Reference Model can be developed with different techniques: DFD, Entity/Relationship model, IDEF0, UML, etc. In addition, the Reference Model and the questionnaire can be useful for determining the critical process in terms of the to-and-fro of paperwork, is the largest, most complicated and important process.

After having chosen the pilot process, the next step is to carry out the **definition of requirements of the process**, which involves collecting the basic information requirements of the life cycle of the product or service that the EDI messages are to deal with.

The next stage in the analysis phase is the **global and partial design of the message processes**. This means that from all the EDI messages that are included in the ANSI X12 or UN/EDIFACT standard we must choose a subset of useful messages and the sequence they must have in the process in question. One of the premises at this stage must be the design of a single ANSI X12 or UN/EDIFACT message or document for each of the documents in the process that has been selected. Furthermore, we must try and make the sequence of messages as standard as possible so that it includes all the problematic issues existing in the sector.

Finally, all the possible combinations of messages expected between the sender and the receiver must be established in order to standardise the transmission of messages from one to the other and to obtain a EDI Diagram. This diagram must reflect the sequences of the messages between sender and receiver. Following the definition of the messages and sequences to be used in the pilot process, the same analysis must be performed for the rest of the processes that were defined at the outset of this phase.

2.3   Design Phase

The objective of this phase is to **design the structure and the format of the EDI messages**. To do so the **EDI work group** must define each of elements of the structure of the EDI messages that are going to be used, and determine their meaning and what they are going to be used for. The EDI messages are structured [1], from the upper level consisting of the messages themselves, down to the data elements. The message is made up of different segments which are in turn composed of data elements, which can be either compound or simple.

In order to carry out the design it is necessary to perform a document-by-document functional analysis. That is to say, the structure of the different ANSI X12 or UN/EDIFACT messages used must be analysed to determine which data elements are going to be useful in the sectoral standard proposed and which are not. Moreover, an attempt must be made to adapt the general meaning these data elements have in the ANSI X12 or UN/EDIFACT standard to the sector in question and to determine the specific needs of the sector in order to decide which data element they can be included in. To do so we must have a thorough knowledge of the data handled by the Information Technology systems in the different enterprises in the sector.

Once the different message profiles have been defined, the ANSI X12 or UN/EDIFACT **implementation guidelines can be drawn up for the sector**. This guide will include all the ANSI X12 or UN/EDIFACT messages, the ones that are used in each of the processes that have been defined for the sector, the sequence in which they must be used and the fields that must be employed, as well as the exactly what they are being used for.

2.4   Development Phase

In this phase the objectives are to use the implementation guidelines drawn up in the previous phase in order **to implement the software** needed to put the EDI into operation and to run a pilot test to check it. It is here that one or several pilot enterprises are chosen from among the group of enterprises taking part in the **EDI work group** so as to be able to conduct the test in an environment that is as real as possible. The enterprises must combine a series of technical requirements that facilitate the technological development of the project.

Although the **EDI work group** also participates, in this phase the fundamental tasks are performed by staff from the selected pilot enterprise or enterprises and from the technological partner. They latter will be in charge of developing the computer applications that allow the EDI messages to be translated and later integrated in the enterprise's Enterprise Resource Planning (ERP). The development strategy can be one of the following:

- Parametrize, according to the needs included in the implementation guidelines, one of the EDI commercial computer software already on the market.

– Develop computer application that allows the EDI messages to be integrated into the different ERPs in each of the enterprises. This alternative can be achieved through the development of web pages, using PHP, XML or ASP pages, depending on the development environment the technological partner believes to be the most suitable [11].

The last activity to be performed in this phase consists of preparing and executing one **battery of partial tests and another made up of global integration tests**. The aim of this is to validate both the parametrized or developed software and its integration with the ERP of the enterprise. Finally, the activity carry out must be the **installation and project launching**.

2.5   Promotion Phase

Once the implementation guidelines and the EDI software have been obtained, the aim of the next phase is **to present, deliver, and disseminate the guidelines**, and the **technical follow-up of the implementation of the EDI**. In order to achieve this, it is useful to hold a series of seminars where the guidelines are given out and made known. Furthermore, it is advisable to run training courses and a technical follow-up of the implementation of the software in the enterprises that decide to adopt the defined sectoral standard and the parametrized or developed EDI software.

2.6   Coordination Phase

This phase involves the whole execution of the project and must be done by the project coordinator. Its main duty consists of acting as a mediator in the problems or conflicts of interests that may arise during the development of the project, but must always think first and foremost of the common welfare of the sector. As we have mentioned above, other duties to be performed by the coordinator would be the organisation and documentation of the meetings held by the **EDI work group**.

3   Case Study: The EDITILE Project

In order to help in the knowledge of the methodology, following a case study is shown. The Spanish tile sector is outstanding in the introduction of tile machinery and technology directly related to the production process, but is somewhat slower when it comes to the implementation of information technologies. In particular, although EDI systems are well consolidated in other sectors such the automotive or the fast moving consumer goods industry and have their own implementation guidelines, the level of implementation in the tile sector is practically zero. It is for this reason that the Association of Tile Manufacturers of Spain started an initiative aimed at carrying out a project to develop an EDI standard in the tile sector following the UN/EDIFACT standard [4]. This project was called EDITILE and was carried out following the methodology described in this paper.

3.1 Preliminary Phase

As the promoter of the initiative, Association of Tile Manufacturers' intention was for the guidelines to become a standard that was followed by the vast majority of enterprises in the sector. In this way, all the enterprises that wanted to interchange information by EDI would have the possibility of doing so under a standard that was defined and accepted by the whole sector. Moreover, by using this particular methodology the idea was that an important kernel of tile enterprises would get involved in the project and would be capable of quickly adopting EDI. The rest of the sector would then be expected to follow suit and large reductions in costs would be achieved.

Following the informative meeting held by Association of Tile Manufacturers, in which the goals, strategy and benefits of the project were explained, the enterprises from the sector were allowed to participate in one of the four work groups that have been defined above in section 2.1. The initial set up of **EDI work group** was as follows: a coordinator who belonged to and represented Association of Tile Manufacturers, different tile manufacturing enterprises integrated in **group 3**, a technological partner, and representatives from university.

3.2 Analysis Phase

The Reference Model of best work practises, and the identification and definition of the requirements was carried out from the tile manufacturer's point of view, but it was also taken into account the problems all the enterprises that take part in the life cycle of tile products had. In other words, not only were the requirements taken into account from the customer's point of view, but also from the perspective of the different suppliers, carriers, distributors, and so on, in the sector. The aim of this was for the processes that had been defined to include the casuistic of the sector, thus enabling the implementation guidelines to become a standard.

This phase was one of the most important since the participants in the **EDI work group** belonged to enterprises with different types of organisation and methods of working, but after several meetings a consensus was reached on the standard processes for the sector as well as their sequence. The processes that were put forward were the following: relation with customers, relation with distributors, relation with carriers, relation with suppliers of raw materials, relation with suppliers of special pieces, and relation with the public administration.

The results obtained in the questionnaire conducted in the enterprises in **group 2** were analysed to determine the highest number of transactions that took place involving documents. The highest number occurred in customers' orders, therefore it was decided that the Relation with customers would be studied as a pilot process.

The messages proposed in the UN/EDIFACT standard for the Relation with customers are the following:

- **PRICAT:** Catalogue of sales prices.
- **ORDERS:** Order from customer.

Table 1. Example of two UN/EDIFACT messages adapted to the tile sector.

Message	Associated document	Origin	Destination
ORDERS	Order from customer	Customer	Manufacturer
ORDRSP	Confirmation of customer's order	Manufacturer	Customer

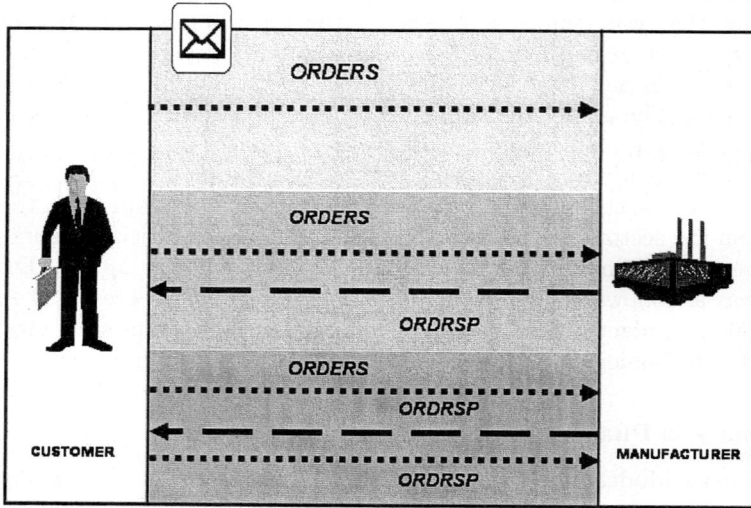

Fig. 2. Possible sequence of EDI messages to carry out Order from customer in the Relation with customers process.

- **ORDRSP:** Answer to the order from customer.
- **ORDCHG:** Request to modify order.
- **OSTRPT:** Report on the state of the order.
- **INSDES:** Shipping instructions.
- **DESADV:** Shipping notification.
- **RECADV:** Notification of reception.
- **INVOIC:** Invoice.

Those chosen for the Order from customer microprocess, which belongs to the Relation with customers process, can be seen in Table 1.

After defining the used messages for this microprocess, their correspondence with the physical documents, their origin and destination, the sequence of messages was determined. This can be seen in Figure 2. The sequence offers three possible ways of interchanging messages in the microprocess Order from customer:

- Sending an ORDERS message that no longer needs confirming. This is the case when all the material requested by the customer can be delivered and on the date proposed by the customer. The manufacturer and the customer have already agreed that sending this message means a enterprise order that does not need confirmation.

- Sending an ORDERS message that needs confirmation from the manufacturer by sending an ORDRSP message. In this case the material requested by the customer is also available but the manufacturer and the customer have agreed that the former should send confirmation.
- Sending an ORDERS message that needs confirming from the manufacturer by sending an ORDRSP message. And this time it is confirmed by the customer sending back another ORDRSP message. This last case occurs when not all the customer's order can be shipped due to problems concerning stock or delivery dates. The manufacturer then sends a confirmation message (ORDRSP) which modifies the order placed by the customer and therefore new confirmation is needed from the customer so that the manufacturer knows whether the modifications have been accepted or not. If the customer accepts the modifications, a message of confirmation (ORDRSP) is sent back. If not, the message is not sent and the customer must go back and start a new sequence of messages if he wishes to make any changes to the modifications stated by the manufacturer. In this way the infinite to-and-fro of messages of confirmation and modification in both directions is avoided.

After defining the messages and sequences that are going to be used in all the microprocesses in the process that was chosen for the pilot test, Relation with customers, the same analysis was performed for the rest of the processes defined at the beginning of this phase.

3.3   Design Phase

In the design phase it is worth discussing the problem of whether the EAN (European Article Numbering Association) [2] International EDI standard coding system, popularly known as the 'bar code', was used to identify customers. If not, if only the **UN/EDIFACT** norm was followed, the use of a unique identifying code, such as the tax identification number (NIF), was enough.

The advantage of the first option was the benefit to be gained from the standardisation and international acceptation of the EAN coding system. On the other hand, the main advantage of the second option was that the tile manufacturers' smaller customers would not have to pay the costs involved in the use of the EAN coding system and this would therefore help the project to expand and become more popular. In the end, the project coordinator acted as a mediator and the pros and cons of both options were studied. It was finally decided that the EAN coding system would be used, since standardisation had been one of the initial points on which the project was based.

3.4   Development Phase

In this phase, the technical partner opted to split the project in two sub-phases, showed in Figure 3, in which the physical implementation of the defined standard

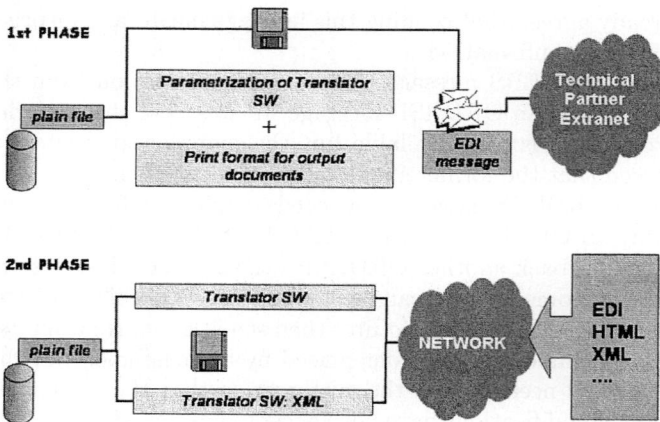

Fig. 3. Technological solutions for the development phase.

was performed using two different technical solutions. In order to improve the profitability of the project, since the use of a value-added network (VAN) is recommended for security and service reasons but the cost of the VAN can be prohibitive [7]:

- **1st Phase (short-term):** for the pilot test a classic EDI system, with private and point-to-point telecommunications networks (VAN), was used as the technical solution. In this phase network access was achieved through the technological partner's extranet. At this time, staff from the technological partner parametrized the EDI translator software application and generated suitable printing formats for the output documents.
- **2nd Phase (mid-term):** an XML-based EDI translator software application was developed which uses public telecommunications networks, as it is performed through the Internet. This system has several advantages, such as the fact that this technology allows images to be attached, for example, to the PRICAT message. Another big advantage is that the final customer can receive the EDI message in the format of its choice: EDI, HTML, XML, etc. In this case it is the manufacturer who has the EDI translator software, while the customer is sent the messages in the suitable format.

In each of the two sub-phases the technological partner took part in the parametrization and development of the EDI software whereas the manufacturing enterprise selected for the pilot test participated in the process of integrating the EDI software in the enterprise's computer system. Figure 4 shows the integration scheme proposed for the manufacturer taking part in the pilot test. In this case, it was necessary the use of data warehouse tools so as to be able to replicate the enterprise's database as a security mechanism. Finally, the battery of partial and overall integration tests had to be prepared with which to check

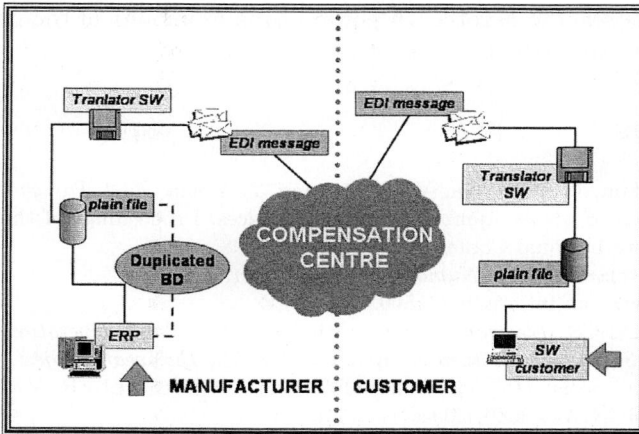

Fig. 4. Diagram showing EDI software integrated in manufacturer's and customer's computer systems.

the functioning of the software applications both alone and after integration in the enterprises' computer systems.

4   Conclusions

From the findings of the work that has been carried out, it can be concluded that the methodology presented in this paper:

- Can be used in sectors with no EDI implementation as a useful guide in the development of a sectoral profile for EDI messages and the corresponding implementation guidelines.
- Propose a way of participating in the project that enables the creation of efficient work groups in terms of the number of participants and representation. Moreover, it requires the involvement of all the enterprises in the sector on different levels so that they participate directly or indirectly in the final outcome of the project. This makes it more easily accepted, therefore allowing a more widespread and faster implementation of the EDI system within the sector.
- Is beneficial for the entire sector that puts it into practice. The enterprises that take part indirectly in the project benefit from the results obtained and can implement an EDI system. This also favours those that have participated directly since it allows them to lower the costs derived from working with partners that have not adopted the EDI system.

Furthermore it must be pointed out that the success of a project of this kind does not depend solely on the technological solutions used, but also on other less tangible aspects such as the culture and the organisational capacity of the enterprises, the negotiation power of the promoters of the project or the structure, size and number of participating enterprises.

References

1. N. R. Adam, I. Adiwijaya, V. Atluri, and Y. Yesha. *EDI Trough A Distributed Information Systems Approach.* System Sciences, Proceedings of the Thirty-First Hawaii International Conference on, vol. 7, 1998.
2. EAN. *European Article Numbering Association.* http://www.ean-int.org/index800.html, 2004.
3. EDIBOLD_SCS. *Electronic Data Interchange For Batch Operations Logistics & Design - Supply Chain System (EDIBOLD-SCS), European Project.* http://dbs.cordis.lu/fep-cgi/srchidadb?ACTION=D&CALLER=PROJ_IST& QM_EP_RCN_A=55086, 2004.
4. C. Huemer. *DIR-XML2 - Unambiguous Access to XML-based Business Documents in B2B E-Commerce.* University of Vienna, Vienna, Austria, Sponsor SIGEcom: ACM Special Interest Group on Electronic Commerce Publisher ACM Press New York, NY, 2001.
5. Ch. L. Iacouvu, I. Benbasat, and A. S. Dexter. *Electronic Data Interchange and Small Organizations: Adoption and Impact of Technology.* MIS Quarterly/December, 1995.
6. R. J. Duff Jr. and M. Jain. *CFO's guide to EDI: How can you control the new paperless environment?* Journal of Corporate Accounting & Finance, vol. 10, issue 1, 1998.
7. K. Ketler, J. Willems, and V. Hampton. *The EDI implementation decision: a small business perspective.* Sponsor SIGCPR: ACM Special Interest Group on Computer Personnel Research Publisher, ACM Press New York, NY, USA, 1997.
8. K. K. Y. Kuan and P. Y. K. Chau. *A perception-based model for EDI adoption in small businesses using a technology-organization-environment framework.* Information and Management, vol. 38, issue 8, 2001.
9. A. J. Marcella and S. Chan. *EDI Security, Control and Audit.* Boston London Artech House cop, 1993.
10. MESSENGER. *Manufacturing, engineering and service status electronically negotiated for greatly enhanced response (MESSENGER).* http://www.isomatic.co.uk/messenger.htm, 2004.
11. T. Weitzel, P. Buxmann, and F. V. Westarp. *A communication architecture for the digital economy 21 st century EDI.* System Sciences, Proceedings of the 33rd Annual Hawaii International Conference on, 2000.

Classification of Business Process Correspondences and Associated Integration Operators*

Georg Grossmann, Michael Schrefl, and Markus Stumptner

University of South Australia, Advanced Computing Research Centre
Mawson Lakes, SA 5095, Adelaide, Australia
{cisgg,cismis,mst}@cs.unisa.edu.au

Abstract. Integration of business processes requires the integration of both the structure and behaviour of the data objects involved. Research in federated information systems has so far mainly addressed integration of object structure or if based on behavior representation, assumed identity of the lower level operations to be integrated. Based on earlier work that showed how a fine differentiation between the possible semantic correspondences could be used to improve the classification process, in this paper we examine business processes from the perspective of a prospective integration tool or designer and likewise provide a classification of different types of similarities between the activities in a process. From this classification, we develop a set of options for integration of business processes and, going to a more detailed level, present the set of choices available depending on particular types of correspondences between the activities of the processes that are to be integrated.

1 Introduction

"Integration" is one of the driving themes in current database and applied computing research in general. A recent special issue of the *Communications of the ACM* [1] and several articles in subsequent issues dealt with integration topics. Whether at the level of classical database applications, web services, workflows, integration of applications is a matter of significant concern.

It is well accepted that, like database design, system integration is an *engineering task* which cannot be fully automated. The methods and techniques developed to be used by a system designer during integration will comprise informal design guidelines as well as formal consistency criteria for checking whether the behaviour of objects in the combined schema is consistent with respect to the behaviour of objects in the originating schemas. This will enable designers to identify and correct problems in the design of such systems before development progresses to the stage of detailed code development, generally resulting in high cost savings in development and later maintenance and extension.

Existing research on the federation or integration of information systems (e.g.[2, 5, 9, 10, 12, 18, 19]) has concentrated almost exclusively on the structural aspects. Integration of object behaviour has received some attention, but only at the level of

---

* This research was partially supported by the Australian Research Council under Discovery Grant DP0210654.

S. Wang et al. (Eds.): ER Workshops 2004, LNCS 3289, pp. 653–666, 2004.

single operations (or "activities" at the conceptual level) [4, 21], despite the fact that in other design areas the results of studying behaviour have generally been found to provide richer outcomes than for structure alone.

Recently, this has been most clearly expressed in work on Web service integration, e.g., [11], where "functionality" of services is expressed in terms of general conditions or even simple predefined scalar types that have to be matched between services. We consider the problem from the perspective of business processes, i.e., sets "of one or more linked procedures or activities which collectively realize a business objective or policy goal, normally within the context of an organisational structure defining functional roles and relationships" [3].

A related area, view integration for object-oriented databases [6], had the opposing assumption that the objects are virtual in the component schemas and real in the integrated schema, whereas in federated databases the situation is just the opposite. Further work [8, 7] was based on the assumption that corresponding activities in different processes are either identical, directly subsume each other, or cannot be examined in more detail. Similar assumptions were made in [15].

It is at this point that our approach diverges. In [19], individual objects were analyzed in terms of a set of semantic correspondences. These correspondences, checked in terms of the *structure* of individual objects in [19], i.e., their instance variables and method interfaces, could then be used to identify a spectrum of possible integration scenarios. Examination of the different combinations of correspondences led to the realization that only a part of the potential combinations of correspondences could actually occur in realistic scenarios, a result that guides us in this paper as well.

In [19], the outcome of the integration process was the definition of a generalized superclass that unified the original objects by "upward inheritance". Here, the goal is the specification of sets of operations that produce correctly integrated business processes. Thus, by following the classification, the choice of correct integration operations is possible rather than relying on informal insights by the developer. The operations are to be based on a description of business process behaviours in terms of UML activity diagrams, thus subsuming and extending the earlier work. Although we intend to eventually provide a metaclass architecture as the framework for inserting such operator definitions, this is not our main goal in this paper; instead we focus on the interesting interrelationships themselves that are made possible by considering different semantic correspondences at the process level.

In [20], we described a generic approach and resulting architecture for the behaviour oriented integration of business processes. It is based on a meta-class architecture that uses inheritance and instantiation relationships to describe high-level integration operators that can adapt and produce individualized integration plans (i.e., groups of operations) for the integration of processes from a particular domain. In this paper, we examine the choices of actual operators and how they are constrained by the properties of the individual processes to be integrated, thus instantiating one of the foundation stones of the scheme from [20].

2   The Integration Process

We explain our approach by the means of integration of business processes from two different domains, i.e., from companies in different business areas. Hence they follow different business goals but they may deal with related objects, e.g., the processes of a car dealer and a car insurance, which refer to a common "car" object. The purpose of this integration could be the sale of cars with a customized insurance.

If each of the domains involved comprises multiple business processes, then processes inside each domain should be integrated first, followed by integration across domains. We expect that corresponding activities are more similar within than across domains. These similarities offer more integration options which are explained later and can be integrated efficiently, e.g., two integrated activities can be executed by one call.

We describe integration for the case of two processes. (Multiple processes can be handled by successive pairwise integration.) Each integration round proceeds in the following steps:(1) *Projection*: If the business processes include activities which affect more than one object (e.g., a car insurance sales process may deal with the objects car, contract, and customer) we have to project the process onto subprocesses consisting of the activities from the original process that are related to a single object. (2) *Identification of relationships and correspondences* between the subprocesses and their activities. (3) *Model transformation* of the subprocesses by using integration operators. Which operators are used and in which order they are applied depends on the relationships that are found and defined in the previous steps. Because of these correspondences several proper integration options may be offered to the process designer who decides which option is the best. (3) *Restructuring the integrated process*: Redundant flows which may result from the model transformation are eliminated. (4) *Merging the object specific integrated processes into one process*. This is the opposite of step 1.

We assume that in a pre-integration phase all heterogeneities (e.g., model conflicts, missing states, naming conflicts) listed in [13, 14] are identified and resolved.

Because of space limitations we describe only steps 2 and 3 in more detail.

3   Classification of Business Process Correspondences

The key phase in the integration process is the model transformation. As a prerequisite for this task we have to find out which specific activities can be integrated by specific integration options. In this chapter we first identify the activities that can be integrated in a reasonable way by examining the relationships of the business processes and their activities. In the next chapter we list integration options in general and in chapter 5 we assign these integration options to the identified relationships of the activities. By following these three steps we achieve the requirement of the model transformation.

In [17, 8, 7, 14] relationships between elements are identified. They are the basis for the transformation of business process models [17] and the integration of statecharts [8, 7] or object lifecycles [14]. In this section we will give an overview of the correspondences we have identified. Instead of analyzing the relationships between whole sub-/processes or statecharts mentioned in [17, 8, 7] we analyze the relationship between the business processes and their activities. The resulting matches are more precise and in addition offer more semantically useful integration options to the process designer.

[14] describes elements in different views as equivalent if both elements model the same real world object, where elements can be activities, states, or labels. For the integration of processes from different domains this definition is not sufficient because they can belong to two different real world objects but still have some common properties. In [19] the semantic relationship between object classes and their attributes and methods are described. This analysis gives the basis for different generalization types of object classes. We extend it to business processes and their activities and explain different integration options on the basis of this analysis in the next section.

Identity Relationship: In [19] the identity relationship between two object classes holds if their real world extensions are identical at all points of time. The real world extension of an object class is defined as all real world objects modelled as instances of that object class. In the business process model we distinguish between two different identity relationships by looking at the extensional and intensional parts of the process. We use the shortcut *Ie* and *Ii*: (a) *Business processes with the same extension (Ie)*: This holds if all their instances are identical at all points of time. (b) *Business processes with disjoint extensions and the same intention (Ii)*: In this case the processes model different real world objects on the same schema. The relation holds between two business processes A and B if the schema of A is identical to the schema of B.

Role Relationship: If two objects model the same real world object in a different situation or context then we speak about a *role relationship*. In [19] the role relationship holds if every instance of an object class that is equivalent to an instance of another class is role-related to this instance. In the case of business process modelling, this applies to two different business processes which deal with the same object but in different situations or contexts, e.g., we have a process involving an employee at a university and another process involving an employee in a company, but the person in both processes is the same. We say that business processes A and B hold a role relationship if every instance a of A that is equivalent to some instance b of B is role-related to b.

History Relationship: In [19] the history relationship describes the relationship between two objects which may represent the same real world object at different times. The history relationship holds between two object classes P and Q if every instance of P that is equivalent to an instance q of Q is history-related to q. This definition also holds in the sense of business process when we say that all real world objects modelled by a business process A are equivalent to all real world objects modelled by another business process B are history-related, e.g., if A deals with activities concerning a person as an applicant and B deals with activities concerning the same person as an employee.

Counterpart Relationship: In [19] two non equivalent objects, i.e. objects not representing the same real world object, are counterpart-related if they share some common properties but represent alternate situations in the real world. The counterpart relationship holds between two object classes if every instance of one class that shares some common properties with some instances of the other class is counterpart related. Two business process models can be counterpart-related if they model two different real world objects which are affected by some common activities but represent alternate situations in the real world, e.g., two business processes of different car insurances A and B, but dealing with the same car. The two processes model some common activities like 'select car manufacturer', 'select usage of the car', and 'calculate the rate' are dealing with the same car. We say that the counterpart relationship holds between business processes A and B if every instance of process A that shares some common activities with some instances of B are counterpart related.

Category Relationship: In [19] two non equivalent objects are category-related if they share some common properties. The category relationship holds between two object classes if every instance of one class is category related to every instance of the other class. Here, this translates to two business processes A and B sharing some common activities, e.g., A deals with house and property insurances and B with car insurances. We say that the category relationship holds between two business processes A and B if every instance of A is category-related to all instances of B.

The difference between the counterpart and the category relationship is that one cannot find pairs of counterparts in category-related objects, e.g., some activities of counterpart related processes can be identical but not the activities of category-related objects.

3.1   Relationships of Activities

We use the classification of business process relationships to get the default relationships of their activities, e.g., if two processes are role-related then normally the activities are role-related as well. However it is possible that the activities show different relationships than their business processes, e.g., activities of role related processes may hold an identity relationship. That is why we have to examine the relationships between individual activities in analogy to the semantic relationships between the attributes and methods of [19]. We use the same classification of relationships between activities as mentioned above. When analyzing the relationships of the activities in the scope of the relationship of their business processes, we find that not all relationship variants can occur with all activity combinations. A table of all possibilities can be seen in [10] and lists the following relationships between pairs of activities:

Activities of identical business processes are in an identity relationship if they model the same functionality of the instances of the process, e.g., two processes 'give details about the car' include the activity 'select manufacturer'. We write *ident_e_rel* and *ident_i_rel* depending on the type of the identity relationship as described in the previous section.

Activities of role-related business process are in an identity relationship (*role_rel*) if they model the same functionality of the instances of the processes but the functionality does not depend on the role in which the processes model their instances, e.g., the activity 'give Social Security Number' exists in two business processes, one concerning an university employee and the other one a company employee. The activities may hold a role relationship if they model a functionality of the instances of the processes in dependence on the specific role of the processes, e.g., the activity 'specify salary' of two business processes concerning an university employee and a company employee.

Activities of History-Related Processes: Such activities are identical if they model the same functionality of the instance of history-related business processes and the functionality does not belong to the points of time at which the objects model their instances, e.g., the activity 'give Social Security Number' of applicant and the activity 'give Social Security Number' of an employee. The activities hold a history relationship (written *hist_rel*) if they model the same functionality and the functionality depends on the points of time at which the processes model their instances, e.g., the activity 'submit CV' of two processes concerning an applicant and an employee.

Activities of Counterpart-Related Processes: Such activities are identical if they specify the same functionality and the functionality is independent from the counterpart relationship of the processes, e.g., the activities 'select manufacturer' for the same car, as part of a service of car insurer A and car insurer B. The activities may be counterpart-related if they model the same functionality but the functionality depends on the counterpart relationship of the processes, e.g., the activity 'calculate the rate' for the same car at car insurer A and car insurer B. The shortcut *count_rel* represents the counterpart relationship of two activities.

Activities of category-related processes correspond if they model functionality which can be perceived to belong to a common category (*cat_rel*), e.g., the activity 'select value of the object which should be insured' of a house/property insurance and a car insurance.

Distinct activities are activities which are not comparable because there exists no equivalent activity in the other process. Distinct activities can be found in identity-, role-, counterpart-, and category-related business processes (expressed by the shortcuts *dis_ident*, *dis_role*, *dis_count*, and *dis_cat*).

4   Integration Options

In this chapter we list integration options which are the basis for the next step in the integration process, the model transformation.

We give an overview of the possible ways to integrate two activities A1 and A2 where A1 belongs to business process BP1 and A2 belongs to business process BP2. We do not look at the relationships and correspondences between BP1, BP2, A1, and A2 earlier in the execution.

In the following list we first classify the options in terms of the possible execution of A1 and A2. Each option has its name listed in parentheses.

4.1   Execute Both Activities

The first possibility is that both activities need to be executed. This execution can be synchronous or asynchronous. If *synchronous* (*sync*), both activities A1 and A2 are merged into one activity and the flows of BP1 and BP2 are synchronized before and after the execution of the merged activity. This can be ensured by the use of synchronization nodes. The execution of BP1 and BP2 can only pass these nodes at the same point of time. If the execution of A1 and A2 is *asynchronous*, i.e., A1 and A2 at different time points, then there are two options: (a) : A1 and A2 are executed in arbitrary order (*anyo*). (b) There is a predetermined sequential order for the execution of A1 and A2, e.g., either A1 can only start when A2 has finished, or vice versa (*seq*). The order of A1 and A2 has to be chosen and will be fixed in the integration process.

Execute Both Activities and Aggregate the Results: An extended case arises if the activities produce results that are important for the intent of the integration, e.g., BP1 is a home and content insurance and BP2 is a car insurance where A1 and A2 return the respective rates which have to be paid twice a year. The intent of integrating both processes might be to get the cumulative rate over all insurance policies. This result can be achieved by adding a new activity to the integrated process which aggregates the results of A1 and A2 to the sum of the rates.

Again, A1 and A2 can be executed synchronously or asynchronously. In the synchronous case (*sync_a*), the flows of BP1 and BP2 are synchronized by control nodes. Between the synchronization nodes A1 and A2 are executed in parallel. The aggregation activity is added after the second control node and receives the results of A1 and A2 at the same point of time. In the asynchronous case, A1 and A2 are executed at different points of time. The aggregation activity can start when both activities have finished. Because time when both activities will be finished is unknown, the results of A1 and A2 have to be stored for later aggregation. We distinguish two possibilities: (a) *Any order* (*anyo_a*): The order of executing A1 and A2 is not important. The aggregation activity will be added in the execution flow after the activity which started later in the execution flow. The result of the last executed activity can be achieved by the aggregate activity immediately. The result of the first executed activity must be loaded for calculating the aggregation function. (b) *Sequential* (*seq_a*): The execution order of A1 and A2 is predetermined. If A1 has to be executed first, then its result will be stored and the aggregation activity will be added to the execution flow after A2. If A2 has to be executed first, it is the other way round.

Execute Both Activities and Process the Results: A third variant computes a new result of the same type out of the results of A1 and A2, e.g., a sum or an average. In this option a new result is generated from A1 and A2 but with a different type. The results of A1 and A2 are related to each other which provides a new conclusion, e.g., comparing the prices of two different insurance companies.

Again, A1 and A2 can be executed synchronously or asynchronously. In the first case (*sync_r*), the flows of BP1 and BP2 are explicitly synchronized. Between the synchronization nodes the activities are executed in parallel. A new activity which relates the results of A1 and A2 is added after the second control node. In the asynchronous case, the results of A1 and A2 have to be stored, to be compared later. This can happen in (a) *Any order* (*anyo_r*): Order of A1 and A2 is not important. A1 is executed before or after A2. The activity which relates the results will be placed after the second activity. Or it can be (b) *Sequential* (*seq_r*): A1 and A2 have to be executed in a specific order. The results of A1 and A2 are compared afterwards.

4.2   Execute One Activity

As opposed to the options described above, where both activities are executed, in this case only one, A1 or A2, is chosen for execution in the integrated process. The problem is how to decide which one. With *static selection* (*stat_s*), the decision which activity is used later in the business process is made before the integration process. This activity replaces the other activity which disappears in the integrated process. With *Dynamic selection*, the decision for an activity cannot be made before the process execution starts. There are three options: (a) *User selection* (*user_s*): Before the execution of A1 and A2 the user is asked which activity should be executed. According to the user selection the activity A1 or A2 is executed. (b) *Conditional expressions* over values of object properties (*cond*): If there is the possibility that the choice of activities may be constrained due to dependencies on the property values of the object which is affected then the selection can be aided by conditional expressions over these values. E.g. a conditional expression is set over the salary of a person which exists in the role 'politician' and 'company employee' to assign him a high or low credit rating. However if there are more than two activities integrated and the constraint leaves a choice between activities then the user decides. (c) *Dynamic binding* (*dyn_b*): If the selection of the activity depends on the type of object which is affected then we use the integration option dynamic binding, e.g., BP1 deals with sports car insurances and BP2 with truck insurances where A1 is 'selecting a manufacturer from sports car manufacturer list' and A2 is 'select a manufacturer from truck manufacturer list'. If BP1 and BP2 are integrated to a vehicle insurance business process then A1 and A2 can be integrated by the dynamic binding. According to the type 'vehicle' for which the integrated process is running, A1 is used for the type sports car and A2 is used for the type truck.

5   Integration Choices for Particular Activity Correspondences

In this step the integration of the two processes takes place. On the basis of the relationships and the integration options which were identified in the previous sections, we propose one or more integration options for each activity relationship. If the activity relationships are not explicitly specified we use the relationship between the business processes. As integration options are independent of the process relationships, the relationships are not mentioned further except in the case of identity-related activities.

For each integration choice we give an example which integrates the business process BP1 with an activity A1 and the business process BP2 with an activity A2.

5.1   Identity-Related Activities

We distinguish between the two relationships *ident_e_rel* and *ident_i_rel*, e.g. 'Two car insurance companies A and B are dealing with the same car C' (*Ie*), and 'Vienna Airport and Adelaide Airport are using the same booking systems for airplanes' (*Ii*).

Furthermore we distinguish between read-only and updating activities and those that cause side effects, i.e., have an effect on objects outside the process (including the real world), e.g., a document is printed out or an order is executed. A side effect happens only in connection with an update function. We use the shortcut *read_only*, *update_side*, and *update* for activities which represent a read-only function, an update function with, and an update function without side effects.

Figure 1(a) gives an example for the relationship *ident_e_rel*, showing a part of two car insurance applications where the customer has to select the car manufacturer. The integration option *sync* for this case is shown in Figure 1(b).

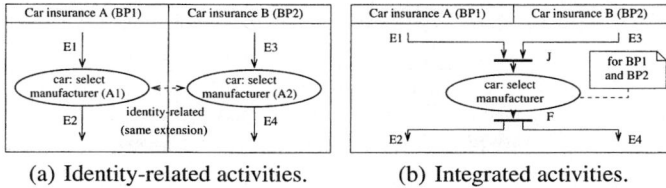

(a) Identity-related activities.   (b) Integrated activities.

Fig. 1. Example for using *sync*.

If the activities A1 and A2 cause side effects then one activity takes the *leader* part, and the other one the *trailer* part. This means that only one activity will be executed and the other activity registers the execution as done. The goal of the leader/trailer split is that side effects are only executed once in the integrated process, e.g. if activities A1 and A2 are 'accept the contract and send it per mail'. Activity A1 is the leader and executes the task, A2, the trailer, marks the insurance application as accepted and sent per mail without executing the task. The integration options for this case are *sync* and *anyo* where the leader activity A1 has to be executed before the trailer activity. For this integration option we consider that there is only contract involving during the execution of the business process, e.g., the customer is choosing the cheapest insurance offer.

For read-only activities (e.g. A1 and A2 are 'list all engine types'), the integration option is *stat_s* because the results are identical and so only one activity needs to be executed.

If the extensions of the business processes are disjoint, e.g. BP1 and BP2 are modelling the same booking service in different cities and their activities hold the relationship *ident_i_rel*, then we take the integration option *dyn_b* for read-only and update functions. The dynamic binding differs between the cities to which an instance belongs and selects the proper activity.

If multiple activities hold a *ident_i_rel* relationship with a read-only function then all activities from one business process will be selected in the static selection option.

If multiple *ident_i_rel* relationships with a read-only function hold between activities of two processes, then selection must be static and for all relationships, the selected activities must come from the same process.

5.2   Role-Related Activities

Figure 2 shows an example of two role-related activities. BP1 deals with a person in the role of a politician and BP2 deals with the same person in the role of a company employee. The activities A1 and A2 are 'output: salary' and print out the salary of the person dependent on his role.

There are 9 integration options for the role relationship. The first four of these are common (and typically preferred) while options 5 to 9 are quite rare.

1. *sync_a*: In this option A1 and A2 are executed synchronously and their results are aggregated, e.g. to obtain the sum of the results. This option is useful if the process designer wants to get the income from a person which consists of the salary as a politician and the salary as a company employee. The result is shown in figure 3(a).

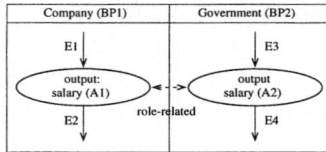

Fig. 2. Role-related activities.

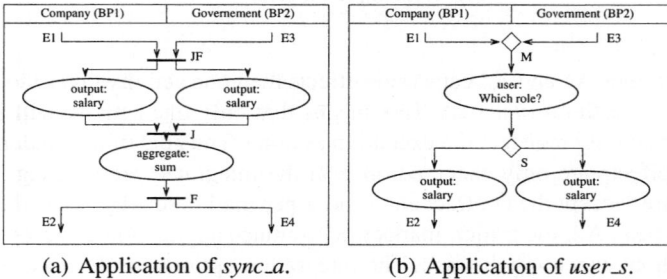

(a) Application of *sync_a*.   (b) Application of *user_s*.

Fig. 3. Application of *sync_a* and *user_s*.

2. *stat_s*: This option is taken if we are interested in one role of a person at a specific activity. During the integration process A1 or A2 is chosen to be always executed. The other activity disappears in the integrated process, e.g. if only the salary of the person in the role of the politician should be taken, then the activity A2 will be deleted.

3. *user_s*: This option offers a dynamic selection and gives the possibility to choose the activities by the user during runtime. The user will be asked before executing role-related activities which role he is interested in. According to the decision one activity will be executed, e.g. see figure 3(b).

4. *cond*: By using a conditional expression over values of business process properties, automated or user supported selection of the activities can be achieved. The defined expression filters the choices of activities by examining the property values of the instances first. Like a guard, the business process offers the remaining activities to the user or decides which activity to execute, e.g. 'if the work address of the person is the parliament, then take the salary of the politician role'.

5. *anyo_r*: Similar to *sync_a* with the difference that both activities are executed asynchronously in any order. The disadvantage of this option is that the result of the activity which is executed first has to be stored for the later aggregation function.

6. *seq_a*: This is the same as *anyo_r* but the A1 and A2 has to be executed in a specific order because they depend on the other, e.g., a politician is allowed to earn a maximum salary as a company employee. If he earns more than the maximum his salary then the salary in the role of a politician will be reduced. In this example the activity A2 is executed first and then A1.

7. *sync_r*: Instead of aggregating the results they are related to each other. A1 and A2 are executed synchronously and the results are related to each other, e.g., 'What is the main income of the person?'.

8. *anyo_r*: This option is similar to *anyo_r* but the results are related to each other.
9. *seq_r*: This option is similar to *seq_a* but the results are related to each other.

5.3 History-Related Activities

If two business processes model the same real world object at different points of time the activities are integrated in a sequence according to the history of the activity execution. The proper integration option is *seq*. In figure 4(a) BP1 models a car dealer with the activity A1 'drive the car' and BP2 models a car insurance with the activity A2 'buy car insurance'. A1 and A2 hold the relationship *hist_rel* where A2 has to be executed before A1 because it is not permitted to drive a car without insurance. The integration option seq of A1 and A2 is shown in figure 4(b).

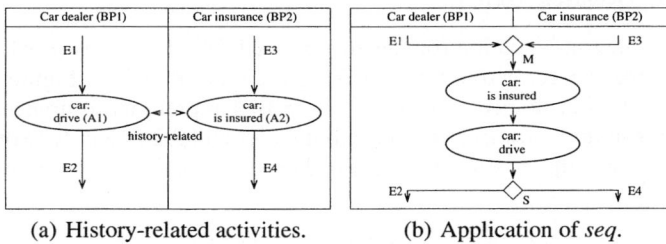

(a) History-related activities.   (b) Application of *seq*.

Fig. 4. Example for integration of history-related activities.

When integrating history-related activities the transitivity of order has to be considered as explained in [16], e.g. a BP1 with the sequence A1, A2 and BP2 with the sequence A3, A4 where A2 and A3 hold a history relationship. If A3 is executed after A2 then the transitivity of order implies that A4 is also executed after A2.

5.4 Counterpart-Related Activities

Figure 5(a) shows an example for a counterpart relationship between A1 and A2. BP1 and BP2 are processes from different car insurance companies, but both deal with the same car. The activities 'calculate monthly rate' share a *count_rel* relationship.

For integrating A1 and A2 the integration options *sync_r*, *anyo_r* and *seq_r* are used. We first explain the preferred option *sync_r* and later the two alternatives.

In *sync_r*, A1 and A2 are executed synchronously and later their results are compared to each other, e.g., to find out which of the insurances is cheaper. The integrated process is shown in figure5(b).

The alternative integration options for relating the results execute A1 and A2 asynchronously in any order or in a sequence. The reason for a sequence could be that insurance company A offers a rate and guarantees the cheapest rate. If the insurance company B is cheaper than the rate of A is changed to the rate of B. However these alternatives have the disadvantage that the result of the first executed activity has to be stored to apply the comparison operator later.

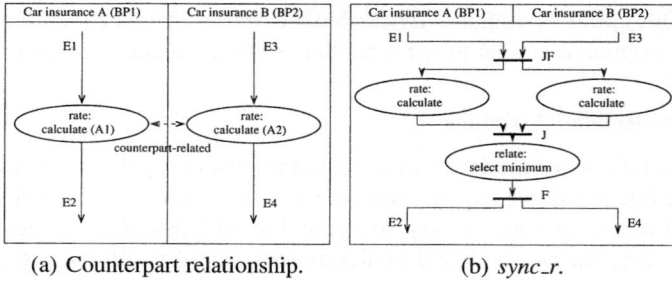

(a) Counterpart relationship.   (b) *sync_r*.

Fig. 5. Example for integration of counterpart-related activities.

5.5   Category-Related Activities

The preferred integration option for category-related activities is *dyn_b* which chooses automatically the correct activity by checking the type of object. Figure 6(a) shows the example of two insurance companies where BP1 represents insurances for sports cars and BP2 insurances for trucks. The activities A1 and A2 'select type of petrol' are holding the relationship *cat_rel*. The integrated process is shown in figure 6(b).

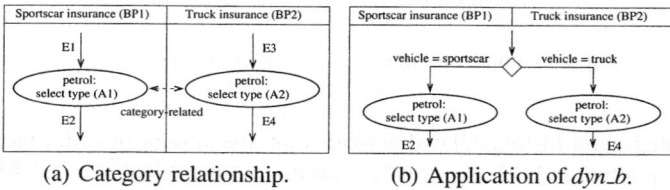

(a) Category relationship.   (b) Application of *dyn_b*.

Fig. 6. Example for integration of category-related activities.

A specific case are category-related activities which start a process. A integrated process can handle different types of object and in the beginning the type of the object is not yet defined. So the business process cannot decide automatically which activity should be executed. We define a relationship between category-related activities which start a process as *cat_rel_start*. The integration options are *stat_s* if the type of object will always be the same and *user_s* if the user should decide which type of object should be handled by the activities.

5.6   Distinct Activities

The way how distinct activities are integrated depends on the relationship of their business processes. We always integrate only one activity because there exists no other activity comparable to it: For identity-related processes, the activity will be integrated by merging the business flows together and leading into the activity. We call the shortcut for this integration option *merg*. For role-related processes, we take the preferred integration option which we used for corresponding role-related activities where only one activity is executed, i.e. *stat_s*, *user_s*, and *cond*. For counterpart-related processes, the integration option is the same as for distinct activities of identity-related processes. The flows of both processes are merged and lead into the activity. For category-related processes, we take the preferred integration option which we used for corresponding category-related activities where only one activity is executed, i.e. *dyn_b* if it is not a starting activity, and *stat_s* or *user_s* if it is a starting activity.

An overview of the integration choices for particular activity correspondences is shown in table 1. The correspondences are identified by the line header and the integration options by the column header. P stands for "preferred integration option" and A for "alternative integration option".

Table 1. Integration choices for particular activity correspondences.

related	sync	anyo	seq	sync_a	anyo_a	seq_a	sync_r	anyo_r	seq_r	stat_s	user_s	cond	dyn_b	merg
ident_e	P	P								P				
ident_i													P	
role				P	A	A	A	A	A	P	P	P		
hist			P											
count							P		A		A			
cat													P	
cat_s										P	P			
d_ident														P
d_role										P	P	P		
d_count														P
d_cat													P	
d_cat_s										P	P			

6 Conclusion

In this paper we have presented an approach to business process integration based on the analysis of the possible semantic relationships in the spirit of [19]. While that work considered the relationships between different object classes, we analyze possible relationships between the different business processes and their constituent activities. Specifically, we have adapted the notions of identity, role, history, counterpart, and category relationships to business processes and then individual activities. We have then examined the different possibilities for integrating multiple activities. These express themselves through different ways of executing the activities, in parallel, sequentially, or alone, with further divisions due to additional boundary conditions, yielding a set of potential integration operators. By examining the different activity and business process correspondences and matching them to the integration operators, we arrived at a classification of the useful combinations of particular activity relationships with execution options. A major outcome is that (as in [19]), a subset of these matches turns out to circumscribe the space of semantically useful integration options.

The classification is a first step towards an automatic integration of business process based on their behavior as described in our conceptual process integration framework [20]. A significant future step will be the automatic detection of relationships and execution of the proper integration options, resulting in reduced development effort and reduced risk for errors during the integration process.

References

1. Special issue on enterprise application integration. 45(10), October 2002.
2. Omran A. Bukhres and Ahmed Elmagarmid. *Object-Oriented Multidatabase Systems: A Solution for Advanced Applications*. Prentice Hall, 1996.
3. Workflow Management Coalition. Terminology and glossary, document number wfmc-tc-1011, document status - issue 3.0, Feb 1999.
4. S. Conrad, B. Eaglestone, W. Hasselbring, M. Roantree, F. Saltor, M. Schonhoff, M. Strassler, and M. W. W. Vermeer. Research issues in federated database systems: Report of EFDBS '97 workshop. *SIGMOD Record*, 26(4):54–56, 1997.
5. Stefan Conrad. *Föderierte Datenbanksysteme. Konzepte der Datenintegration*. Springer Verlag, 1997.
6. H. Frank. *View Integration für objektorientierte Datenbanken*. Reihe "Dissertationen zu Datenbanken und Informationssystemen", DISDBIS 32. Infix-Verlag, 1997.
7. Heinz Frank and Johann Eder. Towards an automatic integration of statecharts. In *ER '99, Paris*, LNCS 1728, pages 430–444. Springer.
8. Heinz Frank and Johann Eder. Integration of statecharts. In *Proc. COOPIS'98, New York City*. IEEE Computer Society, 1998.
9. M. García-Solaco, F. Saltor, and M. Castellanos. A structure based schema integration methodology. In *Proc. ICDE '95, Taipei*. IEEE Computer Society, 1995.
10. W. Klas and M. Schrefl. *Metaclasses and their Applications: Data Model Tailoring and Database Integration*. LNCS 943. Springer-Verlag, Berlin, Heidelberg, 1995.
11. Brahim Medjahed, Athman Bouguettaya, and Ahmed K. Elmagarmid. Composing web services on the semantic web. *The VLDB Journal*, 12(4), 2003.
12. Christine Parent and Stefano Spaccapietra. Issues and approaches of database integration. *Communications of the ACM*, 41(5es):166–178, 1998.
13. G. Preuner and M. Schrefl. Observation consistent integration of views of object life-cycles. In *Proc. of the 16th BNCOD*, pages 32–48, 1998.
14. G. Preuner and S.Conrad. View Integration of Object Life-cycles in Object-oriented Design. LNCS 1728, pages 413–429, 1999.
15. M. Preuner and M. Schrefl. Requester-centered composition of business processes from internal and external services. *DKE*, 2004. To appear.
16. Shazia W. Sadiq. *On dynamically changing workflow processes*. PhD thesis, University of Queensland, Australia, August 2002.
17. Wasim Sadiq and Maria E. Orlowska. On business process model transformations. In *Proc. ER '00: Salt Lake City*. Springer Verlag, October 2000.
18. I. Schmitt. *Schema Integration for the Design of Federated Databases*. Dissertationen zu Datenbanken und Informationssystemen, Vol. 43. infix-Verlag, St. Augustin, 1998.
19. Michael Schrefl and Erich J. Neuhold. Object class definition by generalization using upward inheritance. In *Proc. ICDE '88, Los Angeles*. IEEE Computer Society, 1988.
20. Markus Stumptner, Michael Schrefl, and Georg Grossmann. On the road to behavior-based integration. In *Proc. of the 1st Asia-Pacific Conference on Conceptual Modelling*, 2004.
21. Mark W. W. Vermeer and Peter M. G. Apers. Behaviour specification in database interoperation. In *Proc. CAiSE '97*, pages 61–74, 1997.

On Behavioural Model Transformation in Web Services

Behzad Bordbar and Athanasios Staikopoulos

School of Computer Science, University of Birmingham, Birmingham B15 2TT, UK
{B.Bordbar,A.Staikopoulos}@cs.bham.ac.uk

Abstract. Web Services are seen as one of the most promising solutions for the integration of autonomous, heterogonous e-business systems. Today's e-commerce systems often involve a combination of multiple Web Services, which are implemented via a mix of technologies such as Business Process Markup Language (BPML), Business Process Execution Language for Web Services (BPEL4WS), and Web Service Choreography Interface (WSCI). Recently, the application of Model Driven Architecture (MDA) to Web Services has received considerable attention. However, most of existing literature deals with the static aspects of Web Service modelling. This paper focuses on the behavioral aspect of the composition of Web Services using a Meta Object Facility (MOF) compliant metamodel for BPEL4WS. The paper presents a transformation of the Unified Modelling Language (UML) Activity diagram to the BPEL4WS.

1 Introduction

In recent years, the Internet has evolved from a simple storage of information into a provider of different kind of e-commerce services, ranging from travel booking, shopping to more complex e-business systems involving complex transactions. One of the main challenges of the design of such systems is to integrate autonomous, heterogonous and distributed components [21]. Currently, *Web Services* [26] are seen as one of the most promising approaches to solve the above problems [20][21]. Web Services are a set of technologies that allow applications to communicate with each other in a platform and a programming-language independent manner. Extensible Mark-Up Language (XML) [23], Simple Object Access Protocol (SOAP) [24], and Web Service Description Language (WSDL) [27] are among the technologies used in Web Service. Developing Web Services by applying Model Driven Architecture (MDA) [10][12][17] has recently received considerable attention [4][8][12]. In particular, [4][12] study the transformation of Web Services and present a set of case studies involving the transformation of Web Services models to various implementation platforms such as Java, WSDL and Enterprise Distributed Object Computing (EDOC) [14]. However, most of existing research focuses on the transformation of models that express the static structure of the system, i.e. models describing what the system contains and how various parts are related together. In this paper, we shall study transformations, which deal with the *dynamic* aspects of the system, which are modeled via *behavioral models,* expressing the way the various components collaborate in order to manage a task and fulfill the system functions.

S. Wang et al. (Eds.): ER Workshops 2004, LNCS 3289, pp. 667–678, 2004.

Defining and supporting business processes and collaborations between Web Services is a more complex problem than defining and supporting individual Web Services, see page 43 in [8]. Currently, the composition of Web Services is carried out via a mix of concrete technologies such as BPML [5], BPEL4WS [2], WSCI [25] and other [8]. Considering the existence of various technologies and languages, there is a clear scope for defining the business processes via behavioral Platform Independent Models (PIM) and providing methods of translation of PIM to Platform Specific Models (PSM) [8][10]. In this paper we shall present a method of transformation of the behavioural models from the UML Activity diagram to BPEL4WS.

The paper is organized as follows. We shall start by a brief introduction on Web Services, business processes and the model transformations in the MDA. Section 3 discusses the transformation of Business processes from Activity diagrams to BPEL4WS. First, we shall presents an example of a Stock Quote Web Service, which serves as our running example. The behavioural aspect of the Stock Quote system is modeled as a UML Activity diagram. Next, we shall sketch a metamodel for the UML Activity diagram and BPEL4WS. To translate the Activity diagrams to BPEL4WS, a set of transformations is introduced. The final part of section 3 applies the transformations to the running example. Section 4 reflects on the lessons learnt and discusses some of the issues regarding the model transformation of the behavioural aspects of systems. Finally, section 5 presents a conclusion.

2   Preliminaries

The Web Services introduce a new paradigm for enabling the exchange of information across the Internet. They can be characterised as self contained, self-describing that can be published, located and invoked across the Web. Various e-business applications can be encapsulated and published as Web Services allowing them to interoperate through standard communication and messaging mechanisms [26]. In general, Web Services are based on the Extensible Mark-Up Language (XML) [23] as the fundamental mechanism for describing protocols, structured data and messages, the Web Service Description Language (WSDL) [27] for describing the exposed interfaces and operations of a Web Service and the Simple Object Access Protocol (SOAP) [24] for providing the communication protocol.

Business integration and collaboration requires more than the ability to conduct simple interactions between Web Services. This has resulted in the creation of the notion of *business process*. A business process can be viewed as a composite activity, which defines a complicated behaviour expressed as a workflow [9]. Workflow specifies the control and data flow among sub activities of an activity. There are several different implementations for representing business processes [13], like the Business Process Execution Language for Web Services (BPEL4WS) [2]. The BPEL4WS provides an XML notation and semantics for specifying business process behaviour based on Web Services and defines, how Web Services can be combined to implement a business process.

In MDA each model is based on a specific metamodel, which defines the language that the model is created in. All metamodels are based on a unique metamodel called Meta Object Facility (MOF) [15]. As a result, model transformations can be carried via defining Transformation Rules between two MOF compliant metamodels [4][8][10], the source and the destination. In this paper, the source metamodel is the UML Activity modelling language and the destination is the BPEL4WS. Fig. 1 depicts the use of transformation rules for model transformation [4]. The transformation rules define a mapping between a source and a destination metamodel that preserves equivalent or isodynamic *(similar)* semantics. A transformation engine executes the transformation rules on the source model (acting as the input) in order to generate its equivalent destination model (output).

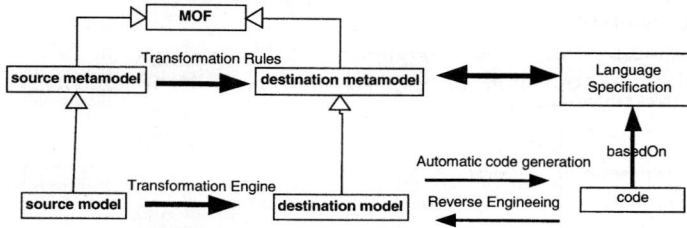

Fig. 1. Transformation in the MDA based on [4]

3 Business Processes Transformation in Web Services

This section demonstrates how a general business process can be modelled as a UML Activity diagram and how it can be mapped to an equivalent MOF based model representing a BPEL4WS. We shall start by presenting an example of a business process, which serves as our running example.

Example: Fig. 2 depicts two UML actors and a business process. The actors, *caller* and *provider*, represent two roles played by Web Services. They are composed together with the help of the *StockQuoteProcess* to create a composite Web Service. The *caller* makes a Stock Quote request to the business process. The process receives the request and forwards it to the *provider*. The *provider* replies with the value of the requested stock quote, which the process sends back to the *caller*.

Fig. 3 depicts a UML Activity Diagram model of the process in terms of workflow, coordination and interaction between the involved Web Services. It can be seen that the three services have been separated via Activity diagram swimlanes [19] and are stereotyped accordingly to their roles. In addition, the service location is indicated by the *external* stereotype regarding the business process's perspective. Next, five tasks *receive request, prepare invocation call, invoke provider service, prepare response message*, and *reply* are modelled as actions and are connected in a sequential order within a process. This order can be regarded as the execution path for handling the *caller*'s request. Furthermore, there are four variables involved; *request, invocation request, response, invocation response*, which are depicted as *object flow*, indicating how objects can be passed around via operation calls.

UML Metamodel for Activity Diagram: Metamodels play an important role within MDA, as they provide the language and the rules for describing models. Fig. 4 depicts a part of the metamodel for Activity diagram, which includes metamodel elements representing *workflow*, *object flow*, *activities*, *actions*, *operation calls* and other various modelling elements expressing control nodes. To create this metamodel, we have abstracted and combined various activities and action modelling elements defined in the UML 2.0 Superstructure Specification [19].

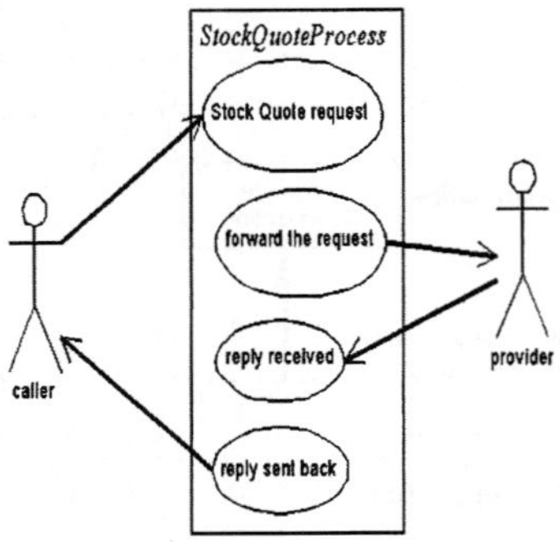

Fig. 2. A stock quote business process

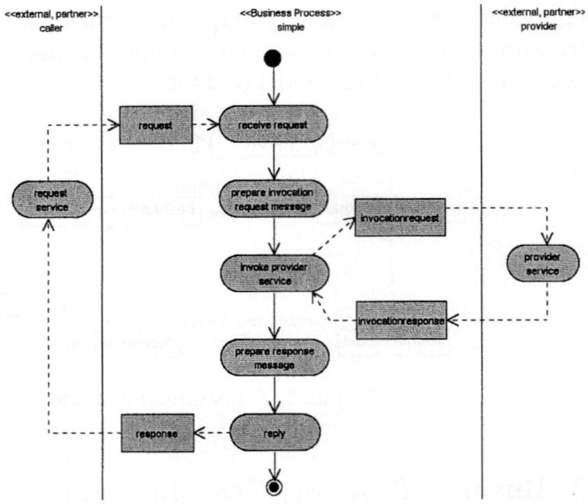

Fig. 3. Activity diagram for the process

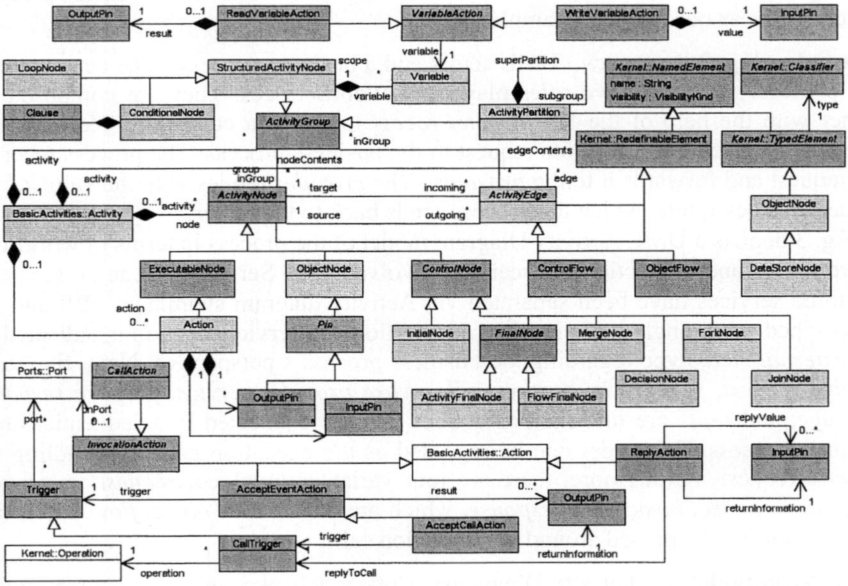

Fig. 4. UML Activity diagram metamodel

Metamodels create a clear view of available model elements. As a result, using the metamodel, it is possible to refine a model and provide a more elaborate representation. For example, consider the *receive request* action of Fig. 3. There are various types of *actions* specified in the metamodel of Fig. 4. For example, *receive request* can be modelled as an *AcceptCallAction* meaning the receipt of a synchronous call request. Such model refinements is a result of implementing new requirements of the system or by making additional assumptions regarding the model elements involved, such as *action types*, *sub activities* and *variables*. Consequently, we have refined the activity diagram of Fig. 3 to the new activity diagram of Fig. 5 by including a set of stereotypes. We have also included *sub activities* to illustrate the internal *variable* manipulation through read and write actions. For more information regarding the Activity model elements refer to chapter 12 of [19].

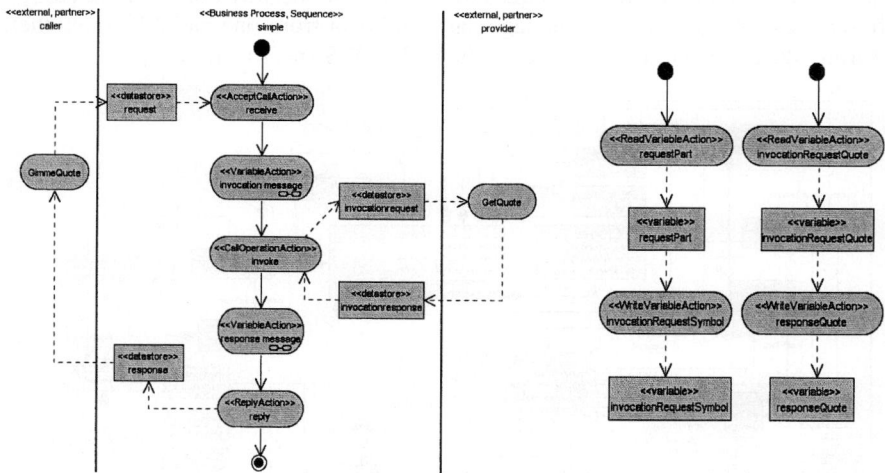

Fig. 5. The refined Activity diagram

BPEL4WS Metamodels: I this section we shall present the metamodel for BPEL4WS; the destination language, see Fig. 1. The BPEL4WS can be seen as an extension of the WSDL [27] supporting the collaboration of Web Services. In other words, the WSDL describes stand alone Web Services and their structure. Thus UML class diagrams, expressing the static aspects, are sufficient to capture and represent the semantics of the WSDL. Currently, there are approaches [4] for specifying and mappings metamodels among UML static aspects with its relevant WSDL elements. Our research can be seen as an extension of the existing work on static modelling by WSDL, to include methods of the mapping of the dynamic aspects.

Fig. 6 depicts the metamodel for BPEL4WS based on BPEL4WS version 1.1 specification and the XML Schema as published in [2]. Since, BPEL4WS is dependant upon a number of WSDL elements, there are references to *port types*, *messages* and *operations*, which are WSDL model elements. The central element of the metamodel is the business *process,* which captures both the structure and the workflow (controlled order of actions) of a business model.

A *process* consists of a number of *variables* for holding messages and representing the states of a process. A *partner links* is used to identify the associated Web Services through their *roles* and their *port types*. A *port type* represents the exposed interfaces of a Web Service via the WSDL specification. The metamodel also includes *fault handlers* for dealing with errors and *event handlers* for reacting to the events triggered. The BPEL4WS also includes *sequence, flow, while, switch and scope* as *structured activities*. *Sequence* activity defines sequential execution of actions. *Flow* provides parallel execution and synchronisation between actions. *While* specifies iterative activities. *Switch* supports conditional behaviour *and Scope* defines the execution context in which local variables and handlers can be defined for an action. Other basic activities for calling Web Service operations are: *Receive* to provide the business process services to Web Service partners. *Reply* to answer an accepted request. *Invoke* to either perform a request/response or a one way operation on partner Web Services. *Assign* to copy message parts and data from one variable to another. For further details, we refer the reader to the BPEL4WS specification [2].

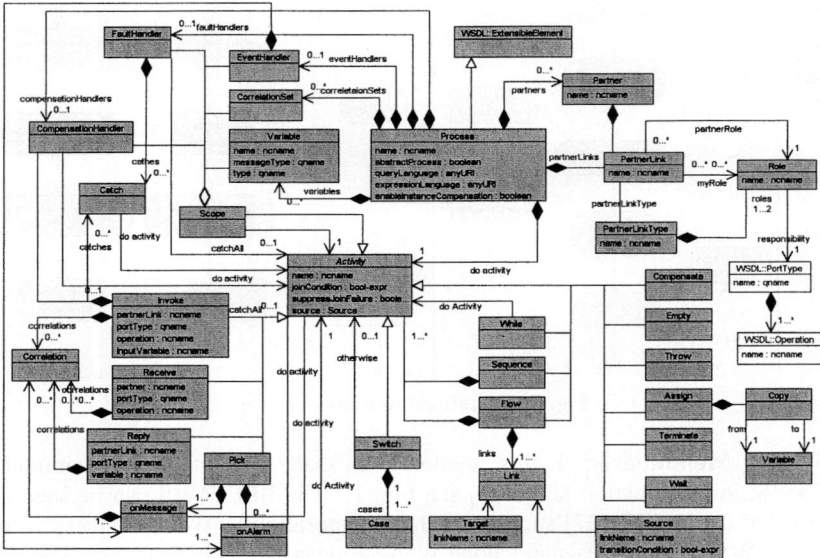

Fig. 6. A metamodel for BPEL4WS

Transformation Mapping: A model transformation, which defines the generation of a target model from a source model, is described by a transformation definition, consisting of a number of transformation rules that are executed by a transformation CASE tool. There are various methods of specifying model transformation [1][4][10]. Currently, there is no standardised language for defining transformations definitions. Consequently, the OMG has issued a request for proposal for a new standard called, *Query Views and Transformations* (QVT) [18], which has received considerable attention [17]. In this paper, we demonstrate the mapping among meta-elements with a table and express their transformation rules with the Object Constraint Language (OCL) [16]. Table 1 depicts a mapping, between the model elements of the metamodel of the Activity diagram and BPEL4WS 1.1. The mapping is based upon the descriptions given in the specifications [19] and [2]. Because of space restrictions, we have only included the relative elements demonstrated in the example

Table 1. Mapping elements of the activity diagram to BPEL4WS

UML 2.0	BPEL 1.1
Class as BehavioredClassifier, Activity, StructuredActivityNode	**\<process\>**
A BehavioredClassifier is a class acting as a container and where behaviour specifications can be defined. It is also an activity that is expressed as a flow of execution via subordinate units.	The BPEL4WS process description.
Datastore, StructuredNode Variable, ObjectNode, Class Attributes	**\<variable\>**
A datastore node is a central buffer node for non-transient information. An object node is an abstract activity node for defining object flow. Variables are elements for passing data between actions indirectly.	They provide the means for holding messages that constitute the state of a business process and the exchanging of messages between partners.
ControlFlow	**\<sequence\>**
It contains one or more activities that are performed sequentially in the order. It completes when the final activity has been completed	An activity that contains one or more other activities and which executes them in a serial order.
AcceptCallAction	**\<receive\>**
Is an accept event action representing the receipt of a synchronous call request. The action produces a token that is needed later to supply the information to the ReplyAction.	A business process provides services to its partners through receive activities and corresponding reply actions.
ReadVariableAction , WriteVariableAction	**\<assign\>**
Variable actions support the reading and writing of variables. The VariableAction metaclass specifies the variable being accessed.	It copies data from one variable to another.
Variable , ObjectNode , OutputPin	**\<from\>**
An output pin is a pin that holds output values produced by an action. Output pins are object nodes that deliver values to other actions .	It specifies the source variable name to be used for an assignment expression.
Variable, ObjectNode , InputPin	**\<to\>**
An input pin is a pin that holds input values to be consumed by an action. They are object nodes that receive values from other actions	It specifies the target variable name to be used for an assignment expression.
CallOperationAction	**\<invoke\>**
An action that transmits an operation call request to the target object. If the action is marked asynchronous, the execution of the call operation waits until the execution completes, otherwise it is completed when the invocation is established.	An activity that is used by a process to make invocation to Web Services provided by partners. Can be either synchronous (request/reply) or asynchronous (one way).
ReplyAction	**\<reply\>**
An action that accepts a set of return values and a token containing return information produced by a previous accept call action.	An action to send a response to a request previously accepted through a receive activity.

Transformation Rules: The transformation method adopted is based on [10] and makes use of the OCL with the following conventions: The *UML* refers to the source language (Activity diagram), the *BPEL4WS* refers to the target language, the *params* refer to any parameters used during the transformation, the *source* and *target* refer to various named source or target meta-language model elements, the source and target conditions refer to source or target language conditions that must hold in order to apply the rule, the *mapping process* performs the mapping among the model elements, and – for various comments.

The transformation process is initiated by performing a mapping between the overall UML Activity diagram and a business process. The transformation *UMLActivity2BusinessProcess* (see Table 2), firstly, maps the *activity name* into a *process name*. Then it performs three nested sub-transformations named as *UML Datastore2BPVariable, ActivityPartition2BPPartnerLink* and *UMLProcessActivity2 BPActivity*. The first sub-transformation maps the *variables* or *datastores* used in the UML activity diagram into the corresponding entities in the business processes. The UML Activity diagram uses *partitions* and *swimlanes* to represent *partners*. The subtransformation *ActivityPartition2BPPartnerLink* maps such *partitions* into to BPEL4WS *partner links*. Finally, the sub-transformation *UMLProcessActivityBP Activity* (see Table 3) maps the UML *Process Activities* into *Business Process Activities*.

Table 2. The *UMLActivity2BusinessProcess* Transformation

Transformation ***UMLActivity2BusinessProcess*** (UML, BPEL4WS)	
params	srcActivity: UML::Activity -- *the UML source Model*
source	srcActNodes: OCL::Set(ActivityNode)
target	trgProcess: BPEL4WS::Process
	trgVariable: BPEL4WS::Variable
	trgPartnerLink: BPEL4WS::PartnerLink
	trgActivity: BPEL4WS::Activity
source cond	srcActNodes = srcActivity.nodes->asSet()->union(
	srcActivity.group->collectNested(ActivityNode))
--mapping *process* srcActivity.name <~> trgProcess.name try **UMLDatastore2BPVariable** on srcActNodes-> collect(DataStore) <~> trgVariable.type try **ActivityPartition2BPPartnerLink** on srcActNodes-> collect(ActivityPartition) <~> trgPartnerLink.type try **UMLProcessActivity2BPActivity** on srcActNodes-> collect(Action) <~> trgActivity.type	

When mapping the *ProcessActivity* to a *BusinessProcess activity* we need to check the *activity* type, which is stereotyped in our model of Fig. 5 in order to trigger the appropriate mapping. In our example, the process activity is of the type *sequence,* therefore the *UMLSequence2BPSequence* (see Table 4) rule is called. If the activity was of type *flow,* then the mapping rule *UMLFlow2BPFlow* would have been applied.

Following, we specify the mapping between *sequential actions* in the UML and BPEL4WS. First the *actions* within a *sequence* activity are passed as parameters. The method adopted involves inductive transformation of the model elements of the

Table 3. The *UMLActivity2BusinessProcess* Transformation

Transformation ***UMLProcessActivity2BPActivity*** (UML, BPEL4WS)	
params	srcActions: OCL::Set(UML::Action)
source	srcActivity: UML::Activity
target	trgSequence: BPEL4WS::Sequence
	trgFlow: BPEL4WS::Flow

```
--mapping process
if srcActivity.oclIsTypeof(Sequence) then
try UMLSequence2BPSequence on scrActions <~> trgSequence.type
elseif srcActivity.oclIsTypeof(Flow) then
try UMLFlow2BPFlow on scrActions <~> trgFlow.type
endif
```

Table 4. The *UMLSequence2BPSequence* Transformation

Transformation ***UMLSequence2BPSequence*** (UML, BPEL4WS)	
params	srcActions: OCL::Set(UML::Action))
source	srcInitActions: OCL::Set(UML::Action) -- *possible set of starting actions*
target	trgActReceive: BPEL4WS::Receive
	trgActInvoke: BPEL4WS::Invoke
	trgActReply: BPEL4WS::Reply
	trgActAssign: BPEL4WS::Assign
source cond	srcInitActions = srcActions.iterate(a:Action acc:Set(ActivityNode))=Set{} | if a.incoming.isEmpty() then acc->including(a)) && srcInitActions.count=1

```
--mapping process
action = srcInitActions
Do while action <> null
If initAction.outgoing.target.oclIsTypeOf(AcceptCallAction)
  try UMLAcceptCallAction2BPReceiveActivity on
action.outgoing.target.type <~> trgActReceive.type
elseif
initActin.outgoing.target.oclIsTypeOf(VariableAction)
  try UMLAssign2BPAssignActivity on
action.outgoing.target.type <~> trgActAssign.type
elseif
initActin.outgoing.target.oclIsTypeOf(CallOperationAction)
  try UMLCallOperationAction2BPInvokeActivity on
action.outgoing.target.type <~> trgActInvoke.type
elseif initActin.outgoing.target.oclIsTypeOf(ReplyAction)
  try UMLReplyAction2BPReplyActivity on
action.outgoing.target.type <~> trgActReply.type
endif
action = action.outgoing.target
loop
```

UML sequence. Starting from the *starting action* (the one that initiates the sequence of actions), each model element is interpreted and transformed into its corresponding model element at the destination. The transformation procedure continues until, the final action is reached. For example, if the type of *outgoing.target,* see the table below, is an *AcceptCallAction*, we have to apply the *AcceptCallAction* to BPEL4WS *ReceiveActivity* mapping rule. Similarly, it is possible to deal with other target types.

The above transformations are samples of the developed rules. Even though, the above sets of rules are sufficient to illustrate the model transformation of our running example. If we apply the transformation rules to the Activity diagram of Fig. 3 or Fig. 5, an equivalent BPEL4WS activity model of Fig. 7 is produced.

4   Discussions and Related Work

The UML Activity diagram of Fig. 5 and Fig. 7 seem very similar, as they represent the same case scenario. From the conceptual point of view, they belong to totally different metamodels and have different properties. In addition, we found that when more complex behaviours are applied (such as representing *scopes* and *repetitions*) the models become more dissimilar.

We have a view of model transformation, which includes the transformation of both static and dynamic aspects of the system. Following the notation of [6][7], suppose that $m_1(s)/f_1$ $(m_2(s)/f_2)$ represent static aspects of the system s as models $m_1(m_2)$ in formalisms f_1 (f_2), respectively. Assume that $m_1(s)/f_1 \to m_2(s)/f_2$ is a transformation that maps static aspects of the system from the formalism f_1, the source, to the formalism f_2, the destination. Suppose that g_1 and g_2 are two formalisms for expressing dynamic aspects of the system s, i.e. they can be used to specify how various components cooperate to manage various tasks and provide functions of the system. As a result, the transformation $m_1(s)/g_1 \to m_2(s)/g_2$ are used to map the dynamic aspects of the system. Since the static and dynamic aspects of a system are closely interrelated, it is naïve to assume that the transformation of the two aspects can be carried out independent of one another. For example, in our case study, the metamodel of BPEL4WS shown in Fig. 6 contains references to WSDL elements; there are references to *port type*, *message* and *operation*, which are static model elements. In other words, since the dynamic aspects represent how entities (defined in the structural models) work together to accomplish a task, it is essential to find a systematic way of integrating the two views together. In our opinion, presenting a general method of integrating static and dynamic aspects of arbitrary systems is a highly non-trivial and challenging task. As a result, our current research focuses only on the transformation of behavioural models of Business Processes, from a UML model to a specific implementation technology (BPEL4WS). In conducting the above case study, we have followed the following trivial rule of thumb:

"If a business process task t1 at the source is transformed to a task t2 at the destination, then t2 and t1 must have the same effect on the corresponding collaborating services."

Would it be possible to formalise the above rule of thumb? Caplat and Sourroulle [6] study a similar question for the static aspects of systems. For example, they show that, similarity of metamodels is not a good criterion for judging the similarity of formalisms. However, answering the above question require further research.

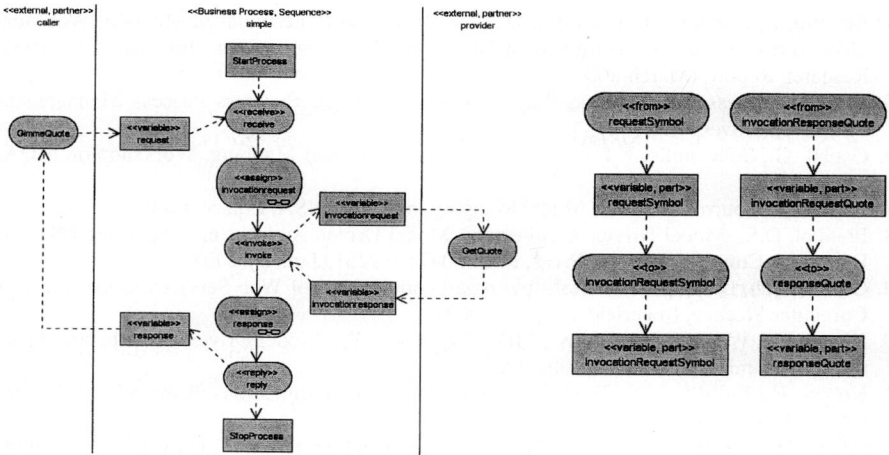

Fig. 7. Activity Diagram for BPEL

The Activity diagram of Fig. 3 is a high level conceptual model, which is too abstract to be translated to an implementation. As a result, we incorporated further information, which required making further design decisions. For example, refine the diagram of Fig. 3 to the Activity diagram of Fig. 5. The metamodel can help us in identifying various possible ways of refining a model. For instance, the metamodel of Fig. 4 specifies that an *ActivityNode* is either a *ControlNode*, *ObjectNode* or *ExecutableNode*, see page 268 in [19].

5   Conclusion

This paper deals with the modelling of the behavioural aspects of composite Web Services. The paper studies a model transformation of business processes from a PIM, created as a UML Activity diagram, into a PSM, modelled via BPEL4WS. We have presented a metamodel for the UML Activity diagram, which can be used for the refinement of the conceptual models of business processes. We have also introduced a MOF compliant metamodel for the BPEL4WS. A mapping between the corresponding model elements in the UML Activity diagram and BPEL4WS is introduced. To translate Activity diagram to BPEL4WS, we have introduced a set of transformation rules, which are specified in the OCL. Finally, we have applied our approach to the model transformation of a Stock Quote Web Service.

References

1. Appukutan, B., Clark, T., Reddy, S., Tratt, T., Venkatesh, R.: A model driven approach to model transformations, Kings College, (2003)
2. BEA, IBM, Microsoft, SAP AG and Siebel Systems: Business Process Execution Language for Web Services, Version 1.1, (May 2003)
3. Bezavin, J., Gerard, S.: A preliminary identification of MDA components, University of Nantes, CEA, (2002)

 4. Bezivin, J., Hammoudi, S., Lopes, D., Jouault, F.: An Experiment in Mapping Web Services to Implementation Platforms, Atlas Group, INRIA and LINA University of Nantes, Research Report, (March 2004)
 5. BPMI: Business Process Modelling Language (BPML), Business Process Management Initiative, (November 2002)
 6. Caplat, G., Sourrouille, J. L.: Considerations about Model Mapping, Workshop on MDA, INSA, (2003)
 7. Caplat, G., Sourrouille, J. L.: Model Mapping in MDA, INSA, France, (2002)
 8. Frankel, D.S.: Model Driven Architecture, Model Driven Architecture: Applying MDA to Enterprise Computing, OMG Press, ISBN: 0471319201,(January 2003)
 9. Ganesarajah, D., Lupu, E.: Workflow-based composition of Web Services, Department of Computer Science, Imperial College, (2002)
10. Kleppe, A., Warmer, J., Bast, W.: MDA Explained. The Model Driven Architecture: Practice and Promise, Addison-Wesley, (April 2003)
11. Kreger, H.: Fulfilling The Web Services Promise, Communications of the ACM, Vol. 46, No. 6, (June 2003)
12. Lopes, D., Hammoudi, S.: Web Services in the Context of MDA, University of Nantes, France, (2003)
13. O' Riordan, D.: Business Process Standards For Web Services, Published by Tect, Chicago, USA
14. OMG: Enterprise Collaboration Architecture (ECA) Specification, Object Management Group, Version 1.0, (February 2004)
15. OMG: Meta Object Facility (MOF) Specification, Object Management Group, Version 1.4, (2002)
16. OMG: Object Constraint Language Specification (OCL), Part of UML 1.4.1 Specification, (2003)
17. OMG: Object Management Group, Available from http://www.omg.com
18. OMG: Request for Proposal: MOF 2.0 Query / Views / Transformations RFP, Object Management Group, (October 2003)
19. OMG: UML 2.0 Superstructure Specification, Object Management Group, Adopted Specification, Version 2, (August 2003)
20. Papazoglou, M.P.,Georgakopoulos, D.: Service Oriented Computing, Communications of the ACM, Vol. 46, No. 10, (October 2003)
21. Siegel, J.: Using OMG's Model Driven Architecture (MDA) to integrate Web Services, Object Management Group White Paper, (November 2001)
22. Tratt, L., Clark, T.: Model Transformation in Converge, Kings College, (2003)
23. W3C: Extensible Markup Language (XML) 1.0, Third Edition, W3C Recommendation, (February 2004)
24. W3C: Simple Object Access Protocol (SOAP), Version 1.2, W3C Recommendation, Available from http://www.w3.org/TR/soap12-part1, (2003)
25. W3C: Web Service Choreography Interface (WSCI) 1.0, W3C Note, Available from http://www.w3.org/TR/wsci, (August 2002)
26. W3C: Web Services Architecture, Working Group Note, (February 2004)
27. W3C: Web Services Description Language (WSDL) Version 2.0, W3C Working Draft, (2003)

Model-Driven Planning and Development of Infrastructures for E-Commerce

Ulrich Frank

University of Koblenz, Universitätsstr. 1, 56070 Koblenz
ulrich.frank@uni-koblenz.de

Abstract. For many companies, especially small and medium sized firms (SME), planning and implementing infrastructures for Electronic Commerce (E-Commerce) that are in line with a convincing long term strategy is too much of a challenge. This paper presents a method that supports companies with re-designing their business processes in order to take advantage of the opportunities offered by E-Commerce. For this purpose, it provides concepts to analyse and design corporate strategies as well as various modelling languages for describing and analysing business processes, resources, IT artifacts etc. In addition to that, the method provides generic strategy networks and a library of reference business processes. They allow prospective users to avoid the effort of developing strategies and new organizational settings from scratch. The modelling language that is used to describe business processes allows for transforming a business process into a corresponding workflow schema, thereby supporting the realisation of a supporting workflow management system.

1 Introduction

E-Commerce, the initiation of, agreement on, and performing of economic transactions, is changing many industries in a fundamental way. This change does not always come as a revolution. But even in industries where it takes place at a slow pace, it is high time to get prepared for the future. Companies that do not care about E-Commerce in time are likely to suffer from the structural changes to come - instead of benefiting from them. However, developing and implementing a successful strategy for E-Commerce is a complex endeavour. Not only that it requires knowledge about traditional methods for strategic planning, it also requires developing appropriate ideas of specific opportunities and risks associated with doing business on the Internet. Furthermore, it will usually be necessary to analyse and eventually redesign existing business processes - sometimes fundamentally. Last, but not least, it has to be taken into account that information systems are the key enabler of E-Commerce. This requires knowledge about appropriate, process-oriented analysis- and design methods. In addition to that, there is a confusing plethora of Internet-specific technologies, such as communication protocols, so called middleware, implementation level languages and standards to promote the exchange of business documents or objects respectively. For many business firms, especially SME, this complexity is too much of

S. Wang et al. (Eds.): ER Workshops 2004, LNCS 3289, pp. 679–690, 2004.

a challenge to cope with. Not only that many companies will lack employees with the required skills. Often, the most competent employees are entirely absorbed by their regular duties. On the other hand, the financial assets of many firms are not sufficient to allow for employing external consultants to a larger extent. At the same time, it may be a question of survival for many firms to act in time. Therefore, there is need for supporting companies effectively with preparing for the transition to E-Commerce. Against this background, the project ECOMOD ('Modelling for Electronic Commerce'), which has been funded by the German National Science Foundation (DFG), was aimed at the development of a method as well as corresponding reference models to reduce both effort and risk of planning and implementing infrastructures for E-Commerce that are in line with an enterprise's long term strategy. In this paper, we will first give an overview of the underlying method, E-MEMO. After that, the generic reference models for guiding strategic planning and designing business processes are described. Finally, it is demonstrated how the reference models in association with E-MEMO can be used to generate workflow schemata.

2   Conceptual Foundation: E-MEMO

The method developed within ECOMOD, called E-MEMO, is based on an already existing method for multi-perspective enterprise modelling, MEMO ([3], [5]). MEMO serves to analyse and design corporate information systems that are balanced with a company's strategy and its organization. For this purpose, it allows for structuring and describing a company from three perspectives: *strategy*, *organisation*, and *information system*. Each perspective can be further differentiated into four aspects: *structure*, *process*, *resources* and *goals*. In order to describe the resulting foci in more detail, MEMO provides a set of modelling languages, such as an object modelling language (MEMO-OML, which is similar to UML, [4]), an organisation modelling language (MEMO-OrgML, [7]) and a strategy modelling language (MEMO-SML). These languages have been adapted to fit the requirements of modelling for E-Commerce. Before we give an overview of the languages' architecture and concepts, we will first focus on the specialized process models that have been adapted from the generic macro- and micro-process models of MEMO.

2.1   Process Models

To guide the long-term transition into E-Commerce as well as particular projects, E-MEMO offers two specialized process models. The macro perspective describes an idealized evolutionary process. To become serious E-Commerce contenders, managers should understand the potential of the Internet. A solid understanding requires a substantial idea of what is possible in the long run - with respect to economic, organisational and technological changes to come. At the same time, investments into E-Commerce recommend a long term perspective. Only then you are able to prepare for a future development step with today's investments

- and only then you are able to protect these investments. The long term development of doing business over the Internet can hardly be predicted, nor is it possible to specify the one best development path. However, one can outline a long term vision that takes into account prototypical development steps. The evolutionary model that has been designed for E-MEMO consists of four prototypical stages. Each stage symbolizes a certain technological and organisational capability. Except for the first stage, every other stage requires the ability of the previous stage to be available. However, the model does not prescribe that a step has to be frozen as soon as it is accomplished. Sometimes it may make sense to take two steps at a time. Also, it will not always be possible to assign a particular state of a company to exactly one evolutionary stage. The purpose of the model is only to give a long term orientation for development in E-Commerce. Such a long term perspective helps to protect investments into first steps, since they can be designed in a way that fosters the realisation of further steps. In the first stage, *presence*, a company uses the Internet only to present itself on one or more web pages. The following stage, *initiation of transactions*, requires to present information about products and, optionally, terms and conditions. With respect to procurement, the initiation of transactions can be supported by making use of suppliers' web pages or electronic market places. *Performing transactions* is at the core of E-Commerce. It includes the entire process from initiating a transaction to after-sales service. The level of automation may vary. However, media clashes should be avoided during the order management process. Similar to the previous stage, it applies to both, sales and procurement. The highest prototypical stage of evolution, cross-organisational value chains, requires not only establishing cross-organisational business processes, which integrate processes and corresponding information systems on buyer and seller sides. In addition to that, it includes the synchronisation of plans, in order to decrease overall cost. If, for instance, a large shipping corporation, orders a part of its trucks a year in advance, the truck producer can decrease its cost by reducing risk. This is also the case for the truck producer's suppliers. Every stage is described in more detail using the following structure: *technological enablers, business opportunities, critical success factors, risks*, and *costs*. The macro process is supplemented by a specialized micro process. It includes five major steps: *analysis of strategic options, design/configuration of a strategy for E-Commerce, analysis of organisational options, selection/specification of business processes* and *implementation of business processes*. Each stage is described using the following structure: *objectives, supporting concepts/artifacts, participants, documentation*.

2.2   Specialized Modelling Languages

The modelling languages that are part of MEMO are aimed at two main objectives. Firstly, they are to provide concepts that help with the analysis and design of complex systems, which include information systems, organisations, corporate strategies or business processes. Hence, they should serve as a tool: "Language is an instrument." (Wittgenstein). On the other hand, a modelling language should foster the communication between the various stakeholders that use corresponding models: "Language is a social art." (Quine) The first aspect requires concepts that are directly related to analysis and design goals. They include abstraction mechanisms, e. g. generalisation or encapsulation, concepts to guide the evaluation of systems, e. g. the concept of a media clash within a business process. In order to cover this aspect, we aimed at reconstructing the concepts being applied in specialized analysis and design methods, as they can be found in software engineering (especially OOA and OOD), in strategic planning or business process re-engineering. The second aspect recommends concepts that are directly related to concepts and visualisations, prospective users are familiar with. Therefore, we tried to take into account methods and terminologies that are widely spread, such as concepts that are common in organisational design and strategic planning or object-oriented concepts. MEMO Organisation Modelling Language (MEMO-OrgML) is a language that fosters the description of a company's organisation, such as business processes or organisation structure. It also includes concepts that allow for modelling resources, e. g. information technology in general, software in particular, machines or human resources. In case there is software to be developed, the MEMO Object Modelling Language (MEMO-OML) supports the design of object models that are tightly integrated with models of business processes or workflows. Note that the MEMO-OML could be replaced by the UML, since both languages are similar. MEMO Strategy Modelling Language (MEMO-SML) provides concepts to describe and analyse corporate strategies. The various models that are part of an enterprise model are interrelated. In order to reduce redundancy and foster integrity, the corresponding modelling languages should be tightly integrated. This is accomplished by a common meta meta model. The language architecture allows for adding further languages. The modelling languages provided by MEMO proved to be well suited for the purpose of ECOMOD. However, in some cases there was need for additional concepts. In other cases, an existing modelling language turned out to be too elaborated and hence to complicated for the intended users of E-MEMO. For these reasons, the MEMO-OrgML was extended with, among other things, concepts to represent specific types of exceptions that may occur during electronic business transactions. Other parts of the MEMO-OrgML, which serve to model resources that are used by business processes, were simplified. With respect to modelling IT resources, the concept of a 'Solution' is of pivotal relevance. It is an abstraction that allows for bundling the applications, middleware, operating system, hardware and human operators, which are required for a particular process type [12]. Fig. 1 shows an excerpt of the corresponding meta model after its modification for E-MEMO.

3 Fostering Analysis and Design Through Reference Models

Separating the overall problem into various perspectives, such as strategy, business process and information system, helps with reducing its complexity. Specialized modelling languages provide further support for conceptualizing the do-

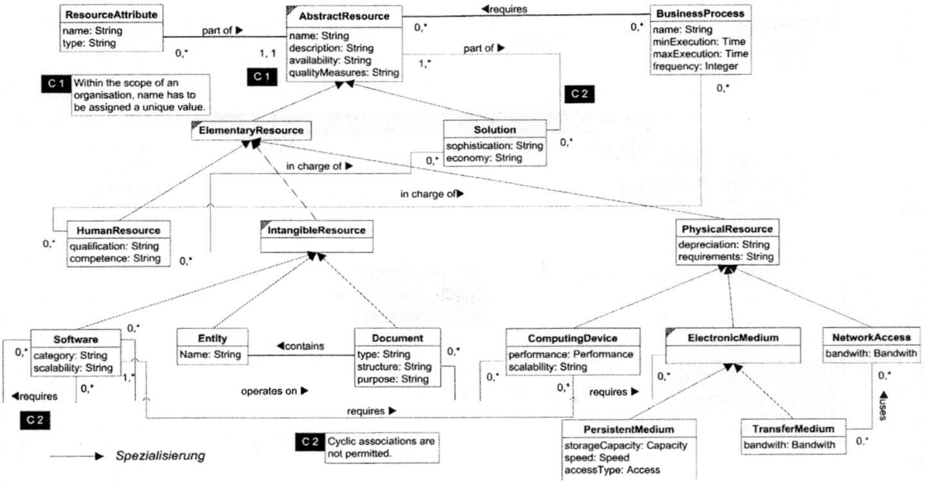

Fig. 1. Excerpt of the meta model for the specification of the MEMO-OrgML

main of interest in an appropriate way. In addition to that, the macro- and micro-process guide managers with long-term planning and project management. However, for many companies, especially for SMEs, this will not be sufficient. In order to provide a more effective support, E-MEMO offers additional reference models that describe strategic options and a library of generic business process models.

3.1   Integrating Reference Models in the Overall Process

The idea of reference models is as attractive as challenging. Conceptual models illustrate prototypical solutions for complex problems. These solutions, e. g. data models or models of business processes, are supposed to be thoroughly developed and generic, i. e. applicable to a wide range of cases. Hence, they promise economic advantages that seem to be irresistible: better quality at less cost. However, within domains that are characterized by a high degree of diversity, finding a proper balance between general applicability and the benefit of re-use in a particular case is a major challenge. E-Commerce is certainly characterized by a tremendous diversity. To cope with this challenge, we took three measures. Firstly, the focus of ECOMOD was narrowed. It is defined by three dimensions: *generic strategy, evolutionary stage, internal value chain.* In his seminal work on strategic planning, Porter differentiates three generic strategies: *focus, cost leader* and *product leader* [16]. In this dimension, the focus of ECOMOD is on cost leader. This is for two reasons. Firstly, we assume that especially with SME, the potential for cost reduction through Internet technologies is of crucial importance for sustainable competitiveness. Secondly, we assume that this generic strategy is better suited to find general patterns both for concrete strategies for

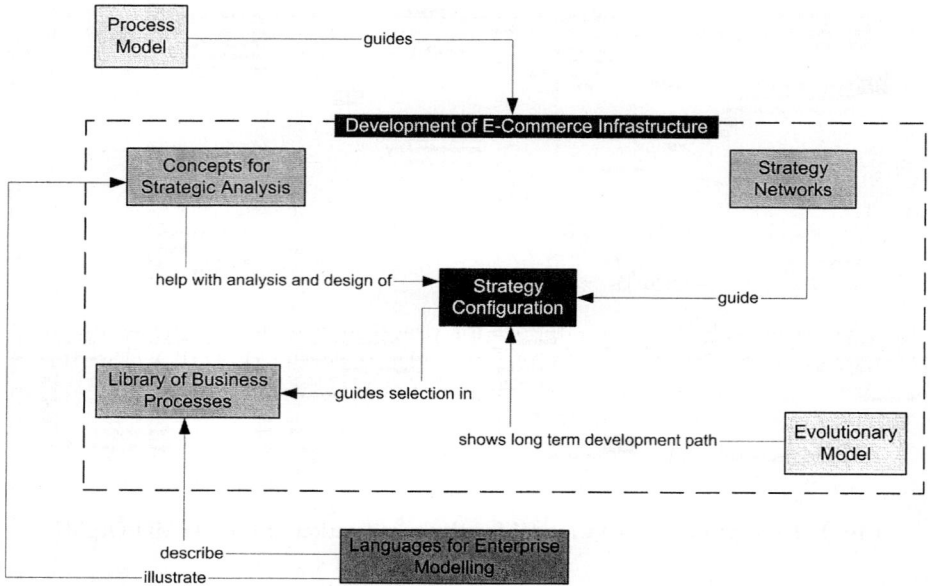

Fig. 2. Components of E-MEMO and their interaction

E-Commerce as well as for corresponding business processes. This is different with focus and product leader, which require taking into account specific peculiarities of customer demands or products. With respect to the dimension *evolutionary stage*, we focus on performing transactions. We assume that this stage is of pivotal relevance for most SME for a number of years to come. Despite its potential, cross-organisational value chain will be too much of a challenge for many SME these days. Finally, there is the internal value chain. Particular aspects of E-Commerce may penetrate the entire value chain. Our focus is on *procurement* and *sales*. We do not take into account operations, such as production, especially because of the tremendous contingency of this activity. Furthermore, the reference models will mostly abstract from peculiarities of particular product types. Secondly, we emphasize that the reference models are not to be mistaken as blueprints. Instead, they provide concrete orientations that will usually require further adaptation to individual needs. For this reason, ECOMOD provides a number of additional analysis instruments that guide the selection and adaptation of reference models. Thirdly, we stress the need for further evaluation and refinement of the reference models offered by ECOMOD. Fig. 2 illustrates, how the various components of E-MEMO are intended to support the design of infrastructures for E-Commerce. Strategic analysis requires identifying strategic options and evaluating corresponding opportunities and threats. E-MEMO provides two kinds of concepts to support strategic analysis: a dictionary of questions and specialised analysis concepts. They are assigned to each stage of a specialized process model. The first stage is aimed at the analysis of customer expectations, technology development as well as the identification and analysis of future competitors. The second stage includes the analysis of sustaining competitive advantages, especially through structural change. The third stage is focussed on the analysis of a firm's value chain with emphasis on approaches to reduce cost in procurement and sales. The last stage is dedicated to finding a proper balance of the various strategic options, i. e. to the final specification of the strategy configuration.

3.2   Strategy Networks

In order to provide effective support for practitioners, ECOMOD is aimed at reconstructing the existing body of knowledge in a way that allows for better accessibility and that gives more concrete guidance for strategic planning in practice. *Strategy networks* are decision networks that provide a 'ballpark view' of strategic options. A strategic option can require other options to be realized and may be refined into other strategic option. Each strategic option is described in detail using the following structure: *investments, required competencies, opportunities, risks, critical success factors* [13]. In addition to that, diagram types or tools can be assigned to each strategic option. So far, we have developed five, partially overlapping strategy networks that are directly or indirectly associated with the generic strategy 'cost leadership'. They focus on sales, procurement, revenues, customer relationships and integration of the internal value chain.

3.3   A Library of Business Processes

Once, the strategy configuration has been completed, the users of E-MEMO are supposed to select business processes from the given library. The selection is supported by keywords that are annotated to process types. Fig. 3 illustrates the integration of strategies and business processes or - in other words - the derivation of appropriate business processes from a given strategy. Currently, the library includes about 80 business process types. They are grouped into various categories, such as *customer relations, reception of delivery, order processing, service, procurement*. A business process type within the library is specified using the MEMO-OrgML [7]. It allows for a graphical description of the control structure of a business process.

Each (sub) process within a process can be assigned resources, roles and textual annotations. In order to support the development of software, it is also possible to refer to services of classes specified in a corresponding object model. We assume, however, that developing software from scratch is no option for prospective users of E-MEMO. Instead, the IT infrastructure will usually be composed of existing applications, components, middleware etc. This can be specified by assigning so called 'solutions' (see fig. 1). The components of a solution are not specific products, but abstractions of systems. Every solution is assigned a description of cost categories, e. g. capital investment, maintenance efforts etc. Each business process type is further characterized by process goals, critical success factors and a list of keywords. The library as well as the

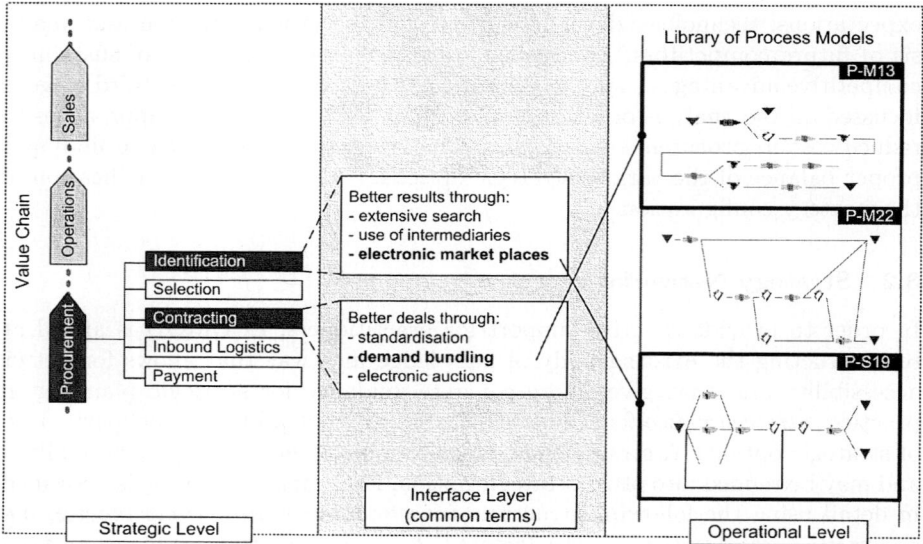

Fig. 3. Integration of Strategic and Operational Level through common Terms

strategy networks are presented on an interactive Web portal (http://www.uni-koblenz.de/ecomod). Navigating through the process library is supported by five types of associations between processes: *interaction, support, aggregation, similarity, specialisation*. Specialisation (or generalisation respectively) is an important concept for maintaining the library, since it implies the propagation of modifications applied to a type to all its subtypes. However, we did not succeed in defining a suitable formal semantics. There are a few suggestions for specifying process specialisation (e. g. [20], [21]), but they remain dissatisfactory, since they fail to guarantee the substitution constraint: Each instance of a specialized type should be suited to replace an instance of its supertype. Nevertheless, the MEMO-OrgML still allows for using specialisation relationships. They serve to document relationships between two types and require further interpretation by a human expert. Adapting a process type to individual requirements can either be done in a non-restricted or in a restricted mode. In the non-restricted mode, a user can change a process simply by modifying it using the MEMO-OrgML. This corresponds very much to copy, paste and change of code. While it allows a high degree of flexibility, it comes with severe disadvantages. It cannot be guaranteed that an adapted process type's documentation is still valid or that the quality of the process type is still satisfactory. In the non-restricted mode, only those modifications are permitted that do not violate integrity constraints. This is especially the case for adapting abstract processes. An abstract process can be assigned constraints that should not be violated in corresponding concrete processes. This may include goals, required resources, information, roles or pre- and postconditions.

4   Generating Workflow Schemata from Business Process Models

In many cases, the technology of choice for implementing process-oriented information systems will be workflow management systems (WMS). A workflow can be regarded as an abstraction of a business process that is focussed on those parts of a process, which can be automated. In other words, a workflow will leave out of account all aspects of a business process that cannot be supported by software. While there are different architectures of WMS, one is of particular appeal and relevance. It is based on a declarative specification of workflow types, called workflow schema, and a workflow engine that instantiates and manages particular workflows. The separation of workflow schema and workflow engine does not only facilitate maintenance, it also fosters the protection of investment, if the language used to specify the schema is standardized. This is case with the XPDL (XML Process Definition Language) proposed by the Workflow Management Coalition (WfMC). XPDL has been adapted from the former Workflow Process Definition Language (WPDL). It is specified as an XML-DTD [23]. Against this background, we decided to define a mapping from MEMO-OrgML to XPDL. It allows for generating XPDL-compliant workflow specifications from the business process models within the ECOMOD reference library. In comparison to MEMO-OrgML, the language concepts offered by XPDL are rather poor. There is, for instance, no concept to directly specify events within the control flow of a process. The specification of resources remains on a superficial level. There is only a few categories of resources (e. g. RESOURCE, ROLE). The resources that are assigned to a workflow type are represented in a list of so called 'participants'. Resources can include internal applications, which are provided by the workflow engine, or external applications. These are not, however, concrete applications, such as a particular text editor. Instead, XPDL allows for assigning application classes (e. g. 'text editor'), which are mapped to particular applications by the workflow engine. In order to demonstrate and test the mapping, we implemented a corresponding tool. For this purpose, we used MetaEdit+, a meta modelling tool, and Shark, a public domain workflow engine. For a detailed description of the mapping see [11]. Since the MEMO-OrgML allows for a more differentiated description of workflows, generating an XPDL document will usually result in the loss of semantics. Nevertheless, the differentiated description of resources within the reference models makes sense, because future releases of the XPDL can be expected to feature more elaborated concepts.

5   Related Work

There are a number of publications that are related to ECOMOD. Some authors analyse the strategic potential of E-Commerce and options to take advantage of it from a management point of view (e g. [1], [2], [9], [14], [17]). They remain on a strategic level and do not get into details of how to implement a strategy on the organisational level. Similar to ECOMOD, the 'handbook of organizational processes' [15] is aimed at reference business processes. However, Malone et al. do not use a modelling language. Therefore, their process descriptions would not allow for generating software artifacts. Also, the process descriptions lack the detail of the ECOMOD process library. EC-Cockpit is a tool that is to support experts with consulting business firms, which want to invest into E-Commerce. It includes a process model to guide structured interviews and a few generic enterprise types. First, a firm is assigned to one of the generic enterprise types. Then, an interview is conducted in order to design an individual value chain, which is the conceptual foundation for evaluating opportunities of E-Commerce. Compared to ECOMOD, EC-Cockpit has a wider scope, but lacks detail. Among other things, it does not include reference processes. Similar to EC-Cockpit, ADONIS provides a tool to support online strategy consulting for E-Commerce ([18], [19]). It suggests identifying the 'business type' of a particular firm. This leads to the selection of questions that help to shape a strategy for E-Commerce. The questions can be answered online. The resulting strategy is generated from the answers by applying a set of rules. ADONIS, too, does not provide reference processes or support for the development of information systems.

6   Conclusions

The second stage of ECOMOD was finished in May 2004. With regard to the objectives of the projects, there are still two challenges to be faced. The first is *dissemination*. Since ECOMOD is aimed at supporting especially SME - or their consultants respectively - academic publications are hardly suited to reach this target group. For this reason, we decided for an interactive, bilingual (English and German) Web portal (http://www.uni-koblenz.de/ecomod). It does not only serve to present the strategy networks and the library of business processes. In addition to that, it provides a comprehensive description of E-MEMO without 'academic overhead'. Users can navigate through the process library and are provided with context sensitive background knowledge. The second challenge is related to *evaluation*. ECOMOD resulted in the development of methods and design artifacts. While such an approach seems to be suitable when it comes to the realisation of future business models and the deployment of future technologies, the evaluation of research results is a severe problem. Within the positivist paradigm, evaluation is typically done by testing a research result against reality using accepted research methods ('correspondence theory of truth'). This kind of evaluation is not possible for evaluating artifacts, unless you are willing to accept that it would take many years - and a huge amount of resources - to apply and test the approach in practice. In a recent article, Hevner et al. emphasize the relevance of what they call design-oriented research in Information Systems [10]. At the same time, they emphasize that the application of "rigorous evaluation methods" is "extremely difficult" (ibid, p. 99). We evaluated E-MEMO primarily against three criteria: *appropriateness*, *comprehensibility*, and *economics*, all of which relate to the method's usability and acceptance. Appropriateness characterizes why and how a method or model contributes to given objectives. This includes the question, whether proper abstractions have been chosen (e. g. no overload with irrelevant aspects). Comprehensibility of a method or visual modelling languages is hard to judge. However, it seems reasonable to assume that a method is the more comprehensible, the more its target group is familiar with the concepts and terminology being used. Therefore we used and reconstructed well known concepts from strategic planning and business process redesign. It is a main objective of E-MEMO to contribute to the economics of the development and implementation of competitive infrastructures for E-Commerce. To accomplish this, we emphasized the construction of re-usable generic domain-level and problem-oriented artifacts. Also, we promoted the use of standards as much as possible. While a discoursive evaluation of these criteria during the research process and its documentation contributes certainly to the overall academic quality, it is not sufficient. There is additional need for empirical evaluation. Therefore, we decided to have prospective users participate in the evaluation process. For this purpose, the interactive Web portal allows users to enter comments and suggestions for refining the models.

References

1. DE FIGUEIREDO, J.M.: *Finding Sustainable Profitability in Electronic Commerce.* In: Brynjolfsson; E.; Urban, G.L. (Eds.): Strategies for e-business success. San Francisco: Jossey Bass 2001, 7–33
2. FEENY, D.: *Making Business Sense of the E-Opportunity.* In: Brynjolfsson, E.; Urban, G.L. (Ed.): Strategies for e-business success. San Francisco: Jossey Bass 2001, 35–60
3. FRANK, U.: *Enriching Object-Oriented Methods with Domain Specific Knowledge: Outline of a Method for Enterprise Modelling.* Arbeitsberichte des Instituts für Wirtschaftsinformatik, University of Koblenz, No. 4, 1997
4. FRANK, U.: *The Memo Object Modelling Language (MEMO-OML).* Arbeitsberichte des Instituts für Wirtschaftsinformatik, University of Koblenz, No. 10, 1998
5. FRANK, U.: *Multi-Perspective Enterprise Modeling (MEMO) - Conceptual Framework and Modeling Languages.* In: Proceedings of the Hawaii International Conference on System Sciences (HICSS-35). Honolulu 2002, 10 pp.
6. FRANK, U.; LANGE, C.: *Corporate Strategies for Electronic Commerce - Stepwise Refinement and Mapping to Generic Business Process Models.* Arbeitsberichte des Instituts für Wirtschaftsinformatik, University of Koblenz, No. 42, 2004
7. FRANK, U.; JUNG, J.: *The MEMO-Organisation Modelling Language (OrgML).* Arbeitsberichte des Instituts für Wirtschaftsinformatik, University of Koblenz, No. 48, 2004
8. FRANK, U.; LANGE, C.: *A Framework to Support the Analysis of Strategic Options for Electronic Commerce.* Arbeitsberichte des Instituts für Wirtschaftsinformatik, University of Koblenz, No. 41, 2004
9. HACKBARTH, G.; KETTINGER, W. J.: *Building an e-business strategy.* In: Information Systems Management, 2000, 78–93
10. HEVNER, A. R.; MARCH, S. T.; PARK, J.; RAM, S.: *Design Science in Information Systems Research.* In: MIS Quarterly, Vol. 28, No. 1, 2004, 75–105

11. JUNG, J.: *Mapping of Business Process Model to Workflow Schemata - an example using MEMO-OrgML and XPDL*. Arbeitsberichte des Instituts für Wirtschaftsinformatik, University of Koblenz, No. 47, 2004

12. JUNG, J.; KIRCHNER, L.: *A Framework for Modelling E-Business Resources*. Arbeitsberichte des Instituts für Wirtschaftsinformatik, University of Koblenz, No. 44, 2004

13. LANGE, C.: *Developing Strategies for Electronic Commerce in Small and Medium Sized Companies - Guidelines for Managers*. Arbeitsberichte des Instituts für Wirtschaftsinformatik, University of Koblenz, No. 39, 2003

14. LUCAS, H.C.: *Strategies for Electronic Commerce and the Internet*. Boston: The MIT Press 2002

15. MALONE, T.W. ET AL.: *Toward a handbook of organizational processes*. In: Management Science, Vol. 45, No. 3, 1999, 425–443

16. PORTER, M. E.: *The Competitive Advantage of Nations*. Basingstoke, Hampshire: MacMillan 1994

17. PORTER, M. E.: *Strategy and the Internet*. In: Harvard Business Review, March 2001, 63–78

18. SCHÜTZ, S.; ZEISSLER, G.: *Ermittlung betriebswirtschaftlicher Anforderungen zur Definition von Geschäftsprozessprofilen*. FWN-Bericht 2001-010 2001

19. THOME, R.; HENNIG, A.; OLLMER, C.: *Kategorisierung von eC-Geschäftsprozessen zur Identifikation geeigneter eC-Komponenten für die organisierte Integration von Standardanwendungssoftware*. FORWIN FWN-2000-011, Bayrischer Forschungsverbund Wirtschaftsinformatik 2000

20. VAN DER AALST, W.M.P.; BASTEN, T.: *Life-cycle Inheritance: A Petri-net-based Approach*. In: Brynjolfsson; E.; Urban, G.L. (Eds.): Strategies for e-business success. San Francisco: Jossey Bass 2001, 7–33

21. VAN DER AALST, W.M. P.; BASTEN, T.: *Inheritance of workflows: an approach to tackling problems related to change*. In: Theoretical Computer Science 270, (1-2), 2002, 125–203

22. WFMC: *Workflow Process Definition Interface - XML Process Definition Language* Doc. No. WFMC-TC-1025, 1.0 Final Draft 2002,
http://www.wfmc.org/standards/docs/TC-1025_10_xpdl_102502.pdf,
checked 04/12/2004

Author Index